Graduate Texts in Mathematics 139

Springer
New York
Berlin
Heidelberg
Barcelona
Hong Kong
London
Milan
Paris
Singapore
Tokyo

Graduate Texts in Mathematics

1 TAKEUTI/ZARING. Introduction to Axiomatic Set Theory. 2nd ed.
2 OXTOBY. Measure and Category. 2nd ed.
3 SCHAEFER. Topological Vector Spaces. 2nd ed.
4 HILTON/STAMMBACH. A Course in Homological Algebra. 2nd ed.
5 MAC LANE. Categories for the Working Mathematician. 2nd ed.
6 HUGHES/PIPER. Projective Planes.
7 SERRE. A Course in Arithmetic.
8 TAKEUTI/ZARING. Axiomatic Set Theory.
9 HUMPHREYS. Introduction to Lie Algebras and Representation Theory.
10 COHEN. A Course in Simple Homotopy Theory.
11 CONWAY. Functions of One Complex Variable I. 2nd ed.
12 BEALS. Advanced Mathematical Analysis.
13 ANDERSON/FULLER. Rings and Categories of Modules. 2nd ed.
14 GOLUBITSKY/GUILLEMIN. Stable Mappings and Their Singularities.
15 BERBERIAN. Lectures in Functional Analysis and Operator Theory.
16 WINTER. The Structure of Fields.
17 ROSENBLATT. Random Processes. 2nd ed.
18 HALMOS. Measure Theory.
19 HALMOS. A Hilbert Space Problem Book. 2nd ed.
20 HUSEMOLLER. Fibre Bundles. 3rd ed.
21 HUMPHREYS. Linear Algebraic Groups.
22 BARNES/MACK. An Algebraic Introduction to Mathematical Logic.
23 GREUB. Linear Algebra. 4th ed.
24 HOLMES. Geometric Functional Analysis and Its Applications.
25 HEWITT/STROMBERG. Real and Abstract Analysis.
26 MANES. Algebraic Theories.
27 KELLEY. General Topology.
28 ZARISKI/SAMUEL. Commutative Algebra. Vol.I.
29 ZARISKI/SAMUEL. Commutative Algebra. Vol.II.
30 JACOBSON. Lectures in Abstract Algebra I. Basic Concepts.
31 JACOBSON. Lectures in Abstract Algebra II. Linear Algebra.
32 JACOBSON. Lectures in Abstract Algebra III. Theory of Fields and Galois Theory.
33 HIRSCH. Differential Topology.
34 SPITZER. Principles of Random Walk. 2nd ed.
35 ALEXANDER/WERMER. Several Complex Variables and Banach Algebras. 3rd ed.
36 KELLEY/NAMIOKA et al. Linear Topological Spaces.
37 MONK. Mathematical Logic.
38 GRAUERT/FRITZSCHE. Several Complex Variables.
39 ARVESON. An Invitation to C^*-Algebras.
40 KEMENY/SNELL/KNAPP. Denumerable Markov Chains. 2nd ed.
41 APOSTOL. Modular Functions and Dirichlet Series in Number Theory. 2nd ed.
42 SERRE. Linear Representations of Finite Groups.
43 GILLMAN/JERISON. Rings of Continuous Functions.
44 KENDIG. Elementary Algebraic Geometry.
45 LOÈVE. Probability Theory I. 4th ed.
46 LOÈVE. Probability Theory II. 4th ed.
47 MOISE. Geometric Topology in Dimensions 2 and 3.
48 SACHS/WU. General Relativity for Mathematicians.
49 GRUENBERG/WEIR. Linear Geometry. 2nd ed.
50 EDWARDS. Fermat's Last Theorem.
51 KLINGENBERG. A Course in Differential Geometry.
52 HARTSHORNE. Algebraic Geometry.
53 MANIN. A Course in Mathematical Logic.
54 GRAVER/WATKINS. Combinatorics with Emphasis on the Theory of Graphs.
55 BROWN/PEARCY. Introduction to Operator Theory I: Elements of Functional Analysis.
56 MASSEY. Algebraic Topology: An Introduction.
57 CROWELL/FOX. Introduction to Knot Theory.
58 KOBLITZ. *p*-adic Numbers, *p*-adic Analysis, and Zeta-Functions. 2nd ed.
59 LANG. Cyclotomic Fields.
60 ARNOLD. Mathematical Methods in Classical Mechanics. 2nd ed.
61 WHITEHEAD. Elements of Homotopy Theory.
62 KARGAPOLOV/MERLZJAKOV. Fundamentals of the Theory of Groups.
63 BOLLOBAS. Graph Theory.

(continued after index)

Glen E. Bredon

Topology and Geometry

With 85 Illustrations

Springer

Glen E. Bredon
Department of Mathematics
Rutgers University
New Brunswick, NJ 08903
USA

Mathematics Subject Classification (2000): 55-01, 58A05

Library of Congress Cataloging-in-Publication Data
Bredon, Glen E.
Topology & geometry/Glen E. Bredon.
p. cm.—(Graduate texts in mathematics; 139)
Includes bibliographical references and indexes.

1. Algebraic topology. I. Title. II. Title: Topology and geometry. III. Series.
QA612.B74 1993
514′.2—dc20 92-31618

Printed on acid-free paper.

Production coordinated by Brian Howe and managed by Francine Sikorski; manufacturing supervised by Vincent Scelta.

9 8 7 6 5

ISBN 978-1-4419-3103-0

Springer-Verlag New York Berlin Heidelberg
A member of BertelsmannSpringer Science+Business Media GmbH

Preface

The golden age of mathematics—that was not
the age of Euclid, it is ours.

C.J. KEYSER

This time of writing is the hundredth anniversary of the publication (1892) of Poincaré's first note on topology, which arguably marks the beginning of the subject of algebraic, or "combinatorial," topology. There was earlier scattered work by Euler, Listing (who coined the word "topology"), Möbius and his band, Riemann, Klein, and Betti. Indeed, even as early as 1679, Leibniz indicated the desirability of creating a geometry of the topological type. The establishment of topology (or "analysis situs" as it was often called at the time) as a coherent theory, however, belongs to Poincaré.

Curiously, the beginning of general topology, also called "point set topology," dates fourteen years later when Fréchet published the first abstract treatment of the subject in 1906.

Since the beginning of time, or at least the era of Archimedes, smooth manifolds (curves, surfaces, mechanical configurations, the universe) have been a central focus in mathematics. They have always been at the core of interest in topology. After the seminal work of Milnor, Smale, and many others, in the last half of this century, the topological aspects of smooth manifolds, as distinct from the differential geometric aspects, became a subject in its own right. While the major portion of this book is devoted to algebraic topology, I attempt to give the reader some glimpses into the beautiful and important realm of smooth manifolds along the way, and to instill the tenet that the algebraic tools are primarily intended for the understanding of the geometric world.

This book is intended as a textbook for a beginning (first-year graduate) course in algebraic topology with a strong flavoring of smooth manifold theory. The choice of topics represents the ideal (to the author) course. In practice, however, most such courses would omit many of the subjects in the book. I would expect that most such courses would assume previous knowledge of general topology and so would skip that chapter, or be limited to a brief run-through of the more important parts of it. The section on homotopy should be covered, however, at some point. I do not go deeply into general topology, but I do believe that I cover the subject as completely as a mathematics student needs unless he or she intends to specialize in that area.

It is hoped that at least the introductory parts of the chapter on differentiable manifolds will be covered. The first section on the Implicit Function Theorem might best be consigned to individual reading. In practice, however, I expect that chapter to be skipped in many cases with that material assumed covered in another course in differential geometry, ideally concurrent. With that possibility in mind, the book was structured so that that material is not essential to the remainder of the book. Those results that use the methods of smooth manifolds and that are crucial to other parts of the book are given separate treatment by other methods. Such duplication is not so large as to be consumptive of time, and, in any case, is desirable from a pedagogic standpoint. Even the material on differential forms and de Rham's Theorem in the chapter on cohomology could be omitted with little impact on the other parts of the book. That would be a great shame, however, since that material is of such interest on its own part as well as serving as a motivation for the introduction of cohomology. The section on the de Rham theory of $\mathbf{CP}^n$ could, however, best be left to assigned reading. Perhaps the main use of the material on differentiable manifolds is its impact on examples and applications of algebraic topology.

As is common practice, the starred sections are those that could be omitted with minimal impact on other nonstarred material, but the starring should not be taken as a recommendation for that aim. In some cases, the starred sections make more demands on mathematical maturity than the others and may contain proofs that are more sketchy than those elsewhere.

This book is not intended as a source book. There is no attempt to present material in the most general form, unless that entails no expense of time or clarity. Exceptions are cases, such as the proof of de Rham's Theorem, where generality actually improves both efficiency and clarity. Treatment of esoteric byways is inappropriate in textbooks and introductory courses. Students are unlikely to retain such material, and less likely to ever need it, if, indeed, they absorb it in the first place.

As mentioned, some important results are given more than one proof, as much for pedagogic reasons as for maintaining accessibility of results essential to algebraic topology for those who choose to skip the geometric treatments of those results. The Fundamental Theorem of Algebra is given no less than four topological proofs (in illustration of various results). In places where choice is necessary between competing approaches to a given topic, preference has been given to the one that leads to the best understanding and intuition.

In the case of homology theory, I first introduce singular homology and derive its simpler properties. Then the axioms of Eilenberg, Steenrod, and Milnor are introduced and used exclusively to derive the computation of the homology groups of cell complexes. I believe that doing this from the axioms, without recourse to singular homology, leads to a better grasp of the functorial nature of the subject. (It also provides a uniqueness proof gratis.) This also leads quickly to the major applications of homology theory. After that point, the difficult and technical parts of showing that singular homology satisfies the axioms are dealt with.

Cohomology is introduced by first treating differential forms on manifolds, introducing the de Rham cohomology and then linking it to singular homology. This leads naturally to singular cohomology. After development of the simple properties of singular cohomology, de Rham cohomology is returned to and de Rham's famous theorem is proved. (This is one place where treatment of a result in generality, for all differentiable manifolds and not just compact ones, actually provides a simpler and cleaner approach.)

Appendix B contains brief background material on "naive" set theory. The other appendices contain ancillary material referred to in the main text, usually in reference to an inessential matter.

There is much more material in this book than can be covered in a one-year course. Indeed, if everything is covered, there is enough for a two-year course. As a suggestion for a one-year course, one could start with Chapter II, assigning Section 1 as individual reading and then covering Sections 2 through 11. Then pick up Section 14 of Chapter I and continue with Chapter III, Sections 1 through 8, and possibly Section 9. Then take Chapter IV except for Section 12 and perhaps omitting some details about CW-complexes. Then cover Chapter V except for the last three sections. Finally, Chapter VI can be covered through Section 10. If there is time, coverage of Hopf's Theorem in Section 11 of Chapter V is recommended. Alternatively to the coverage of Chapter VI, one could cover as much of Chapter VII as is possible, particularly if there is not sufficient time to reach the duality theorems of Chapter VI.

Although I do make occasional historical remarks, I make no attempt at thoroughness in that direction. An excellent history of the subject can be found in Dieudonné [1]. That work is, in fact, much more than a history and deserves to be in every topologist's library.

Most sections of the book end with a group of problems, which are exercises for the reader. Some are harder, or require more "maturity," than others and those are marked with a ◆. Problems marked with a ◇ are those whose results are used elsewhere in the main text of the book, explicitly or implicitly.

Glen E. Bredon

Acknowledgments

It was perfect, it was rounded, symmetrical, complete, colossal.

MARK TWAIN

Unlike the object of Mark Twain's enthusiasm, quoted above (and which has no geometric connection despite the four geometric–topological adjectives), this book is far from perfect. It is simply the best I could manage. My deepest thanks go to Peter Landweber for reading the entire manuscript and for making many corrections and suggestions. Antoni Kosinski also provided some valuable assistance. I also thank the students in my course on this material in the spring of 1992, and previous years, Jin-Yen Tai in particular, for bringing a number of errors to my attention and for providing some valuable pedagogic ideas.

Finally, I dedicate this book to the memory of Deane Montgomery in deep appreciation for his long-term support of my work and of that of many other mathematicians.

Glen E. Bredon

Contents

Preface v

Acknowledgments ix

CHAPTER I
General Topology 1

1. Metric Spaces 1
2. Topological Spaces 3
3. Subspaces 8
4. Connectivity and Components 10
5. Separation Axioms 12
6. Nets (Moore–Smith Convergence) ☼ 14
7. Compactness 18
8. Products 22
9. Metric Spaces Again 25
10. Existence of Real Valued Functions 29
11. Locally Compact Spaces 31
12. Paracompact Spaces 35
13. Quotient Spaces 39
14. Homotopy 44
15. Topological Groups 51
16. Convex Bodies 56
17. The Baire Category Theorem 57

CHAPTER II
Differentiable Manifolds 63

1. The Implicit Function Theorem 63
2. Differentiable Manifolds 68
3. Local Coordinates 71
4. Induced Structures and Examples 72

5. Tangent Vectors and Differentials 76
6. Sard's Theorem and Regular Values 80
7. Local Properties of Immersions and Submersions 82
8. Vector Fields and Flows 86
9. Tangent Bundles 88
10. Embedding in Euclidean Space 89
11. Tubular Neighborhoods and Approximations 92
12. Classical Lie Groups ☼ 101
13. Fiber Bundles ☼ 106
14. Induced Bundles and Whitney Sums ☼ 111
15. Transversality ☼ 114
16. Thom–Pontryagin Theory ☼ 118

CHAPTER III
Fundamental Group 127

1. Homotopy Groups 127
2. The Fundamental Group 132
3. Covering Spaces 138
4. The Lifting Theorem 143
5. The Action of π_1 on the Fiber 146
6. Deck Transformations 147
7. Properly Discontinuous Actions 150
8. Classification of Covering Spaces 154
9. The Seifert–Van Kampen Theorem ☼ 158
10. Remarks on **SO**(3) ☼ 164

CHAPTER IV
Homology Theory 168

1. Homology Groups 168
2. The Zeroth Homology Group 172
3. The First Homology Group 172
4. Functorial Properties 175
5. Homological Algebra 177
6. Axioms for Homology 182
7. Computation of Degrees 190
8. CW-Complexes 194
9. Conventions for CW-Complexes 198
10. Cellular Homology 200
11. Cellular Maps 207
12. Products of CW-Complexes ☼ 211
13. Euler's Formula 215
14. Homology of Real Projective Space 217
15. Singular Homology 219
16. The Cross Product 220
17. Subdivision 223
18. The Mayer–Vietoris Sequence 228
19. The Generalized Jordan Curve Theorem 230
20. The Borsuk–Ulam Theorem 240
21. Simplicial Complexes 245

22. Simplicial Maps 250
23. The Lefschetz–Hopf Fixed Point Theorem 253

CHAPTER V
Cohomology 260

1. Multilinear Algebra 260
2. Differential Forms 261
3. Integration of Forms 265
4. Stokes' Theorem 267
5. Relationship to Singular Homology 269
6. More Homological Algebra 271
7. Universal Coefficient Theorems 281
8. Excision and Homotopy 285
9. de Rham's Theorem 286
10. The de Rham Theory of $\mathbf{CP}^n$ ☼ 292
11. Hopf's Theorem on Maps to Spheres ☼ 297
12. Differential Forms on Compact Lie Groups ☼ 304

CHAPTER VI
Products and Duality 315

1. The Cross Product and the Künneth Theorem 315
2. A Sign Convention 321
3. The Cohomology Cross Product 321
4. The Cup Product 326
5. The Cap Product 334
6. Classical Outlook on Duality ☼ 338
7. The Orientation Bundle 340
8. Duality Theorems 348
9. Duality on Compact Manifolds with Boundary 355
10. Applications of Duality 359
11. Intersection Theory ☼ 366
12. The Euler Class, Lefschetz Numbers, and Vector Fields ☼ 378
13. The Gysin Sequence ☼ 390
14. Lefschetz Coincidence Theory ☼ 393
15. Steenrod Operations ☼ 404
16. Construction of the Steenrod Squares ☼ 412
17. Stiefel–Whitney Classes ☼ 420
18. Plumbing ☼ 426

CHAPTER VII
Homotopy Theory 430

1. Cofibrations 430
2. The Compact-Open Topology 437
3. H-Spaces, H-Groups, and H-Cogroups 441
4. Homotopy Groups 443
5. The Homotopy Sequence of a Pair 445
6. Fiber Spaces 450
7. Free Homotopy 457
8. Classical Groups and Associated Manifolds 463

9. The Homotopy Addition Theorem 469
10. The Hurewicz Theorem 475
11. The Whitehead Theorem 480
12. Eilenberg–Mac Lane Spaces 488
13. Obstruction Theory ☼ 497
14. Obstruction Cochains and Vector Bundles ☼ 511

Appendices

App. A. The Additivity Axiom 519
App. B. Background in Set Theory 522
App. C. Critical Values 531
App. D. Direct Limits 534
App. E. Euclidean Neighborhood Retracts 536

Bibliography 541

Index of Symbols 545

Index 549

CHAPTER I
General Topology

A round man cannot be expected to fit in a square hole right away. He must have time to modify his shape.

MARK TWAIN

1. Metric Spaces

We are all familiar with the notion of distance in euclidean *n*-space: If **x** and **y** are points in $\mathbf{R}^n$ then

$$\operatorname{dist}(\mathbf{x}, \mathbf{y}) = \left(\sum_{i=1}^{n} (x_i - y_i)^2 \right)^{1/2}.$$

This notion of distance permits the definition of continuity of functions from one euclidean space to another by the usual ϵ–δ definition:

$$f: \mathbf{R}^n \to \mathbf{R}^k \text{ is continuous at } \mathbf{x} \in \mathbf{R}^n \text{ if, given } \epsilon > 0,$$
$$\exists \delta > 0 \ni \operatorname{dist}(\mathbf{x}, \mathbf{y}) < \delta \quad \Rightarrow \quad \operatorname{dist}(f(\mathbf{x}), f(\mathbf{y})) < \epsilon.$$

Although the spaces of most interest to us in this book are subsets of euclidean spaces, it is useful to generalize the notion of "space" to get away from such a hypothesis, because it would be very complicated to try to verify that spaces we construct are always of this type. In topology, the central notion is that of continuity. Thus it would usually suffice for us to treat "spaces" for which we can give a workable definition of continuity.

We could define continuity as above for any "space" which has a suitable notion of distance. Such spaces are called "metric spaces."

1.1. Definition. A *metric space* is a set X together with a function

$$\operatorname{dist}: X \times X \to \mathbf{R},$$

called a *metric*, such that the following three laws are satisfied:

(1) (positivity) $\operatorname{dist}(x, y) \geq 0$ with equality $\Leftrightarrow x = y$;
(2) (symmetry) $\operatorname{dist}(x, y) = \operatorname{dist}(y, x)$; and
(3) (triangle inequality) $\operatorname{dist}(x, z) \leq \operatorname{dist}(x, y) + \operatorname{dist}(y, z)$.

In a metric space X we define the "ϵ-ball," $\epsilon > 0$, about a point $x \in X$ to be

$$\boxed{B_\epsilon(x) = \{y \in X \mid \operatorname{dist}(x, y) < \epsilon\}.}$$

Also, a subset $U \subset X$ is said to be "open" if, for each point $x \in U$, there is an ϵ-ball about x completely contained in U. A subset is said to be "closed" if its complement is open. If $y \in B_\epsilon(x)$ and if $\delta = \epsilon - \operatorname{dist}(x, y)$ then $B_\delta(y) \subset B_\epsilon(x)$ by the triangle inequality. This shows that all ϵ-balls are open sets.

It turns out that, for metric spaces, continuity can be expressed completely in terms of open sets:

1.2. Proposition. *A function $f: X \to Y$ between metric spaces is continuous $\Leftrightarrow$ $f^{-1}(U)$ is open in X for each open subset U of Y.*

PROOF. If f is continuous and $U \subset Y$ is open and $f(x) \in U$ then there is an $\epsilon > 0$ such that $B_\epsilon(f(x)) \subset U$. By continuity, there is a $\delta > 0$ such that f maps the δ-ball about x into $B_\epsilon(f(x))$. This means that $B_\delta(x) \subset f^{-1}(U)$. This implies that $f^{-1}(U)$ is open.

Conversely, suppose $f(x) = y$ and that $\epsilon > 0$ is given. By hypothesis, $f^{-1}(B_\epsilon(y))$ is open and contains x. Therefore, by the definition of an open set, there is a $\delta > 0$ such that $B_\delta(x) \subset f^{-1}(B_\epsilon(y))$. It follows that if $\operatorname{dist}(x, x') < \delta$ then $f(x') \in B_\epsilon(y)$, and so $\operatorname{dist}(f(x), f(x')) < \epsilon$, proving continuity in the ϵ–δ sense. □

The only examples of metric spaces we have discussed are euclidean spaces and, of course, subsets of those. Even with those, however, there are other reasonable metrics:

$$\operatorname{dist}_2(\mathbf{x}, \mathbf{y}) = \sum_{i=1}^{n} |x_i - y_i|,$$

$$\operatorname{dist}_3(\mathbf{x}, \mathbf{y}) = \max(|x_i - y_i|).$$

It is not hard to verify, from the following proposition, that these three metrics give the same open sets, and so behave identically with respect to continuity (for maps into or out of them).

1.3. Proposition. *If* dist_1 *and* dist_2 *are metrics on the same set X which satisfy the hypothesis that for any point $x \in X$ and $\epsilon > 0$ there is a $\delta > 0$ such that*

$$\operatorname{dist}_1(x, y) < \delta \quad \Rightarrow \quad \operatorname{dist}_2(x, y) < \epsilon,$$

and

$$\operatorname{dist}_2(x, y) < \delta \quad \Rightarrow \quad \operatorname{dist}_1(x, y) < \epsilon,$$

then these metrics define the same open sets in X.

PROOF. The proof is an easy exercise in the definition of open sets and is left to the reader. □

PROBLEMS

1. Consider the set X of all continuous real valued functions on $[0, 1]$. Show that

$$\operatorname{dist}(f, g) = \int_0^1 |f(x) - g(x)|\, dx$$

defines a metric on X. Is this still the case if continuity is weakened to integrability?

2. -◇- If X is a metric space and x_0 is a given point in X, show that the function $f: X \to \mathbf{R}$ given by $f(x) = \operatorname{dist}(x, x_0)$ is continuous.

3. -◇- If A is a subset of a metric space X then define a real valued function d on X by $d(x) = \operatorname{dist}(x, A) = \inf\{\operatorname{dist}(x, y) \mid y \in A\}$. Show that d is continuous. (*Hint*: Use the triangle inequality to show that $|d(x_1) - d(x_2)| \le \operatorname{dist}(x_1, x_2)$.)

2. Topological Spaces

Although most of the spaces that will interest us in this book are metric spaces, or can be given the structure of metric spaces, we will usually only care about continuity of mappings and not the metrics themselves. Since continuity can be expressed in terms of open sets alone, and since some constructions of spaces of interest to us do not easily yield to construction of metrics on them, it is very useful to discard the idea of metrics and to abstract the basic properties of open sets needed to talk about continuity. This leads us to the notion of a general "topological space."

2.1. Definition. A *topological space* is a set X together with a collection of subsets of X called "*open*" sets such that:

(1) the intersection of two open sets is open;
(2) the union of any collection of open sets is open; and
(3) the empty set $\varnothing$ and whole space X are open.

Additionally, a subset $C \subset X$ is called "*closed*" if its complement $X - C$ is open.

Topological spaces are much more general than metric spaces and the range of difference between them and metric spaces is much wider than that between metric spaces and subspaces of euclidean space. For example, it is possible to talk about convergence of sequences of points in metric spaces with little difference from sequences of real numbers. Continuity of functions can be described in terms of convergence of sequences in metric spaces. One can also talk about convergence of sequences in general topological spaces but that no longer is adequate to describe continuity (as we shall see later). Thus it is necessary to exercise care in developing the theory of general topological spaces. We now begin that development, starting with some further basic definitions.

2.2. Definition. If X and Y are topological spaces and $f: X \to Y$ is a function, then f is said to be *continuous* if $f^{-1}(U)$ is open for each open set $U \subset Y$. A *map* is a continuous function.

Since closed sets are just the complements of open sets and since inverse images preserve complements (i.e., $f^{-1}(Y - B) = X - f^{-1}(B)$), it follows that a function $f: X \to Y$ is continuous $\Leftrightarrow f^{-1}(F)$ is closed for each closed set $F \subset Y$.

2.3. Definition. If X is a topological space and $x \in X$ then a set N is called a *neighborhood* of x in X if there is an open set $U \subset N$ with $x \in U$.

Note that a neighborhood is not necessarily an open set, and, even though one usually thinks of a neighborhood as "small," it need not be: the entire space X is a neighborhood of each of its points.

Note that the intersection of any two neighborhoods of x in X is a neighborhood of x, which follows from the axiom that the intersection of two open sets is open.

The intuitive notion of "smallness" of a neighborhood is given by the concept of a neighborhood basis at a point:

2.4. Definition. If X is a topological space and $x \in X$ then a collection $\mathbf{B}_x$ of subsets of X containing x is called a *neighborhood basis* at x in X if each neighborhood of x in X contains some element of $\mathbf{B}_x$ and each element of $\mathbf{B}_x$ is a neighborhood of x.

Neighborhood bases are sometimes convenient in proving functions to be continuous:

2.5. Definition. A function $f: X \to Y$ between topological spaces is said to be *continuous at* x, where $x \in X$, if, given any neighborhood N of $f(x)$ in Y, there is a neighborhood M of x in X such that $f(M) \subset N$.

Since $f(f^{-1}(N)) \subset N$, this is the same as saying that $f^{-1}(N)$ is a neighborhood of x, for each neighborhood N of $f(x)$. Clearly, this need only be checked for N belonging to some neighborhood basis at $f(x)$.

2.6. Proposition. *A function $f: X \to Y$ between topological spaces is continuous $\Leftrightarrow$ it is continuous at each point $x \in X$.*

PROOF. Suppose that f is continuous, i.e., that $f^{-1}(U)$ is open for each open $U \subset Y$. Let N be a neighborhood of $f(x)$ in Y and let U be an open set such that $f(x) \in U \subset N$ as guaranteed by the definition of neighborhood. Then $x \in f^{-1}(U) \subset f^{-1}(N)$ and $f^{-1}(U)$ is open. It follows that $f^{-1}(N)$ is a neighborhood of x. Thus f is continuous at x.

Conversely, suppose that f is continuous at each point and let $U \subset Y$ be an open set. For any $x \in f^{-1}(U)$, $f^{-1}(U)$ is then a neighborhood of x. Thus there exists an open set V_x in X with $x \in V_x \subset f^{-1}(U)$. Hence $f^{-1}(U)$ is the union of the sets V_x for x ranging over $f^{-1}(U)$. Since the union of any collection of open sets is open, it follows that $f^{-1}(U)$ is open. But U was an arbitrary open set in Y and, consequently, f is continuous. □

2.7. Definition. A function $f: X \to Y$ between topological spaces is called a *homeomorphism* if $f^{-1}: Y \to X$ exists (i.e., f is one–one and onto) and both f and f^{-1} are continuous. The notation $X \approx Y$ means that X is homeomorphic to Y.

Two topological spaces are, then, homeomorphic if there is a one–one correspondence between them as sets which also makes the open sets correspond. Homeomorphic spaces are considered as essentially the same. One of the main problems in topology is to find methods of deciding when two spaces are homeomorphic or not.

To describe a topological space it is not necessary to describe completely the open sets. This can often be done more simply using the notion of a "basis" for the topology:

2.8. Definition. If X is a topological space and **B** is a collection of subsets of X, then **B** is called a *basis* for the topology of X if the open sets are precisely the unions of members of **B**. (In particular, the members of **B** are open.) A collection **S** of subsets of X is called a *subbasis* for the topology of X if the set **B** of *finite* intersections of members of **S** is a basis.

Note that *any* collection **S** of subsets of any set X is a subbasis for *some* topology on X, namely, the topology for which the open sets are the arbitrary unions of the finite intersections of members of **S**. (The empty set and whole set X are taken care of by the convention that an intersection of an empty collection of sets is the whole set and the union of an empty collection of sets is the empty set.) Thus, to define a topology, it suffices to specify some collection of sets as a subbasis. The resulting topology is called the topology "generated" by this subbasis.

In a metric space the collection of ϵ-balls, for all $\epsilon > 0$, is a basis, So is the collection of ϵ-balls for $\epsilon = 1, \frac{1}{2}, \frac{1}{3}, \ldots$.

Here are some examples of topological spaces:

1. (Trivial topology.) Any set X with only the empty set and the whole set X as open.
2. (Discrete topology.) Any set X with all subsets being open.
3. Any set X with open sets being those subsets of X whose complements are finite, together with the empty set. (That is, the closed sets are finite sets and X itself.)

4. $X = \omega \cup \{\omega\}$ with the open sets being all subsets of ω together with complements of finite sets. (Here, ω denotes the set of natural numbers.)
5. Let X be any partially ordered set. For $\alpha \in X$ consider the one-sided intervals $\{\beta \in X \mid \alpha < \beta\}$ and $\{\beta \in X \mid \alpha > \beta\}$. The "order topology" on X is the topology generated by these intervals. The "strong order topology" is the topology generated by these intervals together with the complements of finite sets.
6. Let $X = I \times I$ where I is the unit interval $[0, 1]$. Give this the "dictionary ordering," i.e., $(x, y) < (s, t) \Leftrightarrow$ either $x < s$ or ($x = s$ and $y < t$). Let X have the order topology for this ordering.
7. Let X be the real line but with the topology generated by the "half open intervals" $[x, y)$. This is called the "half open interval topology."
8. Let $X = \Omega \cup \{\Omega\}$ be the set of ordinal numbers up to and including the least uncountable ordinal Ω; see Theorem B.28. Give it the order topology.

2.9. Definition. A topological space is said to be *first countable* if each point has a countable neighborhood basis.

2.10. Definition. A topological space is said to be *second countable* if its topology has a countable basis.

Note that all metric spaces are first countable. Some metric spaces are not second countable, e.g., the space consisting of any uncountable set with the metric $\operatorname{dist}(x, y) = 1$ if $x \neq y$, and $\operatorname{dist}(x, x) = 0$ (which yields the discrete topology).

Euclidean spaces are second countable since the ϵ-balls, with ϵ rational, about the points with all rational coordinates, is easily seen to be a basis.

2.11. Definition. A sequence $f_1, f_2, \ldots$ of functions from a topological space X to a metric space Y is said to *converge uniformly* to a function $f: X \to Y$ if, for each $\epsilon > 0$, there is a number n such that $i > n \Rightarrow \operatorname{dist}(f_i(x), f(x)) < \epsilon$ for all $x \in X$.

2.12. Theorem. *If a sequence $f_1, f_2, \ldots,$ of continuous functions from a topological space X to a metric space Y converges uniformly to a function $f: X \to Y$, then f is continuous.*

PROOF. Given $\epsilon > 0$, let n_0 be such that

$$n \geq n_0 \quad \Rightarrow \quad \operatorname{dist}(f(x), f_n(x)) < \epsilon/3 \qquad \text{for all} \quad x \in X.$$

Given a point x_0, the continuity of f_{n_0} implies that there is a neighborhood N of x_0 such that $x \in N \Rightarrow \operatorname{dist}(f_{n_0}(x), f_{n_0}(x_0)) < \epsilon/3$. Thus, for any $x \in N$ we have

$$\begin{aligned}\operatorname{dist}(f(x), f(x_0)) &\leq \operatorname{dist}(f(x), f_{n_0}(x)) + \operatorname{dist}(f_{n_0}(x), f_{n_0}(x_0)) + \operatorname{dist}(f_{n_0}(x_0), f(x_0)) \\ &< \epsilon/3 + \epsilon/3 + \epsilon/3 = \epsilon.\end{aligned}$$

□

2.13. Definition. A function $f: X \to Y$ between topological spaces is said to be *open* if $f(U)$ is open in Y for all open $U \subset X$. It is said to be *closed* if $f(C)$ is closed in Y for all closed $C \subset X$.

2.14. Definition. If X is a set and some condition is given on subsets of X which may or may not hold for any particular subset, then if there is a topology T whose open sets satisfy the condition, and such that, for any topology T' whose open sets satisfy the condition, then the T-open sets are also T'-open (i.e., $T \subset T'$), then T is called the *smallest* (or *weakest* or *coarsest*) topology satisfying the condition. If, instead, for any topology T' whose open sets satisfy the condition, any T'-open sets are also T-open, then T is called the *largest* (or *strongest* or *finest*) topology satisfying the condition.

The terms "weak" and "strong" are the oldest historically. However, they are used in some places to mean the opposite of the above meaning in general topology. Even some topology books disagree on their meaning. For this reason, the terms "coarse" and "fine" were introduced to rectify the confusion. They are metaphors for thinking of open sets as grains in a rock (the fewer grains, the coarser the rock). The terms "smallest" and "largest" were introduced for the same reason, and they are mathematically more precise as applied to the topologies as collections of open sets. We prefer the latter terms in general.

For example (see Section 13), if $f: X \to Y$ is a function and X is a topological space, then there is a largest topology on Y making f continuous, namely that topology having open sets $\{V \subset Y \mid f^{-1}(V) \text{ is open in } X\}$. There is also a smallest such topology, the trivial topology, but it is not very interesting. Also see Sections 8 and 13 for other examples of this concept.

If a topology is the largest one satisfying some given condition then usually (in fact, always) there is another condition for which the given topology is the smallest one satisfying the new condition. For example, the topology on Y, in the example of the previous paragraph, is the smallest topology satisfying the condition "for all spaces Z and all functions $g: Y \to Z$, $g \circ f$ continuous $\Rightarrow g$ continuous." Thus it is meaningless to argue whether a given topology is "weak" or "strong," etc., unless the defining condition is specified.

PROBLEMS

1. ⌖ Show that in a topological space X:
   (a) the union of two closed sets is closed;
   (b) the intersection of any collection of closed sets is closed; and
   (c) the empty set $\varnothing$ and whole space X are closed.

2. Consider the topology on the real line generated by the half open intervals $[x, y)$ *together with* those of the form $(x, y]$. Show that this coincides with the discrete topology.

3. Show that the space $\Omega \cup \{\Omega\}$ in the order topology cannot be given a metric consistent with its topology.

4. ⌖ If $f:X \to Y$ is a function between topological spaces, and $f^{-1}(U)$ is open for each open U in some subbasis for the topology of Y, show that f is continuous.

5. ⌖ Suppose that S is a set and that we are given, for each $x \in S$, a collection $\mathbf{N}(x)$ of subsets of S satisfying:
(1) $N \in \mathbf{N}(x) \Rightarrow x \in N$;
(2) $N, M \in \mathbf{N}(x) \Rightarrow \exists P \in \mathbf{N}(x) \ni P \subset N \cap M$; and
(3) $x \in S \Rightarrow \mathbf{N}(x) \neq \emptyset$.
Then show that there is a unique topology on S such that $\mathbf{N}(x)$ is a neighborhood basis at x, for each $x \in S$. (Thus a topology can be defined by the specification of such a collection of neighborhoods at each point.)

3. Subspaces

There are several techniques for producing new topological spaces out of old ones. The simplest is the passing to a "subspace," which is merely an arbitrary subset inheriting a topology from the mother space in a quite natural way.

3.1. Definition. If X is a topological space and $A \subset X$ then the *relative topology* or the *subspace topology* on A is the collection of intersections of A with open sets of X. With this topology, A is called a *subspace* of X.

The following propositions are all easy consequences of the definitions and the proofs are left to the reader:

3.2. Proposition. *If Y is a subspace of X then $A \subset Y$ is closed in $Y \Leftrightarrow A = Y \cap B$ for some closed subset B of X.* □

3.3. Proposition. *If X is a topological space and $A \subset X$ then there is a largest open set U with $U \subset A$. This set is called the "interior" of A in X and is denoted by* $\operatorname{int}(A)$. □

3.4. Proposition. *If X is a topological space and $A \subset X$ then there is a smallest closed set F with $A \subset F \subset X$. This set is called the "closure" of A in X and is denoted by $\bar{A}$.* □

If we need to specify the space in which a closure is taken (the X), we shall use the notation $\bar{A}^X$. A consequence of the following fact is that this notation need not be used very often:

3.5. Proposition. *If $A \subset Y \subset X$ then $\bar{A}^Y = \bar{A}^X \cap Y$. Thus, if Y is closed in X then $\bar{A}^Y = \bar{A}^X$.* □

3.6. Definition. If X is a topological space and $A \subset X$ then the *boundary* or *frontier* of A is defined to be $\partial A = \operatorname{bdry}(A) = \bar{A} \cap \overline{X - A}$.

3.7. Proposition. *If $Y \subset X$ then the set of intersections of Y with members of a basis of X is a basis of the relative topology of Y.* □

3.8. Proposition. *If X, Y, Z are topological spaces and Y is a subspace of X and Z is a subspace of Y, then Z is a subspace of X.* □

3.9. Proposition. *If X is a metric space and $A \subset X$ then $\bar{A}$ coincides with the set of limits in X of sequences of points in A.*

PROOF. If x is the limit of a sequence of points in A then any open set about x contains a point of A. Thus $x \notin \text{int}(X - A)$. Since $X - \text{int}(X - A) = \bar{A}$ (see the problems at the end of this section), $x \in \bar{A}$. Conversely, if $x \in \bar{A}$ and $n > 0$ is any integer, then $B_{1/n}(x)$ must contain a point in A because otherwise x would lie in $\text{int}(X - A)$. Take one such point and name it x_n. Then it follows immediately that $x = \lim(x_n)$ is a limit of a sequence of points in A. □

3.10. Definition. A subset A of a topological space X is called *dense* in X if $\bar{A} = X$. A subset A is said to be *nowhere dense* in X if $\text{int}(\bar{A}) = \varnothing$.

PROBLEMS

1. ⌖ Let X be a topological space and $A, B \subset X$.
   (a) Show that
   $$\text{int}(A) = \{a \in X \mid \exists U \text{ open} \ni a \in U \subset A\}$$
   and
   $$\bar{A} = \{x \in X \mid \forall U \text{ open with } x \in U, U \cap A \neq \varnothing\}.$$
   (b) Show that A is open $\Leftrightarrow A = \text{int}(A)$ and that A is closed $\Leftrightarrow A = \bar{A}$.
   (c) Show that $X - \text{int}(A) = \overline{X - A}$ and that $X - \bar{A} = \text{int}(X - A)$.
   (d) Show that $\text{int}(A \cap B) = \text{int}(A) \cap \text{int}(B)$ and that $\overline{A \cup B} = \bar{A} \cup \bar{B}$.
   (e) Show that
   $$\bigcap \text{int}(A_\alpha) \supset \text{int}\left(\bigcap A_\alpha\right) = \text{int}\left(\bigcap \text{int}(A_\alpha)\right),$$
   $$\bigcup \bar{A}_\alpha \subset \text{closure}\left(\bigcup A_\alpha\right) = \text{closure}\left(\bigcup \bar{A}_\alpha\right),$$
   $$\bigcup \text{int}(A_\alpha) \subset \text{int}\left(\bigcup A_\alpha\right),$$
   $$\bigcap \bar{A}_\alpha \supset \text{closure}\left(\bigcap A_\alpha\right),$$
   and give examples showing that these inclusions need not be equalities.
   (f) Show $A \subset B \Rightarrow [\bar{A} \subset \bar{B}$ and $\text{int}(A) \subset \text{int}(B)]$.

2. ⌖ For $A \subset X$, a topological space, show that X is the *disjoint* union of $\text{int}(A)$, $\text{bdry}(A)$, and $X - \bar{A}$.

3. ⌖ Show that a metric space is second countable $\Leftrightarrow$ it has a countable dense set (a countable set whose closure is the whole space). (Such a metric space is called "separable.")

4. -◇- Show that the union of two nowhere dense sets is nowhere dense.

5. A topological space X is said to be "irreducible" if, whenever $X = F \cup G$ with F and G closed, then either $X = F$ or $X = G$. A subspace is irreducible if it is so in the subspace topology. Show that if X is irreducible and $U \subset X$ is open, then U is irreducible.

6. A "Zariski space" is a topological space with the property that every descending chain $F_1 \supset F_2 \supset F_3 \supset \cdots$ of closed sets is eventually constant. Show that every Zariski space can be expressed as a finite union $X = Y_1 \cup Y_2 \cup \cdots \cup Y_n$ where the Y_i are closed and irreducible and $Y_i \not\subset Y_j$ for $i \neq j$. Also show that this decomposition is unique up to order.

7. Let X be the real line with the topology for which the open sets are $\emptyset$ together with the complements of finite subsets. Show that X is an irreducible Zariski space.

8. -◇- Let $X = A \cup B$, where A and B are closed. Let $f: X \to Y$ be a function. If the restrictions of f to A and B are both continuous then show that f is continuous.

4. Connectivity and Components

In a naively intuitive sense, a connected space is a space in which one can move from any point to any other point without jumps, Another way to view it intuitively is as the idea that the space does not fall into two or more pieces which are separated from one another. There are two ways of making these crude ideas precise and both of them will be important to us. One of them, called "connectivity," is the subject of this section, while the other, called "arcwise connectivity," is taken up in the problems at the end.

4.1. Definition. A topological space X is called *connected* if it is not the disjoint union of two nonempty open subsets.

4.2. Definition. A subset A of a topological space X is called *clopen* if it is both open and closed in X.

4.3. Proposition. *A topological space X is connected $\Leftrightarrow$ its only clopen subsets are X and $\emptyset$.* □

4.4. Definition. A *discrete valued map* is a map (continuous) from a topological space X to a discrete space D.

4.5. Proposition. *A topological space X is connected $\Leftrightarrow$ every discrete valued map on X is constant.*

PROOF. If X is connected and $d: X \to D$ is a discrete valued map and if $y \in D$ is in the range of d, then $d^{-1}(y)$ is clopen and nonempty and so must equal X, and so d is constant with only value y.

Conversely, if X is not connected then $X = U \cup V$ for some disjoint clopen sets U and V. Then the map $d: X \to \{0, 1\}$ which is 0 on U and is 1 on V is a nonconstant discrete valued map. □

4.6. Proposition. *If $f: X \to Y$ is continuous and X is connected, then $f(X)$ is connected.*

PROOF. Let $d: f(X) \to D$ be a discrete valued map. Then $d \circ f$ is a discrete valued map on X and hence must be constant. But that implies that d is constant, and hence that $f(X)$ is connected. □

4.7. Proposition. *If $\{Y_i\}$ is a collection of connected sets in a topological space X and if no two of the Y_i are disjoint, then $\bigcup Y_i$ is connected.*

PROOF. Let $d: \bigcup Y_i \to D$ be a discrete valued map. Let p, q be any two points in $\bigcup Y_i$. Suppose $p \in Y_i$ and $q \in Y_j$ and $r \in Y_i \cap Y_j$. Then, since d must be constant on each Y_i, we have $d(p) = d(r) = d(q)$. But p and q were completely arbitrary. Thus d is constant. □

4.8. Corollary. *The relation "p and q belong to a connected subset of X" is an equivalence relation.* □

4.9. Definition. The equivalence classes of the equivalence relation in Corollary 4.8 are called the *components* of X.

4.10. Proposition. *Components of space X are connected and closed. Each connected set is contained in a component. (Thus the components are "maximal connected subsets.") Components are either equal or disjoint, and fill out X.*

PROOF. The last statement follows from the fact that the components are equivalence classes of an equivalence relation. By definition, the component of X containing p is the union of all connected sets containing p, and that is connected by Proposition 4.7. This also implies that a connected set lies in a component. That a component is closed follows from the fact that the closure of a connected set is connected (left to the reader in the problems below). □

4.11. Proposition. *The statement "$d(p) = d(q)$ for every discrete valued map d on X" is an equivalence relation.* □

4.12. Definition. The equivalence classes of the relation in Proposition 4.11 are called the *quasi-components* of X.

4.13. Proposition. *Quasi-components of a space X are closed. Each connected set is contained in a quasi-component. (In particular, each component is contained in a quasi-component.) Quasi-components are either equal or disjoint, and fill out X.*

Proof. If $p \in X$ then the quasi-component containing it is just

$$\{q \in X \mid d(q) = d(p) \text{ for all discrete valued maps } d\}.$$

But this is

$$\bigcap \{d^{-1}(d(p)) \mid d \text{ a discrete valued map}\}$$

which is an intersection of closed sets and hence is closed. The rest is obvious. □

Problems

1. ◇ If A is a connected subset of the topological space X and if $A \subset B \subset \bar{A}$ then show that B is connected.
2. ◇ A space X is said to be "locally connected" if for each $x \in X$ and each neighborhood N of x, there is a connected neighborhood V of x with $V \subset N$. If X is locally connected, show that its components are open and equal its quasi-components.
3. ◇ Show that the unit interval $[0, 1]$ in the real number is connected. (*Hint*: Assume that $[0, 1] = U \cup V$, where U and V are disjoint nonempty open sets, and $1 \in V$. Consider $x = \sup(U)$. Show that $x < 1$ and derive a contradiction.)
4. Consider the subspace X of the unit square in the plane consisting of the vertical line segments $\{1/n\} \times [0, 1]$ for $n = 1, 2, 3, \ldots$, and the two points $(0, 0)$ and $(0, 1)$. Show that the latter two points are components of X but not quasi-components. Show that the two point set $\{(0, 0), (0, 1)\}$ is a quasi-component which is not connected.
5. ◇ A topological space X is said to be "arcwise connected" if for any two points p and q in X there exists a map $\lambda: [0, 1] \to X$ with $\lambda(0) = p$ and $\lambda(1) = q$. A space X is "locally arcwise connected" if every neighborhood of any point contains an arcwise connected neighborhood. An "arc component" is a maximal arcwise connected subset. Show that:
   (a) an arcwise connected space is connected;
   (b) a space is the disjoint union of its arc components;
   (c) an arc component of a space is contained in some component;
   (d) the arc components of a locally arcwise connected space are clopen, and coincide with the components;
   (e) the space with exactly two points p and q and open sets $\varnothing, \{p\}, \{p, q\}$ (only) is arcwise connected; and
   (f) the subspace of the plane consisting of $\{0\} \times [-1, 1] \cup \{(x, \sin(1/x)) \mid x > 0\}$ is connected but not arcwise connected.

5. Separation Axioms

The axioms defining a topological space are extremely general and weak. It should be no surprise that most spaces of interest will have further restrictions on them. We refer here not to structures like a metric, but to conditions completely describable in terms of the topology itself, i.e., in terms of the points and open sets. We begin with the so-called separation axioms.

5.1. Definition. The separation axioms:

(T_0) A topological space X is called a T_0-*space* if for any two points $x \neq y$ there is an open set containing one of them but not the other.

(T_1) A topological space X is called a T_1-*space* if for any two points $x \neq y$ there is an open set containing x but not y and another open set containing y but not x.

(T_2) A topological space X is called a T_2-*space* or *Hausdorff* if for any two points $x \neq y$ there are disjoint open sets U and V with $x \in U$ and $y \in V$.

(T_3) A T_1-space X is called a T_3-*space* or *regular* if for any point x and closed set F not containing x there are disjoint open sets U and V with $x \in U$ and $F \subset V$.

(T_4) A T_1-space X is called a T_4-*space* or *normal* if for any two disjoint closed sets F and G there are disjoint open sets U and V with $F \subset U$ and $G \subset V$.

Axiom T_0 simply says that points can be distinguished by the open sets in which they lie.

Axiom T_1 is the same as saying that one-point sets (singletons) are closed sets, because if we single out a point x and, for each different point y we take U_y to be an open set containing y but not x, then $X - \{x\} = \bigcup U_y$ is the union of open sets and so is open. Conversely, if $\{x\}$ is closed then the open set $X - \{x\}$ can be taken, in the axiom, as the open set containing any other point.

Axiom T_2 is the most important of these axioms and will be assumed in the majority of the text of this book. We shall see later that it essentially means that "limits" are unique.

5.2. Proposition. *A Hausdorff space is regular* $\Leftrightarrow$ *the closed neighborhoods of any point form a neighborhood basis of the point.*

PROOF. Suppose that X is regular, let $x \in V$, with V open, and put $C = X - V$. By regularity there are open sets U, W, with $x \in U$, $C \subset W$, and $U \cap W = \emptyset$. Then $X - W$ is closed, and we have $X - W \subset X - C = V$, so any neighborhood V of x contains a closed neighborhood $X - W$ of x, as was to be shown.

Conversely, suppose that every point has a closed neighborhood basis. Let $x \notin C$ with C closed and put $V = X - C$. By the assumption, there is an open set U with $\bar{U} \subset V = X - C$ and $x \in U$. Then $C \subset X - \bar{U}$, and $U \cap (X - \bar{U}) = \emptyset$. Thus X is regular. □

5.3. Corollary. *A subspace of a regular space is regular.*

PROOF. If $A \subset X$ is a subspace, just intersect a closed neighborhood basis in

X of a point $a \in A$ with A and you get a closed neighborhood basis of a in A. □

PROBLEMS

1. Give an example of a space that is not T_0, and an example of a T_0-space that is not T_1. (*Hint*: Spaces with only two points suffice.)
2. Show that a finite T_1-space is discrete.
3. Consider the set ω of natural numbers together with two other points named x, y. Put a partial ordering on this set which orders ω as usual and makes both x and y greater than any integer, but does not order x against y. Give this the strong order topology. Show it is T_1 but not Hausdorff.
4. Consider the space X whose point set is the plane but whose open sets are given by the basis consisting of the usual open sets in the plane together with the sets $\{(x, y) | x^2 + y^2 < a, y \neq 0\} \cup \{(0, 0)\}$ for all $a > 0$. Show that X is Hausdorff but not regular.
5. ✧ Show that a subspace of a Hausdorff space is Hausdorff.
6. ✧ Show that a Hausdorff space is normal $\Leftrightarrow$ for any sets U open and C closed with $C \subset U$ there is an open set V with $C \subset V \subset \bar{V} \subset U$.
7. Show that there is a smallest topology on the real numbers such that every singleton is closed. Which of the separation axioms does it satisfy?
8. Show that if a Zariski space (see Section 3, Problem 6) is Hausdorff then it is finite.
9. ◆ ✧ Show that a metric space is normal.

6. Nets (Moore–Smith Convergence) ☼

In metric spaces continuity of functions can be expressed in terms of the convergence of sequences. This is not true in general topological spaces. However, there is a generalization of sequences that does work and permits proofs of some things analogously to proofs using sequences in metric spaces. This can be of great help to the intuition. The generalization of a sequence is called a net, and we will develop this subject in this section. Although we will use this concept in proving a couple of important results in subsequent sections, those results will not be used in the main body of the book, and for that reason, this section can be skipped without serious harm to subsequent developments.

6.1. Definition. A *directed set* D is a partially ordered set such that, for any two elements α and β of D, there is a $\tau \in D$ with $\tau \geq \alpha$ and $\tau \geq \beta$.

6.2. Definition. A *net* in a topological space X is a directed set D together with a function $\Phi: D \to X$.

Note that a sequence is simply a net based on the natural numbers as indexing set.

6.3. Definition. If $\Phi: D \to X$ is a net in the topological space X and $A \subset X$ then we say that Φ is *frequently* in A if for any $\alpha \in D$ there is a $\beta \geq \alpha$ such that $\Phi(\beta) \in A$. It is said to be *eventually* in A if there is an $\alpha \in D$ such that $\Phi(\beta) \in A$ for all $\beta \geq \alpha$.

6.4. Definition. A net $\Phi: D \to X$ in a topological space is said to converge to $x \in X$ if, for every neighborhood $U \subset X$ of x, Φ is eventually in U.

Note that if a net Φ is eventually in two sets U and V then it is eventually in $U \cap V$. Also, this is impossible if $U \cap V = \emptyset$. This proves half of the following fact. The remainder of the proof constructs a net which is typical of the nets encountered with general topological spaces.

6.5. Proposition. *A topological space X is Hausdorff $\Leftrightarrow$ any two limits of any convergent net are equal. (Thus one can speak of* the *limit of a net in such a space.)*

Proof. The implication $\Rightarrow$ follows from the preceding discussion. Thus suppose that X is not Hausdorff, and that $x, y \in X$ are two points which cannot be separated by open sets. Consider the directed set whose elements are ordered pairs $\alpha = \langle U, V \rangle$ of open sets where $x \in U$ and $y \in V$ with the ordering $\langle U, V \rangle \geq \langle A, B \rangle \Leftrightarrow (U \subset A$ and $V \subset B)$. For any $\alpha = \langle U, V \rangle$, let $\Phi(\alpha)$ be some point in $U \cap V$. This defines a net Φ which we claim converges to both x and y.

To see this, let W be any neighborhood of x. We claim that Φ is eventually in W. In fact, take any open set V containing y and an open set U with $x \in U \subset W$ and let $\alpha = \langle U, V \rangle$. If $\beta = \langle A, B \rangle \geq \alpha$ then $A \subset U$ and $B \subset V$ so that $\Phi(\beta) \in A \cap B \subset U \subset W$, as claimed. Thus Φ converges to x. Similarly, it converges to y. □

Next we show that nets are "sufficient" to describe continuity.

6.6. Proposition. *A function $f: X \to Y$ between two topological spaces is continuous $\Leftrightarrow$ for every net Φ in X converging to $x \in X$, the net $f \circ \Phi$ in Y converges to $f(x)$.*

Proof. First suppose that f is continuous and let Φ be a net in X converging to x. Let V be any open set in Y containing $f(x)$ and put $U = f^{-1}(V)$, which is a neighborhood of x. By definition of convergence, Φ is eventually in U, and so $f \circ \Phi$ is eventually in V, and thus converges to $f(x)$.

Conversely, suppose that f is not continuous. Then there is an open set $V \subset Y$ such that $K = f^{-1}(V)$ is not open. Let $x \in K - \operatorname{int}(K)$. Consider the directed set consisting of open neighborhoods of x ordered by inclusion, i.e.,

$A \le B$ means $A \supset B$. For any such neighborhood A of x, A cannot be completely inside K, so we can choose a point $w_A \in A - K$. Define the net Φ by putting $\Phi(A) = w_A$. If N is any neighborhood of x and if $B \ge N$ (i.e., $B \subset N$) then $\Phi(B) = w_B \in B - K \subset N$, showing that Φ is eventually in N. Thus Φ converges to x. However $(f \circ \Phi)(A) \notin V$, for any A, so that $f \circ \Phi$ is not eventually in V, and thus does not converge to $f(x)$. □

Given a particular net $\Phi: D \to X$ let $x_\alpha = \Phi(\alpha)$, for $\alpha \in D$. Then it is common to speak of $\{x_\alpha\}$ as being the net in question. This notation makes discussion of nets similar to the notation commonly used with sequences. For example, one can phrase the condition in Proposition 6.6 as

$$f(\lim x_\alpha) = \lim(f(x_\alpha)).$$

6.7. Proposition. *If $A \subset X$ then $\bar{A}$ coincides with the set of limits of nets in A which converge in X.*

PROOF. If $x \in \bar{A}$ then any open neighborhood U of x must intersect A nontrivially. Thus we can base a net on this set of neighborhoods, ordered by inclusion and such points $x_U \in U \cap A$. This clearly converges to x. Conversely, if $\{x_\alpha\}$ is any net of points in A which converges to a point $x \in X$ then, by definition, this net is eventually in any given neighborhood of x. Thus any neighborhood of x contains a point in A and so $x \in \bar{A}$. (Here we are using Problem 1(a) of Section 3.) □

In the case of ordinary sequences, a subsequence can be thought of in two different ways: (1) by discarding elements of the sequence and renumbering, or (2) by composing the sequence, thought of as a function $\mathbf{Z}^+ \to X$, with a function $h: \mathbf{Z}^+ \to \mathbf{Z}^+$, such that $i > j \Rightarrow h(i) > h(j)$. The first of these turns out to be inadequate for nets in general spaces. For the second method, a little thought should convince the reader that the last condition of monotonicity of h is stronger than is necessary for the usual uses of subsequences. Modifying it leads to the more general notion of a "subnet," which we now define.

6.8. Definition. If D and D' are directed sets and $h: D' \to D$ is a function, then h is called *final* if, $\forall \delta \in D, \exists \delta' \in D' \ni (\alpha' \ge \delta' \Rightarrow h(\alpha') \ge \delta)$.

6.9. Definition. A *subnet* of a net $\mu: D \to X$, is the composition $\mu \circ h$ of μ with a final function $h: D' \to D$.

6.10. Proposition. *A net $\{x_\alpha\}$ is frequently in each neighborhood of a given point $x \in X \Leftrightarrow$ it has a subnet which converges to x.*

PROOF. Consider the directed set D' consisting of ordered pairs (α, U) where $\alpha \in D$, U is a neighborhood of x, and $x_\alpha \in U$, ordered by the D ordering and inclusion. If (α, U) and (β, V) are in D' then, since $\{x_\alpha\}$ is frequently in $U \cap V$, there is a $\gamma \geq \alpha, \beta$ with $x_\gamma \in U \cap V$. Thus $(\gamma, U \cap V) \in D'$ and $(\gamma, U \cap V) \geq (\alpha, U)$, (β, V), showing that D' is directed. Map $D' \to D$ by $(\alpha, U) \mapsto \alpha$. For any $\delta \in D$, we have $(\delta, X) \in D'$. Now $(\alpha, U) \geq (\delta, X)$ implies that $\alpha \geq \delta$, which means that $D' \to D$ is final, and so $\{x_{(\alpha,U)}\}$ is a subnet of $\{x_\alpha\}$. We claim that it converges to x. Let N be any neighborhood of x. By assumption, there is some $x_\beta \in N$. If $(\alpha, U) \geq (\beta, N)$ then $x_{(\alpha,U)} = x_\alpha \in U \subset N$. Consequently, $\{x_{(\alpha,U)}\}$ is eventually in N. The converse is immediate. □

Next we treat a powerful concept for nets which has no analogue for sequences.

6.11. Definition. A net in a set X is called *universal* if, for any $A \subset X$, the net is either eventually in A or eventually in $X - A$.

6.12. Proposition. *The composition of a universal net in X with a function $f: X \to Y$ is a universal net in Y.*

Proof. If $A \subset Y$ then the net is eventually in either $f^{-1}(A)$ or $X - f^{-1}(A)$ by definition. But $X - f^{-1}(A) = f^{-1}(Y - A)$ and it follows that the composed net is eventually in either A or $Y - A$, respectively. □

Except for somewhat trivial cases, the definition of a universal net may seem so strong that the reader may reasonably doubt the existence of universal nets. However:

6.13. Theorem. *Every net has a universal subnet.*

Proof. Let $\{x_\alpha | \alpha \in P\}$ be a net in X. Consider all collections $\mathbf{C}$ of subsets of X such that:

(1) $A \in \mathbf{C} \Rightarrow \{x_\alpha\}$ is frequently in A; and
(2) $A, B \in \mathbf{C} \Rightarrow A \cap B \in \mathbf{C}$.

For example, $\mathbf{C} = \{X\}$ is such a collection. Order the family of all such collections $\mathbf{C}$ by inclusion. The union of any simply ordered set of such collections is clearly such a collection, i.e., satisfies (1) and (2). By the Maximality Principle, there is a maximal such collection $\mathbf{C}_0$.

Let $P_0 = \{(A, \alpha) \in \mathbf{C}_0 \times P | x_\alpha \in A\}$ and order P_0 by

$$(B, \beta) \geq (A, \alpha) \quad \Leftrightarrow \quad B \subset A \text{ and } \beta \geq \alpha.$$

This gives a partial order on P_0 making P_0 into a directed set. Map $P_0 \to P$ by taking (A, α) to α. This is clearly final and thus defines a subnet we shall denote by $\{x_{(A,\alpha)}\}$. We claim that this subnet is universal.

Suppose S is any subset of X such that $\{x_{(A,\alpha)}\}$ is frequently in S, Then, for any $(A, \alpha) \in P_0$, there is a $(B, \beta) \geq (A, \alpha)$ in P_0 with $x_\beta = x_{(B,\beta)} \in S$. Then $B \subset A$, $\beta \geq \alpha$, and $x_\beta \in B$. Thus $x_\beta \in S \cap B \subset S \cap A$. We conclude that $\{x_\alpha\}$ is frequently in $S \cap A$, for any $A \in \mathbf{C}_0$. But then we can throw S and all the sets $S \cap A$, for $A \in \mathbf{C}_0$, into $\mathbf{C}_0$ and conditions (1) and (2) will still hold. By maximality, we must have $S \in \mathbf{C}_0$. If $\{x_{(A,\alpha)}\}$ were *also* frequently in $X - S$ then $X - S$ would be in $\mathbf{C}_0$, and so $\emptyset = S \cap (X - S)$ would be in $\mathbf{C}_0$, by (2), and this is contrary to (1). Thus we conclude that $\{x_{(A,\alpha)}\}$ is not frequently in $X - S$, and so is eventually in S.

We have shown that if $\{x_{(A,\alpha)}\}$ is frequently in a set S then, in fact, it is eventually in S. This implies that $\{x_{(A,\alpha)}\}$ is universal. □

Note that this proof uses the Axiom of Choice in the guise of the Maximality Principle. In fact, it can be shown that Theorem 6.13 is equivalent to the Axiom of Choice.

The following fact is immediate from the definitions:

6.14. Proposition. *A subnet of a universal net is universal.* □

PROBLEMS

1. Show that a sequence is a universal net if and only if it is eventually constant.
2. Consider the space $X = \Omega \cup \{\Omega\}$ of ordinals up to and including the first uncountable ordinal Ω with the order topology. Show explicitly that there is a net in Ω which converges to $\{\Omega\}$ but that there is no *sequence* which does so.
3. Prove Proposition 6.14.
4. ◆ Let H be a dense set in the topological space X and let $f: H \to Y$ be a map with Y regular. Let $g: X \to Y$ be a *function*. Suppose that for any net $\{h_\alpha\}$ in H with $h_\alpha \to x \in X$ we have $f(h_\alpha) \to g(x)$. Then show that $g: X \to Y$ is continuous. Also show that the condition of regularity on Y is needed by giving a counterexample without it.

7. Compactness

The notion of compactness is one of the most important ideas in mathematics. The reader has undoubtedly already met it in connection with some of the fundamental facts about the real numbers used in calculus.

7.1. Definition. A *covering* of a topological space X is a collection of sets whose union is X. It is an *open covering* if the sets are open. A *subcover* is a subset of this collection which still covers the space.

If $A \subset X$ then, for convenience, we sometimes use "cover A" for a collection of subsets of X whose union contains A.

7.2. Definition. A topological space X is said to be *compact* if every open covering of X has a finite subcover. (This is sometimes referred to as the *Heine–Borel property.*)

7.3. Definition. A collection C of sets has the *finite intersection property* if the intersection of any finite subcollection is nonempty.

The following fact is just a simple translation of the definition of compactness in terms of open sets to a statement about the (closed) complements of those sets:

7.4. Theorem. *A topological space X is compact $\Leftrightarrow$ for every collection of closed subsets of X which has the finite intersection property, the intersection of the entire collection is nonempty.* □

7.5. Theorem. *If X is a Hausdorff space, then any compact subset of X is closed.*

PROOF. Let $A \subset X$ be compact and suppose $x \in X - A$. For $a \in A$ let $a \in U_a$ and $x \in V_a$ be open sets with $U_a \cap V_a = \emptyset$. Now $A = \bigcup (U_a \cap A)$, which implies, by compactness of A, that there are $a_1, a_2, \ldots, a_n \in A$, such that $A \subset U_{a_1} \cup \cdots \cup U_{a_n} = U$. But $x \in V_{a_1} \cap \cdots \cap V_{a_n} = V$, which is open, and $U \cap V = \emptyset$. Thus $x \in V \subset X - U \subset X - A$ and V is open. Since this is true for any $x \in X - A$, we conclude that $X - A$ is open, and so A is closed. □

7.6. Theorem. *If X is compact and $f: X \to Y$ is continuous, then $f(X)$ is compact.*

PROOF. We may as well replace Y by $f(X)$ and so assume that f is onto. For any open cover of Y look at the inverse images of its sets and apply the compactness of X. □

7.7. Theorem. *If X is compact, and $A \subset X$ is closed, then A is compact.*

PROOF. Cover A by open sets in X, throw in the open set $X - A$ and apply the compactness of X. □

The following fact provides an easy way to check that certain constructions yield homeomorphisms, as we shall see:

7.8. Theorem. *If X is compact and Y is Hausdorff and $f: X \to Y$ is continuous, one–one, and onto, then f is a homeomorphism.*

PROOF. We are to show that f^{-1} is continuous. That is the same as showing that f is a closed mapping (takes closed sets to closed sets). But if $A \subset X$ is closed, then A is compact by Theorem 7.7, so $f(A)$ is compact by Theorem 7.6, whence $f(A)$ is closed by Theorem 7.5. □

7.9. Theorem. *The unit interval* $I = [0, 1]$ *is compact.*

PROOF. Let **U** be an open covering of I. Put

$$S = \{s \in I \mid [0, s] \text{ is covered by a finite subcollection of } \mathbf{U}\}.$$

Let b the least upper bound of S. Clearly S must be an interval of the form $S = [0, b)$ or $S = [0, b]$. In the former case, however, consider a set $U \in \mathbf{U}$ containing the point b. This set must contain an interval of the form $[a, b]$. But then we can throw U in with the hypothesized finite cover of $[0, a]$ to obtain a finite cover of $[0, b]$. Thus we must have that $S = [0, b]$ for some $b \in [0, 1]$. But if $b < 1$, then a similar argument shows that there is a finite cover of $[0, c]$ for some $c > b$, contradicting the choice of b. Thus $b = 1$ and we have found the desired finite cover of $[0, 1]$. □

Note, of course, that any finite closed interval $[a, b]$ of real numbers is homeomorphic to $[0, 1]$ and hence is also compact. Any closed subset of $[a, b]$ is then compact. By looking at the covering of any subset of **R** by the intervals $(-n, n)$, we see that a compact set in **R** must be bounded. Consequently, a subset of **R** is compact $\Leftrightarrow$ it is closed and bounded. The reader is cautioned not to think that this holds in all metric spaces; see Corollary 8.7 and Theorem 9.4.

7.10. Theorem. *A real valued map on a compact space assumes a maximum value.*

PROOF. If $f: X \to \mathbf{R}$ is continuous and X is compact then $f(X)$ is compact by Theorem 7.6. Thus $f(X)$ is closed and bounded. Thus $\sup(f(X))$ exists, is finite, and belongs to $f(X)$ since $f(X)$ is closed. □

7.11. Theorem. *A compact Hausdorff space is normal.*

PROOF. Suppose X is compact Hausdorff. We will first show that X is regular. For this, suppose C is a closed subset and $x \notin C$. Since X is Hausdorff, for any point $y \in C$ there are open sets U_y and V_y with $x \in U_y$, $y \in V_y$ and $U_y \cap V_y = \emptyset$. Since C is closed, it is compact, and the sets V_y cover it. Thus there are points $y_1, \ldots, y_n$, so that $C \subset V_{y_1} \cup \cdots \cup V_{y_n}$. If we put $U = U_{y_1} \cap \cdots \cap U_{y_n}$ and $V = V_{y_1} \cup \cdots \cup V_{y_n}$ then $x \in U, C \subset V$, and $U \cap V = \emptyset$ as desired. The remainder of the proof goes exactly the same way with C playing the role of x and the other closed set playing the role of C. □

The following notion is mainly of use for locally compact spaces X, Y (see Section 11), but makes sense for all topological spaces:

7.12. Definition. A map $f: X \to Y$ between topological spaces is said to be *proper* if $f^{-1}(C)$ is compact for each compact subset C of Y.

7.13. Theorem. *If* $f: X \to Y$ *is a* closed *map and* $f^{-1}(y)$ *is compact for each* $y \in Y$, *then* f *is proper.*

Proof. Let $C \subset Y$ be compact and let $\{U_\alpha | \alpha \in A\}$ be a collection of open sets whose union contains $f^{-1}(C)$. For any $y \in C$ there is a finite subset $A_y \subset A$ such that

$$f^{-1}(y) \subset \bigcup \{U_\alpha | \alpha \in A_y\}.$$

Put

$$W_y = \bigcup \{U_\alpha | \alpha \in A_y\}$$

and

$$V_y = Y - f(X - W_y),$$

which is open. Note that $f^{-1}(V_y) \subset W_y$ and $y \in V_y$. Since C is compact and is covered by the V_y, there are points $y_1, \ldots, y_n$ such that $C \subset V_{y_1} \cup \cdots \cup V_{y_n}$. Thus

$$\begin{aligned} f^{-1}(C) &\subset f^{-1}(V_{y_1}) \cup \cdots \cup f^{-1}(V_{y_n}) \subset W_{y_1} \cup \cdots \cup W_{y_n} \\ &= \bigcup \{U_\alpha | \alpha \in A_{y_i};\ i = 1, 2, \ldots, n\}, \end{aligned}$$

a finite union. □

7.14. Theorem. *For a topological space X the following are equivalent:*

(1) *X is compact.*
(2) *Every collection of closed subsets of X with the finite intersection property has a nonempty intersection.*
(3) *Every universal net in X converges.*
(4) *Every net in X has a convergent subnet.*

Proof. We have already handled the equivalence of (1) and (2). For the rest:

(1) ⇒ (3) Suppose $\{x_\alpha\}$ is a universal net that does not converge. Then given $x \in X$, there is an open neighborhood U_x of x such that x_α is not eventually in U_x. Then x_α is eventually in $X - U_x$ by definition of universal. That is, there is an index β_x such that $\alpha \geq \beta_x \Rightarrow x_\alpha \notin U_x$. Cover X by $U_{x_1} \cup \cdots \cup U_{x_n}$. Let $\alpha \geq \beta_{x_i}$ for all i. Then $x_\alpha \notin U_{x_i}$ for any i, which means that $x_\alpha \notin X$, an absurdity.

(3) ⇒ (4) is clear since every net has a universal subnet.

(4) ⇒ (2) Let $\mathbf{F} = \{C\}$ be a collection of closed sets with the finite intersection propety. We can throw in all finite intersections and so assume that $\mathbf{F}$ is closed under finite intersection. Then $\mathbf{F}$, ordered by $C \geq C' \Leftrightarrow C \subset C'$, is directed. For each $C \in \mathbf{F}$ let $x_C \in C$, defining a net. By assumption, there is a convergent subnet, given by a final map $f: D \to \mathbf{F}$, say. Thus, for $\alpha \in D, f(\alpha) \in \mathbf{F}$ and $x_{f(\alpha)} \in f(\alpha)$. Suppose $x_{f(\alpha)} \to x$. Let $C \in \mathbf{F}$. Then there is a $\beta \in D \ni \alpha \geq \beta \Rightarrow f(\alpha) \subset C$, and so $x_{f(\alpha)} \in f(\alpha) \subset C$. Since C is closed it follows from Proposition 6.7 that $x \in C$. Thus $x \in \bigcap \{C \in \mathbf{F}\}$, proving (2). □

Problems

1. Give a direct proof of (1) ⇒ (4) in Theorem 7.14 without use of universal nets.

2. ⌖ Let X be a compact space and let $\{C_\alpha | \alpha \in A\}$ be a collection of closed sets, closed with respect to finite intersections. Let $C = \bigcap C_\alpha$ and suppose that $C \subset U$ with U open. Show that $C_\alpha \subset U$ for some α.

3. Give an example showing that the hypothesis, in Theorem 7.13, that f is closed, cannot be dropped.

8. Products

Let X and Y be topological spaces. Then we can define a topology (called the "product topology") on $X \times Y$ by taking the collection of sets $U \times V$ to be a subbase, where $U \subset X$ and $V \subset Y$ are open. Since

$$U_1 \times V_1 \cap U_2 \times V_2 = (U_1 \cap U_2) \times (V_1 \cap V_2),$$

this is, in fact, a basis. Therefore the open sets are precisely the arbitrary unions of such "rectangles."

Similarly we can define a product topology on finite products $X_1 \times X_2 \times \cdots \times X_n$ of topological spaces.

For an infinite product $\times\{X_\alpha | \alpha \in A\}$, we define the product topology as the topology with a basis consisting of the sets $\times\{U_\alpha | \alpha \in A\}$ where the U_α are open and where we demand that $U_\alpha = X_\alpha$ for all but a finite number of α's. Note that the collection of sets of the form $U_\alpha \times \times\{X_\beta | \beta \neq \alpha\}$ is a subbasis for the product topology. This topology is also called the "Tychonoff topology."

8.1. Proposition. *The projections $\pi_X: X \times Y \to X$ and $\pi_Y: X \times Y \to Y$ are continuous, and the product topology is the smallest topology for which this is true. Similarly for the case of infinite products.*

PROOF. The subbasis last described consists of exactly those sets which must be open for the projections to be continuous, and the proposition is just expressing that. □

8.2. Proposition. *If X is compact then the projection $\pi_Y: X \times Y \to Y$ is closed.*

PROOF. Let $C \subset X \times Y$ be closed. We are to show that $Y - \pi_Y(C)$ is open. Let $y \notin \pi_Y(C)$, i.e., $\langle x, y\rangle \notin C$ for all $x \in X$. Then, for any $x \in X$, there are open sets $U_x \subset X$ and $V_x \subset Y$ such that $x \in U_x$, $y \in V_x$, and $(U_x \times V_x) \cap C = \emptyset$.

Since X is compact there are points $x_1, \ldots, x_n \in X$ such that $U_{x_1} \cup \cdots \cup U_{x_n} = X$. Let $V = V_{x_1} \cap \cdots \cap V_{x_n}$. Then

$$(X \times V) \cap C = (U_{x_1} \cup \cdots \cup U_{x_n}) \times (V_{x_1} \cap \cdots \cap V_{x_n}) \cap C = \emptyset.$$

Thus, $y \in V \subset Y - \pi_Y(C)$ and V is open. Since y was an arbitrary point of $Y - \pi_Y(C)$ it follows that this set is open, and so its complement $\pi_Y(C)$ is closed. □

8.3. Corollary. *If X is compact then $\pi_Y: X \times Y \to Y$ is proper.*

PROOF. This follows immediately from Theorem 7.13 and Proposition 8.2. □

8.4. Corollary. *If X and Y are both compact, then $X \times Y$ is compact.* □

8.5. Corollary (Tychonoff Theorem for Finite Products). *If the X_i are compact then $X_1 \times \cdots \times X_n$ is compact.* □

8.6. Corollary. *The cube $I^n \subset \mathbf{R}^n$ is compact.* □

8.7. Corollary. *A subspace of $\mathbf{R}^n$ is compact $\Leftrightarrow$ it is closed and bounded.*

PROOF. Let X be the subspace in question.

($\Rightarrow$) Since X is compact, it is closed. Cover X by the open balls of radius k about the origin, $k = 1, 2, \ldots$. Since this has, by hypothesis, a finite subcover, X must be in one of these balls, and hence is bounded.

($\Leftarrow$) If X is closed and bounded, then it is in some ball of radius k about the origin, which in turn is contained in $[-k, k] \times \cdots \times [-k, k]$ (n times), which is compact. Thus X is a closed subset of a compact set and so is compact by Theorem 7.7. □

8.8. Proposition. *A net in a product space $X = \times X_\alpha$ converges to the point $(\ldots, x_\alpha, \ldots) \Leftrightarrow$ its composition with each projection $\pi_\alpha: X \to X_\alpha$ converges to x_α.*

PROOF. This is an easy exercise in the definition of product spaces and of convergence of nets, which will be left to the reader. □

8.9. Theorem (Tychonoff). *The product of an arbitrary collection of compact spaces is compact.*

PROOF. Let $X = \times X_\alpha$ where the X_α are compact. Let $f: D \to X$ be a universal net in X. Then the composition $\pi_\alpha \circ f$ is also a universal net by Proposition 6.12. Therefore this composition converges, say to x_α by Theorem 7.14. But this means that the original net converges to the point whose αth coordinate is x_α by Proposition 8.8 and so X is compact by Theorem 7.14. □

Tychonoff's Theorem has the reputation of being difficult. So, how can we prove it with such ease here? The answer is that the entire difficulty has been subsumed in the results about universal nets. The basic facts about universal nets depend on the axiom of choice, and so it follows that so does the Tychonoff Theorem. In fact, it is known that the Tychonoff Theorem is equivalent to the axiom of choice. That is why we gave a separate treatment of the finite case, which does not depend on the axiom of choice. (Also, the finite case is all that is needed in the main body of this book.)

If X is a space and A is a set, the product of A copies of X is often denoted by X^A and can be thought of as the space of functions $f: A \to X$. In this context, Proposition 8.8 takes the following form:

8.10. Proposition. *A net $\{f_\alpha\}$ in X^A converges to $f \in X^A \Leftrightarrow \forall x \in X, f_\alpha(x) \to f(x)$. In particular,* $\lim(f_\alpha(x)) = (\lim f_\alpha)(x)$. □

When A also has a topology, the notation X^A is often used for the set of all *continuous* functions $f: A \to X$. In that context a topology is often used on this set that differs from the product topology. There are several useful topologies in particular circumstances, and so the context must indicate what topology, if any, is meant by this notation.

8.11. Definition. If X and Y are spaces, then their *topological sum* or *disjoint union* $X + Y$ is the set $X \times \{0\} \cup Y \times \{1\}$ with the topology making $X \times \{0\}$ and $Y \times \{1\}$ clopen and the inclusions $x \mapsto (x, 0)$ of $X \to X + Y$ and $y \mapsto (y, 1)$ of $Y \to X + Y$ homeomorphisms to their images. More generally, if $\{X_\alpha | \alpha \in A\}$ is an indexed family of spaces then their *topological sum* $+_\alpha X_\alpha$ is $\bigcup \{X_\alpha \times \{\alpha\} | \alpha \in A\}$ given the topology making each $X_\alpha \times \{\alpha\}$ clopen and each inclusion $x \mapsto (x, \beta)$ of $X_\beta \to +_\alpha X_\alpha$ a homeomorphism to its image $X_\beta \times \{\beta\}$.

In ordinary parlance, if X and Y are disjoint spaces, one regards $X + Y$ as $X \cup Y$ with the topology making X and Y open subspaces.

PROBLEMS

1. Let X and Y be metric spaces. Define a metric on $X \times Y$ by
$$\operatorname{dist}(\langle x_1, y_1 \rangle, \langle x_2, y_2 \rangle) = (\operatorname{dist}(x_1, x_2)^2 + \operatorname{dist}(y_1, y_2)^2)^{1/2}.$$
Show that the topology induced by this metric is the product topology.

2. Do the same as Problem 1 for the metric:
$$\operatorname{dist}(\langle x_1, y_1 \rangle, \langle x_2, y_2 \rangle) = \max\{\operatorname{dist}(x_1, x_2), \operatorname{dist}(y_1, y_2)\}.$$

3. ◇ For a collection of spaces Y_α show that a function $f: X \to \times \{Y_\alpha\}$ is continuous $\Leftrightarrow$ each composition $X \to \times \{Y_\alpha\} \to Y_\alpha$, with the projection, is continuous.

4. ◇ Show that an arbitrary product of Hausdorff spaces is Hausdorff. Also show that an arbitrary product of regular spaces is regular. (*Hint*: Use Proposition 5.2 for the latter.)

5. ◇ If X is a topological space, the "diagonal" of $X \times X$ is the subspace $\Delta = \{\langle x, x \rangle | x \in X\}$. Show that X is Hausdorff $\Leftrightarrow \Delta$ is closed in $X \times X$.

6. ◇ Let $f, g: X \to Y$ be two maps. If Y is Hausdorff then show that the subspace $A = \{x \in X | f(x) = g(x)\}$ is closed in X.

7. Give an alternative proof of Proposition 8.2 using nets.

8. ◆ Let A be an uncountable set. For each $\alpha \in A$ let $X_\alpha = \{0, 1\}$ with the discrete topology. Put $X = \times_{\alpha \in A} X_\alpha$. (That is, $X = \{0, 1\}^A$.) Let $p \in X$ be the point with all components $p_\alpha = 1$. Let $K = \{q \in X | q_\alpha = 0$ except for a countable number of $\alpha\}$.
   (a) Show that p does not have a countable neighborhood basis.
   (b) Show that there is no neighborhood basis for p simply ordered by inclusion.
   (c) Show that $\bar{K} = X$ but that if H is a countable subset of K then $\bar{H} \subset K$.
   (d) Give an explicit description of a net in K which converges to p.

9. ◆ ⌖ Show that a product of a family of connected spaces is connected. Do the same for arcwise connectivity.

9. Metric Spaces Again

In this section we discuss the central concept of "completeness" of a metric space, which says, intuitively, that sequences that should converge do, in fact, converge. We also show that certain topological conditions on a topological space suffice for the existence of a metric on that space consistent with the given topology.

9.1. Definition. A *Cauchy sequence* in a metric space is a sequence $x_1, x_2, x_3, \ldots$ such that $\forall \epsilon > 0, \exists N > 0 \ni n, m > N \Rightarrow \text{dist}(x_n, x_m) < \epsilon$.

9.2. Definition. A metric space X is called *complete* if every Cauchy sequence in X converges in X.

9.3. Definition. A metric space X is *totally bounded* if, for each $\epsilon > 0$, X can be covered by a finite number of ϵ-balls.

9.4. Theorem. *In a metric space X the following conditions are equivalent:*

(1) *X is compact.*
(2) *Each sequence in X has a convergent subsequence.*
(3) *X is complete and totally bounded.*

PROOF. (1) ⇒ (2) Let $\{x_n\}$ be a sequence. Suppose that x is not a limit of a subsequence. Then there is an open neighborhood U_x of x containing x_n for only a finite number of n. Since X can be covered by a finite number of the U_x, this contradicts the infinitude of indexes n.

(2) ⇒ (3) Let $\{x_n\}$ be a Cauchy sequence. It follows from (2) that some subsequence $x_{n_j} \to x$ for some $x \in X$. The triangle inequality then implies that $x_n \to x$ and hence X is complete. Now suppose that X is not covered by a finite number of ϵ-balls. Then one can choose points $x_1, x_2, \ldots$ such that $\text{dist}(x_i, x_j) > \epsilon$ for all $j < i$. It follows that the distance between any two of these points is greater than ϵ. Such a sequence can have no convergent subsequences, contrary to (2). So, in fact, X must be totally bounded.

(3) ⇒ (2) Let $\{x_n\}$ be an arbitrary sequence in X. Since X is totally bounded by assumption, it can be covered by a finite number of 1-balls. Thus some one of these 1-balls, say B_1, must contain x_n for an infinite number of n. Next, X, and hence B_1, can be covered by a finite number of $\frac{1}{2}$-balls and so one of these balls, say B_2, must be such that $B_1 \cap B_2$ contains x_n for an infinite number of n. Continuing in this way we can find, for $n = 1, 2, 3 \ldots$, a $(1/n)$-ball B_n such that $B_1 \cap B_2 \cap \cdots \cap B_n$ contains x_i for an infinite number of i. Thus we can choose a subsequence $\{x_{n_i}\}$ such that $x_{n_i} \in B_1 \cap \cdots \cap B_i$ for all i. If $i < j$

then it follows that x_{n_i} and x_{n_j} are both in B_i and hence $\text{dist}(x_{n_i}, x_{n_j}) < 1/i$. This implies that this subsequence is Cauchy and so it must converge by completeness.

(2) ⇒ (1) Suppose $\{U_\alpha | \alpha \in A\}$ is an open covering of X. Since X is totally bounded (by (2) ⇒ (3)), we can find a dense sequence of points $x_1, x_2, \ldots$ in X. For each x_i there is a positive integer n such that $B_{1/n}(x_i) \subset U_\alpha$ for some α. Denote one such U_α by $V_{n,i}$. Now, given $x \in X$, there is an n such that $B_{2/n}(x) \subset U_\alpha$ for some α. By density, there is also an i such that $\text{dist}(x_i, x) < 1/n$. Then $B_{1/n}(x_i) \subset B_{2/n}(x) \subset U_\alpha$ so that $V_{n,i}$ is defined. Thus, $x \in B_{1/n}(x_i) \subset V_{n,i}$. Therefore the $V_{n,i}$ cover X and this is a countable subcover of the original cover. Let us rename this countable subcover $\{V_1, V_2, \ldots\}$. If this has a finite subcover then we are done. If not then the closed sets

$$\begin{aligned} C_1 &= X - V_1, \\ C_2 &= X - (V_1 \cup V_2), \\ C_3 &= X - (V_1 \cup V_2 \cup V_3), \\ &\cdots \end{aligned}$$

are all nonempty. Also note that $C_1 \supset C_2 \supset C_3 \supset \cdots$. Choose $x_i \in C_i$ for each i. By our assumption, there is a convergent subsequence $x_{n_i} \to x$, say. Since $x_{n_i} \in C_n$ for all $n_i > n$, and C_n is closed, x must be in C_n, for all n. Thus

$$x \in \bigcap \{C_n\} = X - \bigcup \{V_n\} = \varnothing.$$

This contradiction completes the proof. □

It clearly would be desirable to know when a given topological space can be given the structure of a metric space, in which case the space is called "metrizable." There are several known theorems of this nature. We shall be content with giving one of the simpler criteria. This development will span the rest of this section.

9.5. Definition. A Hausdorff space X is said to be *completely regular*, or $T_{3\frac{1}{2}}$, if, for each point $x \in X$ and closed set $C \subset X$ with $x \notin C$, there is a map $f: X \to [0, 1]$ such that $f(x) = 0$ and $f \equiv 1$ on C.

By following such a function with a map $[0, 1] \to [0, 1]$ which is 0 on $[0, \frac{1}{2}]$ and stretches $[\frac{1}{2}, 1]$ onto $[0, 1]$, we see that the function f in Definition 9.5 can be taken so that it is 0 on a neighborhood of x.

9.6. Proposition. *Suppose X is a metric space. Define:*

$$\text{dist}'(x, y) = \begin{cases} 1 & \text{if } \text{dist}(x, y) > 1, \\ \text{dist}(x, y) & \text{if } \text{dist}(x, y) \le 1. \end{cases}$$

Then dist *and* dist′ *give rise to the same topology on X.*

Proof. It is clear that the topology only depends on the open ϵ-balls for small ϵ, and these are the same in the two metrics. □

9.7. Proposition. *Let $X_i, i = 1, 2, 3, \ldots,$ be a metric space with metric bounded by 1 (see Proposition 9.6). Define a metric on $\times\{X_i\}$ by $\operatorname{dist}\{x, y\} = \sum_i \operatorname{dist}(x_i, y_i)/2^i$, where x_i is the ith coordinate of x, etc. Then this metric gives rise to the product topology.*

PROOF. Let X denote the product space with the product topology, and X' the same set with the metric topology. By Problem 3 of Section 8, to show that $X' \to X$ is continuous, it suffices to show that its composition with the projection to each X_i is continuous. But this projection decreases distance and then multiplies it by the constant 2^i and that clearly implies continuity. For the converse, it suffices to show that for any point $x \in X$, the ϵ-ball about x contains a neighborhood of x in the product topology. Recall that

$$B_\epsilon(x) = \left\{ y \,\middle|\, \sum_i (\operatorname{dist}(x_i, y_i)/2^i) < \epsilon \right\}.$$

Let n be so large that $2^{-n} < \epsilon/4$ and then let $y_i \in X_i$ be such that $\operatorname{dist}(x_i, y_i) < \epsilon/2$ for $i = 1, 2, \ldots, n-1$ and arbitrary for $i \geq n$. Then we compute

$$\begin{aligned}\operatorname{dist}(x, y) &= \sum_{i=1}^{n-1} \operatorname{dist}(x_i, y_i)/2^i + \sum_{i=n}^{\infty} \operatorname{dist}(x_i, y_i)/2^i \\ &< \sum_{i=1}^{n-1} \epsilon/2^{i+1} + \tfrac{1}{4}\epsilon\left(1 + \frac{1}{2} + \frac{1}{2^2} + \cdots\right) \\ &< \epsilon/2 + \epsilon/2 = \epsilon.\end{aligned}$$

Thus

$$x \in B_{\epsilon/2}(x_1) \times \cdots \times B_{\epsilon/2}(x_{n-1}) \times X_n \times X_{n+1} \times \cdots \subset B_\epsilon(x)$$

and the middle term is a basic open set in the product topology. □

9.8. Lemma. *Suppose that X is Hausdorff and that $f_i: X \to [0, 1]$ are maps $(i = 1, 2, 3, \ldots)$ such that, for any point $x \in X$ and any closed set $C \subset X$ with $x \notin C$, there is an index i such that $f_i(x) = 0$ and $f_i \equiv 1$ on C. Define $f: X \to \times\{[0, 1] \mid i = 1, 2, 3, \ldots\}$ by $f(x) = \times\{f_i(x) \mid i = 1, 2, 3, \ldots\}$. Then f is an embedding, i.e., a homeomorphism onto its image.*

PROOF. f is continuous by Problem 3 of Section 8. It is also clear that f is one–one (but not onto). Thus it suffices to show that: $C \subset X$ closed $\Rightarrow f(C)$ is closed in $f(X)$. Suppose we have a sequence $c_i \in C$ such that $f(c_i) \to f(x)$. It then suffices to show that $x \in C$. If not, then there is an index i such that $f_i(x) = 0$ and $f_i \equiv 1$ on C. Then $1 = f_i(c_n) \to f_i(x) = 0$ and this contradiction concludes the proof. □

9.9. Lemma. *Suppose that X is a second countable and completely regular space and let $\mathbf{S}$ be a countable basis for the open sets. For each pair $U, V \in \mathbf{S}$ with $\bar{U} \subset V$, select a map $f: X \to [0, 1]$ which is 0 on U and 1 on $X - V$, provided such a function exists. Call this set of maps $\mathbf{F}$, possibly empty, and note that*

F *is countable. Then for each* $x \in X$ *and each closed set* $C \subset X$ *with* $x \notin C$*, there is an* $f \in \mathbf{F}$ *with* $f \equiv 0$ *on a neighborhood of* x *and* $f \equiv 1$ *on* C.

PROOF. The whole point of the lemma is, of course, that the map f can be chosen from the previously defined countable collection **F**. Given $x \notin C$ as stated, we can find a $V \in \mathbf{S}$ with $x \in V \subset X - C$ (by definition of a basis). Since X is completely regular we can find a map $g: X \to [0, 1]$ which is 0 at x and 1 on $X - V$. As remarked below Definition 9.5, this can be assumed to be 0 on a *neighborhood* of x. This contains a neighborhood $U \in \mathbf{S}$ and so we have provided a triple U, V, g satisfying the initial requirements in the lemma. By assumption, this g can be replaced by another map $f \in \mathbf{F}$ with the same properties and this f clearly satisfies the final requirements. □

9.10. Theorem (Urysohn Metrization Theorem). *If a space* X *is second countable and completely regular then it is metrizable.*

PROOF. Find a countable family **F** of functions satisfying Lemma 9.9. Apply Lemma 9.8 to obtain an embedding of X into a countable (!) product of unit intervals. Finally, apply Proposition 9.7 to see that this countable product of intervals, and hence X, is metrizable. □

The following lemma will be useful to us later on in the book. The diameter, $\text{diam}(A)$, of a subset A of a metric space is $\sup\{\text{dist}(p, q) \mid p, q \in A\}$.

9.11. Lemma (Lebesgue Lemma). *Let* X *be a compact metric space and let* $\{U_\alpha\}$ *be an open covering of* X. *Then there is a* $\delta > 0$ *(a "Lebesgue number" for the covering) such that* $(A \subset X, \text{diam}(A) < \delta) \Rightarrow A \subset U_\alpha$ *for some* α.

PROOF. For each $x \in X$ there is an $\epsilon(x) > 0$ such that $B_{2\epsilon(x)}(x) \subset U_\alpha$ for some α. Then X is covered by a finite number of the balls $B_{\epsilon(x)}(x)$, say for $x = x_1, \ldots, x_n$. Define $\delta = \min\{\epsilon(x_i) \mid i = 1, \ldots, n\}$. Suppose $\text{diam}(A) < \delta$ and pick a point $a_0 \in A$. Then there is an index $1 \leq i \leq n$ such that $\text{dist}(a_0, x_i) < \epsilon(x_i)$. If $a \in A$, then $\text{dist}(a, a_0) < \delta \leq \epsilon(x_i)$. By the triangle inequality, $\text{dist}(a, x_i) < 2\epsilon(x_i)$. Thus $A \subset B_{2\epsilon(x_i)}(x_i) \subset U_\alpha$ for some α. □

PROBLEMS

1. Show that a countable product of copies of the real line is metrizable.
2. ⌖ Show that a subspace of a completely regular space is completely regular.
3. Let X be a metric space. If $\{x_n\}$ and $\{y_n\}$ are Cauchy sequences in X such that $\text{dist}(x_n, y_n) \to 0$ then call $\{x_n\}$ and $\{y_n\}$ "equivalent." Let Y be the set of equivalence classes $[\{x_n\}]$ of Cauchy sequences $\{x_n\}$ in X. Give Y the metric

$$\text{dist}([\{x_n\}], [\{y_n\}]) = \lim \text{dist}(x_n, y_n).$$

   (a) Show that this *is* a metric on Y.

(b) Show that the function $f: X \to Y$ given by $x \mapsto [\{x\}]$ is an isometric embedding of X as a dense subspace of Y. ("Isometric" means "preserving distance.")
(c) Show that Y is complete. (It is called the "completion" of X.)
(d) If $g: X \to Z$ is an isometry (into) and Z is complete then show that there is a unique factorization $X \xrightarrow{f} Y \xrightarrow{h} Z$ of g with h an isometry.
(e) If $g(X)$, in part (d), is dense in Z then show that h is onto.

4. Show that a completely regular space is regular.

5. ◆ Show that an uncountable product of unit intervals is not first countable and hence is not metrizable.

10. Existence of Real Valued Functions

In the metrization theorem of the last section, we gave conditions for metrizability that included complete regularity of the space. This relies on knowing about the existence of sufficiently many, in some sense, continuous real valued functions on the space. That leaves open the question of finding purely *topological* assumptions that will guarantee such functions, and that is what we are going to address in this section.

10.1. Lemma. *Suppose that, on a topological space X, we are given, for each dyadic rational number $r = m/2^n$ $(0 \le m \le 2^n)$, an open set U_r such that $r < s \Rightarrow \bar{U}_r \subset U_s$. Then the function $f: X \to \mathbf{R}$ defined by*

$$f(x) = \begin{cases} \inf\{r \mid x \in U_r\} & \text{if } x \in U_1, \\ 1 & \text{if } x \notin U_1, \end{cases}$$

is continuous.

PROOF. Note that, for r dyadic:

$$\begin{array}{llll} f(x) < r \Rightarrow x \in U_r & \text{hence} & f(x) \ge r \Leftarrow x \notin U_r, & \\ f(x) \le r \Leftarrow x \in U_r & \text{hence} & f(x) > r \Rightarrow x \notin U_r & \Rightarrow x \in X - U_r. \end{array}$$

Thus, for α real,

$$f^{-1}(-\infty, \alpha) = \{x \mid f(x) < \alpha\} = \bigcup\{U_r \mid r < \alpha\}$$

which is open, and

$$f^{-1}(\beta, \infty) = \{x \mid f(x) > \beta\} = \bigcup\{X - U_r \mid r > \beta\} = \bigcup\{X - \bar{U}_s \mid s > \beta\}$$

which is also open. Since these half infinite intervals give a subbasis for the topology of $\mathbf{R}$, f is continuous. (See Problem 4 of Section 2.) □

10.2. Lemma (Urysohn's Lemma). *If X is normal and $F \subset U$ where F is closed and U is open, then there is a map $f: X \to [0, 1]$ which is 0 on F and 1 on $X - U$.*

Proof. Put $U_1 = U$ and use normality to find $F \subset U_0$, $\bar{U}_0 \subset U_1$,

$$\bar{U}_0 \subset U_{1/2} \quad \text{and} \quad \bar{U}_{1/2} \subset U_1,$$
$$\bar{U}_0 \subset U_{1/4} \quad \text{and} \quad \bar{U}_{1/4} \subset U_{1/2} \quad \text{and} \quad \bar{U}_{1/2} \subset U_{3/4} \quad \text{and} \quad \bar{U}_{3/4} \subset U_1,$$

and so on. Apply Lemma 10.1. □

10.3. Corollary. *Normality ⇒ Complete Regularity.* □

10.4. Theorem (Tietze Extension Theorem). *Let X be normal and $F \subset X$ be closed and let $f: F \to \mathbf{R}$ be continuous. Then there is a map $g: X \to \mathbf{R}$ such that $g(x) = f(x)$ for all $x \in F$. Moreover, it can be arranged that*

$$\sup_{x\in F} f(x) = \sup_{x\in X} g(x) \quad \textit{and} \quad \inf_{x\in F} f(x) = \inf_{x\in X} g(x).$$

Proof. First let us take the case in which f is bounded. Without loss of generality, we can assume $0 \le f(x) \le 1$ with infimum 0 and supremum 1. By the Urysohn Lemma (Lemma 10.2), there exists a function $g_1: X \to [0, \frac{1}{3}]$ such that

$$g_1(x) = \begin{cases} 0 & \text{if } x \in F \text{ and } f(x) \le \frac{1}{3}, \\ \frac{1}{3} & \text{if } x \in F \text{ and } f(x) \ge \frac{2}{3}. \end{cases}$$

Put $f_1 = f - g_1$ and note that $0 \le f_1(x) \le \frac{2}{3}$ for all $x \in F$.

Repeating this, find $g_2: X \to [0, \frac{1}{3}\cdot\frac{2}{3}]$ such that

$$g_2(x) = \begin{cases} 0 & \text{if } x \in F \text{ and } f_1(x) \le \frac{1}{3}\cdot\frac{2}{3}, \\ \frac{1}{3} \times \frac{2}{3} & \text{if } x \in F \text{ and } f_1(x) \ge \frac{2}{3}\cdot\frac{2}{3}. \end{cases}$$

Put $f_2 = f_1 - g_2$ and note that $0 \le f_2(x) \le (\frac{2}{3})^2$ for all $x \in F$.

For the inductive step, suppose we have defined a function f_n with $0 \le f_n(x) \le (\frac{2}{3})^n$ for $x \in F$. Then find $g_{n+1}: X \to [0, (\frac{1}{3})(\frac{2}{3})^n]$ such that

$$g_{n+1}(x) = \begin{cases} 0 & \text{if } x \in F \text{ and } f_n(x) \le (\frac{1}{3})(\frac{2}{3})^n, \\ (\frac{1}{3})(\frac{2}{3})^n & \text{if } x \in F \text{ and } f_n(x) \ge (\frac{2}{3})(\frac{2}{3})^n. \end{cases}$$

Put $f_{n+1} = f_n - g_{n+1}$.

Now put $g(x) = \sum g_n(x)$. This series converges uniformly since $0 \le g_n(x) \le (\frac{1}{3})(\frac{2}{3})^{n-1}$. Thus g is continuous, by Theorem 2.12.

For $x \in F$ we have

$$f - g_1 = f_1,$$
$$f_1 - g_2 = f_2,$$
$$\cdots$$

By adding and canceling we get

$$f - (g_1 + g_2 + \cdots + g_n) = f_n \quad \text{and} \quad 0 \le f_n(x) \le (\tfrac{2}{3})^n,$$

and taking the limit gives that $g(x) = f(x)$ on F. Clearly the bounds are also correct.

Now we consider the unbounded cases:

Case I: f is unbounded in both directions.
Case II: f is bounded below by a.
Case III: f is bounded above by b.

Let h be a homeomorphism:

$$\begin{aligned} (-\infty, \infty) &\to (0,1) && \text{in Case I,} \\ [a, \infty) &\to [0,1) && \text{in Case II,} \\ (-\infty, b] &\to (0,1] && \text{in Case III.} \end{aligned}$$

Then $h \circ f$ is bounded by 0, 1 and we can extend it to g_1 say. If we can arrange that $g_1(x)$ is never 0 (resp. 1) if $h \circ f$ is never 0 (resp. 1) then $g = h^{-1} \circ g_1$ would be defined and would extend f.

Thus put

$$\begin{aligned} C &= \{x \mid g_1(x) = 0 \text{ or } 1\} && \text{in Case I,} \\ C &= \{x \mid g_1(x) = 1\} && \text{in Case II,} \\ C &= \{x \mid g_1(x) = 0\} && \text{in Case III.} \end{aligned}$$

Then C is closed and $C \cap F = \emptyset$, so there exists a function $k: X \to [0,1]$ such that $k \equiv 0$ on C and $k \equiv 1$ on F. Put $g_2 = k \cdot g_1 + (1-k) \cdot \frac{1}{2}$. Then g_2 is always between g_1 and $\frac{1}{2}$ with $g_2 \neq g_1$ on C. Also, $g_2 = g_1 = h \circ f$ on F. Thus $g = h^{-1} \circ g_2$ extends f in the desired manner. □

Problems

1. If X is a compact Hausdorff space then show that its quasi-components are connected (and hence that its quasi-components coincide with its components). [*Hint*: If C is a quasi-component, let $C = \bigcap C_\alpha$ where the C_α are the clopen sets containing C. If C is disconnected, then $C = A \cup B$, $A \cap B = \emptyset$, A, B closed. Let $f: X \to [0,1]$ be 0 on A and 1 on B. Put $U = f^{-1}([0, \frac{1}{2}))$ and apply Problem 2 of Section 7.]

2. ⌖ If F is a closed subspace of the normal space X then show that any map $F \to \mathbf{R}^n$ can be extended to X.

11. Locally Compact Spaces

There are many spaces, the most important being euclidean spaces, which are not compact but which contain enough compact subspaces to be important for many properties of the space itself. One class of such spaces is the subject of this section.

11.1. Definition. A topological space is said to be *locally compact* if every point has a compact neighborhood.

11.2. Theorem. *If X is a locally compact Hausdorff space then each neighborhood of a point $x \in X$ contains a compact neighborhood of x. (That is, the compact neighborhoods of x form a neighborhood basis at x.) In particular, X is completely regular.*

PROOF. Let C be a compact neighborhood of x and U an arbitrary neighborhood of x. Let $V \subset C \cap U$ be open with $x \in V$. Then $\bar{V} \subset C$ is compact Hausdorff and therefore regular. Thus there exists a neighborhood $N \subset V$ of x in C which is closed in $\bar{V}$ and hence closed in X. Since N is closed in the compact space C, it is compact by Theorem 7.7. Since N is a neighborhood of x in $\bar{V}$ and since $N = N \cap V$, N is a neighborhood of x in the open set V and hence in X. □

11.3. Theorem. *Let X be a locally compact Hausdorff space. Put $X^+ = X \cup \{\infty\}$ where ∞ just represents some point not in X. Define an open set in X^+ to be either an open set in $X \subset X^+$ or $X^+ - C$ where $C \subset X$ is compact. Then this defines a topology on X^+ which makes X^+ into a compact Hausdorff space called the "one-point compactification" of X. Moreover, this topology on X^+ is the only topology making X^+ a compact Hausdorff space with X as a subspace.*

PROOF. The whole space X^+ and $\emptyset$ are clearly open. If $V \subset X$ is open and $U = X^+ - C$ with $C \subset X$ compact then $U \cap V = V - C$ which is open in X (C being closed in X by Theorem 7.5). The other cases of an intersection of two open sets are trivial.

For arbitrary unions of open sets, let $U = \bigcup\{U_\alpha\}$. If all the U_α are open subsets of X then the union is clearly open. If some $U_\beta = X^+ - C$ then $X^+ - U = \bigcap\{X^+ - U_\alpha\} = C \cap (\bigcap\{X - U_\alpha | \alpha \neq \beta\})$ which is closed in C and therefore compact. Thus, this is a topology.

Suppose that $\{U_\alpha\}$ is an open cover of X^+. One of these sets, say U_β, contains $\{\infty\}$. Then $X - U_\beta$ is compact and hence is covered by a finite subcollection of the other U_α. Therefore X^+ is compact.

To see that X^+ is Hausdorff, it clearly suffices to separate ∞ from any point $x \in X$. Let V be an open neighborhood of x in X such that $\bar{V} \subset X$ is compact. Then $x \in V$ and $\infty \in X - \bar{V}$ provide the required separation.

For uniqueness, let $U \subset X^+$ be an open set in some such topology. Then $C = X^+ - U$ is closed and therefore compact. If $C \subset X$ then U is open in the described topology. If $C \not\subset X$ then $U \subset X$ and must be open in X since X is a subspace. Thus, again U is open in the described topology. It remains to show that we are forced to take the described open sets as open. Since X is a subspace, if $U \subset X$ is open in X then $U = U' \cap X$ for some U' open in X^+. But X is an open subset of X^+ since points are closed in a Hausdorff space, so $U = U' \cap X$ is open in X^+. Next, if C is compact in X then it is compact in X^+, since compactness does not depend on the containing space, and thus C is closed in X^+. It follows that $X^+ - C$ is open in X^+. □

Note that if X is already compact, then ∞ is an isolated point (clopen) in X^+ and X is also clopen in X^+.

11.4. Theorem. *Suppose that X and Y are locally compact, Hausdorff spaces and that $f: X \to Y$ is continuous. Then f is proper $\Leftrightarrow f$ extends to a continuous $f^+: X^+ \to Y^+$ by setting $f^+(\infty_X) = \infty_Y$.*

PROOF. $\Rightarrow$: f^+ exists as a function, so it suffices to check continuity of it. Suppose $U \subset Y^+$ is open. In case $U \subset Y$ then $(f^+)^{-1}(U) = f^{-1}(U)$ is open. In the other case, $U = Y^+ - C$ with $C \subset Y$ compact. Then $(f^+)^{-1}(U) = X^+ - f^{-1}(C)$ is open in X^+ since $f^{-1}(C)$ is compact, and therefore closed, by properness.

$\Leftarrow$: If f^+ exists then $(f^+)^{-1}(\infty_Y) = \{\infty_X\}$ and thus $(f^+)^{-1}(Y) = X$. If $C \subset Y$ is compact then it is closed and so $f^{-1}(C)$ is closed in X^+ and hence is compact and is contained in X. Thus f is proper. □

11.5. Proposition. *If $f: X \to Y$ is a proper map between locally compact Hausdorff spaces, then f is closed.*

PROOF. There is an extension $f^+: X^+ \to Y^+$. If $F \subset X$ is closed in X then $F \cup \{\infty\}$ is closed in X^+ and hence compact. Consequently, $f^+(F \cup \{\infty\})$ is compact by Theorem 7.6 and hence closed in Y^+ by Theorem 7.5. But then $f(F) = f^+(F \cup \{\infty\}) \cap Y$ is closed in Y. □

11.6. Definition. A subspace A of a topological space is said to be *locally closed* if each point $a \in A$ has an open neighborhood U_a such that $U_a \cap A$ is closed in U_a.

11.7. Proposition. *A subspace $A \subset X$ is locally closed $\Leftrightarrow$ it has the form $A = C \cap U$ where U is open in X and C is closed in X.*

PROOF. Put $U = \bigcup\{U_a | a \in A\}$, as in Definition 11.6, which is open, and $C = \bar{A}$ which is closed. Then

$$C \cap U = \bar{A} \cap \left(\bigcup U_a\right) = \bigcup(\bar{A} \cap U_a) = \bigcup(A \cap U_a) = A \cap U = A. \qquad \square$$

11.8. Theorem. *For a Hausdorff space X the following conditions are equivalent:*

(1) *X is locally compact.*
(2) *X is a locally closed subspace of a compact Hausdorff space.*
(3) *X is a locally closed subspace of a locally compact Hausdorff space.*

PROOF. If X is locally compact then it is an open subspace of its one-point compactification. Thus (1)$\Rightarrow$(2). Clearly, (2)$\Rightarrow$(3). If $Y \supset X$ is locally compact and $X = C \cap U$ where $C \subset Y$ is closed and $U \subset Y$ is open, then C is locally compact, and $X = U \cap C$ is open in C and hence is also locally compact. Thus (3)$\Rightarrow$(1). □

The remainder of this section is not used in the remainder of this book and so can be skipped. It assumes knowledge of nets from Section 6.

The preceding results suggest the question of when a topological space X can be embedded in a compact, Hausdorff space Y (as a subspace). Since Y is normal, it is also completely regular. Since a subspace of a completely regular space is completely regular, it follows that X must be completely regular. This turns out to be the precise condition needed.

If X is a completely regular space, consider the set $\mathbf{F}$ of all maps $f: X \to [0, 1]$. Define

$$\Phi: X \to [0, 1]^{\mathbf{F}} = \times\{[0, 1] \mid f \in \mathbf{F}\}$$

by $\Phi(x)(f) = f(x)$. (Here we regard an element of $[0, 1]^{\mathbf{F}}$ as a function $\mathbf{F} \to [0, 1]$.)

11.9. Definition. If X is a completely regular space, and $\Phi: X \to [0, 1]^{\mathbf{F}}$ is defined as above then the closure of $\Phi(X)$ is called the *Stone–Čech compactification* of X and is denoted by $\beta(X)$.

11.10. Theorem. *If X is a completely regular space, then $\beta(X)$ is compact Hausdorff and $\Phi: X \to \beta(X)$ is an embedding.*

PROOF. The function Φ is one-one since, if $\Phi(x) = \Phi(y)$, then $f(x) = f(y)$ for all maps $f: X \to [0, 1]$ and this implies that $x = y$ by complete regularity.

To prove continuity, let x_α be a net in X converging to x. Then

$$\lim(\Phi(x_\alpha)(f)) = \lim(f(x_\alpha)) = f(x) = \Phi(x)(f)$$

for all maps $f: X \to [0, 1]$. This implies that $\lim \Phi(x_\alpha) = \Phi(x)$ by Proposition 8.8.

For continuity of the inverse, suppose that $\{x_\alpha\}$ is a net in X such that $\Phi(x_\alpha)$ converges to $\Phi(x)$. Then, for all maps $f: X \to [0, 1]$,

$$\lim(f(x_\alpha)) = \lim(\Phi(x_\alpha)(f)) = \Phi(x)(f) = f(x).$$

If x_α does not converge to x then there is a neighborhood U of x such that x_α is frequently in $X - U$. But there is a map $f: X \to [0,1]$ which is 0 at x and 1 on $X - U$. Thus $f(x_\alpha)$ is frequently 1, while $f(x) = 0$ contradicting the convergence of $f(x_\alpha)$ to $f(x)$. □

11.11. Theorem. *If X is completely regular and $f: X \to \mathbf{R}$ is a bounded real valued map, then f can be extended uniquely to a map $\beta(X) \to \mathbf{R}$.*

PROOF. It suffices to treat the case in which the image of f is in $[0, 1]$. Consider the function $\bar{f}: [0, 1]^{\mathbf{F}} \to \mathbf{R}$ defined by $\bar{f}(\mu) = \mu(f)$. If $\{\mu_\alpha\}$ is a net in

$[0, 1]^{\mathbf{F}}$ converging to μ then

$$\lim(\bar{f}(\mu_\alpha)) = \lim(\mu_\alpha(f)) = \mu(f) = \bar{f}(\mu),$$

which shows that $\bar{f}$ is continuous.

If $x \in X$ then $\bar{f}(\Phi(x)) = (\Phi(x))(f) = f(x)$, showing that $\bar{f}$ does extend f. □

The problems give other properties of the Stone–Čech compactification.

Problems

1. Show that the Stone–Čech compactification $\beta(\cdot)$ is a functor on completely regular spaces by showing that a map $f\colon X \to Y$ induces a unique commutative diagram

$$\begin{array}{ccc} X & \xrightarrow{f} & Y \\ \downarrow & & \downarrow \\ \beta(X) & \xrightarrow{\beta(f)} & \beta(Y) \end{array}$$

such that $\beta(f \circ g) = \beta(f) \circ \beta(g)$ and $\beta(1_X) = 1_{\beta(X)}$.

2. Show that the Stone–Čech compactification is the "largest" compactification of a completely regular space X by showing that if $g\colon X \hookrightarrow Y$ is any compactification, then there is a unique map $\beta(X) \to Y$ factoring g.

3. Let ω be the set of natural integers and let X be its Stone–Čech compactification. Show that the sequence given by the usual ordering of ω can have no convergent subsequence in X. Conclude that X is not metrizable and not second countable. (Note that this sequence does have a convergent subnet since that is always true in a compact space.)

12. Paracompact Spaces

The notion of "paracompactness" of a space is a type of localization of compactness. It is very different, however, from local compactness. Paracompact spaces are very close to being metrizable, but the concept of paracompactness is sometimes simpler to deal with than that of a metric. In this book, most spaces in which we shall be interested are paracompact. The most important property of paracompact spaces is the existence of "partitions of unity," see Definition 12.7.

12.1. Definition. If $\mathbf{U}$ and $\mathbf{V}$ are open coverings of a space then $\mathbf{U}$ is said to be a *refinement* of $\mathbf{V}$ if each element of $\mathbf{U}$ is a subset of some element of $\mathbf{V}$.

12.2. Definition. A collection $\mathbf{U}$ of subsets of a topological space X is said to be *locally finite* if each point $x \in X$ has a neighborhood N which meets, nontrivially, only a finite number of the members of $\mathbf{U}$.

12.3. Definition. A Hausdorff space X is said to be *paracompact* if every open covering of X has an open, locally finite refinement.

12.4. Proposition. *A closed subspace of a paracompact space is paracompact.*

PROOF. If A is a closed subspace of the paracompact space X, cover A with sets open in X. Throw in the set $X - A$. Take a locally finite refinement of this open covering of X and intersect it with A. This gives a locally finite refinement of the original covering of A. □

12.5. Theorem. *A paracompact space is normal.*

PROOF. We will first show the paracompact space X to be regular. Thus suppose $x \in X$ and $C \subset X$ is closed with $x \notin C$. For each point $y \in C$ there are disjoint open sets U_y, V_y with $x \in U_y$ and $y \in V_y$. Cover X by $X - C$ together with the sets V_y. Then there is an open locally finite refinement, say by sets U_α. Let $U = \bigcup\{U_\alpha | U_\alpha \subset \text{some } V_y\}$ and note that this contains C. Since this is a locally finite collection, its closure $\bar{U}$ is the union of the closures of the same U_α's. But x is not in any of the $\bar{U}_\alpha$ and so $x \notin \bar{U}$. Thus U and $X - \bar{U}$ provide the required separation.

The same argument, with C playing the role of x and the other closed set playing the role of C, shows X to be normal. □

Thus paracompact spaces are close to being metric spaces because all that is needed is second countability. Also, it is known that metric spaces are paracompact. (This is very hard to prove and we will not attempt it.) However, there are paracompact spaces that are not metrizable. There are also examples of paracompact spaces having subspaces which are not paracompact, and, of course, that cannot happen with metrizable spaces.

Normality implies that a paracompact space has many real valued maps, a property that we will now exploit.

12.6. Definition. If f is a real valued map then the *support* of f is

$$\text{support}(f) = \text{closure}\{x | f(x) \neq 0\}.$$

12.7. Definition. Let $\{U_\alpha | \alpha \in A\}$ be an open covering of the space X. Then a *partition of unity* subordinate to this covering is a collection of maps

$$\{f_\beta : X \to [0,1] | \beta \in B\}$$

such that:

(1) There is a locally finite open refinement $\{V_\beta | \beta \in B\}$ such that $\text{support}(f_\beta) \subset V_\beta$ for all $\beta \in B$; and
(2) $\sum_\beta f_\beta(x) = 1$ for each $x \in X$.

12.8. Theorem. *If X is paracompact and* **U** *is an open covering of X then there exists a partition of unity subordinate to* **U**.

PROOF. Without loss of generality, we may assume that the covering $\mathbf{U} = \{U_\alpha | \alpha \in A\}$ is locally finite.

Consider a family $F = \{g_\beta : X \to [0, 1] | \beta \in A_F\}$, where $A_F \subset A$, such that if $W_\beta = \{x | g_\beta(x) > 0\}$, then $\bar{W}_\beta \subset U_\beta$ and $\{W_\beta | \beta \in A_F\} \cup \{U_\alpha | \alpha \in A - A_F\}$ covers X. Let **F** be a collection of such families F which is simply ordered by inclusion and is a maximal such collection; see Theorem B.18(C). Let $G = \bigcup \mathbf{F}$. We claim that $G \in \mathbf{F}$. Now $G = \{g_\beta | \beta \in A_G = \bigcup A_F \text{ for } F \in \mathbf{F}\}$. If $G \notin \mathbf{F}$ then there is a point $x \in X$ not in any W_β for $\beta \in A_G$ or in any U_α for $\alpha \in A - A_G$. Let $\{\alpha_1, \ldots, \alpha_n\}$ be the finite nonempty set of indices for which $x \in U_{\alpha_i}$, all i. Then each $\alpha_i \in A_G$. Since **F** is simply ordered there is an $F \in \mathbf{F}$ for which $\alpha_i \in A_F$ for all $i = 1, \ldots, n$. But then x must be in some W_{α_i}, a contradiction. Therefore $G \in \mathbf{F}$ as claimed.

Next we claim that $A_G = A$. If not then let $\alpha \in A - A_G$ and put

$$D = \bigcup \{W_\beta | \beta \in A_G\} \cup \bigcup \{U_\gamma | \gamma \in A - A_G, \gamma \neq \alpha\}.$$

Then $X = D \cup U_\alpha$. Let $C = X - D$ which is closed and inside U_α. Since X is normal, there exists an open set V with $C \subset V$ and $\bar{V} \subset U_\alpha$. By Urysohn's Lemma there is a map $g_\alpha : X \to [0, 1]$ which is 1 on C and 0 outside V. Then $W_\alpha \supset C$ and so $X = D \cup W_\alpha$, showing that $G \cup \{g_\alpha\}$ is a collection of maps, as above, properly containing G. This contradicts the maximality of **F**, and so $A_G = A$ as claimed.

Thus we now have a collection $\{g_\alpha | \alpha \in A\}$ of maps such that the $W_\alpha = \{x | g_\alpha(x) > 0\}$ cover X and $\bar{W}_\alpha \subset U_\alpha$. Let $g = \sum_\alpha g_\alpha$, which makes sense by the local finiteness, and note that $g(x) > 0$ for all x. Then put $f_\alpha = g_\alpha / g$. This fulfills our requirements. □

Note that we proved a little more than was stated in Theorem 12.8. Namely, if the original covering is already locally finite then it need not be refined as in Definition 12.7. Also note that the sets W_α form a covering and that $\bar{W}_\alpha \subset U_\alpha$. Thus, we have:

12.9. Proposition. *If X is paracompact and $\{U_\alpha\}$ is a locally finite open covering of X then there is an open covering $\{V_\alpha\}$ such that, for each α, $\bar{V}_\alpha \subset U_\alpha$.* □

Generally it is difficult to check that a space is paracompact by just using the definition. Also, we would like to avoid using that metric spaces are paracompact, since we have not proved it. However, the following criterion will apply in most cases of interest to us here.

12.10. Definition. A space is called *σ-compact* if it is the union of countably many compact subspaces.

12.11. Theorem. *A locally compact, Hausdorff space is paracompact $\Leftrightarrow$ it is the disjoint union of open σ-compact subsets.*

PROOF. $\Rightarrow$: Using local compactness, cover the space with open sets U_α such that $\bar{U}_\alpha$ is compact. Using paracompactness it is easy to see that this covering can be replaced by one which is also locally finite, so assume that. We shall inductively construct open sets V_i whose closures are compact. We start with $V_1 = U_\beta$ for some given β. If V_n has been defined then consider all the U_α which intersect $\bar{V}_n$. By compactness of $\bar{V}_n$ and the local finiteness of the cover, this set of U_α's is finite. Let V_{n+1} be the union of these U_α. Then $\bar{V}_{n+1}$ is the union of the closures of this finite set of U_α's and so it is compact. Also $\bar{V}_n \subset V_{n+1}$. Put $V = \bigcup V_n$. Then V is the union of countably many of the U_α's. By local finiteness $\bar{V}$ is the union of the closures of these U_α's but each of these closures is contained in some $\bar{V}_n \subset V_{n+1} \subset V$. Thus $\bar{V} = V$ is clopen and, by construction, is σ-compact. The remainder of the proof of the implication ($\Rightarrow$) is accomplished by a Maximality Principle argument which will be left to the reader since this implication will not be used in this book.

$\Leftarrow$: It is clear that a disjoint union of open paracompact spaces is paracompact, and so we may as well assume the space X to be σ-compact; $X = C_1 \cup C_2 \cup \cdots$ where the C_i are compact. In sequence, alter the C_i by adding to each C_{i+1} (and to the following ones at the same time) a finite union of compact sets (by local compactness) whose interiors cover C_i. In this way, we get $C_i \subset \operatorname{int}(C_{i+1})$ for all i. Define compact sets $A_1 = C_1$, and $A_i = C_i - \operatorname{int} C_{i-1}$ for $i > 1$. (Think of the C_i as concentric disks and the A_i as the rings between them.) Note that each A_i intersects nontrivially with only (at most) A_{i-1} and A_{i+1}. A little more work, using the compactness, hence normality, of the A_i shows that we can enlarge the A_i slightly to provide compact sets B_i whose interiors contain the A_i and which intersect only with the two B_j with adjoining indices. Now, given an open covering, consider the induced covering of each of the compact sets A_i. We can select a finite refinement still covering A_i and with none of the covering sets overflowing from B_i. It is then clear that these finite coverings taken together provide a locally finite refinement of the original cover. □

12.12. Theorem. *If X is locally compact, Hausdorff, and second countable then its one-point compactification X^+ is metrizable and X is σ-compact and paracompact.*

PROOF. Let **B** be a countable basis for X and let $C \subset X$ be compact. If $x \in C$ then x has a compact neighborhood N and there is a member $U_x \in \mathbf{B}$ with $x \in U_x \subset N$. Hence C is covered by a finite union $U_{x_1} \cup \cdots \cup U_{x_n}$ of such sets. Put $V = X - \bigcup \bar{U}_{x_i}$. Then $V \cup \{\infty\}$ is a neighborhood of ∞ in X^+ contained in the arbitrary neighborhood $X^+ - C$. These sets V are indexed by finite subsets of **B** and hence are countable in number; see the remark below Theorem B.27. This shows that X^+ is second countable, and also shows that

X is σ-compact. By Theorem 12.11, X is paracompact. By Theorem 7.11, Corollary 10.3 and Theorem 9.10, X^+ is metrizable. □

PROBLEMS

1. Without using Theorem 12.11 or the fact that metric spaces are paracompact, show that any open subspace of euclidean space is σ-compact, and hence paracompact by Theorem 12.11.
2. Suppose X is paracompact. For any open subset U of $X \times [0, \infty)$ which contains $X \times \{0\}$ show that there is a map $f: X \to (0, \infty)$ such that $(x, y) \in U$ for all $y \leqslant f(x)$.

13. Quotient Spaces

The notion of a quotient space or identification space is of central importance in topology. It gives, for example, a firm foundation for the intuitive idea of the operation of "pasting" spaces together. It also provides many other techniques of producing new spaces out of old ones. It can also be difficult to understand when met for the first time, and the reader is advised to study it fully before going further in this book.

13.1. Definition. Let X be a topological space, Y a set, and $f: X \to Y$ an onto function. Then we define a topology on Y called the *topology induced by* f or the *quotient topology*, by specifying a set $V \subset Y$ to be open $\Leftrightarrow f^{-1}(V)$ is open in X. Note that this is the largest topology on Y which makes f continuous.

13.2. Definition. Let X be a topological space and $\sim$ an equivalence relation on X. Let $Y = X/\sim$ be the set of equivalence classes and $\pi: X \to Y$ the canonical map taking $x \in X$ to its equivalence class $[x] \in X/\sim$. Then Y, with the topology induced by π, is called a *quotient space* of X.

Quotient spaces often have very non-Hausdorff topologies. For example, if X is the real line and $x \sim y \Leftrightarrow x - y$ is rational, then $X/\sim$ is an uncountable set but has the trivial topology, as the reader is asked to verify in Problem 6. We will mostly be concerned with quotient spaces that are better behaved.

The reader can verify the following fact directly from the definition:

13.3. Proposition. *A quotient space of a quotient space of X is a quotient space of X. That is, if $X \to Y \to Z$ are two onto functions and Y is given the quotient topology from X, and Z is given the quotient topology from Y, then Z has the quotient topology from X induced by the composition of the two functions.* □

13.4. Definition. A map $X \to Y$ is called an *identification map* if it is onto and Y has the quotient topology.

13.5. Proposition. *A surjection $f: X \to Y$ is an identification map $\Leftrightarrow$ (for all functions $g: Y \to Z$, ($g \circ f$ is continuous $\Leftrightarrow$ g is continuous)).*

PROOF. The $\Rightarrow$ part is clear from the definitions. For $\Leftarrow$ specialize to the case $Z = Y$ as sets, with the identification topology (on Z) and g the identity function. Then the composition $X \to Y \xrightarrow{g} Z$ is continuous, so the condition demands that g be continuous. But g^{-1} is continuous because the composition $X \to Z \to Y$ is f which is continuous (by the case $Z = Y$ as spaces and $g = 1_Y$) and since Z has the quotient topology. Thus g is a homeomorphism, meaning that $Y = Z$ as *spaces*. □

13.6. Example. The projective plane is often defined as the sphere $\mathbf{S}^2$ with antipodal points identified. That is, it is given the quotient topology from the relation that identifies antipodal points in a sphere. A second description of the projective plane one often sees is that it is the unit disk $\mathbf{D}^2$ with antipodal points on the boundary identified. Regard $\mathbf{D}^2$ as the upper hemisphere and consider the diagram:

$$\begin{array}{ccc} \mathbf{D}^2 & \xrightarrow{i} & \mathbf{S}^2 \\ \downarrow{\scriptstyle f} & & \downarrow{\scriptstyle g} \\ \mathbf{D}^2/\sim & \xrightarrow{k} & \mathbf{S}^2/\sim \end{array}$$

where the maps f and g are the identifications, i is the inclusion, and k is induced (the only function making the diagram commute). If $U \subset \mathbf{S}^2/\sim$ is open then $g^{-1}(U)$ is open so $(gi)^{-1}(U) = i^{-1}(g^{-1}(U))$ is open. But this is the same as $(kf)^{-1}(U) = f^{-1}(k^{-1}(U))$. Thus $k^{-1}(U)$ is open by the definition of the quotient topology. That means that k is continuous. But k is also clearly one–one and onto. Moreover, $\mathbf{D}^2/\sim$ is compact since $\mathbf{D}^2$ is. Also, $\mathbf{S}^2/\sim$ is easily seen to be Hausdorff, and so we finally conclude that k is a homeomorphism from Theorem 7.8. Thus, indeed, these two ways of defining the projective plane as a topological space are equivalent. This is a typical argument involving spaces obtained via identifications.

An often used special case of quotient spaces is the idea of "collapsing" a subspace:

13.7. Definition. If X is a space and $A \subset X$, then X/A denotes the quotient space obtained via the equivalence relation whose equivalence classes are A and the single point sets $\{x\}$, $x \in X - A$.

The following is an easy exercise:

13.8. Proposition. *If X is regular and A is closed then X/A is Hausdorff. If X is normal and A is closed, then X/A is normal.* □

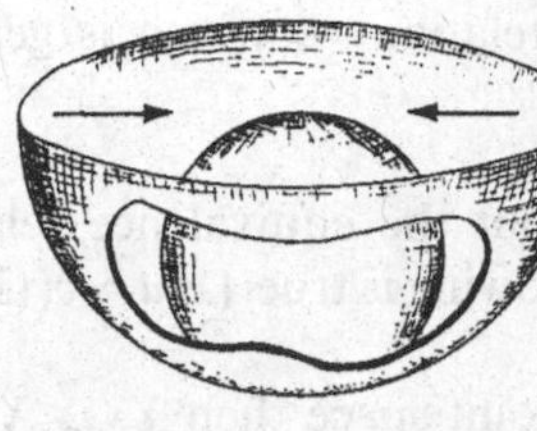

Figure I-1. The sphere as quotient space of a disk.

13.9. Example. Consider the cylinder $\mathbf{S}^n \times I$. Define $f:\mathbf{S}^n \times I \to \mathbf{D}^{n+1}$ by $f(x, t) = tx$. This carries the set $\mathbf{S}^n \times \{0\}$ to the origin and so f factors through $\mathbf{S}^n \times I/\mathbf{S}^n \times \{0\}$. The resulting map $g:\mathbf{S}^n \times I/\mathbf{S}^n \times \{0\} \to \mathbf{D}^{n+1}$ is clearly one–one and onto. Thus it is a homeomorphism by Theorem 7.8.

13.10. Example. Consider the n-disk $\mathbf{D}^n$. This is clearly homeomorphic to the lower n-hemisphere of radius 2 centered at 1 on the "vertical" axis. (See Figure I-1.) We can map this onto the n-sphere $\mathbf{S}^n$ of radius 1 centered at the origin by projection towards the vertical axis. It maps the boundary of the disk to the north pole of the sphere. This function is distance decreasing and hence continuous. Also consider the quotient space $\mathbf{D}^n/\mathbf{S}^{n-1}$. One can factor the projection of the disk to the sphere through this space. By an argument similar to that in Example 13.9 one can show that the resulting map $\mathbf{D}^n/\mathbf{S}^{n-1} \to \mathbf{S}^n$ is a homeomorphism.

If the method in Example 13.9 is not available, the following gives a criterion for deciding the same sort of question.

13.11. Definition. If $A \subset X$ and if $\sim$ is an equivalence relation on X then the *saturation* of A is $\{x \in X \mid x \sim a \text{ for some } a \in A\}$.

13.12. Proposition. *If $A \subset X$ and $\sim$ is an equivalence relation on X such that every equivalence class intersects A nontrivially, then the induced map $k: A/\!\sim \to X/\!\sim$ is a homeomorphism if the saturation of every open (resp. closed) set of A is open (resp. closed) in X.*

PROOF. If $f: A \to A/\!\sim$ and $g: X \to X/\!\sim$ are the canonical maps, and U is an open set in $A/\!\sim$ then $g^{-1}(k(U))$ is the saturation of $f^{-1}(U)$. Moreover, by definition, it is open $\Leftrightarrow k(U)$ is open in $X/\!\sim$. Also k is clearly one–one, onto, and continuous. □

Another common application of the idea of a quotient space is a space obtained by "attaching":

13.13. Definition. Let X and Y be spaces and $A \subset X$ closed. Let $f: A \to Y$ be a map. Then we denote by $Y \cup_f X$, the quotient space of the disjoint union

$X + Y$ by the equivalence relation $\sim$ which is generated by the relations $a \sim f(a)$ for $a \in A$.

(To be more precise about the equivalence relation, for points u, v in $X + Y$, $u \sim v$ if one of the following is true: (1) $u = v$; (2) $u, v \in A$, and $f(u) = f(v)$; (3) $u \in A$ and $v = f(u) \in Y$.)

Note that if Y is a one-point space then $Y \cup_f X = X/A$.

The following is an easy verification left to the reader:

13.14. Proposition. *The canonical map $Y \to Y \cup_f X$ is an embedding onto a closed subspace. The canonical map $X - A \to Y \cup_f X$ is an embedding onto an open subspace.* □

13.15. Definition. If A is a subspace of a space X then a map $f: X \to A$ such that $f(a) = a$ for all points $a \in A$, is called a *retraction*, and A is said to be a *retract* of X.

Special cases of attachments of importance to us are the "mapping cylinder" and the "mapping cone." As is usual, the unit interval $[0, 1]$ will be denoted by I here.

13.16. Definition. If $f: X \to Y$ is a map then the *mapping cylinder* of f is the space $M_f = Y \cup_{f_0} X \times I$ where $f_0: X \times \{0\} \to Y$ is $f_0(x, 0) = f(x)$. See Figure I-2.

Note that $X \approx X \times \{1\}$ is embedded as a closed subset of M_f. By an abuse of notation, we will regard this as an inclusion $X \subset M_f$. Also note then that there is the factorization of f, $X \subset M_f \xrightarrow{r} Y$ where r is the retraction of M_f onto Y induced by the projection $X \times I \to X \times \{0\}$.

13.17. Definition. If $f: X \to Y$ is a map then the *mapping cone* of f is the space $C_f = M_f/(X \times \{1\})$.

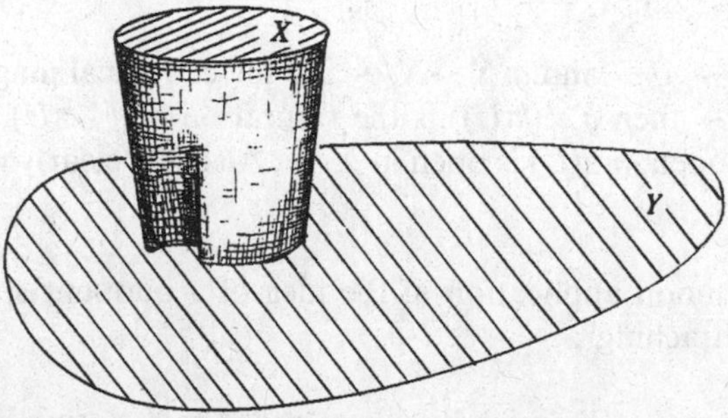

Figure I-2. Mapping cylinder.

It is often of interest to know when a function $M_f \to Z$, taking a mapping cylinder into another space Z, is continuous. This is usually quite easy to check by use of the following simple fact. The proof is an easy application of the definition of the quotient topology on a mapping cylinder, and is left to the reader.

13.18. Proposition. *A function $M_f \to Z$ is continuous $\Leftrightarrow$ the induced functions $X \times I \to Z$ and $Y \to Z$ are both continuous.* □

Here is a result we shall need later.

13.19. Proposition. *If $f: X \to Y$ is an identification map and K is a locally compact Hausdorff space then $f \times 1: X \times K \to Y \times K$ is an identification map.*

PROOF. Suppose that $g: Y \times K \to W$ and let $h = g \circ (f \times 1): X \times K \to W$. Then, by Proposition 13.5, it suffices to prove that h continuous $\Rightarrow g$ continuous. Let $U \subset W$ be open and suppose that $g(y_0, k_0) \in U$. Let $f(x_0) = y_0$. Then $h(x_0, k_0) = g(y_0, k_0) \in U$. Therefore there is a compact neighborhood N of k_0 such that $h(x_0 \times N) \subset U$. Put $A = \{y \in Y \mid g(y \times N) \subset U\}$. Then $y_0 \in A$ and it suffices to show that A is open. Thus it suffices to show that $f^{-1}(A)$ is open. Now

$$f^{-1}(A) = \{x \in X \mid h(x \times N) = g(f(x) \times N) \subset U\}$$

and so $X - f^{-1}(A) = \pi_X(h^{-1}(W - U) \cap (X \times N))$ is closed by Proposition 8.2. □

PROBLEMS

1. ✧ If $f: X \to A$ and $g: Y \to B$ are open identification maps, show that $f \times g: X \times Y \to A \times B$ is also an open identification map.
2. If X, Y are normal, $A \subset X$ is closed, and $f: A \to Y$ is a map, show that $Y \cup_f X$ is normal.
3. If $f: X \to Y$ is a map between Hausdorff spaces, show that M_f and C_f are Hausdorff.
4. There are four common definitions of the torus $\mathbf{T}^2$:
   (1) as $\mathbf{R}^2/\mathbf{Z}^2$, i.e., the plane modulo the equivalence relation $(x, y) \sim (u, w) \Leftrightarrow x - u$ and $y - w$ are both integers;
   (2) as a square with opposite edges identified (see Figure I-3);
   (3) as the product $\mathbf{S}^1 \times \mathbf{S}^1$; and
   (4) as the "anchor ring," the surface of revolution obtained by rotating a circle about an axis in its plane and disjoint from it.
   Show that these are all homeomorphic to one another.
5. The "Klein bottle" $\mathbf{K}^2$ is a square with opposite vertical edges identified in the same direction and opposite horizontal edges identified in the opposite direction (see Figure I-3). Consider the space (denoted by $\mathbf{P}^2 \# \mathbf{P}^2$) resulting from an annulus by identifying antipodal points on the outer circle, and also identifying antipodal points on the inner circle. Show that $\mathbf{K}^2 \approx \mathbf{P}^2 \# \mathbf{P}^2$.

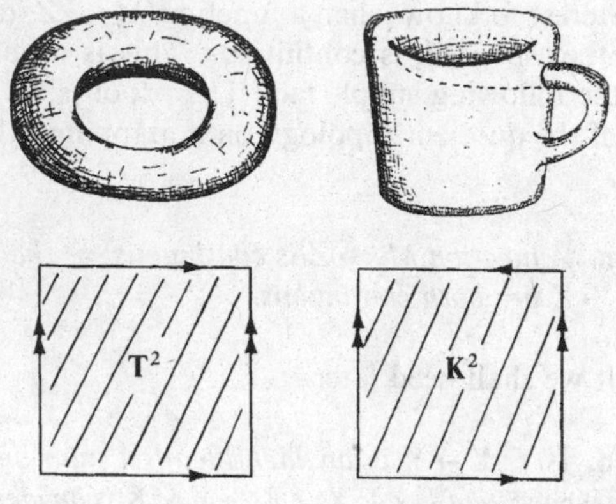

Figure I-3. The torus (left) and Klein bottle (right).

6. Consider the real line $\mathbf{R}$, with the equivalence relation $x \sim y \Leftrightarrow x - y$ is rational. Show that $\mathbf{R}/\sim$ has an uncountable number of points, but its topology is the trivial one.
7. Consider the real line $\mathbf{R}$ and the integers $\mathbf{Z}$. Let $A = \mathbf{R}/\mathbf{Z}$ (the identification of the subspace $\mathbf{Z}$ to a point). Also consider the subspace B of the plane which is the union of the circles of radius $1/n$ $(n = 1, 2, \ldots)$ in the upper half plane all tangent to the real line at the origin. Also consider the subspace C of the plane which is the union of the circles of radius n $(n = 1, 2, \ldots)$ in the upper half plane all tangent to the real line at the origin. Finally, consider the space $D = \mathbf{S}^1/N$ where $\mathbf{S}^1$ is the unit circle in the complex numbers and $N = \{e^{i\pi/n} | n = 1, 2, \ldots\} \cup \{1\}$. Which of these four spaces A, B, C, D are homeomorphic to which others of them?
8. Let (X, x_0) and (Y, y_0) be "pointed spaces," i.e., spaces with distinguished "base" points. Define the "one-point union" $X \vee Y$ to be the quotient space of the topological sum $X + Y$ by the equivalence relation identifying x_0 with y_0. Show that $X \vee Y \approx X \times \{y_0\} \cup \{x_0\} \times Y$, where the latter is regarded as a subspace of $X \times Y$.

14. Homotopy

A homotopy is a family of mappings parametrized by the unit interval. This notion is of central importance in topology. Here we lay down the basic definitions and properties of this concept.

14.1. Definition. If X and Y are spaces then a *homotopy* of maps from X to Y is a map $F: X \times I \to Y$, where $I = [0, 1]$.

Two maps $f_0, f_1: X \to Y$ are said to be *homotopic* if there exists a homotopy $F: X \times I \to Y$ such that $F(x, 0) = f_0(x)$ and $F(x, 1) = f_1(x)$ for all $x \in X$.

The relation "f is homotopic to g" is an equivalence relation on the set of all maps from X to Y (see Definition 14.11) and is denoted by $f \simeq g$. The following is elementary:

14.2. Proposition. *If $f, g: X \to Y, h: X' \to X$ and $k: Y \to Y'$ then*

$$f \simeq g \quad \Rightarrow \quad f \circ h \simeq g \circ h \text{ and } k \circ f \simeq k \circ g. \qquad \square$$

14.3. Definition. A map $f: X \to Y$ is said to be a *homotopy equivalence* with *homotopy inverse* g if there is a map $g: Y \to X$ such that $g \circ f \simeq 1_X$ and $f \circ g \simeq 1_Y$. This relationship is denoted by $X \simeq Y$. One also says, in this case, that X and Y have the *same homotopy type.*

This is an equivalence relation between spaces, since, if $h: Y \to Z$ is another homotopy equivalence with homotopy inverse k then

$$(gk)(hf) = g(kh)f \simeq g1_Yf = gf \simeq 1_X$$

and similarly for the opposite composition.

14.4. Definition. A space is said to be *contractible* if it is homotopy equivalent to the one-point space.

14.5. Proposition. *A space X is contractible $\Leftrightarrow$ the identity map $1_X: X \to X$ is homotopic to a map $r: X \to X$ whose image is a single point.*

PROOF. Let $Y = \{x_0\} = \text{im}(r)$. Then we have the inclusion map $i: Y \to X$ and the retraction $r: X \to Y$. Now $r \circ i = 1_Y$ and $i \circ r \simeq 1_X$ by assumption. The converse is also easy. $\square$

14.6. Example. Consider euclidean space $X = \mathbf{R}^n$ and the homotopy $F: X \times I \to X$ given by $F(x, t) = tx$. This is a homotopy between $f_1 = 1_X$ and f_0, which is the map taking everything to $\{0\}$. Consequently, $\mathbf{R}^n$ is contractible. Note that each f_t is onto for $t > 0$ but that, suddenly, f_0 is far from onto. This may challenge the intuition of some readers.

14.7. Example. Consider the unit sphere $\mathbf{S}^{n-1}$ in $\mathbf{R}^n$ and the punctured euclidean space $\mathbf{R}^n - \{0\}$. Let $i: \mathbf{S}^{n-1} \to \mathbf{R}^n - \{0\}$ be the inclusion and $r: \mathbf{R}^n - \{0\} \to \mathbf{S}^{n-1}$ be the central projection $r(x) = x/\|x\|$. Then $r \circ i = 1$ and $i \circ r \simeq 1$ where the latter homotopy is given by $F: (\mathbf{R}^n - \{0\}) \times I \to \mathbf{R}^n - \{0\}$, where $F(x, t) = tx + (1-t)x/\|x\|$. Thus $\mathbf{S}^{n-1} \simeq \mathbf{R}^n - \{0\}$.

These two examples illustrate and suggest the following:

14.8. Definition. A subspace A of X is called a *strong deformation retract* of

X if there is a homotopy $F: X \times I \to X$ (called a deformation) such that:

$$F(x,0) = x,$$
$$F(x,1) \in A,$$
$$F(a,t) = a \qquad \text{for} \quad a \in A \text{ and all } t \in I.$$

It is just a *deformation retract* if the last equation is required only for $t = 1$.

As in the examples, a deformation retract A of a space X is homotopically equivalent to X.

Is the sphere $\mathbf{S}^n$ contractible? Our intuition tells us the answer is "no" but, in fact, this is quite difficult to prove. This is one type of question which algebraic topology is equipped to answer, and we will answer it, and many more such questions, in later pages.

14.9. Example. If $f: X \to Y$ is a map then the canonical map $r: M_f \to Y$ is a strong deformation retraction, as the reader can verify (see the end of the proof of Theorem 14.18). Hence $M_f \simeq Y$. Thus, the mapping cylinder allows replacing "up to homotopy" the arbitrary map f by the inclusion $X \hookrightarrow M_f$.

14.10. Definition. If $A \subset X$ then a homotopy $F: X \times I \to Y$ is said to be *relative to* A (or rel A) if $F(a,t)$ is independent of t for $a \in A$. A homotopy that is rel X is said to be a *constant* homotopy.

Two homotopies of X into Y can be "concatenated" if the first ends where the second begins, by going through the first at twice the normal speed and then the second at that speed. We will now study this construction. The reader should note the important special case in which X is a single point and so the homotopies are simply paths in Y. It might help the reader's understanding if he draws pictures, in this case, for all the basic homotopies produced in Propositions 14.13, 14.15, and 14.16.

14.11. Definition. If $F: X \times I \to Y$ and $G: X \times I \to Y$ are two homotopies such that $F(x,1) = G(x,0)$ for all x, then define a homotopy $F * G: X \times I \to Y$, which is called *the concatenation of F and G*, by

$$(F * G)(x,t) = \begin{cases} F(x,2t) & \text{if } t \le \frac{1}{2}, \\ G(x,2t-1) & \text{if } t \ge \frac{1}{2}. \end{cases}$$

(See Figure I-4.)

One does not have to combine these homotopies at $t = \frac{1}{2}$. We can do it at any point and with arbitrary speed:

14.12. Lemma (Reparametrization Lemma). *Let ϕ_1 and ϕ_2 be maps $(I, \partial I) \to (I, \partial I)$ which are equal on ∂I. (Note the case where one of these is the*

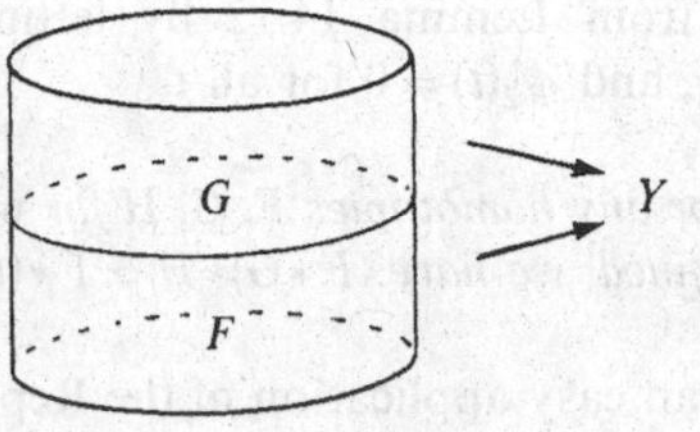

Figure I-4. Concatenation of homotopies.

identity.) Let $F: X \times I \to Y$ *be a homotopy and let* $G_i(x,t) = F(x, \phi_i(t))$ *for* $i = 1, 2$. *Then* $G_1 \simeq G_2 \operatorname{rel} X \times \partial I$.

PROOF. Define $H: X \times I \times I \to Y$ by $H(x,t,s) = F(x, s\phi_2(t) + (1-s)\phi_1(t))$. Then

$$\begin{aligned} H(x,t,0) &= F(x,\phi_1(t)) = G_1(x,t), \\ H(x,t,1) &= F(x,\phi_2(t)) = G_2(x,t), \\ H(x,0,s) &= F(x,\phi_1(0)) = G_1(x,0), \\ H(x,1,s) &= F(x,\phi_2(1)) = G_2(x,1), \end{aligned}$$

with the last two equations coming from $\phi_1(0) = \phi_2(0)$ and $\phi_1(1) = \phi_2(1)$. □

We shall use C to denote a constant homotopy, whichever one makes sense in the current context. For example $F * C$ is concatenation with the constant homotopy C for which $C(x,t) = F(x,1)$, but use of $C * F$ will imply the one for which $C(x,t) = F(x,0)$.

14.13. Proposition. *We have* $F * C \simeq F \operatorname{rel} X \times \partial I$, *and, similarly,* $C * F \simeq F \operatorname{rel} X \times \partial I$.

PROOF. This follows from Lemma 14.12 by letting $\phi_1(t) = 2t$ for $t \le \frac{1}{2}$ and $= 1$ for $t \ge \frac{1}{2}$, and $\phi_2(t) = t$ in the first case, and $\phi_1(t) = 0$ for $t \le \frac{1}{2}$ and $= 2t - 1$ for $t \ge \frac{1}{2}$ and $\phi_2(t) = t$ in the second case. □

We define the "inverse" F^{-1} of a homotopy F to be this homotopy with t running backward. Note that this has nothing to do with the inverse of the map (which probably does not exist anyway).

14.14. Definition. If $F: X \times I \to Y$ is a homotopy, then we define $F^{-1}: X \times I \to Y$ by $F^{-1}(x,t) = F(x, 1-t)$.

14.15. Proposition. *For a homotopy* F *we have* $F * F^{-1} \simeq C \operatorname{rel} X \times \partial I$ *where* $C(x,t) = F(x,0)$ *for all* x *and* t; *i.e.,* C *is a constant homotopy.*

PROOF. This follows from Lemma 14.12 by letting $\phi_1(t)=2t$ for $t\leq\frac{1}{2}$, $\phi_1(t)=2-2t$ for $t\geq\frac{1}{2}$, and $\phi_2(t)=0$ for all t. □

14.16. Proposition. *For any homotopies F, G, H for which the concatenations $F*G$ and $G*H$ are defined, we have* $(F*G)*H\simeq F*(G*H)\operatorname{rel} X\times\partial I$.

PROOF. Again, this is an easy application of the Reparametrization Lemma (Lemma 14.12). □

14.17. Proposition. *For homotopies F_1, F_2, G_1, and G_2, if $F_1\simeq F_2 \operatorname{rel} X\times\partial I$ and $G_1\simeq G_2\operatorname{rel} X\times\partial I$ then $F_1*G_1\simeq F_2*G_2\operatorname{rel} X\times\partial I$.*

PROOF. If $H: X\times I\times I\to Y$ and $K: X\times I\times I\to Y$ are the homotopies giving $F_1\simeq F_2$ and $G_1\simeq G_2$, respectively, then it is easy to check that $H*K$ is the required homotopy of homotopies. (The reader should fill in the details here.) □

Note that all of the discussion of concatenation of homotopies goes through with no difficulties for the cases in which all homotopies are relative to some subspace $A\subset X$ or are homotopies of pairs $(X,A)\to(Y,B)$.

It follows from the stated results that homotopy between maps of pairs $(X,A)\to(Y,B)$ is an equivalence relation. The set of homotopy classes of these maps is commonly denoted by $[X,A;Y,B]$, or just $[X;Y]$ if $A=\emptyset$.

We will now prove that the homotopy type of a mapping cylinder or cone depends only on the homotopy class of the map.

14.18. Theorem. *If $f_0\simeq f_1: X\to Y$ then $M_{f_0}\simeq M_{f_1}\operatorname{rel} X+Y$ and $C_{f_0}\simeq C_{f_1}$ rel Y + vertex.*

PROOF. The part for the cone follows from that for the mapping cylinder. Let $F: X\times I\to Y$ be the given homotopy between f_0 and f_1. Define $h: M_{f_0}\to M_{f_1}$ by $h(y)=y$ for $y\in Y$ and

$$h(x,t)=\begin{cases}F(x,2t) & \text{for } t\leq\frac{1}{2},\\ (x,2t-1) & \text{for } \frac{1}{2}\leq t.\end{cases}$$

Note that $h(x,\frac{1}{2})=F(x,1)=f_1(x)=(x,0)$. To prove continuity, we only have to show that the compositions $Y\to M_{f_1}$ and $X\times I\to M_{f_1}$ are continuous by Proposition 13.18, but this is trivial.

Define $k: M_{f_1}\to M_{f_0}$ in the analogous fashion. Then the composition $kh: M_{f_0}\to M_{f_0}$ is the identity on Y and, on the cylinder portion, it is $F*(F^{-1}*E)$ where $E: X\times I\to M_{f_0}$ is induced by the identity on $X\times I\to X\times I$; see Figure I-5. This is homotopic to the identity $\operatorname{rel} X\times\{1\}+Y$. Similarly for hk. It remains to check the continuity of this homotopy. We have described a homotopy $M_{f_0}\times I\to M_{f_0}$. For continuity, it is sufficient to know that $M_{f_0}\times I\approx M_{f_0\times I}$ because then we only have to check continuity of the

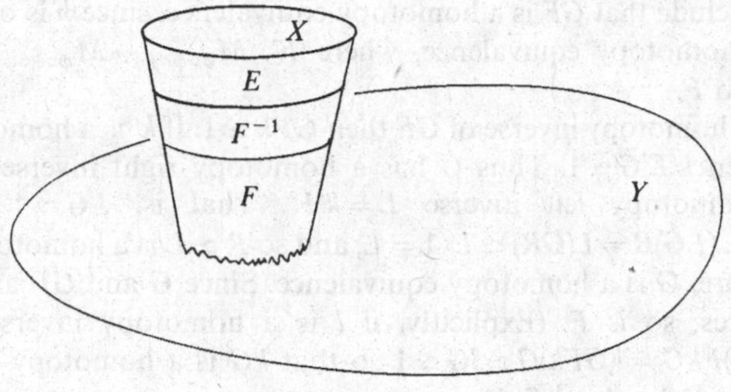

Figure I-5. Deformation of a mapping cylinder.

composition $(X \times I + Y) \times I \to M_{f_0} \times I \to M_{f_0}$, and that is trivial. (On $Y \times I$ it is the constant homotopy and on $X \times I \times I$ it results from $F*(F^{-1}*E) \simeq E \operatorname{rel} X \times \partial I$.) That is, it suffices to show that $M_{f_0} \times I$ has the identification topology from the map $f_0 \times I$. But that is a consequence of Proposition 13.19. □

We conclude this section by studying the effect on mapping cones of changing the target space by a homotopy equivalence.

Let $f: X \to Y$. If $\phi: Y \to Y'$ is a map then there is the induced map $F: M_f \to M_{\phi \circ f}$ induced from ϕ on Y and the identity on $X \times I$.

14.19. Theorem. *If $\phi: Y \to Y'$ is a homotopy equivalence then so is $F: (M_f, X) \to (M_{\phi \circ f}, X)$ and hence so is $F: C_f \to C_{\phi \circ f}$.*

Proof. Let $\psi: Y' \to Y$ be a homotopy inverse of ϕ and let $G: M_{\phi \circ f} \to M_{\psi \circ \phi \circ f}$ be the map induced by ψ on Y' and the identity on $X \times I$. The composition $GF: M_f \to M_{\psi \circ \phi \circ f}$ is induced from $\psi \circ \phi: Y \to Y$ and the identity on $X \times I$. Let $H: Y \times I \to Y$ be a homotopy from 1 to $\psi \circ \phi$; i.e., $H(y, 0) = y$ and $H(y, 1) = \psi\phi(y)$. By the proof of Theorem 14.18 there is the homotopy equivalence $h: M_f \to M_{\psi \circ \phi \circ f}$ rel X given by $h(y) = y$ and

$$h(x, t) = \begin{cases} H(f(x), 2t) & \text{for } t \le \frac{1}{2}, \\ (x, 2t - 1) & \text{for } t \ge \frac{1}{2}. \end{cases}$$

We claim that $h \simeq GF$ rel X. Indeed, the homotopy H can be extended to $M_f \times I \to M_{\psi \circ \phi \circ f}$ by putting

$$H((x, s), t) = \begin{cases} H(f(x), 2s + t) & \text{for } 2s + t \le 1, \\ (x, (2s + t - 1)/(t + 1)) & \text{for } 2s + t \ge 1. \end{cases}$$

Then $H((x, s), 0) = h(x, s)$, $H(y, 0) = y$, and $H((x, s), 1) = (x, s)$, $H(y, 1) = \psi\phi(y)$, so that $H(\cdot, 0) = h$ and $H(\cdot, 1) = GF$.

We conclude that GF is a homotopy equivalence, since h is one. Likewise, $F'G$ is a homotopy equivalence, where $F'\colon M_{\psi\circ\phi\circ f} \to M_{\phi\circ\psi\circ\phi\circ f}$ is defined similarly to F.

If k is a homotopy inverse of GF then $GFk \simeq 1$. If k' is a homotopy inverse of $F'G$ then $k'F'G \simeq 1$. Thus G has a homotopy right inverse $R = Fk$ and also a homotopy left inverse $L = k'F'$. That is, $LG \simeq 1 \simeq GR$. Then $R = 1 \circ R \simeq (LG)R = L(GR) \simeq L \circ 1 = L$, and so $R \simeq L$ is a homotopy inverse of G. Therefore, G is a homotopy equivalence. Since G and GF are homotopy equivalences, so is F. (Explicitly, if l is a homotopy inverse of G then $FkG \simeq (lG)FkG = l(GFk)G \simeq lG \simeq 1$ so that kG is a homotopy inverse of F, since $kGF \simeq 1$ by the definition of k.) □

Problems

1. Let $\mathbf{S}^2 \vee \mathbf{S}^1$ be the "one-point union" of a 2-sphere and a circle; see Problem 8 of Section 13. Let $\mathbf{S}^2 \cup A$ denote the union of the unit 2-sphere and the line segment joining the north and south poles. Show that these spaces are homotopically equivalent.

2. Show that the union of a 2-sphere and a flat unit 2-cell through the origin is homotopically equivalent to the one-point union of two 2-spheres.

3. Show that the union of a standard 2-torus with two disks, one spanning a latitudinal circle and the other spanning a longitudinal circle of the torus, is homotopically equivalent to a 2-sphere.

4. Show that the projective plane is homeomorphic to the mapping cone of the map $z \mapsto z^2$ of the unit circle in the complex numbers to itself.

5. Consider the mapping cone of the map f of the unit circle in the complex numbers to itself, given by $f(z) = z^4$ for z in the upper semicircle and by $f(z) = \bar{z}^2$ for z in the lower semicircle. Show that this space is contractible.

6. The "dunce cap" space is the quotient of a triangle (and interior) obtained by identifying all three edges in an inconsistent manner. That is, if the vertices of the triangle are p, q, r then we identify the line segment (p, q) with (q, r) and with (p, r) in the orientation indicated by the order given of the vertices. (See Figure I-6.) Show that the dunce cap is contractible. (*Hint*: Describe this space as the mapping cone of a certain map from $\mathbf{S}^1$ to itself, and study this map.)

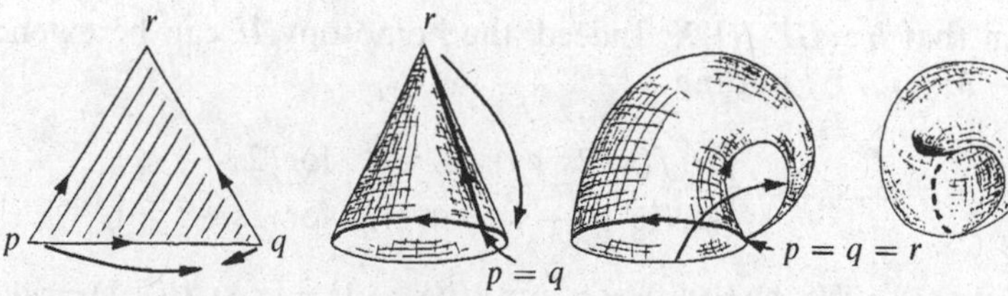

Figure I-6. The dunce cap.

7. If $\phi: X' \to X$ and $f: X \to Y$ are maps, define an induced map $F: C_{f \circ \phi} \to C_f$. If ϕ is a homotopy equivalence then show that F is a homotopy equivalence.

8. Show that a retract of a contractible space is contractible.

9. For any two maps $f, g: X \to \mathbf{S}^n$ such that $f(x) \neq -g(x)$ for all x, show that $f \simeq g$.

15. Topological Groups

Topological groups, spaces which are also groups in the algebraic sense, form a rich territory for important examples in topology and geometry. Here we shall develop the roots of the theory behind them.

15.1. Definition. A *topological group* is a Hausdorff topological space G together with a group structure on G such that:

(1) group multiplication $(g, h) \mapsto gh$ of $G \times G \to G$ is continuous; and
(2) group inversion $g \mapsto g^{-1}$ of $G \to G$ is continuous.

15.2. Definition. A *subgroup* H of a topological group G is a subspace which is also a subgroup in the algebraic sense.

15.3. Definition. If G and G' are topological groups then a *homomorphism* $f: G \to G'$ is a group homomorphism which is also continuous.

15.4. Definition. If G is a topological group and $g \in G$ then *left translation* by g is the map $L_g: G \to G$ given by $L_g(h) = gh$. Similarly *right translation* by g is the map $R_g: G \to G$ given by $R_g(h) = hg^{-1}$.

15.5. Proposition. *In a topological group G we have $L_g \circ L_h = L_{gh}$ and $R_g \circ R_h = R_{gh}$. Moreover, both L_g and R_g are homeomorphisms as is conjugation by g ($h \mapsto ghg^{-1}$) and inversion ($h \mapsto h^{-1}$).*

PROOF. The first statement is a trivial computation, and implies that $L_{g^{-1}} = L_g^{-1}$ and similarly for right translation, and it follows that these are homeomorphisms. Conjugation $h \mapsto ghg^{-1}$ is the same as $R_g \circ L_g$ and so is a homeomorphism. Inversion is continuous by assumption and is its own inverse, and thus is a homeomorphism. □

In a topological group G, if A, B are subsets then we let $AB = \{ab \mid a \in A, b \in B\}$ and $A^{-1} = \{a^{-1} \mid a \in A\}$.

15.6. Definition. A subset A of a topological group is called *symmetric* if $A = A^{-1}$.

15.7. Proposition. *In a topological group G with unity element e, the symmetric neighborhoods of e form a neighborhood basis at e.*

Proof. If U is any neighborhood of e then so is U^{-1}, and hence so is $U \cap U^{-1}$, which is symmetric. □

It is an easy exercise using the continuity of multiplication to see the following two results:

15.8. Proposition. *If G is a topological group and $g \in G$, and U is any neighborhood of g, then there is a symmetric neighborhood V of e such that $VgV^{-1} \subset U$.* □

15.9. Proposition. *If G is a topological group and U is any neighborhood of e and n is any positive integer, then there exists a symmetric neighborhood V of e such that $V^n \subset U$.* □

15.10. Proposition. *If H is any subgroup of a topological group G then $\bar{H}$ is also a subgroup of G. If H is a normal subgroup then so is $\bar{H}$.*

Proof. It follows from continuity of inversion and multiplication that $\bar{H}^{-1} \subset \bar{H}$ and $\bar{H}\bar{H} \subset \bar{H}$ so that $\bar{H}$ is a subgroup. If H is a normal subgroup and $g \in G$ then continuity also implies that $g\bar{H}g^{-1} \subset \bar{H}$ and the opposite inclusion follows by applying this formula to g^{-1}. □

15.11. Proposition. *If G is a topological group and H is a closed subgroup then the space G/H of left cosets of H in G, with the topology induced by the canonical map $\pi: G \to G/H$, is a Hausdorff space. Moreover, π is open and continuous.*

Proof. If $U \subset G$ is open then $\pi^{-1}\pi(U) = UH = \bigcup\{Uh \mid h \in H\}$ is a union of open sets and so is open. By definition of the quotient topology it follows that $\pi(U)$ is open, proving the last statement. To see that G/H is Hausdorff, suppose that $g_1H \neq g_2H$ (hence representing different points in G/H). This is the same as saying that $g_1^{-1}g_2 \notin H$. Since $G-H$ is an open set containing $g_1^{-1}g_2$, Proposition 15.8 implies that there is a symmetric open neighborhood U of e such that $(Ug_1^{-1}g_2U) \cap H = \emptyset$. Thus $g_1^{-1}g_2U \cap UH = \emptyset$ which is the same as $g_2U \cap g_1UH = \emptyset$ which implies in turn that $g_2UH \cap g_1UH = \emptyset$. This shows that $\pi(g_i) \in g_iUH$ which are disjoint open sets in G/H. □

15.12. Proposition. *If H is a closed normal subgroup of the topological group G then G/H, with the quotient topology, is a topological group.*

Proof. G/H is Hausdorff by Proposition 15.11 and it remains to show that the group operations are continuous. Consider the following diagram, where the horizontal arrows are group multiplications:

$$\begin{array}{ccc} G \times G & \longrightarrow & G \\ \downarrow{\scriptstyle \pi \times \pi} & & \downarrow{\scriptstyle \pi} \\ G/H \times G/H & \longrightarrow & G/H. \end{array}$$

An easy consequence of the fact (Proposition 15.11) that π is open is that $\pi \times \pi$ is an identification map (see Section 13, Problem 1). Taking an open set in G/H (the lower right), we must show its inverse image on the lower left is open. But that is the same as showing that its inverse image in $G \times G$ is open. But this inverse image is the same as that via the top and right maps, and that is open since those maps are continuous. A similar argument gives the continuity of inversion in G/H. □

The most important class of topological groups is that of the so-called Lie groups which also carry a differentiable structure. We will discuss differentiable structures in Chapter II. Here, we will only discuss some of the important "classical Lie groups."

The set M_n of $n \times n$-matrices is just a euclidean space of dimension n^2. The determinant function $M_n \to \mathbf{R}$ is continuous since it is just a polynomial in the matrix coefficients. Thus the inverse image of $\{0\}$ is a closed set. Its complement is the set of nonsingular matrices, and this forms a group under multiplication. It is called the "general linear group" and is denoted by $\mathbf{Gl}(n, \mathbf{R})$. It is an open subset of euclidean n^2-space and that is the topology it is given. Matrix multiplication is given by polynomials in the coefficients and so is continuous. Matrix inversion is a rational function of the coefficients by Cramer's rule and so that is continuous. Thus $\mathbf{Gl}(n, \mathbf{R})$ is a topological group.

In the same way, the general linear group $\mathbf{Gl}(n, \mathbf{C})$ over the complex numbers can be seen to be a topological group.

The special linear group $\mathbf{Sl}(n, \mathbf{R})$ is the subgroup of $\mathbf{Gl}(n, \mathbf{R})$ consisting of matrices of determinant 1, and similarly for $\mathbf{Sl}(n, \mathbf{C})$ over the complexes.

Similarly, the general linear group $\mathbf{Gl}(n, \mathbf{H})$ over the quaternions is a topological group, although, in this case, the argument is a little harder since quaternionic matrices lack a determinant function. (See Problem 12.)

The set $\mathbf{O}(n)$ of orthogonal (real) matrices forms a subgroup of $\mathbf{Gl}(n, \mathbf{R})$ and it is a closed subset, since it is defined via continuous relations ($AA^t = I$). Since the coefficients of an orthogonal matrix are bounded by 1 in absolute value, $\mathbf{O}(n)$ is a bounded closed subset of euclidean n^2-space, and hence is compact by Corollary 8.7.

Similarly, the set $\mathbf{U}(n)$ of unitary matrices ($AA^* = I$) is a compact subgroup of $\mathbf{Gl}(n, \mathbf{C})$.

The quaternionic analogue of the orthogonal and unitary groups is called the symplectic group $\mathbf{Sp}(n)$. Its elements are quaternionic matrices A such that $AA^* = I$, where A^* is the quaternionic conjugate transpose of A, conjugation meaning reversal of all three imaginary components. This group is a compact subgroup of $\mathbf{Gl}(n, \mathbf{H})$.

These three classes of examples are called the "classical Lie groups."

Note that the map $\mathbf{Gl}(n, \mathbf{R}) \times \mathbf{R}^n \to \mathbf{R}^n$ is given by polynomials in the coefficients of the matrix and the vector, and so is continuous.

An orthogonal matrix $A \in \mathbf{O}(n)$, as a transformation of euclidean n-space, preserves lengths of vectors, and so it is a map of the sphere $\mathbf{S}^{n-1}$ to itself. We can regard $\mathbf{O}(n-1)$ as the subgroup of $\mathbf{O}(n)$ fixing the last coordinate.

Consider the point $(0,0,\ldots,0,1)$. This point is left fixed by $\mathbf{O}(n-1)$. We can map $\mathbf{O}(n)$ into $\mathbf{S}^{n-1}$ by taking a matrix into where it moves the point $(0,0,\ldots,0,1)$. That is, we define the map

$$f:\mathbf{O}(n)\to\mathbf{S}^{n-1} \qquad \text{by} \qquad f(A)=A(0,0,\ldots,0,1)^t.$$

If $B\in\mathbf{O}(n-1)$ then clearly $f(AB)=f(A)$. This means that the map f factors through the left coset space $\mathbf{O}(n)/\mathbf{O}(n-1)$. A short computation will show that the induced map $\mathbf{O}(n)/\mathbf{O}(n-1)\to\mathbf{S}^{n-1}$ is one–one onto and continuous. Since this is a one–one mapping of a compact space onto a Hausdorff space it is a homeomorphism by Theorem 7.8. Let us abstract these observations.

15.13. Definition. If G is a topological group and X is space, then an *action* of G on X is a map $G\times X\to X$, with the image of (g,x) being denoted by $g(x)$, such that:

(1) $(gh)(x)=g(h(x))$; and
(2) $e(x)=x$.

For a point $x\in X$, the set $G(x)=\{g(x)\mid g\in G\}$ is called the *orbit* of x, and the subgroup $G_x=\{g\in G\mid g(x)=x\}$ is called the *isotropy* or *stability* group at x. The action is said to be *transitive* if there is only one orbit, the whole space X. The action is said to be *effective* if $(g(x)=x$ for all $x)\Rightarrow g=e$, the identity element of G.

Note that, in describing $G\times X\to X$ as a "map," we are assuming it to be continuous. The following, then, is the general setting in which our comments on $\mathbf{O}(n)$ acting on $\mathbf{S}^{n-1}$ lie.

15.14. Proposition. *If G is a compact topological group acting on the Hausdorff space X and G_x is the isotropy group at x, then the map $\phi\colon G/G_x\to G(x)$ given by $gG_x\mapsto g(x)$ is a homeomorphism.*

PROOF. If $g_1(x)=g_2(x)$ then $g_1^{-1}g_2\in G_x$ and so $g_1G_x=g_2G_x$, showing that ϕ is one–one onto $G(x)$. It is continuous by the definition of the quotient topology on G/G_x, and the result then follows from Theorem 7.8. □

Just as with the case of $\mathbf{O}(n)$, $\mathbf{U}(n)$ acts on $\mathbf{S}^{2n-1}$, and it is transitive because one can find a unitary matrix moving any vector of length 1 into any other. The isotropy group at $(0,0,\ldots,0,1)$ is $\mathbf{U}(n-1)$, and so $\mathbf{U}(n)/\mathbf{U}(n-1)\approx\mathbf{S}^{2n-1}$
Similarly $\mathbf{Sp}(n)/\mathbf{Sp}(n-1)\approx\mathbf{S}^{4n-1}$.

More generally, if we let (as is usual) $V_{n,k}$ denote the "Stiefel manifold" of k-frames in n-space (a k-frame being an orthonormal set of k vectors in n-space), then $\mathbf{O}(n)$ acts transitively on $V_{n,k}$ with an isotropy group $\mathbf{O}(n-k)$, and so $\mathbf{O}(n)/\mathbf{O}(n-k)\approx V_{n,k}$. The reader can make analogous observations for the unitary and symplectic cases.

Some other matrix groups are obtained by restriction to matrices of determinant 1:

$$\mathbf{SO}(n) = \{A \in \mathbf{O}(n) | \det(A) = 1\} = \text{the special orthogonal group},$$
$$\mathbf{SU}(n) = \{A \in \mathbf{U}(n) | \det(A) = 1\} = \text{the special unitary group}.$$

There is no analogue in the symplectic case.

With appropriate restrictions, these groups also act transitively on spheres and we get

$$\mathbf{SO}(n)/\mathbf{SO}(n-1) \approx \mathbf{S}^{n-1} \quad \text{for} \quad n \geq 2,$$
$$\mathbf{SU}(n)/\mathbf{SU}(n-1) \approx \mathbf{S}^{2n-1} \quad \text{for} \quad n \geq 2.$$

Similar results can be obtained for the Stiefel manifolds, as the reader can verify.

Problems

1. If G is a topological group and G_0 is the component containing the identity element, then show that G_0 is a closed normal subgroup of G.
2. If $\phi: G \to H$ is an onto homomorphism of topological groups, show that the kernel K of ϕ is a closed normal subgroup of G. If, moreover, G is compact, show that $G/K \approx H$ as topological groups.
3. If $g \in G$, a compact topological group, and $A = \{g^n | n = 0, 1, 2, \ldots\}$, then show that $\bar{A}$ is a subgroup of G. Is this true without compactness of G?
4. If G is a compact topological group, then show that every neighborhood of e contains a neighborhood V which is invariant under conjugation (i.e., $gVg^{-1} = V$ for all $g \in G$).
5. ⌖ If G is a topological group and H is a closed subgroup, show that if H and G/H are both connected then so is G.
6. If G is a topological group acting on the space X and if we put
$$H = \{h \in G | \forall x \in X, h(x) = x\},$$
then show that H is a closed normal subgroup of G.
7. ⌖ Show that $\mathbf{SO}(2) \approx \mathbf{S}^1$, $\mathbf{SU}(2) \approx \mathbf{S}^3$, and $\mathbf{Sp}(1) \approx \mathbf{S}^3$ (as spaces).
8. ⌖ Show that $\mathbf{SO}(n)$ is connected. (*Hint*: Use Problem 5.) Further, show that $\mathbf{SO}(n)$ is the component of $\mathbf{O}(n)$ containing the identity.
9. ⌖ Show that $\mathbf{U}(n)$ and $\mathbf{SU}(n)$ are both connected and that $\mathbf{U}(n)/\mathbf{SU}(n) \approx \mathbf{S}^1$.
10. Show that the center $\{h \in G | \forall g \in G, hg = gh\}$ of a topological group G is closed.
11. Show that real projective n-space $\mathbf{RP}^n \approx \mathbf{O}(n+1)/(\mathbf{O}(n) \times \mathbf{O}(1))$ and that complex projective n-space $\mathbf{CP}^n \approx \mathbf{U}(n+1)/(\mathbf{U}(n) \times \mathbf{U}(1))$.
12. ⌖ Show that $\mathbf{Gl}(n, \mathbf{H})$ is open in $M_n(\mathbf{H})$.
13. Consider the multiplicative group of all upper diagonal 2×2 matrices of deter-

minant 1. Show that the conclusion of Problem 4 is false for this (noncompact) topological group.

14. Show that a topological group is regular. (*Hint*: Let U, V be symmetric open neighborhoods of e such that $V^2 \subset U$ and deduce that $\bar{V} \subset U$.)

15. Prove Propositions 15.8 and 15.9.

16. Convex Bodies

In topology, we often need to know that certain familiar objects are homeomorphic. For example, we shall have reason to want to know that a disk in euclidean space is homeomorphic to a cube, and to a cylinder, and a simplex (the analogue of a tetrahedron), and so on. In this section we give a general result that provides a unified proof of these special cases and many others.

16.1. Definition. A *convex body* in $\mathbf{R}^n$ is a closed set $C \subset \mathbf{R}^n$ with the property that whenever $p, q \in C$ the line segment between p and q is contained in C. The *boundary* of C is $\partial C = C - \text{int}(C)$.

16.2. Proposition. *If $C \subset \mathbf{R}^n$ is a convex body and $0 \in \text{int}(C)$ then any ray from the origin intersects ∂C in at most one point.*

Proof. Suppose R is a ray from the origin and $p, q \in R \cap C$, with neither p nor q being the origin. Suppose q is further from the origin than p. Since the origin is assumed to lie in $\text{int}(C)$ there is a ball B about the origin completely contained in C. Then consider the union of all line segments from points in B to q (the cone on B subtended from q). The point p is clearly in the interior of this cone, and the cone is contained completely in C, since C is convex, and so p must be in $\text{int}(C)$. □

16.3. Proposition. *Let $C \subset \mathbf{R}^n$ be a compact convex body with $0 \in \text{int}(C)$. Then the function $f: \partial C \to \mathbf{S}^{n-1}$ given by $f(x) = x/\|x\|$ is a homeomorphism.*

Proof. Since f is the composition of the inclusion $\partial C \hookrightarrow \mathbf{R}^n - \{0\}$ with the radial retraction $r: \mathbf{R}^n - \{0\} \to \mathbf{S}^{n-1}$, it is continuous. Proposition 16.2 implies that f is one–one, and f is obviously onto. By Theorem 7.8, f is a homeomorphism. □

16.4. Theorem. *A compact convex body C in $\mathbf{R}^n$ with nonempty interior is homeomorphic to the closed n-ball, and $\partial C \approx \mathbf{S}^{n-1}$.*

Proof. By translation, we can assume the origin is in the interior of C. Let $\mathbf{D}^n$ denote the unit disk in $\mathbf{R}^n$ and let f be as in Proposition 16.3. Then the function $k: \mathbf{D}^n \to C$ given by $k(x) = \|x\| f^{-1}(x/\|x\|)$ for $x \neq 0$ and $k(0) = 0$ clearly maps $\mathbf{D}^n$ onto C and is continuous everywhere except possibly at the origin. However, since C is compact, there is a bound M for $\{\|x\| \mid x\in C\}$. Then $\|k(x)\| \leq M\cdot\|x\|$ which implies continuity at the origin. It is also clear that k is one–one, and hence it is a homeomorphism by Theorem 7.8. □

17. The Baire Category Theorem

Often, one is interested in a condition on points of a space that is satisfied by an open dense set of points. For example, if $p(x_1,\ldots,x_n)$ is a polynomial function on $\mathbf{R}^n$ then the condition $p(x)\neq 0$ has this property, and a special case of that is the determinant function on square matrices. If one has two such conditions then the set of points satisfying both conditions is still open and dense. The same, then, is true for any finite number of such conditions. But what of a countably infinite number of such conditions? Certainly, one cannot expect that the set of points satisfying all the conditions is open, but the density of this set does survive for a wide class of spaces, as we show in this section. This fact has many important consequences in analysis as well as in topology.

17.1. Theorem (Baire Category Theorem). *Let X be either a complete metric space or a locally compact Hausdorff space. Then the union of countably many nowhere dense subsets of X has empty interior.*

PROOF. Let U be an open subset of X and suppose that $A_i\subset X$ is nowhere dense $(i=0,1,\ldots)$. Construct a sequence of nonempty open sets V_1, $V_2,\ldots$, such that $\bar{V}_{i+1}\subset V_i-\bar{A}_i$, where $V_0=U$. (In the complete metric case, this can be achieved by taking $V_{i+1}=B_\epsilon(x)$ for some $x\in V_i-\bar{A}_i$ such that $B_{2\epsilon}(x)\subset V_i-\bar{A}_i$.)

If X is locally compact then also construct the V_i so that $\bar{V}_1$, and hence each $\bar{V}_i$, is compact. Then the $\bar{V}_i$ satisfy the finite intersection property and so $\varnothing\neq\bigcap\bar{V}_i\subset U-\bigcup\bar{A}_i$.

If X is complete metric, then also construct the V_i so that $\operatorname{diam}(V_i)<2^{-i}$. Then a sequence of points $x_i\in V_i$ is Cauchy since, for $i<j$, $\operatorname{dist}(x_i,x_j)\leq\operatorname{diam}(V_i)<2^{-i}$. Then $x_n\in\bar{V}_i$ for all $n\geq i$ and so $x=\lim(x_n)\in\bar{V}_i$ for all i. Thus $x\in\bigcap\bar{V}_i\subset U-\bigcup\bar{A}_i$.

In both cases this shows that $U\not\subset\bigcup\bar{A}_i$. Since U is an arbitrary open set, we conclude that $\operatorname{int}(\bigcup\bar{A}_i)=\varnothing$. □

The word "category" in the theorem refers to the following definition:

17.2. Definition. A subset S of a space X is said to be of *first category* if it is the countable union of nowhere dense subsets. Otherwise it is said to be of *second category*. A set of second category is said to be *residual* if its complement is of first category.

Thus Theorem 17.1 can be rephrased: "An open subset of a complete metric space, or a locally compact Hausdorff space, is of second category in itself."

It is also worth while to state the contrapositive of Theorem 17.1:

17.3. Corollary. *Let X be either a complete metric space or a locally compact Hausdorff space. Then the intersection of any countable family of dense open sets in X (i.e., a residual set) is dense.* □

We close this section with some applications of this result. The first application deals with pointwise limits of functions.

17.4. Corollary. *If $\{f_n\}$ is a sequence of continuous functions $f_n: X \to Y$ from a complete metric space X to a metric space Y and if $f(x) = \lim f_n(x)$ exists for each x then the set of points of continuity of f is residual and hence dense.*

PROOF. For positive integers m, k let

$$U_{m,k} = \bigcup_{n \geq m} \{x \mid \text{dist}(f_n(x), f_m(x)) > 1/k\}$$

which is open. Since

$$\bigcap_{m \geq 1} U_{m,k}$$

consists of points where $f_n(x)$ does not converge, it is empty. It follows that

$$\bigcap_{m \geq 1} \bar{U}_{m,k} \subset \bigcup_{m \geq 1} (\bar{U}_{m,k} - U_{m,k})$$

which is a countable union of nowhere dense sets. Therefore

$$\bigcup_{k \geq 1} \bigcap_{m \geq 1} \bar{U}_{m,k}$$

is also a countable union of nowhere dense sets. Thus its complement

$$C = \bigcap_{k \geq 1} \bigcup_{m \geq 1} \text{int}\Big(\bigcap_{n \geq m} \{x \mid \text{dist}(f_n(x), f_m(x)) \leq 1/k\}\Big)$$

is residual. But $y \in C$ means that

$$\forall k \geq 1, \exists m \geq 1 \ni \exists \delta > 0 \ni \text{dist}(x, y) < \delta \quad \Rightarrow \quad \forall n \geq m, \text{dist}(f_n(x), f_m(x)) \leq 1/k.$$

Hence, for such k, m, δ and $\text{dist}(x, y) < \delta$ we have that $\text{dist}(f(x), f_m(x)) \leq 1/k$ and also that $\text{dist}(f(y), f_m(y)) \leq 1/k$. By taking δ smaller, if necessary, we can also assure that $\text{dist}(f_m(x), f_m(y)) \leq 1/k$ by the continuity of f_m. Therefore $\text{dist}(f(x), f(y)) \leq 3/k$ for these choices, showing that f is continuous at y. (We hasten to point out that the set of points of discontinuity of f, while of first category, can well be dense. It is not hard to produce such examples.) □

17.5. Corollary. *There exists a connected 2-manifold (i.e., a Hausdorff space in which each point has a neighborhood homeomorphic to the plane) with the following properties:*

(1) *it has a countable dense set;*
(2) *it has an uncountable discrete subset, and hence is not second countable; and*
(3) *it is not normal, and hence not metrizable.*

PROOF. We will describe a similar manifold M "with boundary." The desired manifold can then be obtained by "doubling" M; i.e., taking two copies of M and identifying their boundaries.

First we describe the point set of M. There are two types of points. The first type consists of the points in the upper half space of the plane, i.e., $\{(x, y) | y > 0\}$. The second type of point is a ray (but we are describing a *single point* of M) from a point of the x-axis pointing into the upper half space.

To describe the topology on this set of points, we shall give a neighborhood basis at each point. For points in the upper half space, we use the usual topology of the plane. For a point corresponding to a ray r from a point x on the x-axis, we take, for a basic neighborhood, the set of points in the upper half plane in the "wedge" between two rays surrounding r and of distance (in the sense of the plane) less than ϵ from $(x, 0)$ together with the points of the second kind consisting of the rays from $(x, 0)$ and lying in the mentioned "wedge." (See Figure I-7.)

To see that this really is a 2-manifold, consider the map ϕ from the upper half plane to itself given by $\phi(x, y) = (x/y, y)$. It is easy to verify that this is a homeomorphism on the upper half plane. Moreover, it maps rays from the origin to vertical lines. Thus the point of M corresponding to a ray from the origin can be thought of as the point on the x-axis attached to the vertical corresponding to the ray under ϕ. Under this correspondence it is evident that the topology becomes the ordinary topology of the closed half plane. This shows that a neighborhood of any point of M given by a ray from the origin, is indeed homeomorphic to an ordinary neighborhood of a boundary point in closed half space, a manifold with boundary. Rays from other points on the original x-axis can be treated similarly.

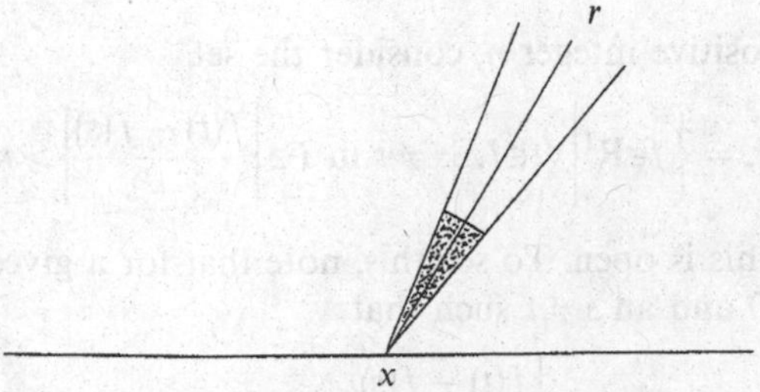

Figure I-7. Construction of a strange manifold.

We now verify the claims for this space. The points in the upper half plane with rational coordinates clearly give a countable dense set, proving (1).

Consider, for each x, the point of the second kind corresponding to a vertical ray from $(x,0)$. Let S be the collection of these points. For any one of them a "wedge" neighborhood of that point intersects S in that point alone. Thus that point is itself an open subset of S. Thus S is discrete, and it is in one–one correspondence with the real axis, and so is uncountable. If M were second countable then any subspace, such as S, would also be second countable (just intersect the basis with the subspace), and that is not true of S.

Finally, we must show that M is not normal. In fact, let A be the subset of S, above, corresponding to rational x, and B that corresponding to irrational x. These are both closed subsets of M and are disjoint. We claim that it is impossible to separate them by disjoint open supersets. Suppose $U \subset M$ is an open set containing B and disjoint from A. For a point $x \in S$, and integer $n > 0$, let $W_n(x)$ be the wedge of angle π/n and radius $1/n$ about the vertical ray from x. Then define

$$T_n = \{x \in B \mid W_n(x) \subset U\}.$$

Then the sets T_n together with the singletons $\{x \in A\}$ comprise a countable collection of sets whose union is $A \cup B$, the real axis. Since the real line in its ordinary topology is complete metric, the closure in $\mathbf{R}$ of one of these sets must contain an interval. This is not true of the singletons, so it must be that $\bar{T}_n \supset (a,b)$ for some n and interval. But then for any "rational" $q \in (a,b) \cap S$ it is clear that every neighborhood in M of q must intersect U. □

The manifold just described is, in fact, a differentiable manifold (see Chapter II) except for failure to be second countable. Indeed, it is the strange properties of examples like this that lead to the restriction to second countable spaces in the definition of a differentiable manifold.

Note that this manifold is a subspace of a normal space, its one-point compactification. Thus, a subspace of a normal space need not be normal.

17.6. Corollary. *In the space $\mathbf{R}^I$ of continuous functions $I \to \mathbf{R}$ in the uniform metric, the set of functions which are nowhere differentiable is dense. Indeed, it is residual in $\mathbf{R}^I$.*

Proof. For a positive integer n, consider the set

$$U_n = \left\{ f \in \mathbf{R}^I \,\middle|\, \forall t \in I, \exists s \neq t \text{ in } I \ni \left| \frac{f(t) - f(s)}{t - s} \right| > n \right\}.$$

We claim that this is open. To see this, note that for a given $f \in U_n$ and $t \in I$, there is an $\epsilon > 0$ and an $s \neq t$ such that

$$\left| \frac{f(t) - f(s)}{t - s} \right| > n + \epsilon.$$

Then, for some such $s = s(t)$ and $\epsilon = \epsilon(t)$, there is an open neighborhood V_t of t such that $s(t) \notin \bar{V}_t$ and such that

$$\left|\frac{f(t') - f(s)}{t' - s}\right| > n + \epsilon$$

for all $t' \in V_t$. The V_t cover I so that some finite union $V_{t_1} \cup \cdots \cup V_{t_k} \supset I$. Let $\epsilon = \min \epsilon(t_i)$, $\delta = \min \operatorname{dist}(s(t_i), \bar{V}_{t_i})$, and suppose that $\| f - g \| < \epsilon\delta/2$. Then, for any $t \in I$, we have $t \in V_{t_i}$ for some i and, for $s = s(t_i)$, we have

$$n + \epsilon < \left|\frac{f(t) - f(s)}{t - s}\right| \le \left|\frac{f(t) - g(t)}{t - s}\right| + \left|\frac{g(t) - g(s)}{t - s}\right| + \left|\frac{g(s) - f(s)}{t - s}\right|.$$

Since $|t - s| \ge \delta$, the first and third terms on the right are each at most $(\epsilon\delta/2)(1/\delta) = \epsilon/2$. It follows that

$$\left|\frac{g(t) - g(s)}{t - s}\right| > n + \epsilon - \epsilon = n,$$

and hence that $g \in U_n$. Therefore, U_n is open as claimed.

Next we claim that each U_n is dense. To see this, let $f \in \mathbf{R}^I$ and $\epsilon > 0$ be given. Let m be so large that $2/m < \epsilon$. By uniform continuity of f there is a k so large that

$$|x - y| \le 1/k \quad \Rightarrow \quad |f(x) - f(y)| \le 1/m.$$

Also, take k so large that $k > nm$. Let $a_i = i/k$, $b_i = a_i + 1/(3k)$, $c_i = a_i + 2/(3k)$, and $y_i = f(a_i)$. Consider the interval $[a_i, a_{i+1}]$. Define a function g on this interval whose graph consists of the three line segments (a_i, y_i) to $(b_i, y_i - (1/m))$ to $(c_i, y_i + (1/m))$ to (a_{i+1}, y_{i+1}); see Figure I-8. These fit together to define g on all of I. By construction, $\| f - g \| \le 2/m < \epsilon$. Let $t \in [a_i, a_{i+1}]$. If $g(t) > y_i$

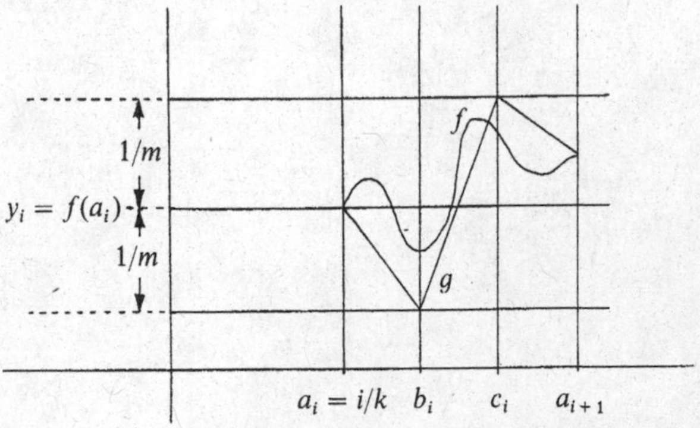

Figure I-8. Creating a nowhere differentiable function.

then take $s = b_i$. Otherwise, take $s = c_i$. Then

$$\left|\frac{g(t) - g(s)}{t - s}\right| \geq \frac{1/m}{1/k} = \frac{k}{m} > \frac{nm}{m} = n.$$

Hence $g \in U_n$ and $\|f - g\| < \epsilon$, concluding the proof that U_n is dense.

Since $\mathbf{R}^I$ is a complete metric space (prove it) we conclude that $A = \bigcap U_n$ is residual. We claim that any function $f \in A$ is nowhere differentiable. Suppose, on the contrary, that f is differentiable at some $t \in I$. Then $|(f(s) - f(t))/(s - t)|$ has a limit as $s \to t$ and so it is bounded for all $s \in I$, $s \neq t$. If n is larger than such a bound then it follows that $f \notin U_n$, a contradiction. □

Problems

1. Below is an outline of a more elegant proof of the fact that U_n is open in the proof of Corollary 17.6. Justify all statements made here.
   (a) The function $\mathbf{R}^I \times I \to \mathbf{R}$ taking $(f, t) \mapsto f(t)$ is continuous.
   (b) For $\Delta = \{(x, x) | x \in I\}$, the function $F: \mathbf{R}^I \times (I \times I - \Delta) \to \mathbf{R}$ taking $(f, s, t) \mapsto |(f(t) - f(s))/(t - s)|$ is continuous.
   (c) The map $\Phi: \mathbf{R}^I \times (I \times I - \Delta) \to \mathbf{R}^I \times (\mathbf{R} - \{0\})$ taking $(f, s, t) \mapsto (f, t - s)$ is closed.
   (d) The projection $p: \mathbf{R}^I \times (\mathbf{R} - \{0\}) \to \mathbf{R}^I$ is open.
   (e) $U_n = p(\mathbf{R}^I \times (\mathbf{R} - \{0\}) - \Phi F^{-1}[0, n])$ which is open.
2. Let X be a complete metric space and let $\mathbf{R}^X$ be the set of continuous functions $X \to \mathbf{R}$. Let $S \subset \mathbf{R}^X$ be a collection of maps $f: X \to \mathbf{R}$ such that $\{f(x) | f \in S\}$ is bounded for each $x \in X$. Show that there is an open set $\emptyset \neq U \subset X$ and a number B such that $|f(x)| \leq B$ for all $x \in U$ and $f \in S$.
3. An upper semicontinuous function of a real variable is a real valued function f on $\mathbf{R}$ such that $f^{-1}(-\infty, r)$ is open for all real r. If $f: \mathbf{R} \to \mathbf{R}$ is upper semicontinuous, show that there is some open interval (a, b) on which f is bounded below.
4. Show that the set of points of continuity of an upper semicontinuous function is residual.

CHAPTER II

Differentiable Manifolds

We have here, in fact, a passage to the limit of unexampled audacity.

F. KLEIN (in reference to Brook Taylor's derivation of his famous theorem)

1. The Implicit Function Theorem

In this section we will prove the Implicit Function Theorem and the Inverse Function Theorem in sufficient generality for our use. Readers who think they already know these theorems, or who are willing to accept them, can skip this section, but they are advised to at least read the statements. This section is self-contained.

A real valued function on an open subset of euclidean space $\mathbf{R}^n$ is said to be C^k if it has continuous partial derivatives through order k (of all orders if $k=\infty$). A function from an open subset of $\mathbf{R}^n$ to an open subset of $\mathbf{R}^m$ is said to be C^k if the m coordinate functions are C^k. A function from an open subset of $\mathbf{R}^n$ to another open subset of $\mathbf{R}^n$ is said to be a "diffeomorphism" if it is C^∞ and has a C^∞ inverse.

1.1. Theorem (The Mean Value Theorem). *Let $f:\mathbf{R}^n \to \mathbf{R}$ be C^1. Let $x=(x_1,\dots,x_n)$ and $\bar{x}=(\bar{x}_1,\dots,\bar{x}_n)$. Then*

$$f(x)-f(\bar{x})=\sum_{i=1}^n \frac{\partial f}{\partial x_i}(\bar{\bar{x}})(x_i-\bar{x}_i)$$

for some point $\bar{\bar{x}}$ on the line segment between x and $\bar{x}$.

PROOF. Apply the Mean Value Theorem found in any freshman calculus book to the function $\mathbf{R}\to\mathbf{R}$ defined by $t\mapsto f(tx+(1-t)\bar{x})$ and use the Chain Rule:

$$\left.\frac{df(tx+(1-t)\bar{x})}{dt}\right|_{t=t_0}=\sum_{i=1}^n \frac{\partial f}{\partial x_i}(\bar{\bar{x}})\left.\frac{d(tx_i+(1-t)\bar{x}_i)}{dt}\right|_{t=t_0}=\sum_{i=1}^n \frac{\partial f}{\partial x_i}(\bar{\bar{x}})(x_i-\bar{x}_i),$$

where $\bar{\bar{x}}=t_0x+(1-t_0)\bar{x}$. □

1.2. Corollary. *Let* $f: \mathbf{R}^k \times \mathbf{R}^m \to \mathbf{R}$ *be* C^1, $x \in \mathbf{R}^k$, $y, \bar{y} \in \mathbf{R}^m$. *Then*

$$f(x,y) - f(x,\bar{y}) = \sum_{i=1}^{m} \frac{\partial f}{\partial y_i}(x, \bar{\bar{y}})(y_i - \bar{y}_i)$$

for some $\bar{\bar{y}}$ *on the line segment between* y *and* $\bar{y}$. □

1.3. Theorem (The Banach Contraction Principle). *Let* X *be a complete metric space and* $T: X \to X$ *a contraction (i.e., for some constant* $K < 1$ *we have* $\operatorname{dist}(Tx, Ty) \le K \cdot \operatorname{dist}(x,y)$ *for all* $x, y \in X$*). Then* T *has a unique fixed point* $\xi \in X$. *Moreover, for any* $x \in X$, $\xi = \lim T^i(x)$.

Proof. Consider a given point $x_0 \in X$ and put $x_1 = Tx_0$, $x_2 = Tx_1$, etc. Then, with $\delta = \operatorname{dist}(x_0, Tx_0) = \operatorname{dist}(x_0, x_1)$, we have

$$\begin{aligned}
\operatorname{dist}(x_0, x_k) &\le \operatorname{dist}(x_0, x_1) + \operatorname{dist}(x_1, x_2) + \cdots + \operatorname{dist}(x_{k-1}, x_k) \\
&\le \operatorname{dist}(x_0, x_1) + K \cdot \operatorname{dist}(x_0, x_1) + K^2 \cdot \operatorname{dist}(x_0, x_1) + \cdots \\
&= \delta(1 + K + K^2 + \cdots) \\
&= \delta/(1-K).
\end{aligned}$$

Also, for $m \ge n$,

$$\begin{aligned}
\operatorname{dist}(x_n, x_m) &\le K \cdot \operatorname{dist}(x_{n-1}, x_{m-1}) \le K^2 \cdot \operatorname{dist}(x_{n-2}, x_{m-2}) \\
&\le \cdots \le K^n \cdot \operatorname{dist}(x_0, x_{m-n}) \le \delta K^n/(1-K)
\end{aligned}$$

which tends to 0 as $n \to \infty$, since $K < 1$.

Thus $x_0, x_1, x_2, \ldots$ is a Cauchy sequence. Put $\xi = \lim(x_i)$. Then we have that $T\xi = \lim Tx_i = \lim x_{i+1} = \xi$.

If x is another fixed point then

$$\operatorname{dist}(x, \xi) = \operatorname{dist}(Tx, T\xi) \le K \cdot \operatorname{dist}(x, \xi).$$

Since $K < 1$, this implies that $\operatorname{dist}(x, \xi) = 0$ and hence that $x = \xi$. □

The following is the basic ingredient going into the Implicit Function Theorem and the Inverse Function Theorem:

1.4. Lemma. *Let* $\xi \in \mathbf{R}^n$ *and* $\eta \in \mathbf{R}^m$ *be given. Let* $f: \mathbf{R}^n \times \mathbf{R}^m \to \mathbf{R}^m$ *be* C^1 *and put* $f = (f_1, \ldots, f_m)$. *(*f *need only be defined in a neighborhood of* (ξ, η).*) Assume that* $f(\xi, \eta) = \eta$ *and that all the following partial derivatives of* f *vanish at* (ξ, η):

$$\frac{\partial f_i}{\partial y_j}(\xi, \eta) = 0,$$

where $x_1, \ldots, x_n$ *are the coordinates in* $\mathbf{R}^n$ *and* $y_1, \ldots, y_m$ *those in* $\mathbf{R}^m$. *Then there exist numbers* $a > 0$ *and* $b > 0$ *such that there exists a unique function* $\phi: A \to B$, *where* $A = \{x \in \mathbf{R}^n \mid \|x - \xi\| \le a\}$ *and* $B = \{y \in \mathbf{R}^m \mid \|y - \eta\| \le b\}$, *such that* $\phi(\xi) = \eta$ *and* $\phi(x) = f(x, \phi(x))$ *for all* $x \in A$. *Moreover,* ϕ *is continuous.*

PROOF. By a transition of coordinates, we may as well assume that $\xi = 0, \eta = 0$, and $f(0,0) = 0$. [Precisely, if $\bar{\phi}(x) = \phi(\xi + x) - \eta$, $\bar{f}(x, y) = f(x + \xi, y + \eta) - \eta$ then $\phi(x) = f(x, \phi(x))$ becomes $\bar{\phi}(x) = \bar{f}(x, \bar{\phi}(x))$ and now the situation is centered at the origin.] We now assume this.

Applying Corollary 1.2 to each coordinate f_i of f and using the assumption that $\partial f_i/\partial y_j = 0$ at the origin, and hence is small in a neighborhood of 0, we can find $a > 0$ and $b > 0$ so that for any preassigned constant $0 < K < 1$,

$$(*) \qquad \|f(x, y) - f(x, \bar{y})\| < K\|y - \bar{y}\|$$

for $\|x\| \le a$, $\|y\| \le b$, and $\|\bar{y}\| \le b$.

Moreover, we can take a to be even smaller so that we also have, for all $\|x\| \le a$,

$$(**) \qquad Kb + \|f(x, 0)\| \le b.$$

Consider the set $\mathbf{F}$ of all functions $\phi: A \to B$ with $\phi(0) = 0$. Give this the "uniform metric": $\operatorname{dist}(\phi, \psi) = \sup\{\|\phi(x) - \psi(x)\| \mid x \in A\}$. This is a complete metric space since $(\{\phi_i\}$ Cauchy$) \Rightarrow (\{\phi_i(x)\}$ Cauchy for all $x) \Rightarrow (\lim \phi_i(x)$ exists for all $x)$. [Note that if we restrict this to continuous ϕ, it is still complete, since uniform limits preserve continuity.]

Define T: $\mathbf{F} \to \mathbf{F}$ by putting $(T\phi)(x) = f(x, \phi(x))$. One must check: (i) $(T\phi)(0) = 0$; and (ii) $(T\phi)(x) \in B$ for $x \in A$. Now $(T\phi)(0) = f(0, \phi(0)) = f(0, 0) = 0$, proving (i). Next we calculate

$$\begin{aligned} \|(T\phi)(x)\| = \|f(x, \phi(x))\| &\le \|f(x, \phi(x)) - f(x, 0)\| + \|f(x, 0)\| \\ &\le K\|\phi(x)\| + \|f(x, 0)\| \quad \text{by } (*) \\ &\le Kb + \|f(x, 0)\| \le b \quad \text{by } (**), \end{aligned}$$

proving (ii).

We claim that T is a contraction. In fact, we compute

$$\begin{aligned} \operatorname{dist}(T\phi, T\psi) &= \sup_{x \in A} (\|(T\phi)(x) - (T\psi)(x)\|) \\ &= \sup_{x \in A} (\|f(x, \phi(x)) - f(x, \psi(x))\|) \\ &\le \sup_{x \in A} K\|\phi(x) - \psi(x)\| \quad \text{by } (*) \\ &= K \cdot \operatorname{dist}(\phi, \psi). \end{aligned}$$

Thus the Banach Principle (Theorem 1.3) states that there is a unique ϕ: $A \to B$ with $\phi(0) = 0$ and $T\phi = \phi$; i.e., $f(x, \phi(x)) = \phi(x)$ for all $x \in A$.

It also states that $\phi = \lim \phi_i$ where ϕ_0 is arbitrary and $\phi_{i+1} = T\phi_i$. Put $\phi_0(x) = 0$ for all x. Then $\phi_1(x) = f(x, \phi_0(x))$ is continuous, $\phi_2(x) = f(x, \phi_1(x))$ is continuous, etc. Hence $\phi = \lim \phi_i$ is a uniform limit of continuous functions, and so it is continuous. □

1.5. Theorem (The Implicit Function Theorem). *Let* $g: \mathbf{R}^n \times \mathbf{R}^m \to \mathbf{R}^m$ *be* C^1

and let $\xi \in \mathbf{R}^n, \eta \in \mathbf{R}^m$ *be given with* $g(\xi, \eta) = 0$. *(g need only be defined in a neighborhood of* (ξ, η).) *Assume that the differential of the composition*

$$\mathbf{R}^m \to \mathbf{R}^n \times \mathbf{R}^m \to \mathbf{R}^m,$$
$$y \mapsto (\xi, y) \mapsto g(\xi, y),$$

is onto at η. [*This is equivalent to the statement that the Jacobian determinant* $J(g_i; y_j) \neq 0$ *at* (ξ, η).] *Then there are numbers* $a > 0$ *and* $b > 0$ *such that there exists a unique function* $\phi: A \to B$ *(with* A, B *as in Lemma 1.4), with* $\phi(\xi) = \eta$, *such that*

$$g(x, \phi(x)) = 0 \qquad \text{for all} \quad x \in A.$$

[*That is,* ϕ *"solves" the implicit relation* $g(x, y) = 0$.]

Moreover, if g *is* C^p *then so is* ϕ *(including the case* $p = \infty$).

PROOF. The differential referred to is the linear map $L: \mathbf{R}^m \to \mathbf{R}^m$ given by

$$L_i(y) = \sum_{j=1}^{m} \frac{\partial g_i}{\partial y_j}(\xi, \eta) y_j,$$

where $y = (y_1, \ldots, y_m)$ and $L(y) = (L_1(y), \ldots, L_m(y))$. That is, it is the linear map represented by the Jacobian matrix $(\partial g_i / \partial y_j)$ at (ξ, η).

The hypothesis says that L is nonsingular, and hence has a linear inverse $L^{-1}: \mathbf{R}^m \to \mathbf{R}^m$.

Let

$$f: \mathbf{R}^n \times \mathbf{R}^m \to \mathbf{R}^m$$

be defined by

$$f(x, y) = y - L^{-1}(g(x, y)).$$

Then $f(\xi, \eta) = \eta - L^{-1}(0) = \eta$. Also, computing differentials at η of $y \mapsto f(\xi, y)$ gives

$$I - L^{-1}L = I - I = 0.$$

Explicitly, this computation is as follows: Let $L = (a_{i,j})$ so that $a_{i,j} = (\partial g_i / \partial y_j)(\xi, \eta)$ and let $L^{-1} = (b_{i,j})$ so that $\sum b_{i,k} a_{k,j} = \delta_{i,j}$. Then

$$\frac{\partial f_i}{\partial y_j}(\xi, \eta) = \delta_{i,j} - \frac{\partial}{\partial y_j}\left[\sum_{k=1}^{m} b_{i,k} g_k(x, y)\right]_{(\xi,\eta)} = \delta_{i,j} - \sum_{k=1}^{m} b_{i,k} \frac{\partial g_k}{\partial y_j}(\xi, \eta)$$
$$= \delta_{i,j} - \sum_{k=1}^{m} b_{i,k} a_{k,j} = 0.$$

Applying Lemma 1.4 to get a, b, and ϕ with $\phi(\xi) = \eta$ and $f(x, \phi(x)) = \phi(x)$ we see that

$$\phi(x) = f(x, \phi(x)) = \phi(x) - L^{-1}(g(x, \phi(x))),$$

which is equivalent to $g(x, \phi(x)) = 0$.

We must now show that ϕ is differentiable. Since the Jacobian $J(g_i; y_j) \neq 0$ at (ξ, η), it is nonzero in a neighborhood, say $A \times B$. To show that ϕ is differentiable at a point $x \in A$ we can use the translation trick in the proof of Lemma 1.4 to reduce the question to the case $x = 0$, and we can also take $\xi = \eta = 0$. With this assumption, which is a minor notational convenience only, apply the Mean Value Theorem (Theorem 1.1) to $g(x, y)$, $g(0, 0) = 0$:

$$0 = g_i(x, \phi(x)) = g_i(x, \phi(x)) - g_i(0, 0)$$
$$= \sum_{j=1}^{n} \frac{\partial g_i}{\partial x_j}(p_i, q_i) \cdot x_j + \sum_{k=1}^{m} \frac{\partial g_i}{\partial y_k}(p_i, q_i) \cdot \phi_k(x),$$

where (p_i, q_i) is some point on the line segment between $(0, 0)$ and $(x, \phi(x))$. Let $h^{(j)}$ denote the point $(0, 0, \ldots, h, 0, \ldots, 0)$, with the h in the jth place, in $\mathbf{R}^n$. Then, putting $h^{(j)}$ in place of x in the above equation and dividing by h, we get

$$0 = \frac{\partial g_i}{\partial x_j}(p_i, q_i) + \sum_{k=1}^{m} \frac{\partial g_i}{\partial y_k}(p_i, q_i) \frac{\phi_k(h^{(j)})}{h}.$$

For j fixed, $i = 1, \ldots, m$ and $k = 1, \ldots, m$ these are m linear equations for the m unknowns

$$\frac{\phi_k(h^{(j)})}{h} = \frac{\phi_k(h^{(j)}) - \phi_k(0)}{h}$$

and they can be solved since the determinant of the coefficient matrix is $J(g_i; y_k) \neq 0$ in $A \times B$. The solution (Cramer's Rule) has a limit as $h \to 0$. Thus $(\partial \phi_k / \partial x_j)(0)$ exists and equals this limit.

We now know that ϕ is differentiable (once) in a neighborhood $A \times B$ of (ξ, η) and thus we can apply standard calculus to compute the derivative of the equations $g(x, \phi(x)) = 0$ (i.e., each $g_i(x, \phi(x)) = 0$) with respect to x_j. The Chain Rule gives

$$0 = \frac{\partial g_i}{\partial x_j}(x, \phi(x)) + \sum_{k=1}^{m} \frac{\partial g_i}{\partial y_k}(x, \phi(x)) \frac{\partial \phi_k}{\partial x_j}(x).$$

Again, these are linear equations ($i = 1, \ldots, m$, and j fixed) with nonzero determinant near (ξ, η) and hence has, by Cramer's Rule, a solution of the form

$$\frac{\partial \phi_k}{\partial x_j}(x) = F_{k,j}(x, \phi(x)),$$

where $F_{k,j}$ is C^{p-1} when g is C^p. ($F_{k,j}$ is just an analytic function of the $\partial g_i / \partial x_j$ and $\partial g_i / \partial y_k$ which are C^{p-1}.)

If ϕ is C^r for $r < p$, then the right-hand side of this equation is also C^r. Thus the left-hand side $\partial \phi_k / \partial x_j$ is C^r and hence the ϕ_k are C^{r+1}. By induction, the ϕ_k are C^p. Consequently, ϕ is C^p when g is C^p, as claimed. □

1.6. Theorem (The Inverse Function Theorem). *Let $\theta: \mathbf{R}^m \to \mathbf{R}^m$ be C^1 with $\theta(\eta) = \xi$ and with differential at η which is nonsingular (i.e., $J(\theta_i; y_j) \neq 0$ at*

$y=\eta$). Then there are numbers $a>0$ and $b>0$ such that there is a function $\phi: A \to B$, where

$$A = \{x \in \mathbf{R}^m \mid \|x-\xi\| \le a\}$$

and

$$B = \{y \in \mathbf{R}^m \mid \|y-\eta\| \le b\},$$

with $\phi(\xi)=\eta$ and with $\theta(\phi(x))=x$ for all $x \in A$.

Moreover, ϕ is C^p when θ is C^p. Furthermore, θ is a diffeomorphism of some neighborhood of η onto a neighborhood of ξ with inverse ϕ, when θ is C^∞.

PROOF. Apply the Implicit Function Theorem (Theorem 1.5) to $g: \mathbf{R}^m \times \mathbf{R}^m \to \mathbf{R}^m$ where $g(x,y)=\theta(y)-x$. We get ϕ with $g(x,\phi(x))=0$; i.e., $0=\theta(\phi(x))-x$, as desired. Now the Jacobian of ϕ is just the inverse of that of θ at $y=\eta$ by the Chain Rule. Thus we can apply the part of Theorem 1.6 now proved, to ϕ in place of θ.

Note that the equation $\theta(\phi(x))=x$ for $x \in A$ shows that ϕ is one–one on A and $\theta|_B$ is *onto* A. The application of the same result to ϕ in place of θ shows that $\phi: A \to B$ is *onto* some neighborhood B' of η in B (hence in $\mathbf{R}^m$).

Thus $\phi: A \to B'$ is one–one onto, and $\theta(\phi(x))=x$ shows that $\theta=\phi^{-1}$ here. If θ is C^∞ then so is ϕ and hence $\theta: B' \to A$ is a diffeomorphism with inverse ϕ. (Technically we should pass to smaller *open* sets here.) □

2. Differentiable Manifolds

A "topological n-manifold" is a Hausdorff space for which each point has a neighborhood homeomorphic to euclidean n-space (or, equivalently, an open subset of euclidean n-space). In each of these euclidean neighborhoods one can introduce a coordinate system. As one travels around the manifold, one must pass from one set of such coordinates to another. This requires a change of coordinates. The changes of coordinates are continuous functions (real functions of several real variables). If one wants to do "calculus" on such a space, however, the changes of coordinates will have to be *differentiable* functions. This leads to the notion of a "smooth" or "differentiable" manifold (or "C^∞-manifold"). We shall now formally define this notion. Actually, we are going to give two quite different definitions of smooth manifolds. The first one is rather traditional. The second one is more elegant and adapts more easily to some more general situations. We will make use of both.

2.1. Definition. An n-dimensional *differentiable manifold* (or *smooth manifold* or *C^∞-manifold*) is a second countable Hausdorff space M^n together with a collection of maps called "charts" such that:

(1) a chart is a homeomorphism $\phi: U \to U' \subset \mathbf{R}^n$ where U is open in M^n and U' is open in $\mathbf{R}^n$;

(2) each point $x \in M$ is in the domain of some chart;
(3) for charts $\phi: U \to U' \subset \mathbf{R}^n$ and $\psi: V \to V' \subset \mathbf{R}^n$ we have that the "change of coordinates" $\phi\psi^{-1}: \psi(U \cap V) \to \phi(U \cap V)$ is C^∞; and
(4) the collection of charts is maximal with properties (1), (2) and (3).

A set of charts satisfying conditions (1), (2), and (3) is often called an "atlas." It should be noted that an atlas can be enlarged *uniquely* to provide a maximal atlas as in the definition. Note that, by Theorem 12.12 of Chapter I, a smooth manifold is paracompact and its one-point compactification is metrizable.

2.2. Definition. Let X be a topological space. A *functional structure* on X is a function F_X defined on the collection of open sets U in X, such that:
(1) $\mathrm{F}_X(U)$ is a subalgebra of the algebra of all continuous real valued functions on U;
(2) $\mathrm{F}_X(U)$ contains all constant functions;
(3) $V \subset U, f \in \mathrm{F}_X(U) \Rightarrow f|_V \in \mathrm{F}_X(V)$; and
(4) $U = \bigcup U_\alpha$ and $f|_{U_\alpha} \in \mathrm{F}_X(U_\alpha)$ for all $\alpha \Rightarrow f \in \mathrm{F}_X(U)$.

The pair (X, F_X) is called a *functionally structured space.* (See Hochschild [1].)

Note: To check item (4) in Definition 2.2 for a particular example, one needs only check it for "small" U_α (diameter $< \varepsilon$ in some metric, for example). To see this, compare U both to the original $\{U_\alpha\}$ and to a "small" covering refining the original.

Some examples of functional structures are $X = \mathbf{R}^n$ with:

(1) $\mathrm{F}_X(U) = C^0(U) =$ all continuous real valued functions on U;
(2) $\mathrm{F}_X(U) = C^k(U) =$ all C^k functions on U;
(3) $\mathrm{F}_X(U) = C^\infty(U) =$ all C^∞ functions on U; and
(4) $\mathrm{F}_X(U) = C^\omega(U) =$ all real analytic functions on U.

2.3. Definition. A *morphism* of functionally structured spaces

$$(X, \mathrm{F}_X) \to (Y, \mathrm{F}_Y)$$

is a map $\phi: X \to Y$ such that composition $f \mapsto f \circ \phi$ carries $\mathrm{F}_Y(U)$ into $\mathrm{F}_X(\phi^{-1}(U))$. An *isomorphism* is a morphism ϕ such that ϕ^{-1} exists as a morphism.

Notation. If (X, F_X) is a given functionally structured space and $U \subset X$ is open, then let $\mathrm{F}_U(V) = \mathrm{F}_X(V)$ for open $V \subset U$. Then (U, F_U) is a functionally structured space. With scant abuse of notation, we shall simply use F to denote F_X, F_U, etc.

We now come to our second definition of a differentiable manifold.

2.4. Definition. An n-dimensional *differentiable manifold* is a second countable functionally structured Hausdorff space (M^n, F) which is locally isomorphic to ($\mathbf{R}^n$, C^∞). That is, each point in M has a neighborhood U such that $(U, \mathrm{F}_U) \approx (V, C_V^\infty)$ for some open $V \subset \mathbf{R}^n$.

In this case a morphism is called a *differentiable* or *smooth* map, an isomorphism is called a *diffeomorphism*, and members of $\mathrm{F}(U)$ are called differentiable (real valued) functions.

We shall now endeavor to show that these two definitions of a smooth manifold are, indeed, equivalent.

(Definition 2.4 $\Rightarrow$ Definition 2.1.) Let (M^n, F) be a given functionally structured space satisfying Definition 2.4. Let a "chart" be a map of an open subset $U \subset M$ to an open set $V \subset \mathbf{R}^n$ which is an isomorphism of functional structures. The domains of the charts cover M by Definition 2.4. Essentially all that needs proving, then, is that the "transition functions" $\theta = \phi\psi^{-1}$ between two charts are C^∞. But θ is an isomorphism of functional structures. Clearly all we need to show, then, is that a morphism $\theta: W \to W'$ of the C^∞ structures on open sets W and W' in $\mathbf{R}^n$ is the same thing as a C^∞ map on such sets.

By definition, θ is a morphism $\Leftrightarrow$ ($f \in C^\infty$ on an open set in $W' \Rightarrow f \circ \theta$ is C^∞). Thus it suffices to show

$$\theta: W \to W' \text{ is } C^\infty \quad \Leftrightarrow \quad f \circ \theta \text{ is } C^\infty \text{ for all } C^\infty f.$$

The implication $\Rightarrow$ is clear. For $\Leftarrow$, let f be a coordinate function (projection to a coordinate axis in $\mathbf{R}^n$) and decompose θ into its coordinate functions

$$\theta(x_1, \ldots, x_n) = (\theta_1(x_1, \ldots, x_n), \ldots, \theta_n(x_1, \ldots, x_n)).$$

Then $f \circ \theta = \theta_i$ for $f = i$th coordinate function. Thus each θ_i is C^∞. But that is exactly what it means for θ to be C^∞.

(Definition 2.1 $\Rightarrow$ Definition 2.4.) Suppose we are given a manifold M^n in the sense of charts. We must define $\mathrm{F}(U)$ for U open. By (4) of Definition 2.2 it suffices to do this for U small, and we shall do so for U the domain of a chart. If $\phi: U \to U' \subset \mathbf{R}^n$ is a chart, put

$$\mathrm{F}(U) = \{f \circ \phi \mid f \in C^\infty(U')\},$$

that is, define F_U such that ϕ is an isomorphism of functional structures. It is then easy to verify that this gives a smooth n-manifold in the sense of Definition 2.4.

2.5. Definition. A map $f: M \to N$ between two smooth manifolds is said to be *smooth* (or *differentiable* or C^∞) if, for any charts ϕ on M and ψ on N, the function $\psi \circ f \circ \phi^{-1}$ is smooth where it is defined. (Also see Problem 3.)

2.6. Definition. An n-manifold together with an atlas such that, for any two charts ϕ, ψ in the atlas, the Jacobian of the change of coordinates function $\phi \circ \psi^{-1}$ has positive determinant at all points in its domain, is called an *oriented manifold*. The particular atlas, maximal with this property, is called an *orientation* of the manifold. An n-manifold having such an atlas is called *orientable*.

Clearly a connected orientable manifold has exactly two orientations. The charts of one have Jacobian determinants that are negative when compared with charts from the opposite orientation. An orientation can be chosen on a connected orientable manifold by the choice of a compatible chart or local coordinates at any one point, which is often the way an orientation is specified.

2.7. Definition. An *n-manifold with boundary* is as in Definition 2.1 except that the target for charts is the half space $\{(x_1,\dots,x_n)\in\mathbf{R}^n \mid x_1 \leq 0\}$. Its *boundary* is the $(n-1)$-manifold consisting of all points mapped to $\{(x_1,\dots,x_n)\in\mathbf{R}^n \mid x_1 = 0\}$ by a chart.

Problems

1. Show that a second countable Hausdorff space X with a functional structure F is an n-manifold $\Leftrightarrow$ every point in X has a neighborhood U such that there are functions $f_1,\dots,f_n \in \mathrm{F}(U)$ such that: a real valued function g on U is in $\mathrm{F}(U) \Leftrightarrow$ there exists a smooth function $h(x_1,\dots,x_n)$ of n real variables $\ni g(p) = h(f_1(p),\dots,f_n(p))$ for all $p \in U$.
2. Complete the discussion of the two definitions of smooth manifold by showing that if one goes from one of the descriptions to the other, as indicated, and then back, one ends up with the same structure as at the start.
3. Show that a map $f: M \to N$ between smooth manifolds, with functional structures F_M and F_N, is smooth in the sense of Definition 2.5 $\Leftrightarrow$ it is smooth in the sense of Definition 2.4 (i.e., $g \in \mathrm{F}_N(U) \Rightarrow g \circ f \in \mathrm{F}_M(f^{-1}(U))$).
4. Let X be the graph of the real valued function $\theta(x) = |x|$ of a real variable x. Define a functional structure on X by taking $f \in \mathrm{F}(U) \Leftrightarrow f$ is the restriction to U of a C^∞ function on some open set V in the plane with $U = V \cap X$. Show that X with this structure is *not* diffeomorphic to the real line with usual C^∞ structure.
5. Consider the half open real line $[0, \infty)$. Define a functional structure F_1 by taking $f \in \mathrm{F}_1(U) \Leftrightarrow f(x) = g(x^2)$ for some C^∞ function g on $\{x \mid x \in U \text{ or } -x \in U\}$. Define another functional structure F_2 by taking $f \in \mathrm{F}_2(U) \Leftrightarrow f$ is the restriction to U of some C^∞ function on an open subset of $\mathbf{R}$. (Note that U is open in $[0, \infty)$ but not necessarily in $\mathbf{R}$.) Convince yourself that it is not unreasonable to believe that these structured spaces are equal, and also try to convince yourself that this is not a triviality; i.e., try to prove it.

3. Local Coordinates

Let M^n be a smooth manifold, let f be a real valued function on M, and let $x: U \to U' \subset \mathbf{R}^n$ be a chart. Let $\bar{f} = f \circ x^{-1}$ which is an ordinary real valued function $\bar{f}(x_1,\dots,x_n)$ of n real variables. Any point p in the domain U of the chart has coordinates

$$x(p) = (x_1(p), \ldots, x_n(p)).$$

Thus

$$f(p) = \bar{f}(x_1(p), \ldots, x_n(p)).$$

By abuse of notation, one often blurs this distinction between f and $\bar{f}$ and, on the domain of the chart, thinks of f as a function of the "local coordinates" $x_1, \ldots, x_n$. One must realize, however, that this representation of f depends on the choice of the chart. In another coordinate system (i.e., chart) this representation would change by the change of coordinates from one chart to another.

4. Induced Structures and Examples

Here we discuss some simple examples of manifolds, mostly with the intention of aiding the reader's understanding of the basic definitions. We also discuss three methods of creating new manifolds from old ones.

4.1. Definition. Suppose F_X is a functional structure on the space X and let $\phi: X \to Y$ be a map. Then the *induced functional structure* on Y is given by

$$F_Y(U) = \{f: U \to \mathbf{R} \mid f \circ \phi \in F_X(\phi^{-1}(U))\}.$$

For example, if $m > n$ and $\phi: \mathbf{R}^m \to \mathbf{R}^n$ is the projection then the induced structure from C^∞ on $\mathbf{R}^m$ is just C^∞ on $\mathbf{R}^n$ as the reader should verify.

We shall now give a number of examples of well-known manifolds defined by both the chart method and the functional structure method.

4.2. Example. Consider the torus $\mathbf{T}^2$ defined as the quotient space of $\mathbf{R}^2$ under the equivalence relation relating points whose coordinates differ by integer amounts. Let $\pi: \mathbf{R}^2 \to \mathbf{T}^2$ be the canonical projection. We wish to give a smooth structure on $\mathbf{T}^2$ by means of charts. This is quite easy, since for a small open disk $U \subset \mathbf{R}^2$ in $\mathbf{R}^2$, π maps U homeomorphically to its image U'. Thus the inverse of this can be taken to be a chart. If ϕ and ψ are two such charts then $\phi\psi^{-1}$ is just a translation and so it is C^∞ (in fact, real analytic), and so this does define a smooth structure on $\mathbf{T}^2$. (See Figure II-1.)

Let us now show how to define the structure by means of functional structures. This is quite trivial in this example, since we can just take the structure induced from the standard one on $\mathbf{R}^2$ by the projection π. In this case, however, we must show that this induced structure is that of a manifold. But this need be done only locally. Let U' be a small open set in $\mathbf{T}^2$ whose inverse image, as above, is the disjoint union of open sets U homeomorphic to U' under π. Let f be a real valued function on U'. If U and V are two of the open sets in $\mathbf{R}^2$ mapping homeomorphically to U' then $f \circ \pi$ is C^∞ on U

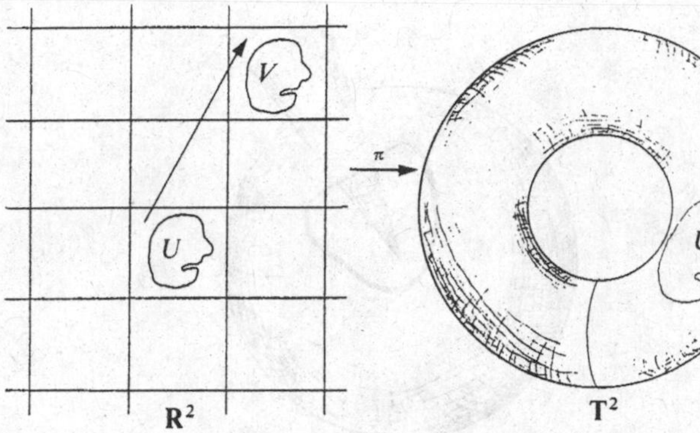

Figure II-1. Differentiable structure on $\mathbf{T}^2$.

if and only if it is on V since the difference is merely a translation $U \to V$, which is smooth. Thus we see that the map $U \to U'$ is an isomorphism of functionally structured spaces where $U \subset \mathbf{R}^2$ has the C^∞ structure. Therefore, $\mathbf{T}^2$ is indeed locally isomorphic to $\mathbf{R}^2$.

4.3. Example. Consider the sphere $\mathbf{S}^2$. As before, we will first give a structure via charts. For this let us take the sphere of radius $\frac{1}{2}$ in the upper half space tangent to the x–y plane at the origin. Thus the north pole is the point $(0, 0, 1)$ and the south pole is the origin. We map $\mathbf{S}^2 - \{(0, 0, 1)\}$ to the plane by "stereographic projection," i.e., we take the line from $(0, 0, 1)$ to another point on the sphere and produce it until it intersects the x–y plane, and the chart ϕ is the map taking that point on the sphere to that intersection point in the plane. (See Figure II-2.) For a second chart ψ, we similarly take the stereographic projection from the origin to the plane $z = 1$ followed by the translation to the x–y plane. The comparison of these charts is $\psi\phi^{-1}: \mathbf{R}^2 - \{0\} \to \mathbf{R}^2 - \{0\}$ and is given by $\mathbf{x} \mapsto \mathbf{x}/\|\mathbf{x}\|^2$ as the reader can

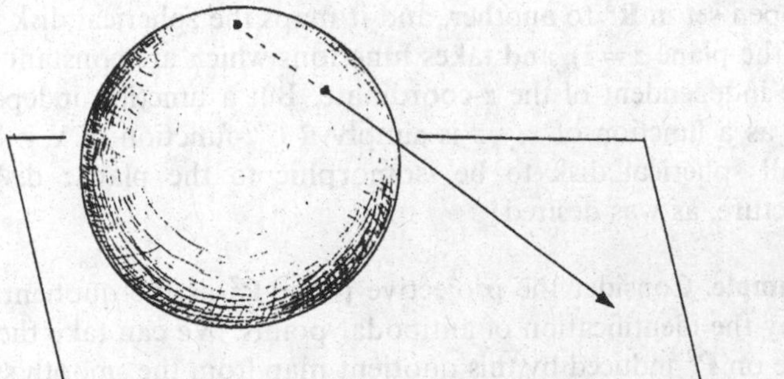

Figure II-2. Stereographic projection.

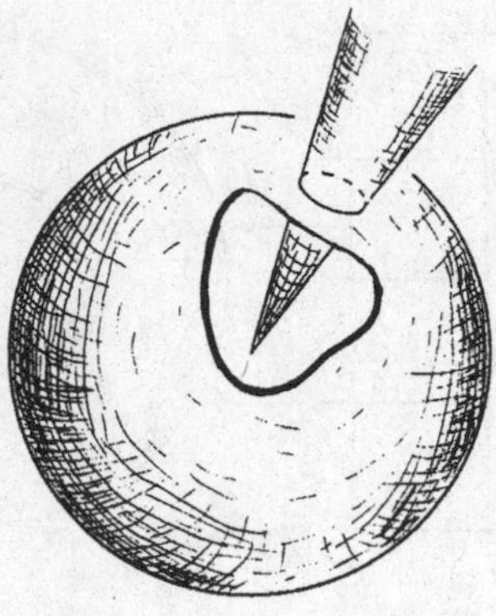

Figure II-3. Defining a functional structure on a sphere.

calculate. Since this map is C^∞ on $\mathbf{R}^2 - \{0\}$, this does define a smooth manifold.

To define the structure on $\mathbf{S}^2$ by means of functional structures we will regard the sphere as the unit sphere in $\mathbf{R}^3$. Consider the radial projection $\pi: \mathbf{R}^3 - \{0\} \to \mathbf{S}^2$. We take the structure induced from C^∞ on $\mathbf{R}^3 - \{0\}$ by this map π.

This is a much simpler description of the structure than that given by charts, but the difficulty is in showing that this does define a C^∞ manifold. To do this, we consider a portion of the sphere cut out by a small circle and the single sheeted open cone through that. (See Figure II-3.) The induced structure on that open "disk" on the sphere simply consists of those functions f such that the function, obtained on the open cone by making f constant along radii, is C^∞. We must show this to be isomorphic to C^∞ on some open set in $\mathbf{R}^2$. To see this, we may as well just consider the case where the disk on the sphere is taken around the north pole, since rotations in $\mathbf{R}^3$ are C^∞. Consider the map of this open cone to an open cylinder, given by $(x, y, z) \mapsto (x/\rho, y/\rho, \rho)$ where $\rho = \sqrt{(x^2 + y^2 + z^2)}$. This map is clearly a diffeomorphism of one open set in $\mathbf{R}^3$ to another, and it maps the spherical disk to a planar disk (in the plane $z = 1$), and takes functions which are constant along radii to those independent of the z-coordinate. But a function independent of z and C^∞ as a function of x, y, z is simply a C^∞-function of x, y. This shows the small spherical disk to be isomorphic to the planar disk with the C^∞-structure, as was desired.

4.4. Example. Consider the projective plane $\mathbf{P}^2$ as the quotient space of a sphere by the identification of antipodal points. We can take the functional structure on $\mathbf{P}^2$ induced by this quotient map from the smooth structure on $\mathbf{S}^2$. Very much in the same way as with the torus in Example 4.2, one simply has to know that the antipodal map is smooth on the sphere, to conclude that this does define a smooth structure on $\mathbf{P}^2$. But that is trivial.

All of these examples could have been done in n dimensions with no further complications. We now discuss induced structures of a type opposite to that above. Although it can be done more generally for maps $X \to Y$, we will confine it to the case of inclusions.

4.5. Definition. Suppose that X is a topological space and that $A \subset X$ is a subspace. Let F be a functional structure on X. We define a functional structure F_A on A by letting $f \in F_A(U \cap A) \Leftrightarrow$ each point of $U \cap A$ has a neighborhood W in $X \ni f$ is the restriction to $W \cap A$ of some function $g \in F(W)$.

Considering $\mathbf{T}^2$ and $\mathbf{S}^2$ to be subspaces of $\mathbf{R}^3$ in the usual ways, this gives another way to define a smooth structure on these spaces. Of course, one would have to show that the structure obtained this way is smooth, but that will follow from general results we will give later on. On the other hand, if we consider the surface of a cone in $\mathbf{R}^3$ then this definition gives a functional structure on the cone. However, this is not a smooth structure. The cone is homeomorphic, but not diffeomorphic to $\mathbf{R}^2$. Thus, functional structures give an easy way to describe "singularities" like the vertex of a cone or of a cusp, which would be much harder to handle with the chart type of definition.

Finally, let us define the product differential structure on a product of two smooth manifolds.

4.6. Definition. Let M^m and N^n be smooth manifolds. If $\phi: U \to \mathbf{R}^m$ is a chart for M and $\psi: V \to \mathbf{R}^n$ is a chart for N, then take $\phi \times \psi: U \times V \to \mathbf{R}^{m+n}$ to be a chart for $M \times N$. This defines a smooth structure on $M \times N$ called the *product structure*.

For example, take the circle $\mathbf{S}^1$ with a smooth structure, then this gives a structure on the tori $\mathbf{T}^2 = \mathbf{S}^1 \times \mathbf{S}^1$, $\mathbf{T}^3 = \mathbf{S}^1 \times \mathbf{S}^1 \times \mathbf{S}^1$, etc. It is not too difficult to show that the smooth manifolds produced this way are diffeomorphic to those described in Example 4.2.

Note that the charts described in Definition 4.6 do not satisfy axiom (4) (maximality) of Definition 2.1, but this is not necessary, as remarked there, since there is a unique maximal atlas containing them.

Note also that, for a product of manifolds, the projections $M \times N \to M$ and $M \times N \to N$ are smooth.

Problems

1. Consider the 3-sphere $\mathbf{S}^3$ as the set of unit quaternions

$$\{x + iy + jz + kw \mid x^2 + y^2 + z^2 + w^2 = 1\}.$$

Let $\phi: U \to \mathbf{R}^3$ be $\phi(x + iy + jz + kw) = (y, z, w) \in \mathbf{R}^3$ where

$$U = \{x + iy + jz + kw \in \mathbf{S}^3 \mid x > 0\}.$$

Consider ϕ as a chart. For each $q \in \mathbf{S}^3$, define a chart ψ_q by $\psi_q(p) = \phi(q^{-1}p)$. Show that this set of charts is an atlas for a smooth structure on $\mathbf{S}^3$.

2. Let X be a copy of the real line $\mathbf{R}$ and let $\phi: X \to \mathbf{R}$ be $\phi(x) = x^3$. Taking ϕ as a chart, this defines a smooth structure on X. Prove or disprove the following statements:
(1) X is diffeomorphic to $\mathbf{R}$;
(2) the identity map $X \to \mathbf{R}$ is a diffeomorphism;
(3) ϕ together with the identity map comprise an atlas;
(4) on the one-point compactification X^+ of X, ϕ and ψ give an atlas, where $\psi(x) = 1/x$, for $x \neq 0, \infty$, and $\psi(\infty) = 0$. (ψ is defined on $X^+ - \{0\}$.)

5. Tangent Vectors and Differentials

All readers are well acquainted with the notion of tangent vectors to curves and surfaces embedded in 3-space. Perhaps, however, many readers are not aware that this notion is intrinsic to the curve or surface and has little to do with the particular embedding in 3-space. It is important to give the notion of a tangent vector an intrinsic setting, not dependent on, or even using, an embedding in euclidean space. One way to do this is to associate the notion of tangent vectors with that of directional derivatives, or derivatives along parametrized curves in the manifold in question. That is the approach we take.

5.1. Definition. Let M be a smooth manifold and $\gamma: \mathbf{R} \to M$ a smooth curve with $\gamma(0) = p$. (γ need only be defined in a neighborhood of 0.) Let $f: U \to \mathbf{R}$ be smooth where U is an open neighborhood of p. Then the *directional derivative* of f along γ at p is

$$D_\gamma(f) = \frac{d}{dt} f(\gamma(t))|_{t=0}.$$

The operator D_γ is called the *tangent vector* to γ at p. For two such curves γ and γ' we regard $D_\gamma = D_{\gamma'}$ if they have the same value at p on each such function f.

5.2. Definition. If M is a smooth manifold and $p \in M$, $T_p(M)$ denotes the vector space of all tangent vectors to M at p. (See below for the fact that this is a vector space.)

5.3. Definition. A *germ* of a smooth real valued function f at $p \in M$ on a smooth manifold M is the equivalence class of f under the equivalence relation $f_1 \sim f_2 \Leftrightarrow f_1(x) = f_2(x)$ for all x in some neighborhood of p.

Note that $D_\gamma(f)$ is defined on the *germ* of f. Letting $D = D_\gamma$, we note two properties of tangent vectors:

(1) $D(af + bg) = aD(f) + bD(g)$ where a and b are constant; and
(2) $D(fg) = f(p)D(g) + D(f)g(p)$.

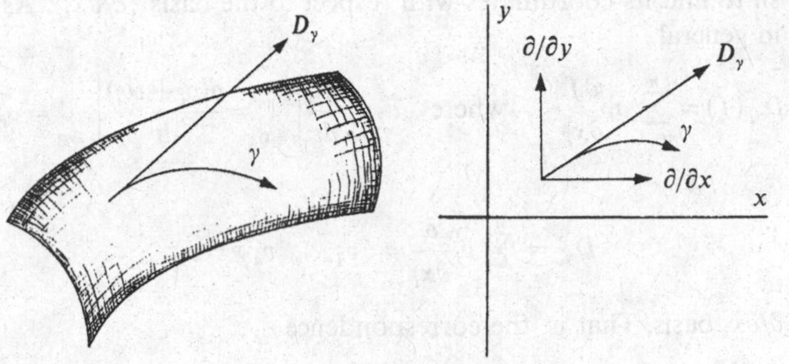

Figure II-4. Tangent vectors.

That is, D is a "derivation" of the algebra of germs of smooth real valued functions on M at p. We remark that one can show (in the C^∞ case only) that any derivation is a tangent vector.

Let us interpret the foregoing in terms of local coordinates. Let $x_1, \ldots, x_n$ be local coordinates at p. Then (by abuse of notation) $\gamma(t) = (\gamma_1(t), \ldots, \gamma_n(t))$ where $\gamma_i(t) = x_i(\gamma(t))$. Then

$$D_\gamma(f) = \frac{d}{dt} f(\gamma_1(t), \ldots, \gamma_n(t))|_{t=0}$$

$$= \sum_{i=1}^{n} \frac{\partial f}{\partial x_i} \frac{d\gamma_i}{dt}\bigg|_{t=0}.$$

Thus

$$D_\gamma = \sum_{i=1}^{n} a_i \frac{\partial}{\partial x_i}\bigg|_p,$$

where $a_i = d\gamma_i/dt$ at $t = 0$.

Now

$$\frac{\partial}{\partial x_i}\bigg|_p = D_{v_i} \qquad \text{where} \quad v_i(t) = (0, \ldots, 0, t, 0, \ldots, 0),$$

where the t is in the ith place. Therefore the set $T_p(M)$ of tangent vectors to M at p is a vector space with basis $\{\partial/\partial x_i | i = 1, \ldots, n\}$. This also shows that, for tangent vectors X and Y at p and scalars $a, b \in \mathbf{R}$, we have

$$\boxed{(aX + bY)(f) = a(X(f)) + b(Y(f)).}$$

5.4. Example. We go through these definitions for the manifold $\mathbf{R}^n$ itself.

Fix a point $p \in \mathbf{R}^n$ and an ordinary n-vector $v = (v_1, \ldots, v_n)$. Let $\gamma_v(t) = p + tv = (p_1 + tv_1, \ldots, p_n + tv_n)$. We then have the tangent vector D_{γ_v}, and wish to find its coordinates with respect to the basis $\{\partial/\partial x_i\}$. As shown above in general,

$$D_{\gamma_v}(f) = \sum_{i=1}^{n} a_i \frac{\partial f}{\partial x_i} \quad \text{where} \quad a_i = \left.\frac{dx_i}{dt}\right|_{t=0} = \left.\frac{d(p_i + tv_i)}{dt}\right|_{t=0} = v_i.$$

Thus

$$D_{\gamma_v} = \sum_{i=1}^{n} v_i \frac{\partial}{\partial x_i} = \langle v_1, \ldots, v_n \rangle = v$$

in the $\partial/\partial x_i$ basis. That is, the correspondence

$$\mathbf{R}^n \leftrightarrow T_p(\mathbf{R}^n),$$
$$v \leftrightarrow D_{\gamma_v},$$

is an isomorphism of vector spaces. By abuse of notation, it is often regarded as an equality.

5.5. Definition. If $\phi: M \to N$ is a smooth map between two smooth manifolds then we define the *differential* of ϕ at $p \in M$ to be the function

$$\phi_*: T_p(M) \to T_{\phi(p)}(N)$$

given by $\phi_*(D_\gamma) = D_{\phi \circ \gamma}$. (The differential ϕ_* is also often denoted by $d\phi$.)

5.6. Proposition. *The differential ϕ_* of a smooth map $\phi: M \to N$ is well defined and linear. It satisfies the equation*

$$(\phi_* D)(g) = D(g \circ \phi).$$

Moreover, $\phi_* \psi_* = (\phi \circ \psi)_*$.

PROOF. Let $g: U \to \mathbf{R}$ be a smooth function where U is a neighborhood of $\phi(p)$. Then

$$\phi_*(D_\gamma)(g) = D_{\phi \circ \gamma}(g) = \frac{d}{dt} g(\phi(\gamma(t)))|_{t=0} = D_\gamma(g \circ \phi).$$

Therefore $(\phi_* D)(g) = D(g \circ \phi)$ which shows that $\phi_* D$ is well defined and satisfies the stated formula.

For tangent vectors X and Y at $p \in M$, we have $\phi_*(aX + bY)(f) = (aX + bY)(f \circ \phi) = aX(f \circ \phi) + bY(f \circ \phi) = (a\phi_*(X) + b\phi_*(Y))(f)$ and so ϕ_* is linear.

Also $(\phi_* \psi_* D)(g) = (\psi_* D)(g \circ \phi) = D(g \circ \phi \circ \psi) = ((\phi \circ \psi)_*(D))(g)$ giving the last formula. □

Note the following special case in which the smooth map ϕ is a curve: Let $\gamma: \mathbf{R} \to M$ be a smooth curve. Then

$$\gamma_*\left(\left.\frac{d}{dt}\right|_{t=0}\right)(f) = \frac{d}{dt}(f \circ \gamma)|_{t=0} = D_\gamma(f)$$

so that

$$\boxed{\gamma_*\left(\frac{d}{dt}\right) = D_\gamma.}$$

Let us now interpret the differential in terms of local coordinates. Let $\phi: M \to N$ be a smooth map and consider local coordinates $x_1, \ldots, x_m$ near the point $p \in M$ and $y_1, \ldots, y_n$ near the point $\phi(p) \in N$. Then we can write

$$\phi(x_1, \ldots, x_m) = (\phi_1(x_1, \ldots, x_m), \ldots, \phi_n(x_1, \ldots, x_m))$$

and we wish to find $\phi_*: T_p(M) \to T_{\phi(p)}(N)$ in terms of the bases $\partial/\partial x_i$ and $\partial/\partial y_j$. The linear map ϕ_* is represented, with respect to these bases, by a matrix $(a_{i,j})$ where

$$\phi_*\left(\frac{\partial}{\partial x_j}\right) = \sum_{i=1}^{n} a_{i,j} \frac{\partial}{\partial y_i}$$

and so

$$\phi_*\left(\frac{\partial}{\partial x_j}\right)(y_k) = \sum_{i=1}^{n} a_{i,j} \frac{\partial}{\partial y_i}(y_k) = a_{k,j}.$$

But the left-hand side of this equation is

$$\frac{\partial}{\partial x_j}(y_k \circ \phi) = \frac{\partial}{\partial x_j} \phi_k.$$

Therefore ϕ_* is represented by the Jacobian matrix

$$\left(\frac{\partial \phi_i}{\partial x_j}\right).$$

5.7. Definition. If $\phi: M \to N$ is smooth then:

(1) if ϕ_* is a monomorphism at all points then ϕ is an *immersion*;
(2) if ϕ_* is onto at all points then ϕ is a *submersion*;
(3) if ϕ is an immersion and one–one then (M, ϕ) is a *submanifold*; and
(4) if (M, ϕ) is a submanifold and $\phi: M \to \phi(M)$ is a homeomorphism for the relative topology on $\phi(M)$, then ϕ is called an *embedding*, and $\phi(M)$ is called an *embedded submanifold* of N.

In case $\phi: M \to N$ is an embedding (or just an immersion) then we may (and often will), without confusion, identify a tangent vector $v \in T_p(M)$ with $\phi_*(v) \in T_{\phi(p)}(N)$.

Similarly, if M and N are smooth manifolds then so is $M \times N$ by taking as charts the products of those for the factors. Clearly then $T_{(p,q)}(M \times N) \approx T_p(M) \times T_q(N)$ and we will often identify them when no confusion can result.

PROBLEMS

1. ♢ If $\phi: \mathbf{R}^m \to \mathbf{R}^n$ is a linear map and we identify $T_p(\mathbf{R}^k)$ with $\mathbf{R}^k$ by identifying $\partial/\partial x_i$ with the ith standard basis vector, show that ϕ_* becomes ϕ.
2. If the curve $\phi: \mathbf{R} \to \mathbf{R}^n$ is an embedding then show that $\phi_*(d/dt)$ coincides with the classical notion of the tangent vector to the curve ϕ under the identification of the tangent space to a euclidean space with the euclidean space.
3. ♢ For a smooth function f defined on a neighborhood of a point $p \in \mathbf{R}^n$, the gradient $\nabla f = \operatorname{grad} f$ of f is the vector
$$\left\langle \frac{\partial f}{\partial x_1}, \ldots, \frac{\partial f}{\partial x_n} \right\rangle.$$
For a vector $v \in \mathbf{R}^n$ show that the directional derivative D_v, denoted by D_{γ_v} in Example 5.4, satisfies the equation
$$\boxed{D_v f = \langle \nabla f, v \rangle,}$$
the standard inner product of ∇f with v in $\mathbf{R}^n$.
4. ♢ If $M^m \subset \mathbf{R}^n$ is a smoothly embedded manifold and f is a smooth real valued function defined on a neighborhood of $p \in M^m$ in $\mathbf{R}^n$ and which is constant on M, show that ∇f is perpendicular to $T_p(M)$ at p.

6. Sard's Theorem and Regular Values

In this section we introduce the notion of "regular value" of a smooth map. This is a type of "general position" concept, and is one of the most useful tools when dealing with smooth manifolds from the topological viewpoint.

6.1. Definition. If $\phi: M^m \to N^n$ is a smooth map then a point $p \in M^m$ is called a *critical point* of ϕ if $\phi_*: T_p(M) \to T_{\phi(p)}(N)$ has rank $< n$. The image in N^n of a critical point is called a *critical value*. A point of N^n which is not a critical value is called a *regular value* (even though it may not be in the image of ϕ).

Note that this means that a point $q \in N^n$ is a regular value provided:
$$m \geq n \quad \Rightarrow \quad \phi_* \text{ is onto at all points } p \in \phi^{-1}(q), \quad \text{and}$$
$$m < n \quad \Rightarrow \quad q \notin \text{Image}(\phi).$$
Any point not in the image is automatically a regular value. It might seem strange to call a point $q \notin \phi(M)$ a "regular value" when it is not even a "value," but this convention makes the statements and proofs of results concerning regular values much simpler than if one were to exclude points not in the image from the definition.

The following is the main result concerning regular values. Its proof requires a small amount of measure theory. The proof contains no ideas of particular interest to us here and so is relegated to Appendix C.

6.2. Theorem (Sard's Theorem). *If $\phi: M^m \to \mathbf{R}^n$ is a smooth map then the set of critical values has measure zero.*

6.3. Corollary (A.B. Brown). *If $\phi: M^m \to N^n$ is smooth then the set of regular values is residual in N^n.*

PROOF. If C is the critical set of ϕ and $K \subset M$ is compact then $\phi(C \cap K)$ is compact and its interior is empty by Theorem 6.2. Therefore $\phi(C \cap K)$ is nowhere dense in N. Since M is covered by a countable number of such sets K, $\phi(C)$ is of first category, and thus its complement is residual. □

Note that, in the case $m = 1$, Sard's Theorem shows that there do not exist smooth "space filling curves," in distinction to the nonsmooth case.

The reader who wishes to examine the proof of Sard's Theorem in Appendix C should first read Section 7 because the proof uses some elementary items from that section.

We shall have many applications of Sard's Theorem later on in this book. It is one of the central tools of differential topology. For now, we will rest content with the following application to a classical result:

6.4. Corollary (Fundamental Theorem of Algebra). *If $p(z)$ is a complex polynomial of positive degree then $p(z)$ has a zero.*

PROOF. (This argument is due to Milnor [3]). Let $p(x + iy) = u(x, y) + iv(x, y)$. Then $p'(z) = (u_x + iv_x) = -i(u_y + iv_y)$ as is seen by differentiating with respect to x and y and comparing the answers (and, of course, is very well known for all complex analytic functions). Thus the Jacobian $J(u, v; x, y) = u_x^2 + u_y^2$ is zero if and only if $p'(z) = 0$. There are only a finite number of points (zeros of p') that can satisfy this. Thus, $p: \mathbf{R}^2 \to \mathbf{R}^2$ has only a finite number of critical points. Let $F \subset \mathbf{R}^2$ be the finite set of critical values.

Letting $p(z) = a_0 z^n + a_1 z^{n-1} + \cdots + a_n$, with $a_0 \neq 0$, the equation

$$|p(z)| = |z|^n |a_0 + a_1 z^{-1} + \cdots + a_n z^{-n}|$$

shows that $|p(z)| \to \infty$ as $|z| \to \infty$. This means that p can be extended continuously to the one-point compactification $\mathbf{S}^2$ of $\mathbf{R}^2$ and hence that p is a proper map. Thus p is a closed mapping by Proposition 11.5 of Chapter I.

For any $c \in \mathbf{C}$, $p^{-1}(c)$ consists of the zeros of the polynomial $p(z) - c$ and so it contains at most n points. Let $k = k(c)$ be the number of points in $p^{-1}(c)$. If c is a regular value and $p^{-1}(c) = \{z_1, \ldots, z_k\}$, then each z_i has a neighborhood U_i mapping diffeomorphically onto a neighborhood $V_i \subset \mathbf{R}^2 - F$ of c. Since $\mathbf{R}^2$ is Hausdorff, we can assume the U_i to be disjoint. There is an open connected neighborhood V of c inside the open set $V_1 \cap \cdots \cap V_k - f(\mathbf{R}^2 - (U_1 \cup \cdots \cup U_k))$. Then $W_i = U_i \cap p^{-1}(V)$ is taken diffeomorphically by p onto V. Moreover, $p^{-1}(V) = W_1 \cup \cdots \cup W_k$ (disjoint) since $p(z) \in V \Rightarrow z \in p^{-1}(V) \cap (U_1 \cup \cdots \cup U_k) = W_1 \cup \cdots \cup W_k$. It follows that $k(c)$ is locally constant on the set $\mathbf{R}^2 - F$ of regular values. Since $\mathbf{R}^2 - F$ is connected, $k(c)$ is constant on

$\mathbf{R}^2 - F$. This constant cannot be zero since that would imply that the image of p is F, and hence that p is constant since its image is connected, but p is not constant. This shows that the image of p contains $(\mathbf{R}^2 - F) \cup F = \mathbf{R}^2$, so that p takes on all values including 0. □

The ideas in the last paragraph of the proof of Corollary 6.4 will be of importance to us in later parts of this chapter; see Theorem 11.6 and Section 16.

PROBLEMS

1. For the map $\phi(x) = x\sin(x)$ of the real line to itself, what are the regular values?
2. For the map $\phi(x, y) = x^2 - y^2$ of the plane to the line, what are the regular values?
3. For the map $\phi(x, y) = \sin(x^2 + y^2)$ of the plane to the line, what are the regular values?
4. Criticize the following "counterexample" of Sard's Theorem: Let M^0 be the real line with the discrete topology. This is a 0-manifold. The canonical map $M^0 \to \mathbf{R}$ then has no regular values.
5. Let $\gamma: \mathbf{R} \to \mathbf{R}^2$ be a smooth curve in the plane. Let K be the set of all $r \in \mathbf{R}$ such that the circle of radius r about the origin is tangent to the curve γ at some point. Show that K has empty interior in $\mathbf{R}$.
6. If C is a circle embedded smoothly in $\mathbf{R}^4$, show that there exists a three-dimensional hyperplane H such that the orthogonal projection of C to H is an embedding.
7. Formulate and prove a "Fundamental Theorem of algebra" for quaternionic polynomials.

7. Local Properties of Immersions and Submersions

This section is mainly a simple generalization of the Inverse Function Theorem from the case of euclidean space to that of general smooth manifolds. There is nothing deep about this generalization and it is mainly a matter of notation.

First let us note that if $\phi: M \to N$ is a smooth map and if $\phi_*: T_p(M) \to T_{\phi(p)}(N)$ is a monomorphism, then it is a monomorphism at any point of some neighborhood of p. Also, if ϕ_* is onto at p then it is onto at any point in some neighborhood of p. The reason for this is that these are the cases for which ϕ_* has maximum possible rank, and the rank of a matrix (the Jacobian in these cases) is the largest size of a square submatrix having nonzero determinant. But the determinant is a continuous function of its entries, and hence of the point p, so it will still be nonzero in some neighborhood of p.

7.1. Theorem. *Let $\theta: M^m \to N^n$ be smooth and assume that $\theta_*: T_p(M^m) \to T_{\theta(p)}(N^n)$ is a monomorphism (at the particular point p). Then there are charts ϕ at p*

and ψ at $\theta(p)$ such that the following diagram commutes:

$$\begin{array}{ccc} M^m & \xrightarrow{\theta} & N^n \\ \uparrow{\scriptstyle \phi^{-1}} & & \uparrow{\scriptstyle \psi^{-1}} \\ \mathbf{R}^m & \xrightarrow{\theta'} & \mathbf{R}^n \end{array}$$

where θ' is the standard inclusion of $\mathbf{R}^m$ in $\mathbf{R}^n$: $(x_1, \ldots, x_m) \mapsto (x_1, \ldots, x_m, 0, \ldots, 0)$. (Accordingly, in these local coordinates, $\theta(M)$ is "flat" in N.)

PROOF. Take arbitrary charts ϕ and ψ such that the origin in euclidean space corresponds to p and $\theta(p)$, respectively. Then $\theta'_*: \mathbf{R}^m = T_0(\mathbf{R}^m) \to T_0(\mathbf{R}^n) = \mathbf{R}^n$ is a monomorphism. By a change of coordinates (a rotation) we can assume that the image of θ'_* is $\mathbf{R}^m \subset \mathbf{R}^n = \mathbf{R}^m \times \mathbf{R}^{n-m}$.

We wish to change coordinates in $\mathbf{R}^n$ by a map $\zeta: \mathbf{R}^n \to \mathbf{R}^n$ so that the diagram

$$\begin{array}{ccc} \mathbf{R}^m & \xrightarrow{\theta'} & \mathbf{R}^n = \mathbf{R}^m \times \mathbf{R}^{n-m} \\ & {\scriptstyle \iota}\searrow & \uparrow{\scriptstyle \zeta} \\ & & \mathbf{R}^n = \mathbf{R}^m \times \mathbf{R}^{n-m} \end{array}$$

commutes, where $\iota(x) = (x, 0)$. We must take $\zeta(x, 0) = \theta'(x)$ and so the obvious candidate for ζ is $\zeta(x, y) = \theta'(x) + y$. With this choice, note that ζ_* takes the tangent space of $\mathbf{R}^m \times \{0\}$ onto itself by θ'_*. Also, it takes the tangent space of $\{0\} \times \mathbf{R}^{n-m}$ onto itself by the identity. Consequently ζ_* is an isomorphism.

By the Inverse Function Theorem (Theorem 1.6), ζ is a diffeomorphism in the neighborhood of the origin. We claim that if we replace the chart ψ with $\zeta^{-1} \circ \psi$ (possibly cutting down on the domain) the new charts satisfy the conclusion of the theorem. But the new θ' is just $\iota = \zeta^{-1} \circ \theta'$ and $\iota(x) = (x, 0)$ as desired. □

7.2. Corollary. *If $M \subset N$ is an embedded submanifold then M has the induced functional structure as a subspace on N.* □

Another way to phrase the corollary is that if f is a smooth real valued function on M, then it extends locally to a smooth function on a neighborhood in N.

7.3. Theorem. *Let $\theta: M^m \to N^n$ be smooth and assume that $\theta_*: T_p(M^m) \to T_{\theta(p)}(N^n)$ is onto (at the particular point p). Then there are charts ϕ at p and ψ at $\theta(p)$ such that the following diagram commutes:*

$$\begin{array}{ccc} M^m & \xrightarrow{\theta} & N^n \\ \uparrow{\scriptstyle \phi^{-1}} & & \uparrow{\scriptstyle \psi^{-1}} \\ \mathbf{R}^m & \xrightarrow{\theta'} & \mathbf{R}^n \end{array}$$

where θ' is the standard projection of $\mathbf{R}^m$ onto $\mathbf{R}^n$: $(x_1, \ldots, x_m) \mapsto (x_1, \ldots, x_n)$.

Proof. Actually, in this proof, we will regard $\mathbf{R}^m$ as $\mathbf{R}^k \times \mathbf{R}^n$ and the projection in question as the map taking $(x, y) \in \mathbf{R}^m$ to $y \in \mathbf{R}^n$. This is of no import. Let ϕ and ψ be charts at p and $\theta(p)$ with $\phi(p) = 0 = \psi(\theta(p))$.

By a rotation of coordinates at $p \in M$ we can assume that $\ker(\theta'_*) = \mathbf{R}^k \times \{0\}$ at 0. We wish to change coordinates in $\mathbf{R}^m$ by a map $\zeta: \mathbf{R}^m \to \mathbf{R}^m$ so that the diagram

$$\begin{array}{ccc} \mathbf{R}^k \times \mathbf{R}^n = \mathbf{R}^m & \xrightarrow{\theta'} & \mathbf{R}^n \\ \zeta \downarrow & \nearrow_{\pi} & \\ \mathbf{R}^k \times \mathbf{R}^n = \mathbf{R}^m & & \end{array}$$

commutes, where $\pi(x, y) = y$. Then the second coordinate of $\zeta(x, y)$ must be $\theta'(x, y)$, so that the obvious candidate for ζ is $\zeta(x, y) = (x, \theta'(x, y))$. With this choice, note that $\ker(\zeta_*) \subset \ker(\theta'_*) = \mathbf{R}^k \times \{0\}$ since $\theta' = \pi \circ \zeta$. But the composition $\mathbf{R}^k \approx \mathbf{R}^k \times \{0\} \hookrightarrow \mathbf{R}^k \times \mathbf{R}^n \xrightarrow{\zeta} \mathbf{R}^k \times \mathbf{R}^n \to \mathbf{R}^k$ is the identity, so that $\ker(\zeta_*) \cap (\mathbf{R}^k \times \{0\}) = \{0\}$. Consequently ζ_* is monomorphic, hence isomorphic, at 0. Therefore, ζ is a diffeomorphism locally at 0. Accordingly we can change the chart ϕ to $\zeta \circ \phi$. Then the new θ' is just $\pi = \theta' \circ \zeta^{-1}$ and $\pi(x, y) = y$ as desired. □

7.4. Corollary. *Suppose that $\theta: M^m \to N^n$ is a smooth map and that $y \in N$ is a regular value of θ. Then $\theta^{-1}(y)$ is an embedded submanifold of M^m of dimension $m - n$.* □

7.5. Example. Consider the map $\theta: \mathbf{R}^n \to \mathbf{R}$ given by $\theta(x_1, \ldots, x_n) = \sum x_i^2$. We claim that 1 is a regular value. To see this, let $p = (x_1, \ldots, x_n)$ where $\sum x_i^2 = 1$. Then some x_i is nonzero, say $x_1 \neq 0$ at p. Then $\partial\theta/\partial x_1 = 2x_1 \neq 0$ at p and so θ_* is onto at p. Hence $\theta^{-1}(1) = \mathbf{S}^{n-1}$ is a submanifold of $\mathbf{R}^n$.

7.6. Definition. Suppose that N_1 and N_2 are embedded submanifolds of M. We say that N_1 intersects N_2 *transversely* (symbolically $N_1 \pitchfork N_2$) if, whenever $p \in N_1 \cap N_2$, we have $T_p(N_1) + T_p(N_2) = T_p(M)$. (The sum is not direct, just the set of sums of vectors, one from each of the two subspaces of $T_p(M)$.)

7.7. Theorem. *If $N_1 \pitchfork N_2$ in M^m then $N_1 \cap N_2$ is a submanifold of M^m of dimension $\dim(N_1 \cap N_2) = \dim(N_1) + \dim(N_2) - \dim(M)$. Moreover, locally in an appropriate coordinate system, we have that $N_1 = \mathbf{R}^{n_1} \times \{0\}$ and $N_2 = \{0\} \times \mathbf{R}^{n_2}$.*

Proof. By taking a chart at p in which N_1 is "flat" (see Theorem 7.1) we can find a coordinate neighborhood U of p and a map $\phi_1: U \to \mathbf{R}^{m-n_1}$ having 0 as a regular value and such that $U \cap N_1 = \phi_1^{-1}(0)$. Similarly, perhaps cutting down U, we can find a map $\phi_2: U \to \mathbf{R}^{m-n_2}$ with 0 as a regular value and such that $U \cap N_2 = \phi_2^{-1}(0)$. Consider $\phi_1 \times \phi_2: U \to \mathbf{R}^{m-n_1} \times \mathbf{R}^{m-n_2}$ taking x to $(\phi_1(x), \phi_2(x))$. We claim that $0 = (0, 0)$ is a regular value. By considering

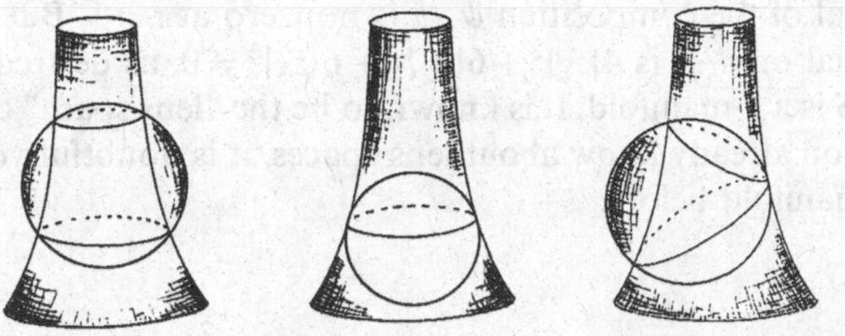

Figure II-5. Intersection of submanifolds.

$\phi_1 \times \phi_2$ followed by the projections we see easily that

$$(\phi_1 \times \phi_2)_*: T_p(M) \to \mathbf{R}^{m-n_1} \times \mathbf{R}^{m-n_2}$$

is given by

$$(\phi_1 \times \phi_2)_*(v) = (\phi_{1_*}(v), \phi_{2_*}(v)).$$

It follows that $\ker(\phi_1 \times \phi_2)_* = \ker(\phi_{1_*}) \cap \ker(\phi_{2_*}) = T_p(N_1) \cap T_p(N_2)$. But the dimension of this is $\dim T_p(N_1) + \dim T_p(N_2) - \dim T_p(M) = n_1 + n_2 - m$. Thus $\dim(\mathrm{im}(\phi_1 \times \phi_2)_*) = m - (n_1 + n_2 - m) = 2m - n_1 - n_2 = \dim(\mathbf{R}^{m-n_1} \times \mathbf{R}^{m-n_2})$ and hence $(\phi_1 \times \phi_2)_*$ is onto at p. Thus 0 is a regular value for $\phi_1 \times \phi_2$ on U and has $(\phi_1 \times \phi_2)^{-1}(0) = N_1 \cap N_2$ (locally), proving the first statement of the theorem.

For the statement about the coordinate system consider the map

$$\theta: U \to \mathbf{R}^{m-n_2} \times \mathbf{R}^{n_1+n_2-m} \times \mathbf{R}^{m-n_1}$$

defined by $\theta(x) = (\phi_2(x), \psi(x), \phi_1(x))$ where ψ is the projection to $N_1 \cap N_2$ in a coordinate system on U in which $N_1 \cap N_2$ is flat.

Now ϕ_{i_*} kills $T_p(N_1 \cap N_2)$ which implies that $\mathrm{im}(\theta_*)$ contains the middle factor (of the tangent space of this product of euclidean spaces). By projection to the first and third factors (together) we see that $\mathrm{im}(\theta_*)$ maps onto $\mathrm{im}(\phi_{2_*} \times \phi_{1_*}) = \mathbf{R}^{m-n_2} \times \{0\} \times \mathbf{R}^{m-n_1}$. It follows that θ_* is onto and hence that θ is a chart (possibly by further restricting its domain) and it clearly satisfies our requirements. □

7.8. Example. Consider $V = \{(z_1, z_2, z_3) \in \mathbf{C}^3 - \{0\} \mid z_1^3 + z_2^2 + z_3^2 = 0\}$. Note that 0 is a regular value of $(z_1, z_2, z_3) \mapsto z_1^3 + z_2^2 + z_3^2$ of $\mathbf{C}^3 - \{0\} \to \mathbf{C}$, so that V is a 4-manifold. Let $S = \mathbf{S}^5 = \{(z_1, z_2, z_3) \in \mathbf{C}^3 \mid |z_1|^2 + |z_2|^2 + |z_3|^2 = 1\}$. Then we claim that $V \pitchfork S$ and hence that $V \cap S$ is a 3-manifold.

To see this, note that we need only show that V has a tangent vector, at any given point $(z_1, z_2, z_3) \in V \cap S$, which is *not* tangent to S. For this, consider the map $\phi: \mathbf{R} \to V \subset \mathbf{C}^3$ given by $\phi(t) = (t^2 z_1, t^3 z_2, t^3 z_3)$ and the map $\psi: \mathbf{C}^3 \to \mathbf{R}$ which is the norm squared. That is, $\psi(z_1, z_2, z_3) = |z_1|^2 + |z_2|^2 + |z_3|^2$. Note that ψ takes vectors tangent to S into 0. Accordingly, it suffices to show that the differential of the composition $\psi \circ \phi$ is nonzero at $t = 1$. But the value of this differential on d/dt is $4|z_1|^2 + 6|z_2|^2 + 6|z_3|^2 \neq 0$, as desired.

Thus $V \cap S$ is a 3-manifold. It is known to be the "lens space" called $L(3, 1)$, and, unless you already know about lens spaces, it is doubtful you have ever seen this 3-manifold before.

Problems

1. Consider the real valued function $f(x, y, z) = (2 - (x^2 + y^2)^{1/2})^2 + z^2$ on $\mathbf{R}^3 - \{(0, 0, z)\}$. Show that 1 is a regular value of f. Identify the manifold $M = f^{-1}(1)$.
2. Show that the manifold M of Problem 1 is transverse to the surface
$$N = \{(x, y, z) \in \mathbf{R}^3 | x^2 + y^2 = 4\}.$$
Identify the manifold $M \cap N$.
3. Show that the manifold M of Problem 1 is *not* transverse to the surface
$$N = \{(x, y, z) \in \mathbf{R}^3 | x^2 + y^2 = 1\}.$$
Is $M \cap N$ a manifold?
4. Show that the manifold M of Problem 1 is *not* transverse to the plane
$$N = \{(x, y, z) \in \mathbf{R}^3 | x = 1\}.$$
Is $M \cap N$ a manifold?
5. Generalize Example 7.8 as far as you can.

8. Vector Fields and Flows

8.1. Definition. A *vector field* on smooth manifold M^n is a function ξ on M^n, such that $\xi(p) \in T_p(M)$ and which is smooth in the following sense: Given local coordinates $x_1, \ldots, x_n$ near $p \in M$, we can write

$$\xi(p) = \sum_{i=1}^{n} a_i(p)\, \partial/\partial x_i$$

and smoothness of ξ means that the a_i are smooth functions.

8.2. Definition. A (smooth) *flow* on a smooth manifold M^n is a smooth map $\theta: \mathbf{R} \times M \to M$ such that:

(1) $\theta(0, x) = x$ for all $x \in M$; and
(2) $\theta(s + t, x) = \theta(s, \theta(t, x))$ for all $x \in M$ and $s, t \in \mathbf{R}$.

It is easy to check that a flow is the same as what we called an "action" of the additive topological group $\mathbf{R}$ of real numbers on the manifold M^n in Definition 15.13 of Chapter I with the addition of smoothness.

A flow generates a vector field by assigning to a point p the vector $\xi(p)$ which is tangent to the curve $\gamma(t) = \theta(t, p)$ at $t = 0$. That is,

$$\xi(p) = \theta_*\left(\frac{d}{dt}\bigg|_{(0,p)}\right) \in T_p(M).$$

This vector field ξ is called the "tangent field" of the flow θ.

Conversely, given a smooth vector field and a coordinate chart in M^n, the field broken up into its n coordinates is just a set of n functions

$$\xi_1(y_1, \ldots, y_n),$$
$$\cdots$$
$$\xi_n(y_1, \ldots, y_n),$$

where the y_i are the local coordinates. A set of "solution curves" for this field are solutions of the first-order system of differential equations

$$dy_1/dt = \xi_1(y_1, \ldots, y_n),$$
$$\cdots$$
$$dy_n/dt = \xi_n(y_1, \ldots, y_n).$$

For the solution going through the point $p = (x_1, \ldots, x_n)$ at time $t = 0$, the existence and uniqueness theorem for first-order differential equations says that there is a solution (smooth) in a neighborhood of $(0, p)$ in $\mathbf{R} \times \mathbf{R}^n$. This means, in the coordinate free notation, that there is a function $\theta: U \to M$, for some neighborhood U of $\{0\} \times M$ in $\mathbf{R} \times M$, whose set of trajectories at $t = 0$ induces the original vector field ξ. By looking at the two functions $\theta(s + t, p)$ and $\theta(s, \theta(t, p))$ as functions of s near $s = 0$, it is not hard to show that they give rise to the same set of differential equations. By the uniqueness theorem, they must coincide for small s, t. Thus a vector field induces a "local" flow (all properties of a flow except that it is defined on some neighborhood of $\{0\} \times M$ and possibly not on all of $\mathbf{R} \times M$). However, if M is compact then one can see that one gets a "global" flow as in Definition 8.2. Since we have

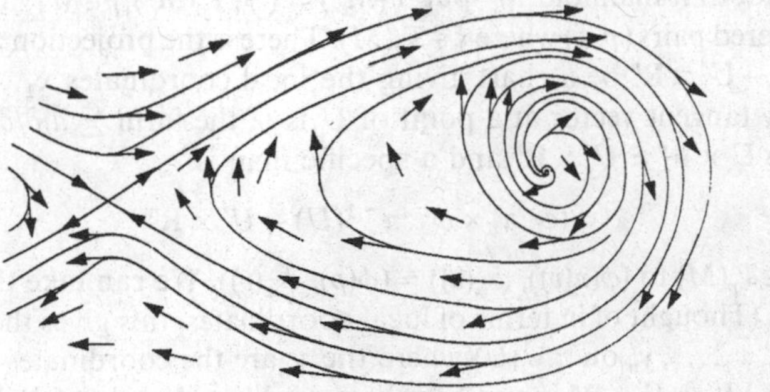

Figure II-6. Vector field and flow.

no great use for these facts, we leave it to the reader to fill in the details of this discussion.

8.3. Example. Consider the sphere $\mathbf{S}^{2n-1} \subset \mathbf{C}^n$ and let $z = (z_1, \ldots, z_n) \in \mathbf{S}^{2n-1}$. Let $\xi(z) = f_*(d/dt)$ where $f(t) = (e^{it}z_1, \ldots, e^{it}z_n)$. Since $(d/dt)(e^{it}z)|_{t=0} = iz$, we see that $\xi(z) = iz$. Consequently, this defines a unit vector field on $\mathbf{S}^{2n-1}$.

It is natural to ask, for a given manifold M, whether or not a vector field exists on M which is everywhere nonzero, like the one just produced on $\mathbf{S}^{2n-1}$. This is a question which algebraic topology is equipped to answer, and it will be taken up again when the needed tools have been developed.

Problems

1. On the 2-sphere, consider the flow

$$\theta(t, \langle x, y, z\rangle) = \langle x, y\cdot\cos(t) - z\cdot\sin(t), y\cdot\sin(t) + z\cdot\cos(t)\rangle.$$

Find the vector field on $\mathbf{S}^2$ induced by this flow.

2. Consider the vector field $\xi(x) = x$ on $\mathbf{R}$. Show that ξ is the tangent field to a flow, and find the flow. (*Hint*: In classical notation, this vector field corresponds to the initial value problem $dy/dt = y$, $y(0) = x$.)

3. Show that the vector field $\xi(x) = x^2$ on $\mathbf{R}$ is *not* the tangent field of any (global) flow.

4. ◇ If X and Y are vector fields on M then XY makes sense as an operator on smooth real valued functions on M. Show that $[X, Y] = XY - YX$ is a vector field. (This is called the "Lie bracket" of X and Y. Sometimes it is defined with the opposite sign.) Also show that XY itself is *not* a vector field.

5. Show that the Klein bottle has an everywhere nonzero vector field. Describe the resulting flow.

9. Tangent Bundles

For a smooth n-manifold M^n put $T(M^n) = \bigcup\{T_p(M^n) | p \in M\}$. This is the set of all ordered pairs (p, ξ) where $\xi \in T_p(M)$. There is the projection $\pi: T(M) \to M$. Let $\phi: U \to U' \subset \mathbf{R}^n$ be a chart giving the local coordinates $x_1, \ldots, x_n$ near p. Then any tangent vector at a point of U is of the form $\sum_i a_i \partial/\partial x_i$. Therefore $\pi^{-1}(U) \approx U \times \mathbf{R}^n \approx U' \times \mathbf{R}^n$ and a specific map is

$$(\phi \circ \pi) \times \phi_*: \pi^{-1}(U) \to U' \times \mathbf{R}^n$$

taking $v \in T_p(M)$ to $(\phi(\pi(v)), \phi_*(v)) = (\phi(p), \phi_*(v))$. We can take this as a chart on $T(M)$. (Thought of in terms of local coordinates, this gives the coordinates $x_1, \ldots, x_n, y_1, \ldots, y_n$ on $\pi^{-1}(U)$ where the y_i are the coordinates of the vector $\sum_i y_i \partial/\partial x_i$.) If $\psi: V \to \mathbf{R}^n$ is another chart on M so that $\theta = \psi\phi^{-1}: \phi(U \cap V) \to \psi(U \cap V)$ is the transition function, then the corresponding transition function on $T(M)$ is clearly $\theta \times \theta_*$. This makes $T(M)$ into a smooth $2n$-manifold, called the "tangent bundle" of M.

A vector field ξ on M is then just a smooth cross section of this bundle. That is, it is a smooth map $\xi: M \to T(M)$ such that $\pi \circ \xi = 1_M$.

A manifold M^n is called "parallelizable" if there is a diffeomorphism θ: $T(M) \to M \times \mathbf{R}^n$ such that each $T_p(M)$ is carried linearly isomorphically onto $\{p\} \times \mathbf{R}^n$. (This is called a "bundle isomorphism" of $T(M)$ to the trivial n-plane bundle.) Clearly, M^n is parallelizable $\Leftrightarrow$ there exist n vector fields on M^n which are independent at each point of M^n.

An example of a parallelizable manifold is the circle $\mathbf{S}^1$, and one way to produce the required nonzero vector field is as the tangent field to the flow $\theta(t, z) = e^{it}z$.

A less trivial example is the sphere $\mathbf{S}^3$. The required three independent vector fields can be obtained by thinking of $\mathbf{S}^3$ as the unit quaternions and the three fields as the tangent fields of the flows $\theta(t, q) = (\cos(t) + i\sin(t))q$, obtaining the other two by replacing i with j and k, respectively.

The sphere $\mathbf{S}^7$ can also be shown to be parallelizable by the same technique using the Cayley numbers. It is known, and nontrivial, that no other spheres are parallelizable. A proof of part of that is given in Corollary 15.16 of Chapter VI.

Problems

1. Show that the n-torus $\mathbf{T}^n = \mathbf{S}^1 \times \cdots \times \mathbf{S}^1$ is parallelizable.
2. Is the Klein bottle parallelizable? (♦ Prove your answer.)
3. Show that the sphere $\mathbf{S}^{4n-1}$ has three vector fields that are everywhere independent.
4. Show that $\mathbf{S}^n \times \mathbf{R}$ is parallelizable for all n.
5. ♦ If n is odd, show that $\mathbf{S}^n \times \mathbf{S}^k$ is parallelizable for all $k \geq 1$.

10. Embedding in Euclidean Space

In this section we prove that every smooth manifold can be smoothly embedded in an appropriate euclidean space.

First we establish a smooth version of the previous result, Theorem 12.8 of Chapter I, on partitions of unity.

10.1. Theorem. *If M is a smooth manifold and $\{U_\alpha\}$ is an open covering then there is a partition of unity $\{f_\beta\}$ subordinate to this covering such that the functions f_β are all smooth.*

Proof. We may as well assume that the original covering is locally finite and also that each member is contained in the domain of a coordinate chart and has compact closure. By Proposition 12.9 of Chapter I, there is a "shrinking" $\{V_\alpha\}$, such that $\bar{V}_\alpha \subset U_\alpha$ for all α.

We shall complete the proof in a sequence of lemmas.

10.2. Lemma. *There exists a smooth function* $B: \mathbf{R} \to \mathbf{R}$ *such that* $B(x) > 0$ *on* $(-1, 1)$, *and* $B(x) = 0$ *for* $|x| \geq 1$.

PROOF. Take $B(x) = e^{-1/(x-1)^2} e^{-1/(x+1)^2}$ for $|x| < 1$ and $B(x) = 0$ otherwise. □

10.3. Lemma. *Let* $p \in U \subset M$ *with* U *open. Then there is a smooth map* $g: M \to [0, \infty)$ *such that* $g(p) > 0$ *and* support$(g) \subset U$. (*That is,* g *vanishes on a neighborhood of* $M - U$.)

PROOF. Take local coordinates $x_1, \ldots, x_n$ at p and assume that the box

$$\{(x_1, \ldots, x_n) \mid \text{all } |x_i| \leq \epsilon\}$$

is contained inside U. Then let $g(x) = B(x_1/\epsilon) \cdot \ldots \cdot B(x_n/\epsilon)$. □

10.4. Lemma. *Let* $K \subset U \subset M$ *with* K *compact and* U *open. Then there is a smooth function* $g: M \to [0, \infty)$ *such that* $g(x) > 0$ *for all* $x \in K$, *and* support(g) $\subset U$.

PROOF. For each $p \in K$, take a function g_p satisfying Lemma 10.3. The sets $\{x \in M \mid g_p(x) > 0\}$ are open and cover K. Thus a finite number of them cover K. Add up the corresponding finite number of g_p, and that clearly works for g. □

Now we shall complete the proof of Theorem 10.1. For each index α use Lemma 10.4 to get a smooth function $g_\alpha: M \to [0, \infty)$ such that $g_\alpha(p) > 0$ if $p \in \bar{V}_\alpha$ and support$(g_\alpha) \subset U_\alpha$. Then put $f_\alpha = g_\alpha / \sum g_\alpha$. □

10.5. Proposition. *Let* U_α *and* f_α *be as in Theorem* 10.1. *Let* $g_\alpha: U_\alpha \to \mathbf{R}$ *be arbitrary smooth functions. Then the function* $g: M \to \mathbf{R}$ *given by* $g(x) = \sum_\alpha f_\alpha(x) g_\alpha(x)$ *is smooth.*

PROOF. Since f_α vanishes on a neighborhood of $M - U_\alpha$, $f_\alpha g_\alpha$ is smooth on U_α and vanishes on a neighborhood of its boundary in M. Thus $f_\alpha g_\alpha$ extends smoothly, by 0, to all of M. Any point $x \in M$ has a neighborhood on which all but a finite number of the $f_\alpha g_\alpha$ vanish and so $\sum_\alpha f_\alpha g_\alpha$ makes sense and is smooth. □

10.6. Theorem. *Let* M *be a smooth manifold and let* $K \subset M$ *be closed. Let* $g: K \to \mathbf{R}$ *be smooth* (*in the induced structure; this means that* g *extends locally at each point* $p \in K$ *to a smooth function on some neighborhood of* p *in* M). *Then* g *extends to a smooth function* $\bar{g}: M \to \mathbf{R}$.

PROOF. Cover K by sets U_α which are open in M and such that there is a smooth function g_α on U_α coinciding with g on $U_\alpha \cap K$. Throw $M-K$ and the zero function into this to get a covering of M. By passing to a refinement we can assume that this is locally finite. By Theorem 10.1 there is a smooth partition of unity $\{f_\alpha\}$ subordinate to this covering. By Proposition 10.5 the function $\bar{g}: M \to \mathbf{R}$ given by $\bar{g}(x) = \sum f_\alpha(x) g_\alpha(x)$ is smooth. For $x \in K$ we have $\bar{g}(x) = \sum (f_\alpha(x) g(x)) = (\sum f_\alpha(x)) g(x) = 1 \cdot g(x) = g(x)$. □

10.7. Theorem (Whitney Embedding Theorem). *If M^n is a compact n-manifold then there exists a smooth embedding $g: M^n \to \mathbf{R}^{2n+1}$.*

PROOF. We can cover M by a finite number of domains U_i of charts ϕ_i, $i = 1, 2, \ldots, k$. We can assume there are sets V_i also covering M such that $\bar{V}_i \subset U_i$ for each i. There are also smooth functions $\lambda_i: M \to \mathbf{R}$ which are 1 on $\bar{V}_i$ and have support in U_i. This follows from Theorem 10.6. Let $\psi_i(p) = \lambda_i(p)\phi_i(p)$ for $p \in U_i$ and 0 otherwise. Then each ψ_i is smooth. Now define $\theta: M \to (\mathbf{R}^n)^k \times \mathbf{R}^k$ by

$$\theta(p) = (\psi_1(p), \ldots, \psi_k(p), \lambda_1(p), \ldots, \lambda_k(p))$$

and note that

$$\theta_* = \psi_{1*} \times \cdots \times \psi_{k*} \times \lambda_{1*} \times \cdots \times \lambda_{k*}.$$

We claim first that θ is an immersion. Look at a point $p \in M$ which must be in some V_i, say $p \in V_j$. Since $\lambda_j = 1$ near p, ψ_j coincides with ϕ_j near p. Thus $\psi_{j*} = \phi_{j*}$ near p, and the latter is monomorphic near p since ϕ_j is a chart. Accordingly, θ_* is monomorphic at p.

Next we claim that θ is one–one. If $\theta(p) = \theta(q)$ then $\lambda_i(p) = \lambda_i(q)$ for each i. Now, $p \in V_i$ for some i and, for that i, $\lambda_i(p) = 1$. Thus $\phi_i(p) = \lambda_i(p)\psi_i(p) = \lambda_i(q)\psi_i(q) = \phi_i(q)$. This means that $p = q$ since ϕ_i is one–one, being a chart.

Since M^n is compact, θ is a homeomorphism onto its image, by Theorem 7.8 of Chapter I. Thus θ is an embedding of M^n into $\mathbf{R}^N$, for some large integer N, which we will regard as an inclusion. It remains to show that we can cut N down to $2n+1$. For this, suppose we can find a vector $w \in \mathbf{R}^N$ such that w is not tangent to M^n at any point, and such that there do not exist points $x, y \in M$ with $x-y$ parallel to w. Then it is clear that the projection of M into the hyperplane $w^\perp$ is still one–one and kills no tangent vector to M. Thus it suffices to show such a vector w exists if $N > 2n+1$. The argument is one of "general position."

Consider the map $\sigma: T(M^n) - M^n \to \mathbf{P}^{N-1}$ (real projective $(N-1)$-space), taking a tangent vector to a vector in $\mathbf{R}^N$ via the embedding and then to its equivalence class in projective $(N-1)$-space. Also consider the map $\tau: M^n \times M^n - \Delta \to \mathbf{P}^{N-1}$ taking a pair (x, y), with $x \neq y$, to the equivalence class of $x-y$. Both of these are smooth maps. The dimensions of both of the source manifolds are $2n$ which is less than the dimension $N-1$ of the target manifold. By Sard's Theorem (Corollary 6.3), it follows that the images of both maps are of first category, and hence their union is also of first category which implies that there must be a vector w satisfying our demands. □

Actually, Whitney [1] proved that M^n can be embedded in $\mathbf{R}^{2n}$, but that is beyond our present capabilities to show.

It takes only a little more argument to remove the compactness restriction from Theorem 10.7 and we now indicate how to do that. We will not use this extension, so the reader can skip it if so desired.

10.8. Theorem. *A smooth manifold M^n can be embedded as a submanifold, and closed subset, of* $\mathbf{R}^{2n+1}$.

PROOF. Cover M^n by open sets with compact closures and take a smooth partition of unity subordinate to a locally finite refinement of this covering. The refinement must be countable and so we can index the partition by the positive integers $\{\lambda_i | i > 0\}$. Let $h(x) = \sum_k k\lambda_k(x)$. This is a smooth *proper* map $M^n \to [1, \infty) \subset \mathbf{R}$. Let $U_i = h^{-1}(i - \frac{1}{4}, i + \frac{5}{4})$ and $C_i = h^{-1}[i - \frac{1}{3}, i + \frac{4}{3}]$. Then U_i is open, C_i is compact, and $\bar{U}_i \subset \operatorname{int} C_i$. Also, all C_{odd} are disjoint as are the C_{even}. For each i, the proof of Theorem 10.7 shows that there exists a smooth map $g_i: M^n \to \mathbf{R}^{2n+1}$ which is an embedding on $\bar{U}_i$ and is 0 outside C_i. By composing this with a diffeomorphism from $\mathbf{R}^{2n+1}$ to an open ball, we can also assume that g_i has bounded image. Let $f_o = \sum g_{\text{odd}}$ and $f_e = \sum g_{\text{even}}$ and $f = \langle f_o, f_e, h \rangle: M^n \to \mathbf{R}^{2n+1} \times \mathbf{R}^{2n+1} \times \mathbf{R}$ and note that $\operatorname{im}(f) \subset K \times \mathbf{R}$ for some compact set $K \subset \mathbf{R}^{2n+1} \times \mathbf{R}^{2n+1}$ since f_o and f_e have bounded images. Then f is proper since h is proper. If $f(x) = f(y)$ then $h(x) = h(y)$ so that x and y are in some common U_i. If i is odd then f_o is an embedding on U_i which implies that $x = y$; and similarly for i even. Hence f is an embedding to a closed subset (by properness). By repeated use of the Sard Theorem argument in the proof of Theorem 10.7 there is a projection p of $\mathbf{R}^{2n+1} \times \mathbf{R}^{2n+1} \times \mathbf{R}$ to a $(2n+1)$-dimensional hyperplane H which is still a one–one immersion on $f(M)$. Moreover, this can be so chosen that the original h coordinate axis is not in $\ker(p)$. That is, if $\pi: \mathbf{R}^{2n+1} \times \mathbf{R}^{2n+1} \times \mathbf{R} \to \mathbf{R}^{2n+1} \times \mathbf{R}^{2n+1}$ is the projection, then $\ker(\pi) \cap \ker(p) = \{0\}$, which implies that $\pi \times p$ is an inclusion, hence proper, where $(\pi \times p)(x) = \pi(x) \times p(x)$. Thus, for $C \subset H$ compact, $K \times \mathbf{R} \cap p^{-1}(C) = (\pi \times p)^{-1}(K \times C)$ is compact, whence p is proper on $f(M) \subset K \times \mathbf{R}$. Therefore $p \circ f$ is an embedding of M^n as a *closed subspace* of $\mathbf{R}^{2n+1}$. □

11. Tubular Neighborhoods and Approximations

In this section we will show that any smoothly embedded manifold has a nice neighborhood analogous to a tube around a curve in 3-space; see Figure II-7. This is then used to prove that continuous maps can be approximated by smooth maps.

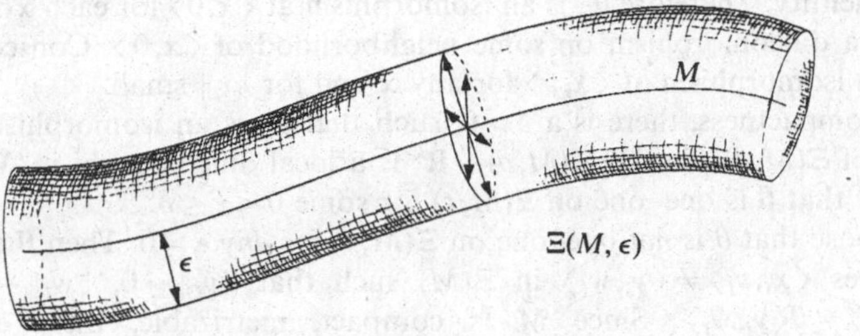

Figure II-7. Tubular neighborhood.

11.1. Definition. Let M^n be a compact smooth manifold embedded in $\mathbf{R}^k$. Then the *normal bundle* of M^n in $\mathbf{R}^k$ is $\Xi(M) = \{\langle x, v\rangle \in M \times \mathbf{R}^k | v \perp T_x(M)\}$. We let $\pi: \Xi(M) \to M$ be the projection $\pi\langle x, v\rangle = x$.

11.2. Proposition. *Each point $x \in M$ has a neighborhood U such that $\pi^{-1}(U) \approx U \times \mathbf{R}^{k-n}$ with the projection $\pi: \pi^{-1}(U) \to U$ corresponding to the canonical projection $U \times \mathbf{R}^{k-n} \to U$.*

PROOF. Let $\phi: V \to \mathbf{R}^{k-n} \times \mathbf{R}^n = \mathbf{R}^k$ be a chart making M flat, i.e., $U = V \cap M$ corresponds to $\{0\} \times \mathbf{R}^n$. Let $\lambda_1, \ldots, \lambda_{k-n}$ be the first $k-n$ coordinate projections $\mathbf{R}^k \to \mathbf{R}$. Then the $\lambda_i \circ \phi$ are constant on U. Thus the vectors $\nabla(\lambda_i \circ \phi)$, at a point of U, give a normal frame to U and provide the splitting $\pi^{-1}(U) \approx U \times \mathbf{R}^{k-n}$; see Problems 3 and 4 of Section 5. (These vectors are independent since they form the first $k-n$ rows of the Jacobian matrix of ϕ.) □

11.3. Definition. In the present situation, let $\theta: \Xi(M) \to \mathbf{R}^k$ be given by $\theta\langle x, v\rangle = x + v$. Also let $\Xi(M, \epsilon) = \{\langle x, v\rangle \in \Xi(M) | \|v\| < \epsilon\}$.

11.4. Theorem (Tubular Neighborhood Theorem). *Let M^n be a compact smooth submanifold of $\mathbf{R}^k$. Then there is an $\epsilon > 0$ such that $\theta: \Xi(M, \epsilon) \to \mathbf{R}^k$ is a diffeomorphism onto the neighborhood $\{y \in \mathbf{R}^k | \text{dist}(M, y) < \epsilon\}$ of M^n in $\mathbf{R}^k$.*

PROOF. First note that there is a canonical splitting $T_{\langle x,0\rangle}(\Xi(M)) \approx T_x(M) \times N_x(M)$ where $N_x(M)$ is the normal space to $T_x(M)$ in $\mathbf{R}^k$.

Now let us compute θ_* at $\langle x, 0\rangle \in \Xi(M)$. Since $\theta\langle x, v\rangle = x + v$ is just a translation for x fixed and v variable, θ_* is the standard inclusion (identity) on $N_x(M) \to \mathbf{R}^k$. Also $\theta_*: T_x(M) \to T_x(\mathbf{R}^k)$ is just the differential of the inclusion of $M \subset \mathbf{R}^k$ and so this part of θ_* is just the standard inclusion of $T_x(M)$ in $T_x(\mathbf{R}^k) = \mathbf{R}^k$. Thus

$$\theta_*: \mathbf{R}^k = T_x(\mathbf{R}^k) = T_x(M) \times N_x(M) \to \mathbf{R}^k$$

is the identity. Therefore θ_* is an isomorphism at $\langle x,0\rangle$ for each $x\in M$, and so θ is a diffeomorphism on some neighborhood of $\langle x,0\rangle$. Consequently, θ_* is an isomorphism at $\langle x,v\rangle$ for any x and for $\|v\|$ small.

By compactness, there is a $\delta>0$, such that θ_* is an isomorphism at all points of $\Xi(M,\delta)$. Thus $\theta:\Xi(M,\delta)\to\mathbf{R}^k$ is a local diffeomorphism. We wish to show that θ is one–one on $\Xi(M,\epsilon)$ for some $0<\epsilon\leq\delta$.

Suppose that θ is *not* one–one on $\Xi(M,\epsilon)$ for *any* $\epsilon>0$. Then there exist sequences $\langle x_i,v_i\rangle\neq\langle y_i,w_i\rangle$ in $\Xi(M)$ such that $\|v_i\|\to 0$, $\|w_i\|\to 0$, and $\theta\langle x_i,v_i\rangle=\theta\langle y_i,w_i\rangle$. Since M is compact, metrizable, there exists a subsequence such that (by reindexing) $x_i\to x$ and $y_i\to y$. Then $\theta\langle x_i,v_i\rangle\to\theta\langle x,0\rangle=x$ and $\theta\langle y_i,w_i\rangle\to\theta\langle y,0\rangle=y$, so that $x=y$. But then, for i large, both $\langle x_i,v_i\rangle$ and $\langle y_i,w_i\rangle$ are close to $\langle x,0\rangle$. Since θ is one–one locally near $\langle x,0\rangle$, this is a contradiction, and thus, as claimed, θ must be one–one on some $\Xi(M,\epsilon)$.

To finish, we must prove the final contention that $\theta(\Xi(M,\epsilon))=\{y\mid\operatorname{dist}(y,M)<\epsilon\}$. The containment $\subset$ is clear, so suppose that y is such that $\operatorname{dist}(y,M)<\epsilon$ and let $x\in M$ be such that $\operatorname{dist}(y,x)$ is a minimum (and hence $<\epsilon$). Then the vector $y-x$ is a normal vector, at the point x, of length $<\epsilon$ and so y does lie in $\theta(\Xi(M,\epsilon))$. □

Note that the map $r=\pi\circ\theta^{-1}:\theta(\Xi(M,\epsilon))\to M^n$ is a smooth retraction of the tubular neighborhood onto M^n. It is also clear that r is homotopic to the identity via a smooth homotopy. That is, r is a smooth "deformation retraction." Also every point of M^n is a regular value of r. We call r the "normal retraction" of the tubular neighborhood onto M^n.

There are more general versions of the Tubular Neighborhood Theorem. Such a theorem can be proved for smooth submanifolds of arbitrary smooth manifolds, and not just euclidean space. This is done in much the same way using geodesics in some Riemannian structure in the same way as straight lines in euclidean space are used in the foregoing proof. A more elementary derivation of that is given at the end of this section. Also, compactness can be removed as a restriction, with a slight modification of the conclusion (the ϵ must be allowed to vary with the point).

11.5. Corollary. *Let M^n be a compact manifold and ξ a nonzero vector field on M^n. Then there is a map $f:M^n\to M^n$ without fixed points, and with $f\simeq 1$. (ξ is assumed to be continuous, but need not be smooth.)*

PROOF. Embed M^n in some $\mathbf{R}^k$. Then ξ is a field of tangent vectors to M^n and, since M^n is compact, there is a constant c such that the vector field $c\xi$ has all vectors of length less than ϵ for any given $\epsilon>0$. Thus each $x+c\xi_x\in\theta(\Xi(M,\epsilon))$. Define $f(x)=r(x+c\xi_x)$. If $f(x)=x$ then $c\xi_x\in N_x(M)\cap T_x(M)=\{0\}$, so that $\xi_x=0$. Therefore, f has no fixed points. The homotopy is given by $F(x,t)=r(x+tc\xi_x)$. □

11.6. Theorem. *Let $f:\mathbf{R}^n\to M^m$ be a smooth map. Assume that $p\in M^m$ is a* *regular value, let $K = f^{-1}\{p\}$, and assume that K is compact. Then there is an open neighborhood N of K inside a tubular neighborhood of K, with normal retraction $r: N \to K$, and an open neighborhood $E \approx \mathbf{R}^m$ of p in M^m such that the map $r \times f: N \to K \times E$ is a diffeomorphism.*

PROOF. Since the critical set of f is closed and disjoint from K, there exists a tubular neighborhood U of K on which f has no critical points. At a point $p \in K$, the tangent space of $\mathbf{R}^n$ at p is the direct sum of the tangent space T_p of K at p and the normal space N_p to T_p in $\mathbf{R}^n$. The differential r_* kills N_p and is an isomorphism (the identity) on T_p. The differential f_* kills T_p and is an isomorphism on N_p since it has maximal rank. It follows that $(r \times f)_*$ is an isomorphism at p. Therefore we can take U small enough so that $(r \times f)_*$ is an isomorphism at each point of U. Then $r \times f$ is an immersion on U. An easy compactness argument, similar to that in the proof of Theorem 11.4, shows that $r \times f$ is one–one on some compact neighborhood $C \subset U$ of K. Then $r \times f$ is a homeomorphism on C to its image by Theorem 7.8 of Chapter I. Another compactness argument shows that, for a sufficiently small open euclidean neighborhood E of p in M, we have that $N = f^{-1}(E) \cap C \subset \operatorname{int} C$. (Just consider the sets $f^{-1}(E) \cap \partial C$ for varying E.)

For $x \in K$, let $f_x: r^{-1}(x) \cap N \to E$ denote the restriction of f to the "fiber" $r^{-1}(x) \cap N$. We claim that each f_x is a diffeomorphism onto E. Since $r \times f$ has no critical points in N and r_* kills tangent vectors to $r^{-1}(x) \cap N$, it follows that f_x has no critical points and hence is a diffeomorphism *into* E. Now $f_x(r^{-1}(x) \cap N) = f(r^{-1}(x) \cap C \cap f^{-1}(E)) = f(r^{-1}(x) \cap C) \cap E$, but $r^{-1}(x) \cap C$ is compact and so this set is closed, as well as open, in E. Consequently, $f_x(r^{-1}(x) \cap N) = E$ since E is connected.

This shows that $r \times f: N \xrightarrow{\approx} K \times E$ is onto and that finishes the proof. □

Figure II-8 illustrates the situation of Theorem 11.6. Let us rephrase Theorem 11.6 slightly. It says that if p is a regular value whose inverse image K is compact then, in a neighborhood of K, f has the form of a smooth projection $K \times E \to E$. Also note that it does *not* say that the product neighborhood is the entire inverse image $f^{-1}(E)$ but only that they coincide

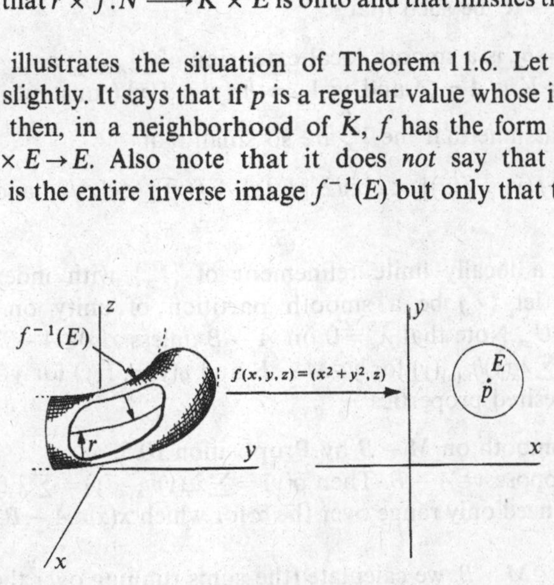

Figure II-8. Regular value and inverse image of its neighborhood.

near K. The projection $(x, y) \mapsto x$ of $\mathbf{S}^1 - \{(0, 1)\} \to \mathbf{R}$ is a counterexample to that. The compactness assumption on K is necessary as is shown by simple examples such as the projection $(x, y) \mapsto x$ of $\{(x, y) | y < |1/x|\} \to \mathbf{R}$.

This result can be generalized to a smooth map from any manifold and not just euclidean space.

The neighborhood $N \approx K \times E$ of Theorem 11.6 is also called a "tubular neighborhood" although it does not quite have the form of a $\Xi(K, \epsilon)$. More generally, any neighborhood of K with the structure of a vector bundle over K (see Section 13) is called a "tubular neighborhood" of K.

We shall now turn to the question of approximating arbitrary continuous maps between smooth manifolds by smooth maps. Note the case $B = \emptyset$ of the following result, which is all that is used in this chapter. The case $B \neq \emptyset$ is needed for an important application in Chapter IV, the Cellular Approximation Theorem (Theorem 11.4 of Chapter IV).

11.7. Theorem. *Let M^n be smooth and $A, B \subset M^n$ closed subsets. Let $f: M^n \to \mathbf{R}^k$ be continuous on M^n and smooth on A (in the induced structure). Then, given $\epsilon > 0$, there exists a map $g: M^n \to \mathbf{R}^k$ which is smooth on $M - B$ and is such that $g(a) = f(a)$ for all $a \in A \cup B$ and such that $\|g(x) - f(x)\| < \epsilon$ for all $x \in M^n$. Moreover, $f \simeq g$* rel *$A \cup B$ via an ϵ-small homotopy.*

PROOF. Let dist be any metric on M; see Theorem 12.12 of Chapter I. For $x \in M$, let $\epsilon(x) = \min(\epsilon, \operatorname{dist}(x, B))$. We remind the reader that "smooth on A" just means that near any point $a \in A$, there is a function defined near a on M which is smooth there and whose restriction to A coincides with f there. Thus, for any $x \in M^n - B$, let $V_x \subset M^n - B$ be a neighborhood of x in $M^n - B$ and let $h_x: V_x \to \mathbf{R}^k$ be such that

(1) $x \in A - B \Rightarrow h_x$ is a smooth local extension of $f|_{A \cap V_x}$; and
(2) $x \notin A \cup B \Rightarrow V_x \cap A = \emptyset$ and $y \in V_x \Rightarrow h_x(y) = f(x)$ (constant in y).

We can also assume that the V_x are so small that

(3) $y \in V_x \Rightarrow \|f(y) - f(x)\| < \epsilon(x)/2$, $\|h_x(y) - f(x)\| < \epsilon(x)/2$ and $\operatorname{dist}(x, y) < \epsilon(x)/2$.

Let $\{U_\alpha\}$ be a locally finite refinement of $\{V_x\}$ with index assignment $\alpha \mapsto x(\alpha)$, and let $\{\lambda_\alpha\}$ be a smooth partition of unity on $M - B$ with $\operatorname{support}(\lambda_\alpha) \subset U_\alpha$. Note that $\lambda_\alpha = 0$ on $A - B$ unless $x(\alpha) \in A - B$.

Put $g(y) = \sum \lambda_\alpha(y) h_{x(\alpha)}(y)$ for $y \in M - B$ and $g(y) = f(y)$ for $y \in B$. We claim this has the desired properties.

First g, is smooth on $M - B$ by Proposition 10.5.

Second, suppose $y \in A - B$. Then $g(y) = \sum \lambda_\alpha(y) h_{x(\alpha)}(y) = \sum \lambda_\alpha(y) f(y) = f(y)$, since the sum need only range over the α for which $x(\alpha) \in A - B$, as remarked above.

Third, for $y \in M - B$, we calculate (the sums running over the α for which $y \in U_\alpha$):

$$\begin{aligned}\|g(y)-f(y)\| &= \|\sum \lambda_\alpha(y)h_{x(\alpha)}(y)-f(y)\|\\ &= \|\sum[\lambda_\alpha(y)(h_{x(\alpha)}(y)-f(x(\alpha)))]+\sum[\lambda_\alpha(y)(f(x(\alpha))-f(y))]\|\\ &\le \|\sum \lambda_\alpha(y)(h_{x(\alpha)}(y)-f(x(\alpha)))\|+\|\sum \lambda_\alpha(y)(f(x(\alpha))-f(y))\|\\ &\le \sum \lambda_\alpha(y)[\|h_{x(\alpha)}(y)-f(x(\alpha))\|+\|f(x(\alpha))-f(y)\|]\\ &< \sum \lambda_\alpha(y)[\epsilon(x(\alpha))/2+\epsilon(x(\alpha))/2]\le \epsilon.\end{aligned}$$

Also,

$$\begin{aligned}y\in U_\alpha \Rightarrow \epsilon(x(\alpha)) &\le \operatorname{dist}(x(\alpha),B)\le \operatorname{dist}(x(\alpha),y)+\operatorname{dist}(y,B)<\\ &\epsilon(x(\alpha))/2+\operatorname{dist}(y,B) \quad\Rightarrow\quad \epsilon(x(\alpha))<2\operatorname{dist}(y,B).\end{aligned}$$

It follows from the displayed inequalities that $\|g(y)-f(y)\|<2\operatorname{dist}(y,B)$. This implies that g is continuous on all of M.

Finally, the standard homotopy $F(x,t)=tf(x)+(1-t)g(x)$ gives the desired ϵ-small homotopy rel $A\cup B$. □

11.8. Theorem (Smooth Approximation Theorem). *Suppose M^m and N^n are smooth manifolds with N^n compact metric. Let $A\subset M^m$ be closed. Let $f\colon M^m\to N^n$ be a map with $f|_A$ smooth. Then for any given $\epsilon>0$, there exists a map $h\colon M^m\to N^n$ such that:*

(1) *h is smooth;*
(2) $\operatorname{dist}(h(x),f(x))<\epsilon$ *for all $x\in M^m$;*
(3) *$h|_A=f|_A$; and*
(4) *$h\simeq f$ by an ϵ-small homotopy* (rel A).

PROOF. Embed N^n in some $\mathbf{R}^k$. By continuity of the inverse of the embedding map of N^n in $\mathbf{R}^k$ and the compactness of N^n, hence uniform continuity, we can find a $\delta>0$ such that $\|p-q\|<\delta\Rightarrow \operatorname{dist}(p,q)<\epsilon$. Thus it will suffice to use the metric in $\mathbf{R}^k$ rather than the one given on N^n.

Take a $\delta/2$-tubular neighborhood U of N^n in $\mathbf{R}^k$ (using a smaller δ if needed), and let $r\colon U\to N^n$ be the related normal retraction map.

Approximate f by a smooth map $g\colon M^m\to\mathbf{R}^k$ within $\delta/2$ using Theorem 11.7. Then $g(M^m)\subset U$. Let $h=r\circ g$. Then:

(a) h is smooth;
(b) $\|h(x)-f(x)\|\le\|r(g(x))-g(x)\|+\|g(x)-f(x)\|<\delta/2+\delta/2=\delta$;
(c) $h|_A=r\circ g|_A=r\circ f|_A=f|_A$; and
(d) $h\simeq f$ rel A by the homotopy $F(x,t)=r(tg(x)+(1-t)f(x))$.

To see that this homotopy is valid, recall that $U=\{y\in\mathbf{R}^k \mid y$ is within $\delta/2$ of $N^n\}$. Since $\|g(x)-f(x)\|<\delta/2$ we have that for $0\le t\le 1$, $tg(x)+(1-t)f(x)\in U$ and is within $\delta/2$ of $f(x)$, showing that the given formula for $F(x,t)$ is valid. It also shows that the homotopy remains within δ of $f(x)$. □

11.9. Corollary. *Suppose that M^m and N^n are smooth manifolds with N^n compact. Then any continuous $f\colon M^m\to N^n$ is homotopic to a smooth map. If f*

and g are smooth and $f \simeq g$, *then* $f \simeq g$ *by a smooth homotopy* $F: I \times M^m \subset \mathbf{R} \times M^m \to N^n$.

PROOF. The first part comes directly from the theorem. Thus suppose that $F: I \times M \to N$ is a given homotopy between two smooth maps. Extend F to $\mathbf{R} \times M \to N$ by making it constant on the ends. Then F is smooth on the subspace $\{0, 1\} \times M$ and so Theorem 11.8 implies that there is a smooth map $G: \mathbf{R} \times M \to N$ which coincides with F on $\{0, 1\} \times M$. □

We now apply the foregoing results of this section to derive some well-known topological facts.

11.10. Theorem. *If* M^m *is a smooth m-manifold and* $m < n$ *then any map* $f: M^m \to \mathbf{S}^n$ *is homotopic to a constant map.*

PROOF. Approximate f by a smooth map g homotopic to it. By Sard's Theorem (Corollary 6.3), there must be a point p which is not in the image of g. But $\mathbf{S}^n - \{p\}$ is homeomorphic to $\mathbf{R}^n$ and so it is contractible. Composing g with such a contraction gives a homotopy of g to a constant map. □

11.11. Theorem. *A sphere* $\mathbf{S}^n$ *is not a retract of the disk* $\mathbf{D}^{n+1}$.

PROOF. (This proof is due to M. Hirsch.) Suppose $f: \mathbf{D}^{n+1} \to \mathbf{S}^n$ is a retraction. We can alter f so that it is the composition of a map $f: \mathbf{R}^{n+1} \to \mathbf{S}^n$ which retracts the disk of radius $\frac{1}{2}$ onto its boundary with one which maps everything outside the disk of radius $\frac{1}{2}$ to the sphere $\mathbf{S}^n$ of radius 1 by radial projection. This makes f smooth on a neighborhood of $\mathbf{S}^n$. Then we can smooth f without changing it near $\mathbf{S}^n$, so we may as well assume that f is smooth and that it is the radial projection near the boundary $\mathbf{S}^n$.

Let $z \in \mathbf{S}^n$ be a regular value of f. Then $f^{-1}(z)$ is a 1-manifold with boundary, and its boundary is the single point $f^{-1}(z) \cap \mathbf{S}^n = \{z\}$. But any compact 1-manifold with boundary is homeomorphic to a disjoint union of circles and closed unit intervals, and hence has an even number of boundary points, a contradiction showing f cannot exist. □

11.12. Corollary (Brouwer's Fixed Point Theorem). *Any map* $f: \mathbf{D}^n \to \mathbf{D}^n$ *has a fixed point* (*i.e., a point* x *such that* $f(x) = x$).

PROOF. If f is such a map and does not have a fixed point then we can define a new map $r: \mathbf{D}^n \to \mathbf{S}^{n-1}$ by letting $r(x)$ be the point where the ray from the point $f(x)$ to the point x passes through $\mathbf{S}^{n-1}$ in the direction indicated. (See Figure II-9.) This is a retraction of $\mathbf{D}^n$ onto $\mathbf{S}^{n-1}$ and hence contradicts Theorem 11.11.

It is not hard to convince oneself that r is continuous, but it is unpleasant to write down a formula that exhibits this. Instead, we will produce another retraction for which the continuity is evident. Consider the disk D of radius

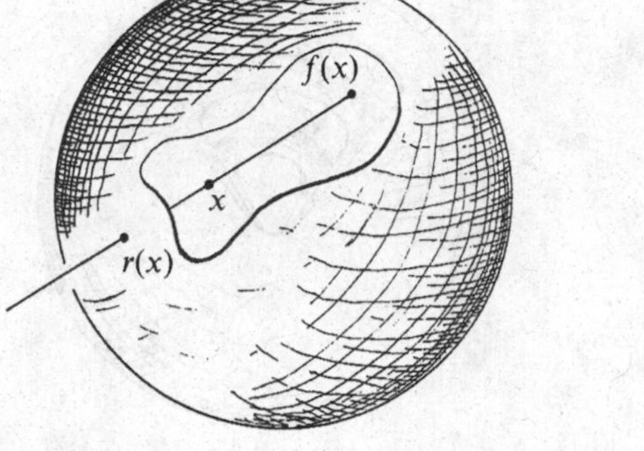

Figure II-9. Retraction of disk.

2 and define a map $g: D \to D$ as follows:

$$g(x) = \begin{cases} (2 - \|x\|)f(x/\|x\|) & \text{if } \|x\| \geq 1, \\ f(x) & \text{if } \|x\| \leq 1. \end{cases}$$

It is clear that g has no fixed points since the image of each point is in the disk of radius 1 where g and f coincide.

Then define $r: D \to D$ by $r(x) = 2(x - g(x))/\|x - g(x)\|$. This is obviously continuous, and, if $\|x\| = 2$, then $g(x) = 0$ whence $r(x) = x$. □

11.13. Corollary. *The sphere* $\mathbf{S}^n$ *is not contractible.*

PROOF. If $\mathbf{S}^n$ is contractible then there is a homotopy $F: \mathbf{S}^n \times \mathbf{I} \to \mathbf{S}^n$ such that $F(x, 1) = x$ and $F(x, 0) = x_0$ for all $x \in \mathbf{S}^n$ and for some point $x_0 \in \mathbf{S}^n$. This factors through the quotient space $\mathbf{S}^n \times I/\mathbf{S}^n \times \{0\} \approx \mathbf{D}^{n+1}$ (by Example 13.9 of Chapter I). The resulting map $\mathbf{D}^{n+1} \to \mathbf{S}^n$ is a retraction of $\mathbf{D}^{n+1}$ onto its boundary $\mathbf{S}^n$, contrary to Theorem 11.11. □

Although the Tubular Neighborhood Theorem (Theorem 11.4) is sufficient for most of our purposes, it will be convenient to have a version for embeddings in arbitrary manifolds, instead of euclidean space. This will be used only in Section 15, which is not used elsewhere in this book.

Let W be a compact manifold smoothly embedded in $\mathbf{R}^n$ and let N be a tubular neighborhood of W with normal retraction $r: N \to W$. Let M be a compact smooth submanifold of W. Define the "normal bundle" of M in W to be

$$\Xi(M, W) = \{\langle x, v\rangle \in M = \mathbf{R}^n | v \in T_x(W) \text{ and } v \perp T_x(M)\}$$

and put

$$\Xi(M, W, \delta) = \{\langle x, v\rangle \in \Xi(M, W) | \ \|v\| < \delta\}.$$

There is a map $\theta: \Xi(M, W) \to \mathbf{R}^n$ given by $\theta(x, v) = x + v$. Then, for δ sufficiently

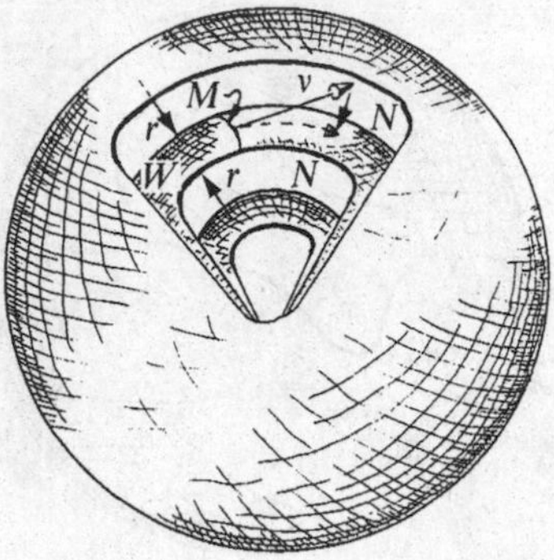

Figure II-10. Normal vector to M in W and normal retraction.

small, θ maps $\Xi(M, W, \delta)$ into N. Consider the composition $\phi = r\circ\theta: \Xi(M, W, \delta) \to W$. At a point $\langle x, 0\rangle$, the differential of θ is monomorphic, as seen before, and maps to the tangent space of W. Since r_* kills only normal vectors, $\phi_* = (r\circ\theta)_*$ is monomorphic at $\langle x, 0\rangle$ and therefore isomorphic by a check of dimensions. As in the proof of Theorem 11.4, there must be an $\epsilon > 0$ such that $\phi: \Xi(M, W, \epsilon) \to W$ is a diffeomorphism to a neighborhood of M in W. (See Figure II-10.)

Although we have assumed that W is compact, that restriction is unnecessary since all manipulations take place in a compact subset, namely, some compact neighborhood of M. We have shown the following generalization of Theorem 11.4.

11.14. Theorem (Tubular Neighborhood Theorem). *Let $M \subset W \subset \mathbf{R}^n$ be a smooth pair of submanifolds with M compact. With the notation above, there is an $\epsilon > 0$ such that the map $\phi: \Xi(M, W, \epsilon) \to W$ is a diffeomorphism of the "ϵ-normal disk bundle to M in W" to a neighborhood V of M in W.* □

In a similar manner, Theorem 11.6 can be generalized so as to replace $\mathbf{R}^n$ by an arbitrary smooth manifold.

The compactness assumptions in Theorems 11.4 and 11.14 can be removed if one allows ϵ to vary with the point. This allows the compactness restrictions in Theorem 11.8 and Corollary 11.9 to be dropped as well. We do not need these results in that generality.

Problems

1. A "probability vector" is a vector in $\mathbf{R}^n$ whose coordinates are all nonnegative and add to 1. A "stochastic matrix" is an $n \times n$ matrix whose columns are probability vectors. Use Corollary 11.12 to show that every stochastic matrix A has a fixed probability vector under the mapping $v \mapsto Av$.

2. Use Corollary 11.12 to show that every $n \times n$ matrix with positive real entries has a positive real eigenvalue.

12. Classical Lie Groups ☼

In this section we develop some elementary properties of the classical groups. This will be used mostly for the study of examples, and so is of minimal importance in understanding the remainder of the book. The reader may want to review parts of Section 15 of Chapter I.

12.1. Lemma. *Let $A = [a_{i,j}]$ be an $n \times n$ matrix over the complex numbers. If c is a bound for the coefficients $|a_{i,j}|$ then $(nc)^k$ is a bound for the coefficients of A^k.*

PROOF. The proof is by an easy induction on k which the reader can supply. □

It follows from Lemma 12.1 that the series of coefficients of

$$I + A + \frac{A^2}{2!} + \frac{A^3}{3!} + \cdots$$

are absolutely dominated by the convergent Taylor's series of e^{nc}. Thus this series of matrices converges to a matrix called e^A. Moreover, the convergence is *uniform* in any compact subset of the space M_n of all $n \times n$ matrices, and so the coefficients of e^A are analytic functions of those of A. That is, the map $\exp\colon M_n \to M_n$ defined by $\exp(A) = e^A$, is analytic. Note that $\exp(0) = I$.

Let us compute the Jacobian of exp at 0, using the usual coordinates (i.e., the matrix coefficients $x_{i,j}$) in M_n. Note that the i, j coordinate of e^A is

$$\delta_{i,j} + x_{i,j} + \text{higher degree terms.}$$

It follows that the Jacobian at $0 \in M_n$ is the $n^2 \times n^2$ identity matrix.

By the Inverse Function Theorem, exp is a diffeomorphism on some neighborhood of $0 \in M_n$ to some neighborhood of $I \in \mathbf{Gl}(n, \mathbf{C})$. (Recall that $\mathbf{Gl}(n, \mathbf{C})$ is an open subset of M_n.)

Also note that for any nonsingular $n \times n$ matrix B,

$$\exp(BAB^{-1}) = I + BAB^{-1} + \frac{BA^2B^{-1}}{2!} = B\left(I + A + \frac{A^2}{2!} + \cdots\right)B^{-1} = Be^AB^{-1}.$$

For any $A \in M_n$, there is a $B \in \mathbf{Gl}(n, \mathbf{C})$ such that BAB^{-1} has super diagonal form with the eigenvalues $\lambda_1, \ldots, \lambda_n$ of A on the diagonal. It follows that $Be^AB^{-1} = \exp(BAB^{-1})$ is also super diagonal with $e^{\lambda_1}, \ldots, e^{\lambda_n}$ on the diagonal. It follows that

$$\det(e^A) = e^{\operatorname{trace}(A)}.$$

In particular, $\det(e^A) \neq 0$ for all A, and so the image of the map exp is in $\mathbf{Gl}(n, \mathbf{C})$.

12.2. Proposition. *If A and B are $n \times n$ matrices that commute then $e^{A+B} = e^A e^B$.*

Proof. For A and B fixed, we shall show that the two functions e^{sA+tB} and $e^{sA}e^{tB}$, of the real variables s and t, are identical. We compute

$$(1) \qquad e^{sA+tB} = \sum_{n=0}^{\infty} \frac{(sA+tB)^n}{n!} = \sum_{n=0}^{\infty} \sum_{j=0}^{n} \frac{(sA)^j (tB)^{n-j}}{j!\,(n-j)!}.$$

On the other hand, an easy computation of partial derivatives in s and t shows that the power series expansion of $e^{sA}e^{tB}$ is

$$e^{sA}e^{tB} = \left(\sum_{i=0}^{\infty} \frac{(sA)^i}{i!} \right) \left(\sum_{j=0}^{\infty} \frac{(tB)^j}{j!} \right)$$

and this is clearly *formally* equally to the right-hand side of (1). Since e^{sA+tB} and $e^{sA}e^{tB}$ are analytic functions of s, t it follows that they are identical. □

12.3. Corollary. *For a given $n \times n$ matrix A, the map $t \mapsto e^{tA}$ is a homomorphism from the additive group of reals into* **Gl**$(n, \mathbf{C})$. □

The homomorphism of Corollary 12.3 is called a "one-parameter subgroup" of **Gl**$(n, \mathbf{C})$. Since **Gl**$(n, \mathbf{C})$ is an open subset of the euclidean space M_n, tangent vectors to **Gl**$(n, \mathbf{C})$ can be regarded as points in M_n. In particular, the tangent vector to the curve $t \mapsto e^{tA}$, at $t = 0$, is

$$\lim_{t \to 0} \frac{e^{tA} - I}{t} = \lim_{t \to 0} \left(A + t\frac{A^2}{2!} + \cdots \right) = A.$$

Consequently, these tangent vectors to **Gl**$(n, \mathbf{C})$ at I fill out the tangent space.

12.4. Proposition. *Let $f: \mathbf{R} \to$ **Gl**$(n, \mathbf{C})$ be a one-parameter subgroup. (That is, f is a continuous homomorphism.) Then there is a matrix $A \in M_n$ such that $f(t) = e^{tA}$ for all $t \in \mathbf{R}$.*

Proof. Consider an open ϵ-neighborhood $U \subset M_n$ of 0 on which the exponential map is one–one onto a neighborhood V of I in **Gl**$(n, \mathbf{C})$. Since $e^{nA} = (e^A)^n$, for integers n, multiplication by n on U, where defined, becomes raising to the nth power on V. It follows that elements in V have *unique* nth roots in V. Let $f, g: \mathbf{R} \to$ **Gl**$(n, \mathbf{C})$ be one-parameter subgroups with $f(t) = g(s)$ for some parameter values t and s. Also assume that $f(t') \in V$ for t' in the interval between 0 and t, and similarly for g and s. Then it follows that $f(t/q) = g(s/q)$ for all integers $q > 0$. In turn, this implies that $f(tp/q) = g(sp/q)$ for all rational numbers p/q. By continuity, we deduce that $f(rt) = g(rs)$ for all real r. Thus $f(u) = g((s/t)u)$ for all real u. Since the one-parameter groups of the form e^{tA} fill out a neighborhood of I in **Gl**$(n, \mathbf{C})$, there is a matrix B and parameters s and t with $f(t) = e^{sB}$ and fulfilling the conditions above for $g(s) = e^{sB}$. Consequently, $f(u) = g(su/t) = e^{(su/t)B} = e^{uA}$, for all u, where $A = (s/t)B$. □

All of the matrix groups discussed in Section 15 of Chapter I are evidently subgroups of **Gl**(n, **C**) except for the ones defined via quaternions. To see this for the latter, note that any quaternionic matrix P can be written as $P = A + Bj$ where A, B are complex matrices. Since $jB = \bar{B}j$, the matrix product is given by

$$(A + Bj)(C + Dj) = (AC - B\bar{D}) + (AD + B\bar{C})j.$$

For $P = A + Bj$, let

$$\phi(P) = \begin{pmatrix} A & B \\ -\bar{B} & \bar{A} \end{pmatrix}. \tag{2}$$

It is easy to calculate that ϕ is an isomorphism of the algebra of quaternionic $n \times n$ matrices into that of $2n \times 2n$ complex matrices. Also $\phi(P^*) = \phi(P)^*$, so that P is symplectic (i.e., $PP^* = I) \Leftrightarrow I = \phi(I) = \phi(PP^*) = \phi(P)\phi(P)^*$, which holds $\Leftrightarrow \phi(P)$ is unitary.

It is elementary to see that a $2n \times 2n$ complex matrix Q has the form (2) $\Leftrightarrow JQJ^{-1} = \bar{Q}$ where

$$J = \begin{pmatrix} 0 & I \\ -I & 0 \end{pmatrix}.$$

If Q is also unitary, then the equation $JQJ^{-1} = \bar{Q}$ can be rewritten as $Q'JQ = J$. Thus the symplectic group **Sp**(n) is isomorphic, via ϕ, to the subgroup of **U**($2n$) consisting of those unitary matrices Q which preserve the bilinear form J (i.e., $Q'JQ = J$). Therefore, we may, and will, consider **Sp**(n) as that subgroup of **U**($2n$).

12.5. Theorem. *For a complex $n \times n$ matrix A the following statements hold:*

(1) *if A is real then e^A is real;*
(2) *if A is skew symmetric ($A' = -A$) and real then $e^A \in$* **O**(n);
(3) *if A is skew Hermitian ($A^* = -A$) then $e^A \in$* **U**(n);
(4) *if A is skew Hermitian, $JA + A'J = 0$, and $n = 2m$ then $e^A =$* **Sp**(m); *and*
(5) *if A has trace 0 then $e^A \in$* **Sl**(n, **C**).

Conversely, there is a neighborhood of 0 *in M_n on which the reverse implications hold.*

Proof. The proofs are all similar and quite easy, so we will only give the argument for (4). If $A^* = -A$ then $(e^A)^* = e^{A^*} = e^{-A} = (e^A)^{-1}$ which implies that e^A is unitary. The equation $JA + A'J = 0$ is equivalent to $JAJ^{-1} = -A'$ and so it implies that $Je^AJ^{-1} = e^{JAJ^{-1}} = e^{-A'} = (e^{-A})' = ((e^A)')^{-1}$. For A skew Hermitian, and hence e^A unitary, this equals the complex conjugate of e^A, implying that $e^A \in$ **Sp**(m).

For the converse, restrict attention to matrices A in a neighborhood of 0 taken by the exponential homeomorphically onto a neighborhood of I. We can assume this neighborhood to be a "ball," and hence invariant under complex conjugation, negation, conjugation by J, and transposition. Then if e^A is unitary, we have that $e^{-A} = (e^A)^{-1} = (e^A)^* = e^{A^*}$. Since the exponential is one–one on a neighborhood of 0 containing both A^* and $-A$, we conclude that $A^* = -A$. If e^A is symplectic then we also have that $\exp(JAJ^{-1}) = J\exp(A)J^{-1}$ is the complex conjugate of e^A. But that is $\exp(\bar{A})$. It follows that $JAJ^{-1} = \bar{A} = -A'$, which is equivalent to $JA + A'J = 0$. □

12.6. Definition. A *Lie group* is a topological group G which also has the structure of a smooth manifold for which the group operations (product and inverse) are smooth maps.

We remark that it can be shown, fairly easily, that a Lie group carries a unique structure as a real analytic manifold for which the group operations are real analytic. It is also known (Hilbert's "fifth problem" proved by A. Gleason, D. Montgomery, and L. Zippin; see Montgomery–Zippin [1]), that a topological group which is locally euclidean (i.e., a topological manifold) carries a unique structure as a Lie group. We shall need neither of these facts.

12.7. Corollary. *Let $G \subset \mathbf{Gl}(n, \mathbf{C})$ be one of the groups $\mathbf{Gl}(n, \mathbf{C})$, $\mathbf{Gl}(n, \mathbf{R})$, $\mathbf{Sl}(n, \mathbf{C})$, $\mathbf{Sl}(n, \mathbf{R})$, $\mathbf{U}(n)$, $\mathbf{SU}(n)$, $\mathbf{O}(n)$, $\mathbf{SO}(n)$, or $\mathbf{Sp}(m)$, where $n = 2m$ in the latter case. Then there is a vector subspace T_G of M_n such that $\exp\colon M_n \to \mathbf{Gl}(n, \mathbf{C})$ maps a neighborhood of 0 in T_G homeomorphically onto a neighborhood of I in G. Also, G is a closed embedded submanifold of $\mathbf{Gl}(n, \mathbf{C})$ and is a Lie group with this structure.*

PROOF. We already have the first statement. The exponential map can be regarded as the inverse of a chart at I in $\mathbf{Gl}(n, \mathbf{C})$, and so the first statement means that G coincides with a plane in this chart near I. By left translation, one sees that this holds at any point of G. The group operations are smooth because they are in the big group. (The restriction of a smooth map to a submanifold is smooth.) □

Corollary 12.7 is a special case of a general statement about closed subgroups of Lie groups. We will state and prove this for the special case of closed subgroups of $\mathbf{Gl}(n, \mathbf{C})$. This is not so special a case, but we won't go into that. Since all the groups of interest to us here (those listed in Corollary 12.7) are themselves closed subgroups of $\mathbf{Gl}(n, \mathbf{C})$, the following development applies to them as to $\mathbf{Gl}(n, \mathbf{C})$. Indeed, the whole development can be done starting with a Lie Group instead of $\mathbf{Gl}(n, \mathbf{C})$. Since our applications will only concern closed subgroups of $\mathbf{Gl}(n, \mathbf{C})$, and are merely examples, we chose this approach to get at the desired items as cheaply as possible.

12.8. Theorem. *Let G be a closed subgroup of $\mathbf{Gl}(n, \mathbf{C})$. Then G is an embedded submanifold of $\mathbf{Gl}(n, \mathbf{C})$ and is a Lie group with this structure.*

PROOF. Consider M_n as a real $4n^2$-dimensional vector space. Let $A_1, \ldots,$

$A_k \in M_n$ be linearly independent and such that $e^{tA_i} \in G$ for all t, and assume that k is maximal with these properties. Let V be the span of $A_1, \ldots, A_k$ and let W be a complementary subspace of V in M_n. Consider the map $\phi: M_n \to \mathbf{Gl}(n, \mathbf{C})$ defined by

$$(*) \qquad \phi(t_1 A_1 + \cdots + t_k A_k + B) = e^{t_1 A_1} e^{t_2 A_2} \cdots e^{t_k A_k} e^B,$$

where $B \in W$. By looking at the differential of ϕ restricted to these k axes and on W, one sees that ϕ_* is the identity. Therefore ϕ is a diffeomorphism on any sufficiently small neighborhood of 0 in M_n to a neighborhood of I in $\mathbf{Gl}(n, \mathbf{C})$. If the right-hand side of $(*)$ is in G then $e^B \in G$ since the other factors are. If we can show that the image under exp of a sufficiently small neighborhood of 0 in W contains no elements of G, other than I, then it follows that a local inverse of ϕ gives a chart on $\mathbf{Gl}(n, \mathbf{C})$ at I with G corresponding to a vector subspace. The group operations on G are smooth near I since they are in $\mathbf{Gl}(n, \mathbf{C})$. Smoothness elsewhere follows from an easy argument using left and right translations.

(This latter is a general fact about Lie groups: one need only have differentiability near the identity, and it follows elsewhere. The proof is more difficult in general than in this special case, much more in case of disconnected groups.) Consequently, the proof will be finished when we establish the following lemma.

12.9. Lemma. *Let $S \subset \mathbf{R}^m$ be a closed nonempty subset. Suppose that $s \in S \Rightarrow ns \in S$ for all integers n. If 0 is not isolated in S then S contains a line through the origin.*

Proof. For each integer $n > 0$ let $x_n \neq 0$ be a point in $B_{1/n}(0) \cap S$. Then there is an integer $k_n \geq n$ such that $k_n x_n \in K$, where $K = \{v \in \mathbf{R}^m \mid 1 \leq \|v\| \leq 2\}$. Let $y_n = k_n x_n$, so that $y_n \in K$. Since K is compact, there is a subsequence $\{y_{n_i}\}$ of $\{y_n\}$ converging to some point $y \in K$. Since the subsequence has all the properties of the original sequence, we may as well assume that $\{y_n\}$ itself converges to y. Let $t \neq 0$ be a real number. Then there is an integer r such that

$$\left|\frac{r}{k_n} - t\right| < \frac{1}{k_n} \leq \frac{1}{n}.$$

Also $(r/k_n) y_n = (r/k_n) k_n x_n = r x_n \in S$. We conclude that

$$\|r x_n - ty\| = \left\|\frac{r}{k_n} y_n - ty\right\| = \left\|\frac{r}{k_n} y_n - t y_n + t y_n - ty\right\|$$

$$\leq \frac{1}{n} \|y_n\| + |t| \, \|y_n - y\|.$$

But this approaches 0 as $n \to \infty$ and so $r x_n \to ty$. Since $r x_n \in S$ and S is closed, we have $ty \in S$ for all real t. □

Although we now know that any closed subgroup G of $\mathbf{Gl}(n, \mathbf{C})$ is a Lie group, we do not have the full force of Corollary 12.7 for it, in that we do not know that G coincides near I with the image under the exponential of a linear subspace of M_n. (The proof of Theorem 12.8 gives G in terms of what is known as "canonical coordinates of the second kind" while Corollary 12.7 corresponds to "canonical coordinates of the first kind." The first kind are natural, while the second kind are not.) If we knew that G has a neighborhood of I that is swept out by one-parameter groups then it would follow from the uniqueness (Proposition 12.4) of one-parameter subgroups that the exponential map takes a neighborhood of 0 in the tangent space of G, thinking of M_n as the tangent space of $\mathbf{Gl}(n, \mathbf{C})$, diffeomorphically onto a neighborhood of I in G, i.e., that Corollary 12.7 holds for any closed subgroup G of $\mathbf{Gl}(n, \mathbf{C})$. For completeness we will indicate the proof of this, but we really don't need it, since we will be concerned only with subgroups that are products of the classical groups, and this fact is immediate (using Corollary 12.7) for those.

12.10. Theorem. *If G is a Lie group and X is a tangent vector to G at the identity $e \in G$, then there is a one-parameter subgroup $f: \mathbf{R} \to G$ whose tangent vector at 0 is taken to X by the differential f_*.*

PROOF. For any $g \in G$ there is the left translation $L_g: G \to G$ defined by $L_g(h) = gh$. This is a diffeomorphism and so the differential of it at e takes X to a tangent vector at g. This gives a tangent vector field on G. By the existence and uniqueness theorem for first-order ordinary differential equations this has, locally at $e \in G$, a unique solution curve $f: \mathbf{R} \to G$ with $f(0) = e$. Suppose that $f(t) = g$ for some t near 0. Then L_g takes e to g and takes the vector field into itself, and hence takes solution curves to solution curves. This implies that, for s near 0, $f(t+s) = L_g(f(s)) = g \cdot f(s) = f(t)f(s)$. Thus f is a homomorphism locally at 0. (Of course, it is only defined near 0 at the moment.) An easy argument, that we leave to the reader, shows that f can be extended, uniquely, to a global homomorphism $\mathbf{R} \to G$. By construction, its tangent vector at 0 is the original vector X. □

13. Fiber Bundles ☼

The theory of fiber bundles, which we have already alluded to and which will be studied in this section, is very well developed. We make no attempt here to go into it deeply, being satisfied to introduce the relevant notions and to discuss some of the examples that will be of use to us later in the book.

13.1. Definition. Let X, B, and F be Hausdorff spaces and $p: X \to B$ a map. Then p is called a *bundle projection* with *fiber* F, if each point of B has a neighborhood U such that there is a homeomorphism $\phi: U \times F \to p^{-1}(U)$ such that $p(\phi\langle b, y\rangle) = b$ for all $b \in U$ and $y \in F$. That is, on $p^{-1}(U)$, p

corresponds to projection $U \times F \to U$. Such a map ϕ is called a *trivialization* of the bundle over U.

13.2. Definition. Let K be a topological group acting effectively on the Hausdorff space F as a group of homeomorphisms. Let X and B be Hausdorff spaces. By a *fiber bundle* over the *base space B* with *total space X*, *fiber F*, and *structure group K*, we mean a bundle projection $p: X \to B$ together with a collection Φ of trivializations $\phi: U \times F \to p^{-1}(U)$, of p over U, called *charts* over U, such that:

(1) each point of B has a neighborhood over which there is a chart in Φ;
(2) if $\phi: U \times F \to p^{-1}(U)$ is in Φ and $V \subset U$ then the restriction of ϕ to $V \times F$ is in Φ;
(3) if ϕ, $\psi \in \Phi$ are charts over U then there is a map $\theta: U \to K$ such that $\psi\langle u, y\rangle = \phi\langle u, \theta(u)(y)\rangle$; and
(4) the set Φ is maximal among collections satisfying (1), (2), and (3).

The bundle is called *smooth* if all these spaces are manifolds and all maps involved are smooth.

Let us investigate the meaning of condition (3) of Definition 13.2. Given charts ϕ and ψ over U, $\phi^{-1}\psi: U \times F \to U \times F$ is a homeomorphism commuting with the projections to U. Thus we can write

$$\phi^{-1}\psi\langle u, y\rangle = \langle u, \mu\langle u, y\rangle\rangle,$$

where

$$\mu: U \times F \to F$$

is the composition $p_F \circ \phi^{-1}\psi$ with the projection $p_F: U \times F \to F$, and hence is continuous. Then $\theta: U \to K$ is given by

$$\theta(u)(y) = \mu\langle u, y\rangle.$$

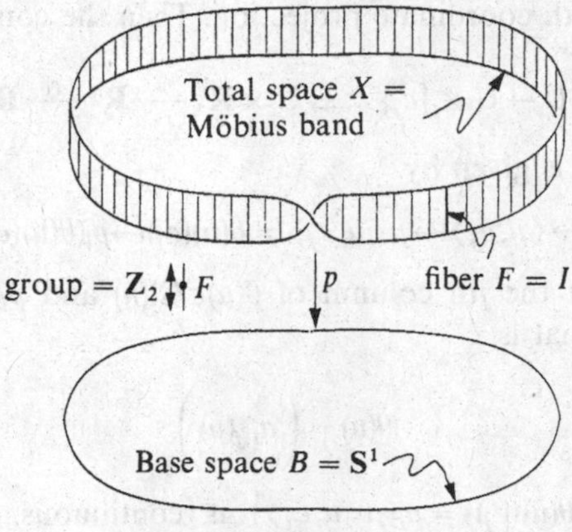

Figure II-11. A fiber bundle.

Therefore θ is completely determined by the charts ϕ and ψ, and it is only a matter of its continuity, and that it takes values in the structure group K, that is of concern in condition (3).

In most cases of interest, the continuity of θ comes free (see Section 2 of Chapter VIII). We shall show that this is the case for "vector bundles" which we now define.

13.3. Definition. A *vector bundle* is a fiber bundle in which the fiber is a euclidean space and the structure group is the general linear group of this euclidean space or some subgroup of that group.

A vector bundle is usually denoted by a Greek letter such as ξ and its total space by $E(\xi)$ and base space by $B(\xi)$. Its fiber projection is denoted by π_ξ or just by π. The following definition, given only for vector bundles, has a fairly obvious generalization to general fiber bundles, but we need it only for vector bundles.

13.4. Definition. If ξ and η are vector bundles then a *bundle map* $\xi \to \eta$ is a map $g: E(\xi) \to E(\eta)$ carrying each fiber of ξ onto some fiber of η isomorphically. (In particular, the fibers have the same dimension and there is an induced map $B(\xi) \to B(\eta)$.) A bundle map g is a *bundle isomorphism* or a *bundle equivalence* if it is a homeomorphism.

13.5. Proposition. *For a vector bundle, condition* (3) *in Definition* 13.2 *can be replaced by the weaker requirement that each* $\theta(u): \mathbf{R}^n \to \mathbf{R}^n$ *be linear. That is, a vector bundle can be defined as a bundle projection* $p: X \to B$ *with fiber* $\mathbf{R}^n$ *such that the changes of coordinates are linear in the fiber.*

Proof. Let $\lambda = \phi^{-1}\psi: U \times \mathbf{R}^n \to U \times \mathbf{R}^n$ be the change of coordinates between any two charts ϕ and ψ over U. The function $\mu: U \times \mathbf{R}^n \to \mathbf{R}^n$ given by $\lambda\langle u, y\rangle = \langle u, \mu\langle u, y\rangle\rangle$ is continuous, as remarked before, and θ is given by $\theta(u)(y) = \mu\langle u, y\rangle$. Let e_j be the jth standard basis vector in $\mathbf{R}^n$ and let $p_i: \mathbf{R}^n \to \mathbf{R}$ be the ith coordinate projection. Then the composition

$$U \to U \times \{e_j\} \hookrightarrow U \times \mathbf{R}^n \xrightarrow{\mu} \mathbf{R}^n \xrightarrow{p_i} \mathbf{R}$$

is continuous and is given by

$$u \mapsto \langle u, e_j\rangle \mapsto \mu\langle u, e_j\rangle = \theta(u)(e_j) \mapsto p_i(\theta(u)(e_j)).$$

But $\theta(u)(e_j)$ is just the jth column of $\theta(u) \in \mathbf{Gl}(n)$ and $p_i(\theta(u)(e_j))$ is the i, j element of $\theta(u)$. That is,

$$\theta(u) = \left(a_{i,j}(u) \right),$$

where $a_{i,j}(u) = p_i(\theta(u)(e_j)) = p_i(\mu\langle u, e_j\rangle)$ is continuous. Hence θ is continuous. □

Note that if we only have the map $\lambda: U \times \mathbf{R}^n \to U \times \mathbf{R}^n$ in the proof of Proposition 13.5 (commuting with the projection to U and an isomorphism on each fiber), then μ is continuous and hence θ is continuous. If we let $\theta'(u) = (\theta(u))^{-1}$ then θ' is also continuous. This implies that the corresponding μ' and λ' are also continuous. But $\lambda' = \lambda^{-1}$ and so λ must, in fact, be a homeomorphism, and a diffeomorphism in the smooth case. This gives the following consequence.

13.6. Corollary. *If ξ and η are vector bundles over B and $f: E(\xi) \to E(\eta)$ is a bundle map inducing the identity on B then f is a bundle isomorphism.* □

13.7. Definition. A *disk bundle*, or *sphere bundle*, is a fiber bundle in which the fiber is a metric disk or sphere in euclidean space and the structure group is the orthogonal group of that space, or a subgroup of that group.

A disk or sphere bundle gives rise to a vector bundle with the orthogonal group as structure group just by replacing the fibers $\mathbf{D}^n$ or $\mathbf{S}^{n-1}$ by $\mathbf{R}^n$ and using the same change of coordinate functions θ. Such a vector bundle is sometimes called a "euclidean bundle." Conversely, every vector bundle over a paracompact base space can be given the structure of a euclidean bundle, meaning that an atlas of charts can be selected for which the changes of coordinates are orthogonal in each fiber. (In Steenrod's terminology, one says that the structure group can be "reduced" to $\mathbf{O}(n)$.) One can see this as follows: the changes of coordinates being orthogonal in each fiber just means that they preserve the quadratic form $\sum x_i^2$ on fibers. That means that there exists a map $q: E(\xi) \to \mathbf{R}$ which is a positive-definite quadratic form on each fiber since, given such a map q, the Gram–Schmidt construction produces an orthonormal framing on the fibers of a coordinate chart and this is equivalent to giving a chart for which the quadratic form is the standard one, $\sum x_i^2$ on each fiber. Thus it suffices to produce such a $q: E(\xi) \to \mathbf{R}$. For any point $x \in B$, one can use any chart to produce such a function q locally about x in B. One can then multiply it by a function $f: B \to [0, \infty)$ which is nonzero at x and vanishes outside the domain of the chart. This gives a $q_x: E(\xi) \to [0, \infty)$ which is a positive semidefinite form on each fiber and is positive definite over a neighborhood of x. Since a positive linear combination of positive definite forms is positive definite, a partition of unity argument then finishes the construction. If the base space is a smooth manifold and ξ is a smooth vector bundle (smooth θ) then one usually wants the quadratic form q to be smooth, and that can be guaranteed by using a smooth partition of unity.

In the case of the tangent bundle of a smooth manifold, such a smooth quadratic form is called a "Riemannian metric" on M.

For another example, suppose M^m is a smooth manifold and that N^n is a submanifold. The tangent bundle of M is a vector bundle over M. Its restriction to N is a vector bundle, or "m-plane bundle" over N. For this bundle over N the structure group can be taken to be (is said to be "reducible" to) the subgroup K of $\mathbf{Gl}(m, \mathbf{R})$ consisting of matrices of the form

$$\begin{pmatrix} A & B \\ 0 & C \end{pmatrix},$$

where A is $n \times n$. This is because the tangent space of M at a point of N has a well-defined subspace, the tangent space of N, and we can restrict our attention to charts preserving this subspace. The matrices A form the transformations of the subspace (i.e., give the transitions for the tangent space of N). This embeds the tangent bundle of N in that of M, restricted to N. One can also form a bundle of the quotients of the tangent spaces to M at points of N by this subbundle, the tangent bundle of N. One gets a vector bundle in which the transition functions are given by the matrices denoted by C above. This vector bundle is called the "normal bundle" of N in M.

For another example, one that will be of use to us in providing examples of later results in algebraic topology, consider a Lie group G and closed subgroup H. Let us assume that G is a closed subgroup of $\mathbf{Gl}(n, \mathbf{C})$. (This assumption is not needed for the truth of what we say below, but is imposed so as to be able to use facts we have previously proved under such an assumption. All our examples will satisfy this.) Then we claim that the canonical map $G \to G/H$ is a bundle projection with fiber H. Recall that from Corollary 12.7 and the discussion above Theorem 12.10, that the exponential map $M_n \to \mathbf{Gl}(n, \mathbf{C})$ takes a linear subspace T_G of M_n smoothly into G and is a diffeomorphism on a neighborhood of 0 in T_G to a neighborhood of the identity $e \in G$. We do not care about the remainder of the exponential map and can forget about $\mathbf{Gl}(n, \mathbf{C})$ now. Similarly, a linear subspace $T_H \subset T_G$ is taken into H by exp, again a diffeomorphism near 0. Let V be a linear subspace of T_G complementary to T_H and consider the map $\phi: V \times H \to G$ defined by $\phi(v, h) = \exp(v)h$. Recalling that the differential of exp at 0 is the identity, we see that the differential ϕ_* of ϕ at $(0, e)$ is the identity on $V \times \{0\} \subset V \times T_H = T_G$. It is also the identity on $\{0\} \times T_H$ since $\phi(0, h) = h$. Thus $\phi_* = 1$ at $(0, e)$ and it follows that ϕ is a diffeomorphism near $(0, e) \in V \times H$ to a neighborhood U of e in G. We claim that ϕ is an embedding on $B \times H$ where $B \subset V$ is a ball about the origin such that $\exp(B)^2 \subset U$. Note that translation by elements of H shows that the differential ϕ_* is an isomorphism at any point of $B \times H$. Consequently, it suffices to show that ϕ is one–one on $B \times H$. Suppose that $\phi(b, h) = \phi(b', h')$ where $b, b' \in B$. This equation is $e^b h = e^{b'} h'$, i.e., $e^{-b'} e^b = h' h^{-1}$. But this is in $\exp(B)^2 \cap H = \{e\}$. Thus, $e^b = e^{b'}$, whence $b = b'$, and $h = h'$.

Clearly, ϕ induces a commutative diagram

$$\begin{array}{ccc} B \times H & \xrightarrow{\phi} & G \\ \big\downarrow{\scriptstyle \text{proj}} & & \big\downarrow \\ B & \xrightarrow{\psi} & G/H \end{array}$$

defining ψ. Obviously ψ maps onto a neighborhood of eH in G/H. If $\psi(v) = \psi(v')$ then $\phi(v, e) = \phi(v', e)h$ for some $h \in H$. But $\phi(v', e)h = \phi(v', h)$, and so this can happen only for $v = v'$ and $h = e$. Thus ψ is one–one. If we take B to be a closed ball then ψ is a one–one map from a compact space to a Hausdorff space and so is a homeomorphism to its image. This provides a trivialization of the map $G \to G/H$ over a neighborhood of eH. Trivializations near other points are obtained simply by left translation in G and G/H. We leave the rest of the easy verification that this is a bundle to the reader. Note that it also follows that G/H is a smooth manifold with the structure induced by the projection $G \to G/H$, and that left translation $G \times G/H \to G/H$ is a smooth action; see Definition 15.13 of Chapter I.

PROBLEMS

1. Finish the proof at the end of this section that $G \to G/H$ is a bundle. Also show that this is a fiber bundle with structure group $N(H)/H$ where $N(H)$ is the normalizer of H in G.

2. ☼ Suppose that G is a Lie group, and assume it is a closed subgroup of $\mathbf{Gl}(n, \mathbf{C})$. Assume that $K \subset H \subset G$ are closed subgroups. Show that the canonical map $G/K \to G/H$ is a fiber bundle with fiber H/K.

3. Show that the Klein bottle is a fiber bundle over $\mathbf{S}^1$ with fiber $\mathbf{S}^1$ and structure group $\mathbf{Z}_2$ acting by a reflection on $\mathbf{S}^1 \subset \mathbf{R}^2$.

4. ☼ A fiber bundle $p: X \to B$ is a "principal bundle" if its fiber is its structure group K acting on itself by right translation. Show that K acts naturally on X (on the left) with orbit space $X/K \approx B$. If p has a continuous cross section then show that it is a trivial bundle.

14. Induced Bundles and Whitney Sums ☼

This section will be used only in Section 15 and in Section 17 of Chapter VI.

If $p: X \to B$ is a fiber bundle with fiber F and if $B' \subset B$, then the restriction of p to $p^{-1}(B') \to B'$ is also a fiber bundle. One can generalize this to arbitrary maps $f: A \to B$ instead of inclusions as follows.

The "induced bundle" $p': f^*X \to A$, induced from p by f, is given by the "pullback"

$$\boxed{f^*(X) = \{\langle a, x\rangle \in A \times X \mid p(x) = f(a)\}}$$

and p' is given by $p'\langle a, x\rangle = a$. With $f': f^*(X) \to X$ given by $f'\langle a, x\rangle = x$, the diagram

$$\begin{array}{ccc} f^*(X) & \xrightarrow{f'} & X \\ {\scriptstyle p'}\downarrow & & \downarrow{\scriptstyle p} \\ A & \xrightarrow{f} & B \end{array}$$

commutes. To see that p' is a bundle projection, let $\phi: U \times F \to p^{-1}(U)$ be a chart over the open set $U \subset X$ and define

$$\phi': f^{-1}(U) \times F \to (p')^{-1} f^{-1}(U) = (fp')^{-1}(U)$$

by

$$\phi'\langle v, y\rangle = \langle v, \phi\langle f(v), y\rangle\rangle.$$

This works since $p\phi\langle f(v), y\rangle = f(v) \in U$ and $fp'\langle v, \phi\langle f(v), y\rangle\rangle = f(v)$. The inverse of ϕ' is

$$\lambda\langle a, x\rangle = \langle a, p_F\phi^{-1}(x)\rangle$$

for $p(x) = f(a) \in U$, and where $p_F: U \times F \to F$ is the projection onto F. To see this, let $p_U: U \times F \to U$ be the projection onto U and compute

$$\begin{aligned} \lambda\phi'\langle v, y\rangle &= \lambda\langle v, \phi\langle f(v), y\rangle\rangle \\ &= \langle v, p_F\phi^{-1}(\phi\langle f(v), y\rangle)\rangle \\ &= \langle v, p_F\langle f(v), y\rangle\rangle \\ &= \langle v, y\rangle \end{aligned}$$

and with $f(a) = p(x)$,

$$\begin{aligned} \phi'\lambda\langle a, x\rangle &= \phi'\langle a, p_F\phi^{-1}(x)\rangle \\ &= \langle a, \phi\langle f(a), p_F\phi^{-1}(x)\rangle\rangle \\ &= \langle a, \phi\langle p(x), p_F\phi^{-1}(x)\rangle\rangle \\ &= \langle a, \phi\langle p_U\phi^{-1}(x), p_F\phi^{-1}(x)\rangle\rangle \\ &= \langle a, \phi\phi^{-1}(x)\rangle \\ &= \langle a, x\rangle. \end{aligned}$$

This shows that p' is a bundle projection with fiber F. If ϕ and ψ are two charts over U for $p: X \to B$ and $\psi\langle u, y\rangle = \phi\langle u, \theta(u)(y)\rangle$ as in (3) of Definition 13.2, then we see that

$$\begin{aligned} \psi'\langle v, y\rangle &= \langle v, \psi\langle f(v), y\rangle\rangle \\ &= \langle v, \phi\langle f(v), \theta(f(v))(y)\rangle\rangle \\ &= \phi'\langle v, \theta(f(v))(y)\rangle \\ &= \phi'\langle v, \theta'(v)(y)\rangle, \end{aligned}$$

where $\theta' = \theta \circ f: f^{-1}(U) \to K$. Therefore θ' is continuous and satisfies (3) of Definition 13.2 for p'. Consequently, p' is a fiber bundle with fiber F and structure group K.

It is clear that if $A' \subset A$ and $B' \subset B$ are subspaces such that $f(A') \subset B'$, then the part of $f^*(X)$ over A' and the part X' of X over B' provide a pullback diagram; i.e., the part of $f^*(X)$ over A' is equivalent to $(f|_{A'})^*(X')$. Applying this to the case of a single point $a \in A$ and its image $b = f(a) \in B$ gives the subdiagram

$$\begin{array}{ccc} f^*(F) & \longrightarrow & F \\ \big\downarrow & & \big\downarrow p \\ \{a\} & \xrightarrow{f} & \{b\}. \end{array}$$

Thus $f^*(F) = \{\langle a, x\rangle \in \{a\} \times F \mid p(x) = f(a)\} = \{a\} \times F$ and this maps to F in the obvious manner.

Now we shall specialize, for the remainder of this section, to vector bundles. As remarked before, one commonly denotes the total space of a vector bundle ξ by $E(\xi)$.

14.1. Theorem. *Suppose ξ is a vector bundle over A and η is a vector bundle over B and let $f: A \to B$. If $\phi: E(\xi) \to E(\eta)$ is a bundle map over f then ξ is isomorphic to $f^*\eta$.*

PROOF. Let $g: E(\xi) \to E(f^*\eta) = \{\langle a, x\rangle \in A \times E(\eta) \mid f(a) = \pi_\eta(x)\}$ be given by $g(v) = \langle \pi_\xi(v), \phi(v)\rangle$. Then g maps the fiber of ξ over $a \in A$ to that of $f^*(\eta)$ over a by the linear isomorphism $v \mapsto \langle a, \phi(v)\rangle$. By Corollary 13.6 this implies that g is a bundle equivalence. □

If ξ_1 and ξ_2 are vector bundles with projections $\pi_1: E(\xi_1) \to B_1$ and $\pi_2: E(\xi_2) \to B_2$ and with fibers $\mathbf{R}^{n_1}$ and $\mathbf{R}^{n_2}$, then $\pi_1 \times \pi_2: E(\xi_1) \times E(\xi_2) \to B_1 \times B_2$, with the obvious charts, is a vector bundle with fiber $\mathbf{R}^{n_1+n_2}$. In the special case for which $B_1 = B = B_2$, the diagonal map $B \to B \times B$ then induces a bundle from $\pi_1 \times \pi_2$. This vector bundle over B with fiber $\mathbf{R}^{n_1+n_2}$ is denoted by $\xi_1 \oplus \xi_2$ and is called the "Whitney sum" of ξ_1 and ξ_2.

For example, if M^m is a smoothly embedded submanifold of $\mathbf{R}^n$ then one has its tangent bundle τ and its normal bundle ν. The fiber of the Whitney sum $\tau \oplus \nu$ is just $\mathbf{R}^n$. Since tangent vectors to M^m can be regarded as vectors in $\mathbf{R}^n$ and similarly for normal vectors, there are maps $E(\tau) \to \mathbf{R}^n$ and $E(\nu) \to \mathbf{R}^n$. This induces a map $E(\tau \oplus \nu) \to \mathbf{R}^n$, by addition of vectors, which is an isomorphism on each fiber. This is a bundle map if $\mathbf{R}^n$ is thought of as the total space of a vector bundle over a point. Therefore $E(\tau \oplus \nu)$ is induced from the trivial bundle over a point and hence is itself a trivial bundle. Thus, in a sense, ν is an additive inverse to τ. That the tangent bundle has an additive inverse in this sense is not a special property of the tangent bundle, as we now show.

14.2. Theorem. *If ξ is a smooth vector bundle over the smooth manifold M^m then there exists a vector bundle η over M^m such that $\xi \oplus \eta$ is isomorphic to a trivial vector buncle over M^m.*

PROOF. We can, and will, regard M^m as the zero section of $E(\xi)$. One can embed some neighborhood $U \subset E(\xi)$ of M^m in some $\mathbf{R}^n$. If τ_U and ν_U are the tangent and normal bundles of U then $\tau_U \oplus \nu_U$ is a trivial bundle over U as seen above. Consider a vector in ξ at $x \in M$. This defines a ray in the fiber of $E(\xi)$ over x. The differential of the embedding $U \to \mathbf{R}^n$ takes the tangent vector to this ray into a tangent vector to U at x that does not lie in the tangent space of M at x. Therefore, it projects nontrivially to the normal space to $T_x(M)$ in $T_x(U)$, regarding both as subspaces of $\mathbf{R}^n$. This induces an isomorphism of ξ with the normal bundle λ to M in U. But $\lambda \oplus \nu_U|_M \approx \nu_M$ and so $\lambda \oplus \nu_U|_M \oplus \tau_M \approx \nu_M \oplus \tau_M$ which is a trivial bundle. This shows that $\xi \approx \lambda$ has an additive inverse, namely, $\nu_U|_M \oplus \tau_M$. □

Remark. The condition in Theorem 14.2, that the base space is a smooth manifold, can be weakened considerably. Indeed, if the base space is any space that can be embedded in euclidean space, then it can be shown that the bundle extends over some neighborhood and can be assumed smooth there. Then the idea of the proof of Theorem 14.2 applies. On the other hand, the canonical line bundle over $\mathbf{P}^\infty$ (essentially the projection $\mathbf{R}^\infty - \{0\} \to \mathbf{P}^\infty$) does not have an inverse. (This follows from the results presented in Section 17 of Chapter VI.)

15. Transversality ☼

Transversality is a central idea in differential topology which allows "general position" arguments. Here we prove some basic facts about it and attempt to indicate its importance via some simple examples. This section will not be used elsewhere in this book except for the isolated result Theorem 11.16 of Chapter VI.

The following is a generalization of the definition of transversality given in Definition 7.6:

15.1. Definition. Let M, X, Y be smooth manifolds and let $f: X \to M$ and $g: Y \to M$ be smooth maps with g an embedding. Then f is said to be transverse to g (denoted by $f \pitchfork g$) if, whenever $f(x) = g(y)$, the images of the differentials $f_*: T_x(X) \to T_{f(x)}(M)$ and $g_*: T_y(Y) \to T_{g(y)}(M) = T_{f(x)}(M)$ span $T_{f(x)}(M)$.

15.2. Theorem. *In the situation of Definition* 15.1, $f^{-1}(f(X) \cap g(Y))$ *is an embedded submanifold of* X *of dimension* $\dim(X) + \dim(Y) - \dim(M)$.

Proof. Let $p \in f(X) \cap g(Y)$ and take local coordinates on the open set $U \subset M$ about p for which $g(Y)$ is flat; i.e., $U \approx \mathbf{R}^n \times \mathbf{R}^k$ with $g(Y)$ corresponding to $\mathbf{R}^n \times \{0\}$. Let $q: U \approx \mathbf{R}^n \times \mathbf{R}^k \to \mathbf{R}^k$ be the projection. Then $q \circ f: f^{-1}(U) \to \mathbf{R}^k$ has 0 as a regular value and so $(q \circ f)^{-1}(0) = f^{-1}(f(X) \cap g(Y)) \cap f^{-1}(U)$ is an embedded submanifold of X of codimension $k = \dim(M) - \dim(Y)$. □

The elegant proof of the following result is due to Kosinski [1].

15.3. Theorem. *Let ξ be a smooth vector bundle over the smooth manifold Y. Let X be a smooth manifold and $f: X \to E(\xi)$ a smooth map. Then there is a smooth cross section $s: Y \to E(\xi)$, as close to the zero section as desired, such that $f \pitchfork s$.*

PROOF. By Theorem 14.2 there is a (smooth) vector bundle η over Y such that $\xi \oplus \eta$ is trivial. Consider the commutative diagram

$$\begin{array}{ccc} E(f^*(\xi \oplus \eta)) & \xrightarrow{f'} & E(\xi \oplus \eta) \\ \downarrow{\scriptstyle \pi'} & & \downarrow{\scriptstyle \pi} \\ X & \xrightarrow{f} & E(\xi) \end{array}$$

defining f'. A trivialization of $\xi \oplus \eta$ provides a diffeomorphism $\phi: Y \times \mathbf{R}^n \xrightarrow{\approx} E(\xi \oplus \eta)$. Let $p: E(\xi \oplus \eta) \to \mathbf{R}^n$ be the resulting projection. Let $z \in \mathbf{R}^n$ be a regular value of the composition $p \circ f': E(f^*(\xi \oplus \eta)) \to E(\xi \oplus \eta) \to \mathbf{R}^n$. By definition of a regular value, the differential $p_* \circ f'_*$ maps the tangent space at any point $v \in E(f^*(\xi \oplus \eta))$ with $pf'(v) = z$ onto that of $\mathbf{R}^n$ at z. This implies that the image of f'_* must span the complement of the tangent space to $Y \times \{z\}$ at $\langle f'(v), z \rangle$ and that simply means that f' is transverse to the section $s': Y \to E(\xi \oplus \eta)$, given, in terms of the trivialization, by $s'(y) = \langle y, z \rangle$. Define the section $s: Y \to E(\xi)$ by $s(y) = \pi s'(y)$; i.e., so that the following diagram commutes:

$$\begin{array}{ccccc} E(f^*(\xi \oplus \eta)) & \xrightarrow{f'} & E(\xi \oplus \eta) & \xleftarrow{s'} & Y \\ \downarrow{\scriptstyle \pi'} & & \downarrow{\scriptstyle \pi} & \swarrow{\scriptstyle s} & \\ X & \xrightarrow{f} & E(\xi). & & \end{array}$$

We claim that $f \pitchfork s$. To see this, let $x \in X$, $y \in Y$ be such that $f(x) = s(y)$. Then $\pi(s'(y)) = s(y) = f(x)$. By the definition of the induced bundle, the point $\langle x, s'(y) \rangle \in E(f^*(\xi \oplus \eta))$ has $f'\langle x, s'(y) \rangle = s'(y)$. Since $f' \pitchfork s'$, the images of the differentials f'_* at $\langle x, s'(y) \rangle$ and s'_* at y span the tangent space to $E(\xi \oplus \eta)$ at $f'\langle x, s'(y) \rangle = s'(y)$. Since π is a submersion, this must also hold for the images of the differentials f_* at x and s_* at y to the tangent space of $E(\xi)$ at $f(x) = s(y)$, but that just means that $f \pitchfork s$. □

For the remainder of this section we will consider submanifolds of some fixed manifold W. We will state the results in general, but will use the Tubular Neighborhood Theorems (Theorems 11.4 and 11.14) which we have proved only in case $W = \mathbf{R}^n$ or W a compact submanifold of $\mathbf{R}^n$. To simplify notation in proofs, the case $W = \mathbf{R}^n$ will be assumed, but there is no real difficulty in extending the proofs to the general case.

15.4. Corollary. *Let M and N be compact smooth manifolds. Let $f: M \to W$ be smooth and let $g_0: N \to W$ be a smooth embedding where W is some smooth manifold. Then there is an arbitrarily small homotopy of g_0 to a smooth* *embedding $g_1: N \to W$ such that $f \pitchfork g_1$. Indeed, the homotopy can be taken to be smooth and such that each $g_t: N \to W$ is an embedding; i.e., it is an "isotopy." M and N can be manifolds with boundary provided the boundaries do not meet.*

PROOF. Let ν be the normal bundle of $g_0(N)$ in W and let $g: E(\nu) \hookrightarrow W$ be a tubular neighborhood. Then $M' = f^{-1}(E(\nu))$ is an open submanifold of M and Theorem 15.3 applies to $f|_{M'}: M' \to E(\nu)$. □

We need to have a version of Corollary 15.4 in which the map f is the one to be deformed. This requires the following fact.

15.5. Lemma. *Let N be a compact smoothly embedded submanifold of W. Let T be a tubular neighborhood of N as constructed in Theorems 11.4 or 11.14. If $s: N \to \operatorname{int} T$ is a section (i.e., $p \circ s = 1$ where $p: T \to N$ is the normal retraction) then there is a homeomorphism $h: T \to T$ which preserves fibers, is the identity on ∂T, and carries s to the zero section N. If s is smooth then h is smooth on int T. Moreover, h is homotopic to the identity through such homeomorphisms.*

PROOF. As stated before, we will do the proof in the case $W = \mathbf{R}^n$ which is notationally, but not conceptually, easier. Let us first remark that a more advanced proof can show that there is such a map h which is smooth on ∂T as well; i.e., such that the extension to $\mathbf{R}^n \to \mathbf{R}^n$ by the identity is smooth. See Kosinski [1]. (It can be seen that the h we produce here is C^1 but not C^2. It is possible to amplify the present proof to produce such a smooth h.)

For our proof, recall that T was constructed as the image of

$$\{(x, v) \in N \times \mathbf{R}^n \mid v \perp T_x(N) \text{ and } \|v\| \le \epsilon\}$$

under the map $(x, v) \mapsto x + v \in \mathbf{R}^n$. The section s can be viewed as a map $x \mapsto s(x) \in \mathbf{R}^n$ where $s(x) \perp T_x(N)$ and $\|s(x)\| < \epsilon$ for all x. Let $\phi: [0, \infty) \to [0, \epsilon)$ be a diffeomorphism which is the identity near 0, and note that $d\phi/dt \to 0$ as $t \to \infty$. If B is the open ϵ-ball about 0 in $\mathbf{R}^n$ then the map $\theta: B \to \mathbf{R}^n$ given by $\theta(v) = \phi^{-1}(\|v\|)v/\|v\|$ for $v \ne 0$ and $\theta(0) = 0$, is a diffeomorphism. Then the map $h: \operatorname{int} T \to \operatorname{int} T$ given by

$$h(x, v) = (x, \theta^{-1}(\theta(v) - \theta(s(x))))$$

is the required map. The reason that it extends continuously to ∂T is as follows. Note that $\|\theta(s(x))\|$ is bounded. As $\|v\| \to \epsilon$ we have that $\|\theta(v)\| \to \infty$. Since $\theta(v)$ and $\theta(v) - \theta(s(x))$ are of bounded distance apart it follows that the distance between $v = \theta^{-1}\theta(v)$ and $\theta^{-1}(\theta(v) - \theta(s(x)))$ approaches 0, basically because $d\phi/dt \to 0$.

Note that the idea of h is that one uses a nice diffeomorphism θ of B to $\mathbf{R}^n$, then translation in $\mathbf{R}^n$ to move the section s to the zero section and then the inverse of θ to bring this back to B from $\mathbf{R}^n$.

The reason that h carries T into itself preserving fibers is that θ carries all vector subspaces of $\mathbf{R}^n$ into themselves, and, in particular, carries the orthogonal complement of $T_x(N)$ into itself.

The statement that $h \simeq 1$ is obvious from the construction. □

15.6. Corollary. *Let M be a compact smooth manifold. Let $f_0: M \to W$ be smooth and let $N \subset W$ be a smooth compact submanifold. Then $f_0 \simeq f_1$ where $f_1 \pitchfork N$. Moreover, the homotopy can be taken to be constant outside $f_0^{-1}(T)$ for a given tubular neighborhood T of N in W.*

PROOF. Theorem 15.3 provides a section s of some tubular neighborhood of N such that $f_0 \pitchfork s(N)$. Composing f_0 with h of Lemma 15.5 (extended by the identity) gives the desired f_1. This f_1, as constructed, is not smooth at the boundary of the tubular neighborhood, but can then be smoothly approximated without changing it near the intersection with N where f_1 is already smooth. □

Although we have only dealt with manifolds without boundary, it is clear that everything goes through for manifolds with boundary as long as the maps do not take any boundary point of one manifold to the image of the other manifold. It is not too hard to weaken that to the case in which transversality already exists near the boundaries. Even that restriction can be dropped, in which case one should also demand that the approximating maps be transverse when *restricted* to either boundary.

15.7. Corollary. *Let M^m be any compact manifold smoothly embedded in $\mathbf{R}^n$ (or $\mathbf{S}^n$). Then any map $f: \mathbf{S}^k \to \mathbf{R}^n - M^m$ can be extended to $\bar{f}: \mathbf{D}^{k+1} \to \mathbf{R}^n - M^m$ provided $k < n - m - 1$.*

PROOF. By a small homotopy (see Theorem 11.7) we can smooth f and then we can extend it to a smooth map $f_0: \mathbf{D}^{k+1} \to \mathbf{R}^n$. By Corollary 15.6, $f_0 \simeq f_1$ rel $\mathbf{S}^k$, with $f_1 \pitchfork M^m$. But, in these dimensions ($k + 1 + m < n$), this implies that $f_1(\mathbf{D}^{k+1}) \cap M^m = \emptyset$ so that $f_1: \mathbf{D}^{k+1} \to \mathbf{R}^n - M^m$ as required. □

As an example, this shows that if $f: \mathbf{S}^1 \to \mathbf{R}^4$ is a smooth embedding, then $\mathbf{R}^4 - f(\mathbf{S}^1)$ is "simply connected," i.e., that any map $\mathbf{S}^1 \to \mathbf{R}^4 - f(\mathbf{S}^1)$ is homotopic to a constant map. This is not true for nonsmooth embeddings in general; see Rushing [1].

As another example, consider embeddings $f: \mathbf{S}^n \to \mathbf{R}^{n+k+1}$ and $g: \mathbf{S}^k \to \mathbf{R}^{n+k+1}$ whose images are disjoint. One can extend f to $f_1: \mathbf{D}^{n+1} \to \mathbf{R}^{n+k+1}$, and, by a homotopy rel $\mathbf{S}^n$, one can assume that $f_1 \pitchfork g$. Then $f_1(\mathbf{D}^{n+1}) \cap g(\mathbf{S}^k)$ is a 0-manifold; a finite set of points where these maps are transverse. Let $K = f_1^{-1}(f_1(\mathbf{D}^{n+1}) \cap g(\mathbf{S}^k))$, again a 0-manifold by Theorem 15.2. At any point $x \in K$, the differential of f_1 induces an $(n+1)$-frame at $f_1(x)$ and that of g induces a k-frame at the same point. Putting these together in this order gives an $(n+k+1)$-frame which may or may not be consistent with the standard $(n+k+1)$-frame of $\mathbf{R}^{n+k+1}$. Assign a plus sign to x if so and a minus sign if not. Then the sum of these signs over all such points x gives an integer called the "linking number" $L(f, g)$ of f and g. To see that this is

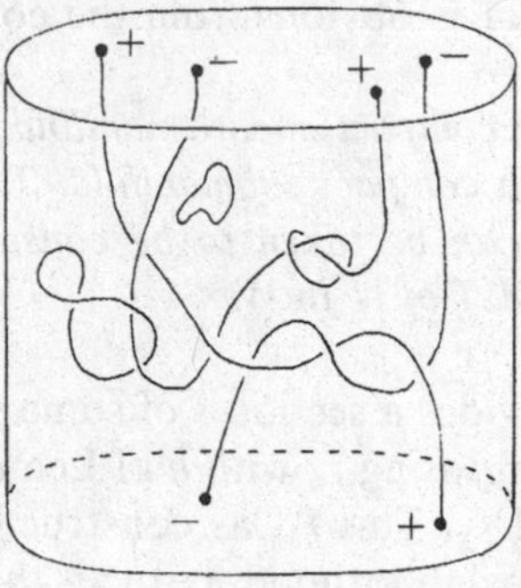

Figure II-12. About linking numbers.

independent of the choice of the approximation f_1, let $f_0: \mathbf{D}^{n+1} \to \mathbf{R}^{n+k+1}$ be another such approximation. Then there is a homotopy $F: \mathbf{D}^{n+1} \times I \to \mathbf{R}^{n+k+1}$ rel $\mathbf{S}^n \times I$ from f_0 to f_1. It can be assumed that F is constant near the two ends, and so it is transverse to g there. Then F can be approximated by some $F_1: \mathbf{D}^{n+1} \times I \to \mathbf{R}^{n+k+1}$ that coincides with F on a neighborhood of $\mathbf{S}^n \times I \cup \mathbf{D}^{n+1} \times \partial I$ and such that $F_1 \pitchfork g$. Then $F_1^{-1}(F_1(\mathbf{D}^{n+1} \times I) \cap g(\mathbf{S}^k))$ is a 1-manifold in $\mathbf{D}^{n+1} \times I$ not meeting $\mathbf{S}^n \times \mathrm{I}$ and which is transverse to the boundary. See Figure II-12. This 1-manifold describes cancellations and creations of pairs of points of opposite sign in its intersections with $\mathbf{D}^{n+1} \times \{0\}$ and $\mathbf{D}^{n+1} \times \{1\}$, which implies the invariance of the linking number. The reader might find it edifying to try to convince himself that $L(f, g)$ differs from $L(g, f)$ only by a sign depending on n and k. There are several other methods of defining $L(f, g)$.

For our last example, let M^m be a compact smooth manifold without boundary and let $\phi: M^m \to \mathbf{R}^n$ be smooth, for some $n \leq m$. Let $x \in \phi(M^m)$ be a regular value (assuming not all of M^m is critical). Then we claim that $K^{m-n} = \phi^{-1}(x)$ is an $(m-n)$-manifold which bounds an $(m-n+1)$-manifold $V \subset M^m$. To see this, let $r \subset \mathbf{R}^n$ be a ray from x of length b where $b > \operatorname{diam} \phi(M^m)$ and let y be its other end. Then the open segment (x, y) has a tubular neighborhood $(x, y) \times \mathbf{R}^{n-1}$ for which the rays are the constant cross-sections. By Theorem 15.3, ϕ is transverse to one of these rays, say s. The required V^{m-n+1} is just $\phi^{-1}(s)$.

16. Thom–Pontryagin Theory ☼

In this section we will investigate pointed homotopy classes of maps $f: \mathbf{S}^{n+k} \to \mathbf{S}^n$. The term "pointed" means that we fix a base point in each space and consider only maps and homotopies taking the base point to the base point. This set of homotopy classes is called the $(n+k)$th "homotopy group" of $\mathbf{S}^n$ and is denoted by $\pi_{n+k}(\mathbf{S}^n)$. The group structure on this set will be defined below, and in more detail in the next chapter.

By composing a pointed map $f:\mathbf{S}^{n+k}\to\mathbf{S}^n$ with the end of a deformation of $\mathbf{S}^n$ collapsing a disk about the base point to the base point, we see that, up to homotopy, we can assume that f takes a neighborhood of the base point of $\mathbf{S}^{n+k}$ to the base point of $\mathbf{S}^n$, and similarly with homotopies of such maps.

Then by removing the base point from $\mathbf{S}^{n+k}$ we can study, instead, maps and homotopies $\mathbf{R}^{n+k}\to\mathbf{S}^n$ which are constant to the base point outside some compact subset of $\mathbf{R}^{n+k}$. By the Smooth Approximation Theorem we can also restrict attention to smooth maps $\mathbf{R}^{n+k}\to\mathbf{S}^n$ and smooth homotopies.

For convenience in notation we shall consider $\mathbf{S}^n$ to be the one-point compactification $\mathbf{R}^n_+=\mathbf{R}^n\cup\{\infty\}$ of euclidean space. Use will be made of some constructions on $\mathbf{S}^n$ which are not smooth at ∞, but this will have no affect on our arguments. For example, a translation of $\mathbf{R}^n$ extends to $\mathbf{R}^n_+$ and is smooth except at ∞.

Suppose given a smooth map $f:\mathbf{R}^{n+k}\to\mathbf{R}^n_+$ as above. Then there is a regular value $p\in\mathbf{R}^n$. By following f by a translation in $\mathbf{R}^n$ (which is, of course, homotopic to the identity as a map of $\mathbf{R}^n_+$ to itself) we can assume that p is the origin $0\in\mathbf{R}^n$. By Theorem 11.6 there is a disk E^n about 0 in $\mathbf{R}^n$ and an embedding $M^k\times E^n\to\mathbf{R}^{n+k}$ onto an open neighborhood N of M^k and whose inverse $N\to M^k\times E^n$ is $r\times f$ where $r:N\to M^k$ is the normal retraction. By another homotopy of f it can and will be assumed that E^n is the open unit disk in $\mathbf{R}^n$ about the origin. In this section, we will refer to such an embedding $g:M^k\times E^n\to\mathbf{R}^{n+k}$, M^k compact, as a "fattened k-manifold."

Now we can follow f by a smooth deformation of $\mathbf{R}^n_+$ starting at the identity and ending with a map $\theta:\mathbf{R}^n_+\to\mathbf{R}^n_+$ which takes E^n diffeomorphically onto $\mathbf{R}^n$ and everything else to ∞. For example, the homotopy

$$\Phi(x,t)=\begin{cases} x/(1-\|x\|^2t^2)^{1/2} & \text{for } \|x\|<1/t,\\ \infty & \text{for } \|x\|\geq 1/t,\end{cases}$$

does this. With this map, $\theta(x)=x/(1-\|x\|^2)^{1/2}$ for $\|x\|<1$. Then the composition $\theta\circ f\simeq f$ can be described as the map taking $N\approx M^k\times E^n\to E^n$ by the projection followed by the diffeomorphism $E^n\xrightarrow{\approx}\mathbf{R}^n$ (the restriction of θ) and taking everything else to ∞. (See Figure II-13.)

Therefore every fattened k-manifold $g:M^k\times E^n\to\mathbf{R}^{n+k}$ gives rise to a map $\phi_g:\mathbf{R}^{n+k}\to\mathbf{R}^n_+$ of this form, and every map $\mathbf{R}^{n+k}\to\mathbf{R}^n_+$, as above, is homotopic to a map arising this way.

Now suppose we are given two fattened k-manifolds $g_0:M^k_0\times E^n\to\mathbf{R}^{n+k}$ and $g_1:M^k_1\times E^n\to\mathbf{R}^{n+k}$ and that the associated maps are homotopic: $\phi_{g_0}\simeq\phi_{g_1}$ via the homotopy $F:\mathbf{R}^{n+k}\times I\to\mathbf{R}^n_+$.

By composing F with a map $\mathbf{R}^{n+k}\times I\to\mathbf{R}^{n+k}\times I$ of the form $1\times\psi$ where $\psi(t)=0$ for t near 0 and $\psi(t)=1$ for t near 1, we can assume that F is a

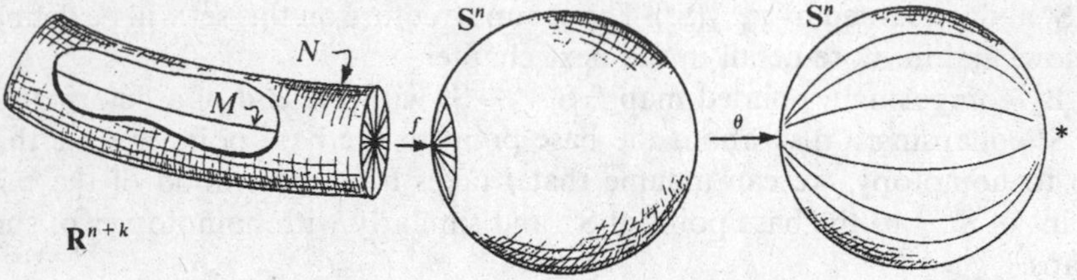

Figure II-13. Thom–Pontryagin construction.

constant homotopy near the two ends. Also, of course, F can be assumed to be smooth away from $F^{-1}(\infty)$.

Let $q \in \mathbf{R}^n$ be a regular value of F and put $V^{k+1} = F^{-1}(\{q\})$. Then there is an open disk D^n about q and an embedding $V^{k+1} \times D^n \to \mathbf{R}^{n+k} \times I$ onto a neighborhood W of V and whose inverse is $r \times F: W \to V^{k+1} \times D^n$, r being the normal retraction. Also, in $\mathbf{R}^{n+1} \times [0, \epsilon]$, for some $\epsilon > 0$, this fattened V^{k+1} has the form of the composition

$$M_0^k \times D^n \times [0,\epsilon] \hookrightarrow M_0^k \times \mathbf{R}^n \times [0,\epsilon] \xrightarrow{\approx} M_0^k \times E^n \times [0,\epsilon] \xrightarrow{g_0 \times 1} \mathbf{R}^{n+k} \times [0,\epsilon]$$

and similarly at the other end. The first inclusion can be replaced by an isotopy (a level preserving embedding $M_0^k \times D^n \times [0,\epsilon] \to M_0^k \times \mathbf{R}^n \times [0,\epsilon]$) which first translates D^n to the origin, then expands it to the unit disk E^n and then expands it to map onto $\mathbf{R}^n$ (essentially the map Φ above with a modification of the parametrization to make it constant near the ends). At the end of this we get the diffeomorphism $M_0^k \times D^n \xrightarrow{\approx} M_0^k \times \mathbf{R}^n \xrightarrow{\approx} M_0^k \times E^n$. We can use the inverse of this to reparametrize the entire fattened V^{k+1} to give a fattened manifold $G: V^{k+1} \times E^n \to \mathbf{R}^{n+k} \times I$ which coincides with g_0 near $\mathbf{R}^{n+k} \times \{0\}$ and with g_1 near $\mathbf{R}^{n+k} \times \{1\}$. This is called a "cobordism" of fattened manifolds in $\mathbf{R}^{n+k}$; see Figure II-14. Since cobordism is taken to be constant near the ends, it is an equivalence relation between fattened k-manifolds in $\mathbf{R}^{n+k}$.

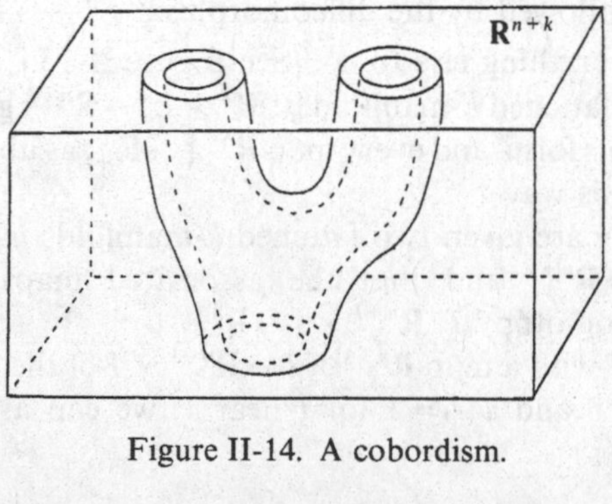

Figure II-14. A cobordism.

Conversely, such a cobordism of fattened manifolds determines a homotopy between the maps $\mathbf{R}^{n+k} \to \mathbf{R}^n_+$ associated with the ends of the cobordism. Thus we have set up a one–one correspondence between pointed homotopy classes of maps $\mathbf{S}^{n+k} \to \mathbf{S}^n$ and cobordism classes of fattened k-manifolds $M^k \times E^n \to \mathbf{R}^{n+k}$. This is close to what we want, but not quite.

A fattened manifold $M^k \times E^n \to \mathbf{R}^{n+k}$ (or $V^{k+1} \times E^n \to \mathbf{R}^{n+k} \times I$) determines a field of normal (meaning here, independent of the tangent space) n-frames on M^k by taking the differentials at points $x \in M^k \subset \mathbf{R}^{n+k}$ of the coordinate axes in $\{x\} \times E^n$. (An n-frame is a set of n independent vectors. We do not assume that they are orthogonal, and they are not in this construction.) Thus we have a "framed manifold" $M^k \subset \mathbf{R}^{n+k}$. Similarly, the fattened V^{k+1} gives a field of normal n-frames to $V^{k+1} \subset \mathbf{R}^{n+k} \times I$, and this is a "framed cobordism."

Conversely, given a (compact) framed manifold $M^k \subset \mathbf{R}^{n+k}$, we can construct a fattening of M^k as follows. Let $\xi_1, \ldots, \xi_n$ be the vector fields in $\mathbf{R}^{n+k}$ defined on M^k and forming an independent set of n normal vectors at each point of M^k. Then define the map $\tau: M^k \times \mathbf{R}^n \to \mathbf{R}^{n+k}$ by

$$\tau(x, t_1, \ldots, t_n) = x + t_1\xi_1(x) + \cdots + t_n\xi_n(x).$$

At any point of M^k, the differential of τ is clearly onto and so, by an argument similar to the proof of Theorem 11.4, there is an $\epsilon > 0$ such that τ maps $M^k \times B_\epsilon(0)$ diffeomorphically onto a neighborhood of M^k in $\mathbf{R}^{n+k}$. By composing with a diffeomorphism $E^n \to B_\epsilon(0)$ which is the identity near 0, we get a fattening $M^k \times E^n \to \mathbf{R}^{n+k}$ of M^k in our sense, and its differential gives back the original n-frame on M^k.

We have almost proved that there is a one–one correspondence between $\pi_{n+k}(\mathbf{S}^n)$ and framed cobordism classes of framed k-manifolds M^k in $\mathbf{R}^{n+k}$. What remains to prove is that if we start with a fattening, pass to the induced framing, and then, by the above construction, to a fattening, we get a fattening which is cobordant to the original one. We shall prove this formally later, in Lemma 16.3.

The group structure on $\pi_{n+k}(\mathbf{S}^n)$ is defined as follows. Use a base point which is in the equator of $\mathbf{S}^{n+k}$. If we collapse the equator to a point, we get a map $\gamma: \mathbf{S}^{n+k} \to \mathbf{S}^{n+k} \vee \mathbf{S}^{n+k}$. If we have pointed maps $f, g: \mathbf{S}^{n+k} \to \mathbf{S}^n$ then we can put f on the first factor of $\mathbf{S}^{n+k} \vee \mathbf{S}^{n+k}$ and g on the second to get a map $\mathbf{S}^{n+k} \vee \mathbf{S}^{n+k} \to \mathbf{S}^n$. Composing this with γ then gives a new map $\mathbf{S}^{n+k} \to \mathbf{S}^n$ called $f * g$. If we use $[f]$ to denote the homotopy class of f then we define $[f] + [g]$ to be $[f * g]$. That this is a group structure will be proved in the next chapter. Looking at the inverse image of a regular value (assuming f and g are smooth) it is clear that the corresponding operation on framed cobordism classes of framed k-manifolds in $\mathbf{R}^{n+k}$ is as follows. Given two framed k-manifolds M^k and N^k, translate M^k in $\mathbf{R}^{n+k}$ until it lies in the lower half space (with respect to the last coordinate, although that does not really matter), and translate N^k to the upper half space. Then M^k and N^k together form a framed k-manifold in $\mathbf{R}^{n+k}$, which we will denote here by $M^k * N^k$. If $[M^k]$ denotes the framed cobordism class of M^k then let $[M^k] + [N^k] =$

$[M^k * N^k]$. It is not hard to see that this latter operation does provide the structure of an abelian group. The identity element is the cobordism class of the empty manifold and the inverse is the class of the mirror image of a framed k-manifold.

Thus, subject to proving the mentioned lemma, we have shown:

16.1. Theorem (Thom–Pontryagin). *The above construction gives an isomorphism of $\pi_{n+k}(\mathbf{S}^n)$ with the group of framed cobordism classes of framed k-manifolds in $\mathbf{R}^{n+k}$.* □

As mentioned, we still must prove Lemma 16.3. By a diffeomorphism of $\mathbf{R}^n$ with E^n which is the identity near the origin, or at least has the identity as differential there, we may replace E^n by $\mathbf{R}^n$ in the definition of "fattening." We need the following definition:

16.2. Definition. Let $\phi, \psi: M^k \times \mathbf{R}^n \to \mathbf{R}^{n+k}$ be two fattenings of the same manifold M^k, i.e., $\phi(x, 0) = \psi(x, 0)$ for all x. Then an *isotopy* between them is an embedding $\Theta: M^k \times \mathbf{R}^n \times I \to \mathbf{R}^{n+k} \times I$ such that $\Theta(x, y, t) \in \mathbf{R}^{n+k} \times \{t\}$, $\Theta(x, 0, t)$ is constant in t, and $\Theta(x, y, 0) = (\phi(x, y), 0)$, and $\Theta(x, y, 1) = (\psi(x, y), 1)$.

Often an isotopy is denoted by $\theta_t(x, y): M^k \times \mathbf{R}^n \to \mathbf{R}^{n+k}$ where

$$\Theta(x, y, t) = (\theta_t(x, y), t).$$

An isotopy can be assumed to be constant near the ends, i.e., θ_t is constant for t near 0 and near 1. Then it is clear that isotopy is an equivalence relation and that it implies cobordism of fattenings. Thus the following lemma suffices to complete the proof of Theorem 16.1.

16.3. Lemma. *If $\phi, \psi: M^k \times \mathbf{R}^n \to \mathbf{R}^{n+k}$ are two fattenings of the same compact manifold M^k and if they induce the same framing of M^k then they are isotopic.*

PROOF. We can shrink the normal disk, keeping a neighborhood of 0 fixed, to make the images of the normal disks as small as we please, and this constitutes an isotopy that does not change the assumptions in the lemma. Thus it is clear that we may assume that the image of ϕ is contained in the image of ψ. Then $\psi^{-1}\phi$ is defined. Therefore we can define $\theta_t(x, y) = \psi(t^{-1}\psi^{-1}\phi(x, tv))$ and we must investigate this as t approaches 0. For simplicity, we can regard M^k as an embedded submanifold of $\mathbf{R}^{n+k}$.

Let $p \in M^k$ and let $x_1, \ldots, x_k$ be local coordinates on an open neighborhood U of p in M^k. Let $y_1, \ldots, y_n$ be coordinates in $\mathbf{R}^n$. Then $x_1, \ldots, x_k, y_1, \ldots, y_n$ can be taken as local coordinates in $\mathbf{R}^{n+k}$ using $\psi: U \times \mathbf{R}^n \to \mathbf{R}^{n+k}$ as a chart. Thus ψ is the identity in these coordinates. We can represent ϕ in these coordinates by

$$\phi(x, y) = (\lambda(x, y), \mu(x, y)),$$

where

$$\lambda(x,y) = (\lambda_1(x,y), \ldots, \lambda_k(x,y)) \in \mathbf{R}^k,$$
$$\mu(x,y) = (\mu_1(x,y), \ldots, \mu_n(x,y)) \in \mathbf{R}^n.$$

Since $\phi(x,0) = (x,0)$ we have that $\lambda(x,0) = x$ and $\mu(x,0) = 0$. Therefore

$$\theta_t(x,y) = \psi(t^{-1}\psi^{-1}\phi(x,ty)) = (\lambda(x,ty), t^{-1}\mu(x,ty))$$

which is defined for sufficiently small $t \neq 0$.

Now $\lambda(x,ty)$ is defined and smooth in x, y, t even at $t = 0$. Since $\mu(x,0) = 0$, we can express

$$\mu_i(x,y) = \sum_j a_{i,j}(x) y_i + \sum_{j,k} b_{i,j,k}(x,y) y_j y_k$$

via Taylor's Theorem, where the $b_{i,j,k}$ are smooth and where

$$a_{i,j}(x) = (\partial \mu_i / \partial y_j)(x,0).$$

By the assumption that ϕ induces the same frame as does ψ, we have that $a_{i,j}(x) = \delta_{i,j}$. Therefore

$$t^{-1}\mu_i(x,ty) = y_i + t\sum_{j,k} b_{i,j,k}(x,ty) y_j y_k,$$

which is defined and smooth in x, y, t even at $t = 0$. Also, putting $t = 0$ gives $\theta_0(x,y) = (x,y)$. This means that the end $t = 0$ of the isotopy $\theta_t(x,y)$ is $\psi(x,y)$. The end $t = 1$ is $\theta_1(x,y) = \phi(x,y)$. □

We will now look at the special case $k = 0$ of maps $\mathbf{S}^n \to \mathbf{S}^n$. By Theorem 16.1 $\pi_n(\mathbf{S}^n)$ is isomorphic to the group of framed cobordism classes of framed 0-manifolds in $\mathbf{R}^n$. A (compact) 0-manifold is just a finite set of points. The framing at each point can be assumed orthonormal by the Gram–Schmidt process, which provides a "homotopy" of the frame, which is a framed cobordism. Also a frame can be rotated so that its first vector agrees with that of $\mathbf{R}^n$, and then a rotation in the orthogonal complement of the first vector can move the second to agree with the second standard basis vector of $\mathbf{R}^n$, if $n > 2$. (This is just a matter of knowing that the special orthogonal group $\mathbf{SO}(n)$ is connected and transitive on the sphere $\mathbf{S}^{n-1}$ if $n > 1$. See Problem 8 of Section 15 of Chapter I.) One can continue this until one gets to the last vector. That finishes it since $\mathbf{SO}(1)$ is not transitive on the 0-sphere. But that leaves all vectors but the last in the standard position and the last is either standard or in the opposite direction from the nth standard vector. One can distinguish these cases simply by the sign of the determinant of the matrix made up of column vectors equal to the original frame, expressed in the standard basis. Thus, we can replace the frame by the sign ± 1, and still have the correspondence. Moreover, one can cancel two opposite signs by a cobordism that is an arc between two such points at the end $t = 0$, and empty at the end $t = 1$. Other points stay constant during the cobordism. Thus, adding up the signs gives an integer, and this integer is a complete invariant for $\pi_n(\mathbf{S}^n)$. This integer is known as the "degree" of the map $f:\mathbf{S}^n \to \mathbf{S}^n$ whose homotopy class is in question. Thus we have:

16.4. Corollary (Hopf). *There is an isomorphism $\pi_n(\mathbf{S}^n) \approx \mathbf{Z}$ which takes a homotopy class $[f]$ to the degree of f.* □

16.5. Corollary (Hopf). *A map $f:\mathbf{S}^n \to \mathbf{S}^n$ of degree 0 is homotopic to a constant.* □

The degree of such a smooth map is, by the discussion, determined by taking a regular value $p \in \mathbf{S}^n$ and adding up the signs of the Jacobians of f at the (finite number of) points in $f^{-1}(p)$.

The method of Pontryagin and Thom was originally intended as an approach to the computation of homotopy groups of spheres. The groups $\pi_{n+1}(\mathbf{S}^n)$ work out fairly easily since they correspond to framed 1-manifolds and 1-manifolds are well known. The groups $\pi_{n+2}(\mathbf{S}^n)$ also work by this method since 2-manifolds are well understood. Even in that case, however, the derivation of $\pi_{n+2}(\mathbf{S}^n)$ is quite difficult. Indeed, Pontryagin originally announced that $\pi_{n+2}(\mathbf{S}^n)$ is trivial; apparently because of a missed framing on the torus. He corrected this shortly thereafter. With great difficulty, the method has been pushed through to compute $\pi_{n+3}(\mathbf{S}^n)$. For higher codimensions, the difficulties become overwhelming. In the meantime, other, algebraic, methods were found for the computation of $\pi_{n+k}(\mathbf{S}^n)$ and many computations have been done, but the complete problem is yet to be cracked. But these results on homotopy groups can be used, through the Thom–Pontryagin construction, to yield information about manifolds, a method that has proved to be highly productive.

Although we have restricted our attention in the discussion to maps from $\mathbf{R}^{n+k}$ to $\mathbf{S}^n$, the only place we used that the maps are from $\mathbf{R}^{n+k}$ was in the definition of the group structure in $\pi_{n+k}(\mathbf{S}^n)$. We mainly made that assumption in order to simplify the argument and aid the reader's intuition. (Also it is by far the most important case.) There is no difficulty in generalizing the results to apply to maps from any compact manifold M^{n+k} to $\mathbf{S}^n$. The upshot of that generalization is the following:

16.6. Theorem. *If M^{n+k} is a compact smooth manifold, then the Thom–Pontryagin construction gives a one–one correspondence between the set $[M^{n+k}; \mathbf{S}^n]$ of homotopy classes of maps $M^{n+k} \to \mathbf{S}^n$ and the set of smooth framed cobordism classes of smooth, compact, normally framed k-submanifolds of M^{n+k}.* □

Returning to smooth maps $f:\mathbf{S}^{n+k} \to \mathbf{S}^n$, note that there is an obvious "suspension" of f to a map $Sf:\mathbf{S}^{n+k+1} \to \mathbf{S}^{n+1}$ induced from $f \times 1:\mathbf{S}^{n+k} \times I \to \mathbf{S}^n \times I$ by passing to the quotient spaces $\mathbf{S}^{n+k+1}$ of $\mathbf{S}^{n+k} \times I$ and $\mathbf{S}^{n+1}$ of $\mathbf{S}^n \times I$ identifying the ends of the cylinders to points. This is not smooth at the poles, but has a regular value on the equator and so it can be smoothed without changing that regular value.

In the viewpoint of the Thom–Pontryagin construction, it is clear that the corresponding operation (at least up to sign) is given by considering a given framed k-manifold M^k in $\mathbf{R}^{n+k}$ as lying in $\mathbf{R}^{n+k} \times \{0\} \subset \mathbf{R}^{n+k+1}$ and adding the new coordinate vector to the frame at each point. Then this defines a homomorphism

$$S: \pi_{n+k}(\mathbf{S}^n) \to \pi_{n+k+1}(\mathbf{S}^{n+1}).$$

We claim that S is an isomorphism for $n > k+1$ and is surjective for $n = k+1$. We will prove this in several steps.

First, note that a given framing can be altered by an isotopy of frames (producing a framed cobordism that is a constant cobordism on the manifold itself) so that the new framing is orthonormal and orthogonal to the manifold. To do that, first project to the normal space. This can be filled in with an isotopy via the standard $tv + (1-t)w$ method. Then use the Gram–Schmidt process to orthogonalize. This also fills in with an isotopy.

Second, for $M^k \subset \mathbf{R}^{n+k+1}$, $n > k$, there is a unit vector v not tangent to M anywhere and not a secant of M; see the proof of the Embedding Theorem (Theorem 10.7). We can rotate $\mathbf{R}^{n+k+1}$ (giving a cobordism) moving v to the last basis vector, and so we can assume that v is this basis vector $e = e_{n+k+1}$. If $p \mapsto (x_1(p), \ldots, x_{n+k+1}(p))$ represents the original embedding of M^k in $\mathbf{R}^{n+k+1}$, then the map $\phi: M^k \times I \to \mathbf{R}^{n+k+1} \times I$ given by $\phi(p,t) = (x_i(p), \ldots, x_{n+k}(p), tx_{n+k+1}(p), t)$ defines a cobordism of M^k to $N^k \subset \mathbf{R}^{n+k} \times \{0\} \subset \mathbf{R}^{n+k+1}$, not yet framed.

To see that ϕ does carry the framing along, note that at each parameter value t, the manifold $p \mapsto \phi(p,t)$ has a tangent space at each p which has an angle $< \pi/2$ with the original tangent space of M at p and hence has trivial intersection with the *original* normal space. Thus the original framing is still a framing of the displaced manifold, even though not orthogonal to it.

Therefore, we may assume that $M^k \subset \mathbf{R}^{n+k} \times \{0\} \subset \mathbf{R}^{n+k+1}$ (if $n > k$) and has an orthonormal framing. Now consider the frame at each point. Referring the vector $e = e_{n+k+1}$ to this frame at a given point $x \in M$ gives a point $\theta(x) \in \mathbf{S}^n$. Since $n > k$, the smooth map $\theta: M^k \to \mathbf{S}^n$ must miss a point. By rotating the framing by an orthogonal transformation constant on M (another cobordism) we can assume that $-v \notin \text{image}(\theta)$, where $v = v_{n+1}$ is the last vector in the given frame $(v_1, \ldots, v_{n+1})$.

Now we claim that we can change the framing via a homotopy so that the last vector $v = v_{n+1}$ of the frame becomes e. This is done by rotating the frame through the 2-plane spanned by e and v moving v to e. This rotation can be described as follows: Let w be the unit vector half way between e and v; i.e. $w = (e+v)/\|e+v\|$. This makes sense since $-v$ is never e. Then the rotation in question is $R_w: \mathbf{R}^{n+1} \to \mathbf{R}^{n+1}$ given by $R_w(u) = T_e(T_w(u))$ where T_w is the reflection in the line $\mathbf{R}w$; whence $T_w(u) = 2\langle u, w\rangle w - u$. The homotopy is given by the family R_{w_t} of rotations where

$$w_t = \frac{te + (1-t)w}{\|te + (1-t)w\|}.$$

Consequently, we can assume that the framing has e as its last vector, but that just means that the new framed k-manifold is in the image of the suspension $S:\pi_{n+k}(\mathbf{S}^n)\to\pi_{n+k+1}(\mathbf{S}^{n+1})$; $n>k$. For a framed cobordism $V_0^{k+1}\subset \mathbf{R}^{n+k+1}\times I$, a similar argument shows that it can be changed into a cobordism $V^{k+1}\subset\mathbf{R}^{n+k}\times\{0\}\times I\subset\mathbf{R}^{n+k+1}\times I$ with the last frame vector e, provided that $n>k+1$. This shows that S is an isomorphism for $n>k+1$. Thus we have proved:

16.7. Theorem (Freudenthal). *For $n\geq 1$, the suspension homomorphism*

$$S:\pi_{n+k}(\mathbf{S}^n)\to\pi_{n+k+1}(\mathbf{S}^{n+1})$$

is an isomorphism for $n>k+1$ and an epimorphism for $n=k+1$. □

Note that this implies that $S:\pi_1(\mathbf{S}^1)\to\pi_2(\mathbf{S}^2)$ is onto and that $\pi_2(\mathbf{S}^2)\to\pi_3(\mathbf{S}^3)\to\cdots$ are all isomorphisms. Thus an alternative to the proof of Corollary 16.4 is to compute $\pi_1(\mathbf{S}^1)\approx\mathbf{Z}$ (done by other means in Chapter III), and to show that $\pi_2(\mathbf{S}^2)$ is infinite (an easy application of Homology Theory in Chapter IV), and to then use Theorem 16.7 to conclude that all these groups are $\mathbf{Z}$.

In Chapter VII we will show that $\pi_3(\mathbf{S}^2)\approx\mathbf{Z}$ and $\pi_4(\mathbf{S}^3)\approx\mathbf{Z}_2$. Thus it will follow from Theorem 16.7 that $\pi_{n+1}(\mathbf{S}^n)\approx\mathbf{Z}_2$ for all $n\geq 3$. It is easy to "explain" (without proof) these facts from the point of view of Thom–Pontryagin. An element of $\pi_{n+1}(\mathbf{S}^n)$ is represented by a framed 1-manifold M^1 in $\mathbf{R}^{n+1}$. It is not hard to see that one can join the components of M via a framed cobordism and similarly one can unknot M. That is, every element of $\pi_{n+1}(\mathbf{S}^n)$ is represented by a framed standardly embedded circle $M=\mathbf{S}^1$ in $\mathbf{R}^{n+1}$. The trivial element is represented by a "trivial" framing: embed $\mathbf{D}^2$ in $\mathbf{R}^2\times I$ meeting $\mathbf{R}^2\times\{1\}$ transversely in $\mathbf{S}^1=\partial\mathbf{D}^2$; then this can be framed and shows that the normal framing of the standard $\mathbf{S}^1\subset\mathbf{R}^2$ is frame cobordant to $\emptyset$ (i.e., $\pi_2(\mathbf{S}^1)=0$; similar arguments show $\pi_k(\mathbf{S}^1)=0$ for all $k>1$). Suspending this adds another normal vector to $\mathbf{S}^1$ in $\mathbf{R}^3$, giving a "trivial" framed $\mathbf{S}^1$ in $\mathbf{R}^3$. Now, given any smooth map $\phi:\mathbf{S}^1\to\mathbf{SO}(2)\approx\mathbf{S}^1$, one can produce a new framing of $\mathbf{S}^1$ by rotating the given frame at $x\in\mathbf{S}^1$ by $\phi(x)\in\mathbf{SO}(2)$. It is clear that all framings come this way. A homotopy of maps $\mathbf{S}^1\times I\to\mathbf{SO}(2)$ gives a cobordism (actually an isotopy) of framings. The homotopy classes of maps $\mathbf{S}^1\to\mathbf{SO}(2)\approx\mathbf{S}^1$ are given by $\pi_1(\mathbf{S}^1)\approx\mathbf{Z}$, by Corollary 16.4, and each of these classes produces, by the frame change construction, an element of $\pi_3(\mathbf{S}^2)$, and it turns out that these elements are all distinct, which explains why $\pi_3(\mathbf{S}^2)\approx\mathbf{Z}$. For framings of $\mathbf{S}^1$ in $\mathbf{R}^4$, one would operate on a trivial normal framing by the maps $\mathbf{S}^1\to\mathbf{SO}(3)$. In Chapter III we will show that there is exactly one nontrivial (up to homotopy) map $\mathbf{S}^1\to\mathbf{SO}(3)$ and this "explains" why $\pi_4(\mathbf{S}^3)\approx\mathbf{Z}_2$. The same fact holds for $\mathbf{S}^1\to\mathbf{SO}(n)$ for all $n\geq 3$, "explaining," without using Theorem 16.7, why $\pi_{n+1}(\mathbf{S}^n)\approx\mathbf{Z}_2$ for all $n\geq 3$.

CHAPTER III

Fundamental Group

Finally, let me propose still another kind of geometry, which, in a sense, is obtained by the most careful sifting process of all, and which, therefore, includes the fewest theorems. This is analysis situs....

F. KLEIN

1. Homotopy Groups

With this chapter we begin the study of *algebraic* topology. The central idea behind algebraic topology is to associate an algebraic situation to a topological situation, and to study the simpler resulting algebraic setup. For example, to each topological space X there could be associated a group $G(X)$, such that homeomorphic (indeed, usually homotopically equivalent) spaces give rise to isomorphic groups. Usually, also, to a map of spaces one associates a homomorphism of the groups attached to those spaces, such that compositions of maps yield compositions of homomorphisms of groups. Then anything one can say about the algebraic situation, gives information about the topological one. For example, if we have two spaces whose associated groups are not isomorphic, then we can conclude that the spaces cannot be homeomorphic.

Many readers will recognize that what we are talking about here is what is known as a "functor" from the category of topological spaces and maps to the category of groups and homomorphisms. Indeed, the whole idea of functors arose out of the field of algebraic topology.

In this chapter we shall study the first and simplest realization of this idea, the fundamental group (or Poincaré group) of a space. This is a special case of so-called homotopy groups, and we shall first define the latter and then specialize, in the rest of the chapter, to the fundamental group, or "first homotopy group."

To define a group, one must define an operation of "multiplication." The reader may recall that we already had such a situation in the first chapter, namely, the concatenation $F*G$ of two homotopies. However, this operation is only defined when the second homotopy starts where the previous one ends. But we can restrict attention to homotopies that all start and end with the same map, and the simplest such map to take is a constant map to a point. Also, one needs certain identities in the definition of a group, such as the associativity of multiplication. But concatenations of homotopies only satisfy the weaker law that $F*(G*H) \simeq (F*G)*H$. This suggests that the objects making up the group should not be homotopies, but equivalence classes of homotopies under some type of equivalence that would make homotopic homotopies equivalent to one another. That is exactly what we are going to do.

First let us recall the notations from Section 14 of Chapter I. If X and Y are spaces, then $[X; Y]$ denotes the set of homotopy classes of maps $X \to Y$. If $A \subset X$ and $B \subset Y$ then $[X, A; Y, B]$ denotes the set of homotopy classes of maps $X \to Y$ carrying A into B (denoted by $(X, A) \to (Y, B)$) such that, moreover, A goes into B during the entire homotopy.

To make a group then, we can select a point $y_0 \in Y$ and consider the set

$$[X \times I, X \times \partial I; Y, \{y_0\}].$$

Here, indeed, one does get a group from the operation of concatenation of homotopies. However, it is technically better to also choose a "base point" $x_0 \in X$ and consider the set

$$[X \times I, \{x_0\} \times I \cup X \times \partial I; Y, \{y_0\}].$$

(Of course, $\partial I = \{0, 1\}$.) For the moment let us set $A = \{x_0\} \times I \cup X \times \partial I$. Then note that maps $X \times I \to Y$ which carry A into $\{y_0\}$ are in one-one correspondence with maps of the quotient space $(X \times I)/A \to Y$ which take the point $\{A\}$ into $\{y_0\}$. Thus we define the space

$$SX = (X \times I)/A = (X \times I)/(\{x_0\} \times I \cup X \times \partial I) \quad \text{with base point } \{A\}.$$

This is called the "reduced suspension" of X.

A space with a base point is often referred to as a "pointed space." We will mostly work in the category of these pointed spaces and pointed maps (maps taking base point to base point). Let us denote the set of homotopy classes of pointed maps of a pointed space X to a pointed space Y, with homotopies preserving the base points, by $[X; Y]_*$. (We use this notation for stress here. In most of the book we will drop the asterisk suffix, depending on the context to make clear what is intended.)

Thus $[SX; Y]_*$ is in a canonical one-one correspondence with $[X \times I, A; Y, \{y_0\}]$.

If f, $g: SX \to Y$ are pointed maps, then they induce homotopies f', g': $X \times I \to Y$ by means of composition with the quotient map $X \times I \to SX$. Then $f' * g': X \times I \to Y$ is defined and factors through SX. The resulting pointed map $SX \to Y$ will be denoted by $f * g$ with little danger of confusion. Note that, geometrically, $f * g$ is obtained by putting f on the bottom and g on the top of the one-point union $SX \vee SX$ and composing the resulting map $SX \vee SX \to Y$ with the map $SX \to SX \vee SX$ obtained by collapsing the middle (parameter value $\frac{1}{2}$) copy of X in SX to the base point. (See Figure III-1.)

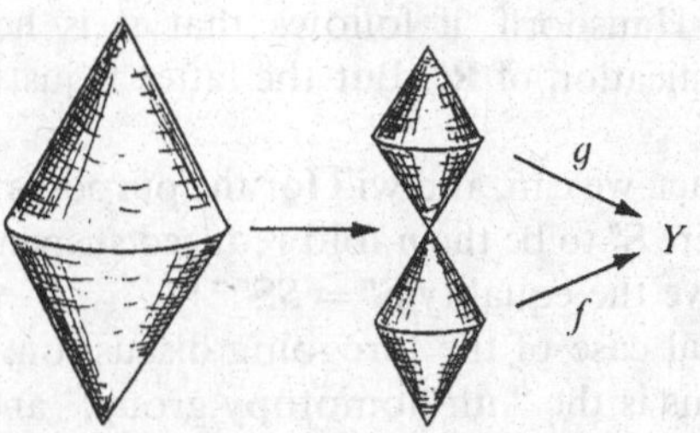

Figure III-1. The product of two map classes $SX \to Y$.

For any map $f:(SX, \{A\}) \to (Y, \{y_0\})$ we denote its homotopy class in $[SX; Y]_*$ by $[f]$. For two such maps f and g we define

$$[f]\cdot[g] = [f * g].$$

Of course, we must check that $[f_1] = [f_2]$ and $[g_1] = [g_2]$ imply that $[f_1 * g_1] = [f_2 * g_2]$, but this follows from Proposition 14.17 of Chapter I.

Let $c_{y_0}: SX \to Y$ be the constant map to the point y_0. Then, from the laws of homotopies developed in Propositions 14.13, 14.15 and 14.16 of Chapter I, we see easily that:

(associativity)	$[f]\cdot([g]\cdot[h]) = ([f]\cdot[g])\cdot[h]$	from I-14.16,
(unity element)	$[c_{y_0}]\cdot[f] = [f] = [f]\cdot[c_{y_0}]$	from I-14.13,
(inverse)	$[f]\cdot[f^{-1}] = [c_{y_0}]$	from I-14.15.

(Recall that f^{-1} stands here for the "inverse" homotopy with time running the opposite way to that in f, and not to an inverse function.)

Thus, under this operation, the set $[SX; Y]_*$ of pointed homotopy classes of pointed maps $SX \to Y$, becomes a group.

Figure III-1 illustrates the group operation. Note that the line segment from the "north pole" in the left-hand side of the illustration to the point x_0 and on to the "south pole" is, in SX, really identified to a point. For the picture it is easier not to try to depict this. One can consider the picture as defining a map on the "unreduced suspension" (the union along X of two cones), which is constant on this line segment. Such a map factors through the reduced suspension, and vice versa, so such illustrations should not lead to problems.

The most important special case of the foregoing is that of suspensions of spheres. To fix the ideas, let $\mathbf{S}^0$ denote the 0-sphere $\{0, 1\}$ with base point $\{0\}$. Pick any base point in the other spheres, say the north pole. We shall use an asterisk "$*$" to denote base points in general.

1.1. Proposition. *The reduced suspension gives* $S\mathbf{S}^{n-1} \approx \mathbf{S}^n$.

Proof. Note that $S\mathbf{S}^{n-1} - \{*\} \approx (\mathbf{S}^{n-1} - \{*\}) \times (0,1) \approx \mathbf{R}^{n-1} \times \mathbf{R} \approx \mathbf{R}^n$. Since

$\mathbf{S}\mathbf{S}^{n-1}$ is compact Hausdorff, if follows that it is homeomorphic to the one-point compactification of $\mathbf{R}^n$. But the latter is just $\mathbf{S}^n$. □

Because of this fact, we can, and will for the purposes of this chapter, *define* the (pointed) n-sphere $\mathbf{S}^n$ to be the n-fold reduced suspension of the two-point set $\mathbf{S}^0$. Then we have the equality $\mathbf{S}^n = S\mathbf{S}^{n-1}$.

Thus, as a special case of the foregoing discussion, the set $[\mathbf{S}^n; Y]_*$ is a group for $n > 0$. This is the "nth homotopy group," and it is denoted by

$$\boxed{\pi_n(Y, y_0) = [\mathbf{S}^n; Y]_*.}$$

If we wish to indicate a homotopy group without specifying the "n" we will just write $\pi_*(Y, y_0)$, or sometimes just $\pi_*(Y)$, the base point being understood.

Of course, the elements of $\pi_n(Y)$ are homotopy classes (rel base point) of maps $\mathbf{S}^n \to Y$. The group operation is easily seen directly. Given maps f, $g: \mathbf{S}^n \to Y$, put them together to give a map of $\mathbf{S}^n \vee \mathbf{S}^n \to Y$, and then compose with the map $\mathbf{S}^n \to \mathbf{S}^n \vee \mathbf{S}^n$ which just collapses the equator (containing the base point) of $\mathbf{S}^n$ to a point. (The reader may well argue here that this description is too vague. For example, on which of the two parts of $\mathbf{S}^n \vee \mathbf{S}^n$ do you put f? As a matter of fact, however, we will see later that this only matters when $n = 1$. Also, it is of little consequence just how one defines $\mathbf{S}^n$ in the first place. Of course, at the present time, we cannot justify these statements, and when proving things we will have to stick to our definition.)

If you think about our definition of $\pi_n(Y)$ for a moment, especially the description of $\mathbf{S}^n$ as a repeated suspension of $\mathbf{S}^0$, you will note that each suspension supplies a parameter in $[0, 1]$ and, in fact, the space $\mathbf{S}^n$ as constructed is actually a quotient space of the cube I^n obtained by collapsing the boundary of the cube to a point (which becomes the base point). Maps $\mathbf{S}^n \to Y$ preserving base points, are in one–one correspondence with maps $I^n \to Y$ which take ∂I^n to the base point of Y. This is a more traditional way of defining $\pi_n(Y)$. This group then becomes the set of homotopy classes of maps $(I^n, \partial I^n) \to (Y, \{y_0\})$. In that context, the group operation is the one coming from the definition:

$$f * g(t_1, \ldots, t_n) = \begin{cases} f(t_1, \ldots, t_{n-1}, 2t_n) & \text{for } t_n \le \frac{1}{2}, \\ g(t_1, \ldots, t_{n-1}, 2t_n - 1) & \text{for } t_n \ge \frac{1}{2}. \end{cases}$$

We will see later that using the last coordinate to do the concatenation of f and g is completely immaterial, and one gets the exact same group operation in $\pi_n(Y)$ by using any other coordinate for the concatenation.

Homotopy groups are very important but they are also very difficult to compute. The most important cases are the groups $\pi_n(\mathbf{S}^k)$. Many of these have been computed, but not all of them. Their study has long been, and continues to be, a very important topic in algebraic topology.

Before returning to the general discussion, let us indicate some of the known facts about these groups. Some of these things will be proved later in this book. The groups $\pi_n(\mathbf{S}^n)$ are known to be infinite cyclic and are generated by the homotopy class of the identity map $\mathbf{S}^n \to \mathbf{S}^n$, a fact that is probably no surprise. Indeed, this was proved in Corollary 16.4 of Chapter II.

The group $\pi_n(\mathbf{S}^k)$ is trivial for $n < k$. In fact, this follows from Theorem 11.10 of Chapter II.

The group $\pi_n(\mathbf{S}^1)$ is trivial for $n > 1$. We will prove this later in this chapter.

The group $\pi_3(\mathbf{S}^2)$ is infinite cyclic. This may well come as a surprise. Consider $\mathbf{S}^3$ as the unit sphere in $\mathbf{C}^2$. Then $(u, v) \mapsto uv^{-1}$ defines a map $\mathbf{S}^3 \to \mathbf{C}^+ \approx \mathbf{S}^2$ and this represents a generator of $\pi_3(\mathbf{S}^2)$. This map, incidentally, is called the "Hopf map" and there are several other ways to define it.

The group $\pi_{n+1}(\mathbf{S}^n)$ is the group $\mathbf{Z}_2$ of two elements for $n > 2$. Going back to maps $\mathbf{S}^3 \to \mathbf{S}^2$, note that one may "suspend" such maps to get maps $\mathbf{S}^4 \to \mathbf{S}^3, \ldots, \mathbf{S}^{n+1} \to \mathbf{S}^n$. (They are obtained from the maps $f \times 1: X \times I \to Y \times I$, for any $f: X \to Y$ by passing to the quotient spaces defining the reduced suspensions.) Starting with the Hopf map described in the last paragraph, these suspensions turn out to yield the generators of all the groups $\pi_{n+1}(\mathbf{S}^n)$ for $n > 2$.

The group $\pi_{n+2}(\mathbf{S}^n)$ is the group $\mathbf{Z}_2$ of two elements for $n \geq 2$. The group $\pi_4(\mathbf{S}^2)$ is generated by the composition of the Hopf map $\mathbf{S}^3 \to \mathbf{S}^2$ with its suspension $\mathbf{S}^4 \to \mathbf{S}^3$, and the higher groups are generated by the suspensions of this.

The group $\pi_{n+3}(\mathbf{S}^n) \approx \mathbf{Z}_{24}$ for $n \geq 5$; also $\pi_5(\mathbf{S}^2) \approx \mathbf{Z}_2$, $\pi_6(\mathbf{S}^3) \approx \mathbf{Z}_{12}$, and $\pi_7(\mathbf{S}^4) \approx \mathbf{Z} \oplus \mathbf{Z}_{12}$.

As might be guessed from some of the stated facts about homotopy groups of spheres, it turns out that $\pi_{n+k}(\mathbf{S}^n)$ is independent of n for n sufficiently large. This is known as "stability." Those who have read Section 16 of Chapter II have already seen a proof of this in Theorem 16.7 of Chapter II.

Enough peeks into the future. Let us resume our general discussion of the groups $[SX; Y]_*$. Let (X, x_0) be a fixed pointed space, and consider maps

$$\phi: (Y, y_0) \to (W, w_0).$$

If $f: (SX, *) \to (Y, y_0)$ is any map then $\phi \circ f: (SX, *) \to (W, w_0)$. Also, if $f \simeq g \operatorname{rel}\{*\}$, then $\phi \circ f \simeq \phi \circ g \operatorname{rel}\{*\}$ so that ϕ induces a function

$$\phi_\#: [SX, *; Y, y_0] \to [SX, *; W, w_0]$$

by $\phi_\#[f] = [\phi \circ f]$. It is clear (see Figure III-2) that $\phi \circ (f * g) = (\phi \circ f) * (\phi \circ g)$ whence $\phi_\#(\alpha\beta) = \phi_\#(\alpha)\phi_\#(\beta)$, i.e., $\phi_\#$ is a homomorphism of groups.

If $\phi: (Y, y_0) \to (W, w_0)$ and $\psi: (W, w_0) \to (Z, z_0)$ then it is clear that

$$\psi_\# \circ \phi_\# = (\psi \circ \phi)_\# \qquad \text{and} \qquad \text{Identity}_\# = \text{Identity},$$

so that $[SX, *; \cdot, \cdot]$ is a *functor*.

Also, if $\phi \simeq \psi: (Y, y_0) \to (Z, z_0)$ then $\phi \circ f \simeq \psi \circ f$ which implies that $\phi_\# = \psi_\#$.

Let us rewrite these observations in terms of the special case of homotopy groups. If $\phi: (Y, y_0) \to (W, w_0)$ and $\psi: (W, w_0) \to (Z, z_0)$ then there are the

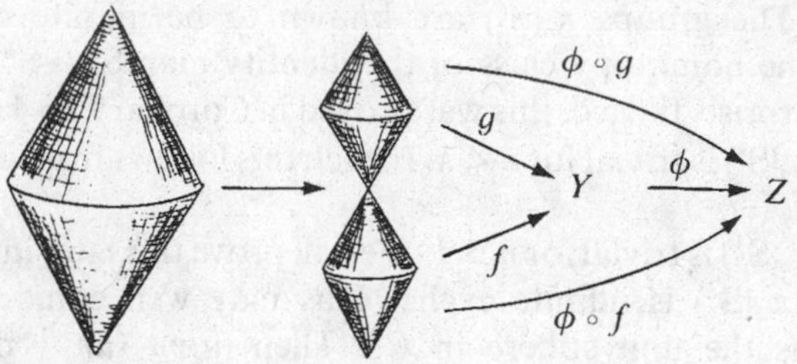

Figure III-2. Functoriality of the product.

homomorphisms

$$\phi_{\#}: \pi_n(Y, y_0) \to \pi_n(W, w_0) \quad \text{and} \quad \psi_{\#}: \pi_n(W, w_0) \to \pi_n(Z, z_0),$$

and we have $(\psi \circ \phi)_{\#} = \psi_{\#} \circ \phi_{\#}$. Also, if $\phi \simeq \psi: (X, x_0) \to (Y, y_0)$ then $\phi_{\#} = \psi_{\#}$.

2. The Fundamental Group

We shall now specialize to the case of $\pi_1(X, x_0)$, the "fundamental" or "Poincaré" group. Via the quotient map $(I, \partial I) \to (\mathbf{S}^1, *)$ we had the one–one correspondence between maps $(I, \partial I) \to (X, x_0)$ and maps $(\mathbf{S}^1, *) \to (X, x_0)$. Thus the fundamental group $\pi_1(X, x_0)$ can be considered as $[I, \partial I; X, x_0]$, i.e., as the set of homotopy classes of closed paths, or "loops," in X at the base point x_0. For loops f, g in X at x_0, $f * g$ is the loop obtained by going along f and then along g. (This is, of course, a special case of the treatment of homotopies. A loop is just a homotopy of maps from a point $\{*\}$ to the space X beginning and ending at the map $\{*\} \to \{x_0\}$.)

Therefore, for each pointed space (X, x_0) there is a group $\pi_1(X, x_0)$, and for each map $\phi: (X, x_0) \to (Y, y_0)$ there is an induced homomorphism $\phi_{\#}: \pi_1(X, x_0) \to \pi_1(Y, y_0)$ such that $(\psi \circ \phi)_{\#} = \psi_{\#} \circ \phi_{\#}$, and $\text{Identity}_{\#} = \text{Identity}$. Finally, if $\phi \simeq \psi: (X, x_0) \to (Y, y_0)$ then $\phi_{\#} = \psi_{\#}$. Moreover, it is clear that $\pi_1(\{x_0\}, x_0) = 1$, the trivial group.

An arcwise connected space X with $\pi_1(X, x_0) = 1$ is called "simply connected." Presently, we will show that this does not depend on the choice of $x_0 \in X$.

As a consequence of this formalism, let us derive an application. Suppose that X is contractible in the strong sense that there exists a homotopy $\phi: I \times (X, x_0) \to (X, x_0)$ of pointed spaces with $\phi(x, 0) = x$ and $\phi(x, 1) = x_0$ for all $x \in X$. The assumption that this is a homotopy of pointed spaces means that $\phi(x_0, t) = x_0$ for all $t \in [0, 1]$. Letting c denote the constant map $X \to X$ with $c(x) = x_0$ for all $x \in X$, we have $\text{Identity} \simeq c$. Thus $\text{Identity}_{\#} = c_{\#}$ on $\pi_1(X, x_0)$. But $\text{Identity}_{\#} = \text{Identity}$, and $c = i \circ k$ where $i: \{x_0\} \to X$ is the inclusion and $k: X \to \{x_0\}$ is the unique map of X into the one point space

$\{x_0\}$. Thus

$$\text{Identity} = \text{Identity}_\# = c_\# = i_\# \circ k_\#.$$

But the right-hand side is a composition through the group $\pi_1(\{x_0\}, x_0)$ which is trivial. The only way this can happen is if $\pi_1(X, x_0) = 1$, the trivial group. (One can give an easier direct proof of this, but we wished to illustrate it as a consequence of the functoriality of the fundamental group.)

Of course, this is not of any use unless we know of spaces X for which the fundamental group $\pi_1(X, x_0)$ is nontrivial. We will find many such spaces, but for the present let us give a proof that this is the case for the circle. We will compute $\pi_1(\mathbf{S}^1)$ later, but for now, we will just show it nontrivial using only methods from the simpler parts of Chapter II.

2.1. Proposition. *The circle $\mathbf{S}^1$ is not simply connected.*

PROOF. Consider the identity map $f: \mathbf{S}^1 \to \mathbf{S}^1$ as a loop in $\mathbf{S}^1$. Thus $[f] \in \pi_1(\mathbf{S}^1, p)$ for some $p \in \mathbf{S}^1$. If $[f] = 1$, the unity element of $\pi_1(\mathbf{S}^1, p)$, then $f \simeq c$ (pointed) where $c: \mathbf{S}^1 \to \mathbf{S}^1$ is the constant map to the base point p. But such a homotopy is a deformation of $\mathbf{S}^1$ to a point and implies that $\mathbf{S}^1$ is contractible, contrary to Corollary 11.13 of Chapter II. □

On the other hand, we have:

2.2. Theorem. *The sphere $\mathbf{S}^n$ is simply connected for $n \geq 2$.*

PROOF. This almost follows from Theorem 11.10 of Chapter II, but to be a correct proof, we have to make sure that the homotopy resulting from the proof of Theorem 11.10 of Chapter II can be taken to preserve the base point. (Later we will show that this is not really necessary.) But the map of the base point to $\mathbf{S}^n$ is smooth, and so the smooth approximations used in Theorem 11.10 of Chapter II can be taken to not move the base point. □

These are important facts, so we will give alternative proofs. We also do this for the reason of illustrating some other approaches to things of this sort. For Proposition 2.1, consider $\mathbf{S}^1$ as the unit circle in the plane and suppose that we have a homotopy $F: \mathbf{S}^1 \times I \to \mathbf{S}^1$ with $F(x, 0) = (1, 0)$ and $F(x, 1) = x$ for all $x \in \mathbf{S}^1$. We can assume this to be a smooth map, since it is already smooth on the ends. Composing this with the quotient map $I \to \mathbf{S}^1$ gives a homotopy $G: I \times I \to \mathbf{S}^1$ with

$$\begin{aligned} G(s, 0) &= (1, 0), \\ G(s, 1) &= (\cos(2\pi s), \sin(2\pi s)), \\ G(0, t) &= (1, 0), \\ G(1, t) &= (1, 0). \end{aligned}$$

Break G into its components $G(s, t) = (x(s, t),\ y(s, t))$ and consider the differential $d\theta = d(\arctan(y/x)) = (x\,dy - y\,dx)/(x^2 + y^2)$. Then consider the line integral

$$f(t) = \oint d\theta = \int_0^1 \frac{d\theta}{ds}(s,t)ds.$$

It is easy to calculate that $f(0) = 0$ and $f(1) = 2\pi$. If you look at the approximating Riemann sums for this integral, you will see that they are just sums of (signed) angles between successive points (x_i, y_i). Any partial sum is the angle from the x-axis to the present point (x_i, y_i). Of course, the angle is determined only up to a multiple of 2π. Since the start and end of each of the parametrized (by s) curves are the same, the approximating Riemann sums must all be multiplies of 2π. It follows that the line integral $f(t) = \oint d\theta$ is itself a multiple of 2π for any t. But $f(t)$ is continuous in t and a continuous function taking values in a discrete set must be constant. This contradiction shows that the homotopy cannot exist. Some readers may be unsure of the rigor of this proof, and they are urged to fill in the details. This argument was one of the precursors to algebraic topology and one of the things we will be doing momentarily is to detail this type of argument, although with different terminology.

Here is another proof of Theorem 2.2: Cover $\mathbf{S}^n$ with open hemispheres. Let $f: I \to \mathbf{S}^n$ be any loop and consider the covering of I by the inverse images under f of the hemispheres. By the Lebesgue Lemma (Lemma 9.11 of Chapter I), there is an integer n such that any interval $[a, b]$ of length $\leq 1/n$ is taken by f into an open hemisphere. Now we will define a homotopy of f rel ∂I. It will be defined as a homotopy of the restriction of f to each interval $[i/n, (i+1)/n]$ rel$\{i/n,\ (i+1)/n\}$. For s in this interval, let $g(s)$ be any parametrization of the line segment in $\mathbf{R}^{n+1}$ from $f(i/n)$ to $f(i+1)/n)$, and note that this does not go through the origin since the end points are in a common open hemisphere. Then let

$$F_i(s,t) = \frac{tg(s) + (1-t)f(s)}{\| tg(s) + (1-t)f(s) \|}.$$

Combining these homotopies then gives a homotopy from f to a loop made up of a finite number of great circle arcs. Such a loop cannot fill up $\mathbf{S}^n$ (prove it) and so there is a point $p \in \mathbf{S}^n$ left over. Thus $[f]$ is in the image of the homomorphism $\pi_1(\mathbf{S}^n - \{p\}) \to \pi_1(\mathbf{S}^n)$. But $\mathbf{S}^n - \{p\}$ is contractible without moving the base point, so its fundamental group is trivial, and $[f]$ must be the trivial class 1. (Where did we use that $n > 1$ in this argument?)

In the things we have done in this section, so far, we had to pay attention to the base point, making sure it did not move during homotopies. This was particularly irksome in the proof that a contractible space has a trivial fundamental group, since we had to assume a stronger type of contractibility, one that does not move the base point. An example of a contractible space that does not satisfy this condition is the "comb space" *Comb* of Figure III-3. Any contraction of *Comb* must move the point x_0. On the other hand,

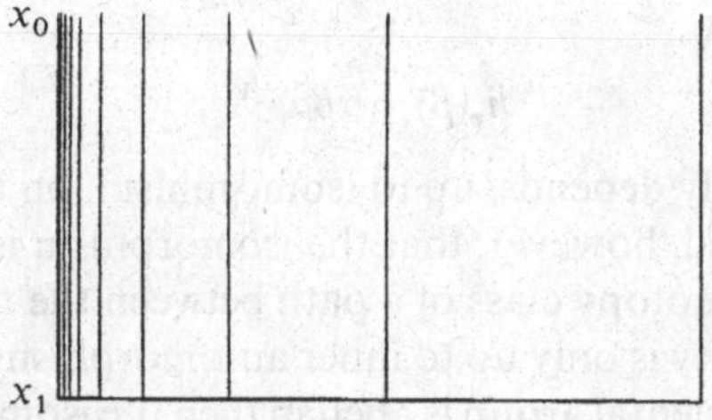

Figure III-3. The comb space $\{0\} \times I \cup I \times (\{0\} \cup \{1/n | n = 1, 2, 3, \ldots\})$.

there is a contraction that does not move the point x_1. Thus we know that $\pi_1(Comb, x_1) = 1$ but nothing we have said tells us anything about $\pi_1(Comb, x_0)$. It can be shown directly that the latter group is trivial, and the reader is encouraged to try to do so.

We now try to rectify this fault by studying change of base point. We will restrict attention at this time to the fundamental group, but a similar treatment can be given for homotopy groups $\pi_n(X)$, in general, and the reader is urged to attempt to give generalizations of the things we do here for the fundamental group. Given a space X, let $p: I \to X$ be a path from $p(0) = x_0$ to $p(1) = x_1$. Then we define a function

$$h_p: \pi_1(X, x_1) \to \pi_1(X, x_0)$$

by

$$h_p[f] = [p * f * p^{-1}].$$

That this is well defined (i.e., depends only on the homotopy class of f) is clear.

This is a homomorphism since $(p * f * p^{-1}) * (p * g * p^{-1}) \simeq p * f * (p^{-1} * p) * g * p^{-1} \simeq p * f * g * p^{-1}$.

Also it is clear that:

(1) $h_p \circ h_q = h_{p*q}$;
(2) $p \simeq q$ rel $\partial I \Rightarrow h_p = h_q$; and
(3) $h_{c_x} = 1$ where c_x is the constant path at x.

Also, using (1) to (3) we get

(4) $h_p \circ h_{p^{-1}} = 1$.

Moreover, if p is a loop then $h_p[f] = [p][f][p]^{-1}$. Thus we have:

2.3. Theorem. *For a path p from x_0 to x_1 in a space X, we have the isomorphism*

$$h_p: \pi_1(X, x_1) \xrightarrow{\approx} \pi_1(X, x_0)$$

with inverse $h_{p^{-1}}$. If p is a loop representing $\alpha = [p]$, then h_p is the inner *automorphism*

$$h_p(\beta) = \alpha\beta\alpha^{-1}. \qquad \square$$

Thus, $\pi_1(X, x_0)$ only depends, up to isomorphism, on the path component of x_0. It must be noted, however, that the isomorphism is *not natural*, in that it depends on the homotopy class of a path between the two base points. The degree of nonnaturality is only up to inner automorphism, however. Thus, for example, if the fundamental group is abelian then the isomorphism connecting different base points *is natural.*

Because of these facts, we sometimes use $\pi_1(X)$ to represent the fundamental group, where the base point taken is immaterial if X is arcwise connected.

Now we take up the study of homotopies of loops which can move the base point, which we shall call "free homotopies." To be more precise, suppose $p: I \to X$ is a path as above. Suppose we have a homotopy $F: I \times I \to X$ such that

$$F(0, s) = F(1, s) = p(s),$$
$$F(t, 0) = f_0(t),$$
$$F(t, 1) = f_1(t).$$

Then we say that f_0 is "freely homotopic" to f_1 along p, and we denote this relationship by $f_0 \simeq_p f_1$.

2.4. Proposition. *In the above situation,* $f_0 \simeq_p f_1 \Leftrightarrow h_p[f_1] = [f_0]$.

Proof. The proof is accomplished by study of the diagrams in Figure III-4. These are pictures of homotopies $I \times I \to X$. In that figure, the cross hatching represents lines along which the maps are constant. The unhatched portion is to be filled in here. The left-hand diagram represents the proof of the $\Rightarrow$ part of the proposition. The unhatched portion can be filled in since $f_0 \simeq_p f_1$. The entire map is then a homotopy showing that $h_p[f_1] = [f_0]$. The right-hand

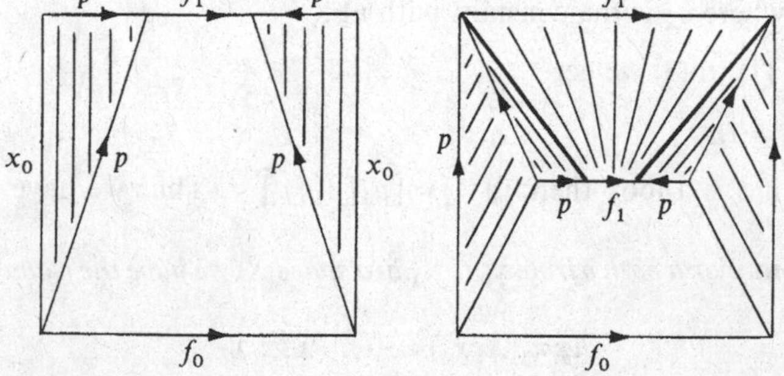

Figure III-4. Basic free homotopy constructions.

diagram represents the proof of $\Leftarrow$. There, the unhatched portion can be filled in since $h_p[f_1] = [f_0]$. The entire map shows that $f_0 \simeq_p f_1$. □

Now we remove the restriction from the proof that contractible spaces have trivial fundamental group. Indeed, we prove a much stronger fact, that homotopically equivalent, arcwise connected spaces have isomorphic fundamental groups.

2.5. Theorem. *If X and Y are arcwise connected spaces and $\phi: X \to Y$ is a homotopy equivalence then $\phi_\#: \pi_1(X, x_0) \to \pi_1(Y, \phi(x_0))$ is an isomorphism.*

PROOF. Let $y_0 = \phi(x_0)$. The only problem with proving this is that we cannot assume that there is a homotopy inverse which takes y_0 to x_0, and we cannot assume that homotopies preserve the base points. Let $\psi: Y \to X$ be a homotopy inverse to ϕ and put $x_1 = \psi(y_0)$. Then we have the homomorphisms

$$\pi_1(X, x_0) \xrightarrow{\phi_\#} \pi_1(Y, y_0) \xrightarrow{\psi_\#} \pi_1(X, x_1)$$

whose composition is $(\psi \circ \phi)_\#$. By assumption $\psi \circ \phi \simeq 1$. During the homotopy the images of the point x_0 sweep out some path, say p, from x_1 to x_0. Composing on the right with a loop f gives $\psi \circ \phi \circ f \simeq_p f$. Putting $\alpha = [f] \in \pi_1(X, x_0)$, we have $(\psi \circ \phi)_\#(\alpha) = h_p(\alpha)$ for all $\alpha \in \pi_1(X, x_0)$. Thus $(\psi \circ \phi)_\#$ is an isomorphism, and it follows that $\phi_\#$ is a monomorphism and $\psi_\#$ is onto. Applying the same discussion but starting with ψ shows that $\psi_\#$ is also a monomorphism. Thus $\psi_\#$ is an isomorphism. Therefore $\phi_\# = \psi_\#^{-1} \circ (\psi \circ \phi)_\#$ is an isomorphism as claimed. □

2.6. Theorem. *Let X and Y be spaces with base points x_0 and y_0 and let $i^X: X \hookrightarrow X \times Y$ and $i^Y: Y \hookrightarrow X \times Y$ be the inclusions $i^X(x) = (x, y_0)$ and $i^Y(y) = (x_0, y)$. Let j^X and j^Y be the projections of $X \times Y$ onto X and Y, respectively. Then the map*

$$i_\#^X \cdot i_\#^Y: \pi_1(X, x_0) \times \pi_1(Y, y_0) \to \pi_1(X \times Y, (x_0, y_0)),$$

given by $i_\#^X \cdot i_\#^Y(\alpha \times \beta) = i_\#^X(\alpha) i_\#^Y(\beta)$, is an isomorphism with inverse

$$j_\#^X \times j_\#^Y: \pi_1(X \times Y, (x_0, y_0)) \to \pi_1(X, x_0) \times \pi_1(Y, y_0).$$

PROOF. Given a loop $f: \mathbf{S}^1 \to X \times Y$, let $f_X = j^X \circ f$ and $f_Y = j^Y \circ f$. Consider the map $f_X \times f_Y: \mathbf{S}^1 \times \mathbf{S}^1 \to X \times Y$ given by $f_X \times f_Y(s, t) = (f_X(s), f_Y(t))$. Also let $\alpha(t) = (t, 0)$, $\beta(t) = (0, t)$, and $\delta(t) = (t, t)$ as maps $\mathbf{S}^1 \to \mathbf{S}^1 \times \mathbf{S}^1$. Clearly $\alpha * \beta \simeq \delta \simeq \beta * \alpha$, which can be seen by thinking of δ as the diagonal of the unit square and α, β as adjacent sides. Thus

$$\begin{aligned} f = (f_X \times f_Y) \circ \delta &\simeq (f_X \times f_Y) \circ (\alpha * \beta) \\ &= \{(f_X \times f_Y) \circ \alpha\} * \{(f_X \times f_Y) \circ \beta\}. \end{aligned}$$

Computing each of these terms at t shows that $(f_X \times f_Y)\circ\alpha = i^X \circ f_X$ and $(f_X \times f_Y)\circ\beta = i^Y \circ f_Y$. Thus $f \simeq (i^X \circ f_X)*(i^Y \circ f_Y)$ which shows that $i^X_\# \cdot i^Y_\#$ is onto.

If $f: \mathbf{S}^1 \to X$ and $g: \mathbf{S}^1 \to Y$ are loops, then it is easy to see that $j^X \circ \{(i^X \circ f)*(i^Y \circ g)\} = f * c_{x_0} \simeq f$ and $j^Y \circ \{(i^X \circ f)*(i^Y \circ g)\} = c_{y_0} * g \simeq g$ from which it follows that $j^X_\# \times j^Y_\#(i^X_\#[f]\, i^Y_\#[g]) = [f] \times [g]$. This means that $(j^X_\# \times j^Y_\#)\circ(i^X_\# \cdot i^Y_\#) = 1$. Therefore $i^X_\# \cdot i^Y_\#$ is one–one onto and $j^X_\# \times j^Y_\#$ is its inverse. Since $j^X_\# \times j^Y_\#$ is a homomorphism, it is an isomorphism and hence so is $i^X_\# \cdot i^Y_\#$. □

We end this section with a simple criterion for the triviality of an element of the fundamental group, which is quite convenient at times.

2.7. Proposition. *Let $f: \mathbf{S}^1 \to X$. Then $[f] = 1 \in \pi_1(X) \Leftrightarrow f$ extends to $\mathbf{D}^2$.*

Proof. If $[f] = 1 \in \pi_1(X)$ then there is a homotopy $\mathbf{S}^1 \times I \to X$ starting with f and ending with the constant map to the base point. We can think of this homotopy as a map from the annulus between the circles of radius 1 and $\frac{1}{2}$ to X, which is f on the outer circle and constant to the base point on the inner circle. But that extends, by a constant map, over the disk of radius $\frac{1}{2}$, giving the desired extension of f to $\mathbf{D}^2$.

Conversely, suppose there is an extension of f to $F: \mathbf{D}^2 \to X$. Compose this with the map $G: I \times I \to \mathbf{D}^2$ given by $G(s,t) = (t\cos(2\pi s), t\sin(2\pi s))$. This is a free homotopy along the path $p(t) = F(G(0, 1-t))$ from f to a constant loop c. Therefore, in the notation above, $[f] = h_p[c] = h_p(1) = 1$. □

Problems

1. Let G be a topological group with unity element e. For loops $f, g: (\mathbf{S}^1, *) \to (G, e)$ define a loop $f \bullet g(t) = f(t)g(t)$ by the pointwise product in G. Show that $f * g \simeq f \bullet g$ rel $*$.

2. Let G be a topological group with unity element e. Show that $\pi_1(G, e)$ is abelian. (*Hint*: Use Problem 1 and the idea of Problem 1 to show that $f \bullet g \simeq g * f$.)

3. If $\mathbf{K}^2$ is Klein bottle, show that $\pi_1(\mathbf{K}^2)$ is generated by two elements, say α and β obtained from the "longitudinal" and "latitudinal" loops. Also show that there is the relation (with proper assignment of α and β) $\alpha\beta\alpha^{-1} = \beta^{-1}$. (You are not asked to show that this is the "only" relation, but, in fact, it is.) (*Hint*: Use the fact that a smooth loop must miss a point.)

3. Covering Spaces

The spaces we shall consider in this section will all be Hausdorff, arcwise connected, and locally arcwise connected.

Note that, in such spaces, every point has a neighborhood basis consisting of arcwise connected sets. In turn, this implies that the arc components of any open subset of such a space are themselves open.

3.1. Definition. A map $p: X \to Y$ is called a *covering map* (and X is called a *covering space* of Y) if X and Y are Hausdorff, arcwise connected, and locally arcwise connected, and if each point $y \in Y$ has an arcwise connected neighborhood U such that $p^{-1}(U)$ is a nonempty disjoint union of sets U_α (which are the arc components of $p^{-1}(U)$) on which $p|_{U_\alpha}$ is a homeomorphism $U_\alpha \xrightarrow{\approx} U$. Such sets U will be called *elementary*, or *evenly covered.*

Note that a covering map must be onto, because that is part of "homeomorphism." Also, it is not enough for a map to be a local homeomorphism (meaning each point of X has a neighborhood mapping homeomorphically onto a neighborhood of the image point). Consider the map $p: (0, 2) \to \mathbf{S}^1$ defined by $p(t) = (\cos(2\pi t), \sin(2\pi t))$. That is a local homeomorphism, but for any small neighborhood U of $1 \in \mathbf{S}^1$, some component of $p^{-1}(U)$ does not map *onto* U.

It is clear that the number of points in the inverse image of a point, under a covering map, is locally constant, and hence constant since the base space is connected. This number is called the "number of sheets" of the covering. Covering maps with two sheets are often called "double coverings" or "two fold" coverings.

Here are some examples of covering maps. Throughout the examples, we will consider $\mathbf{S}^1$ to be the unit circle in the complex numbers $\mathbf{C}$.

(1) The map $\mathbf{R} \to \mathbf{S}^1$ taking $t \mapsto e^{2\pi i t}$ is a covering map with infinitely many sheets.
(2) The map $\mathbf{S}^1 \to \mathbf{S}^1$ taking $z \mapsto z^n$ for a fixed positive integer n, is a covering with n sheets.
(3) The canonical map $\mathbf{S}^2 \to \mathbf{P}^2$, the real projective plane, is a double covering.

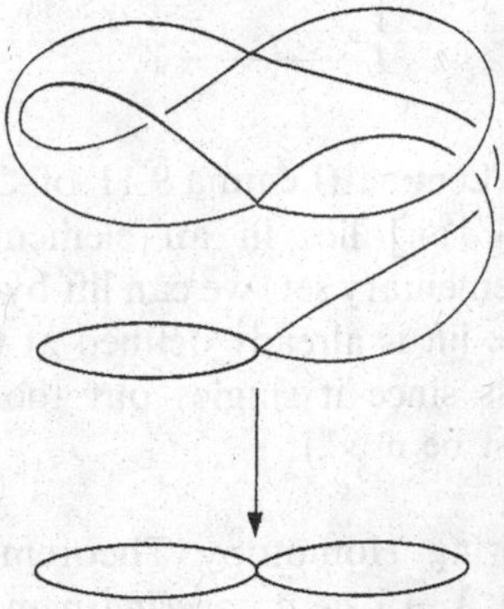

Figure III-5. A threefold covering space.

(4) Consider the equivalence relation on the plane $\mathbf{R}^2$ which is *generated* by the equivalences $(x, y) \sim (x, y+1)$ and $(x, y) \sim (x+1, -y)$. The canonical map $\mathbf{R}^2 \to \mathbf{R}^2/\sim$ is a covering with infinitely many sheets. (The quotient space is the Klein bottle.)

(5) Figure III-5 shows an interesting covering of the "figure eight" space, with three sheets. We shall have reason to refer back to this example later. It would be wise to try out results and proofs on this example.

(6) Let $p(z)$ be a complex polynomial considered as a map $\mathbf{C} \to \mathbf{C}$, and let F be the set of critical values of $p(z)$. Then the induced map $\mathbf{C} - p^{-1}(F) \to \mathbf{C} - F$ is a covering map, as follows from the proof of Corollary 6.4 of Chapter II, and has $\deg(p(z))$ sheets.

3.2. Lemma. *Let W be an arbitrary space and let $\{U_\alpha\}$ be an open covering of $W \times I$. Then for any point $w \in W$ there is a neighborhood N of w in W and a positive integer n such that $N \times [i/n, (i+1)/n] \subset U_\alpha$ for some α, for each $0 \le i < n$.*

Proof. We can cover $\{w\} \times I$ by a refinement of $\{U_\alpha\}$ of the form $N_1 \times V_1, N_2 \times V_2, \ldots, N_k \times V_k$ by compactness of I and the definition of the product topology. The Lebesgue Lemma (Lemma 9.11 of Chapter I) implies that there is an $n > 0$ such that each $[i/n, (i+1)/n]$ is contained in one of the V_j. Just take this n and $N = \bigcap N_i$. □

3.3. Theorem (The Path Lifting Property). *Let $p: X \to Y$ be a covering map and let $f: I \to Y$ be a path. Let $x_0 \in X$ be such that $p(x_0) = f(0)$. Then there exists a unique path $g: I \to X$ such that $p \circ g = f$ and $g(0) = x_0$. This can be summarized by saying that the following commutative diagram can be completed uniquely:*

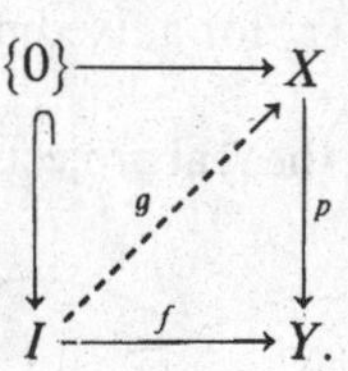

Proof. By the Lebesgue Lemma (Lemma 9.11 of Chapter I), there is an n such that each $f[i/n, (i+1)/n]$ lies in an elementary set. By the local homeomorphisms over elementary sets we can lift by induction on i. (At each stage of the induction, the lift is already defined at the left-hand end point, leading to the uniqueness since it singles out the component above the elementary set which must be used.) □

3.4. Theorem (The Covering Homotopy Theorem). *Let W be a locally connected space and let $p: X \to Y$ be a covering map. Let $F: W \times I \to Y$ be a homotopy and let $f: W \times \{0\} \to X$ be a lifting of the restriction of F to $W \times \{0\}$. Then there is a unique homotopy $G: W \times I \to X$ making the following diagram* *commute*:

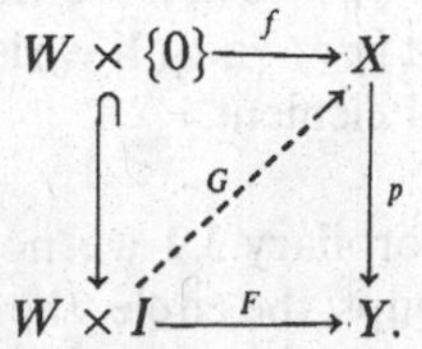

Moreover, if F is a homotopy rel W' *for some* $W' \subset W$, *then so is G.*

Proof. Define G on each $\{w\} \times I$ by Theorem 3.3. This is unique. For continuity, let $w \in W$. By Lemma 3.2 we can find a connected neighborhood N of w in W and integer n so that each $F(N \times [i/n, (i+1)/n])$ is in some elementary set U_i. Assuming that G is continuous on $N \times \{i/n\}$ we see that $G(N \times \{i/n\})$, being connected, must be contained in a single component, say V of $p^{-1}(U_i)$. But then on $N \times [i/n, (i+1)/n]$, the lift G must be F composed with the inverse of the homeomorphism $p|_V: V \to U_i$ (again using connectivity). But that means G is continuous on all of $N \times [i/n, (i+1)/n]$. A finite induction then shows that G is continuous on each $N \times I$, and hence everywhere. The last statement follows from the construction of G. □

The condition that W be locally connected, in Theorem 3.4, can be dropped. The proof is only slightly more difficult. The reader might try proving that.

3.5. Corollary. *Let* $p: X \to Y$ *be a covering map. Let* f_0 *and* f_1 *be paths in* Y *with* $f_0 \simeq f_1$ rel ∂I. *Let* $\tilde{f}_0$ *and* $\tilde{f}_1$ *be liftings of* f_0 *and* f_1 *such that* $\tilde{f}_0(0) = \tilde{f}_1(0)$. *Then* $\tilde{f}_0(1) = \tilde{f}_1(1)$ *and* $\tilde{f}_0 \simeq \tilde{f}_1$ rel ∂I. □

3.6. Corollary. *Let* $p: X \to Y$ *be a covering map. Let* $f:(I, \partial I) \to (Y, y_0)$ *be a loop. If f is homotopic to a constant loop* rel ∂I *then* any *lift of f to a path is a loop and is homotopic to a constant loop* rel ∂I. □

3.7. Corollary. *Let* $p: X \to Y$ *be a covering map and* $p(x_0) = y_0$. *Then*

$$p_{\#}: \pi_1(X, x_0) \to \pi_1(Y, y_0)$$

is a monomorphism whose image consists of the classes of loops at y_0 *in* Y *which lift to loops at* x_0 *in* X. □

3.8. Corollary. *Let* $p: X \to Y$ *be a covering map and* $p(x_0) = y_0$. *If f is a loop in Y at* y_0 *which lifts to a loop in X at* x_0 *then any loop homotopic to f* rel ∂I *also lifts to a loop in X at* x_0. *That is, lifting to a loop is a property of the class* $[f]$. □

3.9. Corollary. *If a Hausdorff, arcwise connected, and locally arcwise connected space Y has a nontrivial covering space then* $\pi_1(Y, y_0) \neq 1$.

PROOF. Take two points $x_0, x_1 \in p^{-1}(y_0)$ and let f be a path between them. Then $p \circ f$ is a loop in Y at y_0 which does not lift to a loop in X at x_0. By Corollary 3.7, it follows that $[p \circ f] \in \pi_1(Y, y_0)$ is not in the image from $\pi_1(X, x_0)$ and hence it is a nontrivial element. □

As a consequence of Corollary 3.9 we now know several spaces having nontrivial fundamental groups: the circle, the Klein bottle, and the projective plane. Later, we will completely calculate these fundamental groups. We shall start with the most important one, the circle.

Consider the exponential map $p: \mathbf{R} \to \mathbf{S}^1$ defined by $p(t) = e^{2\pi i t}$ which is a covering map. Let $f: I \to \mathbf{S}^1$ be any loop at $1 \in \mathbf{S}^1$. Let $\tilde{f}: I \to \mathbf{R}$ be a lifting of f such that $\tilde{f}(0) = 0$. Then $\tilde{f}(1) \in p^{-1}(\{1\}) = \mathbf{Z}$. Let $n = \tilde{f}(1)$. By Corollary 3.5, n depends only on the homotopy class $[f] \in \pi_1(\mathbf{S}^1)$. This integer n is called the "degree" of f, and we write $n = \deg(f)$.

3.10. Theorem. $\deg: \pi_1(\mathbf{S}^1) \to \mathbf{Z}$ *is an isomorphism.*

PROOF. First, we show that deg is a homomorphism. Given loops f, g, and liftings $\tilde{f}, \tilde{g}$, both starting at $0 \in \mathbf{R}$ we have $\tilde{f}(1) = \deg(f) = n$, say, and $\tilde{g}(1) = \deg(g) = m$. Define $\tilde{g}'(t) = \tilde{g}(t) + n$. Then $\tilde{g}'(0) = n = \tilde{f}(1)$, and so $\tilde{f} * \tilde{g}'$ is defined, covers $f * g$ and $\tilde{f} * \tilde{g}'(1) = \tilde{g}'(1) = \tilde{g}(1) + n = m + n = \deg(f) + \deg(g)$, as claimed.

Second, deg is onto since a path from 0 to n in $\mathbf{R}$ maps to a loop in $\mathbf{S}^1$ which has degree n by definition.

Third, we show that deg is a monomorphism by showing its kernel is zero. Suppose $f: I \to \mathbf{S}^1$ has degree 0. Then, for a lifting $\tilde{f}$ of f we have $\tilde{f}(1) = 0 = \tilde{f}(0)$ so that $\tilde{f}$ is a loop and represents an element $[\tilde{f}] \in \pi_1(\mathbf{R}, 0) = 1$, since $\mathbf{R}$ is contractible. Thus $[f] = p_\#[\tilde{f}] = p_\#(1) = 1$. □

3.11. Proposition. *The map* $z \mapsto z^n$ *of* $\mathbf{S}^1 \to \mathbf{S}^1$ *has degree* n. □

3.12. Corollary (Fundamental Theorem of Algebra). *If* $p(z)$ *is a complex polynomial of degree* $n > 0$ *then* $p(z)$ *has a zero.*

PROOF. We may assume that $p(z) = z^n + a_1 z^{n-1} + \cdots + a_n, n > 0$. Assuming p has no zeros, consider the homotopy $F: \mathbf{S}^1 \times I \to \mathbf{S}^1$ defined by

$$F(z, t) = \frac{p((1-t)z/t)}{|p((1-t)z/t)|} = \frac{t^n p((1-t)z/t)}{|t^n p((1-t)z/t)|}.$$

Since

$$t^n p((1-t)z/t) = (1-t)^n z^n + a_1(1-t)^{n-1} z^{n-1} t + \cdots + a_n t^n,$$

F is defined and continuous even at $t = 0$. We have $F(z, 0) = z^n$ and $F(z, 1) = p(0)/|p(0)|$. Therefore the map $z \mapsto z^n$ of $\mathbf{S}^1 \to \mathbf{S}^1$ is freely homotopic to a constant map contrary to Proposition 2.4, Theorem 3.10, and Proposition 3.11. □

PROBLEMS

1. Referring to example (5), find at least two more coverings of the figure eight space with three sheets. Find at least three different double coverings of the figure eight space. Are there any others?
2. Show that the fundamental group of the projective plane is the unique group $\mathbf{Z}_2$ of two elements.
3. Compute the fundamental group of an n-dimensional torus (a product of n circles).
4. Use the covering of the figure eight in example (5) to show that the fundamental group of the figure eight is not abelian. (*Hint*: Consider liftings of loops representing $\alpha\beta$ and $\beta\alpha$, for appropriate classes α and β.)
5. Show that, for maps $\mathbf{S}^1 \to \mathbf{S}^1$, the notion of "degree" in this section coincides with that defined above Corollary 16.4 of Chapter II.

4. The Lifting Theorem

The "lifting problem" in topology is to decide when one can "lift" a map $f: W \to Y$ to a map $g: W \to X$, where $p: X \to Y$ is given. That is, under what conditions can one complete the following diagram (making it commutative):

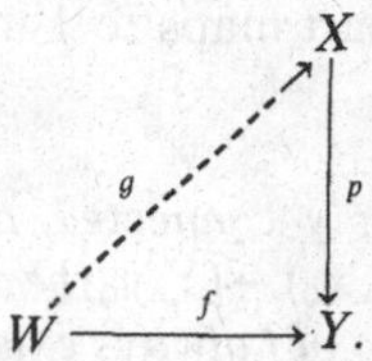

One might also add conditions such as having a lifting already given on some subspace.

This is an important problem in topology, since many topological questions can be phrased in terms of finding such liftings.

If one adds the condition that base points x_0, y_0, and w_0 are given and must correspond under the mappings, and if p is a covering map, then we can answer this question now.

4.1. Theorem (The Lifting Theorem). *Assume that $p: X \to Y$ is a covering mapping with $p(x_0) = y_0$. Assume that W is arcwise connected and locally arcwise connected and that $f: W \to Y$ is a given map with $f(w_0) = y_0$. Then a map $g: (W, w_0) \to (X, x_0)$ such that $p \circ g = f$ exists $\Leftrightarrow f_{\#}\pi_1(W, w_0) \subset p_{\#}\pi_1(X, x_0)$. Moreover, g is unique.*

PROOF. First let us define the function g. Given $w \in W$, let $\lambda: I \to W$ be a path from w_0 to w. Then $f \circ \lambda$ is a path in Y. Lift this to a path $\mu: (I, 0) \to (X, x_0)$ and put $g(w) = \mu(1)$. Then $p \circ g(w) = p(\mu(1)) = f(\lambda(1)) = f(w)$.

To see that g is well defined, suppose λ' is another path in W from w_0 to w and put $\eta = (\lambda')^{-1}$. Then $\lambda * \eta$ is a loop at w_0 in W, so $f \circ \lambda * f \circ \eta$ is a loop at y_0 in Y. Since $[f\circ\lambda * f\circ\eta] = f_\#[\lambda*\eta]\in \operatorname{im}(f_\#)\subset \operatorname{im}(p_\#)$, $f\circ\lambda * f\circ\eta$ lifts to a loop in X at x_0. The reverse of the portion of this lift corresponding to η then is a lift μ' of λ', and $\mu'(1)=\mu(1)$, as required.

Next, we have to show that g is continuous. This is where the condition that W be locally arcwise connected comes in. Let $w\in W$ and put $y=f(w)$. Let $U\subset Y$ be an elementary neighborhood of y, and let V be an arcwise connected neighborhood of w such that $f(V)\subset U$. For any point $w'\in V$, we can construct a path from w_0 to w' by concatenating a given path λ from w_0 to w with a path σ in V from w to w'. Since $f(V)$ is contained in an elementary set, the lift of $f\circ\sigma$ is simply $f\circ\sigma$ composed with the inverse of p taking U to that component of $p^{-1}(U)$ containing $g(w)$. This same component is used for all $w'\in V$ and it follows that g is continuous at w.

The converse is immediate from $f_\# = p_\#\circ g_\#$. □

To see that, in this theorem, the hypothesis that W be locally arcwise connected cannot be dropped, consider the example illustrated in Figure III-6. The map f there is a quotient map that collapses the "sin(1/x)" part of W to a point. Take the point, to which this set is collapsed, as 1 on the circle and let it be the base point. Take 0 as the base point in **R**. If the lift g of f is constructed as in the proof of Theorem 4.1 (which is forced by continuity of path lifting), then the straight part of the "sin(1/x)" set maps to 0 under g and the wiggly part maps to 1 under g, so that g is seen to be discontinuous.

4.2. Corollary. *Let W be simply connected, arcwise connected, and locally arcwise connected, and let $p:(X,x_0)\to(Y,y_0)$ be a covering map. Let $f:(W,w_0)\to(Y,y_0)$ be any map. Then a lift g of f always exists taking w_0 to any given point in $p^{-1}(y_0)$. The lift g is unique if the image of w_0 is specified.* □

4.3. Corollary. *The homotopy group $\pi_n(\mathbf{S}^1)$ is trivial for $n>1$. That is, any map $\mathbf{S}^n\to\mathbf{S}^1$ is homotopically trivial for $n>1$.*

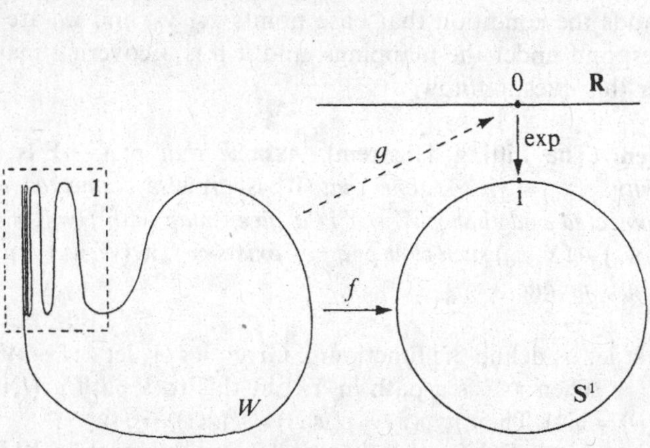

Figure III-6. Discontinuous lifting.

Proof. Any given map $f: \mathbf{S}^n \to \mathbf{S}^1$ lifts to $g: \mathbf{S}^n \to \mathbf{R}$ by Corollary 4.2. But g is homotopically trivial since $\mathbf{R}$ is contractible, and so $f = p \circ g$ is also homotopically trivial. □

4.4. Lemma. *Let W be connected. Let $p: X \to Y$ be a covering map and $f: W \to Y$ a map. Let g_1 and g_2 be maps $W \to X$ both of which are liftings of f. If $g_1(w) = g_2(w)$ for some point $w \in W$ then $g_1 \equiv g_2$.*

Proof. Let $w \in W$ be such that $g_1(w) = g_2(w) = x$, say. Let U be an open elementary neighborhood of $f(w)$ in Y for the covering map p. Let V be the component of $p^{-1}(U)$ containing x. Then $A = g_1^{-1}(V) \cap g_2^{-1}(V)$ is an open set in W and for $a \in A$ we have $g_1(a) = g_2(a)$ since the homeomorphism $p: V \to U$ maps them both to $f(a)$. This shows that the set $\{w \in W \mid g_1(w) = g_2(w)\}$ is open. But this set is also closed since it is the inverse image of the diagonal under the map $g_1 \times g_2: W \to X \times X$, and the diagonal is closed by Problem 5 in Section 8 of Chapter I, since X is Hausdorff. Since W is connected, this set is either empty or all of W. □

4.5. Corollary. *Let $p_i: W_i \to Y$, $i = 1, 2$, be covering maps such that W_1 is simply connected, and let $w_i \in W_i$ and $y \in Y$ be such that $p_i(w_i) = y$. Then there is a unique map $g: W_1 \to W_2$ such that $g(w_1) = w_2$ and $p_2 \circ g = p_1$. Moreover, g is a covering map.*

Proof. This follows directly from Lemma 4.4 except for the addendum that g is a covering map. The latter is a simple exercise in the definition of covering maps and is left to the reader. □

4.6. Corollary. *Let $p_i: W_i \to Y$, $i = 1, 2$, be covering maps such that W_1 and W_2 are both simply connnected. If $w_i \in W_i$ are such that $p_1(w_1) = p_2(w_2)$ then there is a unique map $g: W_1 \to W_2$ such that $p_2 \circ g = p_1$ and $g(w_1) = w_2$. Moreover, g is a homeomorphism.* □

Proof. Use Corollary 4.5 to produce g and to also produce a map $k: W_2 \to W_1$ going the other direction. Then $k \circ g: W_1 \to W_1$ covers the identity map and equals the identity map at w_1. By Lemma 4.4, it equals the identity everywhere. Similarly, with $g \circ k$, so $k = g^{-1}$. □

In the situation of Corollary 4.6, g is called an "equivalence" of covering spaces. Thus all simply connected covering spaces of a given space are equivalent. Such covering spaces are also called "universal" covering spaces. They do not always exist, but they do exist under a mild restriction, as we shall see presently.

Problems

1. Show that $\pi_n(\mathbf{P}^m)$ is trivial for $1 < n < m$.

2. Show that any map of the projective plane to the circle is homotopic to a constant map.

3. Complete the proof of Corollary 4.5.

5. The Action of π_1 on the Fiber

The next few sections are devoted to the classification of all covering spaces of a given space. It should be clear from our previous results that this is closely associated with the study of the fundamental groups of the spaces involved. In this section we define and study an action of the fundamental group of the base space of a covering map, on the "fiber," the set of points mapping to the base point in the base space. This will play an important role in the study of the classification problem.

Throughout this section let $p: X \to Y$ be a given covering space. Also let $y_0 \in Y$ be a fixed base point. To simplify notation, we define

$$J = \pi_1(Y, y_0) \qquad \text{and} \qquad F = p^{-1}(y_0).$$

The discrete set F is called the "fiber" of p. We are going to describe an action of the group J on F as a group of permutations. For convenience the group will act on the *right* of the set. This action is called the "monodromy" action.

Let $x \in F$ and $\alpha \in J$. Represent α by a loop $f: I \to Y$. Lift f to get a path g in X with $g(0) = x$. Then define

$$x \cdot \alpha = g(1).$$

By Corollary 3.5, this does not depend on the choice of f and so it is a well-defined function

$$F \times J \longrightarrow F.$$

Now we shall derive some properties of this function.

(1) $x \cdot 1 = x$.
(2) $(x \cdot \alpha) \cdot \beta = x \cdot (\alpha\beta)$.
These say that J acts as a group of permutations of F. (1) is clear. To prove (2), lift a loop representing α to a path f starting at x. This goes from x to $x \cdot \alpha$. Then lift a loop representing β to a path g starting at $x \cdot \alpha$. This goes from $x \cdot \alpha$ to $(x \cdot \alpha) \cdot \beta$. But then $f * g$ is a lift of a loop representing $\alpha\beta$ and starts at x and hence ends at $x \cdot (\alpha\beta)$, proving (2).
(3) This is a transitive action. That is, given $x, x_0 \in F$, $\exists \alpha \in J \ni x = x_0 \cdot \alpha$.
To see this, merely choose a path in X from x_0 to x. This projects to a loop f in Y. Then $\alpha = [f]$ works.
(4) Put $J_{x_0} = \{\alpha \in J \mid x_0 \cdot \alpha = x_0\}$ (called the "isotropy subgroup" of J at x_0). Then $J_{x_0} = \operatorname{im}\{p_\#: \pi_1(X, x_0) \to \pi_1(Y, y_0) = J\}$.

To see this, note that $\alpha \in J_{x_0} \Leftrightarrow (\alpha = [f]$ and f lifts to a loop at $x_0) \Leftrightarrow \alpha \in \operatorname{im} p_{\#}$ (as shown earlier in Corollary 3.7).

(5) The map $\phi: J_{x_0} \backslash J \to F$ taking the right coset $J_{x_0}\alpha$ to $x_0 \cdot (J_{x_0}\alpha) = x_0 \cdot \alpha$ is a bijection.
This is a simple computation left to the reader.

Summarizing this, we now have:

5.1. Theorem. *Let $p: X \to Y$ be a covering with $p(x_0) = y_0$. Then there is a one–one correspondence between the set $p_{\#}\pi_1(X, x_0) \backslash \pi_1(Y, y_0)$ of right cosets, and the fiber $p^{-1}(y_0)$.* □

Note that $p_{\#}\pi_1(X, x_0) \approx \pi_1(X, x_0)$ since $p_{\#}$ is a monomorphism by Corollary 3.7.

5.2. Corollary. *The number of sheets of a covering map equals the index of $p_{\#}\pi_1(X, x_0)$ in $\pi_1(Y, y_0)$.* □

5.3. Corollary. *If $p: X \to Y$ is a covering with X simply connected, then the number of sheets equals the order of $\pi_1(Y, y_0)$.* □

For example, since $\mathbf{S}^n$ is simply connected for $n > 1$ and $\mathbf{S}^n$ is a double covering of the real projective n-space $\mathbf{P}^n$, it follows that $\pi_1(\mathbf{P}^n) \approx \mathbf{Z}_2$.

PROBLEMS

1. Show that any map of the projective plane to itself which is nontrivial on the fundamental group can be lifted to a map $T: \mathbf{S}^2 \to \mathbf{S}^2$ such that $T(-x) = -T(x)$ for all $x \in \mathbf{S}^2$.

2. Show that a map $f: \mathbf{S}^1 \to \mathbf{S}^1$ of degree 1 is homotopic to the identity. (*Hint*: If $\pi: \mathbf{R}^1 \to \mathbf{S}^1$ is the exponential map, consider the lift of $f \circ \pi$ to a map $\mathbf{R}^1 \to \mathbf{R}^1$.)

6. Deck Transformations

In this section, a covering map $p: X \to Y$ will be fixed once and for all. Also the notation from Section 5 will continue to be used here.

6.1. Definition. Let $p: X \to Y$ be a covering map. A homeomorphism $D: X \to X$ which covers the identity map on Y (i.e., $p \circ D = p$) is called a *deck transformation* or *automorphism* of the covering.

If D is a deck transformation, then D^{-1} is also. Also, the composition of two deck transformations of the same covering is a deck transformation. Thus the deck transformations form a group $\Delta = \Delta(p)$ under composition.

Note that, by Lemma 4.4, if $D \in \Delta$ and $D(x) = x$ for some $x \in X$ then $D = 1$.

6.2. Proposition. *If $D \in \Delta$, $\alpha \in \pi_1(Y, y_0)$, and $x \in p^{-1}(y_0)$ then $(Dx) \cdot \alpha = D(x \cdot \alpha)$.*

Proof. Let f be a loop at y_0 representing α and lift f to a path g starting at x. Then $g(1) = x \cdot \alpha$ by definition. Look at the path $D \circ g$. It is a lift of f and starts at Dx. Thus it ends at $(Dx) \cdot \alpha$ by definition of the latter. But it ends at D of the end of g, i.e., at $D(x \cdot \alpha)$. □

Recall that the "normalizer" of a subgroup H of a group G is

$$\boxed{N(H) = \{n \in G \mid nHn^{-1} = H\}.}$$

6.3. Theorem. *Let $x_0 \in X$ be such that $p(x_0) = y_0$ and let $x \in p^{-1}(y_0)$. Then the following statements are equivalent:*

(1) $\exists D \in \Delta \ni D(x_0) = x$.
(2) $\exists \alpha \in N(p_\# \pi_1(X, x_0)) \ni x = x_0 \cdot \alpha$.
(3) $p_\# \pi_1(X, x_0) = p_\# \pi_1(X, x)$.

Proof. By Theorem 4.1 a map D covering the identity and taking the point x_0 to x exists $\Leftrightarrow p_\# \pi_1(X, x_0) \subset p_\# \pi_1(X, x)$. Similarly, a map D' exists covering the identity and taking x to $x_0 \Leftrightarrow$ the opposite inclusion holds. If both exist then $D \circ D'$ covers the identity and has a point in common with the identity map, so $D \circ D' = 1$ by Lemma 4.4. This proves the equivalence (1)$\Leftrightarrow$(3).

Now compute

$$\begin{aligned} J_{x_0 \cdot \alpha} &= \{\beta \mid (x_0 \cdot \alpha) \cdot \beta = (x_0 \cdot \alpha)\} \\ &= \{\beta \mid x_0 \cdot \alpha\beta\alpha^{-1} = x_0\} \\ &= \{\beta \mid \alpha\beta\alpha^{-1} \in J_{x_0}\}. \end{aligned}$$

Thus

$$\boxed{J_{x_0 \cdot \alpha} = \alpha^{-1} J_{x_0} \alpha.}$$

Next we prove (2)$\Rightarrow$(3): If $x = x_0 \cdot \alpha$ and $\alpha \in N(J_{x_0})$ then $J_x = J_{x_0 \cdot \alpha} = \alpha^{-1} J_{x_0} \alpha = J_{x_0}$, as claimed.

For (3)$\Rightarrow$(2), Suppose $J_{x_0} = J_x$ and $x = x_0 \cdot \alpha$. (Note that such an α exists since J is transitive on F.) Then $J_{x_0} = J_x = J_{x_0 \cdot \alpha} = \alpha^{-1} J_{x_0} \alpha$ which shows that $\alpha \in N(J_{x_0})$. □

From (2)$\Leftrightarrow$(1) of Theorem 6.3, and the last part of its proof, we get:

6.4. Corollary. *The subgroup $p_\# \pi_1(X, x_0)$ is normal in $\pi_1(Y, y_0) \Leftrightarrow \Delta$ is (simply) transitive on $p^{-1}(y_0)$.* □

6.5. Corollary. *If x ranges over $p^{-1}(y_0)$ and x_0 is one such point then $p_{\#}\pi_1(X,x)$ ranges over all conjugates of $p_{\#}\pi_1(X,x_0)$.*

PROOF. This is really a consequence of the *proof* of Theorem 6.3, namely, it is contained in the formula $J_{x_0 \cdot \alpha} = \alpha^{-1} J_{x_0} \alpha$ derived there. □

6.6. Definition. A covering map p is said to be *regular* if Δ is transitive on the fiber $p^{-1}(y_0)$, i.e., if $p_{\#}\pi_1(X,x_0)$ is normal in $\pi_1(Y,y_0)$.

The examples (1) through (4) of Section 3 are all regular. Example (5) is not regular since it is obvious that Δ is not transitive. (Indeed, Δ is clearly the trivial group in that example.)

6.7. Definition. Define a function $\Theta: N(J_{x_0}) \to \Delta$ by $\Theta(\alpha) = D_\alpha$ where D_α is that unique deck transformation such that $D_\alpha(x_0) = x_0 \cdot \alpha$.

6.8. Theorem (Classification of Deck Transformations). *The map $\Theta: N(J_{x_0}) \to \Delta$ is an epimorphism with kernel J_{x_0}. Consequently,*

$$\Delta \approx N(p_{\#}\pi_1(X,x_0))/p_{\#}\pi_1(X,x_0).$$

PROOF. First compute $D_\beta D_\alpha(x_0) = D_\beta(x_0 \cdot \alpha) = (D_\beta(x_0)) \cdot \alpha = (x_0 \cdot \beta) \cdot \alpha = x_0 \cdot (\beta\alpha) = D_{\beta\alpha}(x_0)$. Thus $D_\beta D_\alpha = D_{\beta\alpha}$, i.e., Θ is a homomorphism.

Next note that if $D \in \Delta$ then there is an $\alpha \in N(J_{x_0})$ such that $Dx_0 = x_0 \cdot \alpha = D_\alpha(x_0)$. Therefore $D = D_\alpha$, which shows that Θ is onto.

Finally we compute the kernel of Θ: $D_\alpha = 1 \Leftrightarrow x_0 \cdot \alpha = x_0$ (since $D_\alpha(x_0) = x_0 \cdot \alpha) \Leftrightarrow \alpha \in J_{x_0}$, as claimed. □

6.9. Corollary. *If the covering map $p: X \to Y$ is regular, then*

$$\Delta \approx \pi_1(Y,y_0)/p_{\#}\pi_1(X,x_0).$$ □

6.10. Corollary. *If $p: X \to Y$ is a covering map with X simply connected then*

$$\Delta \approx \pi_1(Y,y_0).$$ □

We will now discuss some examples. The covering $\mathbf{R} \to \mathbf{S}^1$ has, as deck transformations, the translations of $\mathbf{R}$ by integer amounts. Thus $\pi_1(\mathbf{S}^1) \approx \Delta \approx \mathbf{Z}$, as we already know.

Similarly, the covering of the torus by the plane has the translations by integer amounts, in both coordinates, as deck transformations, so that the fundamental group of the torus $\mathbf{S}^1 \times \mathbf{S}^1$ is $\mathbf{Z} \oplus \mathbf{Z}$, as also follows from Theorem 2.6.

Any double covering by a simply connected space has exactly two deck transformations, the identity and one "switching the sheets." Thus the fundamental group of the base space must be $\mathbf{Z}_2$. For example, $\pi_1(\mathbf{P}^n) \approx \mathbf{Z}_2$ for $n \geq 2$.

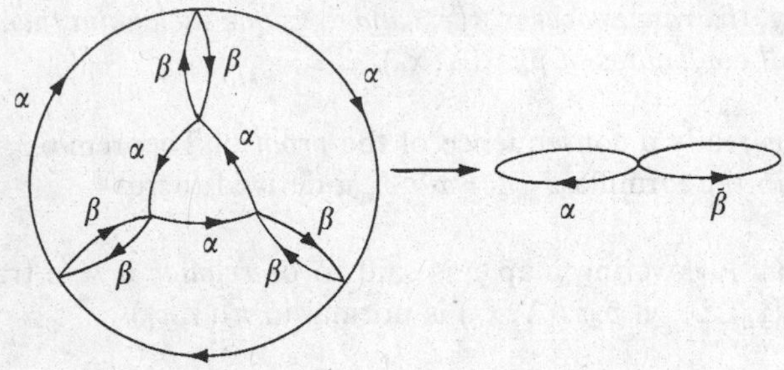

Figure III-7. A sixfold covering of the figure 8.

Note the covering illustrated in Figure III-7. The arcs labeled α, on the left, map to the one in the figure eight at the right with the indicated orientation. Similarly for the ones marked β. It is easy to see, by looking at the figure on the left, that the deck transformation group Δ is σ_3, the permutation group on three letters. If we alter the covering map by reversing the arrows α on the inner circle only, then the deck transformation group becomes $\mathbf{Z}_2 \times \mathbf{Z}_3 \approx \mathbf{Z}_6$. In both cases the covering is regular since Δ is transitive on the fiber (the six vertices on the left).

Problems

1. If $p: X \to Y$ is a covering map with X simply connected then $\pi_1(Y)$ acts on the fiber F in two ways:
   (1) by deck transformations via the isomorphism Θ of Definition 6.7; and
   (2) by the monodromy action.
   Show that these actions coincide$\Leftrightarrow\pi_1(Y)$ is abelian.

7. Properly Discontinuous Actions

Recall, from Section 15.13 of Chapter I, that an "action" of a group G on a space X is a map $G \times X \to X$, where the image of (g, x) will be denoted by gx, such that $(gh)x = g(hx)$ and $ex = x$. In this section G will have the discrete topology, which we mean to imply by calling it a "group" rather than a "topological group."

For $x \in X$, the "orbit" of x is the set $G(x) = \{gx \mid g \in G\}$. It is easy to see that two orbits are either disjoint or identical. Thus they partition the space X. The set of all orbits is denoted by X/G, with the quotient topology from the map $X \to X/G$ taking x to $G(x)$, and is called the "orbit space." Note that the canonical map $p: X \to X/G$ is open since, for $U \subset X$ open and U^* its image in X/G, then

$$p^{-1}(U^*) = \bigcup \{gU \mid g \in G\}$$

which is a union of open sets and hence is open.

7.1. Definition. An action of a group G on a space X is said to be *properly discontinuous* if each point $x \in X$ has a neighborhood U such that $gU \cap U \neq \emptyset \Rightarrow g = e$, the identity element of G.

For example, if $p: X \to Y$ is a covering map, then the group Δ of deck transformations acts properly discontinuously. Moreover, if p is regular then $X/\Delta \approx Y$.

7.2. Proposition. *If G acts properly discontinuously on the arcwise connected and locally arcwise connected Hausdorff space X then $p: X \to X/G$ is a regular covering map with deck transformation group $\Delta = G$.*

PROOF. Let $U \subset X$ be an arcwise connected open set as in Definition 7.1 and put $U^* = p(U)$ which is open as remarked above. Since $U \to U^*$ is continuous, U^* is arcwise connected. Also, the sets gU are the components of $p^{-1}(U^*)$. The maps $gU \to U^*$ are continuous, open, one–one and onto, and hence are homeomorphisms. Thus p is a covering map. Elements of G are deck transformations and act transitively on a fiber. There are no other deck transformations by Lemma 4.4. □

7.3. Corollary. *If X is simply connected and locally arcwise connected and G acts properly discontinuously on X, then $\pi_1(X/G) \approx G$.* □

7.4. Example (Lens Spaces). Consider $\mathbf{S}^{2n-1} \subset \mathbf{C}^n$ as $\{z = (z_1, \ldots, z_n) \mid \|z\| = 1\}$. Let $\epsilon = e^{2\pi i/p}$ be a primitive pth root of unity and let $q_1, \ldots, q_n$ be integers relatively prime to p. Consider $\mathbf{Z}_p = \{1, \epsilon, \epsilon^2, \ldots, \epsilon^{p-1}\}$ and let it act on $\mathbf{S}^{2n-1}$ by $\epsilon(z_1, \ldots, z_n) = (\epsilon^{q_1} z_1, \ldots, \epsilon^{q_n} z_n)$. This is properly discontinuous, as is any action by a finite group such that $gx = x \Rightarrow g = e$ (and which the reader should check). The orbit space is denoted by $L(p; q_1, \ldots, q_n)$ and is called a "lens space." By Corollary 7.3, the fundamental group of any of these spaces is $\mathbf{Z}_p$. For the classical case $n = 2$, $L(p; 1, q)$ is commonly denoted by $L(p, q)$.

7.5. Example (Klein Bottle). Consider the group of transformations of the plane generated by α and β, where $\alpha(x, y) = (x + 1, y)$ and $\beta(x, y) = (1 - x, y + 1)$. A close inspection of this action should convince the reader that the orbit space $\mathbf{R}^2/G$ is the Klein bottle. The group is the group abstractly defined as generated by elements α an β and having the single relation $\beta^{-1}\alpha\beta = \alpha^{-1}$. (This is easily checked geometrically.) This group is nonabelian, has a normal infinite cyclic subgroup (generated by α) with a quotient group also infinite cyclic generated by the image of β. By Corollary 7.3, this group is the fundamental group of the Klein bottle.

7.6. Example (Figure Eight). Let G be the free group on two letters α and β. Define a graph $X = \text{Graph}(G, \alpha, \beta)$ as follows: The vertices of X are the elements of G, so they are reduced words in α and β. The edges are of two types $(g, g\alpha)$ and $(g, g\beta)$. (Note then that there are exactly four edges abutting the vertex g, namely, $(g, g\alpha)$, $(g, g\beta)$, $(g\alpha^{-1}, g)$, and $(g\beta^{-1}, g)$.) The group element $h \in G$ acts on X by taking an edge $(g, g\alpha)$ to $(hg, hg\alpha)$ and $(g, g\beta)$ to $(hg, hg\beta)$. That is, it is the obvious action on X induced by left translation on G, the vertex set. It is clear that this action is properly discontinuous and that X/G is the figure eight space, whose two loops are the images of edges $(g, g\alpha)$ for one loop and $(g, g\beta)$ for the other loop.

(The precise description of the space X is as $(G \times V)/\sim$ where V is the graph with the three vertices e, a, and b and two edges (e, a) and (e, b) and where $\sim$ is the equivalence relation generated by $g \times a \sim g\alpha \times e$ and $g \times b \sim g\beta \times e$.)

We claim that this space X is simply connected. It suffices to show that any finite connected subgraph is contractible (since the image of a loop is in a finite connected subgraph). The proof will be done via the following two lemmas.

7.7. Lemma. *A finite connected graph with no cycles (a finite "tree") is contractible.*

PROOF. Such a graph must have a vertex which is on only one edge (or the graph is a single vertex). If the graph obtained by deleting that vertex and edge (but not the other vertex of this edge) is contractible then clearly the original graph is contractible. Thus the result follows by induction on the number of vertices. (We remark that this lemma is true without the word "finite." The reader might attempt to prove this.) □

7.8. Lemma. *The graph X of Example 7.6 has no cycles.*

PROOF. Start constructing a cycle beginning at the vertex g. The vertices one visits have to be of the form $g, g\alpha, g\alpha\beta, g\alpha\beta^2, \ldots$ That is, it is g followed by a growing *reduced* word in α, β. Thus, upon return to the vertex g the vertex we stop at is gw where w is a reduced word. Thus $g = gw$, so $w = 1$. But w is a reduced word, and in a free group no nontrivial reduced word equals 1. □

Thus, finally, we see that the fundamental group of the figure eight is the free group on two generators. The reader can prove similar results for more than two circles attached at a common point. We simply state the final result:

7.9. Theorem. *If X is the one-point union of n circles then $\pi_1(X)$ is a free group on n generators.* □

Theorem 7.9 holds for infinite n provided the correct topology is used on the union. (It should be a CW-complex, see Chapter IV.)

We shall go on to find the fundamental group of any finite connected graph in the following sequence of lemmas.

7.10. Lemma. *A finite connected graph G contains a maximal tree. Any such tree $T \subset G$ contains all the vertices of G.*

PROOF. The existence of T is obvious. If it does not contain all vertices of G then there must be an edge of G one of whose vertices is in T and the other not. But then addition of this edge to T still makes a tree, contradicting maximality. □

7.11. Definition. If G is a finite connected graph with V vertices and E edges then its *Euler characteristic* $\chi(G)$ is defined to be the integer $V - E$.

7.12. Lemma. *If T is a finite tree then $\chi(T) = 1$. If T is a maximal tree in the finite connected graph G then $\chi(G) = 1 - n$, where n is the number of edges of G not in T.*

PROOF. The first statement is an easy induction on the number of edges using the fact that a tree that is more than a single vertex has an edge with a vertex on no other edge. Removing such an edge leaves a tree with one less edge and one less vertex. The second statement is even more trivial. □

7.13. Lemma. *If G is a finite connected graph then G is homotopy equivalent to the one-point union of n circles where $n = 1 - \chi(G)$.*

PROOF. The graph G is obtained from one of its maximal trees T by attaching edges. Each of these attachments is just the mapping cone of a map of $\{0, 1\}$ to the pair of vertices of that edge. Since the vertices are in T and T is contractible by Lemma 7.7, this mapping cone is homotopy equivalent to the mapping cone of the map of $\{0, 1\}$ to a single vertex (any vertex), by Theorem 14.18 of Chapter I. By Theorem 14.19 of Chapter I, this argument can be repeated for subsequent attachments and so G is homotopy equivalent to T with n circles attached to any vertex. Since T is contractible this space is homotopy equivalent to the subspace consisting of the n circles joined at a vertex by Theorem 14.19 of Chapter I again. □

The following theorem is a direct consequence of the foregoing results:

7.14. Theorem. *If G is a finite connected graph then $\pi_1(G)$ is a free group on $1 - \chi(G)$ generators.* □

PROBLEMS

1. ⌖ Suppose that G is a finite group acting on the Hausdorff space X in such a way that $g(x) = x$, for some $x \in X$, $\Rightarrow g = e$, the identity element of G. (Such an action is called "free.") Then show that G acts properly discontinuously.

2. If $G \neq \{e\}$ acts properly discontinuously on $\mathbf{R}$, show that $G \approx \mathbf{Z}$.

3. Find an example of a free, but not properly discontinuous, action of some infinite group on some space.

4. ◆ Either prove or find a counterexample to the following statement:
If $p: X \to Y$ is a covering map and if $\phi: X \to X$ is a map such that $p \circ \phi = p$, then ϕ is a deck transformation.

8. Classification of Covering Spaces

Recall that two covering spaces X and X' of a given space Y are called equivalent if there is a homeomorphism $X \to X'$ covering the identity on Y. There is also a stronger form of equivalence, that one for which one specifies base points, which, of course, must correspond under the mappings.

8.1. Theorem. *Let Y be arcwise connected and locally arcwise connected, and suppose that Y has a simply connected covering space $\tilde{Y}$. Then the equivalence classes of covering spaces of Y with base points mapping to $y_0 \in Y$, are in one–one correspondence with subgroups of $\pi_1(Y, y_0)$. Equivalence classes without base points are in one–one correspondence with conjugacy classes of such subgroups. The correspondence is given by $X \leftrightarrow p_\# \pi_1(X)$ where $p: X \to Y$ is the covering map.*

PROOF. The second statement follows from the first and Corollary 6.5, so we shall restrict our attention to the first. We are to show the function taking a covering map, with base point, $p: (X, x_0) \to (Y, y_0)$, into the subgroup $p_\# \pi_1(X, x_0)$ of $\pi_1(Y, y_0)$ is a one–one correspondence. This function is one–one by the Lifting Theorem (Theorem 4.1). To see that it is onto, suppose that $H \subset \pi_1(Y, y_0) = J$ is an arbitrary subgroup. Since $\tilde{Y}$ is simply connected, Theorem 6.8 gives the isomorphism $\Theta: J \to \Delta$, where $\Theta(\alpha) = D_\alpha$. Under this map, the subgroup H goes to a subgroup $\Delta_H \subset \Delta$. Put $X = \tilde{Y}/\Delta_H$ which maps to Y canonically. Let x_0 be the image of the base point $\tilde{y}_0$ of $\tilde{Y}$. We wish to identify $p_\# \pi_1(X, x_0)$. Let f be a loop in X at x_0. Lifting this to $\tilde{Y}$ at $\tilde{y}_0$ gives the same path as a lifting of the projection of f to a loop at y_0 in Y. Thus the lift ends at the point $D_\alpha(\tilde{y}_0)$, where $\alpha \in \pi_1(Y, y_0)$ is the homotopy class of the projection of f to Y. But for f to be a loop in $X = \tilde{Y}/\Delta_H$, we must have that $\tilde{y}_0$ and $D_\alpha(\tilde{y}_0)$ are in the same orbit of Δ_H. This is true if and only if $D_\alpha \in \Delta_H$, and this holds if and only if $\alpha \in H$. But α is an arbitrary element of $p_\# \pi_1(X, x_0)$. Thus $H = p_\# \pi_1(X, x_0)$. □

8.2. Corollary. *If G is a free group on n generators and H is a subgroup of index p in G then H is a free group on $pn-p+1$ generators.*

PROOF. From Theorem 7.9 we know that G is the fundamental group of the bouquet Y of n circles. By Theorem 8.1 and Corollary 5.2, H is isomorphic to the fundamental group of a p-fold covering X of Y. Then $\chi(X)=p\chi(Y)=p(1-n)$. By Theorem 7.14, H is a free group on $1-\chi(X)=1-p(1-n)=pn-p+1$ generators. □

Now we turn our attention to the question: "When does an arcwise connected and locally arcwise connected Hausdorff space X have a simply connected covering space?" A necessary condition is readily at hand: If a loop is in an evenly covered subspace of X then the loop lifts to the covering space, and if that is simply connected, the loop must be homotopically trivial in the covering space. The homotopy can be composed with the map to X and so the original loop must be homotopically trivial in X. That is, "small" loops in X must be homotopically trivial in X. This leads to:

8.3. Definition. A space X is said to be *semilocally* 1*-connected* or *locally relatively simply connected* if each point $x\in X$ has a neighborhood U such that all loops in U are homotopically trivial in X (i.e., for any $u\in U$, the homomorphism $\pi_1(U,u)\to\pi_1(X,u)$ is trivial).

It turns out that this condition is also sufficient, as we now show.

8.4. Theorem. *If Y is arcwise connected and locally arcwise connected, then Y has a simply connected covering space $\Leftrightarrow$ Y is locally relatively simply connected.*

PROOF. Only the $\Leftarrow$ part is left to be proved. We must construct the simply connected covering space $\tilde{Y}$. Choose a base point $y_0\in Y$ once and for all and let

$$\tilde{Y}=\{[f]\operatorname{rel}\partial I\,|\,f\text{ is a path in }Y\text{ with }f(0)=y_0\}$$

and let $p\colon\tilde{Y}\to Y$ be $p([f])=f(1)$. We are going to topologize $\tilde{Y}$ and show that p is then the desired covering map.

Let $\mathbf{B}=\{U\subset Y\,|\,U$ is open, arcwise connnected, and relatively simply connected$\}$ and note that this is a basis for the topology of Y. If $f(1)\in U\in\mathbf{B}$ let

$$U_{[f]}=\{[g]\in p^{-1}(U)\,|\,g\simeq f*\alpha\operatorname{rel}\partial I,\text{ for some path }\alpha\text{ in }U\}$$

which is a subset of $\tilde{Y}$. (See Figure III-8.)

We shall now prove a succession of properties of these definitions, culminating in the proof of the theorem. In the discussion, all homotopies

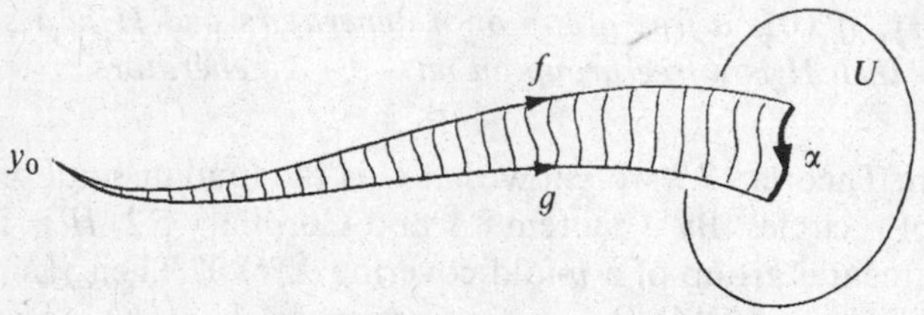

Figure III-8. The set $U_{[f]}$.

of paths starting at y_0 will be assumed, to simplify notation, to be rel ∂I unless otherwise indicated.

(1) $[g] \in U_{[f]} \Rightarrow U_{[g]} = U_{[f]}$.

To prove this, let $[h] \in U_{[g]}$. Then $h \simeq g * \beta$, for some path β in U. Since $g \simeq f * \alpha$ we conclude that $h \simeq (f * \alpha) * \beta \simeq f * (\alpha * \beta)$ showing that $[h] \in U_{[f]}$. Thus $U_{[g]} \subset U_{[f]}$. But $g \simeq f * \alpha \Rightarrow g * \alpha^{-1} \simeq f * \alpha * \alpha^{-1} \simeq f$, so $[f] \in U_{[g]}$. Consequently, $U_{[f]} \subset U_{[g]}$, proving the claim.

(2) p maps $U_{[f]}$ one–one onto U.

That this is onto is clear since U and Y are arcwise connected. To show that it is one–one, let $[g]$, $[g'] \in U_{[f]}$, which we now know is the same as $U_{[g]}$ and $U_{[g']}$. Suppose that $g(1) = g'(1)$. Since $[g'] \in U_{[g]}$ we have that $g' \simeq g * \alpha$ for some loop α in U. But then α is homotopically trivial in Y since U is relatively simply connected in Y. Thus $g' \simeq g * \alpha \simeq g * constant \simeq g$. Therefore $[g'] = [g]$, showing the map in question to be one–one.

(3) $U, V \in \mathbf{B}, U \subset V, f(1) \in U \Rightarrow U_{[f]} \subset V_{[f]}$.

This is obvious.

(4) The $U_{[f]}$ for $U \in \mathbf{B}$ and $f(1) \in U$, form a basis for a topology on $\tilde{Y}$.

Suppose $[f] \in U_{[g]} \cap V_{[g']} = U_{[f]} \cap V_{[f]}$. Let $W \subset U \cap V$ be in $\mathbf{B}$ with $f(1) \in W$. Then $[f] \in W_{[f]} \subset U_{[f]} \cap V_{[f]}$ yielding the claim.

(5) p is open and continuous.

We have $p(U_{[f]}) = U$ by (2), and these sets form bases, so it follows that p is open. Also, $p^{-1}(U) = \bigcup \{U_{[f]} | f \in p^{-1}(U)\}$ which is open for $U \in \mathbf{B}$, so p is continuous.

(6) $p: U_{[f]} \xrightarrow{\approx} U$.

This is because p is one–one, continuous, and open.

Now we have shown that p satisfies all the requirements to be a covering map except for showing that the space $\tilde{Y}$ is arcwise connected. To do this, we need the next claim.

(7) Let $F: I \times I \to Y$ be a homotopy with $F(0, t) = y_0$. Put $f_t(s) = F(s, t)$ which

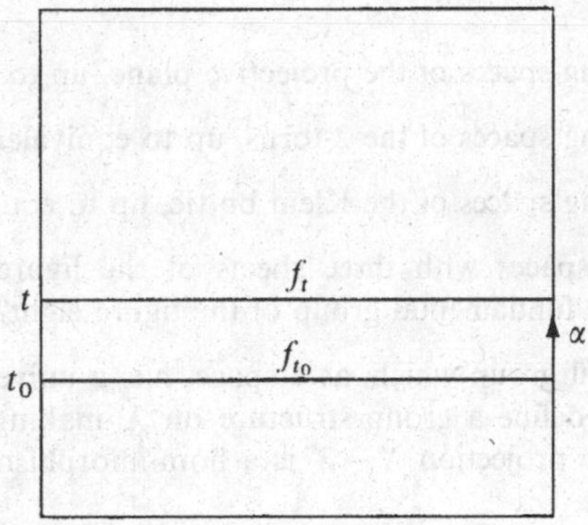

Figure III-9. Proof of item (7).

is a path starting at y_0. Let $\tilde{f}(t) = [f_t] \in \tilde{Y}$. Then $\tilde{f}$ is a path in $\tilde{Y}$ covering the path $f_t(1) = F(1, t)$ in Y.

The only thing that needs proving here is that $\tilde{f}$ is continuous. Let $t_0 \in I$. We shall prove continuity at t_0. Let $U \in \mathbf{B}$ be a neighborhood of $f_{t_0}(1)$. For t near t_0, $f_t(1) \in U$. Thus

$$\tilde{f}(t) = [f_t] \in U_{[f_{t_0}]} \qquad \text{for } t \text{ near } t_0$$

because the portion of $F(\cdot, t)$ for t in a small interval near t_0 is a homotopy rel$\{0\}$ between f_{t_0} and f_t with the right end of the homotopy describing a path α in U, i.e., $f_t \simeq f_{t_0} * \alpha$; see Figure III-9. Since $U_{[f_{t_0}]}$ maps homeomorphically to U it follows that $\tilde{f}(t)$ is continuous at t_0 because it maps to the continuous function $F(1, t)$ in U, for t near t_0.

(8) $\tilde{Y}$ is arcwise connected. (Hence p is a covering map.)

For $[f] \in \tilde{Y}$, put $F(s, t) = f(st)$. By (7) this yields a path in $\tilde{Y}$ from $\tilde{y}_0 = [1_{y_0}]$ to the arbitrary point $[f] \in \tilde{Y}$.

(9) $\tilde{Y}$ is simply connected.

Let $\alpha \in \pi_1(Y, y_0)$ and let f be a loop in Y representing α. Let $F(s, t) = f(st)$ and let $f_t(s) = F(s, t)$. Then we have the path $\tilde{f}$, where $\tilde{f}(t) = [f_t]$. This path covers f since $p(\tilde{f}(t)) = p[f_t] = f_t(1) = f(t)$.

Now $\tilde{f}(0) = [f_0] = \tilde{y}_0$. Also, by definition. $\tilde{y}_0 \cdot \alpha = \tilde{f}(1) = [f_1] = [f]$.
If $\tilde{y}_0 \cdot \alpha = \tilde{y}_0$ then

$$1 = [1_{y_0}] = \tilde{y}_0 = \tilde{y}_0 \cdot \alpha = [f] = \alpha$$

so that $\alpha = 1$ in $\pi_1(Y, y_0)$. By (4) of Section 5, we conclude that

$$\{1\} = \{\alpha \mid \tilde{y}_0 \cdot \alpha = \tilde{y}_0\} = J_{\tilde{y}_0} = p_{\#}\pi_1(\tilde{Y}, \tilde{y}_0).$$

Since $p_{\#}$ is a monomorphism by Corollary 3.7, $\tilde{Y}$ must be simply connected. $\square$

PROBLEMS

1. Describe all covering spaces of the projective plane, up to equivalence.
2. Describe all covering spaces of the 2-torus, up to equivalence.
3. Describe all covering spaces of the Klein bottle, up to equivalence.
4. Find all covering spaces with three sheets of the figure eight. What are the implications for the fundamental group of the figure eight?
5. If X is a topological group which, as a space, has a universal covering space $\tilde{X}$ then show how to define a group structure on $\tilde{X}$ making it into a topological group such that the projection $\tilde{X} \to X$ is a homomorphism. Also show that this is essentially unique.
6. If Y satisfies the hypotheses of Theorem 8.1, show that the equivalence classes, ignoring base points, of k-fold covering spaces of Y are in one–one correspondence with the equivalence classes of representations of $\pi_1(Y)$ as a transitive permutation group of $\{1, 2, \ldots, k\}$ modulo the equivalence relation induced by renumbering; i.e., modulo inner automorphisms of the symmetric group on k letters. Also interpret the correspondence geometrically.
7. The connected sum M of $\mathbf{T}^2$ and $\mathbf{P}^2$ has fundamental group $\{a, b, c \mid aba^{-1}b^{-1}c^2 = 1\}$. Find the number of regular 3-fold covering spaces of M up to equivalence.

9. The Seifert–Van Kampen Theorem ☼

In this optional section we shall prove a powerful result about the fundamental group of a union of two spaces. First, we provide some needed group-theoretic background material on free products of groups.

Let $\{G_\alpha \mid \alpha \in S\}$ be a disjoint collection of groups. Then the "free product" of these groups is denoted by $G = \bigstar\{G_\alpha \mid \alpha \in S\}$ (or by $G_1 * G_2 * \cdots$, etc.). It is defined to be the set of "reduced words"

$$w = x_1 x_2 \cdots x_n$$

where each x_i is in some G_α, no $x_i = 1$ and adjacent x_i are in different G_α's. These are multiplied by juxtaposition and then reduction (after juxtaposition, the last x_i in the first word may be in the same group as the first x_i in the second word, and they must be combined; this combination may cancel those x's out, etc.). The unity element is the empty word which we just denote by 1. The proof that this does, in fact, define a group is messy but straightforward and intuitive, so we shall omit it.

There are the canonical monomorphisms $i_\alpha: G_\alpha \to G$ whose images generate G.

9.1. Proposition. *The free product $G = \bigstar\{G_\alpha \mid \alpha \in S\}$ is characterized by the "universal property" that if H is any group and $\psi_\alpha: G_\alpha \to H$ are homomorphisms then there is a unique homomorphism $f: G \to H$ such that $f \circ i_\alpha = \psi_\alpha$ for all $\alpha \in S$.*

PROOF. For a reduced word $w = x_1 \cdots x_n$ we must have $f(w) = f(x_1) \cdots f(x_n) = \psi_{\alpha_1}(x_1) \cdots \psi_{\alpha_n}(x_n)$ and this serves as the definition of f. An easy induction on the length of the word proves f to be a homomorphism. $\square$

The "free group" on one generator x is $F_x = \{\ldots, x^{-2}, x^{-1}, 1, x, x^2, \ldots\}$. The "free group" on a set S is $F_S = \bigstar\{F_x \mid x \in S\}$ and S generates it, since F_S is the group of words in S. The reader may supply the proof to:

9.2. Proposition. *The free group F_S on a set S satisfies the "universal property" that if H is any group and $g: S \to H$ is any function then there is a unique homomorphism $f: F_S \to H$ such that $f|_S = g$.* $\square$

Suppose we are given groups G_1, G_2, and A and homomorphisms $\phi_1: A \to G_1$ and $\phi_2: A \to G_2$. Then we define the "free product with amalgamation" $G_1 *_A G_2$ as $(G_1 * G_2)/N$ where N is the normal subgroup generated by the words $\phi_1(a)\phi_2(a)^{-1}$ for $a \in A$. Otherwise stated, this consists of the words in G_1 and G_2 with the relations $\phi_1(a) = \phi_2(a)$. There is the commutative diagram:

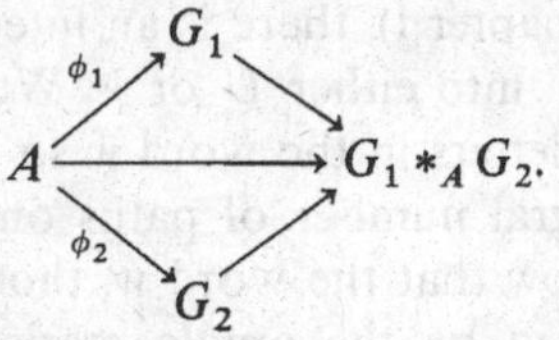

The notation $G_1 *_A G_2$ leaves something to the imagination since it does not indicate the homomorphisms ϕ_1 and ϕ_2 explicitly, and they are, of course, important to the construction.

9.3. Proposition. *The free product with amalgamation satisfies the universal property that a commutative diagram*

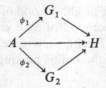

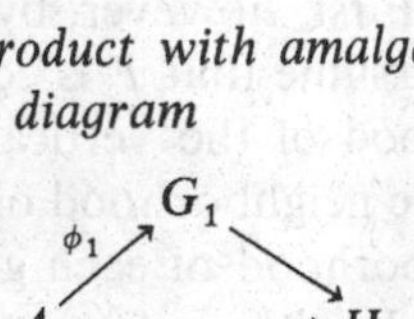

induces a unique homomorphism $G_1 *_A G_2 \to H$ through which the homomorphisms from G_1, G_2, and A to H factor.

PROOF. There is a unique extension to a homomorphism $G_1 * G_2 \to H$. Words of the form $\phi_1(a)\phi_2(a)^{-1}$ are in the kernel, and so it factors through $G_1 *_A G_2$. $\square$

9.4. Theorem (Seifert–Van Kampen). *Let $X = U \cup V$ with U, V, and $U \cap V$ all open, nonempty, and arcwise connected. Let the base point of all these be some point $x_0 \in U \cap V$. Then the canonical maps of the fundamental groups of U, V, and $U \cap V$ into that of X induce an isomorphism $\Theta: \pi_1(U) *_{\pi_1(U \cap V)} \pi_1(V) \xrightarrow{\approx} \pi_1(X)$.*

Proof. The homomorphism Θ is defined and it is a matter of showing it to be one–one and onto. To show it is onto, let f be a loop in X (at the base point in $U \cap V$). By the Lebesgue Lemma (Lemma 9.11 of Chapter I) there is an integer n such that f takes each subinterval $[i/n, (i+1)/n]$ into either U or V. If one of these is in U and the next one is in V, or vice versa, then the common point must map to $U \cap V$. Insert into the path at that point a path through $U \cap V$ running to the base point and then back again. This does not change the homotopy class of f. When that is done at all the division points i/n where it is appropriate, we then have a path that can be broken up into parts that are *loops* in U or in V. Thus the images in $\pi_1(X)$ from $\pi_1(U)$ and $\pi_1(V)$ generate the whole fundamental group, proving that Θ is onto.

Now we will prove that Θ is one–one. Suppose we have a word $w = \alpha_1\beta_1\alpha_2\cdots$ where $\alpha_1 \in \pi_1(U)$ and $\beta_i \in \pi_1(V)$. Let α_i be represented by the loop f_i in U and β_i by the loop g_i in V. Suppose that this word is in the kernel of Θ. That means that there is a homotopy $F: I \times I \to X$ with $F(s,0) = f_1 * g_1 * f_2 * \cdots$, $F(s,1) = x_0$, $F(0,t) = x_0$ and $F(1,t) = x_0$. By the Lebesgue Lemma (Lemma 9.11 of Chapter I), there is an integer n so that any square of side $1/n$ is mapped by F into either U or V. We can assume that n is a multiple of the number of letters in the word w so that each of the loops f_i and g_i consists of an integral number of paths on intervals of length $1/n$. What we must do is to show that the word w, thought of as an element of $\pi_1(U) * \pi_1(V)$, can be altered by the amalgamation relations (passing an element of $\pi_1(U \cap V)$ from one element in the word to the next), so as to end up with the trivial word. The procedure will be, after some preliminaries, to show that one can thus pass from the word represented by $F(s, i/n)$ to that represented by $F(s, (i+1)/n)$. First, however, by a homotopy not changing the word w illegally, we can assume that F is constant along the horizontal (s direction) on a neighborhood of the verticals $\{i/n\} \times I$. Then it can be made constant vertically in the neighborhood of the horizontals $I \times \{i/n\}$. It is then constant in the neighborhood of each grid point $(i/n, j/n)$. Then one can use paths in U, V, or $U \cap V$, with preference to the latter, to replace the constant disks by functions of the radius making the radii into paths to the

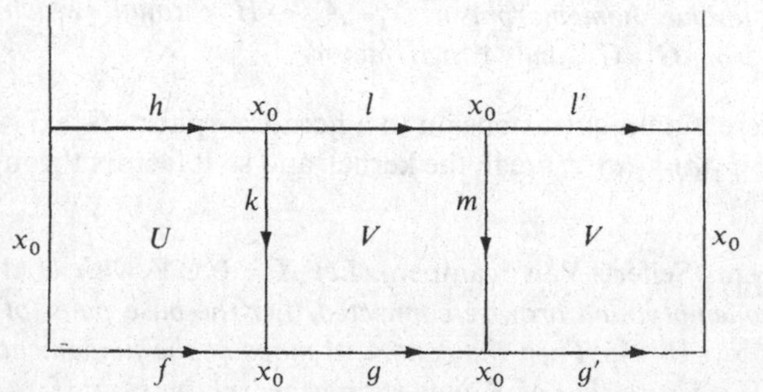

Figure III-10. Used in the proof of Theorem 9.4.

base point so that the grid points all map to the base point x_0. Now we will indicate the passage from one horizontal to the next by means of a typical example indicated in Figure III-10. In that figure f and h represent loops in U and should be thought of as giving elements in $\pi_1(U)$ of the word along the horizontal. Similarly, g, g', l, and l' represent loops in V. Now we give the manipulation of the bottom word to the top one, taking place in $\pi_1(U) *_{\pi_1(U \cap V)} \pi_1(V)$. Subscripts show what group the indicated homotopy class is meant to be in:

$$\begin{aligned}
[f]_U[g]_V[g']_V &= [hk]_U[k^{-1}lm]_V[m^{-1}l']_V \\
&= [hk]_U[k^{-1}ll']_V && \text{(multiplication in } \pi_1(V)) \\
&= ([h]_U[k]_U)[k^{-1}ll']_V \\
&= [h]_U([k]_V[k^{-1}ll']_V) && \text{(by amalgamation of } [k] \in \pi_1(U \cap V)) \\
&= [h]_U[ll']_V && \text{(multiplication in } \pi_1(V)).
\end{aligned}$$

Thus we have passed from the bottom word to the top word in Figure III-10. In such a way, we change the original word to the word along the top of F. But the latter is the trivial word, so we have shown that the original word w represents 1 in the amalgamated product. These remarks should make it clear how to complete a formal proof and the details of that will be left to the reader. □

9.5. Corollary. *If $X = U \cup V$ with U, V, and $U \cap V$ open and arcwise connected and with $U \cap V \neq \emptyset$ and simply connected, then $\pi_1(X) \approx \pi_1(U) * \pi_1(V)$.* □

For example, the figure eight is the union of two circles with whiskers (making them open sets) and with contractible intersection. Thus its fundamental group is the free product of $\mathbf{Z}$ with itself, i.e., it is the free group on two letters (as we know by different means). Similarly, we can add another circle making a three-petaled rose, and it follows that its fundamental group is the free group on three letters, etc.

9.6. Corollary. *Suppose $X = U \cup V$ with U, V, and $U \cap V \neq \emptyset$ open and arcwise connected, and with V simply connected. Then $\pi_1(X) = \pi_1(U)/N$ where N is the normal subgroup of $\pi_1(U)$ generated by the image of $\pi_1(U \cap V)$.* □

For example, consider Figure III-11, which illustrates the Klein bottle $\mathbf{K}^2$. If we remove a small disk from the center of the square, the result contracts onto the "boundary," which, under the equivalences, is the figure eight. One gets the Klein bottle back from this by pasting on a slightly larger (open) disk, which is contractible. The fundamental group of the intersection (an annulus) is $\mathbf{Z}$ generated by the obvious circle. This circle deforms to the figure eight and represents the word $aba^{-1}b$ there. Then it follows from Corollary 9.6 that $\pi_1(\mathbf{K}^2)$ is $\{a, b \mid aba^{-1}b = 1\}$. This agrees with our earlier calculation using a universal covering space.

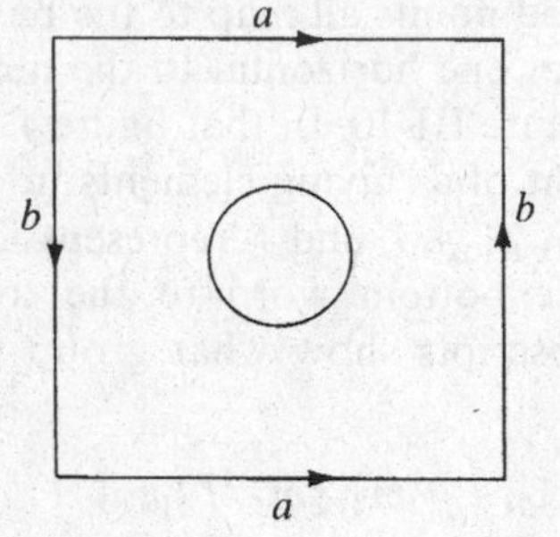

Figure III-11. The Klein bottle and a disk in it.

Problems

1. The surface H_n obtained from the sphere $\mathbf{S}^2$ by attaching n handles is homeomorphic to the space obtained from a $4n$-gon by identifications on the boundary according to the word $a_1b_1a_1^{-1}b_1^{-1}a_2b_2a_2^{-1}b_2^{-1}\cdots a_nb_na_n^{-1}b_n^{-1}$. Similarly, the surface H_n' obtained from the projective plane by attaching n handles is homeomorphic to the space obtained from a $(4n+2)$-gon by identifications on the boundary according to the above word with c^2 added. See Figure III-12. Calculate the fundamental groups of these surfaces. Also calculate the abelianized fundamental groups $\pi_1/[\pi_1, \pi_1]$. From the latter, show that none of these surfaces are homeomorphic to any of the others. (H_n is called the surface of "genus" n.)
2. Consider the space which is obtained from a disjoint union of countably many circles by identifying one point from each of them to a common base point (an infinite petaled rose). Show that the fundamental group of this is the free group on a countable set of letters. (The topology is the quotient topology with respect to the indicated identifications and is important to the validity of the result.) (*Hint*: Use the known result for a finite rose to a deduce this result directly, instead of trying to cite a theorem.)
3. Consider an annulus. Identify antipodal points on the outer circle. Also identify antipodal points on the inner circle. Calculate the fundamental group of this

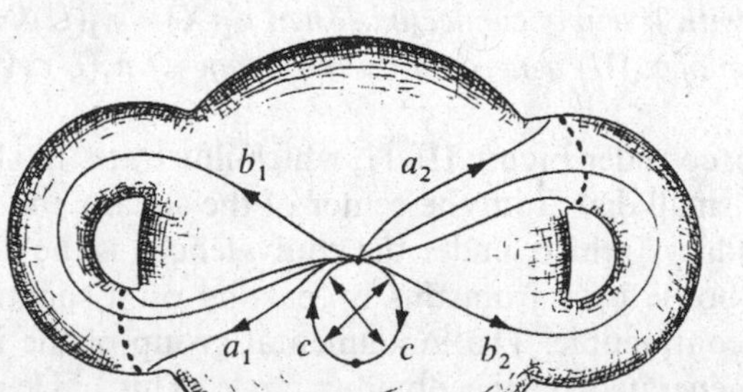

Figure III-12. Projective plane with two handles.

surface. Can this surface be homeomorphic to one of those in Problem 1? If not, prove it. If so, which one?

4. Let H''_n be the surface resulting from the Klein bottle by attaching n handles, $n \geq 0$. It is known that the H''_n together with the H_n and H'_n of Problem 1 form a complete list without repetitions of compact surfaces up to homeomorphism. Let $C_n, n \geq 1$, be the surface obtained from $\mathbf{S}^2$ by removing n disjoint open disks and identifying antipodal points on each remaining boundary circle, giving what are called "crosscaps." It is known that the H_n and the C_n form a complete list, without repetition, of all compact surfaces up to homeomorphism. Granting this, C_n must be homeomorphic to some H'_k or H''_k. Determine this correspondence.

5. Let X be the union of the unit sphere in 3-space with the straight line segment from the north pole to the south pole. Find $\pi_1(X)$.

6. Let X be the union of the unit sphere in 3-space with the unit disk in the x–y plane. Find $\pi_1(X)$.

7. Let X be the quotient space of $\mathbf{D}^2$ obtained by identifying points on the boundary that are 120° apart. Find $\pi_1(X)$.

8. Let X be the quotient space of an annulus obtained by identifying antipodal points on the outer circle, and identifying points on the inner circle which are 120° apart. Find $\pi_1(X)$.

9. Let $\mathbf{P}^2 = U_1 \cup \cdots \cup U_n$ where each U_i is homeomorphic to the plane. Put $V_i = U_1 \cup \cdots \cup U_i$ for $1 \leq i < n$. Show that there is an $i \leq n$ such that $U_i \cap V_{i-1}$ is disconnected or empty.

10. Let $X = \mathbf{D}^2 \times \mathbf{S}^1 \cup_f \mathbf{S}^1 \times \mathbf{D}^2$ where $f: \mathbf{S}^1 \times \mathbf{S}^1 \to \mathbf{S}^1 \times \mathbf{S}^1$ is the map induced by the linear map $\mathbf{R}^2 \to \mathbf{R}^2$ given by the matrix

$$\begin{pmatrix} a & b \\ c & d \end{pmatrix}$$

Compute $\pi_1(X)$ in terms of the integers a, b, c, d.

11. ◆ Consider $\mathbf{S}^3 = \mathbf{S}^1 \times \mathbf{D}^2 \cup \mathbf{D}^2 \times \mathbf{S}^1$. The intersection of these sets is the torus $\mathbf{S}^1 \times \mathbf{S}^1$. The "torus knot" $K_{p,q}$, p and q relatively prime, is the curve $px = qy$ in $\mathbf{R}^2$ projected to this torus and considered as a closed curve in $\mathbf{S}^3$. Show that

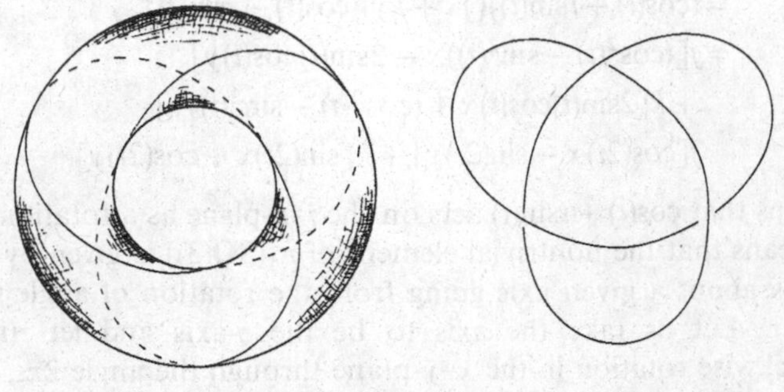

Figure III-13. Cloverleaf knot as the torus knot $K_{2,3}$.

$\pi_1(S^3 - K_{p,q}) = \{\alpha, \beta \mid \alpha^p\beta^q = 1\}$. The knot $K_{2,3}$ is also known as the "cloverleaf knot." Show that the "group $\pi_1(S^3 - K_{2,3})$ of the knot" is nonabelian by exhibiting an epimorphism of it to the dihedral group of order 6 (the symmetry group of a regular triangle), from which it follows that $K_{2,3}$ is really knotted. (It can be shown that the $K_{p,q}$ all have distinct groups, except, of course, that $K_{p,q} = K_{q,p}$.) See Figure III-13.

12. Construct a compact 4-manifold whose fundamental group is the free group on k generators, $k = 1, 2, 3, \ldots$.

13. ◆ If G is finitely presented (i.e., a group defined by a finite number of generators and relations) then show how to construct a compact 4-manifold M with $\pi_1(M) \approx G$.

10. Remarks on SO(3) ☼

The special orthogonal group **SO**(3) in three variables is homeomorphic to real projective 3-space $\mathbf{P}^3$. One way to see this is to consider the action of $\mathbf{S}^3 \approx \mathbf{Sp}(1)$ on itself by conjugation. This action is linear and preserves norm and so is an action by orthogonal transformations. It leaves the real axis fixed and so operates on its perpendicular complement, the span of i, j, and k. Thus it defines a homomorphism $\mathbf{Sp}(1) \to \mathbf{SO}(3)$. It is easy to check that the kernel of this map is $\{\pm 1\}$. It can be seen that this map is onto (an easy proof is given in Section 8 of Chapter VII, relying on some results presented in Chapter IV), and it follows that $\mathbf{SO}(3) \approx \mathbf{Sp}(1)/\{\pm 1\} \approx \mathbf{P}^3$. Therefore $\pi_1(\mathbf{SO}(3)) \approx \mathbf{Z}_2$.

Consider the semicircle

$$\{\cos(t) + i\sin(t) \in \mathbf{Sp}(1) \mid 0 \le t \le \pi\}.$$

This maps to a loop in **SO**(3) and so this projection represents the nontrivial element of $\pi_1(\mathbf{SO}(3))$. When it acts by conjugation on the quaternions we see that it fixes the complex plane and so acts only on the j–k-plane. For $z = jx + ky$ we have

$$\begin{aligned}
&(\cos(t) + i\sin(t))(jx + ky)(\cos(t) + i\sin(t))^{-1} \\
&= (\cos(t) + i\sin(t))(jx + ky)(\cos(t) - i\sin(t)) \\
&= j[(\cos^2(t) - \sin^2(t))x - 2\sin(t)\cos(t)y] \\
&\quad + k[2\sin(t)\cos(t)x + (\cos^2(t) - \sin^2(t))y] \\
&= j[\cos(2t)x - \sin(2t)y] + k[\sin(2t)x + \cos(2t)y],
\end{aligned}$$

which means that $\cos(t) + i\sin(t)$ acts on the j–k-plane as a rotation through $2t$. This means that the nontrivial element of $\pi_1(\mathbf{SO}(3))$ is given by the path of rotations about a given axis going from the rotation of angle 0 to that of angle 2π. Let us take the axis to be the z-axis and let $r(t)$ be the counterclockwise rotation in the x–y-plane through the angle $2\pi t$, so that r is a loop in **SO**(3) representing the nontrivial class of $\pi_1(\mathbf{SO}(3))$. This group is the cyclic group of order 2, so we have that the loop $r*r$ is homotopic to the constant loop at $e \in \mathbf{SO}(3)$. This loop $\lambda = r*r$ is the loop whose value $\lambda(t)$ is the rotation of angle $4\pi t$.

The fact that λ is homotopically trivial means that there is a homotopy

$$F: I \times I \to \mathbf{SO}(3)$$

with

$$\begin{aligned} F(s,0) &= e, \\ F(s,1) &= \lambda(s) = \text{rotation by } 4\pi s, \\ F(0,t) &= e, \\ F(1,t) &= e. \end{aligned}$$

Consider the map $\Phi: \mathbf{R}^3 \times I \to \mathbf{R}^3$ given by

$$\Phi(x,t) = \begin{cases} F(\rho - 1, 1 - t)\cdot x & \text{for } 1 \leq \rho \leq 2, \\ x & \text{otherwise,} \end{cases}$$

where ρ is the distance from x to the center. Thus

$$\begin{aligned} \Phi(x,0) &= \lambda(\rho - 1) && \text{for } 1 \leq \rho \leq 2, \\ \Phi(x,0) &= x && \text{for } \rho \leq 1 \text{ or } \rho \geq 2, \\ \Phi(x,1) &= x && \text{for all } x. \end{aligned}$$

Think of the parameter t as time and, for given t, think of $\Phi(\cdot, t): \mathbf{R}^3 \to \mathbf{R}^3$ as representing a contortion of space. At time $t = 0$, the contortion is given by the rotation of the sphere of radius ρ through the angle $4\pi(\rho - 1)$. At time $t = 1$, there is no longer any contortion. For $\rho \leq 1$ and $\rho \geq 2$, there is no contortion at any time. This can be illustrated by the following thought experiment. Suppose we have a steel ball of radius 1. Suppose it is placed in the center of a cavity of radius 2 in a block of steel. In the region between, put some very flexible, but not liquid, Jello attached firmly to both ball and block. Now rotate the ball about some axis until it has made two full rotations. During this rotation, the Jello must move, but we suppose it does not break. At the end, the ball is in its original position, but the Jello is all wound around. Now clamp the ball so it cannot move and give the block a kick. The Jello will move through the region between the stable ball and stable block, and it will totally untangle itself. (This action is what is described by the map $\Phi(x, t)$ where t is representing time. The wound up position is $\Phi(x, 0)$ and the unwound position is $\Phi(x, 1) = x$.)

One can actually carry out such an experiment with some minor changes. Take some strings and attach one end of them to some small object. Attach the other ends to some fixed objects in the room, say some chairs and tables and chandeliers. Do this carefully so that the strings are not tangled. Then rotate the small object one complete turn. Then try to untangle the string keeping the object stationary. You will not be able to do that. But rotate the object one more time in the same direction. You would think this would just

Figure III-14. Two full rotations are homotopic to the identity.

tangle the strings more thoroughly, but, in fact, you can now untangle them, while holding stationary the small object to which they are attached.

The photographs in Figure III-14 show another illustration of this phenomenon. In them a hand holding a cup is rotated about the vertical axis two complete turns, with palm up at all times, and at the end the arm has untwisted itself.

This phenomenon has been used in mechanical devices. It has also been cited by Dirac to explain why a rotating body can have an angular momentum of half a quantum, but of no other fraction.

CHAPTER IV
Homology Theory

Others [topological invariants] were discovered by Poincaré. They are all tied up with his homology theory which is perhaps the most profound and far reaching creation in all topology.

S. LEFSCHETZ

1. Homology Groups

One necessary annoyance when dealing with the fundamental group is keeping the base point under control. Let us discuss another approach that does not require base points, but which leads necessarily to something other than the fundamental group.

Instead of loops, consider paths $I \to X$ and "sums" of paths in a formal sense. The sums of paths are called chains, more precisely 1-chains. We are mostly interested in "closed" chains, defined as follows: If σ is a path then put $\partial\sigma = \sigma(1) - \sigma(0)$, a formal sum of signed points. Then for a chain $c = \sum \sigma_i$ let $\partial c = \sum \partial\sigma_i$. A "closed" chain is a chain c with $\partial c = 0$. A closed chain is more commonly called a "cycle." The left part of Figure IV-1 shows a 1-cycle. Instead of using homotopies to identify different cycles, we use the relation arising from regarding the boundary of a triangle to be trivial. (The boundary consists of three paths oriented consistently.) The right side of Figure IV-1 indicates the sum of the boundaries of several triangles and, allowing for cancellations of paths identical except for having the opposite orientation, the boundary is the indicated inner cycle c_1 minus the outer cycle c_0, if both of those are oriented in the clockwise direction. Accordingly, $c_1 - c_0$ is regarded as zero, i.e., c_1 and c_0 give the same equivalence class. The equivalence relation here, resulting from making boundaries zero, is called "homology."

Note that if the paths in these chains are joined to make loops then the loops are homotopic in this example. It is important to understand that this need not be the case in general. For example, if the torus is divided into triangles (it is "triangulated") and if the interior of one of the triangles is removed, then its boundary (a chain of three paths) is still homologically trivial since it bounds the sum of all the remaining triangles, oriented consistently, on the torus. However, regarded as a loop, it is *not* homotopically

Figure IV-1. Examples of chains.

trivial. Thus the relation of homology is weaker than that of homotopy. This has the disadvantage that it yields an invariant of spaces which is less sharp than that of homotopy, but the advantage, as we shall see, that it gives invariants which are easier to compute and work with.

The foregoing was somewhat vague. We shall soon give a more precise description of homology. Historically, the first method of introducing homology was to restrict attention to spaces having a "triangulation." Then the invariants (homology groups) one gets must be shown to be independent of the triangulation, a difficult matter. Also, one is restricted to dealing with spaces with such a structure. There are two ways of generalizing this to general topological spaces: One way of generalizing is to discard the idea of physical triangles, edges, etc., and substitute *maps* of a standard triangle into the space. This leads to what is known as "singular homology" and it is the approach we will take. The other way to generalize is to approximate the space, in some sense, with triangulated spaces. This leads to what is known as "Čech homology."

We now embark on the detailed description of "singular homology."

1.1. Definition. Let $\mathbf{R}^\infty$ have the standard basis $e_0, e_1, \ldots$. Then the *standard p-simplex* is $\Delta_p = \{x = \sum_{i=0}^p \lambda_i e_i \mid \sum \lambda_i = 1, 0 \le \lambda_i \le 1\}$. The λ_i are called *barycentric coordinates.*

1.2. Definition. Given points $v_0, \ldots, v_n$ in $\mathbf{R}^N$, let $[v_0, \ldots, v_n]$ denote the map $\Delta_n \to \mathbf{R}^N$ taking $\sum_i \lambda_i e_i \to \sum_i \lambda_i v_i$. This is called an *affine singular n-simplex.*

Note that the image of $[v_0, \ldots, v_n]$ is the convex span of the v_i. The v_i are *not* assumed to be independent, so this convex object may be degenerate. For example, the image of $[e_0, e_1, e_2, e_2]$ is a triangle instead of a tetrahedron.

The notation of putting a hat over one of a group of similar symbols, indicates that that one is omitted. Thus $[e_0, \ldots, \hat{e}_i, \ldots, e_p]$ denotes the affine singular $(p-1)$-simplex obtained by dropping the ith vertex (counting from 0). Note that the image of this is in Δ_p.

1.3. Definition. The affine singular simplex $[e_0, \ldots, \hat{e}_i, \ldots, e_p]$: $\Delta_{p-1} \to \Delta_p$ is called the ith *face map* and is denoted by F_i^p.

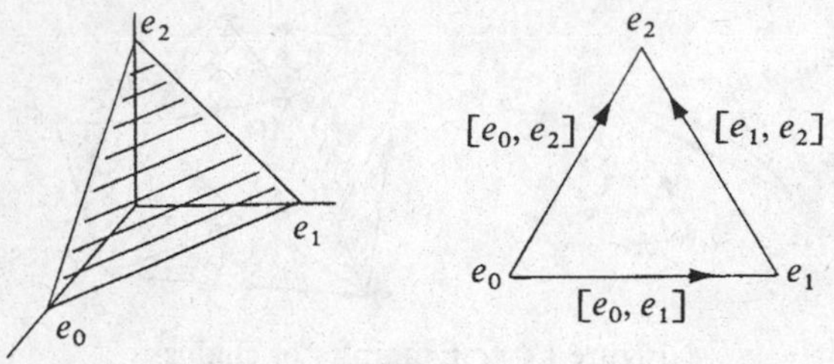

Figure IV-2. Standard 2-simplex and its faces.

Note that the ith face referred to here is the face opposite vertex number i. Figure IV-2 indicates the standard 2-simplex and its faces. Note that the faces in the figure are not oriented consistently. That is something that will have to be dealt with.

We have that $F_i^p(e_j) = e_j$ for $j < i$, and $F_i^p(e_j) = e_{j+1}$ for $j \geq i$. It is easy to see that

$$j > i \quad \Rightarrow \quad F_j^{p+1} \circ F_i^p = [e_0, \ldots, \hat{e}_i, \ldots, \hat{e}_j, \ldots, e_p],$$
$$j \leq i \quad \Rightarrow \quad F_j^{p+1} \circ F_i^p = [e_0, \ldots, \hat{e}_j, \ldots, \hat{e}_{i+1}, \ldots, e_p].$$

1.4. Definition. If X is a topological space then a *singular p-simplex* of X is a map $\sigma_p \colon \Delta_p \to X$. The *singular p-chain group* $\Delta_p(X)$ is the free abelian group based on the singular p-simplices.

Thus a p-chain (for short) in X is a formal finite sum $c = \sum_\sigma n_\sigma \sigma$ of p-simplices σ with integer coefficients n_σ. Or you can consider this a sum over *all* singular p-simplices with all but a finite number of the coefficients zero.

1.5. Definition. If $\sigma \colon \Delta_p \to X$ is a singular p-simplex, then the ith *face* of σ is $\sigma^{(i)} = \sigma \circ F_i^p$. The *boundary* of σ is $\partial_p \sigma = \sum_{i=0}^p (-1)^i \sigma^{(i)}$, a $(p-1)$-chain. If $c = \sum_\sigma n_\sigma \sigma$ is a p-chain, then we put $\partial_p c = \partial_p(\sum_\sigma n_\sigma \sigma) = \sum_\sigma n_\sigma \partial_p \sigma$. That is, ∂_p is extended to $\Delta_p(X)$ so as to be a homomorphism

$$\partial_p \colon \Delta_p(X) \to \Delta_{p-1}(X).$$

1.6. Lemma. *The composition* $\partial_p \partial_{p+1} = 0$.

PROOF. This is a simple calculation as follows:

$$\partial_p \partial_{p+1} \sigma = \partial_p \sum_{j=0}^{p+1} (-1)^j (\sigma \circ F_j^{p+1}) = \sum_{j=0}^{p+1} (-1)^j \sum_{i=0}^{p} (-1)^i (\sigma \circ F_j^{p+1}) \circ F_i^p$$
$$= \sum_{j=0}^{p+1} \sum_{i=0}^{p} (-1)^{i+j} \sigma \circ F_j^{p+1} \circ F_i^p$$

$$= \sum_{0 \le i < j \le p+1} (-1)^{i+j} \sigma \circ F_j^{p+1} \circ F_i^p + \sum_{0 \le j \le i \le p} (-1)^{i+j} \sigma \circ F_j^{p+1} \circ F_i^p$$

$$= \sum_{0 \le i < j \le p+1} (-1)^{i+j} \sigma \circ F_j^{p+1} \circ F_i^p + \sum_{0 \le j \le i \le p} (-1)^{i+j} \sigma \circ F_{i+1}^{p+1} \circ F_j^p.$$

But, if we replace $i+1$ by i, the second sum becomes the negative of the first, so this comes out zero as claimed. □

For convenience we put $\Delta_p(X) = 0$ for $p < 0$, and $\partial_p = 0$ for $p \le 0$. Thus the composition

$$\Delta_{p+1}(X) \xrightarrow{\partial_{p+1}} \Delta_p(X) \xrightarrow{\partial_p} \Delta_{p-1}(X)$$

is always zero. Chains in the kernel of ∂_p are called "p-cycles," and those in the image of ∂_{p+1} are called "p-boundaries." That is,

$$p\text{-cycles} = \ker \partial_p = Z_p(X),$$
$$p\text{-boundaries} = \operatorname{im} \partial_{p+1} = B_p(X).$$

1.7. Definition. The pth *singular homology group* of a space X is

$$H_p(X) = Z_p(X)/B_p(X) = (\ker \partial_p)/(\operatorname{im} \partial_{p+1}).$$

As said before, a p-chain c is called a "cycle" if $\partial c = 0$. Two chains c_1 and c_2 are said to be "homologous" if $c_1 - c_2 = \partial d$ for some $(p+1)$-chain d, and this is indicated by $c_1 \sim c_2$. The equivalence class of a cycle c under the relation of homology is denoted by $[\![c]\!] \in H_p(X)$.

Homology groups are obviously invariant under homeomorphism. In fact, even the chain groups are invariant, but they are too large to be useful invariants.

Clearly the chain group $\Delta_p(X)$ is the direct sum of the chain groups of the arc components of X. The boundary operator ∂ preserves this and so $H_p(X) = \bigoplus H_p(X_\alpha)$, where the X_α are the arc components of X. Accordingly, if suffices to study arcwise connected spaces.

Let us compute the homology groups of a one-point space $*$. This is not a complete triviality since there is a singular simplex in each dimension $p \ge 0$. A p-simplex σ has $p+1$ faces, all of which are the unique $(p-1)$-simplex. But they have alternating signs in $\partial_p \sigma$ starting with $+$. Thus $\partial_p \sigma = 0$ when p is odd, and ∂_p is an isomorphism $\Delta_p \to \Delta_{p-1}$ when p is even, $p \ne 0$. It follows that $H_p(*) = 0$ for $p \ne 0$ and $H_0(*) \approx \mathbf{Z}$ and is generated by the homology class of the unique 0-simplex σ of $*$. One of the problems with the definition of singular homology is that the one-point space is about the only space for which the homology groups can be computed directly out of the definition. Rest assured, however, that we shall develop tools which will make computation usually fairly easy, at least for simple spaces.

For a space X, the sequence of groups $\Delta_i(X)$ and homomorphisms $\partial: \Delta_i(X) \to \Delta_{i-1}(X)$ is called the "singular chain complex" of X.

If $H_i(X)$ is finitely generated then its rank is called the ith "Betti number" of X.

2. The Zeroth Homology Group

In this section we shall calculate $H_0(X)$ for any space X. A 0-simplex σ in X is a map from Δ_0 to X. But Δ_0 is a single point, so a 0-simplex in X is essentially the same thing as a point in X. By agreement, $\partial_0 = 0$.

Thus a 0-chain can be regarded as a formal sum $c = \sum_x n_x x$ over points $x \in X$, where $n_x = 0$ except for a finite number of x. A 0-chain is automatically a 0-cycle.

Let us define, for $c = \sum_x n_x x$, $\epsilon(c) = \sum_x n_x \in \mathbf{Z}$. Then $\epsilon: \Delta_0(X) \to \mathbf{Z}$ is a homomorphism.

If σ is a singular 1-simplex then $\partial\sigma$ is the difference of two points. Thus $\epsilon(\partial\sigma) = 0$. Since ϵ is a homomorphism and any 1-chain is a sum of 1-simplices, it follows that $\epsilon(\partial d) = 0$ for any 1-chain d.

It follows that ϵ induces a homomorphism

$$\boxed{\epsilon_*: H_0(X) \to \mathbf{Z}.}$$

Both ϵ and ϵ_* are called the "augmentation."

2.1. Theorem. *If $X \neq \varnothing$ is arcwise connected, then $\epsilon_*: H_0(X) \to \mathbf{Z}$ is an isomorphism.*

PROOF. That ϵ_* is onto is clear. Choose a point $x_0 \in X$. For any $x \in X$, let λ_x be a path from x_0 to x. This is, of course, a 1-simplex with $\partial\lambda_x = x - x_0$. Suppose $c = \sum_x n_x x$ is a 0-chain with $\epsilon_*[\![c]\!] = \epsilon(c) = \sum_x n_x = 0$, i.e., $c \in \ker \epsilon$. Then $c - \partial \sum_x n_x \lambda_x = c - \sum_x n_x \partial\lambda_x = \sum_x n_x x - \sum_x n_x(x - x_0) = (\sum_x n_x) x_0 = 0$.

Therefore $c = \partial \sum_x n_x \lambda_x$, so that $[\![c]\!] = 0$. □

2.2. Corollary. *If X is arcwise connected then $H_0(X) \approx \mathbf{Z}$ and is generated by $[\![x]\!]$ for any $x \in X$.* □

2.3. Corollary. *$H_0(X)$ is canonically isomorphic to the free abelian group based on the arc components of X.* □

3. The First Homology Group

We now give ourselves the task of "finding" the first homology group $H_1(X)$ of any space, something a good deal more difficult than for H_0. As should be suspected, the answer is in terms of the fundamental group of the space, and provides a definite link between the present subject of homology and our previous discussion of homotopy.

It suffices, of course, to treat the case in which X is arcwise connected. As in the last section, we pick a point $x_0 \in X$ once and for all. For any point $x \in X$, we again let λ_x denote some path from x_0 to x. We shall take λ_{x_0} to be the constant path.

Let $\tilde{\pi}_1$ be the abelianized fundamental group of X at x_0. That is, $\tilde{\pi}_1 = \pi_1(X, x_0)/[\pi_1, \pi_1]$, where the denominator stands for the normal subgroup generated by the commutators of elements in $\pi_1(X, x_0)$. Recall that this has the universal property that any homomorphism from $\pi_1(X, x_0)$ to any abelian group can be factored through $\tilde{\pi}_1$.

Since we shall be dealing with both homotopy classes and homology classes in this section, we remind the reader that $[f]$ stands for a homotopy class, while $[\![f]\!]$ stands for a homology class. Also recall that $f \simeq g$ stands for "f is homotopic to g," while $f \sim g$ stands for "f is homologous to g."

3.1. Lemma. *If f and g are paths in X such that $f(1) = g(0)$ then the 1-chain $f * g - f - g$ is a boundary.*

PROOF. On the standard 2-complex Δ_2 put f on the edge (e_0, e_1) and g on the edge (e_1, e_2). Then define a singular 2-simplex $\sigma: \Delta_2 \to X$ to be constant on the lines perpendicular to the edge (e_0, e_2). This results in the path $f * g$ being on the edge (e_0, e_2). Therefore $\partial\sigma = g - (f * g) + f$ as claimed. □

Note that this lemma implies that one can replace the 1-simplex $f * g$ by the 1-chain $f + g$ modulo boundaries.

3.2. Lemma. *If f is a path in X then $f + f^{-1}$ is a boundary. Also a constant path is a boundary.*

PROOF. The boundary of a constant 2-simplex is a constant 1-simplex since two of the faces cancel. If we put f on the edge (e_0, e_1) and then define a 2-simplex $\sigma: \Delta_2 \to X$ by making it constant on lines parallel to the edge (e_0, e_2), then the edge (e_1, e_2) carries f^{-1} and we have that $\partial\sigma = f + f^{-1} - \text{constant}$. Since the constant edge is a boundary, so is $f + f^{-1}$. □

3.3. Lemma. *If f and g are paths then $f \simeq g$ rel $\partial I \Rightarrow f \sim g$.*

PROOF. If $F: I \times I \to X$ is a homotopy from f to g then, since the edge $\{0\} \times I$ maps to a single point, F factors through the map $I \times I \to \Delta_2$ which collapses that edge to the vertex e_0. This provides a singular simplex σ which is f on the edge (e_0, e_1) and g on (e_0, e_2) and is constant on (e_1, e_2). Then $\partial\sigma = f - g + \text{constant}$. Since the constant edge is a boundary, so is $f - g$. □

Now if $f: I \to X$ is a loop then f, as a 1-chain, is a cycle. Thus, by Lemma 3.3, we have a well-defined function

$$\phi: \pi_1(X, x_0) \to H_1(X),$$

taking $[f]$ to $[\![f]\!]$. We claim that this is a homomorphism. To see this, let f and g be loops and note that $\phi([f][g]) = \phi[f*g] = [\![f*g]\!] = [\![f]\!] + [\![g]\!]$, by Lemma 3.1, as claimed.

Consequently, ϕ induces a homomorphism

$$\boxed{\phi_*: \tilde{\pi}_1(X, x_0) \to H_1(X).}$$

3.4. Theorem (Hurewicz). *The homomorphism ϕ_* is an isomorphism if X is arcwise connected.*

PROOF. First, we will define the function that will be the inverse of ϕ_*. Let $f \in \Delta_1(X)$ be a path. Then put $\hat{f} = \lambda_{f(0)} * f * \lambda_{f(1)}^{-1}$ which is a loop at x_0. Define $\psi(f) = [\hat{f}] \in \tilde{\pi}_1(X)$. This extends to a homomorphism

$$\psi: \Delta_1(X) \to \tilde{\pi}_1(X).$$

(Note that this is defined since we are going into $\tilde{\pi}_1(X)$ instead of $\pi_1(X)$.) We will need two further lemmas before completing the proof of Theorem 3.4.

3.5. Lemma. *The map ψ takes the group $B_1(X)$ of 1-boundaries into $1 \in \tilde{\pi}_1(X)$.*

PROOF. Let $\sigma: \Delta_2 \to X$ be a 2-simplex. Put $\sigma(e_i) = y_i$ and $f = \sigma^{(2)}$, $g = \sigma^{(0)}$ and $h = (\sigma^{(1)})^{-1}$. Then

$$\begin{aligned}
\psi(\partial\sigma) &= \psi(\sigma^{(0)} - \sigma^{(1)} + \sigma^{(2)}) \\
&= \psi(g - h^{-1} + f) = \psi(f + g - h^{-1}) \\
&= \psi(f)\psi(g)\psi(h^{-1})^{-1} \\
&= [\hat{f}][\hat{g}][((h^{-1})\hat{\ })^{-1}] \\
&= [\hat{f} * \hat{g} * ((h^{-1})\hat{\ })^{-1}] \\
&= [\lambda_{y_0} * f * \lambda_{y_1}^{-1} * \lambda_{y_1} * g * \lambda_{y_2}^{-1} * (\lambda_{y_0} * h^{-1} * \lambda_{y_2}^{-1})^{-1}] \\
&= [\lambda_{y_0} * f * \lambda_{y_1}^{-1} * \lambda_{y_1} * g * \lambda_{y_2}^{-1} * \lambda_{y_2} * h * \lambda_{y_0}^{-1}] \\
&= [\lambda_{y_0} * f * g * h * \lambda_{y_0}^{-1}] = [\mathit{constant}] = 1,
\end{aligned}$$

since $f * g * h \simeq \mathit{constant}$. (See Figure IV-3.) □

Thus ψ induces the homomorphism $\psi_*: H_1(X) \to \tilde{\pi}_1(X)$. If f is a loop then $\psi_*\phi_*[f] = \psi_*[\![f]\!] = [\lambda_{x_0} * f * \lambda_{x_0}^{-1}] = [f]$ since λ_{x_0} was chosen to be a constant path.

Thus we have shown that $\psi_*\phi_* = 1$, and it remains to show that the opposite composition is also 1.

The assignment $x \mapsto \lambda_x$ takes 0-simplices into 1-simplices and thus extends to a homomorphism $\lambda: \Delta_0(X) \to \Delta_1(X)$ by $\lambda_{\Sigma n_x x} = \lambda(\sum_x n_x x) = \sum_x n_x \lambda_x$.

3.6. Lemma. *If σ is a 1-simplex in X then the class $\phi_*\psi(\sigma)$ is represented by*

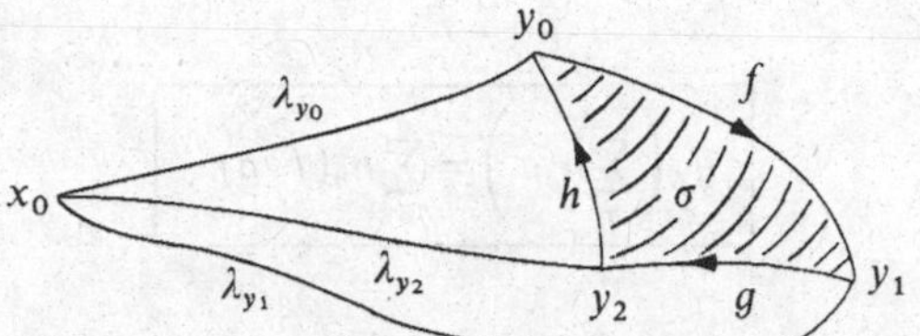

Figure IV-3. Inverse of the Hurewicz homomorphism.

the cycle $\sigma + \lambda_{\sigma(0)} - \lambda_{\sigma(1)} = \sigma - \lambda_{\partial\sigma}$. *Also, if c is a* 1*-chain then* $\phi_*\psi(c) = [\![c - \lambda_{\partial c}]\!]$. *If c is a* 1*-cycle then* $\phi_*\psi(c) = [\![c]\!]$.

PROOF. We compute

$$\begin{aligned}\phi_*\psi(\sigma) &= \phi_*[\lambda_{\sigma(0)} * \sigma * \lambda_{\sigma(1)}^{-1}] \\ &= [\![\lambda_{\sigma(0)} * \sigma * \lambda_{\sigma(1)}^{-1}]\!] \\ &= [\![\lambda_{\sigma(0)} + \sigma + \lambda_{\sigma(1)}^{-1}]\!] \\ &= [\![\lambda_{\sigma(0)} + \sigma - \lambda_{\sigma(1)}]\!]\end{aligned}$$

by Lemmas 3.1 and 3.2. The last two statements follow immediately. □

If c is a 1-cycle, then, by Lemma 3.6, $\phi_*\psi_* [\![c]\!] = [\![c]\!]$, finishing the proof of Theorem 3.4. □

Using Theorem 3.4, we can now calculate $H_1(X)$ for a large number of spaces. Here are a few such results:

3.7. Corollary. *The following isomorphisms hold, where* $\mathbf{P}^n$ *is real projective n-space, and* $\mathbf{T}^n$ *is the n-dimensional torus (the product of n circles):*

$$\begin{aligned}H_1(\mathbf{S}^1) &\approx \mathbf{Z}, \\ H_1(\mathbf{S}^n) &\approx 0 \quad \textit{for} \quad n > 1, \\ H_1(\mathbf{P}^n) &\approx \mathbf{Z}_2 \quad \textit{for} \quad n \geq 2, \\ H_1(\mathbf{T}^n) &\approx \mathbf{Z}^n, \\ H_1(\textit{Klein bottle}) &\approx \mathbf{Z} \oplus \mathbf{Z}_2, \\ H_1(\textit{Figure eight}) &\approx \mathbf{Z} \oplus \mathbf{Z}.\end{aligned}$$ □

4. Functorial Properties

Suppose that $f: X \to Y$ is a map. For any singular p-simplex $\sigma: \Delta_p \to X$ in X, the composition $f \circ \sigma: \Delta_p \to Y$ is a singular p-simplex in Y. This extends uniquely to a homomorphism:

$$\boxed{f_\Delta: \Delta_p(X) \to \Delta_p(Y)}$$

by

$$f_\Delta\left(\sum_\sigma n_\sigma \sigma\right) = \sum_\sigma n_\sigma (f \circ \sigma).$$

4.1. Proposition. *For a map $f: X \to Y$, the induced homomorphism $f_\Delta: \Delta_*(X) \to \Delta_*(Y)$ is a "chain map." That is, $f_\Delta \circ \partial = \partial \circ f_\Delta$.*

PROOF. We compute

$$\begin{aligned} f_\Delta(\partial\sigma) &= f_\Delta\left(\sum_i (-1)^i \sigma^{(i)}\right) \\ &= \sum_i (-1)^i f \circ \sigma^{(i)} \\ &= \sum_i (-1)^i f \circ \sigma \circ F_i \\ &= \sum_i (-1)^i (f \circ \sigma)^{(i)} \\ &= \partial(f \circ \sigma) = \partial(f_\Delta(\sigma)). \end{aligned}$$

This shows that $f_\Delta \circ \partial = \partial \circ f_\Delta$ on generators, and thus these homomorphisms must also be identical on the entire group $\Delta_*(X)$. □

4.2. Corollary. *A map $f: X \to Y$ induces homomorphisms $f_*: H_p(X) \to H_p(Y)$ defined by $f_*[\![c]\!] = [\![f_\Delta(c)]\!]$. Moreover*

$$(f \circ g)_* = f_* \circ g_*$$

and

$$1_* = 1.$$

That is, H_* is a "functor."

PROOF. If $c \in \Delta_p(X)$ is a cycle then so is $f_\Delta(c)$, since $\partial f_\Delta(c) = f_\Delta(\partial c) = f_\Delta(0) = 0$. If $c \sim c'$ then $c' = c + \partial d$ for some chain $d \in \Delta_{p+1}(X)$ and so $f_\Delta(c') = f_\Delta(c + \partial d) = f_\Delta(c) + f_\Delta(\partial d) = f_\Delta(c) + \partial f_\Delta(d)$. This shows that $f_\Delta(c') \sim f_\Delta(c)$ so that $[\![f_\Delta(c')]\!] = [\![f_\Delta(c)]\!]$. Thus f_* is defined. The two equations follow immediately from the definition. □

4.3. Corollary. *If $f: X \to Y$ is a homeomorphism then $f_*: H_p(X) \to H_p(Y)$ is an isomorphism.* □

PROBLEMS

1. ⌖ If X is arcwise connected and $f: X \to X$ is any map, show that $f_*: H_0(X) \to H_0(X)$ is the identity.

2. ◇ If $f: X \to Y$ is a map and $f(x_0) = y_0$ then show that the diagram

$$\begin{array}{ccc} \pi_1(X, x_0) & \xrightarrow{f_\#} & \pi_1(Y, y_0) \\ \downarrow \phi_X & & \downarrow \phi_Y \\ H_1(X) & \xrightarrow{f_*} & H_1(Y) \end{array}$$

commutes, where ϕ_X and ϕ_Y are the Hurewicz homomorphisms.

3. If $f: X \to Y$ is a covering map then $f_\#: \pi_1(X, x_0) \to \pi_1(Y, y_0)$ is a monomorphism by covering space theory. Is it true that $f_*: H_1(X) \to H_1(Y)$ is a monomorphism? Give either a proof or a counterexample.

5. Homological Algebra

For a space X we defined the "singular chain complex" $(\Delta_*(X), \partial)$ of X. From that point, the definition of the homology groups $H_p(X)$ and some simple properties were derived completely algebraically. Such "chain complexes" can, and will, occur in other contexts. Accordingly, it is very useful to abstract the algebraic part of the process, in order to apply it to future situations. We will begin that in this section. At the end of the section are some applications to singular homology.

5.1. Definition. A *graded group* is a collection of abelian groups C_i indexed by the integers.

5.2. Definition. A *chain complex* is a graded group $\{C_i\}$ together with a sequence of homomorphisms $\partial: C_i \to C_{i-1}$ such that $\partial^2: C_i \to C_{i-2}$ is zero. The operator ∂ is called a *boundary operator* or *differential.*

Our only example, so far, is the singular chain complex $C_i = \Delta_i(X)$ for $i \geq 0$, and $C_i = 0$ for $i < 0$.

5.3. Definition. If $C_* = (\{C_i\}, \partial)$ is a chain complex, then we define its *homology* to be the graded group

$$H_i(C_*) = \frac{\ker \partial: C_i \to C_{i-1}}{\operatorname{im} \partial: C_{i+1} \to C_i}.$$

Thus $H_i(X) = H_i(\Delta_*(X))$.

5.4. Definition. If A_* and B_* are chain complexes then a *chain map* $f: A_* \to B_*$ is a collection of homomorphisms $f: A_i \to B_i$ such that $f \circ \partial = \partial \circ f$.

In other words, a chain map is a "ladder" of homomorphisms which commutes:

$$\begin{array}{ccccccccc} \cdots & \xrightarrow{\partial} & A_{i+1} & \xrightarrow{\partial} & A_i & \xrightarrow{\partial} & A_{i-1} & \xrightarrow{\partial} & \cdots \\ & & \downarrow f & & \downarrow f & & \downarrow f & & \\ \cdots & \xrightarrow{\partial} & B_{i+1} & \xrightarrow{\partial} & B_i & \xrightarrow{\partial} & B_{i-1} & \xrightarrow{\partial} & \cdots. \end{array}$$

A chain map $f: A_* \to B_*$ induces a homomorphism of graded groups $f_*: H_*(A_*) \to H_*(B_*)$ by $f_*[\![a]\!] = [\![f(a)]\!]$, such that $(f \circ g)_* = f_* \circ g_*$ and $1_* = 1$.

5.5. Definition. A sequence of groups $A \xrightarrow{i} B \xrightarrow{j} C$ is called *exact* if $\text{im}(i) = \ker(j)$.

Exact sequences are common and fundamental in algebraic topology. Note that an exact sequence of the form $0 \to A \xrightarrow{i} B \xrightarrow{j} C \to 0$ means that i is an isomorphism of A onto a subgroup of B and j induces an isomorphism of $B/i(A)$ onto C. Also note that to say that $k: A \to B$ is an isomorphism (onto), is the same as to say that $0 \to A \xrightarrow{k} B \to 0$ is exact.

5.6. Theorem. *A "short" exact sequence* $0 \to A_* \xrightarrow{i} B_* \xrightarrow{j} C_* \to 0$ *of chain complexes and chain maps induces a "long" exact sequence*

$$\cdots \xrightarrow{\partial_*} H_p(A_*) \xrightarrow{i_*} H_p(B_*) \xrightarrow{j_*} H_p(C_*) \xrightarrow{\partial_*} H_{p-1}(A_*) \xrightarrow{i_*} \cdots,$$

where $\partial_*[\![c]\!] = [\![i^{-1} \circ \partial \circ j^{-1}(c)]\!]$ *and is called the "connecting homomorphism".*

PROOF. The arguments in this proof are of a type called "diagram chasing" consisting, in this case, of carrying elements around in the diagram

$$\begin{array}{ccccccccc}
0 \to & A_{p+1} & \xrightarrow{i} & B_{p+1} & \xrightarrow{j} & C_{p+1} & \to 0 \\
& \downarrow \partial & & \downarrow \partial & & \downarrow \partial & \\
0 \to & A_p & \xrightarrow{i} & B_p & \xrightarrow{j} & C_p & \to 0 \\
& \downarrow \partial & & \downarrow \partial & & \downarrow \partial & \\
0 \to & A_{p-1} & \xrightarrow{i} & B_{p-1} & \xrightarrow{j} & C_{p-1} & \to 0.
\end{array}$$

We will see this one through in detail, but later such arguments will be abbreviated, since they are almost always straightforward to one with previous experience. We begin by checking that the definition given for ∂_* really does define a unique homomorphism $H_p(C_*) \to H_{p-1}(A_*)$. Suppose given $c \in C_p$ such that $\partial c = 0$. Since j is onto there is a $b \in B_p$ with $c = j(b)$. Then $j(\partial b) = \partial(j(b)) = \partial(c) = 0$. By exactness there is a unique element $a \in A_{p-1}$ such that $i(a) = \partial b$. Then $i(\partial a) = \partial(i(a)) = \partial\partial b = 0$. Thus $\partial a = 0$ since i is a monomorphism. Therefore $[\![a]\!] \in H_{p-1}(A_*)$ is defined. As indicated $\partial_*[\![c]\!]$ is defined to be $[\![a]\!]$.

We must show that this does not depend on the choices of b and of c within its homology class. First suppose $c = j(b')$, so that $j(b - b') = 0$. Then $b - b' = i(a_0)$ for some $a_0 \in A_p$. Thus $\partial b - \partial b' = \partial(i(a_0)) = i(\partial a_0)$. But the left-hand side of this equation is $\partial b - \partial b' = i(a) - i(a') = i(a - a')$. It follows that $a - a' = \partial a_0$ and so $a \sim a'$ as desired.

We now consider the effect of changing c within its homology class. Let $c' = c + \partial c''$. Then we can set $c = j(b)$ and $c'' = j(b'')$. Let $b' = b + \partial b''$. We calculate $j(b') = j(b) + j(\partial b'') = c + \partial c'' = c'$. But $\partial b' = \partial b + \partial\partial b'' = \partial b$ and so ∂b and $\partial b'$, being equal, pull back to the same thing under i^{-1}.

One also must show that ∂_* is a homomorphism. But for two classes c and c', we can trace the definition back for both and, at any stage, the addition of the elements going into the definition work for the sum $c + c'$, which proves this contention.

Finally, we must show that the indicated "long" sequence is exact. First, we show it is of "order two" (i.e., the composition of adjoining homomorphisms is zero). There are three cases. First, $j_* i_* = (j \circ i)_* = 0_* = 0$.

Second, consider $\partial_* j_* [\![b]\!]$, where $\partial b = 0$. By definition of ∂_*, this is obtained by taking b, then applying ∂ to it, giving $\partial b = 0$, and pulling this back (to 0) to A_*.

Third, consider $i_* \partial_*$. This is the result of taking an element of C_*, pulling it back to B_*, taking ∂ of it, pulling that back to A_* (this being the ∂_* part) and then pushing this out to B_* again. But this element of B_* is, by construction, ∂ of something, which has homology class 0, as claimed.

Now we must show that an element in the kernel of one of the maps i_*, j_*, or ∂_* is in the image of the preceding one. Again the proof of this has three cases.

First, we show the exactness at $H_*(B_*)$. Suppose that $j_* [\![b]\!] = 0$. This means that $j(b) = \partial c$ for some $c \in C_*$. Let $b' \in B_*$ be such that $j(b') = c$. Then

$$j(b - \partial b') = j(b) - j(\partial b') = \partial c - \partial(j(b')) = \partial c - \partial c = 0.$$

This shows that we could have taken the representative b of its homology class to be such that $j(b) = 0$. For this choice, then, $b = i(a)$ for some $a \in A_*$ (and $\partial a = 0$ since it maps, by the monomorphism i, into $\partial b = 0$). Thus $[\![b]\!] = i_* [\![a]\!]$ as claimed.

Second, for the exactness at $H_*(A_*)$, suppose that $i_* [\![a]\!] = 0$. Then $i(a) = \partial b$ for some $b \in B_*$. Then put $c = j(b)$. We have $\partial c = \partial j(b) = j(\partial b) = j(i(a)) = 0$. Thus c represents a homology class, and by construction of ∂_*, $\partial_* [\![c]\!] = [\![a]\!]$.

Third, for the exactness at $H_*(C_*)$, suppose that $\partial_* [\![c]\!] = 0$. Then for an element $b \in B_*$ for which $j(b) = c$, there is an $a \in A_*$ such that $i(a) = \partial b$, by the construction of ∂_*, and a must be a boundary since it represents $\partial_* [\![c]\!] = 0$. Thus let $a = \partial a'$. Then $\partial i(a') = i(\partial a') = i(a) = \partial b$. Accordingly, $\partial(b - i(a')) = 0$ and $j(b - i(a')) = c - 0 = c$. Therefore $j_* [\![b - i(a')]\!] = [\![c]\!]$ as required. □

A short exact sequence $0 \to A \xrightarrow{i} B \xrightarrow{j} C \to 0$ is said to "split" if there exists an idempotent ($\phi^2 = \phi$) endomorphism $\phi: B \to B$ whose kernel *or* image equals $\operatorname{im}(i) = \ker(j)$. Then $1 - \phi$ is also idempotent and $\ker(1 - \phi) = \{b \mid b = \phi(b)\} = \{b \mid b = \phi(b')\} = \operatorname{im}(\phi)$, since, if $b = \phi(b')$, then $\phi(b) = \phi^2(b') = \phi(b') = b$. Also $B = \ker(\phi) \oplus \operatorname{im}(\phi)$ given by $b = (1 - \phi)(b) + \phi(b)$. If $\ker(\phi) = \operatorname{im}(i)$ (otherwise use $1 - \phi$ instead) then we have $\{0\} = \operatorname{im}(\phi) \cap \ker(\phi) = \operatorname{im}(\phi) \cap \ker(j)$ so that $j: \operatorname{im}(\phi) \to C$ is injective. But $j: \operatorname{im}(\phi) \to C$ is also surjective since $j: B \to C$ is surjective and kills the summand $\ker(\phi) = \ker(j)$ complementary to $\operatorname{im}(\phi)$. Therefore $B = \ker(\phi) \oplus \operatorname{im}(\phi) = \operatorname{im}(i) \oplus \operatorname{im}(\phi) \approx \operatorname{im}(i) \oplus C$ via $1 \oplus j$.

Conversely, if $\operatorname{im}(i)$ is a direct summand of B then the projection $\phi: B \to \operatorname{im}(i)$ is a splitting.

If $0 \to A \xrightarrow{i} B \xrightarrow{j} C \to 0$ is exact and if $s: B \to A$ satisfies $s \circ i = 1_A$ then $\phi = i \circ s$ is idempotent and $\operatorname{im}(\phi) = \operatorname{im}(i)$ so that the sequence splits. Conversely, a splitting provides such a left inverse s of i.

If $0 \to A \xrightarrow{i} B \xrightarrow{j} C \to 0$ is exact and if $t: C \to B$ satisfies $j \circ t = 1_C$ then $\phi = t \circ j$ is idempotent and $\ker(\phi) = \ker(j) = \operatorname{im}(i)$ so that the sequence splits. Conversely, a splitting provides such a right inverse t of j.

Therefore, to specify a splitting, either of the maps s or t, as above, will do. Both are called "splitting maps."

If $0 \to A \xrightarrow{i} B \xrightarrow{j} C \to 0$ is exact and if C is free abelian, then the sequence splits since a splitting $t: C \to B$ can be defined by just specifying its value on a basis element to be any preimage under j.

Similar remarks also hold in the noncommutative case as the reader should show.

5.7. Example. Let $A \subset X$ be a pair of spaces. Clearly $\Delta_i(A)$ is a subgroup of $\Delta_i(X)$ and the inclusion is a chain map. Let $\Delta_i(X, A) = \Delta_i(X)/\Delta_i(A)$. Then

$$0 \to \Delta_*(A) \to \Delta_*(X) \to \Delta_*(X, A) \to 0$$

is an exact sequence of chain complexes and chain maps. We define the "relative homology" of (X, A) to be

$$\boxed{H_p(X, A) = H_p(\Delta_*(X, A)).}$$

Then we have an induced "exact homology sequence of the pair (X, A)":

$$\boxed{\cdots \xrightarrow{\partial_*} H_p(A) \xrightarrow{i_*} H_p(X) \xrightarrow{j_*} H_p(X, A) \xrightarrow{\partial_*} H_{p-1}(A) \xrightarrow{i_*} \cdots.}$$

Note that the group $\Delta_p(X, A) = \Delta_p(X)/\Delta_p(A)$ is free abelian since it can be seen to be isomorphic to the free group generated by the singular p-simplices of X whose images are not completely in A. This gives a splitting $\Delta_*(X, A) \to \Delta_*(X)$. It is important to realize, however, that this splitting is *not* a chain map (why not?) and so it *does not* induce a map in homology.

5.8. Example (Homology with Coefficients). Let G be an abelian group. Then the tensor product $\Delta_*(X) \otimes G$ is a chain complex with the differential $\partial \otimes 1$.

[Those not familiar with tensor products can regard $\Delta_p \otimes G$ as the group of finite formal sums $\sum_\sigma g_\sigma \sigma$ with $g_\sigma \in G$. All the properties of the tensor product we will use here are easily verified for this definition. Tensor products are discussed in detail in Section 6 of Chapter V, and are not really needed until then.]

We define homology groups with coefficients in G by

$$\boxed{H_p(X;G) = H_p(\Delta_*(X) \otimes G).}$$

For $A \subset X$, the sequence

$$0 \to \Delta_*(A) \otimes G \to \Delta_*(X) \otimes G \to \Delta_*(X,A) \otimes G \to 0 \tag{1}$$

is exact because of the splitting map $\Delta_*(X,A) \to \Delta_*(X)$. Define

$$\boxed{H_p(X,A;G) = H_p(\Delta_*(X,A) \otimes G).}$$

Then (1) induces the long exact sequence

$$\boxed{\cdots \to H_p(A;G) \to H_p(X;G) \to H_p(X,A;G) \to H_{p-1}(A;G) \to \cdots.}$$

If $0 \to G' \to G \to G'' \to 0$ is an exact sequence of abelian groups then

$$0 \to \Delta_*(X) \otimes G' \to \Delta_*(X) \otimes G \to \Delta_*(X) \otimes G'' \to 0$$

is exact, since $\Delta_*(X)$ is free abelian. Consequently, there is the long exact sequence

$$\boxed{\cdots \to H_p(X;G') \to H_p(X;G) \to H_p(X;G'') \to H_{p-1}(X;G') \to \cdots}$$

and similarly for the relative groups $H_*(X,A;\cdot)$. The connecting homomorphism $H_p(X;G'') \to H_{p-1}(X;G')$ is sometimes called the "Bockstein" homomorphism in this case, although that appellation is usually reserved for the special cases of the coefficient sequences $0 \to \mathbf{Z} \to \mathbf{Z} \to \mathbf{Z}_p \to 0$ and $0 \to \mathbf{Z}_p \to \mathbf{Z}_{p^2} \to \mathbf{Z}_p \to 0$.

Especially note the case of coefficients in a field F. In this case all the groups such as $H_p(X,A;F)$ are vector spaces over F. More generally, if R is a commutative ring, then $H_p(X,A;R)$ is an R-module.

5.9. Example (Reduced Homology). Consider the chain complex C_* where $C_i = \Delta_i(X)$ for $i \geq 0$, $C_{-1} = \mathbf{Z}$, and $C_i = 0$ for $i < -1$, and where the differential $C_0 \to C_{-1}$ is the augmentation $\epsilon: \Delta_0(X) \to \mathbf{Z}$. The homology of this complex is called the "reduced homology" of X and is denoted by $\tilde{H}_*(X)$. This differs from $H_*(X)$ only in degree zero where $\tilde{H}_0(X)$ can easily be seen to be the kernel of the map $H_0(X) \to H_0(point)$ induced by the map of X to the one-point space. (For $X = \emptyset$ it also differs in degree -1 since $\tilde{H}_{-1}(\emptyset) \approx \mathbf{Z}$. However, one usually does not talk of reduced homology in this case.)

The following algebraic lemma is useful throughout algebraic topology:

5.10. Lemma (The 5-Lemma). *If the following diagram is commutative and has exact rows, and if f_1, f_2, f_4, and f_5 are isomorphisms then f_3 is also an* *isomorphism*:

$$\begin{array}{ccccccccc} A_1 & \longrightarrow & A_2 & \longrightarrow & A_3 & \longrightarrow & A_4 & \longrightarrow & A_5 \\ \approx\downarrow f_1 & & \approx\downarrow f_2 & & \downarrow f_3 & & \approx\downarrow f_4 & & \approx\downarrow f_5 \\ B_1 & \longrightarrow & B_2 & \longrightarrow & B_3 & \longrightarrow & B_4 & \longrightarrow & B_5. \end{array}$$

PROOF. This is a fairly straightforward diagram chase. First suppose $a_3 \in \ker(f_3)$. Then a_3 maps into $\ker(f_4) = 0$, so that a_3 comes from some $a_2 \in A_2$ by exactness. If we push a_2 to $b_2 \in B_2$ then that goes to 0 in B_3 and thus comes from some $b_1 \in B_1$, and in turn that can be lifted to $a_1 \in A_1$. But a_1 maps to a_2, since the images of these in B_2 are equal. But then a_1 maps to a_3 and so the latter is 0 by exactness. This shows that f_3 is a monomorphism.

Now, forgetting the above notation, let $b_3 \in B_3$. Map this to $b_4 \in B_4$ and pull it up to $a_4 \in A_4$. This must map to 0 in A_5 since it goes to 0 in B_5. By exactness, there is an $a_3 \in A_3$ mapping to a_4. If we map this to B_3 and subtract it from the original b_3 we conclude that this goes to 0 in B_4. Accordingly, we may as well assume that the original b_3 maps to 0 in B_4, and hence comes from some $b_2 \in B_2$. Pulling this up to A_2 and pushing it into A_3 gives us an element that maps to b_3, showing that f_3 is onto. □

5.11. Definition. A space X is said to be *acyclic* if $\tilde{H}_*(X) = 0$.

Note that $\emptyset$ is not acyclic since $\tilde{H}_{-1}(\emptyset) \neq 0$.

PROBLEMS

1. Multiplication by the prime $p: \mathbf{Z} \to \mathbf{Z}$ fits in a short exact sequence

$$0 \to \mathbf{Z} \xrightarrow{p} \mathbf{Z} \to \mathbf{Z}_p \to 0.$$

Use this to derive the natural split exact sequence

$$0 \to \frac{H_n(X)}{pH_n(X)} \to H_n(X; \mathbf{Z}_p) \to \ker\{p: H_{n-1}(X) \to H_{n-1}(X)\} \to 0.$$

(The splitting is not natural.)

2. ◇ The proof of the 5-lemma did not really use the full strength of the assumptions. Find the minimal assumptions on the homomorphisms f_1, f_2, f_4, and f_5 needed to prove that f_3 is a monomorphism (resp., onto). Give examples showing that further weakening of the assumptions on these maps is not possible.

3. If $\emptyset \neq A \subset X$ and A is acyclic then show that $H_*(X, A) \approx \tilde{H}_*(X)$.

6. Axioms for Homology

We shall now temporarily abandon singular homology in order to present the Eilenberg–Steenrod–Milnor axioms for homology. Then we shall derive consequences and applications directly out of the axioms. Only later will we return to singular homology and show that it satisfies the axioms. This illogical approach is used in order to get as quickly as possible to some of the main applications of homology theory. It is useful, also, to force ourselves to derive some of the main consequences of homology in the "right way" instead of making ad hoc arguments with singular chains.

The axioms are not presented in their most general form, as we prefer to state them for the full category of topological spaces and maps.

6.1. Definition. A *homology theory* (on the category of all pairs of topological spaces and continuous maps) is a functor H assigning to each pair (X, A) of spaces, a graded (abelian) group $\{H_p(X, A)\}$, and to each map $f:(X, A) \to (Y, B)$, homomorphisms $f_*: H_p(X, A) \to H_p(Y, B)$, together with a natural transformation of functors $\partial_*: H_p(X, A) \to H_{p-1}(A)$, called the *connecting homomorphism* (where we use $H_*(A)$ to denote $H_*(A, \emptyset)$, etc.), such that the following five axioms are satisfied:

(1) (Homotopy axiom.)

$$f \simeq g:(X, A) \to (Y, B) \quad \Rightarrow \quad f_* = g_*: H_*(X, A) \to H_*(Y, B).$$

(2) (Exactness axiom.) For the inclusions $i: A \hookrightarrow X$ and $j: X \hookrightarrow (X, A)$ the sequence

$$\cdots \xrightarrow{\partial_*} H_p(A) \xrightarrow{i_*} H_p(X) \xrightarrow{j_*} H_p(X, A) \xrightarrow{\partial_*} H_{p-1}(A) \xrightarrow{i_*} \cdots$$

is exact.

(3) (Excision axiom.) Given the pair (X, A) and an open set $U \subset X$ such that $\bar{U} \subset \operatorname{int}(A)$ then the inclusion $k:(X - U, A - U) \hookrightarrow (X, A)$ induces an isomorphism

$$k_*: H_*(X - U, A - U) \xrightarrow{\approx} H_*(X, A).$$

(4) (Dimension axiom.) For a one-point space P, $H_i(P) = 0$ for all $i \neq 0$.

(5) (Additivity axiom.) For a topological sum $X = +_\alpha X_\alpha$ the homomorphism

$$\bigoplus (i_\alpha)_*: \bigoplus H_n(X_\alpha) \to H_n(X)$$

is an isomorphism, where $i_\alpha: X_\alpha \hookrightarrow X$ is the inclusion.

The statement that ∂_* is a "natural transformation" means that for any map $f:(X, A) \to (Y, B)$, the diagram

$$\begin{array}{ccc} H_p(X, A) & \xrightarrow{\partial_*} & H_{p-1}(A) \\ \downarrow f_* & & \downarrow f_* \\ H_p(Y, B) & \xrightarrow{\partial_*} & H_{p-1}(B) \end{array}$$

is commutative. The statement that H is a functor means that for maps $f:(X, A) \to (Y, B)$ and $g:(Y, B) \to (Z, C)$ we have $(g \circ f)_* = g_* \circ f_*$, and also $1_* = 1$, where 1 stands for any identity mapping.

6.2. Definition. For a homology theory, $H_0(P) = G$ is called the *coefficient group* of the theory, where P is a one-point space.

Note that singular theory $H_*(\Delta_*(\cdot))$ has coefficients $\mathbf{Z}$ and $H_*(\Delta_*(\cdot)\otimes G)$ has coefficient group G.

Singular homology clearly satisfies the additivity axiom.

It should be noted that there are important "homology theories" that do not satisfy all of these axioms. ("Čech homology" does not satisfy the exactness axiom, and "bordism" and "K-theories" do not satisfy the dimension axiom.)

So far, we have proved that singular homology is a functor with the natural transformation ∂_* that satisfies the exactness, dimension, and additivity axioms. Later, we shall prove the more difficult homotopy and excision axioms. For now, we will assume we have a theory that does satisfy all these axioms and shall derive many consequences from this assumption.

6.3. Proposition.

$$(X, A) \simeq (Y, B) \quad \Rightarrow \quad H_*(X, A) \approx H_*(Y, B).$$

Proof. If $f:(X, A) \to (Y, B)$ is a homotopy equivalence with homotopy inverse g then $g \circ f \simeq 1_{(X,A)}$, so that $g_* \circ f_* = 1_* = 1$. Similarly, $f_* \circ g_* = 1$. Thus $f_*: H_*(X, A) \to H_*(Y, B)$ is an isomorphism with inverse g_*. □

The following shows that the additivity axiom is needed only for infinite disjoint unions. (However, the axiom has implications for spaces which are not disjoint unions.)

6.4. Proposition. *If $i_X: X \hookrightarrow X + Y$ and $i_Y: Y \hookrightarrow X + Y$ are the inclusions, where $X + Y$ is the topological sum of X and Y, then $i_{X*} \oplus i_{Y*}: H_*(X) \oplus H_*(Y) \to H_*(X + Y)$ is an isomorphism for any homology theory satisfying axioms (1)–(4).*

Proof. Consider the exact homology sequence for the pair $(X + Y, X)$:

$$\cdots \xrightarrow{\partial_*} H_p(X) \xrightarrow{i_{X*}} H_p(X + Y) \xrightarrow{j_*} H_p(X + Y, X) \xrightarrow{\partial_*} H_{p-1}(X) \xrightarrow{i_*} \cdots.$$

The inclusion map $k: (Y, \varnothing) \hookrightarrow (X + Y, X)$ is an excision map (with $U = X$) and so there is the isomorphism $k_*: H_p(Y) \xrightarrow{\approx} H_p(X + Y, X)$. But $k = j \circ i_Y$. Accordingly, $i_{Y*} \circ k_*^{-1}$ is a splitting of the long exact sequence, as claimed. □

Reduced homology was previously defined for singular homology, but it is easy to define it from the axioms as follows: Assume that $X \neq \varnothing$ and let $\epsilon: X \to P$ be the unique map to a one-point space P. This induces $\epsilon_*: H_*(X) \to H_*(P)$. For any map $i: P \to X$ we have $\epsilon \circ i = 1$ so that ϵ_* is onto. We define $\tilde{H}_0(X) = \ker(\epsilon_*)$. That is, the reduced homology group is defined by exactness of the sequence

$$0 \to \tilde{H}_0(X) \to H_0(X) \xrightarrow{\epsilon_*} H_0(P) \to 0.$$

The group on the right is, of course, the coefficient group G. This sequence splits via i_*, but this splitting is not natural, as it depends on the inclusion i chosen. Accordingly,

$$H_0(X) \approx \tilde{H}_0(X) \oplus G \quad \text{(not natural)}.$$

We set $\tilde{H}_p(X) = H_p(X)$ for $p \neq 0$. We also set $\tilde{H}_p(X, A) = H_p(X, A)$ if $A \neq \varnothing$. Then the map $(X, A) \to (P, P)$ induces the commutative diagram:

$$\begin{array}{ccccccccc} & & \tilde{H}_0(A) & \to & \tilde{H}_0(X) & & & & \\ & \nearrow & \downarrow & & \downarrow & \searrow & & & \\ H_1(X, A) & \to & H_0(A) & \to & H_0(X) & \to & H_0(X, A) & \to & H_{-1}(A) \\ \downarrow & & \downarrow & & \downarrow & & \downarrow & & \\ 0 = H_1(P, P) & \to & H_0(P) & \to & H_0(P) & \to & H_0(P, P) = 0. & & \end{array}$$

A diagram chase shows the *reduced* homology sequence is exact. The following result is immediate from Proposition 6.3 and the definition of reduced homology.

6.5. Theorem. *If X is contractible then $\tilde{H}_*(X) = 0$.* □

Note that this is just a convenient way of saying that $H_p(X) = 0$ for $p \neq 0$ and $H_0(X) = G$.

Let us denote by $\mathbf{D}^n_+$ the closed upper hemisphere of the sphere $\mathbf{S}^n$.

6.6. Theorem. *For $n \geq 0$, we have*

(S_n)
$$\tilde{H}_i(\mathbf{S}^n) \approx \begin{cases} G & \text{if } i = n, \\ 0 & \text{if } i \neq n, \end{cases}$$

(D_n)
$$H_i(\mathbf{D}^n, \mathbf{S}^{n-1}) \approx \begin{cases} G & \text{if } i = n, \\ 0 & \text{if } i \neq n, \end{cases}$$

(R_n)
$$H_i(\mathbf{S}^n, \mathbf{D}^n_+) \approx \begin{cases} G & \text{if } i = n, \\ 0 & \text{if } i \neq n. \end{cases}$$

Proof. The labels (S_n), (D_n), and (R_n) stand for these statements which will all be proved by a recursive procedure. First we shall prove (R_0). This follows from the excision and dimension axioms giving $H_i(\mathbf{S}^0, \mathbf{D}^0_+) \approx H_i(P) \approx G$ for $i = 0$, and 0 otherwise.

Next we shall show that $(R_n) \Leftrightarrow (S_n)$. This follows from the exact homology sequence of the inclusion map $\mathbf{D}^n_+ \to \mathbf{S}^n$:

$$0 = \tilde{H}_i(\mathbf{D}^n_+) \to \tilde{H}_i(\mathbf{S}^n) \to H_i(\mathbf{S}^n, \mathbf{D}^n_+) \to \tilde{H}_{i-1}(\mathbf{D}^n_+) = 0.$$

Similarly, $(D_n) \Leftrightarrow (R_n)$ follows from

$$H_i(\mathbf{S}^n, \mathbf{D}^n_+) \approx H_i(\mathbf{S}^n - U, \mathbf{D}^n_+ - U) \approx H_i(\mathbf{D}^n_-, \mathbf{S}^{n-1}),$$

where U is a small disk neighborhood of the north pole of $\mathbf{S}^n$, and the first isomorphism is by the excision axiom and the second by Proposition 6.3.

Next, the exact sequence of the pair $(\mathbf{D}^n, \mathbf{S}^{n-1})$ in its reduced form is

$$\tilde{H}_i(\mathbf{D}^n) \to H_i(\mathbf{D}^n, \mathbf{S}^{n-1}) \to \tilde{H}_{i-1}(\mathbf{S}^{n-1}) \to \tilde{H}_{i-1}(\mathbf{D}^n).$$

The two groups on the ends are zero since $\mathbf{D}^n$ is contractible, and so the middle map is an isomorphism, proving $(D_n) \Leftrightarrow (S_{n-1})$.

Finally, having proved (R_0) and the following implications, all claims are proved:

$$(D_0) \Leftrightarrow (R_0) \Leftrightarrow (S_0) \Rightarrow (D_1) \Rightarrow (R_1) \Rightarrow (S_1) \Rightarrow (D_2) \Rightarrow \cdots. \qquad \square$$

6.7. Corollary. *The sphere* $\mathbf{S}^{n-1}$ *is not a retract of the disk* $\mathbf{D}^n$.

PROOF. (This was proved by smooth manifold methods in Theorem 11.11 of Chapter II, but we now give a totally independent, and very simple, proof via homology.) If $r: \mathbf{D}^n \to \mathbf{S}^{n-1}$ is a retraction map, and $i: \mathbf{S}^{n-1} \to \mathbf{D}^n$ is the inclusion, then $r \circ i = 1$. Thus the composition

$$G = \tilde{H}_{n-1}(\mathbf{S}^{n-1}) \xrightarrow{i_*} \tilde{H}_{n-1}(\mathbf{D}^n) \xrightarrow{r_*} \tilde{H}_{n-1}(\mathbf{S}^{n-1}) = G$$

factors the identity map $1 = r_* \circ i_*: G \to G$ through the middle group which is 0. This implies that the coefficient group G is zero. Consequently, any homology theory with nonzero coefficients, such as the integers, gives a contradiction. $\square$

The Brouwer Fixed Point Theorem (Theorem 11.12 of Chapter II), follows from this by the same simple geometric argument given in Chapter II.

The reader should note the form of the proof of Corollary 6.7. The geometric assumption that there is a retraction can be stated in terms of maps and their compositions (the identity map of the sphere to itself factors through the disk). This statement translates, by applying the homology functor, to an analogous statement about groups and homomorphisms and their compositions. Since the resulting statement about groups is obviously false, the original one about spaces must also be false. This type of argument is typical of applications of homology theory.

6.8. Definition. If $f: \mathbf{S}^n \to \mathbf{S}^n$ is a map, we let $\deg(f)$ be that integer such that $f_*(a) = (\deg(f))a$ for all $a \in \tilde{H}_n(\mathbf{S}^n; \mathbf{Z}) \approx \mathbf{Z}$.

The following is a triviality:

6.9. Proposition. *If* $f, g\colon \mathbf{S}^n \to \mathbf{S}^n$ *then* $\deg(f \circ g) = \deg(f)\deg(g)$. □

6.10. Proposition. *For* $\mathbf{S}^n \subset \mathbf{R}^{n+1}$ *with coordinates* $x_0, \ldots, x_n$, *let* $f\colon \mathbf{S}^n \to \mathbf{S}^n$ *be given by* $f(x_0, x_1, \ldots, x_n) = (-x_0, x_1, \ldots, x_n)$, *the reversal of the first coordinate only. Then* $\deg(f) = -1$.

PROOF. First we prove the case $n = 0$. The zero sphere $\mathbf{S}^0$ is just two points, say $\mathbf{S}^0 = \{x, y\}$, where $x = 1$ and $y = -1$ on the real axis. Note that f interchanges x and y. By Proposition 6.4, the inclusions induce the isomorphism

$$H_0(\{x\}) \oplus H_0(\{y\}) \xrightarrow{\approx} H_0(\mathbf{S}^0)$$

and f_* becomes, on the direct sum, the interchange $(a, b) \mapsto (b, a)$, where we identify the homology of all one point spaces via the unique maps between them. The map $\mathbf{S}^0 \to P$, to a one-point space, is, in homology, $(a, b) \mapsto a + b$. Hence, under this isomorphism with the direct sum, $\tilde{H}_0(\mathbf{S}^0)$ becomes $\{(a, -a) \in H_0(P) \oplus H_0(P) \approx H_0(\mathbf{S}^0)\}$. Thus, in this representation, f_* takes $(a, -a)$ to $(-a, a) = -(a, -a)$. Accordingly, f has degree -1 in this case as claimed.

Suppose we know the result on $\mathbf{S}^k$ for $k < n$. Let $\mathbf{D}^n_+ = \{(x_0, \ldots, x_n) \in \mathbf{S}^n | x_n \geq 0\}$ and $D^n_- = \{(x_0, \ldots, x_n) \in \mathbf{S}^n | x_n \leq 0\}$. Note that these are preserved by f. Consider the commutative diagram

$$\begin{array}{ccccccc}
H_n(\mathbf{S}^n) & \xrightarrow{\approx} & H_n(\mathbf{S}^n, \mathbf{D}^n_+) & \xleftarrow{\approx} & H_n(\mathbf{D}^n_-, \mathbf{S}^{n-1}) & \xrightarrow{\approx} & \tilde{H}_{n-1}(\mathbf{S}^{n-1}) \\
\downarrow f_* & & \downarrow f_* & & \downarrow f_* & & \downarrow f_* = -1 \\
H_n(\mathbf{S}^n) & \xrightarrow{\approx} & H_n(\mathbf{S}^n, \mathbf{D}^n_+) & \xleftarrow{\approx} & H_n(\mathbf{D}^n_-, \mathbf{S}^{n-1}) & \xrightarrow{\approx} & \tilde{H}_{n-1}(\mathbf{S}^{n-1})
\end{array}$$

in which the horizontal isomorphisms are from the proof of Theorem 6.6. This shows that all of the vertical maps f_* are multiplication by -1. The one on the left establishes the induction. □

The map f in Proposition 6.10 is topologically no different from the map reversing any other single coordinate. So all these maps have degree -1. On $\mathbf{S}^n \subset \mathbf{R}^{n+1}$ there are $n + 1$ coordinates. If we compose all these $n + 1$ maps, we get the antipodal map, and so that composition must have degree $(-1)^{n+1}$. Therefore we have proved:

6.11. Corollary. *The antipodal map* $-1\colon \mathbf{S}^n \to \mathbf{S}^n$ *has degree* $(-1)^{n+1}$. □

6.12. Corollary. *For n even, the antipodal map on* $\mathbf{S}^n$ *is not homotopic to the identity map.* □

6.13. Corollary. *If n is even and* $f\colon \mathbf{S}^n \to \mathbf{S}^n$ *is any map then there is a point* $x \in \mathbf{S}^n$ *such that* $f(x) = \pm x$.

PROOF. Suppose not. That is, suppose there is a map $f\colon \mathbf{S}^n \to \mathbf{S}^n$ such that

$f(x)$ is never $\pm x$. Define two homotopies F and G as follows:

$$F(x,t) = \frac{tf(x) + (1-t)x}{\|tf(x) + (1-t)x\|}, \qquad G(x,t) = \frac{-tx + (1-t)f(x)}{\|-tx + (1-t)f(x)\|}.$$

Now, $F(x,0) = x$, $F(x,1) = f(x)$, $G(x,0) = f(x)$, and $G(x,1) = -x$. Consequently, $F * G$ is a homotopy between the identity and the antipodal map on $\mathbf{S}^n$, contrary to Corollary 6.12. □

6.14. Corollary. *The sphere $\mathbf{S}^n$, for n even, does not have an everywhere nonzero (continuous) tangent vector field.*

Proof. If ξ_x is a tangent vector at $x \in \mathbf{S}^n$ which is nonzero and continuous, we can divide by its length (as a vector in $\mathbf{R}^{n+1}$) to get one of length one everywhere. Consequently, putting $f(x) = \xi_x / \|\xi_x\|$ gives a map $\mathbf{S}^n \to \mathbf{S}^n$ for which $f(x)$ is never $\pm x$ (since it is $\perp x$). This contradicts Corollary 6.13. □

It follows, for example, that at any given time there is a point on the surface of the earth where the wind velocity is zero.

Note that an odd-dimensional sphere $\mathbf{S}^{2n-1}$ does have a nonzero vector field, the field assigning to $(x_1, \ldots, x_{2n}) \in \mathbf{R}^{2n}$, the vector $(x_2, -x_1, x_4, -x_3, \ldots, x_{2n}, -x_{2n-1})$.

We end this section by proving a stronger form of the exactness axiom, namely, exactness for a triple (X, A, B).

6.15. Theorem. *If $B \subset A \subset X$ and we let $\partial_* : H_i(X,A) \to H_{i-1}(A,B)$ be the composition of $\partial_* : H_i(X,A) \to H_{i-1}(A)$ with the map $H_{i-1}(A) \to H_{i-1}(A,B)$ induced by inclusion, then the following sequence is exact, where the maps other than ∂_* come from inclusions:*

$$\cdots \xrightarrow{\partial_*} H_p(A,B) \xrightarrow{i_*} H_p(X,B) \xrightarrow{j_*} H_p(X,A) \xrightarrow{\partial_*} H_{p-1}(A,B) \xrightarrow{i_*} \cdots.$$

Proof. There is the following commutative diagram:

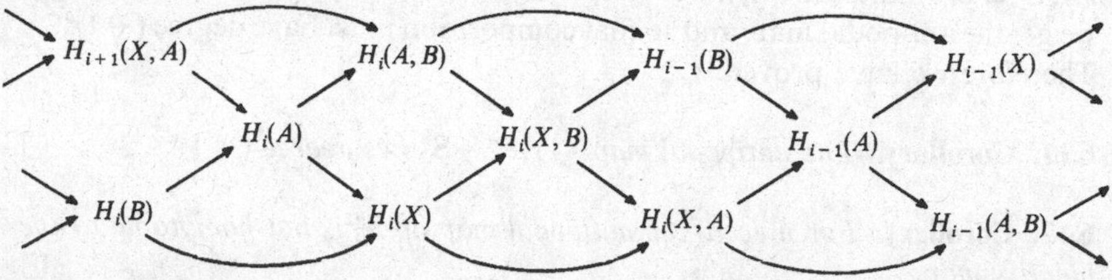

This is called a "braid" diagram. (This kind of diagram is due to Wall [1] and Kervaire.) There are four braids and three of them are exact (the exact sequences for the pairs (X,A), (X,B), and (A,B)). The fourth is the sequence of the triple (X,A,B) which we wish to prove exact. This sequence is easily seen from the commutativity to be of order two except for the composition

$$H_i(A,B)\to H_i(X,B)\to H_i(X,A).$$

But this composition factors through $H_i(A,A)=0$ (see Problem 5), so the entire sequence is of order two. The theorem now follows from the following completely algebraic fact.

6.16. Lemma (Wall). *Consider the following commutative braid diagram:*

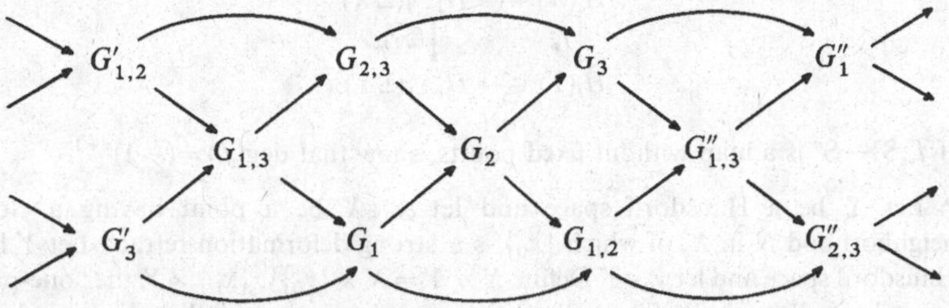

If all the (3) *braids except the one with single subscripts are exact and the fourth is of order two, then the fourth one is also exact.*

PROOF. The proof is by a diagram chase. For simplicity in doing the chase we shall introduce some special notation for it. Elements of, for example, $G'_{1,2}$ will be denoted by $a'_{1,2}$, $b'_{1,2}$, etc. To indicate that an element $a_1\in G_1$ comes from an element $b_{1,3}$, not yet defined, we just write $\exists b_{1,3}\to a_1$. To indicate that an element $a_1\in G_1$ goes to $b_{1,2}\in G_{1,2}$ we just write $a_1\to b_{1,2}$, and if $b_{1,2}$ has not yet been defined, this does so. The notation $a_2\to 0_3$ means that the element $a_2\in G_2$ goes to 0 in G_3. Now we begin the chase. In following the arguments the reader will find it helpful to diagram the elements and relations as they arise.

First we prove exactness at G_2 for the composition $G_1\to G_2\to G_3$. Thus suppose $a_2\to 0_3$. Then if $a_2\to a_{1,2}$, $a_{1,2}\to 0''_{1,3}$ by commutativity. By exactness, $\exists a_1\to a_{1,2}$. Let $a_1\to b_2$. Then $a_2\to a_{1,2}$ and $b_2\to a_{1,2}$ imply $a_2-b_2\to 0_{1,2}$. Thus $\exists a_{2,3}\to a_2-b_2$. Since $a_1\to b_2$ we have $b_2\to 0_3$. Thus $a_2-b_2\to 0_3$, and so $a_{2,3}\to 0_3$. Therefore $\exists a_{1,3}\to a_{2,3}$. Let $a_{1,3}\to b_1$. Since $a_{1,3}\to a_{2,3}\to a_2-b_2$ and $a_{1,3}\to b_1$ it follows from commutativity that $b_1\to a_2-b_2$. But $a_1\to b_2$, so $a_1+b_1\to b_2+(a_2-b_2)=a_2$, as desired.

Next we prove exactness at G_3. Thus let $a_3\to 0''_1$. Define $a_3\to a''_{1,3}$. Then $a''_{1,3}\to 0''_1$, so that $\exists a_{1,2}\to a''_{1,3}$. Since $a_3\to a''_{1,3}\to 0''_{2,3}$ we have $a_{1,2}\to 0''_{2,3}$. Consequently, $\exists a_2\to a_{1,2}$. Let $a_2\to b_3$. Then $a_3-b_3\to 0''_{1,3}$. Thus $\exists a_{2,3}\to a_3-b_3$. Let $a_{2,3}\to c_2$. Then $a_2+c_2\to b_3+(a_3-b_3)=a_3$ as claimed.

Finally we prove exactness at G_1. Thus suppose $a_1\to 0_2$. Then $a_1\to 0_{1,2}$ so $\exists a_{1,3}\to a_1$. Let $a_{1,3}\to a_{2,3}$. Then $a_{2,3}\to 0_2$ so that $\exists a'_{1,2}\to a_{2,3}$. Let $a'_{1,2}\to b_{1,3}$. Then note $b_{1,3}\to 0_1$. Then we have $a_{1,3}-b_{1,3}\to 0_{2,3}$ so $\exists a'_3\to a_{1,3}-b_{1,3}$. Now $a_{1,3}-b_{1,3}\to a_1-0_1=a_1$, so that $a'_3\to a_1$, as claimed. □

PROBLEMS

1. ◇ Define the "unreduced suspension" ΣX of a space X to be the quotient space of $I \times X$ obtained by identifying $\{0\} \times X$ and $\{1\} \times X$ to points. (This is the union of two cones on X.) For any homology theory (satisfying the axioms) show that there is a natural isomorphism $\tilde{H}_i(X) \xrightarrow{\approx} \tilde{H}_{i+1}(\Sigma X)$. Here "natural" means that for a map $f: X \to Y$, and its suspension $\Sigma f: \Sigma X \to \Sigma Y$, the following diagram commutes:

$$\begin{array}{ccc} \tilde{H}_i(X) & \xrightarrow{\approx} & \tilde{H}_{i+1}(\Sigma X) \\ \downarrow f_* & & \downarrow (\Sigma f)_* \\ \tilde{H}_i(Y) & \xrightarrow{\approx} & \tilde{H}_{i+1}(\Sigma Y). \end{array}$$

2. If $f: \mathbf{S}^n \to \mathbf{S}^n$ is a map without fixed points, show that $\deg(f) = (-1)^{n+1}$.

3. ◇ Let X be a Hausdorff space and let $x_0 \in X$ be a point having a closed neighborhood N in X, of which $\{x_0\}$ is a strong deformation retract. Let Y be a Hausdorff space and let $y_0 \in Y$. Define $X \vee Y = X \times \{y_0\} \cup \{x_0\} \times Y$, the "one-point union" of X and Y. Show that the inclusion maps induce isomorphisms $\tilde{H}_i(X) \oplus \tilde{H}_i(Y) \xrightarrow{\approx} \tilde{H}_i(X \vee Y)$, for any homology theory, whose inverse is induced by the projections of $X \vee Y$ to X and Y.

4. If n is even, show that any map $f: \mathbf{P}^n \to \mathbf{P}^n$ (real projective n-space) has a fixed point. (*Hint*: Use Corollary 6.13.)

5. ◇ Prove from the axioms that $H_i(\varnothing) = 0$ for all i, and that $H_i(X, X) = 0$ for all i and all spaces X.

7. Computation of Degrees

In the next few sections it will be important for us to be able to compute the degree of a map $\mathbf{S}^n \to \mathbf{S}^n$. We will give a method for doing that in this section for any homology theory satisfying the axioms. (Degree was defined in Definition 6.8.)

In this section, and in the remainder of the book, the coefficient group for homology is assumed to be $\mathbf{Z}$ if not otherwise indicated.

7.1. Proposition. *Let A be a nonsingular $n \times n$ matrix. As a map $\mathbf{R}^n \to \mathbf{R}^n$ it can be extended, by adding the point at infinity, to a map $\mathbf{S}^n \to \mathbf{S}^n$, and as such it has degree equal to the sign of the determinant $|A|$.*

PROOF. Since A is a homeomorphism it does extend to the one-point compactification. Clearly, if the statement on degree holds for matrices A and B then it holds for AB. Consequently, it suffices to prove it for the elementary matrices. The elementary matrix which is diagonal with one term on the diagonal different from 1 (and from 0) is homotopic, as a map, through such matrices to the identity matrix or to the matrix differing from the identity by the sign of one of the diagonal entries. These cases are taken care of by Proposition 6.10. The elementary matrix differing from the identity by having one off-diagonal entry is homotopic to the identity, taking care of that case. The remaining case of an elementary matrix obtained by interchange of two rows of the identity is topologically a reflection through a codimension-one hyperplane, so this is also taken care of by Proposition 6.10. □

7.2. Proposition. *Suppose that $f: \mathbf{S}^n \to \mathbf{S}^n$ is differentiable and that the north pole p is a regular value such that $f^{-1}(p)$ is exactly one point q. Then* $\deg(f)$ *is the sign of the Jacobian determinant at q, computed from coordinate systems at p and q that differ by a rotation of* $\mathbf{S}^n$, *i.e., operation by an element of* $\mathbf{SO}(n+1)$.

PROOF. By a rotation (and using that $\mathbf{SO}(n+1)$ is arcwise connected) we may as well assume that $q = p$. Using Proposition 7.1 we can compose f with a (one-point compactified) linear map so that the differential at p of the new map is the identity. Thus we can assume that to be the case for f, and our task becomes to show that the new f has degree one. By Taylor's Theorem, we can write $f(x) = x + g(x)$ in some local coordinates $x = (x_1, \dots, x_n)$ about p, where, for some $\epsilon > 0$, $\|g(x)\| < \frac{1}{2}\|f(x)\|$ for $0 < \|x\| \le 2\epsilon$. Define $F: \mathbf{S}^n \times I \to \mathbf{S}^n$ by

$$F(x,t) = \begin{cases} f(x) & \text{for } 2\epsilon \le \|x\|, \\ f(x) - t(2 - \|x\|/\epsilon)g(x) & \text{for } \epsilon \le \|x\| \le 2\epsilon, \\ f(x) - tg(x) & \text{for } \|x\| \le \epsilon, \end{cases}$$

and put $f_1(x) = F(x, 1)$. Then f_1 is the identity near p and maps no other point to p because of the stipulation that $\|g(x)\| < \frac{1}{2}\|f(x)\|$ for $\|x\| \le 2\epsilon$. Then there is an open disk about p so small that if D is the complementary disk in $\mathbf{S}^n$ then $f_1(D) \subset D$ and f_1 is the identity on ∂D. Such a map on a disk is homotopic to the identity rel the boundary. (Considering D as $\mathbf{D}^n$, the homotopy $G(x,t) = tx + (1-t)f_1(x)$ does this.) This homotopy extends to give a homotopy of f_1 to the identity, and so $\deg(f) = \deg(f_1) = \deg(1_{\mathbf{S}^n}) = 1$. □

7.3. Proposition. *Let $X = \mathbf{S}_1^n \vee \cdots \vee \mathbf{S}_k^n$ be the one-point union of k copies of the n-sphere, for $n > 0$. Then the homomorphism $H_n(\mathbf{S}_1^n) \oplus \cdots \oplus H_n(\mathbf{S}_k^n) \to H_n(X)$ induced by the inclusion maps, is an isomorphism whose inverse is induced by the projections $X \to \mathbf{S}_i^n$ (obtained by collapsing all the other spheres to the base point).*

PROOF. This follows from Problem 3 of Section 6 by an induction on k. □

To state our next, and main, result, we first need some notation and a description of the situation we will be concerned with. Let Y be a space with a base point y_0. Let $E_1, \dots, E_k$ be disjoint open sets in $\mathbf{S}^n$ each homeomorphic

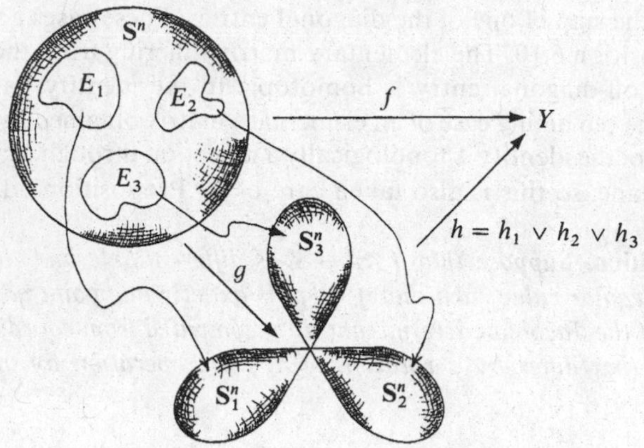

Figure IV-4. Setup for Theorem 7.4.

to $\mathbf{R}^n$, $n>0$. Let $f:\mathbf{S}^n \to Y$ be a map which takes $\mathbf{S}^n - \bigcup E_i$ to y_0. Then f factors through the space $\mathbf{S}^n/(\mathbf{S}^n - \bigcup E_i) \approx \mathbf{S}_1^n \vee \cdots \vee \mathbf{S}_k^n$ where $\mathbf{S}_i^n = \mathbf{S}^n/(\mathbf{S}^n - E_i)$:

$$f:\mathbf{S}^n \xrightarrow{g} \mathbf{S}_1^n \vee \cdots \vee \mathbf{S}_k^n \xrightarrow{h} Y.$$

Let $i_j:\mathbf{S}_j^n \hookrightarrow \mathbf{S}_1^n \vee \cdots \vee \mathbf{S}_k^n$ be the jth inclusion and let $p_j:\mathbf{S}_1^n \vee \cdots \vee \mathbf{S}_k^n \to \mathbf{S}_j^n$ be the jth projection. Then Proposition 7.3 implies that $\sum_j i_{j*}p_{j*} = 1$ on $H_n(\mathbf{S}_1^n \vee \cdots \vee \mathbf{S}_k^n)$. Let $g_j = p_j \circ g:\mathbf{S}^n \to \mathbf{S}_j^n$, and let $h_j = h \circ i_j:\mathbf{S}_j^n \to Y$, and put $f_j = h_j \circ g_j:\mathbf{S}^n \to Y$. See Figure IV-4. Note that f_j is just the map which is f on E_j and maps the complement of E_j to the base point y_0.

7.4. Theorem. *In the above situation we have* $f_* = \sum_{j=1}^k f_{j*}: H_n(\mathbf{S}^n) \to H_n(Y)$.

Proof. For $\alpha \in H_n(\mathbf{S}^n)$ we have $g_*(\alpha) = \sum_j i_{j*}p_{j*}g_*(\alpha) = \sum_j i_{j*}g_{j*}(\alpha)$ and so $f_*(\alpha) = h_*g_*(\alpha) = \sum_j h_* i_{j*}g_{j*}(\alpha) = \sum_j h_{j*}g_{j*}(\alpha) = \sum_j f_{j*}(\alpha)$. □

7.5. Corollary. *Let* $f:\mathbf{S}^n \to \mathbf{S}^n$ *be differentiable and let* $p \in \mathbf{S}^n$ *be a regular value. Let* $f^{-1}(p) = \{q_1, \ldots, q_k\}$. *Let* $d_i = \pm 1$ *be the "local degree" of* f *at* q_i, *computed as the sign of the Jacobian at* q_i *using a coordinate system at* q_i *which is a rotation of the coordinate system at* p. *Let* $d = \deg(f)$. *Then* $d = \sum_{i=1}^k d_i$. *(More generally,* f *need only be smooth on* $f^{-1}(U)$ *for some neighborhood* U *of* p.)

Proof. By definition of regular value there is a disk D about p such that $f^{-1}(D)$ is a union of disjoint disks about the q_i which map diffeomorphically to D. We can follow f by the end of a deformation which stretches D so that its interior maps diffeomorphically onto the complement of the antipode of p and the rest of $\mathbf{S}^n$ goes to this antipode. This composition has the same degree as f and it falls under the situation of Theorem 7.4. The result then follows from Theorem 7.4 and Proposition 7.2. □

7.6. Corollary. *The degree of a map $f: \mathbf{S}^n \to \mathbf{S}^n$ is independent of the homology theory (integer coefficients) used to define it.* □

7.7. Example. Consider the circle $\mathbf{S}^1$ as the unit circle in the complex plane and let $\phi_k: \mathbf{S}^1 \to \mathbf{S}^1$ be $\phi_k(z) = z^k$. This is differentiable and all points are regular values. The inverse image of a point is k equally spaced points around $\mathbf{S}^1$ and the local degree at each of them is clearly 1. Consequently, $\deg(\phi_k) = k$.

One can give another argument for this directly from Theorem 7.4: Divide $\mathbf{S}^1$ into k equal arcs, which are the E_i of Theorem 7.4. Let $f = \phi_k$. The map f_i just takes one of these arcs and wraps it around $\mathbf{S}^1$, in the same direction, and maps the complement to a point. It is clear that f_i is homotopic to the identity. Accordingly, $\deg(f_i) = 1$ and it follows from Theorem 7.4 that $\deg(\phi_k) = \sum_i \deg(f_i) = k$.

7.8. Example. We shall calculate $H_*(\mathbf{P}^2)$, where $\mathbf{P}^2$ is the projective plane, even though this computation will follow easily from things we shall develop later. Consider $P = \mathbf{P}^2$ as the attachment of $\mathbf{D}^2$ to $\mathbf{S}^1$ by a map $f: \mathbf{S}^1 \to \mathbf{S}^1$ which is ϕ_2 of Example 7.7. Let D denote the disk of radius $\frac{1}{2}$ in $\mathbf{D}^2$ and let C be its boundary. Let $S = \mathbf{S}^1$, the circle the disk is being attached to. Let $U = P - \{p\}$ where p is the center of $\mathbf{D}^2$. Note that S is a deformation retract of U, so that these have the same homology. Consider the exact sequence of the pair (P, U):

$$\cdots \to H_{k+1}(P, U) \to H_k(U) \to H_k(P) \to H_k(P, U) \to \cdots.$$

By excision and a homotopy $H_i(P, U) \approx H_i(P - S, U - S) \approx H_i(D, C)$ which, by Theorem 6.6, is $\mathbf{Z}$ for $i = 2$ and is 0 otherwise. Thus $H_k(U) \to H_k(P)$ is an isomorphism if $k \neq 1, 2$. Since U is homotopy equivalent to $\mathbf{S}^1$ it follows that $H_k(P) = \mathbf{Z}$ for $k = 0$ and is 0 for $k > 2$. It remains to compute $H_1(P)$ and $H_2(P)$.

The pair (D, C) maps to the pair (P, U) by inclusion. Since this is an excision and $H_2(U) = 0$, $H_1(P, U) = 0$, $H_i(D) = 0$ for $i = 1, 2$, there is the following commutative diagram with exact rows:

$$\begin{array}{ccccccccc} & & 0 & \to & H_2(D, C) & \to & H_1(C) & \to & 0 \\ & & & & \downarrow\approx & & \downarrow & & \\ 0 & \to & H_2(P) & \to & H_2(P, U) & \to & H_1(U) & \to & H_1(P) \to 0. \end{array}$$

Thus the map $H_2(P, U) \to H_1(U)$ is essentially the same as $H_1(C) \to H_1(U) \xrightarrow{\approx} H_1(S)$, and this is induced by the inclusion $C \hookrightarrow U$ followed by the retraction of U onto S. But this map is just ϕ_2 and so has degree 2. Thus the sequence looks like $0 \to H_2(P) \to \mathbf{Z} \xrightarrow{2} \mathbf{Z} \to H_1(P) \to 0$. Accordingly, $H_2(P) = 0$ and $H_1(P) \approx \mathbf{Z}_2$.

Note that we could have deduced that $H_1(P) \approx \mathbf{Z}_2$ for singular theory from the connection between it and $\pi_1(P)$, and then the rest follows from the exact sequence of (P, U) alone. However, the argument we gave is the type of idea we are going to generalize.

7.9. Corollary (Fundamental Theorem of Algebra). *If $p(z)$ is a complex polynomial of positive degree then $p(z)$ has a zero.*

Proof. There was another proof in Corollary 6.4 of Chapter II and we shall use parts of that proof here. The function p can be extended to $\mathbf{S}^2 \to \mathbf{S}^2$ and there are only a finite number of points which are not regular values. Since the image of p is connected and p is not constant, there must be a point in the image which is a regular value. But the Jacobian (which was computed to be $u_x^2 + u_y^2$) is always positive. Thus the local degree at a point mapping into a regular value must be 1. Therefore the degree of $p: \mathbf{S}^2 \to \mathbf{S}^2$ is strictly positive by Corollary 7.5. Hence p is not homotopic to a constant map and so its image cannot miss a point. □

Problems

1. Give another proof of Example 7.7 for singular homology using the fact that ϕ_k is a k-fold covering map, and using the relationship between $\pi_1(\mathbf{S}^1)$ and $H_1(\mathbf{S}^1)$.
2. Consider the surface S of a tetrahedron. Number its vertices in any order and number the three vertices of a triangle 1, 2, 3. Let S' be the triangle with the sides collapsed to a point. Note that both S and S' are homeomorphic to $\mathbf{S}^2$. Define a map $S \to S'$ by taking a triangle T in S and mapping its vertices to the points 1, 2, 3 of the triangle in the order of the numbering of the vertices of T. Extend that to an affine map of T to the triangle and then collapse to get a map to S'. Find the degree of this map up to sign. Do the same for all regular triangulated polyhedra. Does the answer depend on the numbering?
3. If $X_k = \mathbf{S}^1 \cup_{\phi_k} \mathbf{D}^2$, where ϕ_k is the map of Example 7.7, compute $H_*(X_k)$.

8. CW-Complexes

A CW-complex is a space made up of "cells" attached in a nice way. The "C" in "CW" stands for "closure finite" and the "W" stands for "weak topology." It is possible to define these spaces intrinsically, but we prefer to do it by describing the process by which they are constructed. For the most part, we will be concerned only with "finite" complexes, meaning complexes having a finite number of cells, but we shall give the definition in general and we will make use of infinite complexes in parts of Chapter VII.

Let $K^{(0)}$ be a discrete set of points. These points are the 0-cells.

If $K^{(n-1)}$ has been defined, let $\{f_{\partial\sigma}\}$ be a collection of maps $f_{\partial\sigma}: \mathbf{S}^{n-1} \to K^{(n-1)}$ where σ ranges over some indexing set. Let Y be the disjoint union of copies $\mathbf{D}_\sigma^n$ of $\mathbf{D}^n$, one for each σ, let B be the corresponding union of the boundaries $\mathbf{S}_\sigma^{n-1}$ of these disks, and put together the maps $f_{\partial\sigma}$ to produce a map $f: B \to K^{(n-1)}$. Then define

$$K^{(n)} = K^{(n-1)} \cup_f Y.$$

The map $f_{\partial\sigma}$ is called the "attaching map" for the cell σ.

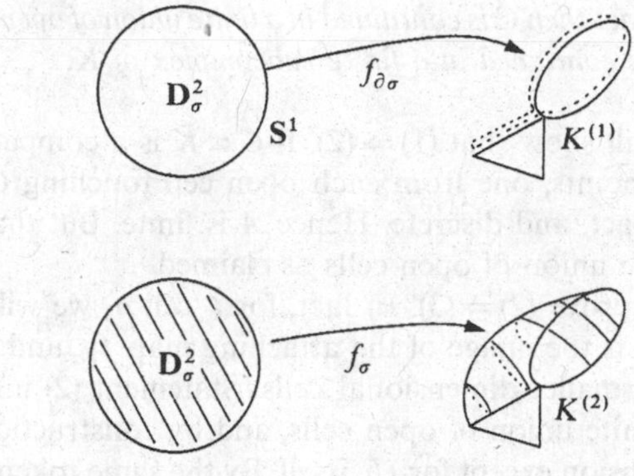

Figure IV-5. A two-dimensional CW-complex.

If $K^{(n)}$ has been defined for all integers $n \geq 0$, let $K = \bigcup K^{(n)}$ with the "weak" topology that specifies that a set is open $\Leftrightarrow$ its intersection with each $K^{(n)}$ is open in $K^{(n)}$. (It follows that a set is closed $\Leftrightarrow$ its intersection with each $K^{(n)}$ is closed.)

For each σ let $f_\sigma: \mathbf{D}_\sigma^n \to K$ be the canonical map given by the attaching of the cell σ. This map is called the "characteristic map" of the cell σ. Let K_σ be the image of f_σ. This is a compact subset of K which will be called a "closed cell." (Note, however, that this is generally not homeomorphic to $\mathbf{D}^n$ since there are identifications on the boundary.) Denote by U_σ the image in K of the open disk $\mathbf{D}_\sigma^n - \mathbf{S}_\sigma^{n-1}$. This is homeomorphic to the interior of the standard n-disk (i.e., to $\mathbf{R}^n$). We shall refer to U_σ as an "open cell," but remember that this is usually *not* an open subset of K. It is open in $K^{(n)}$. See Figure IV-5.

It is clear that the topology of each $K^{(n)}$, and hence of K itself, is characterized by the statement that a subset is open (closed) $\Leftrightarrow$ its inverse image under each f_σ is open (closed) $\Leftrightarrow$ its intersection with each K_σ is open (closed) in K_σ where the topology of the latter is the topology of the quotient of $\mathbf{D}^n$ by the identifications made by the attaching map $f_{\partial\sigma}$.

Also note that a function $g: K \to X$, for any space X, is continuous if and only if each $g \circ f_\sigma: \mathbf{D}_\sigma^n \to X$ is continuous.

By a "subcomplex" we mean a union of some of the closed cells which is itself a CW-complex with the same attaching maps. (Thus if the cell σ is in the subcomplex then K_σ is contained in the subcomplex, as well as all the open cells it touches.) It is clear that an intersection of any collection of subcomplexes of K is a subcomplex of K. Consequently, there is usually a minimal subcomplex satisfying any given condition.

The subcomplex $K^{(n)}$ of K is called the "n-skeleton" of K.

8.1. Proposition. *If K is a CW-complex then the following statements all hold:*

(1) *if $A \subset K$ has no two points in the same open cell, then A is closed and discrete;*

(2) *if $C \subset K$ is compact then C is contained in a finite union of open cells; and*
(3) *each cell of K is contained in a finite subcomplex of K.*

PROOF. First, we will show that (1) $\Rightarrow$ (2). If $C \subset K$ is a compact set then let $A \subset C$ be a set of points, one from each open cell touching C. By (1) A is closed, hence compact, and discrete. Hence A is finite. But that means C is contained in a finite union of open cells as claimed.

Second, we will prove (2) $\Rightarrow$ (3). In fact, for a cell σ, we will only use (2) for the set C which is the image of the attaching map $f_{\partial\sigma}$ and the images of attaching maps of smaller-dimensional cells. Statement (2) implies that K_σ is contained in a finite union of open cells, and by construction all of these are of smaller dimension except for U_σ itself. By the same token each of these lower-dimensional closed cells is contained in a finite union of open cells of even smaller dimension (except for the cell itself), and so on. This reasoning obviously must come to an end with 0-cells in a finite number of steps, and, at that stage, the union of the cells produced is a finite subcomplex.

Third, we will prove (3) $\Rightarrow$ (1). Consider the intersection of A with a closed cell. By (3) this is contained in a finite subcomplex. Since A has at most one point in common with any open cell this intersection must be finite, and hence closed. For any point $x \in A$, the set $A - \{x\}$ satisfies the hypothesis for A, and so it must also be closed. Hence $\{x\}$ is open in A, so A is discrete.

Finally, we put all this together. Statement (1) clearly holds for $K^{(0)}$. Suppose we know (1) for the n-skeleton $K^{(n)}$. Then we also have (2) for $K^{(n)}$. In turn, we get (3) for $K^{(n)}$. But the proof of (2) $\Rightarrow$ (3) for a particular k-cell only used (2) for subsets of $K^{(k-1)}$. Thus we actually have (3) for $K^{(n+1)}$. Thus we also have (1) for $K^{(n+1)}$. Consequently, we have all three statements for $K^{(n)}$ for all n. But any cell is in some $K^{(n)}$, and so we know (3) for K itself, and this implies (1) then (2) for K. □

8.2. Theorem. *If K is a CW-complex then any compact subset of K is contained in a finite subcomplex.*

PROOF. Let $C \subset K$ be compact. By (2) of Proposition 8.1, C is contained in a union of a finite number of open cells. By (3) of Proposition 8.1, each of these is contained in a finite subcomplex. The union of this finite number of finite subcomplexes is a finite subcomplex which contains C. □

Note that it follows that the attaching map $f_{\partial\sigma}$ for any cell σ is an attachment onto a finite subcomplex. Part (3) of Proposition 8.1 is the origin of the term "closure finite." The notion of a CW-complex is due to J.H.C. Whitehead [2].

Here are some examples of finite CW-complexes.

8.3. Example. The n-sphere is a CW-complex with one 0-cell and one n-cell, where, of course, the only attaching map is the unique map $\mathbf{S}^{n-1} \to$ *point*. There are no cells at all in other dimensions.

8.4. Example. Another CW-structure on the n-sphere is obtained by starting with two 0-cells, making $K^{(0)} \approx \mathbf{S}^0$. To this, attach two 1-cells attached by homeomorphisms $\mathbf{S}^0 \to K^{(0)}$. This makes $K^{(1)} \approx \mathbf{S}^1$. Then attach two 2-cells (the north and south hemispheres) again by homeomorphisms $\mathbf{S}^1 \to K^{(1)}$. This gives $K^{(2)} \approx \mathbf{S}^2$. The pattern is now clear. This complex has exactly two cells in each dimension from 0 to n.

8.5. Example. In Example 8.4, we can take the two cells in dimension i, for each i, to be

$$D^i_+ = \{(x_1,\dots,x_{i+1},0,\dots,0)\in\mathbf{S}^n \mid x_{i+1} \geq 0\}$$

and

$$D^i_- = \{(x_1,\dots,x_{i+1},0,\dots,0)\in\mathbf{S}^n \mid x_{i+1} \leq 0\}.$$

Then the antipodal map interchanges D^i_+ and D^i_-. If we identify these two i-cells by the antipodal map for each $i = 0,\dots,n$ then we get a CW-complex structure on $\mathbf{P}^n$, the real projective n-space. (See Example 8.9 for the complex analogue giving another approach to the definition of this structure.)

8.6. Example. The 2-sphere has a CW-structure given by its dissection as a dodecahedron. This has twenty 0-cells, thirty 1-cells, and twelve 2-cells.

8.7. Example. The torus $\mathbf{T}^2$ can be considered as the space resulting from a square by identifying opposite sides. Under this identification all four vertices of the square are identified. Thus one can consider this as a CW-complex with one 0-cell, two 1-cells, each corresponding to one of the two pairs of identified sides, and one 2-cell. The attaching maps for the 1-cells are unique since the 0-skeleton is a single point. The 1-skeleton $K^{(1)}$ is a figure eight. If the two loops of the figure eight are named α and β then the attaching map of the 2-cell is given by the word $\alpha\beta\alpha^{-1}\beta^{-1}$.

8.9. Example. Consider complex projective n-space $\mathbf{CP}^n$. Let $(z_0 : z_1 : \cdots : z_n)$ from Example 8.7 is that the attaching of the 2-cell is by the word $\alpha\beta\alpha^{-1}\beta$.

8.9. Example. Consider complex projective n-space $\mathbf{CP}^n$. Let $(z_0 : z_1 : \cdots z_n)$ denote the homogeneous coordinates of a point in $\mathbf{CP}^n$. Let $f: \mathbf{D}^{2n} \to \mathbf{CP}^n$ be given by $f(z_0,\dots,z_{n-1}) = (z_0:\cdots:z_{n-1}:(1-\sum_{i<n}|z_i^2|)^{1/2})$. Then f takes $\partial\mathbf{D}^{2n}$ into the points with $z_n = 0$, i.e., into $\mathbf{CP}^{n-1}$. If f_0 denotes the restriction of f to $\mathbf{S}^{2n-1}$ then f factors through $\mathbf{CP}^{n-1} \cup_{f_0} \mathbf{D}^{2n}$. The resulting map $\mathbf{CP}^{n-1} \cup_{f_0} \mathbf{D}^{2n} \to \mathbf{CP}^n$ is easily seen to be one–one and onto. Since it is from a compact space to a Hausdorff space it is a homeomorphism by Theorem 7.8 of Chapter I. Therefore, this gives a structure as a CW-complex on $\mathbf{CP}^n$ with exactly one i-cell in each *even* dimension $0 \leq i \leq 2n$. There are no odd dimensional cells.

We conclude this section by showing that a covering space of a CW-complex has a canonical structure as a CW-complex. This is rather a triviality, but perhaps it is useful to illustrate a direct argument concerning the weak topology. Let $p: X \to Y$ be a covering map and assume that Y is a CW-complex with characteristic maps $f_\alpha: \mathbf{D}^n \to Y$. Since $\mathbf{D}^n$ is simply connected, each f_α lifts to maps $f_{\tilde{\alpha}}: \mathbf{D}^n \to X$ which are unique upon specification of the image of any point. Take the collection of all such liftings of all f_α to define a cell structure on X.

8.10. Theorem. *With the above notation, the covering space X of the CW-complex Y is a CW-complex with the $f_{\tilde{\alpha}}$ as its characteristic maps.*

Proof. The only thing that really needs proving is that X has the weak topology. That is, we must show that a set $A \subset X$ is open $\Leftrightarrow$ each $f_{\tilde{\alpha}}^{-1}(A)$ is open in the disk which is the domain of $f_{\tilde{\alpha}}$. The implication $\Rightarrow$ is trivial since $f_{\tilde{\alpha}}$ is continuous. Thus we must show that if $A \subset X$ has each $f_{\tilde{\alpha}}^{-1}(A)$ open, then A is open. If U ranges over all components of $p^{-1}(V)$ where V ranges over all connected evenly covered open sets in Y, then $A = \bigcup (A \cap U)$ and $f_{\tilde{\alpha}}^{-1}(A \cap U) = f_{\tilde{\alpha}}^{-1}(A) \cap f_{\tilde{\alpha}}^{-1}(U)$. This shows that it suffices to consider the case in which $A \subset U$ for some such U.

We claim that

$$f_\alpha^{-1}(p(A)) = \bigcup \{f_{\tilde{\alpha}}^{-1}(A) \mid f_{\tilde{\alpha}} \text{ a lift of } f_\alpha\}.$$

Indeed, if $x \in f_\alpha^{-1}(p(A))$ then $f_\alpha(x) = p(a)$ for some $a \in A$ and there exists a lifting $f_{\tilde{\alpha}}$ of f_α such that $f_{\tilde{\alpha}}(x) = a$. Thus $x \in f_{\tilde{\alpha}}^{-1}(a) \subset f_{\tilde{\alpha}}^{-1}(A)$. Conversely, if $x \in f_{\tilde{\alpha}}^{-1}(A)$ then $f_{\tilde{\alpha}}(x) = a \in A$ and so $f_\alpha(x) = (p \circ f_{\tilde{\alpha}})(x) = p(a) \in p(A)$, giving that $x \in f_\alpha^{-1}(p(A))$, as claimed.

Therefore, if $f_{\tilde{\alpha}}^{-1}(A)$ is open for all $\tilde{\alpha}$, then the union above is open and so $f_\alpha^{-1}(p(A))$ is open for all α. Since Y has the weak topology by definition, $p(A)$ is open. But $A \subset U$ and $p: U \to p(U) = V$ is a homeomorphism by the assumption that U is a component of $p^{-1}(V)$ for the evenly covered open set V. Therefore, A is open in U and hence in X. □

Problems

1. Fill in the details of Example 8.9.
2. ◇ Give the analogue of Example 8.9 for quaternionic projective spaces.
3. Let X be the union in $\mathbf{R}^n$ of infinitely many copies of the circle which all go through the origin, but are otherwise disjoint. Show that X does *not* have the structure of a CW-complex.

9. Conventions for CW-Complexes

For the purposes of computing the homology of CW-complexes in the following sections we need some notational conventions. For technical reasons we will consider characteristic maps as being defined on the cubes $\mathbf{I}^n$ rather than on $\mathbf{D}^n$.

For spaces X, Y with base points (pointed spaces), and using $*$ to denote all base points, the one-point union $X \vee Y$ is the disjoint union $X + Y$ with base points identified. The "smash product" is $X \wedge Y = X \times Y/(X \times \{*\} \cup \{*\} \times Y) = X \times Y/X \vee Y$. If X and Y are compact then $X \wedge Y$ is the one-point compactification of $(X - \{*\}) \times (Y - \{*\})$. Thus, for example, $\mathbf{S}^p \wedge \mathbf{S}^q \approx \mathbf{S}^{p+q}$.

Let $I = [0, 1]$ with base point $\{0\}$. Let $\mathbf{I}^n = I \times \cdots \times I$ (n times). Thus $\mathbf{I}^1 = I$. Let $\mathbf{S}^1 = \mathbf{I}^1/\partial\mathbf{I}^1 = I/\{0, 1\}$, a *pointed* space. Let $\gamma_1: \mathbf{I}^1 \to \mathbf{S}^1$ be the quotient map. Let $\mathbf{S}^p = \mathbf{S}^1 \wedge \cdots \wedge \mathbf{S}^1$ (p times) and $\gamma_p: \mathbf{I}^p = I \times \cdots \times I \to \mathbf{S}^1 \wedge \cdots \wedge \mathbf{S}^1 = \mathbf{S}^p$ be $\gamma_p = \gamma_1 \wedge \cdots \wedge \gamma_1$, and note that this factors through $I \wedge \cdots \wedge I$. Note that γ_p collapses $\partial\mathbf{I}^p$ to the base point and is a homeomorphism on the interior. ($\partial\mathbf{I}^p \subset \mathbf{I}^p$ is the set of points with some coordinate 0 or 1.) Also note that

$$\gamma_{p+q} = \gamma_p \wedge \gamma_q: \mathbf{I}^{p+q} = \mathbf{I}^p \times \mathbf{I}^q \to \mathbf{S}^p \wedge \mathbf{S}^q = \mathbf{S}^{p+q}.$$

The reason for using $\mathbf{I}^n$ rather than $\mathbf{D}^n$ is that we have $\mathbf{I}^{p+q} = \mathbf{I}^p \times \mathbf{I}^q$, not just homeomorphic. This will be very convenient when considering product complexes. Note that, with these conventions $\partial\mathbf{I}^p$ is *homeomorphic* to $\mathbf{S}^{p-1}$ but *not* equal to it.

Now let H_* be a homology theory satisfying the axioms. We will assume it to have integer coefficients, but this is not really necessary. It could have any coefficient group G but statements about "generators" would have to be replaced with statements about "given isomorphisms with G."

Let $\mathbf{I}^0$ be a singleton, so that $H_0(\mathbf{I}^0) \approx \mathbf{Z}$, the coefficients of the theory, and take, once and for all, a generator $[\mathbf{I}^0] \in H_0(\mathbf{I}^0)$. This element $[\mathbf{I}^0]$ is often called a "fundamental class," or an "orientation."

Take $\gamma_0: \mathbf{I}^0 \to \mathbf{S}^0 = \{0, 1\}$ to be the map to $\{1\}$, $\{0\}$ being the base point $*$ of $\mathbf{S}^0$. (For any space X, $X/\varnothing$ should be regarded as the disjoint union of X and a base point. Thus $\gamma_0: \mathbf{I}^0 \to \mathbf{S}^0$ is the collapse $\mathbf{I}^0 \to \mathbf{I}^0/\partial\mathbf{I}^0 \approx \mathbf{S}^0$, where the homeomorphism is the unique *pointed* map. Note that $\partial\mathbf{I}^0 = \varnothing$.)

Now we shall orient $\mathbf{I}^p$, $\mathbf{S}^p$, and $\partial\mathbf{I}^p$ (which, here, is *not* $\mathbf{S}^{p-1}$) inductively as follows. If $\mathbf{I}^p$ has been oriented by choice of a generator $[\mathbf{I}^p] \in H_p(\mathbf{I}^p, \partial\mathbf{I}^p)$ then orient $\mathbf{S}^p$ by taking $[\mathbf{S}^p] \in H_p(\mathbf{S}^p, *)$ to correspond to $[\mathbf{I}^p]$ under the isomorphism

$$\gamma_{p*}: H_p(\mathbf{I}^p, \partial\mathbf{I}^p) \xrightarrow{\approx} H_p(\mathbf{S}^p, *).$$

That is, take $[\mathbf{S}^p] = \gamma_{p*}[\mathbf{I}^p]$. That γ_{p*} is an isomorphism follows from the fact that, up to a homeomorphism, γ_p is the collapsing map γ in the diagram

$$\begin{array}{ccc} (\mathbf{D}^p_+, \mathbf{S}^{p-1}) & \hookrightarrow & (\mathbf{S}^p, \mathbf{D}^p_-) \\ {\scriptstyle\gamma}\downarrow & \swarrow {\scriptstyle h} & \\ (\mathbf{S}^p, *) & & \end{array}$$

where h is a homotopy equivalence (stretching $\mathbf{D}^p_+$ to cover $\mathbf{S}^p$) and the inclusion induces an isomorphism in homology by the excision and homotopy axioms (see Theorem 6.6).

Now we orient $\partial\mathbf{I}^{p+1}$: Consider the map $1 \wedge \gamma_p: \mathbf{I}^{p+1} = \mathbf{I}^1 \times \mathbf{I}^p \to \mathbf{I}^1 \wedge \mathbf{S}^p$. This *restricts* to $\partial\mathbf{I}^{p+1} = (\partial\mathbf{I}^1) \times \mathbf{I}^p \cup \mathbf{I}^1 \times (\partial\mathbf{I}^p) \to \partial\mathbf{I}^1 \times \mathbf{S}^p \cup I \times \{*\} \to \mathbf{S}^0 \wedge \mathbf{S}^p = \mathbf{S}^p$. (The latter equality is the canonical homeomorphism $X \to \mathbf{S}^0 \wedge X$

taking x to $\{1\} \wedge x$, for any space X.) This is a map which collapses all but the face $\{1\} \times \mathbf{I}^p \subset \mathbf{I}^1 \times \mathbf{I}^p$ to the base point and is essentially γ_p on this face. Clearly it is a homotopy equivalence. (Indeed, it is homotopic to a homeomorphism.) This map $1 \wedge \gamma_p: \partial\mathbf{I}^{p+1} \to \mathbf{S}^p$ then induces an isomorphism

$$(1 \wedge \gamma_p)_*: H_p(\partial\mathbf{I}^{p+1}, *) \xrightarrow{\approx} H_p(\mathbf{S}^p, *)$$

and we take $[\partial\mathbf{I}^{p+1}] \in H_p(\partial\mathbf{I}^{p+1}, *)$ to go into $[\mathbf{S}^p]$.

Finally, we orient $\mathbf{I}^{p+1}$ via the isomorphism

$$\partial_*: H_{p+1}(\mathbf{I}^{p+1}, \partial\mathbf{I}^{p+1}) \xrightarrow{\approx} H_p(\partial\mathbf{I}^{p+1}, *),$$

choosing, of course, $[\mathbf{I}^{p+1}]$ to go into $[\partial\mathbf{I}^{p+1}]$. This completes the inductive definitions.

Now, suppose given a CW-complex K. For each n-cell σ of K we choose, once and for all, a characteristic map

$$f_\sigma: \mathbf{I}^n \to K^{(n)}.$$

Its restriction to $\partial\mathbf{I}^n$ is the "attaching map"

$$f_{\partial\sigma}: \partial\mathbf{I}^n \to K^{(n-1)}.$$

Consider $K^{(n)}/K^{(n-1)}$ (for $n = 0$ this is $K^0 + \{*\}$) as a pointed space. It is the one-point union

$$K^{(n)}/K^{(n-1)} \approx \bigvee \mathbf{S}^n \quad \text{(one copy for each } n\text{-cell } \sigma).$$

There is a "projection" to the σth summand $\mathbf{S}^n$ and the composition

$$\mathbf{I}^n \xrightarrow{f_\sigma} K^{(n)} \to K^{(n)}/K^{(n-1)} \xrightarrow{\sigma\text{th projection}} \mathbf{S}^n$$

collapses $\partial\mathbf{I}^n$ to the base point and is otherwise a homeomorphism. We are free to change this by a self-homeomorphism of $\mathbf{S}^n$ and thus, without loss of generality, this composition can be *assumed* to be the given collapse γ_n.

With this assumption, the σth projection $K^{(n)} \to K^{(n)}/K^{(n-1)} \to \mathbf{S}^n$ will be denoted by p_σ. Thus $p_\sigma: K^{(n)} \to \mathbf{S}^n$ is that *unique* map such that

$$\boxed{p_\sigma f_\sigma = \gamma_n: \mathbf{I}^n \to \mathbf{S}^n}$$

and

$$\boxed{p_\sigma f_{\sigma'} = \text{constant map to base point, for } n\text{-cells } \sigma' \neq \sigma.}$$

10. Cellular Homology

We will now show how to compute the homology of a CW-complex using only information about the degrees of certain maps arising from the attaching maps. Throughout this section let K be a CW-complex and A a subcomplex. Let $K^{(n)}$ denote the *union* of the n-skeleton of K with the entire space A.

Let $+_\sigma \mathbf{I}^n$ be the disjoint union of n-disks over the n-cells σ of K not in A. The characteristic maps f_σ fit together to give a map $f\colon +_\sigma \mathbf{I}^n \to K^{(n)}$ which takes $+_\sigma \partial\mathbf{I}^n \to K^{(n-1)}$. Additivity gives

$$H_n(+_\sigma(\mathbf{I}^n, \partial\mathbf{I}^n)) \approx \bigoplus_\sigma H_n(\mathbf{I}^n, \partial\mathbf{I}^n)$$

which is a free abelian group on the n-cells of K not in A.

10.1. Lemma. *The map $\bigoplus_\sigma f_{\sigma *}\colon \bigoplus_\sigma H_n(\mathbf{I}^n, \partial\mathbf{I}^n) \to H_n(K^{(n)}, K^{(n-1)})$ is an isomorphism. Also $H_i(K^{(n)}, K^{(n-1)}) = 0$ for $i \neq n$.*

PROOF. This is equivalent to the map $f_*\colon H_*(+_\sigma(\mathbf{I}^n, \partial\mathbf{I}^n)) \to H_*(K^{(n)}, K^{(n-1)})$ being an isomorphism. In turn, this follows from the diagram:

$$\begin{array}{ccc}
H_*(+(\mathbf{I}^n, \partial\mathbf{I}^n)) & \xrightarrow{f_*} & H_*(K^{(n)}, K^{(n-1)}) \\
\downarrow \approx & & \downarrow \approx \text{ (homotopy axiom and 5-lemma)} \\
H_*(+_\sigma(\mathbf{I}^n, \mathbf{I}^n - \{0\})) & \longrightarrow & H_*(K^{(n)}, K^{(n)} - \bigcup f_\sigma(0)) \\
\uparrow \approx & & \uparrow \approx \text{ (excision)} \\
H_*(+_\sigma(\operatorname{int}\mathbf{I}^n, \operatorname{int}\mathbf{I}^n - \{0\})) & \xrightarrow{\approx} & H_*(K^{(n)} - K^{(n-1)}, K^{(n)} - K^{(n-1)} - \bigcup f_\sigma(0)).
\end{array}$$

The isomorphism on the bottom holds because the map of spaces is a homeomorphism. □

By the naturality of the homology sequence of a triple we get the following result.

10.2. Corollary. *The following diagram commutes:*

$$\begin{array}{ccc}
H_n(K^{(n)}, K^{(n-1)}) & \xrightarrow{\partial_*} & \tilde{H}_{n-1}(K^{(n-1)}, A) \\
\uparrow \bigoplus_\sigma f_{\sigma *} & & \uparrow \bigoplus_\sigma f_{\partial\sigma *} \\
\bigoplus_\sigma H_n(\mathbf{I}^n, \partial\mathbf{I}^n) & \xrightarrow{\bigoplus \partial_*} & \bigoplus_\sigma \tilde{H}_{n-1}(\partial\mathbf{I}^n).
\end{array}$$ □

We have seen that $H_n(K^{(n)}, K^{(n-1)})$ is a free abelian group on the n-cells of K not in A, and that $H_i(K^{(n)}, K^{(n-1)}) = 0$ for $i \neq n$. The exact homology sequence of $(K^{(n)}, K^{(n-1)})$ then implies that $H_i(K^{(n-1)}, A) \to H_i(K^{(n)}, A)$ is onto for $i \neq n$, is one–one for $i \neq n - 1$, and thus is an isomorphism for $i \neq n, n - 1$. By induction on n with i fixed we conclude that

$$H_i(K^{(n)}, A) = 0 \quad \text{for} \quad n < i \quad (\text{and for } i < 0),$$

and we have the exact sequence

$$0 \to H_n(K^{(n)}, A) \to H_n(K^{(n)}, K^{(n-1)}) \xrightarrow{\partial_*} H_{n-1}(K^{(n-1)}, A) \to H_{n-1}(K^{(n)}, A) \to 0.$$

Consider the following diagram with exact row and column, defining β_{n+1} by commutativity:

$$\begin{array}{ccccccc}
 & & & & 0 & & \\
 & & & & \downarrow & & \\
H_{n+1}(K^{(n+1)}, A) & \xrightarrow{j_{n+1}} & H_{n+1}(K^{(n+1)}, K^{(n)}) & \xrightarrow{\partial_{n+1}} & H_n(K^{(n)}, A) & \xrightarrow{i_n} & H_n(K^{(n+1)}, A) \to 0 \\
 & & & \searrow^{\beta_{n+1}} & \downarrow j_n & & \\
 & & & & H_n(K^{(n)}, K^{(n-1)}) & & \\
 & & & & \downarrow \partial_n & & \\
 & & & & \tilde{H}_{n-1}(K^{(n-1)}, A). & &
\end{array}$$

Accordingly, $\beta_{n+1} = j_n \circ \partial_{n+1}$. Since $\beta_n \beta_{n+1} = j_{n-1} \partial_n j_n \partial_{n+1} = j_{n-1} \circ 0 \circ \partial_{n+1} = 0$, this gives a boundary operator β. Note that

$$\ker \beta_{n+1} = \ker \partial_{n+1} = \operatorname{im} j_{n+1}.$$

Hence

$$\ker \beta_n = \ker \partial_n = \operatorname{im} j_n,$$
$$\operatorname{im} \beta_{n+1} = j_n(\operatorname{im} \partial_{n+1})$$

and j_n is a monomorphism on $H_n(K^{(n)})$. Consequently, j_n induces isomorphisms

$$\begin{array}{ccc}
H_n(K^{(n)}, A) & \xrightarrow{\approx} & \ker \beta_n \\
\cup & & \cup \\
\operatorname{im} \partial_{n+1} & \xrightarrow{\approx} & \operatorname{im} \beta_{n+1}
\end{array}$$

and hence induces

$$H_n(K^{(n+1)}, A) \approx \operatorname{coker} \partial_{n+1} \xrightarrow{\approx} \ker \beta_n / \operatorname{im} \beta_{n+1}.$$

But

$$H_n(K^{(n+1)}, A) \xrightarrow{\approx} H_n(K^{(n+2)}, A) \xrightarrow{\approx} H_n(K^{(n+3)}, A) \xrightarrow{\approx} \cdots$$

since $H_n(K^{(i)}, K^{(i-1)}) = 0$ for $i > n$, and by the exact sequences. Thus, if $\dim(K) < \infty$, we get

$$H_n(K, A) \approx H_n(K^{(n+1)}, A) \approx \ker \beta_n / \operatorname{im} \beta_{n+1}.$$

For arbitrary K, and for singular homology this also follows from the fact that any singular chain has compact image and hence is inside some finite subcomplex. It is proved in general from the axioms in Appendix A but the case of finite-dimensional cell complexes is all we really need here.

The above facts then show that $H_n(K, A)$ can be computed from the chain complex whose nth term is $H_n(K^{(n)}, K^{(n-1)})$ and whose boundary operator is β. We now look more closely at this chain complex.

Let $C_n(K, A)$ be the free abelian group based on the n-cells of K not in A. Define the homomorphisms

$$C_n(K, A) \underset{\Phi}{\overset{\Psi}{\rightleftarrows}} H_n(K^{(n)}, K^{(n-1)})$$

given by

$$\Psi\left(\sum_\sigma n_\sigma \sigma\right) = \sum_\sigma n_\sigma f_{\sigma*}[\mathbf{I}^n],$$

$$\Phi(\alpha) = \sum_\sigma \phi_n(p_{\sigma*}(\alpha))\sigma,$$

where $\phi_n: H_n(\mathbf{S}^n, *) \to \mathbf{Z}$ is the unique homomorphism such that $\phi_n[\mathbf{S}^n] = 1$. Note that Ψ has already been shown to be an isomorphism. We *claim* that Φ is its inverse. To show this, it suffices to show that $\Phi \circ \Psi = 1$. Thus start with some n-cell σ, i.e., a basis element of $C_n(K, A)$. The composition in question is then $\Phi(\Psi(\sigma)) = \Phi(f_{\sigma*}[\mathbf{I}^n]) = \sum_\tau \phi_n(p_{\tau*} f_{\sigma*}[\mathbf{I}^n])\tau$. Using that $p_\tau f_\sigma = \gamma_n$ for $\tau = \sigma$ and is constant otherwise, we see that this is

$$\phi_n(p_{\sigma*} f_{\sigma*}[\mathbf{I}^n])\sigma = \phi_n(\gamma_{n*}[\mathbf{I}^n])\sigma = \phi_n([\mathbf{S}^n])\sigma = 1 \cdot \sigma = \sigma,$$

as claimed.

These explicit maps allow us to "compute" the homology as follows: We define, of course, the boundary operator ∂ on $C_*(K, A)$ by commutativity of

$$
\begin{array}{ccc}
C_{n+1}(K, A) & \xrightarrow{\;\;\partial_{n+1}\;\;} & C_n(K, A) \\
\big\uparrow{\scriptstyle \approx} & & \big\uparrow{\scriptstyle \approx} \\
H_{n+1}(K^{(n+1)}, K^{(n)}) \xrightarrow{\partial_{n+1}} H_n(K^{(n)}) & \xrightarrow{\;j_n\;} & H_n(K^{(n)}, K^{(n-1)})
\end{array}
$$

$$\beta_{n+1}$$

where the verticals are the isomorphisms given above.

For a generator $\sigma \in C_{n+1}(K, A)$ (an $n+1$ cell) we trace this diagram as follows (where the $[\tau : \sigma] \in \mathbf{Z}$ are the coefficients that make the diagram commutative):

$$
\begin{array}{ccc}
\sigma & \longmapsto & \sum_\tau [\tau : \sigma]\tau \\
\big\uparrow & & \big\uparrow \\
(f_\sigma)_*[\mathbf{I}^{n+1}] \xmapsto{\;\partial_*\;} (f_{\partial\sigma})_*[\partial\mathbf{I}^{n+1}] & \longmapsto & (f_{\partial\sigma})_*[\partial\mathbf{I}^{n+1}].
\end{array}
$$

Looking at our previous description of the right-hand vertical map we conclude that the "incidence number" $[\tau:\sigma]$ is

$$\begin{aligned}[\tau:\sigma] &= \phi_n(p_{\tau*}(f_{\partial\sigma*}[\partial\mathbf{I}^{n+1}]))\\ &= \phi_n((p_\tau f_{\partial\sigma})_*[\partial\mathbf{I}^{n+1}])\\ &= \phi_n(\deg(p_\tau f_{\partial\sigma})\cdot[\mathbf{S}^n])\\ &= \deg(p_\tau f_{\partial\sigma})\end{aligned}$$

by the definition of degree and the choice of the generators $[\partial\mathbf{I}^{n+1}]\in H_n(\partial\mathbf{I}^{n+1})$ and $[\mathbf{S}^n]\in H_n(\mathbf{S}^n)$. Consequently, we have shown:

10.3. Theorem. *$H_*(K, A)$ is isomorphic to the homology of the chain complex $C_*(K, A)$, where the boundary operator $\partial: C_{n+1}(K, A)\to C_n(K, A)$ is given by*

$$\partial\sigma = \sum_\tau [\tau:\sigma]\tau$$

where

$$[\tau:\sigma] = \deg(p_\tau f_{\partial\sigma}).$$

□

The boundary operator ∂: $C_1(K, A)\to C_0(K, A)$ is the least important one to understand since the singular homology in this area is easy to understand. (We already computed it in Sections 2 and 3.) However, it is perhaps the most difficult for the student to understand from the point of view of the present section. Thus we will work that out here. Suppose that x and y are 0-cells in K and that a is a 1-cell attached by the map $f_{\partial a}$: $\{0, 1\} = \partial\mathbf{I}^1\to\{x, y\}\subset K^{(0)}$ taking 0 to x and 1 to y. We wish to compute $\partial a = [y: a]y + [x: a]x$, where $[x: a] = \deg(p_x f_{\partial a})$ and $[y: a] = \deg(p_y f_{\partial a})$. The projections p_x and p_y are both compositions of the form (disregarding 0-cells other than x and y)

$$K^{(0)}\to K^{(0)}/K^{(-1)} = K^{(0)}/A = (K^{(0)} - A) + \{*\} = \{*, x, y\}\to\mathbf{S}^0$$

and the only difference is that p_x takes x to 1 and y to 0, while p_y does the opposite. Thus $p_x f_{\partial a}$ takes 1 to 0 and 0 to 1, while $p_y f_{\partial a}$ is the identity on $\mathbf{S}^0 = \{0, 1\}$. Therefore $[x: a] = \deg(p_x f_{\partial a}) = -1$ (the reader may wish to verify the details of this) and $[y: a] = \deg(p_y f_{\partial a}) = 1$. Hence, we can now compute $\partial a = [y: a]y + [x: a]x = y - x$. (If either x or y is in A, just discard it.)

We shall now illustrate these results by doing some simple examples.

10.4. Example. Consider the real projective plane $K = \mathbf{P}^2$ as a CW-complex with exactly one 0-cell x, one 1-cell a, and one 2-cell σ. There is only one way to attach the 1-cell to the 0-cell, and it makes $K^{(1)}$ into a circle. The 2-cell is attached to this by the map that is the quotient map of identifying antipodal points on $\mathbf{S}^1$. The chain complex $C_*(K)$ is

$$\cdots\to 0\to C_2(K)\to C_1(K)\to C_0(K)\to 0\to\cdots$$

which is

$$\cdots \to 0 \to \mathbf{Z} \to \mathbf{Z} \to \mathbf{Z} \to 0 \to \cdots.$$

For the 1-cell a we have that $\partial a = x - x = 0$, by the remarks above. For the 2-cell σ, the attaching map is just the map that can be thought of as $z \mapsto z^2$ of the unit complex numbers. This is the map $p_a f_{\partial\sigma}$ up to a difference of orientation (which only affects the sign of the incidence number $[a{:}\,\sigma]$). By Example 7.7 this map has degree 2. Our neglect of "orientation" here means that we cannot depend on the sign, but anyway, $[a{:}\,\sigma] = \pm 2$. Accordingly, the chain complex looks like

$$\cdots \to 0 \to \mathbf{Z} \xrightarrow{\pm 2} \mathbf{Z} \xrightarrow{0} \mathbf{Z} \to 0 \to \cdots$$

so that $H_2(\mathbf{P}^2) = 0$, $H_1(\mathbf{P}^2) \approx \mathbf{Z}_2$, and $H_0(\mathbf{P}^2) \approx \mathbf{Z}$.

10.5. Example. Consider the torus $K = \mathbf{T}^2$ as the space arising from the unit square by identifying opposite sides. The obvious structure as a CW-complex is to let the corners be the unique 0-cell x, the edges of the square giving two 1-cells a, b and the interior giving the single 2-cell σ. Thus $C_0(\mathbf{T}^2) \approx \mathbf{Z}$, generated by x, $C_1(\mathbf{T}^2) \approx \mathbf{Z} \oplus \mathbf{Z}$, generated by a and b, and $C_2(\mathbf{T}^2) \approx \mathbf{Z}$, generated by σ. By the above remarks $\partial a = 0$ and $\partial b = 0$. The attaching map for σ is the loop in the figure eight (formed by a and b) running around a then b then a^{-1} then b^{-1}. The composition $p_a f_{\partial\sigma}$ is this map followed by mapping the a part of the figure eight around the circle by degree ± 1, and the b part to the base point. This is just the loop $a * \text{constant} * a^{-1} * \text{constant} \simeq \text{constant}$. Hence it has degree 0, and similarly for $p_b f_{\partial\sigma}$. Consequently, $[a{:}\,\sigma] = 0$ and $[b{:}\,\sigma] = 0$ so that $\partial = 0$ in degree 2, and thus in all degrees. Therefore $H_*(\mathbf{T}^2) \approx C_*(\mathbf{T}^2) \approx \mathbf{Z}, \mathbf{Z} \oplus \mathbf{Z}, \mathbf{Z}$ in degrees 0, 1, 2, respectively.

10.6. Example. The Klein bottle $\mathbf{K}^2$ can be constructed similarly to the torus but with the "b" edges identified with a flip. Thus the attaching map, up to orientation of cells, would be the loop $a * b * a^{-1} * b$. In this case the composition $p_b f_{\partial\sigma}$ becomes the loop, in $\mathbf{S}^1$, $\text{constant} * b * \text{constant} * b$ which has degree ± 2. Thus the boundary map computation will give $[b{:}\,\sigma] = \pm 2$, where the sign depends on how one defines the attaching maps in detail. Thus $\partial\sigma = [a{:}\,\sigma]a + [b{:}\,\sigma]b = 0a \pm 2b = \pm 2b$. It then follows that $H_2(\mathbf{K}^2) = 0$, $H_1(\mathbf{K}^2) \approx \mathbf{Z} \oplus \mathbf{Z}_2$, and $H_0(\mathbf{K}^2) \approx \mathbf{Z}$.

10.7. Example. We will define a space L as a quotient space of $\mathbf{D}^3$ as follows. The only identifications will be on the boundary $\mathbf{S}^2$. Let relatively prime integers $p > 1$ and q be given. Identify a point on the closed upper hemisphere of $\mathbf{S}^2$ with the point on the lower hemisphere obtained by rotating clockwise about the vertical axis through an angle of $2\pi q/p$ and then reflecting through the equator. Then a point in the open upper hemisphere is identified to exactly one other point which is in the open lower hemisphere. However, a point on the equator is identified with all points on the equator making an angle with it of any multiple of $2\pi q/p$. Since q is prime to p, this implies that points on the equator $2\pi n/p$ apart are identified for all $n = 1, 2, \ldots, p-1$. We take the class of some point on the equator to define the unique 0-cell x of L. The equivalents of this point divide the equator into p intervals of angle $2\pi/p$ all equivalent to one another. Take such a (clockwise) interval to define the unique 1-cell a of L. The upper hemisphere is then taken to define the unique 2-cell b of L and $\mathbf{D}^3$ itself is taken to define the unique 3-cell c of L. Then $\partial a = x - x = 0$ and $\partial b = pa$ (or $-pa$ with the opposite choice of orientation). It is clear that ∂c is either $\pm(b-b) = 0$ or $\pm(b+b) = \pm 2b$. But $\pm 2b$ is not a cycle and so we must have $\partial c = 0$. Consequently, $H_0(L) \approx \mathbf{Z}$, $H_1(L) \approx \mathbf{Z}_p$, $H_2(L) = 0$ and $H_3(L) \approx \mathbf{Z}$. This space can be seen to be the lens space $L(p, q)$; see Example 7.4 of Chapter III. Indeed, this is the classic description of the lens space (except for our use of a simple CW-decomposition instead of a much more complicated simplicial decomposition).

Problems

1. Give a CW-structure on the 3-torus $\mathbf{T}^3$ and use it to compute the homology.
2. Consider the space X which is the union of the unit sphere $\mathbf{S}^2$ in $\mathbf{R}^3$ and the line segment between the north and south poles. Give it a CW-structure and compute its homology.
3. Show that the space X in Problem 2 is homotopy equivalent to the one-point union $\mathbf{S}^2 \vee \mathbf{S}^1$ of a 2-sphere and a circle. Use this to give an easier computation of $H_*(X)$.
4. Compute the homology of the real projective 3-space. (*Hint*: Try to use information you already have from the computation for projective 2-space.)
5. The "dunce cap" space is the space resulting from a triangle (a, b, c) and its interior by identifying the three edges by $(a, b) \sim (b, c) \sim (a, c)$. Compute its homology using the induced CW-structure.
6. Compute the homology of the space obtained from a circle by attaching a 2-cell by a map of degree 2, and another 2-cell by a map of degree 3. Generalize.
7. ◇ Show that
$$H_i(\mathbf{CP}^n) \approx \begin{cases} \mathbf{Z} & \text{for } i \text{ even}, 0 \le i \le 2n, \\ 0 & \text{otherwise.} \end{cases}$$
8. ◇ Show that
$$H_i(\mathbf{QP}^n) \approx \begin{cases} \mathbf{Z} & \text{for } i \text{ divisible by } 4, 0 \le i \le 4n, \\ 0 & \text{otherwise.} \end{cases}$$
9. Let K be the quotient space of the cube $\{(x, y, z) \mid |x| \le 1, |y| \le 1, |z| \le 1\}$ by the identifications $(x, y, 1) \sim (-y, x, -1)$, $(x, 1, z) \sim (z, -1, -x)$, and $(1, y, z) \sim (-1, -z, y)$; i.e., by identifying each face with the opposite face by a counter clockwise rotation through 90°. Compute $H_*(K; \mathbf{Z})$ and $H_*(K; \mathbf{Z}_2)$.
10. Let X result from $\mathbf{D}^3$ by identifying points on $\mathbf{S}^2$ taken into one another by the

180° rotation about the vertical axis. Give X the structure of a CW-complex and compute its homology.

11. Let C be the circle on the torus $\mathbf{T}^2$ which is the image, under the covering map $\mathbf{R}^2 \to \mathbf{T}^2$, of the line $px = qy$. Let $X = \mathbf{T}^2/C$, the quotient space obtained by identifying C to a point. Compute $H_*(X)$.

12. For a CW-complex, show that $\sum_\tau [\omega:\tau][\tau:\sigma] = 0$ for all $(n+1)$-cells σ and $(n-1)$-cells ω, and with τ ranging over all n-cells.

11. Cellular Maps

In the previous section we showed how to compute $H_*(K)$ for a CW-complex K. In this section we show how to compute $f_*: H_*(K) \to H_*(L)$ for a map $f: K \to L$ between CW-complexes. We shall use that $K \times I$ has the structure of a CW-complex, where I is regarded as the 3-cell complex $\{I, \{0\}, \{1\}\}$. The cell structure of $K \times I$ is obvious (if not, see Section 12). That the product topology coincides with the CW-topology is easy to see; a proof is given in Theorem 12.3, not requiring the reading of the rest of Section 12.

11.1. Definition. A map $f: K \to L$ between CW-complexes is said to be *cellular* if $f(K^{(n)}) \subset L^{(n)}$ for all n.

11.2. Lemma. *Let K be a CW-complex. Then any map ϕ: $\mathbf{D}^n \to K$ such that $\phi(\mathbf{S}^{n-1}) \subset K^{(n-1)}$ is homotopic* rel $\mathbf{S}^{n-1}$, *to a map into $K^{(n)}$. The same is true for $\mathbf{D}^n$ replaced by a quotient space resulting from $\mathbf{D}^n$ by some identifications on the boundary. (Accordingly, $(K, K^{(n)})$ is "n-connected.")*

Proof. Since any compact subset of K is contained in a finite subcomplex, it suffices to show that if (X, Y) is a pair of spaces with $X = Y \cup_\phi \mathbf{I}^m$, where $\phi: \partial\mathbf{I}^m \to Y$, for some $m > n$, and if $f: (\mathbf{D}^n, \mathbf{S}^{n-1}) \to (X, Y)$ then f is homotopic, rel $\mathbf{S}^{n-1}$, to a map into Y. Let $U = X - Y \approx \mathbf{R}^m$ and put $M^n = f^{-1}(U)$ which is an open set in $\mathbf{R}^n$. Let $E \neq \emptyset$ be an open set with $\bar{E} \subset U$ compact. Put $B = M - f^{-1}(E)$ which is a closed subset of M. By the smooth approximation Theorem (Theorem 11.7 of Chapter II), there is a map $g: M \to U$ which equals f on B, is smooth on $M - B$, and is homotopic to f rel B. Since $g = f$ on B, and since $B \cup f^{-1}(Y)$ is a neighborhood of $f^{-1}(Y)$, g extends to a map $g: \mathbf{D}^n \to X$ by taking $g(x) = f(x)$ for $x \in \mathbf{D}^n - M = f^{-1}(Y)$. Let $p \in E$ be a regular value of $g|_{M-B}$. Since $m > n$, this just means that $p \notin g(M - B)$. But $p \notin g(B) = f(B)$, and so $p \notin g(\mathbf{D}^n)$. However, $X - \{p\}$ deforms to Y and so there is a homotopy rel $\mathbf{S}^{n-1}$ of g, and hence of f, to a map into Y.

For the case of a map from a space Q which is $\mathbf{D}^n$ with some identifications on the boundary, just apply what we now have to the composition $\mathbf{D}^n \to Q \to K$. Since the resulting homotopy is rel $\mathbf{S}^{n-1}$, it factors through $Q \times I$ and the induced function $Q \times I \to K$ is continuous because $Q \times I$ has the quotient topology by Proposition 13.19 of Chapter I. □

11.3. Corollary. *If K is a CW-complex then any map $\mathbf{D}^n \times \{0\} \cup \mathbf{S}^{n-1} \times I \to K$ such that $f(\mathbf{S}^{n-1} \times \{1\}) \subset K^{(n-1)}$ extends to a map $\mathbf{D}^n \times I \to K$ taking $\mathbf{D}^n \times \{1\}$ into $K^{(n)}$. This also holds for maps from $Q \times I$ where Q is a quotient space of $\mathbf{D}^n$ resulting from identifications on $\mathbf{S}^{n-1}$.*

Proof. This follows from Lemma 11.2 and the evident fact that the pair $(\mathbf{D}^n \times I, \mathbf{D}^n \times \{0\} \cup \mathbf{S}^{n-1} \times I)$ is homeomorphic to the pair $(\mathbf{D}^n \times I, \mathbf{D}^n \times \{0\})$; see Figure VII-6 on p. 451. □

11.4. Theorem (Cellular Approximation Theorem). *Let K and Y be CW-complexes and $L \subset K$ a subcomplex. Suppose $\phi: K \to Y$ is a map such that $\phi|_L$ is cellular. Then ϕ is homotopic,* rel L, *to a cellular map $\psi: K \to Y$.*

Proof. Extend $\phi \cup (\phi \circ \pi_L)$: $K \times \{0\} \cup L \times I \to Y$ by induction over the skeletons of $K - L$ using Corollary 11.3. Continuity of the result is a consequence of having the weak topology on $K \times I$. □

11.5. Corollary. *If ϕ, ψ: $K \to Y$ are cellular maps which are homotopic, then they are homotopic via a cellular homotopy $K \times I \to Y$. Also, if the original homotopy is already cellular on $L \times I$, for some subcomplex L of K, then the new homotopy can be taken to be identical to the old one on $L \times I$.*

Proof. Just apply Theorem 11.4 to $(K \times I, K \times \partial I \cup L \times I)$. □

In the remainder of this section, we shall show, for a cellular map g: $(K, A) \to (L, B)$ of pairs of CW-complexes, how to compute the induced map g_*: $H_i(K, A) \to H_i(L, B)$ in homology. As in the previous section, we shall simplify the notation by letting $K^{(n)}$ denote the *union* of the n-skeleton of K with all of A and similarly for (L, B).

Since g is cellular, it induces the commutative diagram:

$$\begin{array}{ccccc}
 & & \xrightarrow{\qquad\qquad\qquad \beta_n^K \qquad\qquad\qquad} & & \\
H_n(K^{(n)}, K^{(n-1)}) & \xrightarrow{\partial_*} & H_{n-1}(K^{(n-1)}) & \xrightarrow{j_*} & H_{n-1}(K^{(n-1)}, K^{(n-2)}) \\
\downarrow g_* & & \downarrow g_* & & \downarrow g_* \\
H_n(L^{(n)}, L^{(n-1)}) & \xrightarrow{\partial_*} & H_{n-1}(L^{(n-1)}) & \xrightarrow{j_*} & H_{n-1}(L^{(n-1)}, L^{(n-2)}) \\
 & & \xrightarrow{\qquad\qquad\qquad \beta_n^L \qquad\qquad\qquad} & &
\end{array}$$

Thus $g_\Delta: C_*(K, A) \to C_*(L, B)$, induced by this diagram, is a chain map. (Although the notation g_Δ for an induced chain map is derived from the notation for the singular chain complex, we use it to denote any chain map induced from a map. This should not cause confusion since cellular chain maps and singular chain maps are never used at the same time in this book.)

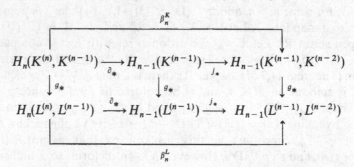

We have shown that there is the following commutative diagram with exact rows:

$$\begin{array}{ccccccccc}
0 & \to & \operatorname{im}\partial_{n+1} & \to & H_n(K^{(n)}) & \xrightarrow{i_n^K} & H_n(K,A) & \to & 0 \\
 & & \approx\downarrow j_n^K & & \approx\downarrow j_n^K & & \vdots\downarrow & & \\
0 & \to & \operatorname{im}\beta_{n+1} & \to & \ker\beta_n & \longrightarrow & \dfrac{\ker\beta_n}{\operatorname{im}\beta_{n+1}} & \to & 0
\end{array}$$

where the dotted arrow is the resulting isomorphism with which we are concerned. (Recall that $H_n(K^{(n+1)}, A) \approx H_n(K, A)$.) There is the same diagram for (L, B) and its obvious naturality shows that g_* induces a map from the diagram for K to that for L. Accordingly, the following diagram commutes:

$$\begin{array}{ccc}
H_n(K,A) & \xrightarrow{g_*} & H_n(L,B) \\
\downarrow\approx & & \downarrow\approx \\
\dfrac{\ker\beta_n^K}{\operatorname{im}\beta_{n+1}^K} & \xrightarrow{g_*} & \dfrac{\ker\beta_n^L}{\operatorname{im}\beta_{n+1}^L}.
\end{array}$$

The bottom of this diagram is simply the map $g_*: H_n(C_*(K, A)) \to H_n(C_*(L, B))$ induced by the chain map $g_\Delta: C_*(K, A) \to C_*(L, B)$. This shows that $g_*: H_n(K, A) \to H_n(L, B)$ can be "computed" from the chain map $g_\Delta: C_*(K, A) \to C_*(L, B)$. It remains to get a formula for this chain map.

Consider the following commutative diagram, where the barred maps are uniquely defined by commutativity, and $g_{\tau,\sigma}$ is defined to be the composition along the bottom. (Here σ denotes an n-cell of $K - A$, τ an n-cell of $L - B$, and $g: K \to L$ is cellular.)

$$\begin{array}{ccccccc}
\mathbf{D}^n & \xrightarrow{f_\sigma} & K^{(n)} & \xrightarrow{g} & L^{(n)} & \xrightarrow{p_\tau} & \mathbf{S}^n \\
\downarrow\gamma_n & & \downarrow & & \downarrow & & \| \\
\mathbf{S}^n & \xrightarrow{\bar f_\sigma} & K^{(n)}/K^{(n-1)} & \xrightarrow{\bar g} & L^{(n)}/L^{(n-1)} & \xrightarrow{\bar p_\tau} & \mathbf{S}^n \\
\end{array}$$
$$\mathbf{S}^n \xrightarrow{\quad g_{\tau,\sigma}\quad} \mathbf{S}^n$$

By the discussion above, the diagram

$$\begin{array}{ccc}
C_n(K,A) & \xrightarrow{g_\Delta} & C_n(L,B) \\
\updownarrow\approx & & \updownarrow\approx \\
H_n(K^{(n)},K^{(n-1)}) & \xrightarrow{g_*} & H_n(L^{(n)},L^{(n-1)})
\end{array}$$

commutes (which is the definition of the chain map g_Δ on the top). For an n-cell σ of $K - A$, we can then compute $g_\Delta(\sigma)$ by going down, right, and up in this diagram as follows:

$$g_\Delta(\sigma) = \sum_\tau \phi_n(p_{\tau*}(g_* f_{\sigma*}[\mathbf{I}^n]))\tau$$

$$= \sum_\tau \phi_n((p_\tau g f_\sigma)_*[\mathbf{I}^n])\tau$$

$$= \sum_\tau \phi_n((\bar{p}_\tau \bar{g} \bar{f}_\sigma \gamma_n)_*[\mathbf{I}^n])\tau$$

$$= \sum_\tau \phi_n((g_{\tau,\sigma})_*[\mathbf{S}^n])\tau$$

$$= \sum_\tau \deg(g_{\tau,\sigma})\tau$$

the sums ranging over n-cells τ of $L - B$. We have shown:

11.6. Theorem. *If $g: K \to L$ is cellular, and $A \subset K$ is a subcomplex of K, and B, one of L, such that $g(A) \subset B$, then the induced chain map $g_\Delta: C_*(K, A) \to C_*(L, B)$ is given by*

$$g_\Delta(\sigma) = \sum_\tau \deg(g_{\tau,\sigma})\tau. \qquad \square$$

We have now shown how to compute the homology groups of CW-complexes and the induced homomorphisms in homology from cellular maps. Also, the computations depend only on the degrees of maps of spheres associated to the attaching maps. These degrees can be found from smooth approximations and signs of determinants of certain Jacobian matrices. None of this data for the computations depends at all on the particular homology theory used. Thus we have a sort of uniqueness theorem for homology theories satisfying the axioms as far as their values on CW-complexes is concerned. This is not quite a proper uniqueness theorem since we should also show how to compute the "boundary" operator $\partial_*: H_n(K,A) \to H_{n-1}(A)$ from the same sort of data. This can be done by similar arguments as for the rest, but we don't need it and so leave it as an exercise for the interested reader.

It is worthwhile noting here that although we have this uniqueness result, we still do not have a corresponding existence theorem. That is, we do not know yet that there is any homology theory that does satisfy the axioms. In later sections we shall remedy this and show that singular theory does satisfy the two axioms (excision and homotopy) that we have not already verified. Only at that point will the applications we have already given be fully justified.

Problems

1. Consider the torus $\mathbf{T}^2$ as the quotient space of $\mathbf{R}^2$ by identifying points, both of whose coordinates differ by integers. In $\mathbf{R}^2$ also consider the line segment from $(0, 0)$ to $(2, 3)$. In the torus this gets embedded as a circle C. Compute, with the methods of this section, the map $H_1(C) \to H_1(\mathbf{T}^2)$ induced by the inclusion $C \hookrightarrow \mathbf{T}^2$.

2. Consider the unit sphere $\mathbf{S}^2$ in 3-space. Attach two handles (connected sum with

two tori) away from the z-axis in such a way that the 180° rotation about the z-axis takes one into the other. Call this space X. Define a quotient space Y of this by identifying points that are symmetric with respect to the z-axis. (Note that Y is homeomorphic to the torus.) Compute the homology of these spaces and the map $\pi_*: H_i(X) \to H_i(Y)$ induced by the quotient map $\pi: X \to Y$.

3. ♦ Let L be a subcomplex of the CW-complex K.
(a) If L is contractible, show that L is a retract of K.
(b) If L and K are both contractible, show that L is a deformation retract of K.
(*Hint*: Construct the retractions by induction over skeletons.)

4. Show that the fundamental group of a finite CW-complex is finitely generated.

5. Show that a connected finite CW-complex has the homotopy type of a CW-complex with a single 0-cell.

12. Products of CW-Complexes ☼

Let K and L be CW-complexes and $K \times L$ the product CW-complex. This need not have the product topology unless one of K and L is locally finite, but it is the case $L = I$ (a complex with cells $\{0\}$, $\{1\}$, and $[0, 1]$) in which we are mainly interested. The attaching maps for $K \times L$ will be described in a moment, and that is how this space is defined. (The fact that the topology may not be the product topology is of very little significance, since it can be shown that the singular homology and homotopy groups are the same, in a strong sense; see Problem 5.)

Since $\mathbf{I}^p \times \mathbf{I}^q = \mathbf{I}^{p+q}$ and $\mathbf{S}^p \wedge \mathbf{S}^q = \mathbf{S}^{p+q}$ by our conventions, we can define, for a p-cell σ of K and a q-cell τ of L, the $p+q$-cell $\sigma \times \tau$ of $K \times L$ having characteristic map

$$f_{\sigma \times \tau} = f_\sigma \times f_\tau: \mathbf{I}^{p+q} = \mathbf{I}^p \times \mathbf{I}^q \to K \times L.$$

Also

$$p_\sigma \wedge p_\tau: K^{(p)} \times K^{(q)} \to \mathbf{S}^p \wedge \mathbf{S}^q = \mathbf{S}^{p+q}$$

extends to $(K \times L)^{(p+q)} = \bigcup \{K^{(i)} \times L^{(j)} | i + j = p + q\}$. If $x \in \mathbf{I}^p$ and $y \in \mathbf{I}^q$ then

$$\begin{aligned}(p_\sigma \wedge p_\tau)(f_{\sigma \times \tau}(x, y)) &= (p_\sigma \wedge p_\tau)(f_\sigma(x), f_\tau(y)) = p_\sigma f_\sigma(x) \wedge p_\tau f_\tau(y) \\ &= \gamma_p(x) \wedge \gamma_q(y) = \gamma_{p+q}(x, y).\end{aligned}$$

Thus, $p_{\sigma \times \tau} = p_\sigma \wedge p_\tau$ extended trivially to $(K \times L)^{(p+q)}$.

We wish to compute the boundary operator of the chain complex $C_*(K \times L)$ which is the free abelian group on the cells $\sigma \times \tau$, in terms of those on $C_*(K)$ and $C_*(L)$.

First, let σ be a p-cell of K, τ a $(p-1)$-cell of K and μ a q-cell of L. To find $\partial(\sigma \times \mu)$ we need to know, in particular, the value of the incidence number

$$[\tau \times \mu : \sigma \times \mu] = \deg(p_{\tau \times \mu} \circ f_{\partial(\sigma \times \mu)}).$$

For $(x, y) \in \partial \mathbf{I}^{p+q} = \partial \mathbf{I}^p \times \mathbf{I}^q \cup \mathbf{I}^p \times \partial \mathbf{I}^q$, we have

$$\begin{aligned} p_{\tau \times \mu} f_{\partial(\sigma \times \mu)}(x, y) &= (p_\tau \wedge p_\mu)(f_\sigma \times f_\mu)(x, y) \\ &= (p_\tau \wedge p_\mu)(f_\sigma(x), f_\mu(y)) \\ &= p_\tau f_\sigma(x) \wedge \gamma_q(y). \end{aligned}$$

If one looks at this, one sees that it is the composition (1_p denoting the identity on $\mathbf{I}^p$ or $\mathbf{S}^p$)

$$\mathbf{I}^p \times \partial \mathbf{I}^q \cup \partial \mathbf{I}^p \times \mathbf{I}^q = \partial(\mathbf{I}^p \times \mathbf{I}^q) \xrightarrow{1_p \wedge \gamma_q} \partial \mathbf{I}^p \wedge \mathbf{S}^q \xrightarrow{p_\tau f_{\partial\sigma} \wedge 1_q} \mathbf{S}^{p-1} \wedge \mathbf{S}^q = \mathbf{S}^{p+q-1}$$

where the first map is a restriction of $1_p \wedge \gamma_q \colon \mathbf{I}^p \times \mathbf{I}^q \to \mathbf{I}^p \wedge \mathbf{S}^q$.

12.1. Lemma. $[\tau \times \mu : \sigma \times \mu] = [\tau : \sigma]$.

PROOF. By the above remarks, we need to show that

$$\deg((p_\tau f_{\partial\sigma} \wedge 1_q) \circ (1_p \wedge \gamma_q)) = \deg(p_\tau f_{\partial\sigma}).$$

In general, for a map $g \colon \partial \mathbf{I}^p \to \mathbf{S}^{p-1}$ we will show that the composition

$$\partial(\mathbf{I}^p \times \mathbf{I}^q) \xrightarrow{1_p \wedge \gamma_q} \partial \mathbf{I}^p \wedge \mathbf{S}^q \xrightarrow{g \wedge 1_q} \mathbf{S}^{p-1} \wedge \mathbf{S}^q = \mathbf{S}^{p+q-1}$$

has the same degree as does g. That is, $\deg(g \wedge 1_q \circ 1_p \wedge \gamma_q) = \deg g$.

Recall from Problem 1 of Section 6 that there is a natural "suspension" isomorphism $\tilde{H}_k(X) \xrightarrow{\approx} \tilde{H}_{k+1}(\Sigma X)$. Also, for pointed X, there is a canonical map $\Sigma X \to X \wedge \mathbf{S}^1$ collapsing the arc between the poles through the base point. For most X this is a homotopy equivalence, and that is true when $X \approx \mathbf{S}^k$. (We only need that $H_{k+1}(\Sigma X) \xrightarrow{\approx} H_{k+1}(X \wedge \mathbf{S}^1)$ when $X \approx \mathbf{S}^k$ and that is an easy consequence of Theorem 7.4 by comparing both spaces to the sphere obtained by collapsing the complement of a nice disk away from the "base point arc.") Therefore, the composition $\tilde{H}_k(X) \to \tilde{H}_{k+1}(X \wedge \mathbf{S}^1)$, of these two maps, is a natural homomorphism which is an isomorphism when $X \approx \mathbf{S}^k$. Iterating this q times and using that $\mathbf{S}^q = \mathbf{S}^1 \wedge \cdots \wedge \mathbf{S}^1$, q times, gives the natural homomorphism $\phi \colon \tilde{H}_k(X) \to \tilde{H}_{k+q}(X \wedge \mathbf{S}^q)$ which is an isomorphism for $X \approx \mathbf{S}^k$. (Incidentally, a general condition on X for ϕ to be an isomorphism is that X be "well-pointed"; see Theorem 1.9 of Chapter VII.) Thus there is the commutative diagram

$$\begin{array}{ccc} H_{p-1}(\partial \mathbf{I}^p) & \xrightarrow[\approx]{\phi} & H_{p+q-1}(\partial \mathbf{I}^p \wedge \mathbf{S}^q) \\ \downarrow g_* & & \downarrow (g \wedge 1_q)_* \\ H_{p-1}(\mathbf{S}^{p-1}) & \xrightarrow[\approx]{\phi} & H_{p+q-1}(\mathbf{S}^{p-1} \wedge \mathbf{S}^q) = H_{p+q-1}(\mathbf{S}^{p+q-1}) \end{array}$$

for *any* g.

Define $\epsilon = \pm 1$ by $\phi[\mathbf{S}^{p-1}] = \epsilon[\mathbf{S}^{p+q-1}]$. By definition, $g_*[\partial \mathbf{I}^p] = (\deg g)[\mathbf{S}^{p-1}]$. Hence

$$(g \wedge 1_q)_* \phi[\partial \mathbf{I}^p] = \phi g_*[\partial \mathbf{I}^p] = \epsilon \deg(g)[\mathbf{S}^{p+q-1}].$$

In the special case where g is $1_1 \wedge \gamma_{p-1}: \partial \mathbf{I}^p \to \mathbf{S}^{p-1}$, our orientation conventions specify that $\deg(1_1 \wedge \gamma_{p-1}) = 1$. Substituting $1_1 \wedge \gamma_{p-1}$ for g in the above equation gives

$$(1_1 \wedge \gamma_{p-1} \wedge 1_q)_* \phi[\partial \mathbf{I}^p] = \epsilon[\mathbf{S}^{p+q-1}].$$

For general g then, these two equations show that

$$\deg(g \wedge 1_q) = \deg(g) \cdot \deg(1_1 \wedge \gamma_{p-1} \wedge 1_q).$$

(The degrees make sense only after orienting $\partial \mathbf{I}^p \wedge \mathbf{S}^q$, but the *equation* is not affected by that choice.) Composing this with $1_p \wedge \gamma_q$, and noting that $(1_1 \wedge \gamma_{p-1} \wedge 1_q) \circ (1_p \wedge \gamma_q) = 1_1 \wedge \gamma_{p+q-1}$ in the diagram

$$\begin{array}{ccc} \partial(\mathbf{I}^p \times \mathbf{I}^q) & \xrightarrow{1_p \wedge \gamma_q} & \partial \mathbf{I}^p \wedge \mathbf{S}^q \\ {\scriptstyle 1_1 \wedge \gamma_{p+q-1}} \downarrow & & \downarrow {\scriptstyle 1_1 \wedge \gamma_{p-1} \wedge 1_q} \\ \mathbf{S}^{p+q-1} & = & \mathbf{S}^{p-1} \wedge \mathbf{S}^q \end{array}$$

we conclude that

$$\begin{aligned} \deg(g \wedge 1_q \circ 1_p \wedge \gamma_q) &= \deg(g) \cdot \deg((1_1 \wedge \gamma_{p-1} \wedge 1_q) \circ (1_p \wedge \gamma_q)) \\ &= \deg(g) \cdot \deg(1_1 \wedge \gamma_{p+q-1}) = \deg(g), \end{aligned}$$

where the last equation is by our orientation conventions. □

12.2. Theorem. *For a product of two CW-complexes with our orientation conventions, the boundary operators satisfy the equation*

$$\partial(\sigma \times \mu) = \partial\sigma \times \mu + (-1)^{\deg \sigma} \sigma \times \partial \mu.$$

PROOF. It is clear, by remarks similar to those above, that $[\sigma \times \phi : \sigma \times \mu] = \pm[\phi : \mu]$ where the sign ± 1 depends only on the dimensions of σ and μ. For a 0-cell σ, it is clear that the sign is $+1$. Direct arguments can determine this sign, as in the proof of Lemma 12.1. But this is difficult and the following approach is much easier.

Let $\epsilon_{p,q} = \pm 1$ be the sign defined by the equation

$$[\sigma \times \phi : \sigma \times \mu] = \epsilon_{p,q}[\phi : \mu]$$

where σ is a p-cell of K, μ is a q-cell of L, and ϕ is a $(q-1)$-cell of L. We have that $\epsilon_{0,q} = 1$ for all q. We compute

$$\begin{aligned} \partial(\sigma \times \mu) &= \sum_\tau [\tau \times \mu : \sigma \times \mu] \tau \times \mu + \sum_\phi [\sigma \times \phi : \sigma \times \mu] \sigma \times \phi \\ &= \sum_\tau [\tau : \sigma] \tau \times \mu + \sum_\phi \epsilon_{p,q}[\phi : \mu] \sigma \times \phi \\ &= (\partial\sigma) \times \mu + \epsilon_{p,q} \sigma \times (\partial \mu). \end{aligned}$$

But we *know* that $\partial^2 = 0$, and so

$$\begin{aligned}0 = \partial\partial(\sigma \times \mu) &= (\partial\partial\sigma) \times \mu + \epsilon_{p-1,q}(\partial\sigma) \times (\partial\mu)\\ &\quad + \epsilon_{p,q}(\partial\sigma) \times (\partial\mu) + \epsilon_{p,q-1}\epsilon_{p,q}\sigma \times (\partial\partial\mu)\\ &= \epsilon_{p-1,q}(\partial\sigma) \times (\partial\mu) + \epsilon_{p,q}(\partial\sigma) \times (\partial\mu).\end{aligned}$$

For this to always hold (for all K and L) we clearly must have $\epsilon_{p,q} = -\epsilon_{p-1,q}$. Since $\epsilon_{0,q} = 1$, we conclude that $\epsilon_{p,q} = (-1)^p$. □

For the sake of completeness, we now prove the fact, about the topology of a product complex, alluded to at the beginning of this section.

12.3. Theorem. *If K and L are CW-complexes and if L is locally finite then the weak topology coincides with the product topology on $K \times L$.*

Proof. For each n-cell σ of K, let D_σ be a copy of $\mathbf{I}^n$ and consider the characteristic map f_σ as defined on D_σ, and similarly for cells τ of L. Then, for the product topology on $K \times L$, there is the map

$$+D_\sigma \times D_\tau \to K \times L,$$

where $+$ denotes disjoint union over all cells σ of K and τ of L. It suffices to show that this map is an identification since that is exactly what the weak topology on $K \times L$ is. We can factor this as the composition

$$+D_\sigma \times D_\tau \to +D_\sigma \times L \to K \times L.$$

The first map is an identification by Proposition 13.19 of Chapter I since $+D_\sigma$ is locally compact. The second map is an identification since L is locally compact (which follows easily from local finiteness). Therefore the composition is an identification by Proposition 13.3 of Chapter I. □

Problems

1. Compute $H_i(\mathbf{S}^p \times \mathbf{S}^q)$.
2. Compute $H_i(\mathbf{P}^2 \times \mathbf{P}^2)$.
3. Compute the homology of the product of a Klein bottle and a real projective plane.
4. Let $X = \mathbf{S}^1 \cup_f \mathbf{D}^2$ and $Y = \mathbf{S}^1 \cup_g \mathbf{D}^2$ where $f\colon \mathbf{S}^1 \to \mathbf{S}^1$ has degree p and $g\colon \mathbf{S}^1 \to \mathbf{S}^1$ has degree q. Compute $H_i(X \times Y)$.
5. Let K and L be CW-complexes. Let $K \times L$ denote the product space with the product topology and let $K \bigcirc L$ denote $K \times L$ with the CW topology. Show that the canonical function $\phi\colon K \bigcirc L \to K \times L$ is continuous. Also show that if X is compact then a function $f\colon X \to K \bigcirc L$ is continuous $\Leftrightarrow \phi \circ f\colon X \to K \times L$ is continuous. Conclude that $\phi_*\colon H_*(K \bigcirc L) \to H_*(K \times L)$ (singular homology) and $\phi_\#\colon \pi_*(K \bigcirc L) \to \pi_*(K \times L)$ are isomorphisms.

13. Euler's Formula

Let us recall, without proof, the Fundamental Theorem of Abelian Groups. A finitely generated free abelian group A is isomorphic to $\mathbf{Z}^r$ for some r. Suppose that $B \subset A$ is a subgroup. Then there exists a basis $a_1, \ldots, a_r$ of A and nonzero integers $n_1 | n_2 | \cdots | n_s$ (each dividing the next) with $s \le r$ such that $n_1 a_1, \ldots, n_s a_s$ is a basis for B. In particular, B is free abelian of rank s and

$$A/B \approx \mathbf{Z}_{n_1} \oplus \cdots \oplus \mathbf{Z}_{n_s} \oplus \mathbf{Z}^{r-s}. \tag{1}$$

The integer $r - s \ge 0$ is called the rank of A/B. Note that it is the dimension of the rational vector space $(A/B) \otimes \mathbf{Q}$ where $\mathbf{Q}$ is the rationals.

Thus any finitely generated abelian group has the form of (1) and if $0 \to B \to A \to C \to 0$ is an exact sequence of finitely generated abelian groups then $\operatorname{rank}(A) = \operatorname{rank}(B) + \operatorname{rank}(C)$.

13.1. Definition. A space X is said to be of *finite type* if $H_i(X)$ is finitely generated for each i. It is of *bounded finite type* if $H_i(X)$ is also zero for all but a finite number of i.

13.2. Definition. If X is a space of bounded finite type then its *Euler characteristic* is

$$\chi(X) = \sum_i (-1)^i \operatorname{rank} H_i(X).$$

Note then that $\chi(X)$ is a topological invariant of X.

13.3. Theorem (Euler–Poincaré). *Let X be a finite CW-complex and let a_i be the number of i-cells in X. Then $\chi(X)$ is defined and*

$$\chi(X) = \sum_i (-1)^i a_i.$$

Proof. Note that $a_i = \operatorname{rank} C_i(X)$. Let $Z_i \subset C_i = C_i(X)$ be the group of i-cycles, $B_i = \partial C_{i+1}$, the group of i-boundaries, and $H_i = H_i(C_*(X)) \approx H_i(X)$. Thus $H_i = Z_i / B_i$.

The exact sequence

$$0 \to Z_i \to C_i \xrightarrow{\partial} B_{i-1} \to 0$$

shows that

$$\operatorname{rank}(C_i) = \operatorname{rank}(Z_i) + \operatorname{rank}(B_{i-1}).$$

Similarly the exact sequence $0 \to B_i \to Z_i \to H_i \to 0$ shows that

$$\operatorname{rank}(Z_i) = \operatorname{rank}(B_i) + \operatorname{rank}(H_i).$$

Adding the last two equations with signs $(-1)^i$ gives

$$\sum_i (-1)^i(\operatorname{rank}(H_i) + \operatorname{rank}(B_i)) = \sum_i (-1)^i \operatorname{rank}(Z_i)$$

$$= \sum_i (-1)^i(\operatorname{rank}(C_i) - \operatorname{rank}(B_{i-1})).$$

The terms in B_* cancel, leaving $\chi(X) = \sum_i (-1)^i \operatorname{rank}(H_i) = \sum_i (-1)^i \operatorname{rank}(C_i) = \sum_i (-1)^i a_i$. □

13.4. Corollary (Euler). *For any CW-complex structure on the 2-sphere with F 2-cells, E 1-cells and V 0-cells, we have $F - E + V = 2$.* □

13.5. Proposition. *If $X \to Y$ is a covering map with k sheets (k finite) and Y is a finite CW-complex then X is also a CW-complex and $\chi(X) = k\chi(Y)$.*

PROOF. Since the characteristic maps $\mathbf{D}^i \to Y$ are maps from a simply connected space, they lift to X in exactly k ways. This gives the structure of a CW-complex on X with the number of i-cells exactly k times that number for Y. (Also see Theorem 8.10.) Thus the alternating sum of these for X is k times the same thing for Y. □

13.6. Corollary. *If $\mathbf{S}^{2n} \to Y$ is a covering map and Y is CW then the number of sheets is either* 1 *or* 2. □

13.7. Corollary. *The Euler characteristic of real projective $2n$-space $\mathbf{P}^{2n}$ is* 1. □

13.8. Corollary. *If $f: \mathbf{P}^{2n} \to Y$ is a covering map and Y is a CW-complex then f is a homeomorphism.* □

The hypothesis that Y is a CW-complex in Corollaries 13.6 and 13.8 can be dropped, but we do not now have the machinery to prove that.

PROBLEMS

1. Use the knowledge of the covering spaces of the torus, but do not use the knowledge of its homology groups, to show that its Euler characteristic is zero.

2. If X is a finite CW-complex of dimension two, and if X is simply connected then show that $\chi(X)$ determines $H_2(X)$ completely. What are the possible values for $\chi(X)$ in this situation?

3. Let

$$A(t) = \sum_{i=0}^{\infty} a_i t^i \quad \text{and} \quad B(t) = \sum_{i=0}^{\infty} b_i t^i$$

be formal power series. Define a relation $A(t) \gg B(t)$ to mean that there is a formal power series

$$C(t) = \sum_{i=0}^{\infty} c_i t^i$$

with all $c_i \geq 0$ such that

$$A(t) - B(t) = (1 + t)C(t).$$

(a) Show that $A(t) \gg B(t)$ is equivalent to the "Morse inequalities":

$$a_0 \geq b_0,$$
$$a_1 - a_0 \geq b_1 - b_0,$$
$$a_2 - a_1 + a_0 \geq b_2 - b_1 + b_0,$$
$$\cdots$$

(b) If C_* is a chain complex of finite type over a field Λ with $C_i = 0$ for $i < 0$, let $C(t) = \sum_i \operatorname{rank}(C_i)t^i$ and $H(t) = \sum_i \operatorname{rank}(H_i(C_*))t^i$. Show that $C(t) \gg H(t)$.
(c) Derive the Euler–Poincaré formula from (b).

4. If X and Y are finite CW-complexes, show that $\chi(X \times Y) = \chi(X)\chi(Y)$.

14. Homology of Real Projective Space

Construct a CW-complex structure on $\mathbf{S}^n$ with exactly two cells in each dimension i, $0 \leq i \leq n$, by letting the i-cells be two hemispheres of $\mathbf{S}^i \subset \mathbf{S}^n$ for each i. For $k \leq n$ denote the two k-cells by σ_k and $T\sigma_k$. We can take the characteristic map of the latter to be

$$f_{T\sigma_k} = T \circ f_{\sigma_k}$$

where $T: \mathbf{S}^n \to \mathbf{S}^n$ is the antipodal map. Note that $p_{T\sigma} f_{T\sigma} = p_\sigma f_\sigma$ since both are equal to γ_k. Also note that the first of these equals $p_{T\sigma} T f_\sigma$. Since the equation $p_\sigma f_\sigma = \gamma_k$ characterizes the projection p_σ we conclude that

$$p_{T\sigma} T = p_\sigma.$$

Now the composition $p_{\sigma_{k-1}} f_{\partial\sigma_k}: \mathbf{S}^{k-1} \to \mathbf{S}^{k-1}$ collapses a hemisphere to a point and is otherwise a homeomorphism. This is clearly homotopic to a homeomorphism and thus has degree ± 1. (This also follows directly from Proposition 7.2 or Corollary 7.5.)

We can choose the characteristic maps f_{σ_k} inductively so that

$$\deg(p_{\sigma_{k-1}} f_{\partial\sigma_k}) = 1.$$

We also have that $\deg(p_{T\sigma_{k-1}} f_{\partial\sigma_k}) = \pm 1$. Thus

$$[\sigma_{k-1} : \sigma_k] = 1$$
$$[T\sigma_{k-1} : \sigma_k] = \pm 1.$$

We will determine the correct sign in a moment. It follows that

(1) $$\partial\sigma_k = \sigma_{k-1} \pm T\sigma_{k-1}.$$

Now

$$p_{T\sigma_{k-1}} f_{\partial T\sigma_k} = p_{T\sigma_{k-1}} T f_{\partial\sigma_k} = p_{\sigma_{k-1}} f_{\partial\sigma_k}$$

has degree 1. Similarly

$$p_{\sigma_{k-1}} f_{\partial T\sigma_k} = p_{\sigma_{k-1}} T f_{\partial\sigma_k} = p_{T\sigma_{k-1}} f_{\partial\sigma_k}$$

has the same degree as the sign in (1). Thus

(2) $$\partial(T\sigma_k) = T\sigma_{k-1} \pm \sigma_{k-1}$$

where the sign is the same as in (1). From (1) and (2) we see that $T: C_* \to C_*$ is a chain map.

Now $\partial\sigma_1 = \sigma_0 \pm T\sigma_0$, but the sign here must be "$-$" in order for H_0 to come out correctly. Thus $\partial\sigma_1 = \sigma_0 - T\sigma_0 = (1-T)\sigma_0$.

Similarly, $\partial\sigma_2 = (1 \pm T)\sigma_1$. But we calculate

$$0 = \partial\partial\sigma_2 = (1 \pm T)\partial\sigma_1 = (1 \pm T)(1-T)\sigma_0.$$

If the sign here is "$-$" then the result is $0 = 2(1-T)\sigma_0$, which is false. Thus the sign must be "$+$." Similar arguments establish that these signs must alternate in order that $\partial\partial = 0$ on all the σ_k. Thus

$$\partial\sigma_k = (1 + (-1)^k T)\sigma_{k-1}.$$

Now the map $\pi: \mathbf{S}^n \to \mathbf{P}^n$ induces a cell structure on $\mathbf{P}^n$ with a single cell τ_k for each $k \le n$ and with characteristic map $f_{\tau_k} = \pi \circ f_{\sigma_k} = \pi T f_{\sigma_k} = \pi f_{T\sigma_k}$.

Thus the map $\pi: \mathbf{S}^n \to \mathbf{P}^n$ induces the chain map $\pi_\Delta: C_k(\mathbf{S}^n) \to C_k(\mathbf{P}^n)$ taking both σ_k and $T\sigma_k$ to τ_k. Therefore

$$\partial\tau_k = \partial\pi_\Delta\sigma_k = \pi_\Delta\partial\sigma_k = \pi_\Delta(1 + (-1)^k T)\sigma_{k-1} = (1 + (-1)^k)\tau_{k-1},$$

which is 0 if k is odd and $2\tau_{k-1}$ if k is even. Thus we can compute the homology groups of $\mathbf{P}^n$ from this, with the final result:

$$n \text{ even} \Rightarrow H_i(\mathbf{P}^n) = \begin{cases} \mathbf{Z} & \text{for } i = 0, \\ \mathbf{Z}_2 & \text{for } i \text{ odd}, 0 < i < n, \\ 0 & \text{otherwise,} \end{cases}$$

$$n \text{ odd} \Rightarrow H_i(\mathbf{P}^n) = \begin{cases} \mathbf{Z} & \text{for } i = 0, n, \\ \mathbf{Z}_2 & \text{for } i \text{ odd}, 0 < i < n, \\ 0 & \text{otherwise.} \end{cases}$$

Problems

1. Use the formulas developed in this section for the CW-structure on $\mathbf{S}^n$ to rederive the degree of the antipodal map.

2. ◆ Let $\mu = e^{2\pi i/p}$. The map $(z_1, \ldots, z_n) \mapsto (\mu z_1, \ldots, \mu z_n)$ is a map of period p on the

sphere $\mathbf{S}^{2n-1}$. It defines a properly discontinuous action of $\mathbf{Z}_p$ on $\mathbf{S}^{2n-1}$ and so gives a p-fold covering $\mathbf{S}^{2n-1} \to L$ of its orbit space L (which is a special case of what is called a "lens space"). Compute the homology groups of L.

15. Singular Homology

We now return to the study of the singular complex of a space. The reader may wish to review the definitions and elementary properties given in Section 1.

First we shall give some further material on homological algebra.

15.1. Definition. Let ϕ and ψ be chain maps $A_* \to B_*$ between chain complexes. Then we say that they are *chain homotopic*, $\phi \simeq \psi$ in symbols, if there exists a sequence of homomorphisms $D: A_i \to B_{i+1}$ such that $\partial D + D\partial = \phi - \psi$.

15.2. Proposition. *If* $\phi \simeq \psi: A_* \to B_*$ *then* $\phi_* = \psi_*: H_*(A_*) \to H_*(B_*)$.

Proof. If $\partial a = 0$ then $\phi_*[\![a]\!] - \psi_*[\![a]\!] = [\![\phi(a) - \psi(a)]\!] = [\![(\partial D + D\partial)a]\!] = [\![\partial D(a)]\!] = 0$. □

15.3. Proposition. *Chain homotopy is an equivalence relation.*

Proof. If $\phi - \psi = \partial D + D\partial$ and $\psi - \eta = \partial D' + D'\partial$ then $\phi - \eta = \partial(D + D') + (D + D')\partial$. □

15.4. Definition. A chain map $\phi: A_* \to B_*$ is called a *chain equivalence* if there is a chain map $\psi: B_* \to A_*$ such that $\phi\psi \simeq 1_{B_*}$ and $\psi\phi \simeq 1_{A_*}$.

Note that in the situation of Definition 15.4, the induced maps satisfy $\phi_*\psi_* = 1$ on $H_*(B_*)$ and $\psi_*\phi_* = 1$ on $H_*(A_*)$, and so ϕ_* and ψ_* are both isomorphisms which are inverses of one another. Clearly the relation of chain equivalence is an equivalence relation on chain complexes.

The following theorem is basic to singular homology theory. Remember that we do not yet know that singular theory satisfies the axioms.

15.5. Theorem. *If* X *is contractible then* $H_i(X) = 0$ *for all* $i \neq 0$.

Proof. Let $F: X \times I \to X$ with $F(x, 0) = x$ and $F(x, 1) = x_0$ for all $x \in X$ and for some base point $x_0 \in X$. Define $D\sigma: \Delta_n \to X$, for each singular simplex $\sigma: \Delta_{n-1} \to X$ of X, by

$$(D\sigma)\left(\sum_{i=0}^{n} \lambda_i e_i\right) = F\left(\sigma\left(\sum_{i=1}^{n} \frac{\lambda_i}{\lambda} e_{i-1}\right), \lambda_0\right)$$

where $\sum_0^n \lambda_i = 1$ and $\lambda = \sum_1^n \lambda_i = 1 - \lambda_0$. (See Figure IV-6.) This is called the "cone construction."

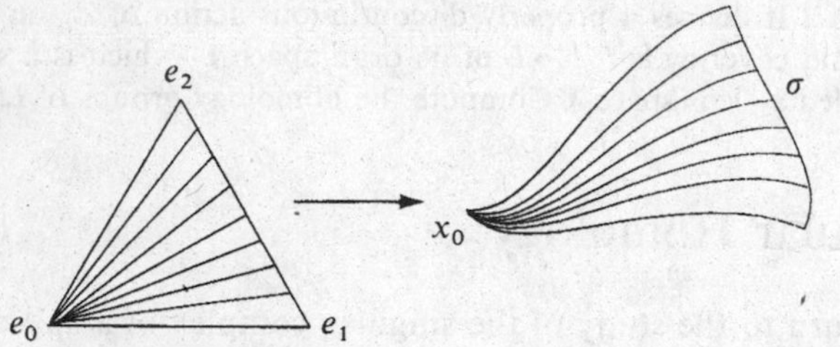

Figure IV-6. A chain homotopy.

Since D has been defined on a basis of the free abelian group $\Delta_{n-1}(X)$ it extends uniquely to a homomorphism $D: \Delta_{n-1}(X) \to \Delta_n(X)$.

To compute the ith face of the singular simplex $D\sigma$ for $i > 0$ just put $\lambda_i = 0$ or eliminate the λ_i terms, and you get

$$(D\sigma)^{(i)} = D(\sigma^{(i-1)}).$$

Clearly also

$$(D\sigma)^{(0)} = \sigma.$$

Let us compute $\partial(D\sigma)$. For $n > 1$ we find

$$\partial(D\sigma) = (D\sigma)^{(0)} - \sum_{i=1}^{n} (-1)^{i-1}(D\sigma)^{(i)} = (D\sigma)^{(0)} - \sum_{j=0}^{n-1} (-1)^j D(\sigma^{(j)}) = \sigma - D(\partial\sigma).$$

For $n = 1$ (i.e., σ a 0-simplex) we have

$$\partial(D\sigma) = \sigma - \sigma_0 \qquad \text{where} \quad \sigma_0: \Delta_0 \to \{x_0\},$$

and $D(\partial\sigma) = 0$ by definition. Thus $\partial D + D\partial = 1 - \epsilon$ where $\epsilon: \Delta_i(X) \to \Delta_i(X)$ is given by $\epsilon = 0$ for $i \neq 0$ and $\epsilon(\sum n_\sigma \sigma) = (\sum n_\sigma)\sigma_0$ for $i = 0$ and where σ_0 is the 0-simplex at x_0. That is, ϵ is the "augmentation." Thus, in homology $1 = 1_* = \epsilon_*$ which is 0 in nonzero dimensions. □

16. The Cross Product

We would like to define a bilinear map (for each $p, q \geq 0$)

$$\times: \Delta_p(X) \times \Delta_q(Y) \to \Delta_{p+q}(X \times Y).$$

The image of $\langle a, b \rangle$ will be denoted by $a \times b$. The geometric idea here is easy to describe. If σ is a singular p-simplex in X and τ is a singular q-simplex in Y then $\langle \sigma, \tau \rangle$ is a map from the prism $\Delta_p \times \Delta_q$ to $X \times Y$. Thus if we somehow subdivide the prism into simplices we can regard that map as defining a $(p+q)$-chain in $X \times Y$. This is essentially what we do, except that the process of subdivision is replaced by an algebraic analogue that is easier to implement.

First let us establish some notation. If $x \in X$ then we will use x to also stand for the singular 0-simplex taking the vertex e_0 into x. That is, $x(e_0) = x$. For $\sigma: \Delta_q \to Y$ we let $x \times \sigma$ be the singular q-simplex of $X \times Y$ taking $w \mapsto (x, \sigma(w))$ and similarly, for $y \in Y$ and $\tau: \Delta_p \to X$, we let $\tau \times y$ denote the singular p-simplex of $X \times Y$ taking $w \mapsto (\tau(w), y)$. This defines $\times$ on $\Delta_0(X) \times \Delta_q(Y)$ and on $\Delta_p(X) \times \Delta_0(Y)$.

16.1. Theorem. *There exist bilinear maps* $\times : \Delta_p(X) \times \Delta_q(Y) \to \Delta_{p+q}(X \times Y)$ *such that:*

(1) *for* $x \in X$, $y \in Y$, $\sigma: \Delta_q \to Y$, *and* $\tau: \Delta_p \to X$, $x \times \sigma$ *and* $\tau \times y$ *are as described above;*

(2) *(naturality) if* $f: X \to X'$ *and* $g: Y \to Y'$ *and if* $\langle f, g \rangle: X \times Y \to X' \times Y'$ *denotes the product map, then*
$$\langle f, g \rangle_\Delta(a \times b) = f_\Delta(a) \times g_\Delta(b); \quad \textit{and}$$

(3) *(boundary formula)* $\partial(a \times b) = \partial a \times b + (-1)^{\deg a} a \times \partial b$.

PROOF. Note that (3) holds when either p or q is zero. The method of proof here goes by the name of "acyclic models." One can give a general form of it allowing the plugging in of specific situations like the present one. We much prefer, however, to just use it as a method, as that gives one a better idea of just what is going on.

Let $\iota_p: \Delta_p \to \Delta_p$ be the identity map thought of as a singular p-simplex of the space Δ_p.

Let $p > 0$ and $q > 0$ be given and assume that $\times$ has been defined for smaller $p + q$ satisfying (1), (2), and (3). The idea is this: first try to define $\iota_p \times \iota_q$ (on the "models"). To do that, write down what its boundary would have to be by (3). Then compute the boundary of that to see that it is a cycle. Since the space $\Delta_p \times \Delta_q$ is contractible, and hence "acyclic," this cycle must be a boundary. What it is a boundary of is then taken to be $\iota_p \times \iota_q$. Next we define $\sigma \times \tau$ in general by applying (2) to the maps $\sigma: \Delta_p \to X$ and $\tau: \Delta_q \to Y$. We now carry this plan out.

If $\iota_p \times \iota_q$ were defined then, by (3), its boundary would be
$$\text{“}\partial(\iota_p \times \iota_q)\text{”} = \partial\iota_p \times \iota_q + (-1)^p \iota_p \times \partial\iota_q$$
in $\Delta_{p+q-1}(\Delta_p \times \Delta_q)$. We compute the boundary of the right-hand side
$$\partial(\text{rhs}) = \partial\partial\iota_p \times \iota_q + (-1)^{p-1}\partial\iota_p \times \partial\iota_q + (-1)^p \partial\iota_p \times \partial\iota_q + \iota_p \times \partial\partial\iota_q = 0.$$

Thus the rhs is a $(p+q-1)$-cycle in $\Delta_p \times \Delta_q$. Since this space is contractible and $p + q > 1$, its homology is zero by Theorem 15.5. Thus the rhs is a boundary of some chain. Choose any such chain to be $\iota_p \times \iota_q$.

Now, if $\sigma: \Delta_p \to X$ and $\tau: \Delta_q \to Y$ are arbitrary singular simplices then, regarding them also as maps which then induce homomorphisms of chain groups, we have $\sigma = \sigma_\Delta(\iota_p)$ and $\tau = \tau_\Delta(\iota_q)$. As before, let $\langle \sigma, \tau \rangle: \Delta_p \times \Delta_q \to$

$X \times Y$ denote the product map. By (2) we *must* define $\sigma \times \tau = \sigma_\Delta(\iota_p) \times \tau_\Delta(\iota_q) = \langle \sigma, \tau \rangle_\Delta(\iota_p \times \iota_q)$. Then it is clear that (2) holds in general in these dimensions.

To verify property (3) we compute

$$\begin{aligned}\partial(\sigma \times \tau) &= \partial(\langle \sigma, \tau \rangle_\Delta(\iota_p \times \iota_q))\\ &= \langle \sigma, \tau \rangle_\Delta(\partial(\iota_p \times \iota_q))\\ &= \langle \sigma, \tau \rangle_\Delta(\partial\iota_p \times \iota_q + (-1)^p \iota_p \times \partial\iota_q)\\ &= \sigma_\Delta(\partial\iota_p) \times \tau_\Delta(\iota_q) + (-1)^p \sigma_\Delta(\iota_p) \times \tau_\Delta(\partial\iota_q)\\ &= \partial\sigma_\Delta(\iota_p) \times \tau_\Delta(\iota_q) + (-1)^p \sigma_\Delta(\iota_p) \times \partial\tau_\Delta(\iota_q)\\ &= \partial\sigma \times \tau + (-1)^p \sigma \times \partial\tau.\end{aligned}$$

This extends to all chains by bilinearity. □

16.2. Definition. If (X, A) and (Y, B) are pairs of spaces then $(X, A) \times (Y, B)$ denotes the pair $(X \times Y, X \times B \cup A \times Y)$.

16.3. Proposition. *The cross product $\Delta_p(X) \times \Delta_q(Y) \to \Delta_{p+q}(X \times Y)$ induces a bilinear map $\times: H_p(X, A) \times H_q(Y, B) \to H_{p+q}((X, A) \times (Y, B))$ defined by $[\![a]\!] \times [\![b]\!] = [\![a \times b]\!]$.*

PROOF. If $a \in \Delta_p(X)$ with $\partial a \in \Delta_{p-1}(A)$ (i.e., if a represents a cycle of (X, A)) and $b \in \Delta_q(Y)$ with $\partial b \in \Delta_{q-1}(B)$ then

$$\partial(a \times b) = \partial a \times b + (-1)^p a \times \partial b \in \Delta_{p+q-1}((A \times Y) \cup (X \times B))$$

so that $a \times b$ does represent a cycle of $(X, A) \times (Y, B)$. We must show that it does not depend on the choices of representatives a and b of the (relative) homology classes. This is clear if they are changed by adding chains in A and B respectively. Also, $(a + \partial a') \times (b + \partial b') = a \times b + a \times \partial b' + \partial a' \times b + \partial a' \times \partial b' = a \times b \pm \partial(a \times b') + \partial(a' \times b) + \partial(a' \times \partial b') +$ (a chain in $A \times Y \cup X \times B$) when a and b are relative cycles. □

The most important case of this is $X = I = [0, 1]$. We will also use I to denote the affine simplex $[\{0\}, \{1\}]: \Delta_1 \to I$ and we let ϵ_0 and ϵ_1 be the 0-simplices $\epsilon_0(e_0) = \{0\}$ and $\epsilon_1(e_0) = \{1\}$ of I. Thus $\partial I = \epsilon_1 - \epsilon_0$.

For a chain $c \in \Delta_q(X)$ we have $I \times c \in \Delta_{q+1}(I \times X)$, and

$$\partial(I \times c) = \epsilon_1 \times c - \epsilon_0 \times c - I \times \partial c.$$

Define $D: \Delta_q(X) \to \Delta_{p+1}(I \times X)$ by $D(c) = I \times c$. Then

$$(\partial D + D\partial)(c) = \partial(D(c)) + D(\partial c) = \epsilon_1 \times c - \epsilon_0 \times c.$$

Let η_0 and η_1 be the maps $X \to I \times X$ given by $\eta_0(x) = (0, x)$ and $\eta_1(x) = (1, x)$. Then $\eta_{i_\Delta}(c) = \epsilon_i \times c$. Thus we have

$$\partial D + D\partial = \eta_{1_\Delta} - \eta_{0_\Delta}.$$

16.4. Theorem. *If* $f_0 \simeq f_1: (X, A) \to (Y, B)$ *then* $f_{0_\Delta} \simeq f_{1_\Delta}: \Delta_*(X, A) \to \Delta_*(Y, B)$ *and therefore* $f_{0*} = f_{1*}: H_*(X, A) \to H_*(Y, B)$.

PROOF. The second statement follows immediately from the first. If $F: I \times (X, A) \to (Y, B)$ is a homotopy between f_0 and f_1 (so that $F \circ \eta_0 = f_0$ and $F \circ \eta_1 = f_1$) then it induces $F_\Delta: \Delta_*(I \times X, I \times A) \to \Delta_*(Y, B)$. If we compose this with $\partial D + D\partial = \eta_{1_\Delta} - \eta_{0_\Delta}$ we obtain

$$\partial(F_\Delta \circ D) + (F_\Delta \circ D)\partial = F_\Delta \circ (\partial D + D\partial) = F_\Delta \circ \eta_{1_\Delta} - F_\Delta \circ \eta_{0_\Delta} = f_{1_\Delta} - f_{0_\Delta},$$

which shows that $F_\Delta \circ D$ is the desired chain homotopy. □

16.5. Corollary. *Singular homology satisfies the Homotopy Axiom.* □

17. Subdivision

We wish to prove that singular homology satisfies the Excision Axiom. First we will indicate the difficulty with doing this, and outline the remedy, and finally we go into the detailed proof.

Suppose, for example, $U \subset A \subset X$ with $\bar{U} \subset \text{int}(A)$ and we wish to "excise" U. If all singular simplices which are not completely within A miss U completely, then we could just discard any simplex in A. Thus the problem is with "large" singular simplices, those touching both $X - A$ and U. These sets are "separated," i.e., their closures do not meet. Thus if we could somehow "subdivide" a singular simplex into smaller simplices (a chain) which satisfy the above condition then we might be able to make excision work.

We are going to define an operator Υ called "subdivision" on $\Delta_i(X)$ and a chain homotopy T from Υ to the identity.

Please recall the standard q-simplex $\Delta_q \subset \mathbf{R}^{q+1}$. Let $L_*(\Delta_q)$ be the subcomplex of $\Delta_*(\Delta_q)$ generated by the affine singular simplices, i.e., singular simplices of the form $\sigma: \Delta_p \to \Delta_q$ such that $\sigma(\sum_i \lambda_i e_i) = \sum_i \lambda_i v_i$ where $\sum_i \lambda_i = 1$ and $v_i = \sigma(e_i)$. We denote such affine singular simplices by $\sigma = [v_0, \ldots, v_p]$.

Now we define the "cone operator" which takes an affine simplex and forms the "cone" on it from some point, producing a simplex of one higher dimension. Let $v \in \Delta_q$ and let $\sigma = [v_0, \ldots, v_p]: \Delta_p \to \Delta_q$ be affine. The cone on σ from v is then defined to be $v\sigma = [v, v_0, \ldots, v_p]: \Delta_{p+1} \to \Delta_q$. For a chain $c = \sum_\sigma n_\sigma \sigma \in L_p(\Delta_q)$, let $vc = \sum_\sigma n_\sigma v\sigma \in L_{p+1}(\Delta_q)$. Taking $c \mapsto vc$ gives a homomorphism

$$L_p(\Delta_q) \to L_{p+1}(\Delta_q).$$

(By definition, $v0 = 0$.) If $p > 0$ then we compute

$$\partial[v, v_0, \ldots, v_p] = [v_0, \ldots, v_p] - \sum_i (-1)^i [v, v_0, \ldots, \hat{v}_i, \ldots, v_p]$$

$$= [v_0, \ldots, v_p] - v(\partial[v_0, \ldots, v_p]).$$

If $p=0$ then $\partial v\sigma = \sigma - [v]$. Thus, for a 0-chain c, $\partial vc = c - \epsilon(c)[v]$ where ϵ is the augmentation, of Section 2, assigning to a 0-chain the sum of its coefficients. Thus we have that

$$\partial(vc) = \begin{cases} c - v(\partial c) & \text{if } \deg(c) > 0, \\ c - \epsilon(c)[v] & \text{if } \deg(c) = 0. \end{cases}$$

We now define the "barycentric subdivision" operator $\Upsilon: L_p(\Delta_q) \to L_p(\Delta_q)$ inductively by

$$\Upsilon(\sigma) = \begin{cases} \underline{\sigma}(\Upsilon(\partial\sigma)) & \text{for } p > 0, \\ \sigma & \text{for } p = 0, \end{cases}$$

where $\underline{\sigma}$ denotes the "barycenter" of the affine simplex σ, i.e., $\underline{\sigma} = (\sum_{i=0}^{p} v_i)/(p+1)$ for $\sigma = [v_0, \ldots, v_p]$. This defines Υ on a basis of $L_p(\Delta_q)$ and thus we extend it linearly to be a homomorphism. See Figure IV-7.

17.1. Lemma. $\Upsilon: L_p(\Delta_q) \to L_p(\Delta_q)$ *is a chain map.*

Proof. We shall show that $\Upsilon(\partial\sigma) = \partial(\Upsilon(\sigma))$ inductively on p where σ is an affine p-simplex. If $p=0$ then $\Upsilon(\partial\sigma) = \Upsilon(0) = 0$, while $\partial(\Upsilon(\sigma)) = \partial\sigma = 0$, since there are no (-1)-chains. If $p=1$ then $\Upsilon(\partial\sigma) = \partial\sigma$ while $\partial(\Upsilon(\sigma)) = \partial(\underline{\sigma}(\Upsilon(\partial\sigma))) = \partial(\underline{\sigma}(\partial\sigma)) = \partial\sigma - \epsilon(\partial\sigma)[\underline{\sigma}] = \partial\sigma$.

For $p > 1$ and assuming that the formula is true for chains of degree $< p$, we have $\partial(\Upsilon(\sigma)) = \partial(\underline{\sigma}\Upsilon(\partial\sigma)) = \Upsilon(\partial\sigma) - \underline{\sigma}(\partial\Upsilon\partial\sigma) = \Upsilon(\partial\sigma)$ since $\partial\Upsilon\partial\sigma = \Upsilon\partial\partial\sigma = 0$ by the inductive assumption. □

Now we define $T: L_p(\Delta_q) \to L_{p+1}(\Delta_q)$ by induction on the formula

$$T\sigma = \underline{\sigma}(\Upsilon\sigma - \sigma - T(\partial\sigma)),$$

and $T=0$ for $p=0$.

We wish to show that $\partial T + T\partial = \Upsilon - 1$. For $p=0$ we compute

$$\partial T\sigma + T\partial\sigma = \partial(\underline{\sigma}(\Upsilon\sigma - \sigma)) = 0$$

since $\Upsilon\sigma = \sigma$ for $p=0$. For the same reason $(\Upsilon - 1)\sigma = 0$.

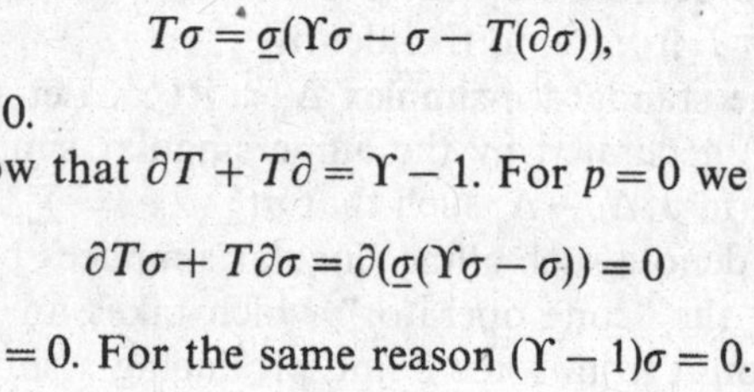

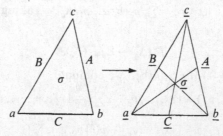

Figure IV-7. Barycentric subdivision.

For $p > 0$ we compute

$$(*) \qquad \partial T\sigma = (\Upsilon\sigma - \sigma - T\partial\sigma) - \underline{\sigma}(\partial\Upsilon\sigma - \partial\sigma - \partial T\partial\sigma).$$

The term $\partial T\partial\sigma = (\Upsilon - 1 - T\partial)(\partial\sigma) = \Upsilon\partial\sigma - \partial\sigma$ so that the entire right-hand term of (*) vanishes, which yields the claimed formula. Thus T is a chain homotopy from 1 to Υ.

We are now done for affine chains in Δ_q. We now transfer these results to general singular chains of X.

We wish to define $\Upsilon: \Delta_p(X) \to \Delta_p(X)$ and $T: \Delta_p(X) \to \Delta_{p+1}(X)$ such that:

(1) (naturality) $\Upsilon f_\Delta(c) = f_\Delta(\Upsilon c)$ and $T(f_\Delta(c)) = f_\Delta(T(c))$ for $f: X \to Y$;
(2) Υ is a chain map and $\partial T + T\partial = \Upsilon - 1$;
(3) Υ and T extend the previous definition on affine chains; and
(4) $\Upsilon\sigma$ and $T\sigma$ are chains in image(σ).

Note that (4) follows from (1). We list it for stress.

Thus let $\sigma: \Delta_p \to X$. Then we have $\sigma = \sigma_\Delta(\iota_p)$ and, of course, $\iota_p \in L_p(\Delta_p)$. We define

$$\Upsilon\sigma = \sigma_\Delta(\Upsilon\iota_p),$$
$$T\sigma = \sigma_\Delta(T\iota_p).$$

Of course, one must check that these coincide with the previous definitions when σ is affine, but this is obvious because Υ and T were defined on affine simplices using only affine operations. Property (4) is also clear, so this settles (3) and (4).

To show naturality (1) we compute $\Upsilon f_\Delta\sigma = \Upsilon(f\circ\sigma)_\Delta(\iota_p) = (f\circ\sigma)_\Delta(\Upsilon\iota_p) = f_\Delta(\sigma_\Delta(\Upsilon\iota_p)) = f_\Delta(\Upsilon\sigma)$, and similarly for T.

It remains to prove property (2). To show that Υ is a chain map, we compute

$$\begin{aligned}
\Upsilon\,\partial\sigma &= \Upsilon(\partial(\sigma_\Delta(\iota_p))) \\
&= \Upsilon(\sigma_\Delta(\partial\iota_p)) && \text{(since } \sigma_\Delta \text{ is a chain map)} \\
&= \sigma_\Delta(\Upsilon(\partial\iota_p)) && \text{(naturality)} \\
&= \sigma_\Delta(\partial(\Upsilon\iota_p)) && \text{(since } \iota_p \text{ is affine)} \\
&= \partial(\sigma_\Delta(\Upsilon\iota_p)) && \text{(since } \sigma_\Delta \text{ is a chain map)} \\
&= \partial(\Upsilon\sigma) && \text{(by definition).}
\end{aligned}$$

Similarly, for the formula involving T we compute

$$T\partial\sigma = T(\sigma_\Delta(\partial\iota_p)) = \sigma_\Delta(T\partial\iota_p)$$

and

$$\partial T\sigma = \partial\sigma_\Delta(T\iota_p) = \sigma_\Delta(\partial T\iota_p)$$

so that

$$(T\partial + \partial T)(\sigma) = \sigma_\Delta((T\partial + \partial T)\iota_p) = \sigma_\Delta((\Upsilon - 1)\iota_p) = (\Upsilon - 1)\sigma_\Delta(\iota_p) = \Upsilon\sigma - \sigma. \quad \square$$

17.2. Corollary. *For $k \geq 1$, $\Upsilon^k: \Delta_p(X) \to \Delta_p(X)$ is chain homotopic to the identity.*

PROOF. This follows from $\Upsilon^2 \simeq \Upsilon \circ 1 \simeq 1 \circ 1 = 1$, etc. Another way to show it, which displays the chain homotopy explicitly is to note that

$$\begin{aligned}\Upsilon^k - 1 &= \Upsilon^k - \Upsilon^{k-1} + \Upsilon^{k-1} - \cdots - 1 = (\Upsilon^{k-1} + \Upsilon^{k-2} + \cdots + \Upsilon + 1)(\Upsilon - 1) \\ &= G(\Upsilon - 1) = G(T\partial + \partial T) = (GT)\partial + \partial(GT). \qquad \square\end{aligned}$$

Let us denote this chain homotopy (GT in the second proof) by T_k and note that it is natural.

17.3. Lemma. *If $\sigma = [v_0, \ldots, v_p]$ is an affine simplex of Δ_q then any simplex in the chain $\Upsilon\sigma$ has diameter at most $(p/(p+1))\text{diam}(\sigma)$.*

PROOF. A simplex in $\Upsilon\sigma$ has the form $\underline{\sigma}\tau$ where τ is a simplex of $\Upsilon(\partial\sigma)$, i.e., τ is a simplex of $\Upsilon(\sigma^{(i)})$ for some i. Thus a simplex of $\Upsilon\sigma$ has the form $[\underline{\sigma}_0, \underline{\sigma}_1, \underline{\sigma}_2, \ldots]$ where $\sigma = \sigma_0 > \sigma_1 > \sigma_2 \ldots$, using $\alpha > \beta$ to mean that β is a proper face of α. Each of the barycenters $\underline{\sigma}_i$ is the average of *some* of the v_k. If $j > i$ then $\underline{\sigma}_j$ is the average of some of *these* v_k. Thus by reordering the vertices, the lemma comes down to the following:

If $w_1, \ldots, w_k \in \mathbf{R}^q$ with $m < k \le p + 1$ then

$$\left\| \frac{1}{k} \sum_{i=1}^{k} w_i - \frac{1}{m} \sum_{i=1}^{m} w_i \right\| \le \frac{p}{p+1} \max \| w_i - w_j \|.$$

Since $x/(x+1)$ is an increasing function and $m < k \le p + 1$, it suffices to show that the left-hand side of this inequality is at most $((k-1)/k) \max \| w_i - w_j \|$. We calculate

$$\begin{aligned}\left\| \frac{1}{m} \sum_{i=1}^{m} w_i - \frac{1}{k} \sum_{i=1}^{k} w_i \right\| &= \left\| \frac{1}{m} \sum_{i=1}^{m} w_i - \frac{1}{k} \sum_{i=1}^{m} w_i - \frac{1}{k} \sum_{i=m+1}^{k} w_i \right\| \\ &= \left\| \frac{k-m}{km} \sum_{i=1}^{m} w_i - \frac{1}{k} \sum_{i=m+1}^{k} w_i \right\| \\ &= \frac{k-m}{k} \left\| \frac{1}{m} \sum_{i=1}^{m} w_i - \frac{1}{k-m} \sum_{i=m+1}^{k} w_i \right\|.\end{aligned}$$

Both the terms in the norm of the last expression are in the convex span of the w_i and so this entire expression is at most $((k-m)/k) \max \| w_i - w_j \| \le ((k-1)/k) \max \| w_i - w_j \|$. $\square$

17.4. Corollary. *Each affine simplex in $\Upsilon^k(\iota_p) \in L_p(\Delta_q)$ has a diameter of at most $(p/(p+1))^k \text{diam}(\Delta_q)$, which approaches 0 as $k \to \infty$.* $\square$

17.5. Corollary. *Let X be a space and $\mathbf{U} = \{U_\alpha\}$ an open covering of X. Let σ be a singular p-simplex of X. Then $\exists k \ni \Upsilon^k(\sigma)$ is $\mathbf{U}$-small. That is, each simplex in $\Upsilon^k(\sigma)$ has image in some U_α.*

PROOF. This is an easy consequence of Corollary 17.4 and the Lebesgue Lemma (Lemma 9.11 of Chapter I). □

17.6. Definition. Let **U** be a collection of subsets of X whose interiors cover X. Let $\Delta_*^{\mathbf{U}}(X) \subset \Delta_*(X)$ be the subcomplex generated by the **U**-small singular simplices and let $H_*^{\mathbf{U}}(X) = H_*(\Delta_*^{\mathbf{U}}(X))$.

17.7. Theorem. *The map $H_*^{\mathbf{U}}(X) \to H_*(X)$ generated by inclusion is an isomorphism.*

PROOF. First we show the map to be a monomorphism. Let $c \in \Delta_p^{\mathbf{U}}(X)$ with $\partial c = 0$. Suppose that $c = \partial e$ for some $e \in \Delta_{p+1}(X)$. We must show that $c = \partial e'$ for some $e' \in \Delta_{p+1}^{\mathbf{U}}(X)$. There is a k such that $\Upsilon^k(e) \in \Delta_{p+1}^{\mathbf{U}}(X)$ and

$$\Upsilon^k(e) - e = T_k(\partial e) + \partial T_k(e) = T_k(c) + \partial T_k(e).$$

Thus

$$\partial \Upsilon^k(e) - \partial e = \partial T_k(c),$$

so that

$$c = \partial e = \partial(\Upsilon^k(e) - T_k(c)) \in \partial(\Delta_*^{\mathbf{U}}(X))$$

by the naturality of T_k.

Now we shall show the map to be onto. Let $c \in \Delta_p(X)$ with $\partial c = 0$. We must show that there is a $c' \in \Delta_p^{\mathbf{U}}(X)$ such that $c \sim c'$. There is a k such that $\Upsilon^k(c) \in \Delta_p^{\mathbf{U}}(X)$. Then

$$\Upsilon^k(c) - c = T_k(\partial c) + \partial T_k(c) = \partial T_k(c).$$

Thus $c' = \Upsilon^k(c)$ works. □

We remark that it can be shown that the isomorphism of Theorem 17.7 is induced by a chain equivalence.

To discuss the relative case of this result, put

$$\Delta_*^{\mathbf{U}}(X, A) = \Delta_*^{\mathbf{U}}(X) / \Delta_*^{\mathbf{U} \cap A}(A)$$

where $\mathbf{U} \cap A$ is the set of intersections of members of **U** with A. We have the commutative diagram

$$\begin{array}{ccccccccc} 0 \to & \Delta_*^{\mathbf{U}\cap A}(A) & \to & \Delta_*^{\mathbf{U}}(X) & \to & \Delta_*^{\mathbf{U}}(X, A) & \to 0 \\ & \downarrow & & \downarrow & & \downarrow & \\ 0 \to & \Delta_*(A) & \to & \Delta_*(X) & \to & \Delta_*(X, A) & \to 0. \end{array}$$

This induces a commutative "ladder" in homology

$$\begin{array}{cccccccccc} \to & H_i^{\mathbf{U}\cap A}(A) & \to & H_i^{\mathbf{U}}(X) & \to & H_i^{\mathbf{U}}(X, A) & \to & H_{i-1}^{\mathbf{U}\cap A}(A) & \to & H_{i-1}^{\mathbf{U}}(X) \to \\ & \downarrow \approx & & \downarrow \approx & & \downarrow & & \downarrow \approx & & \downarrow \approx \\ \to & H_i(A) & \to & H_i(X) & \to & H_i(X, A) & \to & H_{i-1}(A) & \to & H_{i-1}(X) \to . \end{array}$$

Thus $H_i^{\mathbf{U}}(X, A) \xrightarrow{\approx} H_i(X, A)$ follows from the 5-lemma.

Now we are prepared to prove the Excision Axiom. Note that the following statement of it is slightly stronger than the axiom itself.

17.8. Theorem (Excision). *If* $B \subset A \subset X$ *with* $\bar{B} \subset \operatorname{int}(A)$ *then the inclusion* $(X - B, A - B) \hookrightarrow (X, A)$ *induces an isomorphism* $H_*(X - B,\ A - B) \xrightarrow{\approx} H_*(X, A)$.

PROOF. Let $\mathbf{U} = \{A, X - B\}$. Then $X = \operatorname{int}(A) \cup (X - \bar{B}) = \operatorname{int}(A) \cup \operatorname{int}(X - B)$. Thus we have $H_*^{\mathbf{U}}(X, A) \xrightarrow{\approx} H_*(X, A)$. Note that

$$\Delta_*^{\mathbf{U}}(X) = \Delta_*(A) + \Delta_*(X - B)$$

as a subgroup of $\Delta_*(X)$. (The sum is *not* direct.) Also

$$\Delta_*(A - B) = \Delta_*(A) \cap \Delta_*(X - B).$$

By one of the Noetherian isomorphisms it follows that inclusion induces the isomorphism

$$\Delta_*(X - B)/\Delta_*(A - B) \xrightarrow{\approx} \Delta_*^{\mathbf{U}}(X)/\Delta_*(A).$$

Thus the inclusion maps induce

$$\begin{array}{ccc} \dfrac{\Delta_*(X - B)}{\Delta_*(A - B)} & \xrightarrow{\approx} & \dfrac{\Delta_*^{\mathbf{U}}(X)}{\Delta_*(A)} \\ & \searrow \quad \swarrow & \\ & \dfrac{\Delta_*(X)}{\Delta_*(A)} & \end{array}.$$

This diagram of chain complexes and chain maps induces the following diagram in homology:

$$\begin{array}{ccc} H_*(X - B, A - B) & \xrightarrow{\approx} & H_*^{\mathbf{U}}(X, A) \\ {\scriptstyle \mathrm{incl}_*} \searrow & & \swarrow {\scriptstyle \approx} \\ & H_*(X, A) & \end{array}.$$

It follows that the map marked incl_* is an isomorphism. □

This concludes the demonstration that singular theory does satisfy all the axioms.

18. The Mayer–Vietoris Sequence

In this section we derive an exact sequence that links the homology of A, B, $A \cap B$, and $A \cup B$ directly. It is often useful in doing computations of homology groups. It will also play a major role in the proofs of some important results.

18.1. Theorem (Mayer–Vietoris). *Let $A, B \subset X$ and suppose that $X = \mathrm{int}(A) \cup \mathrm{int}(B)$. Let $\mathrm{U} = \{A, B\}$. Let $i^A: A \cap B \hookrightarrow A$, $i^B: A \cap B \hookrightarrow B$, $j^A: A \hookrightarrow A \cup B$, and $j^B: B \hookrightarrow A \cup B$ be the inclusions. Then the sequence*

$$0 \to \Delta_p(A \cap B) \xrightarrow{i^A_\Delta \oplus i^B_\Delta} \Delta_p(A) \oplus \Delta_p(B) \xrightarrow{j^A_\Delta - j^B_\Delta} \Delta^{\mathrm{U}}_p(A \cup B) \to 0$$

is exact and so induces the long exact sequence

$$\cdots \to H_p(A \cap B) \xrightarrow{i^A_* \oplus i^B_*} H_p(A) \oplus H_p(B) \xrightarrow{j^A_* - j^B_*} H_p(A \cup B)$$

$$\to H_{p-1}(A \cap B) \to \cdots$$

called the "Mayer–Vietoris" sequence. If $A \cap B \neq \emptyset$ then the reduced sequence is also exact.

Proof. This is all straightforward and elementary, given our present machinery. □

The reader should note that we could just as well have taken the difference instead of the sum in the first map and the sum in the second.

We wish to give a version of this for relative homology. Consider the diagram

$$\begin{array}{ccccccccc}
 & & 0 & & 0 & & 0 & & \\
 & & \downarrow & & \downarrow & & \downarrow & & \\
0 & \to & \Delta_p(A \cap B) & \to & \Delta_p(A) \oplus \Delta_p(B) & \to & \Delta^{\mathrm{U}}_p(A \cup B) & \to & 0 \\
 & & \downarrow & & \downarrow & & \downarrow & & \\
0 & \to & \Delta_p(X) & \to & \Delta_p(X) \oplus \Delta_p(X) & \to & \Delta_p(X) & \to & 0 \\
 & & \downarrow & & \downarrow & & \downarrow & & \\
0 & \to & \dfrac{\Delta_p(X)}{\Delta_p(A \cap B)} & \to & \dfrac{\Delta_p(X)}{\Delta_p(A)} \oplus \dfrac{\Delta_p(X)}{\Delta_p(B)} & \to & \dfrac{\Delta_p(X)}{\Delta^{\mathrm{U}}_p(A \cup B)} & \to & 0 \\
 & & \downarrow & & \downarrow & & \downarrow & & \\
 & & 0 & & 0 & & 0 & &
\end{array}$$

with exact columns. The first two nontrivial rows are exact, and an easy diagram chase shows the third to be exact. (Another way to see that is to regard the rows as chain complexes. The columns then provide a short exact sequence of chain complexes and chain maps. Thus there is an induced long exact homology sequence with every two out of three terms zero, and so the remaining terms are also zero; which is equivalent to the third row being exact.)

Also consider the diagram

$$\begin{array}{ccccccccc}
0 & \to & \Delta^{\mathrm{U}}_p(A \cup B) & \to & \Delta_p(X) & \to & \dfrac{\Delta_p(X)}{\Delta^{\mathrm{U}}_p(A \cup B)} & \to & 0 \\
 & & \downarrow & & \downarrow & & \downarrow & & \\
0 & \to & \Delta_p(A \cup B) & \to & \Delta_p(X) & \to & \dfrac{\Delta_p(X)}{\Delta_p(A \cup B)} & \to & 0
\end{array}$$

with exact rows. The first two vertical maps induce isomorphisms and so the third one must also induce an isomorphism in homology by the 5-lemma. Putting this together gives:

18.2. Theorem. *If A, B are open in $A \cup B \subset X$ then there is the long exact Mayer–Vietoris sequence*

$$\cdots \to H_p(X, A \cap B) \to H_p(X, A) \oplus H_p(X, B) \to H_p(X, A \cup B)$$
$$\to H_{p-1}(X, A \cap B) \to \cdots,$$

where the map to the direct sum is induced by the sum of the inclusions and the map from the direct sum is the difference of those induced by the inclusions. □

PROBLEMS

1. Use the Mayer–Vietoris sequence to give another derivation of the homology groups of spheres (of all dimensions).
2. Use the Mayer–Vietoris sequence to compute the homology of the space which is the union of three n-disks along their common boundaries.
3. Use the Mayer–Vietoris sequence to give another derivation of the homology groups of the projective plane.
4. ♢ For $X = \text{int}(A) \cup \text{int}(B)$, consider the following commutative braid diagram:

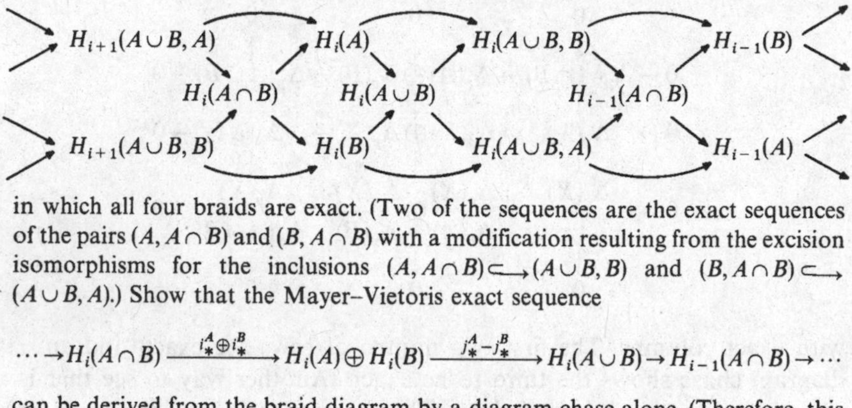

in which all four braids are exact. (Two of the sequences are the exact sequences of the pairs $(A, A \cap B)$ and $(B, A \cap B)$ with a modification resulting from the excision isomorphisms for the inclusions $(A, A \cap B) \hookrightarrow (A \cup B, B)$ and $(B, A \cap B) \hookrightarrow (A \cup B, A)$.) Show that the Mayer–Vietoris exact sequence

$$\cdots \to H_i(A \cap B) \xrightarrow{i_*^A \oplus i_*^B} H_i(A) \oplus H_i(B) \xrightarrow{j_*^A - j_*^B} H_i(A \cup B) \to H_{i-1}(A \cap B) \to \cdots$$

can be derived from the braid diagram by a diagram chase alone. (Therefore, this Mayer–Vietoris sequence follows from the axioms alone.)

19. The Generalized Jordan Curve Theorem

The Jordan Curve Theorem states that a circle in the plane divides the plane into two parts. The Generalized Jordan Curve Theorem proved below is the analogue in higher dimensions. As consequences of this theorem we also prove two classic theorems called Invariance of Domain and Invariance of Dimension. These show that for an open set in euclidean n-space the dimension n is a topological invariant. That is, if an open set in n-space is homeomorphic to one in m-space then $n = m$. This intuitively "obvious" fact is, in fact, rather difficult to prove without the sort of advanced tools we now have at our disposal.

First, however, we prove a result saying, in effect, that the homology of each piece of an increasing union of open sets determines that of the union.

Although we use the terminology "direct limit" ($\varinjlim$) in this section, it is not really necessary to know what a direct limit is, since we only use the two properties listed below in Theorem 19.1. However, the reader can consult Appendix D for this definition and its simple properties, if so desired.

19.1. Theorem. *Let X be a space and let $X = \bigcup\{U_i | i = 1, 2, 3, \ldots\}$ where the U_i are open and $U_i \subset U_{i+1}$, for all i. Let $i_n: U_n \hookrightarrow X$ and $i_{m,n}: U_n \hookrightarrow U_m$ be the inclusions. Then $H_p(X) = \varinjlim H_p(U_i)$. That is:*

(1) *each $\alpha \in H_p(X)$ is in $\operatorname{im}(i_{n*})$ for some n; and*
(2) *if $\alpha_n \in H_p(U_n)$ and $i_{n*}(\alpha_n) = 0$ then $i_{m,n*}(\alpha_n) = 0$ for m sufficiently large.*

PROOF. Note that if $C \subset X$ is compact then $C \subset U_n$ for some n. To prove (1), let α be represented by the singular chain $a \in \Delta_p(X)$. Since a involves only a finite number of singular simplices and the image of each of them is compact we must have that $a \in \Delta_p(U_n)$ for some n and this implies (1). Part (2) is proved in the same way. □

We wish to show that if $f: \mathbf{D}^r \to \mathbf{S}^n$ is an embedding then $\tilde{H}_*(\mathbf{S}^n - f(\mathbf{D}^r)) = 0$. The next theorem will be a generalization of this.

First, let us point out that the complement of the image of this r-disk in $\mathbf{S}^n$ need not be contractible. (If it were the result would be trivial.) One counterexample of this is the "wild arc" of Fox and Artin illustrated in Figure IV-8. It can be shown that the complement of this arc in $\mathbf{R}^3$ is not simply connected. Although this is quite plausible, a proof is difficult, and we will not attempt it here.

Another counterexample is the Alexander horned disk depicted in Figure IV-9. This is an inductively constructed object in 3-space. We will briefly describe it with no attempt at proving anything about it. One starts with an ordinary disk (right-hand side of the figure). Embed that in 3-space in the standard way. Then take two disks inside it and map them to "horns" on

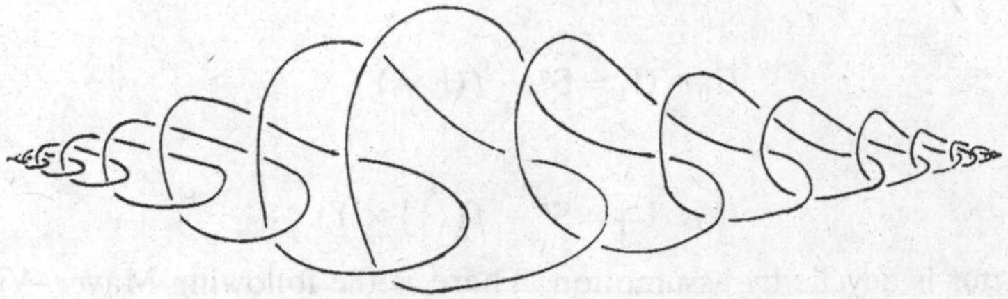

Figure IV-8. Wild arc of Fox and Artin.

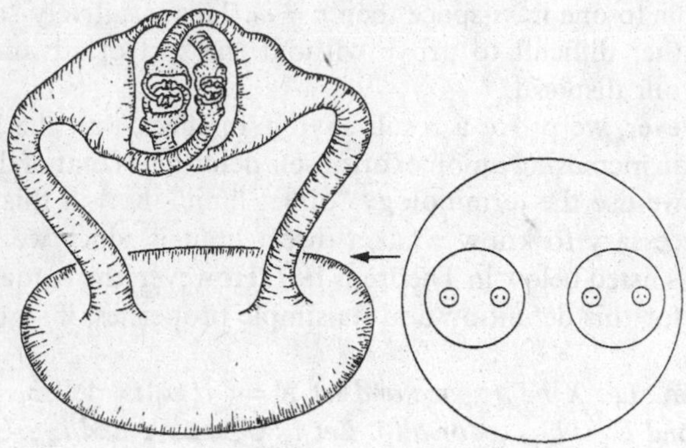

Figure IV-9. The Alexander horned disk.

the embedded disk. You now have an embedded disk shaped with horns and a map of the standard disk to the embedded disk. In each of these two subdisks, now take two more disks and use them to put horns on the horns in such a way that the two pairs of horns seem to link like a chain. This process is continued ad infinitum. If done carefully you will end up with a one–one map (hence a homeomorphism) of the standard disk to an embedded disk in 3-space whose complement is not simply connected. For details about these, and similar, examples see Rushing [1].

The following theorem, and its proof, is essentially due to Alexander (see Dieudonné [1], p. 57):

19.2. Theorem. *Let n be fixed. Suppose that Y is a compact space with the property that* $\tilde{H}_*(\mathbf{S}^n - f(Y)) = 0$ *for* every *embedding* $f: Y \to \mathbf{S}^n$. *Then* $I \times Y$ *also has this property.*

Proof Let $f: I \times Y \to \mathbf{S}^n$ be an embedding and suppose that $0 \neq \alpha \in \tilde{H}_i(\mathbf{S}^n - f(I \times Y))$. Put

$$U_0 = \mathbf{S}^n - f([0, \tfrac{1}{2}] \times Y)$$

and

$$U_1 = \mathbf{S}^n - f([\tfrac{1}{2}, 1] \times Y).$$

Then

$$U_0 \cap U_1 = \mathbf{S}^n - f(I \times Y)$$

and

$$U_0 \cup U_1 = \mathbf{S}^n - f(\{\tfrac{1}{2}\} \times Y).$$

The latter is acyclic by assumption. There is the following Mayer–Vietoris exact sequence:

$$\cdots \to \tilde{H}_{i+1}(U_0 \cup U_1) \to \tilde{H}_i(U_0 \cap U_1) \to \tilde{H}_i(U_0) \oplus \tilde{H}_i(U_1) \to \tilde{H}_i(U_0 \cup U_1) \to \cdots.$$

The terms involving $U_0 \cup U_1$ are 0 by assumption. Our element $0 \neq \alpha \in \tilde{H}_i(U_0 \cap U_1)$ must then map nontrivially into at least one of $\tilde{H}_i(U_0)$ and $\tilde{H}_i(U_1)$ and it is no loss of generality to suppose it is U_0. Now replace the interval I by $[0, \frac{1}{2}]$ and repeat the argument.

In this way we get a sequence of intervals $I \supset I_1 \supset I_2 \supset \cdots$ where I_k has length 2^{-k} and α maps to $\alpha_k \neq 0$ in $\tilde{H}_i(\mathbf{S}^n - f(I_k \times Y))$. Let $V_k = \mathbf{S}^n - f(I_k \times Y)$, an open set. Then we have $V_0 \subset V_1 \subset V_2 \subset \cdots$, $\bigcap I_k = \{x\}$ for some $x \in I$, and $\bigcup V_k = \mathbf{S}^n - \bigcap f(I_k \times Y) = \mathbf{S}^n - f(\{x\} \times Y)$.

By Theorem 19.1, we have $0 \neq \{\alpha_k\} \in \varinjlim \tilde{H}_i(V_k) = \tilde{H}_i(\bigcup V_k) = \tilde{H}_i(\mathbf{S}^n - f(\{x\} \times Y)) = 0$, (i.e., α maps to $\alpha_k \neq 0$ for all k and so it must map to a nonzero element of $\tilde{H}_i(\bigcup V_k)$ by Theorem 19.1), a contradiction. □

19.3. Corollary. *If $f: \mathbf{D}^r \to \mathbf{S}^n$ is an embedding then $\tilde{H}_*(\mathbf{S}^n - f(\mathbf{D}^r)) = 0$. In particular $\mathbf{S}^n - f(\mathbf{D}^r)$ is connected.*

Proof. This is clearly true for $r = 0$. By Theorem 19.2, it follows for $r = 1, 2, 3, \ldots$. □

19.4. Theorem (The Generalized Jordan Curve Theorem). *If $f: \mathbf{S}^r \to \mathbf{S}^n$ is an embedding then*

$$\tilde{H}_i(\mathbf{S}^n - f(\mathbf{S}^r)) \approx \begin{cases} \mathbf{Z} & \text{for } i = n - r - 1, \\ 0 & \text{for } i \neq n - r - 1. \end{cases}$$

That is, $\tilde{H}_i(\mathbf{S}^n - f(\mathbf{S}^r)) \approx \tilde{H}_i(\mathbf{S}^{n-r-1})$.

Proof. We use induction on r. For $r = 0$ we have $\mathbf{S}^n - f(\mathbf{S}^0) \approx \mathbf{R}^n - \{0\} \simeq \mathbf{S}^{n-1}$. Suppose we know the result for $r - 1$, i.e., that $\tilde{H}_i(\mathbf{S}^n - f(\mathbf{S}^{r-1})) \approx \mathbf{Z}$ for $i = n - r$ and is 0 otherwise. Then for an embedding f of $\mathbf{S}^r$ in $\mathbf{S}^n$ put

$$U_+ = \mathbf{S}^n - f(\mathbf{D}^r_+) \qquad \text{and} \qquad U_- = \mathbf{S}^n - f(\mathbf{D}^r_-),$$

so that

$$U_+ \cap U_- = \mathbf{S}^n - f(\mathbf{S}^r) \qquad \text{and} \qquad U_+ \cup U_- = \mathbf{S}^n - f(\mathbf{S}^{r-1}).$$

In the Mayer–Vietoris sequence for the pair U_+ and U_-,

$$\cdots \longrightarrow \tilde{H}_{i+1}(U_+) \oplus \tilde{H}_{i+1}(U_-) \longrightarrow \tilde{H}_{i+1}(\mathbf{S}^n - f(\mathbf{S}^{r-1})) \longrightarrow \tilde{H}_i(\mathbf{S}^n - f(\mathbf{S}^r))$$
$$\longrightarrow \tilde{H}_i(U_+) \oplus \tilde{H}_i(U_-) \longrightarrow \cdots$$

the direct sum terms are 0 by Corollary 19.3. Thus the terms between are isomorphic and this implies the result. □

Now we shall apply the foregoing results in a sequence of propositions culminating in two classic facts called Invariance of Domain and Invariance of Dimension. These results are so intuitively plausible that it is very easy to allow some unproved intuition to creep into the proofs. The reader should keep this in mind when going through the demonstrations.

19.5. Corollary (Jordan–Brouwer Separation Theorem). *If $f: \mathbf{S}^{n-1} \to \mathbf{S}^n$ is an embedding then $\mathbf{S}^n - f(\mathbf{S}^{n-1})$ consists of exactly two components and both of them are acyclic. Moreover, $f(\mathbf{S}^{n-1})$ is the topological boundary of each of these components.*

Proof. By Theorem 19.4, $H_0(\mathbf{S}^n - f(\mathbf{S}^{n-1})) \approx \mathbf{Z} \oplus \mathbf{Z}$ and all other homology groups are zero. This is equivalent to the first statement of the corollary. For the last statement, let U and V be the two components of $\mathbf{S}^n - f(\mathbf{S}^{n-1})$. Note that since $f(\mathbf{S}^{n-1})$ is compact, its complement is open and any point in it is in an open ϵ-ball completely contained in U or V. Thus such a point is not in the boundary of either U or V. Therefore $\partial U \subset f(\mathbf{S}^{n-1})$ and $\partial V \subset f(\mathbf{S}^{n-1})$. Let $p \in \mathbf{S}^{n-1}$ and suppose that $f(p) \notin \partial U$. Then some open neighborhood N of $f(p)$ does not touch U. Let W be an open ball in $\mathbf{S}^{n-1}$ about p so small that $f(W) \subset N$. Since $\mathbf{S}^{n-1} - W \approx \mathbf{D}^{n-1}$, we have that $Y = \mathbf{S}^n - f(\mathbf{S}^{n-1} - W)$ is an open connected set by 19.3. Now $Y = U \cup V \cup f(W) \subset U \cup V \cup N$. But $U \cap (V \cup N) = \varnothing$ and so $Y = (U \cap Y) \cup ((V \cup N) \cap Y)$ is a disjoint union of two nonempty open sets, a contradiction. Thus $f(p) \in \partial U$, and a similar argument shows that $f(p) \in \partial V$. □

19.6. Corollary. *If $f: \mathbf{S}^{n-1} \to \mathbf{R}^n$, $n \geq 2$, is an embedding then $\mathbf{R}^n - f(\mathbf{S}^{n-1})$ has exactly two components, one of which is bounded and the other not. The bounded component is acyclic and the other has the homology of $\mathbf{S}^{n-1}$.*

Proof. Let V be a component of $\mathbf{S}^n - f(\mathbf{S}^{n-1})$ and let $x \in V$. Then consider the exact sequence of the pair $(V, V - \{x\})$:

$$\cdots \to \tilde{H}_{i+1}(V) \to H_{i+1}(V, V - \{x\}) \to \tilde{H}_i(V - \{x\}) \to \tilde{H}_i(V) \to \cdots.$$

The homology of V is trivial by Corollary 19.5, so

$$\begin{aligned} \tilde{H}_i(V - \{x\}) &\approx H_{i+1}(V, V - \{x\}) \approx H_{i+1}(\mathbf{D}^n, \mathbf{D}^n - \{0\}) \\ &\approx \tilde{H}_i(\mathbf{D}^n - \{0\}) \approx \tilde{H}_i(\mathbf{S}^{n-1}), \end{aligned}$$

where the first isomorphism is from the sequence for $(V, V - \{x\})$, the second is by excision, the third is from the exact sequence for $(\mathbf{D}^n, \mathbf{D}^n - \{0\})$, and the fourth is by homotopy. □

19.7. Corollary. *If f is an embedding of $\mathbf{D}^n$ in $\mathbf{S}^n$ (or in $\mathbf{R}^n$) then $f(\mathbf{D}^n - \mathbf{S}^{n-1}) = f(\mathbf{D}^n) - f(\mathbf{S}^{n-1})$ is open in $\mathbf{S}^n$ (or $\mathbf{R}^n$) and equals a component of $\mathbf{S}^n - f(\mathbf{S}^{n-1})$ (or $\mathbf{R}^n - f(\mathbf{S}^{n-1})$).*

Proof. Let U_0 and U_1 be the two components of $\mathbf{S}^n - f(\mathbf{S}^{n-1})$. Then $\mathbf{S}^n - f(\mathbf{S}^{n-1})$ is the disjoint union $U_0 + U_1$. But it is also the disjoint union $(\mathbf{S}^n - f(\mathbf{D}^n)) + f(\mathbf{D}^n - \mathbf{S}^{n-1})$. By Corollary 19.3, $A_0 = \mathbf{S}^n - f(\mathbf{D}^n)$ is open and acyclic and therefore connected. Also $A_1 = f(\mathbf{D}^n - \mathbf{S}^{n-1})$ is connected. Now $A_0 \subset$ some U_i, say U_0. Then $A_1 \subset U_1$. Since $A_0 \cup A_1 = U_0 \cup U_1$ we must have $A_0 = U_0$ and $A_1 = U_1$. □

19.8. Corollary. *If M^n is a topological n-manifold and $f: M^n \to \mathbf{R}^n$ is one–one and continuous, then f is an open mapping.*

PROOF. Let $V \subset M^n$ be open and $x \in V$. Then there is an open neighborhood $U \subset V$ of x and a homeomorphism $g: \mathbf{R}^n \to U$ such that $g(0) = x$. Then $g(\mathbf{D}^n - \mathbf{S}^{n-1})$ is open in U, by Corollary 19.7, and therefore is open in M. Consider the composition $f \circ g|_{\mathbf{D}^n}: \mathbf{D}^n \to \mathbf{R}^n$, which is an embedding by Theorem 7.8 of Chapter I. Now $(f \circ g)(\mathbf{D}^n - \mathbf{S}^{n-1})$ is open by Corollary 19.7 and contains $f(x)$. But this set is $f(g(\mathbf{D}^n - \mathbf{S}^{n-1})) \subset f(U)$ and so $f(U)$ is a neighborhood of $f(x)$. Therefore $f(V)$ contains a neighborhood of each of its points and so is open. □

19.9. Corollary (Invariance of Domain). *If M^n and N^n are topological n-manifolds and $f: M^n \to N^n$ is one–one and continuous, then f is open.*

PROOF. It suffices to show that the image under f of a neighborhood of x is a neighborhood of $f(x)$. Let $f(x) \in V \subset N^n$ be an open set homeomorphic to $\mathbf{R}^n$. Then the restriction of f to $f^{-1}(V) \to V \approx \mathbf{R}^n$ is one–one, continuous and hence open by Corollary 19.8. Now, if U is an open neighborhood of x then so is $U \cap f^{-1}(V)$. Therefore f takes this into an open neighborhood of $f(x)$. Hence $f(U)$ is a neighborhood of $f(x)$. □

19.10. Corollary (Invariance of Dimension). *If an m-manifold M^m is homeomorphic to an n-manifold N^n then $n = m$.*

PROOF. If $m < n$ then one can embed $\mathbf{R}^m$ in $N^n \approx M^m$ in such a way that $\mathbf{R}^m$ is *not* open in N^n. But then this $\mathbf{R}^m$ is not open in M^m contrary to Corollary 19.9.

There is also an easy direct proof of Corollary 19.10 as follows: By embedding an m-disk $\mathbf{D}^m$ in M^m with 0 corresponding to x we compute

$$\begin{aligned} H_i(M, M - \{x\}) &\approx H_i(\mathbf{D}^m, \mathbf{D}^m - \{0\}) && \text{(by excision)} \\ &\approx \tilde{H}_{i-1}(\mathbf{D}^m - \{0\}) && \text{(by the exact sequence)} \\ &\approx \tilde{H}_{i-1}(\mathbf{S}^{m-1}) && \text{(by homotopy)} \end{aligned}$$

and we know this group to be $\mathbf{Z}$ for $i = m$ and 0 otherwise. It follows that m is a topological invariant of M^m since the groups $H_i(M, M - \{x\})$ are topological invariants. □

In the classical situation of embeddings $f: \mathbf{S}^1 \to \mathbf{S}^2$, it is also known that the closure of each of the two components of $\mathbf{S}^2 - f(\mathbf{S}^1)$ is homeomorphic to $\mathbf{D}^2$. This is called the Schoenflies Theorem and follows from the strong form of the Riemann Mapping Theorem in complex analysis. (See, for example, Nehari [1].)

In higher dimensions, the analogue of this is false, in general, as is shown by the Alexander horned sphere. However, by assuming a "bicollaring" on the embedding, one can prove the higher-dimensional case. The rest of this section is devoted to just that. It will not be used elsewhere in this book and so it can be skipped. However, it does provide an exposure to some interesting topological techniques quite different from others studied in this book.

This higher-dimensional case was first proved by Mazur [1] with an ingenious argument. His argument assumed an additional niceness hypothesis that was later removed by Morse [1]. Independently of Morse, M. Brown [1] gave an elegant new proof that also avoided the additional condition. Our treatment is based on Brown's proof.

19.11. Theorem (Generalized Schoenflies Theorem). *Suppose* $f: \mathbf{S}^{n-1} \times [-1,1] \to \mathbf{S}^n$ *is an embedding. Then the closure of each component of* $\mathbf{S}^n - f(\mathbf{S}^{n-1} \times \{0\})$ *is homeomorphic to* $\mathbf{D}^n$.

PROOF. First we need the following four lemmas:

19.12. Lemma. *Let* $K_1, \ldots, K_k$ *be compact sets in* $\mathbf{R}^n$, *let* $K = \bigcup K_i$ *and let* $\mathbf{R}^n/\{K_i\}$ *denote the quotient space obtained by identifying each* K_i *to a point. If* $g: \mathbf{R}^n/\{K_i\} \to \mathbf{R}^n$ *is a one–one map then it is an embedding onto an open set.*

PROOF. We must show that g is open. Let $f: \mathbf{R}^n \to \mathbf{R}^n$ be the composition of g with the quotient map $\mathbf{R}^n \to \mathbf{R}^n/\{K_i\}$. Let D be an n-disk with $K \subset \operatorname{int} D$. Then $f(D - \partial D)$ is contained in a component W of $\mathbf{R}^n - f(\partial D)$. Note that $f(D - \partial D) = f(D) \cap W$. By Invariance of Domain, $f(D - \partial D - K)$ is open. But $f(D)$ is compact and so $f(D - \partial D - K) = f(D) \cap (W - f(K))$ is closed in $W - f(K)$. Since $f(K)$ is a finite, $W - f(K)$ is connected and we conclude that $f(D - \partial D) = W$, which is open. Now g is an embedding on $D/\{K_i\}$ by Theorem 7.8 of Chapter I and so $g: (\operatorname{int} D)/\{K_i\} \to W$ is a homeomorphism. Thus $(\operatorname{int} D)/\{K_i\}$, and hence $\mathbf{R}^n/\{K_i\}$, is an n-manifold and so g is open by Invariance of Domain. □

19.13. Lemma. *Let A and B be disjoint compact subsets of* $\mathbf{R}^n$. *Suppose that*

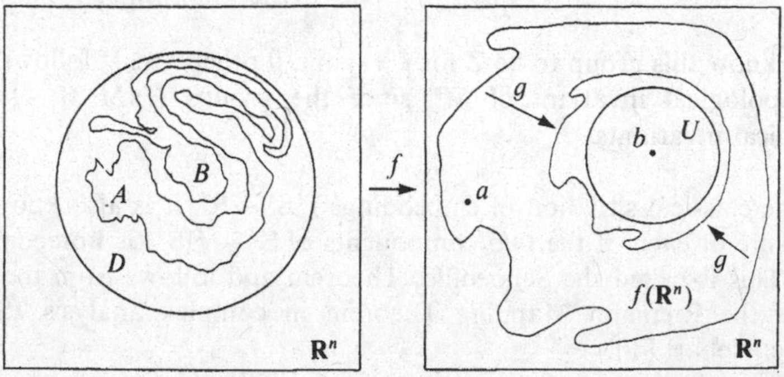

Figure IV-10. Avoiding B.

the quotient space $\mathbf{R}^n/\{A, B\}$ obtained by identifying A to a point and B to another point can be embedded in $\mathbf{R}^n$. Then $\mathbf{R}^n/A$ and $\mathbf{R}^n/B$ can be embedded in $\mathbf{R}^n$ as open sets.

PROOF. (See Figure IV-10.) Let $f: \mathbf{R}^n \to \mathbf{R}^n$ be the composition of the quotient map $\mathbf{R}^n \to \mathbf{R}^n/\{A, B\}$ with the embedding of the latter in $\mathbf{R}^n$. Let $f(A) = \{a\}$ and $f(B) = \{b\}$. By Lemma 19.12, $f(\mathbf{R}^n)$ is open in $\mathbf{R}^n$.

Let $U \subset f(\mathbf{R}^n)$ be an open disk with $b \in U$ and $a \notin \bar{U}$. Let $g: \mathbf{R}^n \to \mathbf{R}^n$ be an embedding which is the identity on U and whose image does not contain a. (The image can be taken to be any open disk slightly larger than $\bar{U}$.) On $f^{-1}(\mathbf{R}^n - \{b\})$ the map $f^{-1}gf$ is defined since $a \notin gf(\mathbf{R}^n)$. It is the identity on $f^{-1}(U) \supset B$. Thus the function $h: \mathbf{R}^n \to \mathbf{R}^n$, defined by

$$h(x) = \begin{cases} x & \text{for } x \in B, \\ f^{-1}gf(x) & \text{for } x \notin B, \end{cases}$$

is continuous, collapses A to a point, and is an embedding on $\mathbf{R}^n - A$. The induced map $\bar{h}: \mathbf{R}^n/A \to \mathbf{R}^n$ is one–one and hence an embedding onto an open set in $\mathbf{R}^n$ by Lemma 19.12. □

19.14. Lemma. *If A is a compact subset of $\mathbf{R}^n$ and $\mathbf{R}^n/A$ can be embedded as an open set in $\mathbf{R}^n$ then A is "cellular," meaning that each neighborhood of A contains a neighborhood homeomorphic to $\mathbf{D}^n$.*

PROOF. (See Figure IV-11.) We may assume that $A \subset \operatorname{int} \mathbf{D}^n$. Let $f: \mathbf{R}^n \to \mathbf{R}^n$ be the composition of the quotient map $\mathbf{R}^n \to \mathbf{R}^n/A$ with the hypothesized embedding $\mathbf{R}^n/A \to \mathbf{R}^n$. Let $U \supset A$ be open. Let $f(A) = \{a\}$, so $a \in f(U)$.

Let V be a small open disk about a with $\bar{V} \subset f(U)$. Let $g: \mathbf{R}^n \to \mathbf{R}^n$ be an embedding such that

$$g(x) = x \quad \text{for} \quad x \in V,$$
$$\operatorname{im}(g) \subset f(U).$$

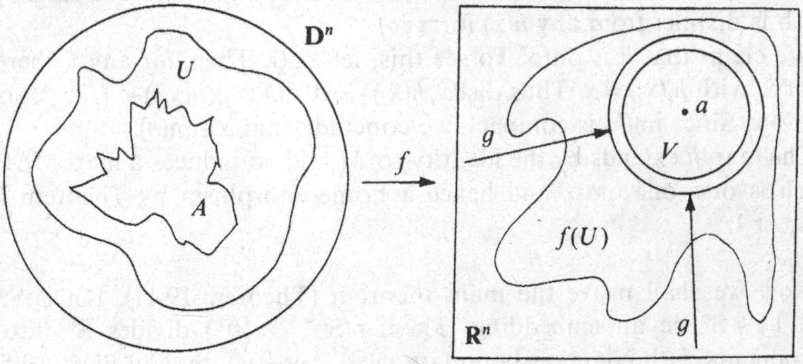

Figure IV-11. Showing A to be cellular.

Then $f^{-1}gf$ is an embedding on $\mathbf{R}^n - A$ and equals the identity on $f^{-1}(V - \{a\})$. Thus the function $h: \mathbf{R}^n \to \mathbf{R}^n$ defined by

$$h(x) = \begin{cases} x & \text{for } x \in A, \\ f^{-1}gf(x) & \text{for } x \notin A, \end{cases}$$

is an embedding. Then $h(\mathbf{D}^n) \approx \mathbf{D}^n$ and $A \subset h(\mathbf{D}^n) \subset U$, as required. □

19.15. Lemma. *Let M^n be a compact topological n-manifold, possibly with boundary. Let $A \subset M - \partial M$ be cellular. Then $M \approx M/A$.*

PROOF. Give M a metric. (For our use we could assume further that $M \subset \mathbf{R}^n$.) Since A is cellular, there exist, in $M - \partial M$, n-cells (homeomorphs of $\mathbf{D}^n$) $Q = Q_0 \supset Q_1 \supset Q_2 \cdots$ with $Q_{i+1} \subset \operatorname{int}(Q_i)$ and $A = \bigcap Q_i$. Let $h_0: Q \to Q$ be a homeomorphism which is the identity on ∂Q and is such that diam $(h_0(Q_1)) < 1$. Then let $h_1: Q \to Q$ be a homeomorphism which equals h_0 on $Q - Q_1$ and has diam $(h_1(Q_2)) < 1/2$. Continue inductively to construct homeomorphisms

$$h_i: Q \xrightarrow{\approx} Q$$

such that

$$h_i(x) = h_{i-1}(x) \qquad \text{on} \quad Q - Q_i$$

and

$$\operatorname{diam}(h_i(Q_{i+1})) < 1/2^i.$$

For any $x \in Q$, we have $\operatorname{dist}(h_{i+1}(x), h_i(x)) \le \operatorname{diam}(h_i(Q_{i+1})) < 1/2^i$ since $h_{i+1} = h_i$ outside Q_{i+1}, and (hence) $h_{i+1}(Q_{i+1}) \subset h_i(Q_{i+1})$.

Therefore the h_i converge uniformly to a map $Q \to Q$ which is the identity on ∂Q. Also

$$\operatorname{dist}(h_i(x), h(x)) \le \operatorname{dist}(h_i(x), h_{i+1}(x)) + \operatorname{dist}(h_{i+1}(x), h_{i+2}(x)) + \cdots < 1/2^{i-1}.$$

Now $h = h_i$ on $Q - Q_{i+1}$. Therefore h is one–one outside A. Since $h(Q_{i+1}) = Q - h(Q - Q_{i+1}) = Q - h_i(Q - Q_{i+1}) = h_i(Q_{i+1})$, we have $h(A) = \bigcap h_i(Q_{i+1})$ which has diameter at most $1/2^i$ for all i. It follows that $h(A)$ is a single point which is distinct from any $h(x)$ for $x \notin A$.

We claim that h is onto. To see this, let $x \in Q$. Then, for any i, there is a point x_i with $h_i(x_i) = x$. Thus $\operatorname{dist}(x, h(x_i)) = \operatorname{dist}(h_i(x_i), h(x_i)) < 1/2^{i-1}$, so that $h(x_i) \to x$. Since $\operatorname{im}(h)$ is compact, we conclude that $x \in \operatorname{im}(h)$.

The map h extends by the identity to M and so induces a map $M/A \to M$ which is one–one onto, and hence a homeomorphism by Theorem 7.8 of Chapter I. □

Now we shall prove the main theorem (Theorem 19.11). Let $\phi: \mathbf{S}^{n-1} \times [-1, 1] \to \mathbf{S}^n$ be an embedding. Then $\phi(\mathbf{S}^{n-1} \times \{0\})$ divides $\mathbf{S}^n$ into two components with common boundary $\phi(\mathbf{S}^{n-1} \times \{0\})$, by Corollary 19.5. Let M_+ be the closure of the one of these components containing $\phi(\mathbf{S}^{n-1} \times \{1\})$

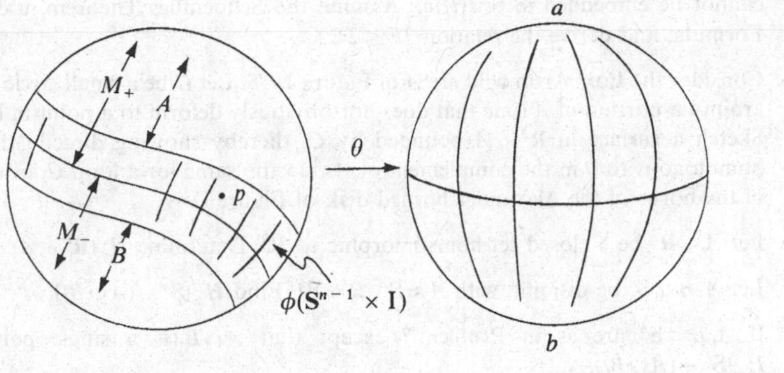

Figure IV-12. Final proof of the Generalized Schoenflies Theorem.

and M_- the closure of the one containing $\phi(\mathbf{S}^{n-1} \times \{-1\})$. Put

$$A = (\mathbf{S}^n - \phi(\mathbf{S}^{n-1} \times (-1, 1))) \cap M_+,$$
$$B = (\mathbf{S}^n - \phi(\mathbf{S}^{n-1} \times (-1, 1))) \cap M_-.$$

(See Figure IV-12.) Consider the map $\theta: \mathbf{S}^{n-1} \times [-1, 1] \to \mathbf{S}^n$ given by

$$\theta(x, t) = \langle x\cos(\pi t/2), \sin(\pi t/2)\rangle \in \mathbf{R}^n \times \mathbf{R} = \mathbf{R}^{n+1}.$$

Then $\theta \circ \phi^{-1}$ extends to a map $f: \mathbf{S}^n \to \mathbf{S}^n$ taking A to the north pole, B to the south pole, M_+ onto $\mathbf{D}^n_+$ and M_- onto $\mathbf{D}^n_-$. In particular, f induces homeomorphisms $M_+/A \xrightarrow{\approx} \mathbf{D}^n_+$ and $M_-/B \xrightarrow{\approx} \mathbf{D}^n_-$.

If $p \in \mathbf{S}^n - (A \cup B)$ then f induces a homeomorphism $(\mathbf{S}^n - \{p\})/\{A, B\} \to \mathbf{S}^n - f(p)$. Regard $\mathbf{R}^n$ as $\mathbf{S}^n - \{p\}$. By Lemma 19.13, $\mathbf{R}^n/A$ and $\mathbf{R}^n/B$ can be embedded as open sets in $\mathbf{R}^n$. By Lemma 19.14, A and B are cellular. By Lemma 19.15, $M_+ \approx M_+/A \approx \mathbf{D}^n_+$ and $M_- \approx M_-/B \approx \mathbf{D}^n_-$. □

Problems

1. Let $Y = A \cup B$ where A, B, and $A \cap B$ are homeomorphic to closed disks. Let X_r be a union of three r-dimensional disks along their common boundary; i.e., $X_r = \{(x_0, \ldots, x_r) \in \mathbf{R}^{r+1} | \sum_i x_i^2 = 1$, or $x_0 = 0$ and $\sum_i x_i^2 \le 1\}$.
   (a) For any embedding $f: Y \to \mathbf{S}^n$ show that $\tilde{H}_*(\mathbf{S}^n - f(Y)) = 0$.
   (b) For any embedding $g: X_r \to \mathbf{S}^n$ find $\tilde{H}_*(\mathbf{S}^n - g(X_r))$.
   (*Hint*: X_r is a union of two sets each of which is homeomorphic to some Y, as above, and with intersection X_{r-1}.)

2. Find a cellular set in $\mathbf{R}^2$ which is not arcwise connected.

3. Find a manifold M^2 and a compact set $A \subset M^2$ such that M/A is a 2-manifold but A is not cellular.

4. Consider the graph G consisting of the vertices and edges of Δ_4. Show that G

cannot be embedded in $\mathbf{S}^2$. (*Hint*: Assume the Schoenflies Theorem, use Euler's Formula, and derive the relation $3F \le 2E$.)

5. Consider the Fox–Artin wild arc A of Figure IV-8. Let C be a small circle looping around a portion of A (one that does not obviously deform to a point in $\mathbf{R}^3 - A$). Sketch a surface in $\mathbf{R}^3 - A$ bounded by C, thereby showing directly that C is homologous to 0 in the complement of A. Do the same for a loop C around one of the horns of the Alexander horned disk of Figure IV-9.
6. Let $A \subset \mathbf{R}^n$ be a closed set homeomorphic to $\mathbf{R}^k$. Determine $H_*(\mathbf{R}^n - A)$.
7. Let $A, B \subset \mathbf{S}^n$ be disjoint, with $A \approx \mathbf{S}^p$, $B \approx \mathbf{S}^q$. Find $H_*(\mathbf{S}^n - (A \cup B))$.
8. If $A, B \subset \mathbf{S}^n$ are as in Problem 7 except that $A \cap B$ is a single point, find $H_*(\mathbf{S}^n - (A \cup B))$.
9. Let $G \subset \mathbf{S}^n$ be a finite connected graph. Find $H_*(\mathbf{S}^n - G)$. Also try this for a disconnected graph.

20. The Borsuk–Ulam Theorem

Let $\pi: X \to Y$ be a two-sheeted covering map. Let $g: X \to X$ be the unique nontrivial deck transformation. We have $g^2 = 1$. Also $g(x) \ne x$ for all $x \in X$.

Throughout this section we will use singular homology with coefficients in $\mathbf{Z}_2$. This can be regarded as the homology based on the chain complex $\Delta_*(X; \mathbf{Z}_2) = \Delta_*(X) \otimes \mathbf{Z}_2$ or more simply as the complex of chains $\sum_\sigma n_\sigma \sigma$ where $n_\sigma \in \mathbf{Z}_2$.

For any simplex $\sigma: \Delta_p \to X$ the simplex $g \circ \sigma$ is distinct from σ. Thus the simplices of X fall into a set of pairs $(\sigma, g \circ \sigma)$. Any simplex τ of Y can be lifted to a simplex of X since the standard simplex is simply connected. There are exactly two such liftings of the form σ and $g \circ \sigma$. Define a chain map $t: \Delta_*(Y; \mathbf{Z}_2) \to \Delta_*(X; \mathbf{Z}_2)$ by taking τ into $\sigma + g \circ \sigma = (1 + g_\Delta)\sigma$. Clearly $\pi_\Delta \circ t = 0$ since we are using mod 2 coefficients. Suppose a chain c of X is in the kernel of $\pi_\Delta: \Delta_*(X; \mathbf{Z}_2) \to \Delta_*(Y; \mathbf{Z}_2)$. That means that c contains the simplex $g \circ \sigma$ whenever it contains σ and that means that it is in the image of t. Thus we have shown that there is the short exact sequence of chain complexes

$$0 \longrightarrow \Delta_*(Y; \mathbf{Z}_2) \xrightarrow{t} \Delta_*(X; \mathbf{Z}_2) \xrightarrow{\pi_\Delta} \Delta_*(Y; \mathbf{Z}_2) \longrightarrow 0.$$

The chain map t (and its induced map t_* in homology) is called the "transfer." We have the induced long exact sequence in homology:

$$\cdots \longrightarrow H_p(Y; \mathbf{Z}_2) \xrightarrow{t_*} H_p(X; \mathbf{Z}_2) \xrightarrow{\pi_*} H_p(Y; \mathbf{Z}_2) \xrightarrow{\partial_*} H_{p-1}(Y; \mathbf{Z}_2) \longrightarrow \cdots.$$

(This sequence is a rather trivial special case of an extensive theory due to P.A. Smith; see Bredon [4].) The composition $t \circ \pi_\Delta$ takes a simplex σ to

$(1+g_\Delta)\sigma$ and so

$$t_* \circ \pi_* = (1+g_*): H_p(X;\mathbf{Z}_2) \to H_p(X;\mathbf{Z}_2).$$

In particular, if $g_* = 1$ then $t_* \circ \pi_* = 0$. Note that this is always the case when we have $H_p(X;\mathbf{Z}_2) \approx \mathbf{Z}_2$.

The exact sequences we have derived are obviously natural with respect to equivariant maps, i.e., maps $f: X \to X'$ between the total spaces of two double sheeted coverings such that $f \circ g = g' \circ f$ (using g' to denote the nontrivial deck transformation of X').

20.1. Theorem. *Let $g: \mathbf{S}^n \to \mathbf{S}^n$ be the antipodal map (and similarly for g' on $\mathbf{S}^m$) and let $\phi: \mathbf{S}^n \to \mathbf{S}^m$ be equivariant. Then $n \le m$.*

Proof. Suppose that $n > m$. The coverings in question are the double coverings of the real projective spaces by the spheres. We claim that the exact sequence for $\mathbf{S}^m$ must look like this:

$$0 \longrightarrow H_m(\mathbf{P}^m;\mathbf{Z}_2) \xrightarrow{\approx} H_m(\mathbf{S}^m;\mathbf{Z}_2) \xrightarrow{0} H_m(\mathbf{P}^m;\mathbf{Z}_2) \xrightarrow{\approx} H_{m-1}(\mathbf{P}^m;\mathbf{Z}_2) \xrightarrow{0} \cdots$$
$$\cdots \xrightarrow{0} H_1(\mathbf{P}^m;\mathbf{Z}_2) \xrightarrow{\approx} H_0(\mathbf{P}^m;\mathbf{Z}_2) \xrightarrow{0} H_0(\mathbf{S}^m;\mathbf{Z}_2) \xrightarrow{\approx} H_0(\mathbf{P}^m;\mathbf{Z}_2) \longrightarrow 0.$$

To see this, we do not need to know the homology of $\mathbf{P}^m$. If the first map marked 0 were nontrivial then the composition $H_m(\mathbf{S}^m;\mathbf{Z}_2) \to H_m(\mathbf{P}^m;\mathbf{Z}_2) \to H_m(\mathbf{S}^m;\mathbf{Z}_2)$ would be nontrivial since the second map is a monomorphism. But that composition is trivial by the remarks above (it is $1+g'_*$). Thus $H_m(\mathbf{P}^m;\mathbf{Z}_2) \approx \mathbf{Z}_2$ and the first two markings are correct. The second 0 is correct since $H_{m-1}(\mathbf{S}^m;\mathbf{Z}_2) = 0$ and the same reasoning applies to the rest.

By naturality we get the commutative diagram (where $\psi: \mathbf{P}^n \to \mathbf{P}^m$ is the map induced by ϕ):

$$\begin{array}{ccc}
H_i(\mathbf{P}^n;\mathbf{Z}_2) & \xrightarrow[\approx]{\partial_*} & H_{i-1}(\mathbf{P}^n;\mathbf{Z}_2) \\
\downarrow{\scriptstyle \psi_*} & & {\scriptstyle \approx}\downarrow{\scriptstyle \psi_*} \\
H_i(\mathbf{P}^m;\mathbf{Z}_2) & \xrightarrow[\approx]{\partial_*} & H_{i-1}(\mathbf{P}^m;\mathbf{Z}_2).
\end{array}$$

The isomorphisms shown are true for $i-1=0$. It follows that ψ_* on the left is also an isomorphism. Then we can look at the same diagram for $i=2,3$ and so on until we no longer know that the horizontal maps are isomorphisms, i.e., until, and including, $i=m$. Now look at the commutative diagram:

$$\begin{array}{ccc}
H_m(\mathbf{P}^n;\mathbf{Z}_2) & \xrightarrow{t_*} & H_m(\mathbf{S}^n;\mathbf{Z}_2) = 0 \\
{\scriptstyle \approx}\downarrow{\scriptstyle \psi_*} & & \downarrow{\scriptstyle \phi_*} \\
H_m(\mathbf{P}^m;\mathbf{Z}_2) & \xrightarrow[\approx]{t_*} & H_m(\mathbf{S}^m;\mathbf{Z}_2).
\end{array}$$

This diagram is self-contradictory, and so we have reached the desired contradiction. □

20.2. Theorem (Borsuk–Ulam Theorem). *If $f: \mathbf{S}^n \to \mathbf{R}^n$ is a map then there exists a point $x \in \mathbf{S}^n$ such that $f(x) = f(-x)$.*

Proof. If not then consider the map $\phi(x) = (f(x) - f(-x))/\| f(x) - f(-x) \| \in \mathbf{S}^{n-1}$. This satisfies $\phi(-x) = -\phi(x)$ so that it contradicts Theorem 20.1. □

20.3. Corollary. *At any particular time there is a place on the earth where the barometric pressure and the temperature are both equal to those at the antipodal point.* □

A ham sandwich consists of two pieces of bread and one piece of ham. The following corollary says that one can always slice it with a straight cut of a knife so as to cut each slice of bread exactly in two and the same for the ham.

20.4. Corollary (The Ham Sandwich Theorem). *Let $A_1, \ldots, A_m$ be Lebesgue measurable bounded subsets of $\mathbf{R}^m$. Then there exists an affine $(m-1)$-plane $H \subset \mathbf{R}^m$ which divides each A_i into pieces of equal measure.*

Proof. The proof is illustrated in Figure IV-13. Regard $\mathbf{R}^m$ as $\mathbf{R}^m \times \{1\} \subset \mathbf{R}^{m+1}$, i.e., the subset $\{(x_1, \ldots, x_{m+1}) | x_{m+1} = 1\}$. For a unit vector $x \in \mathbf{R}^{m+1}$, let

$$V_x = \mathbf{R}^m \times \{1\} \cap \{y \in \mathbf{R}^{m+1} | \langle x, y \rangle \geq 0\}$$

and

$$H_x = \mathbf{R}^m \times \{1\} \cap \{y \in \mathbf{R}^{m+1} | \langle x, y \rangle = 0\}.$$

Let $f_i = \text{measure}(V_x \cap A_i)$ which is continuous since A_i is bounded. Then put $f = (f_1, \ldots, f_m): \mathbf{S}^m \to \mathbf{R}^m$. By the Borsuk–Ulam Theorem (Theorem 20.2),

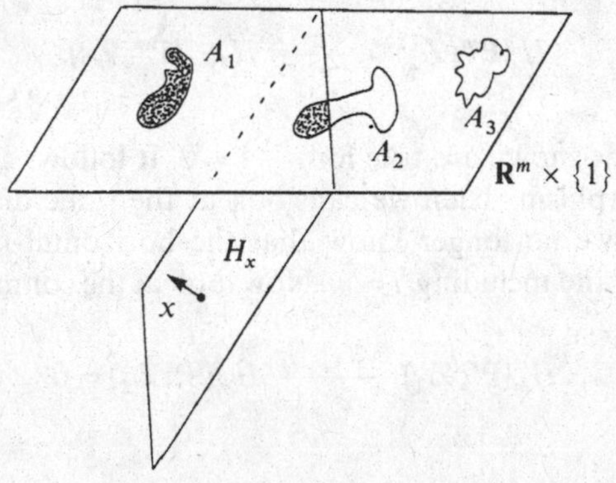

Figure IV-13. Ham sandwich.

there is a vector x_0 such that $f(x_0) = f(-x_0)$. Then H_{x_0} is the desired hyperplane. □

Let us discuss briefly a generalization of the Ham Sandwich Theorem due to Stone and Tukey [1].

Suppose we are given maps $f_0, \ldots, f_n: \mathbf{R}^k \to \mathbf{R}$ and bounded measurable sets $A_1, \ldots, A_n \subset \mathbf{R}^k$. Suppose that the f_i are independent in the sense that $\{x \in \mathbf{R}^k \mid \sum_{i=0}^n \lambda_i f_i(x) = 0\}$ has measure zero unless all the λ_i are zero. For $\lambda = (\lambda_0, \ldots, \lambda_n) \in \mathbf{S}^n$ define

$$V_\lambda^+ = \left\{ x \in \mathbf{R}^k \mid \sum_{j=0}^n \lambda_j f_j(x) \geq 0 \right\},$$

$$V_\lambda^- = \left\{ x \in \mathbf{R}^k \mid \sum_{j=0}^n \lambda_j f_j(x) \leq 0 \right\},$$

$$V_\lambda^0 = \left\{ x \in \mathbf{R}^k \mid \sum_{j=0}^n \lambda_j f_j(x) = 0 \right\},$$

and note that $V_\lambda^- = V_{-\lambda}^+$. Define

$$g: \mathbf{S}^n \to \mathbf{R}^n$$

by $g = (g_1, \ldots, g_n)$ where $g_i(\lambda) = \text{measure}(V_\lambda^+ \cap A_i)$. That g is continuous is not hard to prove, but requires more measure theory than we wish to assume. It will be clear in the examples we give. Then by Theorem 20.2, there exists a $\lambda \in \mathbf{S}^n$ with $g(\lambda) = g(-\lambda)$. That is,

$$\text{measure}(V_\lambda^+ \cap A_i) = \text{measure}(V_\lambda^- \cap A_i) \qquad \text{for all } i,$$

so that the solution set $V_\lambda^0 = \{x \mid \sum_{j=0}^n \lambda_j f_j(x) = 0\}$ cuts each A_i in half. For example, if $f_0 = 1, f_1 = x_1, \ldots, f_n = x_n$, this is the Ham Sandwich Theorem. For $f_0 = x^2 + y^2$, $f_1 = x$, $f_2 = y$, $f_3 = 1$, we conclude that, given three sets A_1, A_2, and A_3 in $\mathbf{R}^2$, there is a circle or a line (the solution set $\{(x, y) \mid \lambda_0(x^2 + y^2) + \lambda_1 x + \lambda_2 y + \lambda_3 = 0\}$) cutting each A_i in half. That "or a line" cannot be omitted is shown by Problem 5. Similarly, the functions $f_0 = x^2$, $f_1 = xy$, $f_2 = y^2$, $f_3 = x$, $f_4 = y$, $f_5 = 1$ show that any five such sets $A_1, \ldots, A_5$ in $\mathbf{R}^2$ can be cut in half by some conic section.

20.5. Theorem (Lusternik–Schnirelmann). *If $\mathbf{S}^n$ is covered by $n+1$ closed sets $A_1, \ldots, A_{n+1}$ then at least one of the A_i contains an antipodal pair of points.*

Proof. Suppose that A_i is disjoint from $-A_i$ for $i = 1, \ldots, n$. By Urysohn's Lemma (Lemma 10.2 of Chapter I), there is a map $f_i: \mathbf{S}^n \to [0, 1]$ which is 0 on A_i and 1 on $-A_i$, for all $i = 1, \ldots, n$. Let $f = (f_1, \ldots, f_n): \mathbf{S}^n \to \mathbf{R}^n$. By Theorem 20.2 there exists a point $x_0 \in \mathbf{S}^n$ such that $f(x_0) = f(-x_0)$. If $x_0 \in A_i$ for some $i \leq n$ then $f_i(x_0) = 0$ while $f_i(-x_0) = 1$ and so $f(x_0) \neq f(-x_0)$. Thus $x_0 \notin A_i$ for any $i \leq n$. Similarly $-x_0 \notin A_i$ for any $i \leq n$. Since the A_i cover $\mathbf{S}^n$, both x_0 and $-x_0$ must lie in A_{n+1}. □

We conclude this section by giving an alternative proof of Theorem 20.1 using a combination of smooth methods and homology. If $\phi: \mathbf{S}^n \to \mathbf{S}^m$ satisfies $\phi(-x) = -\phi(x)$ with $m < n$ then composing ϕ with the inclusion $\mathbf{S}^m \hookrightarrow \mathbf{S}^n$ gives an equivariant map $f: \mathbf{S}^n \to \mathbf{S}^n$ of degree zero. Thus Theorem 20.1 follows from:

20.6. Theorem. *If* $f: \mathbf{S}^n \to \mathbf{S}^n$ *satisfies* $f(-x) = -f(x)$ *then* $\deg(f)$ *is odd.*

PROOF. Let $g: \mathbf{S}^n \to \mathbf{S}^n$ be a smooth approximation to f so close that $\| f(x) - g(x) \| < 1$ for all $x \in \mathbf{S}^n$. Put $h(x) = (g(x) - g(-x))/2 \in \mathbf{R}^{n+1} - \{0\}$ and $k(x) = h(x)/\|h(x)\|$. Then $2\|h(x) - f(x)\| = \|g(x) - g(-x) - f(x) + f(-x)\| \leq \|g(x) - f(x)\| + \|g(-x) - f(-x)\| < 2$ so that $\|h(x) - f(x)\| < 1$. This implies that $h(x)$, and hence $k(x)$, is in the open half space in $\mathbf{R}^{n+1}$ with $f(x)$ as pole and this implies, in turn, that $k: \mathbf{S}^n \to \mathbf{S}^n$ is homotopic to f via the homotopy $F(x, t) = (tk(x) + (1-t)f(x))/\|tk(x) + (1-t)f(x)\|$. Also k is smooth and equivariant. This shows that we may assume that f is smooth. Now discard the notation g, h, k.

Let $v \in \mathbf{S}^n$ be a regular value of f and put $f^{-1}(v) = \{x_1, \ldots, x_d\}$; see Figure IV-14. Note that $d \equiv \deg(f) \pmod 2$ by Corollary 7.5. There is a great sphere $\mathbf{S}^{n-1} \subset \mathbf{S}^n$ not containing any of the points x_i and hence not containing any of the $-x_i$. Let $D_+ \approx \mathbf{D}^n$ be the closure of one of the components of $\mathbf{S}^n - \mathbf{S}^{n-1}$ and D_- the closure of the other component. Let $\{y_1, \ldots, y_d\}$ be those points among the $\pm x_i$ contained in D_+. Let $r: \mathbf{S}^n \to \mathbf{S}^n$ be the "folding" map that is the identity on the hemisphere with center v and the reflection taking $-v$ to v on the other hemisphere. Since $f(\mathbf{S}^{n-1}) \subset \mathbf{S}^n - \{v, -v\}$, and hence $rf(\mathbf{S}^{n-1}) \subset \mathbf{S}^n - \{v\} \approx \mathbf{R}^n$, the restriction $rf|_{D_+}$ extends to $D_- \to \mathbf{S}^n - \{v\}$. Let $g: \mathbf{S}^n \to \mathbf{S}^n$ be rf on D_+ and this extension on D_-. Then $g^{-1}(v) = \{y_1, \ldots, y_d\}$ and so $\deg(g) \equiv d \pmod 2$ by Corollary 7.5. Note that $-v \notin g(D_+)$. Let $k: \mathbf{S}^n - \{v, -v\} \to \mathbf{S}^{n-1}$ be the obvious equivariant map, which is a homotopy equivalence, and put $h = k \circ g: \mathbf{S}^{n-1} \to \mathbf{S}^{n-1}$ which is equivariant. The commutative diagram

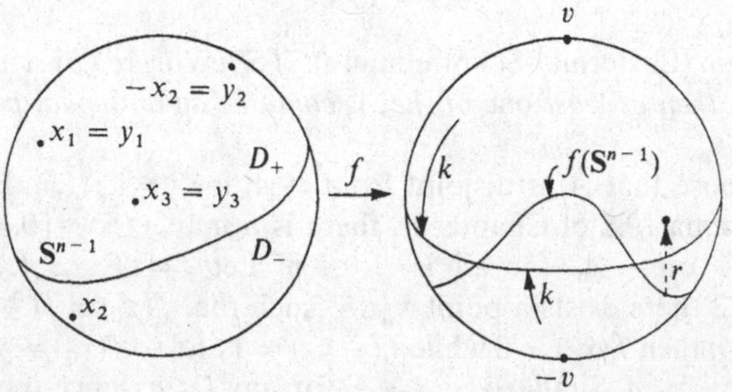

Figure IV-14. Proof of Theorem 20.6.

$$\begin{array}{ccccccccc}
H_n(\mathbf{S}^n) & \xrightarrow[\approx]{} & H_n(\mathbf{S}^n, D_-) & \xleftarrow[\approx]{} & H_n(D_+, \mathbf{S}^{n-1}) & \xrightarrow[\approx]{\partial_*} & H_{n-1}(\mathbf{S}^{n-1}) & \xrightarrow[\approx]{} & H_{n-1}(\mathbf{S}^{n-1}) \\
\downarrow g_* & & \downarrow g_* & & \downarrow g_* & & \downarrow g_* & & \downarrow h_* \\
H_n(\mathbf{S}^n) & \xrightarrow[\approx]{} & H_n(\mathbf{S}^n, \mathbf{S}^n - \{v\}) & \xleftarrow[\approx]{} & H_n(\mathbf{S}^n - \{-v\}, \mathbf{S}^n - \{v, -v\}) & \xrightarrow[\approx]{\partial_*} & H_{n-1}(\mathbf{S}^n - \{v, -v\}) & \xrightarrow[\approx]{k_*} & H_{n-1}(\mathbf{S}^{n-1})
\end{array}$$

shows that $\deg(h) = \pm\deg(g) \equiv d \equiv \deg(f) \pmod 2$. Since the result is clear for $n = 0$, it follows from this by induction on n. □

Problems

1. Let f be a map $\mathbf{P}^n \to \mathbf{P}^m$ with $n > m > 0$. Show that $f_\#: \pi_1(\mathbf{P}^n) \to \pi_1(\mathbf{P}^m)$ is trivial.
2. Show that $\mathbf{P}^2$ is not a retract of $\mathbf{P}^3$.
3. Show that $\mathbf{S}^n$ can be covered by $n+2$ closed sets, none containing an antipodal pair.
4. If $A_1, \ldots, A_n$ are measurable subsets of $\mathbf{S}^n$, show that there is a great $\mathbf{S}^{n-1}$ cutting each A_i exactly in half.
5. It has been stated in the popular literature that, given three regions in the plane, there is a circle dividing each into equal pieces. Show by example that this is false; indeed, give an example of just two regions for which no such circle exists. However, show that for any three regions on $\mathbf{S}^2$ there is a circle on $\mathbf{S}^2$ dividing each region into equal pieces.
6. Give a simple derivation of Theorem 20.1 out of Theorem 20.2.
7. Derive Theorem 20.6 from Theorem 20.1. (*Hint*: From an equivariant map $\mathbf{S}^n \to \mathbf{S}^n$ of even degree, show how to construct one of degree zero, and from such a map of degree zero construct an equivariant map $\mathbf{S}^{n+1} \to \mathbf{S}^n$. There is also an easy proof out of the *proof* of Theorem 20.1.)
8. Justify the statement, in the proof of Theorem 20.6, that there exists a great sphere $\mathbf{S}^{n-1}$ not containing any of the x_i.

21. Simplicial Complexes

Points $v_0, \ldots, v_n \in \mathbf{R}^\infty$ are "affinely independent" if they span an affine n-plane, i.e., if

$$\left(\sum_{i=0}^{n} \lambda_i v_i = 0, \sum_{i=0}^{n} \lambda_i = 0\right) \quad \Rightarrow \quad \text{all } \lambda_i = 0.$$

If this is not the case then if, for example, $\lambda_0 \neq 0$ we can assume $\lambda_0 = -1$ and solve the equations to get

$$v_0 = \sum_{i=1}^{n} \lambda_i v_i, \qquad \sum_{i=1}^{n} \lambda_i = 1,$$

which means that v_0 is in the affine space spanned by $v_1, \ldots, v_n$.

If $v_0, \ldots, v_n$ are affinely independent then we put

$$\sigma = (v_0, \ldots, v_n) = \left\{ \sum_{i=0}^{n} \lambda_i v_i \,\middle|\, \sum_{i=0}^{n} \lambda_i = 1, \lambda_i \geq 0 \right\}$$

which is the "affine simplex" spanned by the v_i. Note that this is the convex hull of the v_i. For $k \leq n$, a k-face of σ is any affine simplex of the form $(v_{i_0}, \ldots, v_{i_k})$ where these vertices are all distinct and so are affinely independent.

21.1. Definition. A (*geometric*) *simplicial complex* K is a collection of affine simplices such that:

(1) $\sigma \in K \Rightarrow$ any face of σ is in K; and
(2) $\sigma, \tau \in K \Rightarrow \sigma \cap \tau$ is a face of both σ and τ, or is empty.

Figure IV-15 illustrates two noncomplexes, and they are not complexes because the intersection of two simplices is not a face of at least one of those simplices. Both of these could be "subdivided" to form simplicial complexes.

If K is a simplicial complex then we put $|K| = \bigcup \{\sigma \mid \sigma \in K\}$. This is called the "polyhedron" of K.

21.2. Definition. A space X is a *polyhedron* if there exists a homeomorphism $h: |K| \xrightarrow{\approx} X$ for some simplicial complex K. The map h, together with the complex K, is called a *triangulation* of X.

Let K be a finite simplicial complex and choose an ordering of the vertices $v_0, v_1, \ldots$ of K. *If* $\sigma = (v_{\sigma_0}, \ldots, v_{\sigma_n})$ is a simplex of K, where $\sigma_0 < \cdots < \sigma_n$ then let $f_\sigma: \Delta_n \to |K|$ be

$$f_\sigma = [v_{\sigma 0}, \ldots, v_{\sigma_n}],$$

in the notation of Definition 1.2. Then this gives a CW-complex structure on $|K|$ with the f_σ as characteristic maps.

In this situation we will use $\langle v_{\sigma_0}, \ldots, v_{\sigma_n} \rangle = \sigma$ to denote σ as an n-cell of this CW-complex.

Recall the boundary formula (Theorem 10.3) for cellular homology of a CW-complex: If σ is an n-cell then $\partial\sigma = \sum_\tau [\tau : \sigma]\tau$, where $[\tau : \sigma] = \deg(p_\tau f_{\partial\sigma})$.

Figure IV-15. Not simplicial complexes.

In the present situation it is clear that

$$p_\tau \circ f_{\partial\sigma} \text{ is } \begin{cases} \text{constant if } \tau \text{ is not a face of } \sigma, \\ \text{of degree } \pm 1 \text{ if } \tau \text{ is a face of } \sigma, \end{cases}$$

because, in the latter case, this map is homotopic to a homeomorphism since it maps a face of the boundary of a simplex homeomorphically and collapses all other faces. Therefore we get the boundary formula

$$\partial\langle v_{\sigma_0}, \ldots, v_{\sigma_n}\rangle = \sum_{i=0}^{n} \epsilon_{n,i}\langle v_{\sigma_0}, \ldots, \hat{v}_{\sigma_i}, \ldots, v_{\sigma_n}\rangle$$

where $\epsilon_{n,i} = \pm 1$. We wish to determine this sign. The equation $\partial\partial = 0$ implies that $\epsilon_{n,i} = -\epsilon_{n,i+1}$. Thus $\epsilon_{n,i} = \epsilon_n(-1)^i$ for some signs $\epsilon_n = \pm 1$.

But, for homology calculations, it does not matter what the signs ϵ_n are, since the cycles and boundaries do not depend on that. Thus, we get the correct homology if we just take $\epsilon_n = 1$. (Technically, to work this out, one must produce a fixed homeomorphism of the n-simplex Δ_n with the n-cube $\mathbf{I}^n$. It is clear that one can make those choices so that the sign ϵ_n comes out any way one wants.) Thus we can arrange so that the boundary formula for a simplicial complex is

$$(*) \qquad \boxed{\partial\langle v_{\sigma_0}, \ldots, v_{\sigma_n}\rangle = \sum_{i=0}^{n} (-1)^i\langle v_{\sigma_0}, \ldots, \hat{v}_{\sigma_i}, \ldots, v_{\sigma_n}\rangle.}$$

This describes a chain complex we shall, temporarily, name $C_n^0(K)$. That is, we order the vertices of K (which we do here by naming them $v_0, v_1, \ldots$) and we let $C_n^0(K)$ be the free abelian group on the n-simplices $\langle v_{\sigma_0}, \ldots, v_{\sigma_n}\rangle$, where $\sigma_0 < \cdots < \sigma_n$, and we take the boundary formula $(*)$. We have shown, by way of considering this as a CW-complex, that

$$H_*(|K|) \approx H_*(C_*^0(K)).$$

There is another way of describing this chain complex. It is called the "oriented simplicial chain complex" and it is defined as follows: Let $C_n(K)$ be the abelian group generated by the symbols $\langle v_{\sigma_0}, \ldots, v_{\sigma_n}\rangle$ where the v_i are distinct vertices of a simplex of K, subject to the relations that make these "alternating" symbols, i.e., we regard

$$\langle v_{\sigma_0}, \ldots, v_{\sigma_n}\rangle = \operatorname{sign}(p)\langle v_{\sigma_{p(0)}}, \ldots, v_{\sigma_{p(n)}}\rangle$$

for any permutation p of $0, \ldots, n$. The boundary operator is still defined by $(*)$.

If we take an ordering of the vertices of K then any chain in $C_n^0(K)$ can be regarded as an oriented chain, and, in fact, we get all oriented chains this way. (For example, $\langle v_1, v_0\rangle$ is an oriented chain, but it is equal, as an oriented chain, to $-\langle v_0, v_1\rangle$, and the latter is a chain in $C_1^0(K)$.) It is clear that this correspondence preserves the boundary operator and so it is an isomorphism of chain complexes. Thus, we also have that

$$H_*(|K|) \approx H_*(C_*(K)),$$

and this does not depend on any ordering of the vertices.

It is easy to generalize these considerations to the case of infinite simplicial complexes as long as they are locally finite in the sense that each vertex is the vertex of at most a finite number of simplices. The only difficulty with general infinite complexes that are not locally finite is in comparing them to a CW-complex, and there it is just a matter of changing the topology to the weak topology to make the discussion go through.

The formula for the boundary operator in simplicial theory is very nice, but it is almost always much more difficult to make computations with it than it is for the cellular homology since there are usually many more simplices in a triangulation than there are cells needed in a CW-complex structure for the same space. For example, the real projective plane $\mathbf{P}^2$ requires only a single cell in dimensions 0, 1, and 2, while Figure IV-16 shows the simplest triangulation of this space. It has 6 vertices, 15 edges, and 10 faces. It is the image of the triangulation of $\mathbf{S}^2$ as an icosahedron.

Of course, Figure IV-16 does not describe a complex in the sense of our definition of a simplicial complex, since we have given no embedding in euclidean space. This can be done by taking those faces of the standard simplex $\Delta_5 \subset \mathbf{R}^6$ which correspond to simplices in the figure. (For example, include the simplex $[e_1, e_3, e_4]$ since $\langle 1, 3, 4\rangle$ is a simplex in the figure.) Obviously, this works for any finite complex, and that suffices for our purposes.

There is another important variety of simplicial complex called an "abstract simplicial complex." We will not make use of this concept in this book, but briefly give its definition. An abstract simplicial complex is a set V, whose elements are called vertices, and a collection K of *finite* subsets of V, called simplices, such that $\sigma \in K$, $\varnothing \neq \tau \subset \sigma \Rightarrow \tau \in K$, and $v \in V \Rightarrow \{v\} \in K$. It is possible to assign a topological space $|K|$ to such an abstract simplicial complex which, for finite complexes, is equivalent to a geometric simplicial

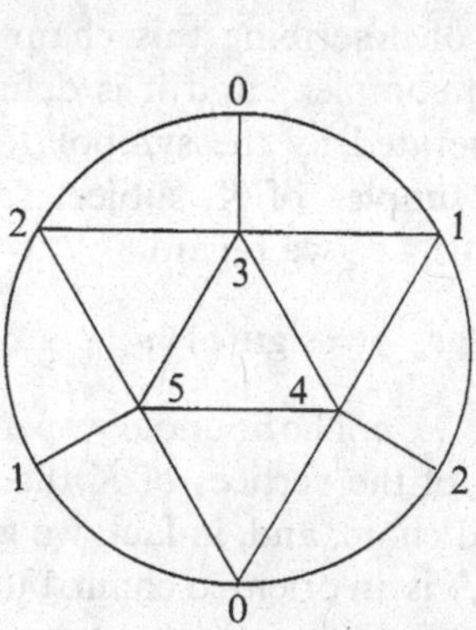

Figure IV-16. Triangulation of the projective plane.

complex. The notion of an abstract simplicial complex is useful in a number of ways, but we shall not have a need for it in this book.

Perhaps it is useful to do one calculation using simplicial theory, and we shall do so with the projective plane as triangulated as in Figure IV-16. We leave the discussion of the zeroth homology to the reader, and shall do the first and second (oriented) homology groups. We start with the easier of the two, the second homology group. Thus suppose we have a 2-cycle c. If c contains a term $n\langle 0,1,3\rangle$ then, this term contributes $\partial n\langle 0,1,3\rangle = n\langle 1,3\rangle - n\langle 0,3\rangle + n\langle 0,1\rangle$ to the expression for ∂c. For c to be a cycle, these have to be cancelled out. But, for example, the only way to cancel out $n\langle 0,3\rangle$ would be to have a term $n\langle 0,3,2\rangle$ in c because that is the only 2-simplex besides the original $\langle 0,1,3\rangle$ that has $\langle 0,3\rangle$ as an edge. Reasoning in the same way for all the "interior" 1-simplices, we see that the cycle c must have the same coefficient on all of the 2-simplices $\langle 0,1,3\rangle$, $\langle 0,3,2\rangle$, $\langle 1,4,3\rangle$, $\langle 3,4,5\rangle$, $\langle 2,3,5\rangle$, $\langle 2,5,1\rangle$ $\langle 1,2,4\rangle$, $\langle 0,4,2\rangle$, $\langle 0,5,4\rangle$, and $\langle 0,1,5\rangle$. These are just all ten 2-simplices written with the vertices in such an order that they go clockwise in the figure. Thus c must be a multiple of the chain which is the sum of these 10 simplices. Computing the boundary of this chain, and just looking at the contribution to the term in $\langle 0,1\rangle$, we see that that is $2\langle 0,1\rangle$. This is not zero, and so this 2-chain cannot be a cycle, nor can any nonzero multiple of it. Thus, $H_2(\mathbf{P}^2) = 0$. (If we compute with mod 2 coefficients, however, note that this 2-chain *would* be a cycle, the only one, and so $H_2(\mathbf{P}^2; \mathbf{Z}_2) \approx \mathbf{Z}_2$.)

Now let us try to compute the first homology group. We shall not try to find all 1-cycles (there are many) but shall look for representatives within a homology class. That is, starting with an unknown 1-cycle c we allow ourselves to change c by adding boundaries of 2-simplices, and thus will try to get c into some standard form. In fact, let us try to get c into a form only involving the 1-simplices $\langle 0,1\rangle$, $\langle 1,2\rangle$, and $\langle 2,0\rangle$. We will first work on the edges around the vertex $\langle 3\rangle$. Clearly we can add a multiple, perhaps, 0, of $\partial\langle 0,1,3\rangle$ to c to get rid of any term involving $\langle 0,3\rangle$. Then we can add a multiple $\partial\langle 3,1,4\rangle$ to get rid of a term in $\langle 1,3\rangle$ (without ruining previous work, of course). Similarly, use $\langle 3,4,5\rangle$ to get rid of $\langle 3,4\rangle$, then use $\langle 3,5,2\rangle$ to get rid of $\langle 3,5\rangle$. At this point, we have gotten rid of all the edges around the vertex $\langle 3\rangle$ except for $\langle 2,3\rangle$. But the resulting coefficient on $\langle 2,3\rangle$ must be 0 or else the boundary of this 1-cycle would have a nonzero term $n\langle 3\rangle$. Working in the same way, we can use $\langle 1,2,4\rangle$ to get rid of $\langle 1,4\rangle$; $\langle 0,4,5\rangle$ to get rid of $\langle 4,5\rangle$; $\langle 0,2,4\rangle$ to get rid of $\langle 0,4\rangle$ and we get rid of $\langle 2,4\rangle$ for free, because of the vertex $\langle 4\rangle$. Then use $\langle 1,5,2\rangle$ to get rid of $\langle 2,5\rangle$, and $\langle 0,5,1\rangle$ to get rid of $\langle 1,5\rangle$, and $\langle 0,5\rangle$ disappears automatically because of $\langle 5\rangle$. Now we have altered c, within its homology class, into the form $c = i\langle 0,1\rangle + j\langle 1,2\rangle + k\langle 2,0\rangle$. Then we calculate $\partial c = i(\langle 1\rangle - \langle 0\rangle) + j(\langle 2\rangle - \langle 1\rangle) + k(\langle 0\rangle - \langle 2\rangle) = (k-i)\langle 0\rangle + (i-j)\langle 1\rangle + (j-k)\langle 2\rangle$. For this to be 0 we must have $i = j = k$. Thus there is an n such that $c \sim n(\langle 0,1\rangle + \langle 1,2\rangle + \langle 2,0\rangle)$, and this is, indeed, a 1-cycle. However, we showed above that there is a 2-chain a (the sum of the 10 clockwise 2-simplices) with $\partial a = 2(\langle 0,1\rangle + \langle 1,2\rangle + \langle 2,0\rangle)$. Thus the homology class $\alpha = [\![\langle 0,1\rangle +$

$\langle 1,2\rangle + \langle 2,0\rangle]\!]$ generates $H_1(\mathbf{P}^2)$ and $2\alpha = 0$. Hence $H_1(\mathbf{P}^2) \approx \mathbf{Z}_2$. Mod 2 coefficients give the same result.

Problems

1. Triangulate the torus and use simplicial theory to compute its homology.
2. Triangulate the Klein bottle and use simplicial theory to compute its homology in both integer and mod 2 coefficients.
3. Consider the triangulation of $\mathbf{S}^2$ as an octahedron. This is invariant under the antipodal map and so gives a CW-decomposition of $\mathbf{P}^2$ into four triangles. Show that this is *not* a triangulation of $\mathbf{P}^2$.

22. Simplicial Maps

For simplicial complexes K and L there is a class of mappings $|K| \to |L|$ of the associated polyhedra which are closely related to the triangulations. These are the "simplicial maps" studied in this section. A simplicial map induces a homomorphism in homology that is very easy to describe. We will prove an important result called the "Simplicial Approximation Theorem" which says that any map $|K| \to |L|$ is homotopic to a simplicial map after, perhaps, subdividing K.

22.1. Definition. If K and L are simplicial complexes then a *simplicial map* $f: K \to L$ is a map $f: |K| \to |L|$ which takes vertices of any given simplex of K into vertices of some simplex of L and is affine on each simplex, i.e., $f(\sum_i \lambda_i v_{\sigma_i}) = \sum_i \lambda_i f(v_{\sigma_i})$.

Note that such a map is characterized by its values on vertices. Also a function f from the vertex set of K to that of L will define a simplicial map if and only if

$$\langle v_{\sigma_0}, \ldots, v_{\sigma_k} \rangle \in K \Rightarrow f(v_{\sigma_0}), \ldots, f(v_{\sigma_k}) \quad \text{are vertices of some simplex of } L$$

(not necessarily distinct). It follows that $f: |K| \to |L|$ is cellular and the induced chain map $f_\Delta: C_*(K) \to C_*(L)$ has values

$$f_\Delta \langle v_{\sigma_0}, \ldots, v_{\sigma_n} \rangle = \begin{cases} \langle f(v_{\sigma_0}), \ldots, f(v_{\sigma_n}) \rangle & \text{if the latter vertices are distinct,} \\ 0 & \text{if they are not distinct.} \end{cases}$$

(To see this, choose an ordering of the vertices of L such that these are in the correct order.) Thus this formula serves to calculate the homomorphism in oriented simplicial homology induced by any simplicial map f.

To generalize this to arbitrary maps between polyhedra we need a way to replace an arbitrary map by a simplicial one. This is called "simplicial approximation." It also requires subdivision of one of the complexes involved, and we shall deal with that first. For the rest of this section we shall deal exclusively with finite simplicial complexes.

Suppose K to be a simplicial complex. Let K' be its "barycentric subdivision," i.e., the complex whose vertices are the barycenters $\underline{\sigma}$ of the simplices σ of K and the simplices are of the form $\langle\underline{\sigma}_0,\ldots,\underline{\sigma}_k\rangle$ where $\sigma_0<\cdots<\sigma_k$. (Recall that the barycenter of $\langle v_0,\ldots,v_n\rangle$ is the average of the v_i, and that the "<" relation between simplices means the first is a proper face of the second.) Also, let $K^{[r]}$ denote the r-fold iterated barycentric subdivision of K.

We have that $|K|=|K'|$ (proof left to the reader).

Define mesh(K) to be the maximum diameter of a simplex of K. In Section 17 it was shown that

$$\text{mesh}(K')\le\frac{n}{n+1}\text{mesh}(K),$$

where $n=\dim K$, and hence that $\text{mesh}(K^{[r]})\to 0$ as $r\to\infty$.

22.2. Definition. For $x\in|K|$, for a simplicial complex K, we define the *carrier* of x (carr(x)) to be the smallest simplex of K containing x.

Note that

$$y\in\text{carr}(x)\quad\Leftrightarrow\quad\text{carr}(y)\subset\text{carr}(x).$$

22.3. Definition. Let $f:|K|\to|L|$ be continuous. A *simplicial approximation* to f is a simplicial map $g:K\to L$ such that $g(x)\in\text{carr}(f(x))$ for each $x\in|K|$.

Note that this condition is equivalent to $\text{carr}(g(x))\subset\text{carr}(f(x))$.

22.4. Corollary. *If g is a simplicial approximation to f then $f\simeq g$.* □

Proof. Since $f(x)$ and $g(x)$ are both in the simplex carr($f(x)$), the line segment between them is in carr($f(x)$). Thus the homotopy $F(x,t)=tg(x)+(1-t)f(x)$ works. □

22.5. Lemma. *If $g:K\to L$ is simplicial then* $g(\text{carr}(x))=\text{carr}(g(x))$.

Proof. Let $(v_0,\ldots,v_p)\in K$ and $x=\sum_{i=0}^p\lambda_iv_i$ with all $\lambda_i>0$. Then $\text{carr}(x)=(v_0,\ldots,v_p)$. Since g is simplicial, $g(x)=\sum_{i=0}^p\lambda_ig(v_i)$. Note that some of the $g(v_i)$ may be equal. Thus $\text{carr}(g(x))=\text{convex hull of }\{g(v_0),\ldots,g(v_p)\}=g(\text{carr}(x))$. □

22.6. Corollary. *If $f_1:|K|\to|L|$ and $f_2:|L|\to|M|$ are continuous and g_i is a simplicial approximation to f_i then $g_2\circ g_1$ is a simplicial approximation to $f_2\circ f_1$.*

Proof. $g_2g_1(x)\in g_2(\text{carr}(f_1(x)))=\text{carr}(g_2(f_1(x)))\subset\text{carr}(f_2(f_1(x)))$. □

22.7. Definition. If v is a vertex of the simplicial complex K then the *open star* of v is

$$\text{St}_k(v) = \{x \in |K| \mid v \in \text{carr}(x)\}.$$

Note that

$$\boxed{x \in \text{St}_K(v) \quad \Leftrightarrow \quad v \in \text{carr}(x).}$$

22.8. Proposition. *For vertices v of K, each $\text{St}_K(v)$ is open, and these cover $|K|$.*

PROOF. Let $\text{St} = \text{St}_K$. We will show that $|K| - \text{St}(v)$ is a union of simplices, and hence is closed. It suffices to show that $x \notin \text{St}(v) \Rightarrow \text{carr}(x) \cap \text{St}(v) = \varnothing$.

Thus suppose $x \notin \text{St}(v)$, i.e., $v \notin \text{carr}(x)$. If $y \in \text{carr}(x)$ then $\text{carr}(y) \subset \text{carr}(x)$ and hence $v \notin \text{carr}(y)$. Therefore $y \notin \text{St}(v)$. That is, $\text{carr}(x) \cap \text{St}(v) = \varnothing$. □

22.9. Proposition. *If $v_0, \dots, v_r$ are vertices of the simplicial complex K then*

$$\bigcap \text{St}_K(v_i) \neq 0 \quad \Leftrightarrow \quad (v_0, \dots, v_r) \in K.$$

PROOF. A point $x \in \bigcap \text{St}(v_i) \Leftrightarrow (v_i \in \text{carr}(x) \text{ for all } i) \Leftrightarrow (v_0, \dots, v_r)$ is a face of $\text{carr}(x)$, and the result follows. □

22.10. Theorem (The Simplicial Approximation Theorem). *Let K and L be finite simplicial complexes and let $f: |K| \to |L|$ be continuous. Then there is an integer $r \geq 0$ such that there is a simplicial approximation $g: K^{[r]} \to L$ to f.*

PROOF. Let $\{w_i\}$ be the set of vertices of L. Let $U_i = f^{-1}(\text{St}_L(w_i))$ and note that these give an open covering of $|K|$. Let δ be a Lebesgue number of this covering. Then $\text{diam}(A) < \delta \Rightarrow A \subset U_i$ for some i. Take r such that $\text{mesh}(K^{[r]}) < \delta/2$. For purposes of the proof we can now forget r and assume that $\text{mesh}(K) < \delta/2$.

Let v be a vertex of K. If $x \in \text{St}_K(v)$ then $v \in \text{carr}(x)$ and so

$$\text{dist}(v, x) \leq \text{mesh}(K) < \delta/2.$$

Therefore $\text{diam}(\text{St}_K(v)) < 2 \cdot \delta/2 = \delta$.

Thus $\text{St}_K(v) \subset U_i = f^{-1}(\text{St}_L(w_i))$ for some i. That is, $f(\text{St}_K(v)) \subset \text{St}_L(w_i)$ for some i. Define $g(v)$ to be one such vertex w_i. That is

$$f(\text{St}_K(v)) \subset \text{St}_L(g(v)),$$

and this is the defining relation for g.

Next we show that g is simplicial. Let $(v_0, \dots, v_k) \in K$. Then $\bigcap \{\text{St}_K(v_i)\} \neq \varnothing$. Thus, $\varnothing \neq f(\bigcap \text{St}_K(v_i)) \subset \bigcap f(\text{St}_K(v_i)) \subset \bigcap \text{St}_L(g(v_i))$. Therefore $\bigcap \text{St}_L(g(v_i)) \neq \varnothing$, and it follows from Proposition 22.9 that $(g(v_0), \dots, g(v_k))$ is a simplex of L, showing that g is simplicial.

Finally, we must show that g is a simplicial approximation to f. That is, we must show that $g(x) \in \text{carr}(f(x))$ for all x. Let $x = \sum_{i=0}^{k} \lambda_i v_i$, where $(v_0, \dots, v_k) =$

carr(x). Then $v_i \in \text{carr}(x)$, so that $x \in \text{St}_K(v_i)$. Thus $f(x) \in f(\text{St}_K(v_i)) \subset \text{St}_L(g(v_i))$. This implies that $g(v_i) \in \text{carr}(f(x))$. Therefore $g(x) = \sum_{i=0}^k \lambda_i g(v_i) \in \text{carr}(f(x))$. □

The Simplicial Approximation Theorem can be used to prove some things that we have used smooth approximation to prove previously. For example, it can be used to prove that any map $\mathbf{S}^k \to \mathbf{S}^n$ is homotopic to a constant if $k < n$. (Because it is homotopic to such a simplicial map, and that map must miss a point in $\mathbf{S}^n$.)

PROBLEMS

1. Let K be any subdivison of Δ_n and let $g: K \to \Delta_n$ be a simplicial approximation to the identity. Show that the number of simplices of K that map onto Δ_n is odd. (This is known as "Sperner's Lemma.") (*Hint*: g must carry $\partial\Delta_n$ into itself.)
2. Triangulate $\mathbf{S}^2$ via the central projection of a regular tetrahedron T inscribed in $\mathbf{S}^2$. Show that there is no simplicial approximation $T \to T$ (without subdividing) to the antipodal map.
3. Let K be a simplicial complex and K' its barycentric subdivision. Let $\Upsilon: C_*(K) \to C_*(K')$ be the subdivision chain map and let $\phi: K' \to K$ be a simplicial approximation to the identity. Show that $\phi_\Delta \circ \Upsilon = 1: C_*(K) \to C_*(K)$.

23. The Lefschetz–Hopf Fixed Point Theorem

Let W be a finite-dimensional vector space over the field Λ and let $f: W \to W$ be a linear transformation. Then the "trace" of f is the sum of the diagonal entries in a matrix representation of f, which is the sum, with multiplicities, of the eigenvalues of f, and is denoted by $\text{tr}(f)$. The reader can supply the proof to the following elementary fact.

23.1. Lemma. *If $f: W \to W$ is linear and $V \subset W$ is a subspace such that $f(V) \subset V$ then f induces a linear map $f_{W/V}$ on W/V and we have* $\text{tr}(f_W) = \text{tr}(f_V) + \text{tr}(f_{W/V})$. □

Let C_* be any chain complex of bounded finite type over a field Λ (i.e., each C_i is finite dimensional and all but a finite number of them are zero), and let Z_i be the i-cycles, B_i the i-boundaries, and H_i the i-homology, all of which are Λ-vector spaces. Let $f: C_* \to C_*$ be a chain map. Then f induces the endomorphisms f_{Z_i}, f_{B_i}, and f_{H_i} on Z_i, B_i, and H_i, respectively. We have the commutative diagram

$$\begin{array}{ccccccccc} 0 & \longrightarrow & Z_i & \longrightarrow & C_i & \xrightarrow{\partial} & B_{i-1} & \longrightarrow & 0 \\ & & \downarrow{\scriptstyle f_{Z_i}} & & \downarrow{\scriptstyle f_{C_i}} & & \downarrow{\scriptstyle f_{B_{i-1}}} & & \\ 0 & \longrightarrow & Z_i & \longrightarrow & C_i & \xrightarrow{\partial} & B_{i-1} & \longrightarrow & 0 \end{array}$$

and it follows from Lemma 23.1 that

$$\operatorname{tr}(f_{C_i}) = \operatorname{tr}(f_{Z_i}) + \operatorname{tr}(f_{B_{i-1}}).$$

Similarly

$$\operatorname{tr}(f_{Z_i}) = \operatorname{tr}(f_{B_i}) + \operatorname{tr}(f_{H_i}).$$

Thus $\sum_i(-1)^i(\operatorname{tr}(f_{C_i}) - \operatorname{tr}(f_{B_{i-1}})) = \sum_i(-1)^i(\operatorname{tr}(f_{B_i}) + \operatorname{tr}(f_{H_i}))$. The terms involving B_i cancel out and so we get the following fact:

23.2. Theorem (Hopf Trace Formula). *If C_* is a chain complex of bounded finite type over a field Λ and $f: C_* \to C_*$ is a Λ-linear chain map then*

$$\sum_i(-1)^i \operatorname{tr}(f_{C_i}) = \sum_i(-1)^i \operatorname{tr}(f_{H_i}) \in \Lambda,$$

where f_{H_i} is the induced map on homology. □

Note that for $f = 1$ and $C_* = C_*(K)$, for a CW-complex K, this is just the Euler–Poincaré formula (Theorem 13.3).

23.3. Definition. Let K be a finite simplicial complex, and consider homology with coefficients in a field Λ. Let $f: |K| \to |K|$ be a map and $f_*: H_i(|K|; \Lambda) \to H_i(|K|; \Lambda)$ the induced map in homology and $\operatorname{tr}_i(f_*)$ its trace in degree i. Then we define the *Lefschetz number* of f to be

$$L_\Lambda(f) = \sum_i(-1)^i \operatorname{tr}_i(f_*) \in \Lambda.$$

23.4. Theorem (Lefschetz–Hopf Fixed Point Theorem). *Let K be a finite simplicial complex and $f: |K| \to |K|$ a map. If $L_\Lambda(f) \neq 0$ then f has a fixed point.*

PROOF. Suppose that f has no fixed points. Then we must show that $L(f) = 0$. Since $\operatorname{dist}(x, f(x))$ is a continuous function there is a number δ such that $\operatorname{dist}(x, f(x)) \geq \delta$ for all $x \in |K|$. Subdivide K so that $\operatorname{mesh}(K) < \delta/2$. We may as well assume this holds for K itself. That is, we have

$$\operatorname{dist}(x, f(x)) > 2 \cdot \operatorname{mesh}(K).$$

Let $g: K^{[n]} \to K$ be a simplicial approximation to f. Thus $x \in |K| \Rightarrow g(x) \in \operatorname{carr}_K(f(x))$. It follows that $\operatorname{dist}(g(x), f(x)) \leq \operatorname{mesh}(K)$. We conclude that

$$\operatorname{dist}(x, g(x)) > \operatorname{mesh}(K).$$

Therefore $x \in \sigma \in K \Rightarrow g(x) \notin \sigma$. This shows that

$$\sigma \cap g(\sigma) = \varnothing \qquad \text{for all simplices } \sigma \text{ of } K.$$

Now g may be regarded as a *cellular* map $K^{[n]} \to K^{[n]}$ even though it is not simplicial. Recall that the induced cellular chain map is given by $g_\Delta(\sigma) = \sum_\tau \deg(g_{\tau,\sigma})\tau$ where, for k-cells σ and τ, $g_{\tau,\sigma}: \mathbf{S}^k \to \mathbf{S}^k$ is induced by the composition of the characteristic map for σ with the collapse to $\bigvee_\tau \mathbf{S}^k$ followed by the projection to the τth sphere. Since $g(|\sigma|)$ has empty intersection with

$|\sigma|$ we have $\deg(g_{\sigma,\sigma}) = 0$. Therefore $g_\Delta: C_k(K^{[n]}) \to C_k(K^{[n]})$ has trace 0. Since $f \simeq g$, we deduce that

$$\begin{aligned} L(f) &= \sum(-1)^i \operatorname{tr}_i(g_*) && \text{(since } g_* = f_*) \\ &= \sum(-1)^i \operatorname{tr}_i(g_\Delta) && \text{(by Hopf)} \\ &= 0 && \text{(since all terms are 0).} \end{aligned}$$ □

Now we will discuss the case of integer coefficients. Let A be an abelian group. Then the tensor product $A \otimes \mathbf{Q}$ of A with the rationals $\mathbf{Q}$ is the abelian group generated by the symbols $a \otimes q$ for $a \in A$ and $q \in \mathbf{Q}$ subject to the relations of being bilinear ($(a+b) \otimes q = a \otimes q + b \otimes q$ and $a \otimes (p+q) = a \otimes p + a \otimes q$). This is a rational vector space (by $r(a \otimes q) = a \otimes rq$ for $r \in \mathbf{Q}$).

If A is finitely generated, as is the case with the applications in this section, then $A \otimes \mathbf{Q}$ is easy to understand. The torsion part $T(A)$ of A disappears in $A \otimes \mathbf{Q}$ because if $na = 0$ for some integer n then $a \otimes q = a \otimes (n)(q/n) = na \otimes (q/n) = 0 \otimes (q/n) = 0$. But $A/T(A)$ is free abelian and it is clear that $(A/T(A)) \otimes \mathbf{Q}$ is just the rational vector space on a basis of this free abelian group, so the same is true of $A \otimes \mathbf{Q}$. Unfortunately, this description is not "natural" and so masks the fact that, for example, tensoring a commutative diagram with $\mathbf{Q}$ gives a commutative diagram.

If $0 \to A \to B \to C \to 0$ is a short exact sequence of abelian groups then

$$0 \to A \otimes \mathbf{Q} \to B \otimes \mathbf{Q} \to C \otimes \mathbf{Q} \to 0$$

is also exact. (That is, $\otimes \mathbf{Q}$ is an exact functor.) We only need this for finitely generated groups and the reader should be able to prove that case with ease. (If not, then refer to Chapter V, Section 6, where general tensor products are discussed in detail.) It follows that for any chain complex C_*, we have $H_*(C_* \otimes \mathbf{Q}) \approx H_*(C_*) \otimes \mathbf{Q}$.

Suppose now that A is finitely generated and that $f: A \to A$ is an endomorphism. Then f induces an endomorphism $A/T(A) \to A/T(A)$. Since $A/T(A)$ is free abelian with a finite basis, this has a matrix representation over the integers, and therefore has a trace in $\mathbf{Z}$. We denote this trace by $\operatorname{tr}(f)$.

Then $f \otimes 1$ is defined on $A \otimes \mathbf{Q} = A/T(A) \otimes \mathbf{Q}$, which, as mentioned above, is a vector space over $\mathbf{Q}$ with basis consisting of a basis over $\mathbf{Z}$ of $A/T(A)$ tensored with 1. Thus $f \otimes 1$ has the same matrix representation as does f. Therefore $\operatorname{tr}_{\mathbf{Q}}(f \otimes 1) = \operatorname{tr}(f)$ and is an integer.

Thus if $f: |K| \to |K|$ is a map of a finite polyhedron to itself then one can define the integral Lefschetz number

$$L(f) = \sum_i (-1)^i \operatorname{tr}(f_{(H_i(|K|)/\text{torsion})}) \in \mathbf{Z}$$

but we have $L(f) = L_{\mathbf{Q}}(f)$ by our remarks. Therefore $L(f)$ can always be computed using rational coefficients, and the result is always an integer.

We shall now make a slight generalization of Theorem 23.4 to the case of a compact "euclidean neighborhood retract" (ENR). An ENR is a space which is homeomorphic to a retract of some open set in some $\mathbf{R}^n$. If X is a compact ENR then X is a retract of a finite polyhedron K because it is the retract of an open set U in some $\mathbf{R}^n$ and one can triangulate $\mathbf{R}^n$ so finely that any simplex touching X is inside U, and the union of those simplices can be taken as K. Conversely, a retract of a finite polyhedron is an ENR since the polyhedron is an ENR as is shown in Appendix E (not needed here). It follows easily from the Tubular Neighborhood Theorem that any smooth compact manifold is a compact ENR. In Appendix E it is shown that any topological manifold embeddable in euclidean space, and hence any compact topological manifold, is an ENR (not needed here).

23.5. Corollary. *Let X be a compact* ENR. *If $f: X \to X$ has $L(f) \neq 0$ then f has a fixed point.*

PROOF. By the above remarks X is a retract of some finite polyhedron K. Let $i: X \hookrightarrow K$ be the inclusion and $r: K \to X$ the retraction, so that $r \circ i = 1$. It follows from this that $H_*(X)$ has bounded finite type since the identity map on it factors though $H_*(K)$ which has bounded finite type. Therefore $L(f)$ is defined. We cannot apply Theorem 23.4 directly because we do not assume that X can be triangulated. Put $f^K = i \circ f \circ r: K \to K$. On $H_p(\cdot; \mathbf{Q})$, and recalling that $\operatorname{tr}(AB) = \operatorname{tr}(BA)$, we have

$$\operatorname{tr}_p(f^K_*) = \operatorname{tr}_p(i_* f_* r_*) = \operatorname{tr}_p(f_* r_* i_*) = \operatorname{tr}_p(f_* \circ 1) = \operatorname{tr}_p(f_*),$$

and so $L(f) = L(f^K)$. If $L(f) \neq 0$ then $L(f^K) \neq 0$ and so f^K has a fixed point $x \in K$ by Theorem 23.4. But $x = f^K(x) = ifr(x)$ shows that $x \in X$ and so $f(x) = x$. □

Let us now examine a few examples and corollaries. First note that if X is arcwise connected then, for any map $f: X \to X$, f_* is the identity on $H_0(X)$. That is so because $H_0(X)$ is generated by a singular 0-simplex, which is essentially just a point in X, and any two of these are homologous because there is a singular 1-simplex having the difference of the points as boundary (a path from one to the other). Consequently, $[\![x]\!] = [\![f(x)]\!] = f_*[\![x]\!]$.

The following result is an immediate consequence of Corollary 23.5 and Corollary 11.5 of Chapter II. Note that $L(1_K) = \chi(K)$.

23.6. Corollary. *If M is a compact smooth manifold with Euler characteristic $\chi(M) \neq 0$ then any tangent vector field on M has a zero.* □

The converse of Corollary 23.6 is also true but cannot be shown here. It is proved in Corollary 14.5 of Chapter VII.

23.7. Corollary. *If $f: \mathbf{S}^n \to \mathbf{S}^n$ has* $\deg(f) \neq (-1)^{n+1}$ *then f has a fixed point.* □

This implies, of course, that the degree of the antipodal map on $\mathbf{S}^n$ is $(-1)^{n+1}$ as seen before.

23.8. Definition. A space X is said to have the *fixed point property* if every map $f: X \to X$ has a fixed point.

For example, any polyhedron X which is acyclic ($\tilde{H}_*(X;\Lambda) = 0$) over any field Λ has the fixed point property. Thus the "dunce cap" space and the real projective plane have the fixed point property. In fact, the real projective space of even dimension $\mathbf{P}^{2m}$ is acyclic over $\mathbf{Q}$ and so has the fixed point property.

We shall give some further examples based on some results we shall have to state without proof (all but one of which will be proved later in the book).

Consider a map $f: \mathbf{CP}^n \to \mathbf{CP}^n$ on complex projective n-space. Recall from Problem 7 of Section 10 that $H_i(\mathbf{CP}^n) \approx \mathbf{Z}$ for i even with $0 \le i \le 2n$. It is known (shown in Proposition 10.2 of Chapter VI, also see Section 10 of Chapter V), that if $0 \ne \alpha \in H_2(\mathbf{CP}^n)$ and $f(\alpha) = k\alpha$ then f_* is multiplication by k^i on $H_{2i}(\mathbf{CP}^n)$. Thus $L(f) = 1 + k + k^2 + \cdots + k^n$. This is not zero for $k = 1$. For $k \ne 1$ we have $L(f) = (1 - k^{n+1})/(1 - k)$. But this has no integer k as a root if n is even. (For n odd, there is the single root -1.) Consequently, $\mathbf{CP}^n$ has the fixed point property for n even.

A similar argument also applies to quaternionic projective space $\mathbf{QP}^n$. The homology of this space is $H_i(\mathbf{QP}^n) \approx \mathbf{Z}$ for i divisible by 4 and $0 \le i \le 4n$. There is also the added fact about these spaces that for $n \ge 2$, it can be shown that there is no map $f: \mathbf{QP}^n \to \mathbf{QP}^n$ which is multiplication by -1 on $H_4(\mathbf{QP}^n)$, i.e., the analogue of k here cannot be -1. It follows that $\mathbf{QP}^n$ has the fixed point property for all $n \ge 2$.

The reduced suspension $S\mathbf{CP}^2$ is a CW-complex with single cells in dimensions 0, 3, and 5. It follows that its mod 2 homology is $\mathbf{Z}_2$ in these dimensions and 0 otherwise. It is a fact that if f is a self-map in $S\mathbf{CP}^2$ then f_* is nontrivial in dimension 3 if and only if it is nontrivial in dimension 5. Thus $L_{\mathbf{Z}_2}(f) = 1$ in $\mathbf{Z}_2$ always. Therefore $S\mathbf{CP}^2$ has the fixed point property. (Proofs of the foregoing facts about $\mathbf{QP}^n$ and $S\mathbf{CP}^n$ can be found in Section 15 of Chapter VI.)

A known theorem in the other direction states that if K is a simplicial complex of dimension at least 3, with no local cut points (i.e., no point of $|K|$ disconnects all small connected neighborhoods), and Euler characteristic $\chi(K) = 0$, then there is a self-map of $|K|$ without fixed points and homotopic to the identity. (See R.F. Brown [1].)

23.9. Proposition. *A retract A of a space X with the fixed point property has the fixed point property.*

Proof. This follows from the last part of the proof of Corollary 23.5. □

23.10. Proposition. *The one-point union $A \vee B$ has the fixed point property $\Leftrightarrow$ both A and B have the fixed point property.*

Proof. Let x_0 stand for the common base point. For ($\Leftarrow$) suppose

$f: A \vee B \to A \vee B$ and let us assume that $f(x_0) \in A$ and is not x_0. Let $g: A \to A$ be $p_A \circ f$ where p_A is the projection of $A \vee B$ onto A. Then $g(a_0) = a_0$ for some $a_0 \in A$.

If $f(a_0) \in A$ then $a_0 = g(a_0) = p_A f(a_0) = f(a_0)$ and we are done.

If $f(a_0) \in B$ then $a_0 = g(a_0) = p_A f(a_0) = x_0$ contrary to the assumption that $f(x_0) \in A - B$.

The implication ($\Rightarrow$) follows from Proposition 23.9. □

Putting together several of the last "results," we see that the space $X = \mathbf{QP}^3 \vee S\mathbf{CP}^2 \vee S\mathbf{CP}^2$ has the fixed point property, but $I \times X$ does not. This is because all three "factors" of X have the fixed point property by the above remarks, but $\chi(I \times X) = \chi(X) = 1 + (3) - (2) - (2) = 0$, so that there is a map without fixed points on $I \times X$ by the theorem cited above. (In this computation, the 1 represents the unique 0-cell, the others are contributions from higher cells in the three factors.) Examples like this are rather hard to come by. They are useful to prevent hard work being expended trying to prove intuitively plausible guesses such as "$I \times X$ has the fixed point property if X does" which are, in fact, false. There are more sophisticated examples of polyhedra K and L such that $K \times L$ has a fixed point free map but every polyhedron homotopy equivalent to either K or L has the fixed point property; see Bredon [3].

Lefschetz originally proved his fixed point theorem for manifolds only. That is the form appearing here in Theorem 12.6 of Chapter VI. The version (Theorem 23.4) in the present section is due to Hopf.

Remarks on computation of homology groups. Let K be a finite CW-complex. Since $B_{p-1} \subset C_{p-1}$ is free abelian by the Fundamental Theorem of Abelian Groups (see Section 13), the exact sequence $0 \to Z_p \to C_p \to B_{p-1} \to 0$ splits. Thus $C_p \approx Z_p \oplus B_{p-1}$ and $C_{p+1} \approx Z_{p+1} \oplus B_p$. In this representation, ∂_{p+1} is just the inclusion $B_p \hookrightarrow Z_p$. By the Fundamental Theorem of Abelian Groups there are bases $\{z_1, \dots, z_n, b_1, \dots, b_r\}$ of C_p and $\{z'_1, \dots, z'_{n'}, b'_1, \dots, b'_{r'}\}$ of C_{p+1}, such that $\partial b'_i = k_i z_i$ where $k_1 | k_2 | \cdots | k_{r'}$. It follows that

$$H_p(K) \approx \mathbf{Z}^{b_p} \oplus \mathbf{Z}_{k_1} \oplus \cdots \oplus \mathbf{Z}_{k_{r'}}$$

where the pth Betti number $b_p = n - r'$. (Some of the k_i may be 1 in which case $\mathbf{Z}_{k_i}$ is trivial.) Thus

$$\begin{aligned} b_p &= n - r' = \operatorname{nullity}(\partial_p) - \operatorname{rank}(\partial_{p+1}) \\ &= \operatorname{rank}(C_p) - \operatorname{rank}(\partial_p) - \operatorname{rank}(\partial_{p+1}). \end{aligned}$$

The matrix A of ∂_{p+1} in these bases is $\operatorname{diag}(k_1, \dots, k_{r'})$ padded with some zero rows and/or columns.

A change of bases is given by $B \mapsto UBV$ where U and V are unimodular (integer matrices with determinant ± 1). This can be achieved by a sequence of row and column operations which are addition of an integer multiple of a row or column to another, interchange of rows or columns, and change of the sign of a row or column. Therefore the matrix A, and hence the k_i, can be obtained, via these operations, from the matrix of ∂_{p+1} in its canonical basis. The latter is just the matrix

$$([\tau_i:\sigma_j])$$

of incidence numbers, where the τ_i are the p-cells of K and the σ_j are the $(p+1)$-cells of K. Such matrix reductions can be given in algorithmic form and hence can (easily) be programmed on a computer.

Problems

1. Let $f:\mathbf{R}^2 \to \mathbf{R}^2$ be the map taking $(x, y) \mapsto (-x-y, x)$. Since this is linear and preserves the integral lattice $\mathbf{Z}^2$, it induces a map $f:\mathbf{T}^2 \to \mathbf{T}^2$ on the torus. Let $w_0 \in \mathbf{T}^2$ be the point corresponding to the origin $\{0\}$ of $\mathbf{R}^2$.
   (a) Find $f_\#: \pi_1(\mathbf{T}^2, w_0) \to \pi_1(\mathbf{T}^2, w_0)$ and use this to find $f_*: H_1(\mathbf{T}^2) \to H_1(\mathbf{T}^2)$.
   (b) Find $f_*: H_2(\mathbf{T}^2) \to H_2(\mathbf{T}^2)$.
   (c) Show that any self-map on $\mathbf{T}^2$ homotopic to f has a fixed point.

2. Let Y be the union of the cylinder $\{(x, y, z) \in \mathbf{R}^3 \mid x^2 + y^2 = 1, |z| \le 1\}$ with the disk $\{(x, y, z) \in \mathbf{R}^3 \mid x^2 + y^2 \le 1, z = 0\}$. Let X be the quotient space of Y obtained by identifying antipodal points on the circle $\{z = 1\}$ in Y and also identifying antipodal points on the circle $\{z = -1\}$ in Y. Show that X has the fixed point property.

3. If X is a polyhedron with the property that any map $f: X \to X$ has $L(f) \ne 0$, then show that any finite polyhedron Y homotopy equivalent to X has the fixed point property. Also show that this is false if Y is allowed to be any topological space homotopy equivalent to X.

4. Consider the subset X of the plane consisting of the two circles A and B of radius 1 centered at $(0, \pm 1)$. Let $f: X \to X$ be the reflection in the y-axis on A and a map $B \to B$ of degree 2 on B. Show that $L(f) = 0$ but that any map $g: X \to X$ homotopic to f has a fixed point.

5. Let $X = A \cup B \cup C$ be the subset of $\mathbf{R}^3$ where $A = \{(1 + 1/t)e^{2\pi i t} \mid 1 \le t < \infty\}$ (a spiral), $B = \{(x, 0, z) \in \mathbf{R}^3 \mid (x-1)^2 + z^2 = 1, z \ge 0\}$ (a semicircle), and $C = \{z \in \mathbf{C} \mid |z| \le 1\}$ (a disk). Show that X is compact and acyclic (whence $L(f) = 1$ for all $f: X \to X$) but there exists a map $f: X \to X$ without fixed points. Why does this not contradict Corollary 23.5?

6. ◆ Let $K = \{(x, y) \in \mathbf{R}^2 \mid x^2 + y^2 \le 16, (x-2)^2 + y^2 \ge 1, (x+2)^2 + y^2 \ge 1\}$ which is a disk with two small separated disks removed; i.e., the simplest surface with three boundary components. If $f: K \to K$ is a homeomorphism without fixed points, show that f must cyclically permute the three boundary components and must reverse orientation. Also, construct such a homeomorphism without fixed points.

CHAPTER V
Cohomology

I shall allow myself free excursions as the changing stimulus of surroundings may lead me to do....

F. KLEIN

1. Multilinear Algebra

Let V be a vector space over the real numbers. Denote by $A^p(V)$ the vector space of all alternating multilinear p-forms on V. That is, $\omega \in A^p(V)$ is a function which assigns to each p-tuple $\langle X_1, \ldots, X_p \rangle$ of vectors in V, a real number $\omega(X_1, \ldots, X_p)$ such that

$$\omega(X_{\sigma_1}, \ldots, X_{\sigma_p}) = \operatorname{sgn}(\sigma)\omega(X_1, \ldots, X_p)$$

for any permutation σ of $1, 2, \ldots, p$, and such that ω is linear in each variable.

There is a product, the "wedge" product or "exterior" product, $A^p \times A^q \to A^{p+q}$ defined by

$$\boxed{\omega \wedge \eta(X_1, \ldots, X_{p+q}) = \sum_{\sigma} \operatorname{sgn}(\sigma)\omega(X_{\sigma_1}, \ldots, X_{\sigma_p})\eta(X_{\sigma_{p+1}}, \ldots, X_{\sigma_{p+q}}),}$$

where the sum is over all permutations σ of $1, 2, \ldots, p+q$ such that $\sigma_1 < \cdots < \sigma_p$ and $\sigma_{p+1} < \cdots < \sigma_{p+q}$. Note that this is the same as

$$\omega \wedge \eta(X_1, \ldots, X_{p+q}) = \frac{1}{p!\,q!} \sum_{\sigma} \operatorname{sgn}(\sigma)\omega(X_{\sigma_1}, \ldots, X_{\sigma_p})\eta(X_{\sigma_{p+1}}, \ldots, X_{\sigma_{p+q}}),$$

where the sum is over all permutations.

The wedge product is bilinear and associative.

For $p = 0$ we define $A^0(V) = \mathbf{R}$. Note that $A^1(V) = V^*$, the dual space to V. It is easy to see that if $\omega_1, \ldots, \omega_p$ are 1-forms then

$$\boxed{(\omega_1 \wedge \cdots \wedge \omega_p)(X_1, \ldots, X_p) = \det[\omega_i(X_j)].}$$

Indeed, that is essentially the definition of a determinant.

1.1. Proposition. *If $\omega_1, \ldots, \omega_n$ is a basis of V^* then $\{\omega_{i_1} \wedge \cdots \wedge \omega_{i_p} | i_1 < \cdots < i_p\}$ is a basis of $A^p(V)$ over* $\mathbf{R}$.

Proof. Let $X_1, \ldots, X_n$ be the dual basis of V, so that $\omega_i(X_j) = \delta_{i,j}$. Let ω be a p-form. For indices $i_1 < \cdots < i_p$ let $a_{i_1,\ldots,i_p} = \omega(X_{i_1}, \ldots, X_{i_p})$. Then we see that

$$\omega(X_{j_1}, \ldots, X_{j_p}) = \sum_{i_1 < \cdots < i_p} a_{i_1,\ldots,i_p} \omega_{i_1} \wedge \cdots \wedge \omega_{i_p}(X_{j_1}, \ldots, X_{j_p})$$

for all $j_1 < \cdots < j_p$ since $\omega_{i_1} \wedge \cdots \wedge \omega_{i_p}(X_{j_1}, \ldots, X_{j_p}) = 1$ when the j_k equal the i_k, and is zero otherwise. It follows that these forms are identical. Thus the forms in question span $A^p(V)$. If the above sum is zero, i.e. if $\omega = 0$, then each of the coefficients $a_{i_1,\ldots,i_p}$ vanishes and so the forms in question are independent. □

1.2. Corollary.

$$\dim A^p(V) = \binom{n}{p} = n!/(p!\,(n-p)!) \quad \textit{where} \quad n = \dim V. \qquad \square$$

1.3. Corollary.

$$\omega \in A^p(V), \quad \eta \in A^q(V) \quad \Rightarrow \quad \omega \wedge \eta = (-1)^{pq} \eta \wedge \omega.$$

Proof. Both sides of the formula are bilinear. They coincide, by direct examination, when ω and η are from the basis given in Proposition 1.1. Thus they coincide for all forms ω and η. □

1.4. Corollary.

$$\omega \in A^p(V), p \textit{ odd} \quad \Rightarrow \quad \omega \wedge \omega = 0. \qquad \square$$

2. Differential Forms

Let M^n be a smooth manifold. Recall that $T_x(M)$ denotes the tangent space of M at $x \in M$.

2.1. Definition. *A differential p-form ω on M is a differentiable function which assigns to each point $x \in M$, a member $\omega_x \in A^p(T_x(M))$.*

In this definition, "differentiable" means the following: In local coordinates $x_1, \ldots, x_n$ about x, $T_x^*(M)$ has the basis $dx_1, \ldots, dx_n$ where $dx_i(\partial/\partial x_j) = \partial x_i/\partial x_j = \delta_{i,j}$. Thus ω can be written as

$$\omega = \sum f_{i_1,\ldots,i_p} dx_{i_1} \wedge \cdots \wedge dx_{i_p}$$

and the $f_{i_1,\ldots,i_p}$ are required to be smooth.

We let $\Omega^p(M)$ be the vector space of all smooth p-forms on M. Note that $\Omega^0(M)$ is the space of all real valued smooth functions on M.

One has the wedge product on $\Omega^p(M)$ by operating pointwise. In the case of a function $f \in \Omega^0(M)$ and a form $\omega \in \Omega^p(M)$ we use the notation $f\omega = f \wedge \omega$ for short.

Note that if $X_1, \ldots, X_p$ are vector fields on M and ω is a p-form then $\omega(X_1, \ldots, X_p)$ is a smooth real valued function on M. (It can be shown that smoothness of ω is equivalent to the smoothness of the functions obtained this way.)

In general, if f is smooth real valued function on M, and X is a smooth vector field, then we define $df(X) = X(f)$. Note that $df = \sum_i (\partial f/\partial x_i)\, dx_i$.

If $\omega \in \Omega^p(M)$ then we define its "exterior derivative" $d\omega \in \Omega^{p+1}(M)$ as follows. If $\omega = f\, dx_{i_1} \wedge \cdots \wedge dx_{i_p}$ then we put $d\omega = df \wedge dx_{i_1} \wedge \cdots \wedge dx_{i_p}$. This is then extended linearly to all forms.

Of course, it must be shown that this definition is independent of the local coordinates used to define it. This is straightforward and will be left to the reader. A nice way to do this is indicated in Problem 1.

Note then that the exterior derivative gives a linear map

$$d: \Omega^p(M) \to \Omega^{p+1}(M).$$

2.2. Proposition. *We have that* $dd = 0$.

PROOF. It is sufficient to check this on $\omega = f\, dx_1 \wedge \cdots \wedge dx_p$. We calculate

$$d\omega = df \wedge dx_1 \wedge \cdots \wedge dx_p = \sum_{i=1}^{n} \frac{\partial f}{\partial x_i} dx_i \wedge dx_1 \wedge \cdots \wedge dx_p.$$

Thus

$$dd\omega = \sum_{j=1}^{n} \sum_{i=1}^{n} \frac{\partial^2 f}{\partial x_j \partial x_i} dx_j \wedge dx_i \wedge dx_1 \wedge \cdots \wedge dx_p.$$

If we rearrange the double sum so that it ranges over $j < i$ then we get two terms for each pair i, j. These terms are identical except for a $dx_j \wedge dx_i$ in one and $dx_i \wedge dx_j$ in the other. Thus it all cancels out. □

2.3. Proposition. *If ω is a p-form and η is a q-form then*

$$d(\omega \wedge \eta) = d\omega \wedge \eta + (-1)^p \omega \wedge d\eta.$$

PROOF. One need only check this when $\omega = f\, dx_1 \wedge \cdots \wedge dx_p$ and $\eta = g\, dx_{i_1} \wedge \cdots \wedge dx_{i_q}$. This is straightforward. □

One can prove (but we will not do so here) the following invariant formula for $d\omega$:

$$d\omega(X_0,\ldots,X_p)=\sum_{i=0}^{p}(-1)^i X_i(\omega(X_0,\ldots,\hat{X}_i,\ldots,X_p))$$
$$+\sum_{i<j}(-1)^{i+j}\omega([X_i,X_j],X_0,\ldots,\hat{X}_i,\ldots,\hat{X}_j,\ldots,X_p),$$

where $[X, Y] = XY - YX$, the so-called Lie bracket.

2.4. Definition. A p-form ω is *closed* if $d\omega = 0$. It is *exact* if $\omega = d\eta$ for some $(p-1)$-form η.

Note that the sequence $\cdots \to \Omega^{p-1}(M) \to \Omega^p(M) \to \Omega^{p+1}(M) \to \cdots$ is of order 2 and so it is a chain complex except that the operator d raises degree instead of lowering it. Thus what we get by taking its "homology" is called "cohomology." more precisely:

2.5. Definition. The pth *de Rham cohomology* group of M is the real vector space

$$H^p_\Omega(M)=\frac{\ker(d\colon\Omega^p(M)\to\Omega^{p+1}(M))}{\operatorname{im}(d\colon\Omega^{p-1}(M)\to\Omega^p(M))}=\frac{\text{closed } p\text{-forms}}{\text{exact } p\text{-forms}}.$$

It is a classical problem in analysis to decide when a given form is exact. It is easy to check whether or not a form is closed since that is just a computation. Thus, if $H^p_\Omega(M)=0$, then exact is the same as closed and so the problem is solved in that case.

Note that there is a product $H^p_\Omega(M)\times H^q_\Omega(M)\to H^{p+q}_\Omega(M)$ given by $[\![\omega]\!]\times[\![\eta]\!]\mapsto[\![\omega\wedge\eta]\!]$. This is well defined by Proposition 2.3 and it is bilinear and associative. Moreover, the constant function 1 is a unity element. Thus $H^*_\Omega(M)$ is a graded (signed) commutative $\mathbf{R}$-algebra.

Let us compute $H^0_\Omega(M)$. This is the kernel of $d\colon\Omega^0(M)\to\Omega^1(M)$. That is, it is $\{f\colon M\to\mathbf{R}\mid df=0\}$. But $df=0$ means all partials of f vanish and that means that f is locally constant. Thus $H^0_\Omega(M)$ is the cartesian product of copies of $\mathbf{R}$, one copy for each component of M. In particular, it is $\mathbf{R}$ if M is connected.

Now we will study "induced forms." Let $\theta\colon M\to N$ be a smooth map. Recall that the differential θ_* of θ at a point $x\in M$ is defined by $\theta_*(X)(f)=X(f\circ\theta)$ where X is a tangent vector at x of M.

2.6. Definition. If $\theta\colon M\to N$ is smooth then $\theta^*\colon\Omega^p(N)\to\Omega^p(M)$ is defined by

$$\theta^*(\omega)(X_{1_x},\ldots,X_{p_x})=\omega(\theta_*(X_{1_x}),\ldots,\theta_*(X_{p_x})),$$

where $X_{1_x},\ldots,X_{p_x}$ are tangent vectors to M at x. On functions $\theta^*(f)=f\circ\theta$.

Note that for a smooth real valued function f there are the following identities which follow immediately from the definitions:

$$d\theta^*(f)(X) = d(f \circ \theta)(X) = X(f \circ \theta) = \theta_*(X)(f) = df(\theta_*(X)) = \theta^*(df)(X).$$

Of course, we must show that the form $\theta^*(\omega)$ is smooth. This follows immediately from part (3) of the following result:

2.7. Proposition. *If θ and ψ are smooth mappings of manifolds then:*

(1) $(\theta \circ \psi)^* = \psi^* \circ \theta^*$;
(2) $\theta^*(\omega \wedge \eta) = \theta^*(\omega) \wedge \theta^*(\eta)$;
(3) $\theta^*(f\,dy_1 \wedge \cdots \wedge dy_p) = f \circ \theta\, d(y_1 \circ \theta) \wedge \cdots \wedge d(y_p \circ \theta)$; *and*
(4) $\theta^*(d\omega) = d(\theta^*(\omega))$.

PROOF. The first two follow directly from the definitions. For the third, we compute

$$\begin{aligned}\theta^*(f\,dy_1 \wedge \cdots \wedge dy_p) &= \theta^*(f) \wedge \theta^*(dy_1) \wedge \cdots \wedge \theta^*(dy_p) \\ &= f \circ \theta\, d(y_1 \circ \theta) \wedge \cdots \wedge d(y_p \circ \theta).\end{aligned}$$

For part (4), and for $\omega = f\,dy_1 \wedge \cdots \wedge dy_p$, we have

$$\begin{aligned}\theta^*(d\omega) &= \theta^*(df) \wedge \theta^*(dy_1 \wedge \cdots \wedge dy_p) \\ &= d(f \circ \theta) \wedge d(y_1 \circ \theta) \wedge \cdots \wedge d(y_p \circ \theta) \\ &= d(f \circ \theta\, d(y_1 \circ \theta) \wedge \cdots \wedge d(y_p \circ \theta)) \\ &= d(\theta^* \omega).\end{aligned}$$ □

It follows from Proposition 2.7 that θ^* induces a ring homomorphism

$$H^p_\Omega(N) \to H^p_\Omega(M),$$

taking $[\![\omega]\!]$ to $[\![\theta^*\omega]\!]$.

PROBLEMS

1. On an open set $U \subset \mathbf{R}^n$ show that the exterior derivative d is the only operator $d\colon \Omega^p(U) \to \Omega^{p+1}(U)$ satisfying:
   (a) $d(\omega + \eta) = d\omega + d\eta$;
   (b) $\omega \in \Omega^p(U)$, $\eta \in \Omega^q(U) \Rightarrow d(\omega \wedge \eta) = d\omega \wedge \eta + (-1)^p \omega \wedge d\eta$;
   (c) $f \in \Omega^0(U) \Rightarrow df(X) = X(f)$; and
   (d) $f \in \Omega^0(U) \Rightarrow d(df) = 0$.
   Deduce that d is independent of the coordinate system used to define it.

2. On the unit circle $\mathbf{S}^1$ in the plane, let $\theta = \arctan(y/x)$ be the usual polar coordinate. Show that $d\theta$ makes sense on $\mathbf{S}^1$ and is a closed 1-form which is not exact.

3. For $\omega \in \Omega^1(M)$, verify the special case $d\omega(X, Y) = X(\omega(Y)) - Y(\omega(X)) - \omega([X, Y])$ of the invariant formula mentioned above Definition 2.4.

3. Integration of Forms

We wish to define the "integration" of an n-form on an n-manifold. First we consider the case of an open subset $U \subset \mathbf{R}^n$ with coordinates $x_1, \ldots, x_n$.

For a form ω, define the support of ω, support(ω), to be the *closure* of $\{x \mid \omega_x \neq 0\}$.

Let ω be an n-form on $U \subset \mathbf{R}^n$ with *compact* support. Note then that ω extends to all of $\mathbf{R}^n$ by 0 with support in some cube. Now ω can be written

$$\omega = f(x_1, \ldots, x_n)\, dx_1 \wedge \cdots \wedge dx_n$$

and $f = 0$ outside some compact set. Thus we can define

$$\int_U \omega = \int_{\mathbf{R}^n} \omega = \int\int \cdots \int_{\mathbf{R}^n} f(x_1, \ldots, x_n)\, dx_1\, dx_2 \cdots dx_n,$$

where the right-hand integral is an ordinary Riemann integral. In fact, f need only be Riemann integrable with compact support for this to make sense.

Now suppose that W is another open set in $\mathbf{R}^n$ and let $\theta: W \to U$ be a diffeomorphism. Then we have the n-form $\theta^*(\omega)$ on W where, by definition,

$$\theta^*(\omega) = (f \circ \theta)\, d(x_1 \circ \theta) \wedge \cdots \wedge d(x_n \circ \theta).$$

Now

$$d(x_i \circ \theta) = \sum_{j=1}^{n} \frac{\partial(x_i \circ \theta)}{\partial x_j} dx_j = \sum_{j=1}^{n} J_{i,j}(\theta) dx_j.$$

where $J_{i,j}(\theta)$ is the i,j entry of the Jacobian matrix of θ. Thus

$$\begin{aligned} d(x_1 \circ \theta) \wedge \cdots \wedge d(x_n \circ \theta) &= \left(\sum_{j_1} J_{1,j_1}(\theta) dx_{j_1} \right) \wedge \cdots \wedge \left(\sum_{j_n} J_{n,j_n}(\theta) dx_{j_n} \right) \\ &= \sum_{s} J_{1,s_1}(\theta) \cdots J_{n,s_n}(\theta) dx_{s_1} \wedge \cdots \wedge dx_{s_n} \\ &= \sum_{s} \operatorname{sgn}(s) (J_{1,s_1}(\theta) \cdots J_{n,s_n}(\theta))\, dx_1 \wedge \cdots \wedge dx_n \\ &= \det(J_{i,j}(\theta)) dx_1 \wedge \cdots \wedge dx_n. \end{aligned}$$

Therefore $\theta^*(\omega) = (f \circ \theta)\, \det(J_{i,j}(\theta))\, dx_1 \wedge \cdots \wedge dx_n$ and so

$$\begin{aligned} \int \theta^*(\omega) &= \int\int \cdots \int_{\mathbf{R}^n} f(\theta(x_1, \ldots, x_n)) \det(J_{i,j}(\theta))\, dx_1 \cdots dx_n \\ &= \pm \int\int \cdots \int f(x_1, \ldots, x_n)\, dx_1 \cdots dx_n = \pm \int \omega \end{aligned}$$

by the standard Riemann change of variables rule, where the sign is the sign of $\det(J_{i,j}(\theta))$. If U is not connected then we are assuming here the same sign on all components.

Now assume that M^n is an *oriented* manifold and use only charts consistent with its orientation. Let ω be an n-form on M^n whose support is contained in the open set U where U is the domain of a chart $\phi: U \to W \subset \mathbf{R}^n$.

Then $(\phi^{-1})^*\omega$ is an n-form on $W \subset \mathbf{R}^n$. Thus we *define*

$$\int_M \omega = \int_{\mathbf{R}^n} (\phi^{-1})^*\omega.$$

To show that this is independent of the choice of ϕ let ψ be another such map and let $\theta = \psi \circ \phi^{-1}$. Then $\psi^{-1} \circ \theta = \phi^{-1}$ so $(\phi^{-1})^* = \theta^* \circ (\psi^{-1})^*$. Thus

$$\int (\phi^{-1})^*\omega = \int \theta^*(\psi^{-1})^*\omega = \int (\psi^{-1})^*\omega$$

which proves this independence.

Now let ω be an arbitrary n-form on M^n with compact support K. Let $f_i: M \to \mathbf{R}$ give a smooth partition of unity so that:

(a) $f_i \geq 0$;
(b) support$(f_i) \subset U_i$ where $\{U_i\}$ is a locally finite covering of M with each U_i the domain of a chart; and
(c) $\sum_i f_i = 1$.

Note that the compact set K touches only a finite number of the U_i. We define

$$\int \omega = \sum_i \int f_i \omega.$$

We must show that this is well defined. Thus suppose $\{g_j\}$ to be another such partition. Then $1 = \sum_{i,j} f_i g_j$. Hence $f_i = f_i \sum_j g_j = \sum_j f_i g_j$. It follows that

$$\int f_i \omega = \sum_j \int f_i g_j \omega.$$

Therefore

$$\int \omega = \sum_i \int f_i \omega = \sum_i \sum_j \int f_i g_j \omega = \sum_j \int g_j \omega,$$

as claimed.

We already had a local change of variables rule which now generalizes easily to the global fact that if $\theta: M^n \to N^n$ is a diffeomorphism of *oriented* n-manifolds, which preserves orientation, then

$$\boxed{\int_M \theta^*\omega = \int_N \omega}$$

for any n-form ω on N^n with compact support.

4. Stokes' Theorem

Consider an n-manifold M^n with boundary. It can be shown, in a similar manner to the Tubular Neighborhood Theorem, that such a manifold can be embedded in an n-manifold without boundary by adding an outward "collar" to ∂M. For technical reasons we shall assume this to be done. (In this way, we do not have to redefine forms, etc.)

We also assume M^n to be oriented. Given, at a boundary point, a system of local coordinates $x_1,..,x_n$ such that M^n is defined by $x_1 \leq 0$ (as in Definition 2.7 of Chapter II), we have that $x_2,\ldots,x_n$ form local coordinates for the boundary. (See Figure V-1.) We take these coordinates, renumbering in the same order, to define an orientation on the boundary ∂M. (The reader should convince himself that the boundary is, indeed, orientable.) Let $B = \partial M$ denote the boundary and $i: B \to M$ the inclusion map.

Let ω be an $(n-1)$-form on M^n. Then $i^*\omega$ is an $(n-1)$-form on B^{n-1}. We will denote this by $\omega|_B = i^*\omega$ or simply by ω when it is clear we are talking about B^{n-1} and not M^n.

Thus, if $\omega = \sum_i f_i \, dx_1 \wedge \cdots \wedge \widehat{dx_i} \wedge \cdots \wedge dx_n$, then

$$\omega|_B = i^*\omega = \sum_i f_i \, d(x_1 \circ i) \wedge \cdots \wedge \widehat{d(x_i \circ i)} \wedge \cdots \wedge d(x_n \circ i) = f_1 \, dx_2 \wedge \cdots \wedge dx_n,$$

since B is given by $x_1 = 0$, whence $d(x_1 \circ i) = 0$.

4.1. Theorem (Stokes' Theorem). *If M^n is an oriented n-manifold with boundary ∂M^n and ω is an $(n-1)$-form on M^n with compact support then*

$$\int_M d\omega = \int_{\partial M} \omega.$$

Proof. We first prove this in the case for which the support of ω is in a "cubic coordinate neighborhood" $\{(x_1,\ldots,x_n) \mid |x_i| \leq a\}$. (At a point on the boundary, we assume as usual that $x_1 \leq 0$ defines M.) We can take

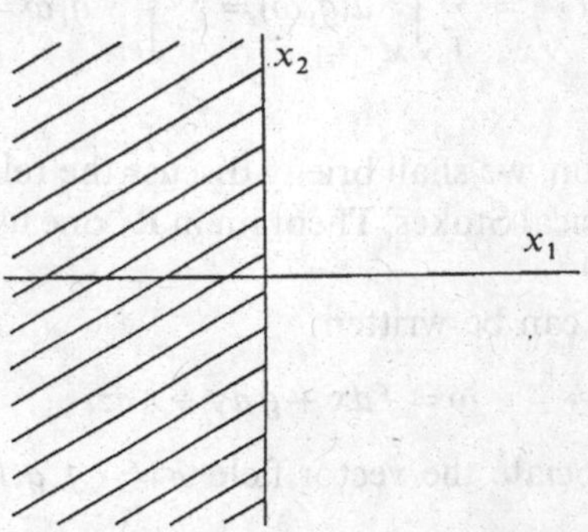

Figure V-1. Coordinate patch at a boundary point.

$\omega = \sum_i f_i \, dx_1 \wedge \cdots \wedge \widehat{dx_i} \wedge \cdots \wedge dx_n$ where support(f_i) is in the coordinate neighborhood. Because of the linearity of both sides of the equation in Theorem 4.1, it suffices to prove Theorem 4.1 for the individual terms

$$\omega = f \, dx_1 \wedge \cdots \wedge \widehat{dx_i} \wedge \cdots \wedge dx_n.$$

Then

$$d\omega = (-1)^{i-1}(\partial f/\partial x_i) \, dx_1 \wedge \cdots \wedge dx_i \wedge \cdots \wedge dx_n.$$

Accordingly, we compute

$$\begin{aligned} \int_M d\omega &= (-1)^{i-1} \int \cdots \int (\partial f/\partial x_i) \, dx_1 \, dx_2 \cdots dx_n \\ &= (-1)^{i-1} \int \cdots \int f(x_1, \ldots, x_n) \Big|_{x_i=-a}^{x_i=a} dx_1 \cdots \widehat{dx_i} \cdots dx_n, \end{aligned}$$

where $x_i = a$ should be replaced with $x_i = 0$ in the case of coordinates at a boundary point *and* $i = 1$. For an interior point this last term is zero since f vanishes where it is being evaluated. At a boundary point, with $i \neq 1$ we also get 0. In both cases $\int_{\partial M} \omega = 0$ since $\omega|_{\partial M} = 0$.

The only case remaining is that of a boundary point, and $i = 1$. In this case,

$$\begin{aligned} \int_M d\omega &= \int \cdots \int f(x_1, \ldots, x_n) \Big|_{x_1=-a}^{x_1=0} dx_2 \cdots dx_n \\ &= \int \cdots \int f \Big|_{\partial M} dx_2 \cdots dx_n \\ &= \int_{\partial M} \omega, \end{aligned}$$

as was to be shown. Now we take the general case. Let $\{g_i\}$ be a smooth partition of unity with the support of each g_i being compact and inside a cubic coordinate neighborhood as above. Then

$$\int_M d\omega = \int_M d\Big(\sum_i g_i \omega\Big) = \sum_i \int_M d(g_i \omega) = \sum_i \int_{\partial M} g_i \omega = \int_{\partial M} \sum_i g_i \omega = \int_{\partial M} \omega.$$

□

To close this section, we shall briefly discuss the relationship between this material and the classical Stokes' Theorem in $\mathbf{R}^3$ one usually sees in a calculus course.

In $\mathbf{R}^3$ a 1-form ω can be written

$$\omega = f \, dx + g \, dy + h \, dz.$$

To this, one can associate the vector field $\vec{F} = \langle f, g, h \rangle$. A 2-form η can be written

$$\eta = f \, dy \wedge dz + g \, dz \wedge dx + h \, dx \wedge dy$$

and this too can be associated with the same vector field $\vec{F} = \langle f, g, h \rangle$. The 3-form $f\,dx \wedge dy \wedge dz$ can be regarded as the function f.

Then one can compute that

$$d\omega = (g_x - f_y)\,dx \wedge dy + (f_z - h_x)dz \wedge dx + (h_y - g_z)\,dy \wedge dz,$$

so that the form $d\omega$ corresponds to the vector field $\operatorname{curl} \vec{F}$.

Similarly, for a 2-form η, as above, we have

$$d\eta = (f_x + g_y + h_z)\,dx \wedge dy \wedge dz$$

which is identified with the function $\operatorname{div} \vec{F}$.

For a function f, the vector field $\operatorname{grad} f$ corresponds to the form df.

On a (compact smooth) solid $M^3 \subset \mathbf{R}^3$ one can see that the 2-form induced from η by the inclusion map $\partial M^3 \hookrightarrow \mathbf{R}^3$, corresponds to $\vec{F} \cdot \vec{n}\,d\sigma$ on ∂M^3 where $\vec{n}$ is the outward normal and $d\sigma$ is the element of surface area. This is essentially the same calculation that is done in calculus books. In this case, then, the formula $\int_M d\eta = \int_{\partial M} \eta$ of Theorem 4.1 becomes

$$\iiint_M \operatorname{div} \vec{F}\,dV = \iint_{\partial M} \vec{F} \cdot \vec{n}\,d\sigma,$$

which is the "Divergence Theorem" or "Gauss' Theorem" in calculus books.

Similarly, if M^2 is a surface with a boundary in $\mathbf{R}^3$ and ω is a 1-form as above, then the restriction of ω to ∂M^2 can be computed to correspond to $\vec{F} \cdot \vec{T}\,ds$ where $\vec{T}$ is the unit tangent vector to the curve ∂M^2 appropriately oriented, and ds is the element of arc length. Thus, the formula $\int_M d\omega = \int_{\partial M} \omega$ of Theorem 4.1 becomes

$$\iint_M \operatorname{curl} \vec{F} \cdot \vec{n}\,d\sigma = \oint_{\partial M} \vec{F} \cdot \vec{T}\,ds$$

which is the "Stokes' Theorem" in calculus books, and becomes, in case M^2 is planar, what is variously called "Green's Theorem" or "Gauss' Theorem" in the plane.

Also, of course, the formula $d^2 = 0$ becomes, in the classical notation, $\operatorname{div} \operatorname{curl} F = 0$ when applied to 1-forms, and $\operatorname{curl} \operatorname{grad} f = 0$ when applied to 0-forms.

5. Relationship to Singular Homology

We wish to relate de Rham cohomology to singular homology. In order to do that we must make a slight modification to singular homology (which will be shown later to give the same groups). We will restrict attention to *smooth* singular simplices $\sigma: \Delta_p \to M^n$. (Actually we require σ to be defined and smooth in a neighborhood of Δ_p in its p-plane, but we don't care about its values there. This is just to avoid a separate definition of "smooth" for nonmanifolds like Δ_p.)

Since we are going to integrate a p-form over the standard p-simplex, we must choose an orientation for Δ_p. Do this by taking the positive orientation on the 0-simplex Δ_0 and, if an orientation has been chosen for Δ_{p-1}, choose the one on $\partial\Delta_p$ making the face map $F_0\colon \Delta_{p-1} \to \partial\Delta_p$ preserve orientation. (Ignore the $(p-2)$-skeleton so you can think of Δ_p as a manifold with boundary.) Then orient Δ_p consistently with its boundary, in the way described in the previous section. Since the face maps F_i satisfy $F_i = [e_i, e_0, e_1, \ldots, \hat{e}_i, \ldots, e_p] \circ F_0$ it follows that F_i preserves orientation if and only if i is even.

Now suppose we have a p-form ω on M^n and a smooth singular simplex $\sigma\colon \Delta_p \to M^n$. Then $\sigma^*\omega$ is a p-form on Δ_p. Ignoring the $(p-2)$-skeleton of Δ_p, we can integrate $\sigma^*\omega$ over Δ_p. (To be precise, one must do an approximation in order to avoid the $(p-2)$-skeleton, or expand the treatment of integration a bit. One could, for example, multiply the form by a smooth function to $[0, 1]$ which is 1 outside a small neighborhood of the $(p-2)$-skeleton and 0 inside a smaller neighborhood, and then pass to a limit as these neighborhoods get smaller. The result of all that is clear, however, so we shall leave the details alone.)

Thus we define

$$\int_\sigma \omega = \int_{\Delta_p} \sigma^*\omega.$$

(Note that M^n need not be oriented for this. It is the orientation of Δ_p that matters.) Define, for a p-chain $c = \sum_\sigma n_\sigma \sigma$,

$$\int_c \omega = \sum_\sigma n_\sigma \int_\sigma \omega.$$

This provides a homomorphism

$$\Psi(\omega)\colon \Delta_p(M) \to \mathbf{R}$$

given by $\Psi(\omega)(c) = \int_c \omega$. Note that this is linear in ω as well. Therefore we have the homomorphism

$$\Psi\colon \Omega^p(M) \to \mathrm{Hom}(\Delta_p(M), \mathbf{R}),$$

where $\mathrm{Hom}(A, B)$ denotes the group of homomorphisms $A \to B$.

Now let ω be a $(p-1)$-form and σ a (smooth) singular p-simplex in M^n. Applying Stokes' Theorem, we obtain

$$\begin{aligned}\Psi(d\omega)(\sigma) &= \int_\sigma d\omega = \int_{\Delta_p} \sigma^*(d\omega)\\ &= \int_{\Delta_p} d(\sigma^*\omega) = \int_{\partial\Delta_p} \sigma^*\omega\\ &= \sum_i (-1)^i \int_{\Delta_{p-1}} F_i^*\sigma^*\omega = \sum_i (-1)^i \int_{\Delta_{p-1}} (\sigma\circ F_i)^*\omega\end{aligned}$$

$$= \sum_i (-1)^i \int_{\sigma \circ F_i} \omega = \int_{\partial \sigma} \omega = \Psi(\omega)(\partial \sigma)$$

$$= \delta(\Psi(\omega))(\sigma),$$

where

$$\delta : \mathrm{Hom}(\Delta_{p-1}(M), \mathbf{R}) \to \mathrm{Hom}(\Delta_p(M), \mathbf{R})$$

is the transpose of ∂, i.e., $(\delta f)(c) = f(\partial c)$.

Thus we have the commutative diagram

$$\begin{array}{ccc} \Omega^{p-1}(M) & \xrightarrow{\Psi} & \mathrm{Hom}(\Delta_{p-1}(M), \mathbf{R}) \\ \downarrow d & & \downarrow \delta = \mathrm{Hom}(\partial, 1) \\ \Omega^p(M) & \xrightarrow{\Psi} & \mathrm{Hom}(\Delta_p(M), \mathbf{R}). \end{array}$$

That is, Ψ is a chain map. We shall study it.

The groups on the right (the duals of the chain groups) are called "cochain groups," their elements being "cochains." The (smooth) singular "cochain complex" is

$$\boxed{\Delta^p(M; \mathbf{R}) = \mathrm{Hom}(\Delta_p(M), \mathbf{R})}$$

together with the transpose δ of ∂. The singular "cohomology" of M is

$$\boxed{H^p(M; \mathbf{R}) = H^p(\Delta^*(M; \mathbf{R})),}$$

that is, the kernel of δ on Δ^p modulo the image of δ from Δ^{p-1}. The chain map Ψ, above, then induces a homomorphism

$$\boxed{\Psi^* : H^p_\Omega(M) \longrightarrow H^p(M; \mathbf{R}).}$$

The "de Rham Theorem" (originally conjectured by E. Cartan) states that Ψ^* is an isomorphism for all M. We will prove that theorem in Section 9. First we must develop the elements of the theory of singular cohomology, and related things.

6. More Homological Algebra

We briefly discussed the tensor product $A \otimes B$ of two abelian groups A and B before. Let us take it up again a little more fully. The group $A \otimes B$ is defined as the abelian group generated by the symbols $a \otimes b$, $a \in A$ and $b \in B$ subject to the relations that make this bilinear, i.e., $(a_1 + a_2) \otimes b = a_1 \otimes b + a_2 \otimes b$ and $a \otimes (b_1 + b_2) = a \otimes b_1 + a \otimes b_2$. It is characterized by the "universal property" that a bilinear map $A \times B \to C$ into an abelian group C factors uniquely through $A \otimes B$ where $A \times B \to A \otimes B$ taking $a \times b$ to $a \otimes b$ is bilinear and the induced map $A \otimes B \to C$ is a homomorphism.

The tensor product is a covariant functor of both variables. That is, homomorphisms $f: A' \to A$ and $g: B' \to B$ induce a homomorphism $f \otimes g: A' \otimes B' \to A \otimes B$ by $(f \otimes g)(a \otimes b) = f(a) \otimes g(b)$.

If $A' \subset A$ and $K \subset A \otimes B$ is the image of $i \otimes 1: A' \otimes B \to A \otimes B$, then there is the homomorphism $(A/A') \otimes B \to (A \otimes B)/K$ taking $[a] \otimes b$ into $[a \otimes b]$. There is also the homomorphism $(A \otimes B)/K \to (A/A') \otimes B$ induced from the homomorphism $A \otimes B \to (A/A') \otimes B$ and taking $[a \otimes b] \mapsto [a] \otimes b$. These two homomorphisms are mutually inverse, and so define an isomorphism $(A/A') \otimes B \approx (A \otimes B)/K$. This is equivalent to saying that the short exact sequence

$$0 \to A' \to A \to A'' \to 0$$

induces the sequence

$$A' \otimes B \to A \otimes B \to A'' \otimes B \to 0$$

which is exact at the $A \otimes B$ term. It is also obviously exact at the $A'' \otimes B$ term, and so it is exact. That means that $\otimes$ is a "right exact functor."

There is the isomorphism $\mathbf{Z} \otimes B \approx B$ making $1 \otimes b$ correspond to b. This covers all of $\mathbf{Z} \otimes B$ because of the identity $n \otimes b = (1 + 1 + \cdots + 1) \otimes b = 1 \otimes b + \cdots + 1 \otimes b = 1 \otimes nb$.

If $\{A_\alpha\}$ is a collection of abelian groups then $(\bigoplus A_\alpha) \otimes B$ is generated by the elements $a_\alpha \otimes b$. The group $\bigoplus (A_\alpha \otimes B)$ is also generated by elements $a_\alpha \otimes b$. This symbol means different things in the two groups, but the correspondence between them generates an isomorphism

$$\left(\bigoplus A_\alpha\right) \otimes B \approx \bigoplus (A_\alpha \otimes B).$$

Similarly, if A and B are abelian groups then $\mathrm{Hom}(A, B)$ denotes the abelian group of all homomorphism $A \to B$. If $0 \to A' \to A \to A'' \to 0$ is exact then it is very easy to see that the induced sequence

$$0 \to \mathrm{Hom}(A'', B) \to \mathrm{Hom}(A, B) \to \mathrm{Hom}(A', B)$$

is exact. Thus $\mathrm{Hom}(\cdot, \cdot)$ is a "left exact" contravariant functor of the first variable. It is easy to see that it is also a left exact covariant functor of the second variable.

Now let (C_*, ∂) be a chain complex and let G be an abelian group. Then $C_* \otimes G$ is a chain complex with the differential $\partial \otimes 1$. Also $\mathrm{Hom}(C_*, G)$ is a chain complex with the differential $\delta = \mathrm{Hom}(\partial, 1)$, but note that this raises degree by one instead of lowering it. (Some people call this a "cochain complex" but we will usually not do that unless we want to stress that the differential is of degree $+1$ instead of -1.)

If A_* and B_* are chain complexes and $\phi: A_* \to B_*$ is a chain map then, for any abelian group G, $\phi \otimes 1: A_* \otimes G \to B_* \otimes G$ is a chain map, and $\phi' = \mathrm{Hom}(\phi, 1): \mathrm{Hom}(B_*, G) \to \mathrm{Hom}(A_*, G)$, given by $\phi'(f)(a) = f(\phi(a))$, is also a chain map, i.e., $\delta\phi' = \phi'\delta$, by

$$\delta(\phi'(f))(a) = \phi'(f)(\partial a) = f(\phi(\partial a)) = f(\partial(\phi(a))) = (\delta f)(\phi(a)) = \phi'(\delta f)(a).$$

Thus there is an induced homomorphism

$$\phi^*: H^*(\mathrm{Hom}(B_*, G)) \to H^*(\mathrm{Hom}(A_*, G)),$$

and similarly

$$\phi_*: H_*(A_* \otimes G) \to H_*(B_* \otimes G).$$

Let $0 \to A_* \to B_* \to C_* \to 0$ be an exact sequence of chain complexes and chain maps which is *split*, by a splitting $C_* \to B_*$ which is not assumed to be a chain map (and usually is not). (For example, the splitting exists if each C_i is free abelian, e.g., $C_* = \Delta_*(X)$.) In this case we get the short exact sequence

$$0 \to \mathrm{Hom}(C_*, G) \to \mathrm{Hom}(B_*, G) \to \mathrm{Hom}(A_*, G) \to 0$$

and therefore the induced long exact sequence

$$\cdots \to H^i(\mathrm{Hom}(C_*, G)) \to H^i(\mathrm{Hom}(B_*, G)) \to H^i(\mathrm{Hom}(A_*, G))$$
$$\to H^{i+1}(\mathrm{Hom}(C_*, G)) \to \cdots$$

and similarly for the tensor product.

For example, in singular theory, if $A \subset X$ then there is the split exact sequence

$$0 \to \Delta_*(A) \to \Delta_*(X) \to \Delta_*(X, A) \to 0$$

which induces

$$0 \to \Delta^*(X, A; G) \to \Delta^*(X; G) \to \Delta^*(A; G) \to 0$$

where, for instance, $\Delta^p(X; G) = \mathrm{Hom}(\Delta_p(X), G)$. Note that this "cochain group" can be thought of as the group of *functions* from the set of singular simplices to G, since these simplices are a set of free generators of the chain group. By the exact sequence, it follows that we can similarly regard $\Delta^p(X, A; G)$ as the set of functions on singular p-simplices of X to G which vanish on simplices totally in A. Of course, the singular cohomology group $H^p(X, A; G)$ with coefficients in G is defined to be

$$\boxed{H^p(X, A; G) = H^p(\Delta^*(X, A; G)).}$$

The last short exact sequence then induces the long exact sequence in singular cohomology with coefficients in G:

$$\boxed{\cdots \to H^i(X, A; G) \to H^i(X; G) \to H^i(A; G) \to H^{i+1}(X, A; G) \to \cdots.}$$

For another example, but one involving changing the other variable in Hom, suppose that $0 \to G' \to G \to G'' \to 0$ is a short exact sequence of abelian groups, and that C_* is a free chain complex. Then this induces the short exact sequence

$$0 \to \mathrm{Hom}(C_*, G') \to \mathrm{Hom}(C_*, G) \to \mathrm{Hom}(C_*, G'') \to 0,$$

the exactness on the right following from the assumption that C_* is free, and hence the long exact sequence

$$\cdots \to H^i(\mathrm{Hom}(C_*, G')) \to H^i(\mathrm{Hom}(C_*, G)) \to H^i(\mathrm{Hom}(C_*, G''))$$
$$\to H^{i+1}(\mathrm{Hom}(C_*, G')) \to \cdots.$$

For singular theory, this gives, for example, the exact sequence

$$\boxed{\cdots \to H^i(X, A; G') \to H^i(X, A; G) \to H^i(X, A; G'') \to H^{i+1}(X, A; G') \to \cdots.}$$

In singular *homology*, the tensor product gives, similarly, the long exact sequences

$$\boxed{\cdots \to H_i(A; G) \to H_i(X; G) \to H_i(X, A; G) \to H_{i-1}(A; G) \to \cdots}$$

and

$$\boxed{\cdots \to H_i(X, A; G') \to H_i(X, A; G) \to H_i(X, A; G'') \to H_{i-1}(X, A; G') \to \cdots.}$$

We shall now introduce and study the "derived functors" Ext and Tor of Hom and $\otimes$, respectively, in the category of abelian groups.

6.1. Definition. An abelian group I is called *injective* if, whenever $G' \subset G$, any homomorphism $G' \to I$ can be extended to a homomorphism $G \to I$. An abelian group D is said to be *divisible* if, for any element $d \in D$ and any integer $n \neq 0$, there is an element $d' \in D$ with $nd' = d$. An abelian group P is called *projective* if, for any surjection $G \to G''$ of abelian groups, any homomorphism $P \to G''$, can be factored via a homomorphism $P \to G$.

The group I being injective then means that the following diagram with exact row can always be completed to be commutative:

$$\begin{array}{ccccc} 0 & \longrightarrow & G' & \longrightarrow & G \\ & & \downarrow & \swarrow & \\ & & I. & & \end{array}$$

Similarly, the group P being projective then means that the following diagram with exact row can be completed to be commutative:

$$\begin{array}{ccccc} & & P & & \\ & \swarrow & \downarrow & & \\ G & \longrightarrow & G'' & \longrightarrow & 0. \end{array}$$

6.2. Proposition. *An abelian group I is injective $\Leftrightarrow$ it is divisible.*

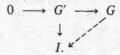

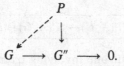

PROOF. If I is injective, then applying the definition of injective to the inclusion $n\mathbf{Z} \hookrightarrow \mathbf{Z}$ and the homomorphism $n\mathbf{Z} \to I$ taking $nk \mapsto kd$ shows that I is divisible. If D is divisible and $h': G' \to D$ is a homomorphism and $G' \subset G$ then consider all pairs (f, H) where H is a subgroup of G containing G' and $f: H \to D$ is a homomorphism extending h'. Order these pairs by inclusion (of homomorphism as well as of group) and note that the union of any chain of this partially ordered set is a member of the set. Thus, by the Maximality Principle, there is a maximal such pair. We may as well suppose that (h', G') is already maximal and must then show that $G' = G$. Suppose $g \in G - G'$. If $ng \notin G'$ for any integer $n > 1$ then we can extend h' by putting $h'(g) = 0$. Thus also define $h'(g' + kg) = h'(g')$, for any $g' \in G'$ and $k \in \mathbf{Z}$. This is a proper extension contradicting maximality. Next suppose that $ng \in G'$ for some minimal integer $n > 1$. Since D is divisible, there is a $d \in D$ such that $nd = h'(ng)$. Then, extend h' by putting $h'(g' + kg) = h'(g') + kd$. This is a well-defined proper extension of h' and so this again contradicts maximality. □

6.3. Proposition. *Any abelian group G is a subgroup of an injective group.*

Proof. There exists an epimorphism $F \to G$ for some free abelian group F; e.g., F could be taken as the free abelian group on the elements of G. Thus G is isomorphic to F/R for some $R \subset F$. If we take D to be the rational vector space on a basis of F then $F \subset D$, so that $G \approx F/R \subset D/R$ and D/R is divisible and hence injective. □

Since the quotient of a divisible group is divisible and hence injective, it follows that, for any abelian group G, there is an exact sequence

$$0 \to G \to I \to J \to 0$$

with both I and J injective. (This is called an "injective resolution of G.") Applying the functor $\mathrm{Hom}(A, \cdot)$ we get an exact sequence

$$0 \to \mathrm{Hom}(A, G) \to \mathrm{Hom}(A, I) \to \mathrm{Hom}(A, J).$$

The group $\mathrm{Ext}(A, G)$ is defined to be the cokernel of the last map. That is, it is defined so that the following sequence is exact:

$$\boxed{0 \to \mathrm{Hom}(A, G) \to \mathrm{Hom}(A, I) \to \mathrm{Hom}(A, J) \to \mathrm{Ext}(A, G) \to 0.}$$

Letting $I^0 = I$ and $I^1 = J$, the resolution can be regarded as a small cochain complex I^* with augmentation $G \to I^*$. Then $C^* = \mathrm{Hom}(A, I^*)$ is also a cochain complex and the above exact sequence amounts to saying that $H^0(C^*) = \mathrm{Hom}(A, G)$ and $H^1(C^*) = \mathrm{Ext}(A, G)$.

6.4. Proposition. *If $G \to I^*$ and $H \to J^*$ are injective resolutions of the abelian groups G and H respectively, and if $h: G \to H$ is a given homomorphism then there exists a chain map $h_*: I^* \to J^*$ extending h. Moreover, any two such extensions are chain homotopic.*

Proof. We will prove this for the resolutions of length two that we have in the case of abelian groups, but it would extend easily to those of any length.

We are to show that there is a completion to a commutative diagram (with exact rows) as shown:

$$\begin{array}{ccccccc} 0 & \longrightarrow & G & \xrightarrow{i} & I^0 & \xrightarrow{i_0} & I^1 \\ & & \Big\downarrow h & & \vdots h_0 & & \vdots h_1 \\ 0 & \longrightarrow & H & \xrightarrow{j} & J^0 & \xrightarrow{j_0} & J^1. \end{array}$$

The composition $j \circ h: G \to J^0$ extends to the map h_0 by injectivity of J^0. This makes the first square commute. Regarding, as we can, i, j as inclusions, there is the induced diagram with exact rows

$$\begin{array}{ccccc} 0 & \longrightarrow & I^0/G & \xrightarrow{i_0'} & I^1 \\ & & \Big\downarrow h_0' & & \vdots h_1 \\ 0 & \longrightarrow & J^0/H & \xrightarrow{j_0'} & J^1. \end{array}$$

The map $j_0' \circ h_0'$ extends to $h_1: I^1 \to J^1$ by injectivity of J^1. This completes the construction of h_0 and h_1, the desired commutativity being clear from the construction.

Now suppose that $h_i': I^i \to J^i$, $i = 0, 1$, give another extension of h. The desired chain homotopy will be a map $D: I^1 \to J^0$ (since it vanishes in other degrees) such that $D \circ i_0 = h_0' - h_0$ and $j_0 \circ D = h_1' - h_1$. Since $h_0' - h_0$ vanishes on G it induces a map $I^0/G \to J^0$ and this extends to a map $D: I^1 \to J^0$ because J^0 is injective. Thus we have the commutative diagram:

$$\begin{array}{ccccccccc} 0 & \longrightarrow & G & \xrightarrow{i} & I^0 & \xrightarrow{i_0} & I^1 & \longrightarrow & 0 \\ & & \Big\downarrow 0 & & \Big\downarrow h_0' - h_0 & \swarrow D & \Big\downarrow h_1' - h_1 & & \\ 0 & \longrightarrow & H & \xrightarrow{j} & J^0 & \xrightarrow{j_0} & J^1 & & \end{array}$$

which is exactly what we were after. (The commutativity of the lower triangle follows from the surjectivity of i_0, the only place we are using that these are resolutions of length 2.) □

This shows that there is a canonical isomorphism between the versions of $\operatorname{Ext}(A, G)$ obtained from two injective resolutions of G and also shows that $\operatorname{Ext}(A, G)$ is a covariant functor of G. Naturality in A of the sequence defining $\operatorname{Ext}(A, G)$ shows that $\operatorname{Ext}(A, G)$ is a contravariant functor of A.

If $0 \to A' \to A \to A'' \to 0$ is exact then so is

$$0 \to \operatorname{Hom}(A'', I) \to \operatorname{Hom}(A, I) \to \operatorname{Hom}(A', I) \to 0,$$

for I injective, the exactness on the right being precisely the definition of "injective." Thus

$$0 \to \operatorname{Hom}(A'', I^*) \to \operatorname{Hom}(A, I^*) \to \operatorname{Hom}(A', I^*) \to 0$$

is a short exact sequence of cochain complexes, and hence induces the (not very) long exact sequence in homology:

$$0 \to \operatorname{Hom}(A'', G) \to \operatorname{Hom}(A, G) \to \operatorname{Hom}(A', G) \to \operatorname{Ext}(A'', G) \\ \to \operatorname{Ext}(A, G) \to \operatorname{Ext}(A', G) \to 0.$$

If P is projective and $0 \to G \to I \to J \to 0$ is an injective resolution of G then $\operatorname{Hom}(P, I) \to \operatorname{Hom}(P, J)$ is onto, by definition of "projective," and so $\operatorname{Ext}(P, G) = 0$. Suppose that $0 \to R \to F \to A \to 0$ is exact with F projective. (Here "F" stands for "free" since a free abelian group is obviously projective, and such a sequence always exists for any A with F free.) This gives the exact sequence

$$0 \to \operatorname{Hom}(A, G) \to \operatorname{Hom}(F, G) \to \operatorname{Hom}(R, G) \to \operatorname{Ext}(A, G) \to 0$$

since $\operatorname{Ext}(F, G) = 0$. Thus $\operatorname{Ext}(A, G)$ could also be defined via this exact sequence.

Suppose now that A is such that $\operatorname{Ext}(A, G) = 0$ for all G. Let $0 \to R \to F \to A \to 0$ be exact with F projective and let $0 \to G_2 \to G_1 \to G_0 \to 0$ be exact. Then

$$0 \to \operatorname{Hom}(A, G_*) \to \operatorname{Hom}(F, G_*) \to \operatorname{Hom}(R, G_*) \to 0$$

is a short exact sequence of chain complexes. Its induced homology sequence contains the segment

$$H_1(\operatorname{Hom}(R, G_*)) \to H_0(\operatorname{Hom}(A, G_*)) \to H_0(\operatorname{Hom}(F, G_*)).$$

But the right-hand term is zero since $\operatorname{Hom}(F, \cdot)$ is exact because F is projective. The left-hand term is also zero since $0 \to \operatorname{Hom}(R, G_2) \to \operatorname{Hom}(R, G_1) \to \operatorname{Hom}(R, G_0)$ is exact. Therefore the middle term is zero which means that $\operatorname{Hom}(A, G_1) \to \operatorname{Hom}(A, G_0)$ is onto. Since this is true for all surjections $G_1 \to G_0$, it follows that A is projective. This shows that $\operatorname{Ext}(A, \cdot) = 0 \Leftrightarrow A$ is projective.

For general A, and for an exact sequence $0 \to R \to F \to A \to 0$ with F projective, the exact sequence

$$0 \to \operatorname{Hom}(A, G) \to \operatorname{Hom}(F, G) \to \operatorname{Hom}(R, G) \to \operatorname{Ext}(A, G) \\ \to \operatorname{Ext}(F, G) \to \operatorname{Ext}(R, G) \to 0$$

shows that $\operatorname{Ext}(R, G) = 0$ for all G, since $\operatorname{Ext}(F, G) = 0$. Therefore R is automatically projective, so that we really have a projective resolution of A. Since R is an arbitrary subgroup of F we conclude that any subgroup of a projective group is projective.

[Of course, it is well known that a subgroup R of a free abelian group F is free abelian. It follows easily that "projective" is the same as "free" for abelian groups. However, we do not need these facts.]

Let $0 \to P_1 \to P_0 \to A \to 0$ be a "projective resolution" of A; e.g., take $P_0 = F$, $P_1 = R$ as above. Then $\operatorname{Hom}(P_*, G)$ is a cochain complex with $H^0(\operatorname{Hom}(P_*, G)) = \operatorname{Hom}(A, G)$ and $H^1(\operatorname{Hom}(P_*, G)) = \operatorname{Ext}(A, G)$ for all G. If

$$0 \to G' \to G \to G'' \to 0$$

is exact then so is

$$0 \to \operatorname{Hom}(P_*, G') \to \operatorname{Hom}(P_*, G) \to \operatorname{Hom}(P_*, G'') \to 0$$

since P_* is projective. The induced homology sequence of this gives the exact sequence

$$\boxed{\begin{aligned}&0 \to \operatorname{Hom}(A, G') \to \operatorname{Hom}(A, G) \to \operatorname{Hom}(A, G'') \to \operatorname{Ext}(A, G')\\ &\to \operatorname{Ext}(A, G) \to \operatorname{Ext}(A, G'') \to 0.\end{aligned}}$$

Similar considerations apply to the tensor product in place of Hom, as we will briefly describe. For any abelian group A, let $0 \to R \to F \to A \to 0$ be a projective resolution. Then there is the exact sequence

$$0 \to A * B \to R \otimes B \to F \otimes B \to A \otimes B \to 0$$

which defines the "torsion product" $A * B$. This is often denoted by $\operatorname{Tor}(A, B)$ or $\operatorname{Tor}_1(A, B)$. It is symmetric in the sense that $A * B \approx B * A$ (in a canonical way). (To see this, chase the diagram made up of the tensor product of a projective resolution of A with one of B.) Also, any exact sequence $0 \to A' \to A \to A'' \to 0$ induces the (somewhat) long exact sequence

$$\boxed{0 \to A' * B \to A * B \to A'' * B \to A' \otimes B \to A \otimes B \to A'' \otimes B \to 0.}$$

Clearly $A * B = 0$ if either A or B is projective since, then, $0 \to R \to F \to A \to 0$ splits.

Now we shall compute Ext and Tor for finitely generated groups. Since they commute with *finite* direct sums, as is easily seen by looking at split exact sequences, it suffices to compute them for cyclic groups.

Since $\mathbf{Z}$ is projective we have

$$\boxed{A * \mathbf{Z} = 0 \qquad \text{and} \qquad \operatorname{Ext}(\mathbf{Z}, A) = 0 \qquad \text{for all } A.}$$

From the exact sequence $0 \to \mathbf{Z} \xrightarrow{n} \mathbf{Z} \to \mathbf{Z}_n \to 0$ we have, for any G, the exact sequence

$$0 \to \operatorname{Hom}(\mathbf{Z}_n, G) \to \operatorname{Hom}(\mathbf{Z}, G) \xrightarrow{n} \operatorname{Hom}(\mathbf{Z}, G) \to \operatorname{Ext}(\mathbf{Z}_n, G) \to 0.$$

Since $\operatorname{Hom}(\mathbf{Z}, G) \approx G$, we conclude that

$$\boxed{\operatorname{Ext}(\mathbf{Z}_n, G) \approx G/nG.}$$

In particular,

$$\boxed{\operatorname{Ext}(\mathbf{Z}_n, \mathbf{Z}) \approx \mathbf{Z}_n.}$$

To compute $\operatorname{Ext}(\mathbf{Z}_n, \mathbf{Z}_m)$ consider the exact sequence $0 \to \mathbf{Z}_m \to \mathbf{Q}/\mathbf{Z} \xrightarrow{m} \mathbf{Q}/\mathbf{Z} \to 0$ where the map on the right is multiplication by m. This induces the exact sequence

$$0 \to \operatorname{Hom}(\mathbf{Z}_n, \mathbf{Z}_m) \to \operatorname{Hom}(\mathbf{Z}_n, \mathbf{Q}/\mathbf{Z}) \xrightarrow{m} \operatorname{Hom}(\mathbf{Z}_n, \mathbf{Q}/\mathbf{Z}) \to \operatorname{Ext}(\mathbf{Z}_n, \mathbf{Z}_m) \to 0$$

where the middle map is multiplication by m. The groups on the ends must be cyclic, and counting orders using exactness shows they must both be $\mathbf{Z}_d$ for some d. Thus it suffices to compute the kernel of multiplication by m on $\operatorname{Hom}(\mathbf{Z}_n, \mathbf{Q}/\mathbf{Z}) \approx \mathbf{Z}_n$. We claim the kernel is the (cyclic) subgroup of order $d = \gcd(n, m)$. To see this let $n = dp$ and $m = dq$ where p and q are relatively prime. Then $k \in \ker(m) \Leftrightarrow n \mid km \Leftrightarrow dp \mid kdq \Leftrightarrow p \mid kq \Leftrightarrow p \mid k$, whence $k = p$ works and is the smallest such, proving the claim. Therefore we have

$$\boxed{\operatorname{Ext}(\mathbf{Z}_n, \mathbf{Z}_m) \approx \mathbf{Z}_d \approx \operatorname{Hom}(\mathbf{Z}_n, \mathbf{Z}_m) \qquad \text{where} \quad d = \gcd(n, m).}$$

The exact sequence $0 \to \mathbf{Z} \xrightarrow{n} \mathbf{Z} \to \mathbf{Z}_n \to 0$ induces the exact sequence

$$0 \to \mathbf{Z} * \mathbf{Z}_m \to \mathbf{Z} * \mathbf{Z}_m \to \mathbf{Z}_n * \mathbf{Z}_m \to \mathbf{Z} \otimes \mathbf{Z}_m \to \mathbf{Z} \otimes \mathbf{Z}_m \to \mathbf{Z}_n \otimes \mathbf{Z}_m \to 0.$$

Since $\mathbf{Z}$ is projective, this reduces to

$$0 \to \mathbf{Z}_n * \mathbf{Z}_m \to \mathbf{Z}_m \xrightarrow{n} \mathbf{Z}_m \to \mathbf{Z}_n \otimes \mathbf{Z}_m \to 0$$

where the middle map is multiplication by n. As above this implies that the groups on the ends are cyclic of order $d = \gcd(n, m)$. That is,

$$\boxed{\mathbf{Z}_n * \mathbf{Z}_m \approx \mathbf{Z}_d \approx \mathbf{Z}_n \otimes \mathbf{Z}_m \qquad \text{where} \quad d = \gcd(n, m).}$$

Recall that an abelian group A is said to be "torsion free" if $na = 0$ for some $0 \neq n \in \mathbf{Z}$ implies that $a = 0$.

6.5. Proposition. *The abelian group A is torsion free $\Leftrightarrow$ $A * B = 0$ for all B.*

Proof. $\Leftarrow$: Suppose A has n-torsion and consider the exact sequence $0 \to \mathbf{Z} \xrightarrow{n} \mathbf{Z} \to \mathbf{Z}_n \to 0$. This induces the exact sequence

$$0 \to A * \mathbf{Z}_n \to A \otimes \mathbf{Z} \xrightarrow{n} A \otimes \mathbf{Z} \to A \otimes \mathbf{Z}_n \to 0$$

where the middle map is multiplication by n. Then $A * \mathbf{Z}_n \approx \ker\{n\colon A \to A\} \neq 0$ as claimed.

$\Rightarrow$: For arbitrary B suppose $A * B \neq 0$, with A torsion free. Let $0 \to R \to F \to B \to 0$ be a projective resolution of B. This induces the exact sequence

$$0 \to A * B \to A \otimes R \to A \otimes F \to A \otimes B \to 0.$$

Let $0 \neq \alpha \in A * B$. Then, regarding α as an element of $A \otimes R$, α can be written $\alpha = \sum a_i \otimes r_i$. Let $G \subset A$ be the subgroup generated by the a_i. Consider the commutative diagram

$$\begin{array}{ccccccccc} & & & & A/G*R & \to & A/G*F & & \\ & & & & \downarrow & & \downarrow & & \\ 0 \to & G*B & \to & G \otimes R & \to & G \otimes F & \to & G \otimes B & \to 0 \\ & \downarrow & & \downarrow & & \downarrow & & \downarrow & \\ 0 \to & A*B & \to & A \otimes R & \to & A \otimes F & \to & A \otimes B & \to 0 \end{array}$$

with exact rows and columns. The element $\alpha \in A * B$ taken into $A \otimes R$ comes from some $\beta \in G \otimes R$ by construction. The image of β in $G \otimes F$ maps to 0 in $A \otimes F$, and so comes from $(A/G) * F$. But $(A/G) * F = 0$ since F is projective. Thus β comes from $G * B$. But G is finitely generated and torsion free, hence free, so $G * B = 0$, a contradiction. □

6.6. Proposition.

(1) *A is projective* $\Leftrightarrow \operatorname{Ext}(A, G) = 0$ *for all G.*
(2) *G is injective* $\Leftrightarrow \operatorname{Ext}(A, G) = 0$ *for all A.*

PROOF. Part (1) has already been proved. For part (2), if G is injective then $0 \to G \to G \to 0 \to 0$ is an injective resolution and $\operatorname{Ext}(A, G) = 0$ from its definition. Conversely, suppose that $\operatorname{Ext}(A, G) = 0$ for all A. Then, for an exact sequence $0 \to A' \to A \to A'' \to 0$, the induced sequence

$$0 \to \operatorname{Hom}(A'', G) \to \operatorname{Hom}(A; G) \to \operatorname{Hom}(A', G) \to 0$$

is exact. The surjectivity on the right is precisely the definition of G being injective. □

We conclude this section with some remarks on other "base rings." If Λ is a commutative ring with unity and A, B are Λ-modules then one can define the tensor product $A \otimes_\Lambda B$ and the module of Λ-homomorphisms $\operatorname{Hom}_\Lambda(A, B)$. The only differences are the addition of the relation $\lambda a \otimes b = a \otimes \lambda b$ for the tensor product and the requirement that $h \in \operatorname{Hom}_\Lambda(A, B)$ be Λ-linear; i.e., $h(\lambda a) = \lambda h(a)$. Then $A \otimes_\Lambda B$ and $\operatorname{Hom}_\Lambda(A, B)$ are Λ-modules. If Λ is a principal ideal domain then everything done in this section, and most of the rest of the book, goes through with no major differences. In fact, however, the only major use made of such generalizations in this book is in the case of a field Λ. Even in that case, we use only the fields $\mathbf{Q}$, $\mathbf{R}$, and $\mathbf{Z}_p$, p prime. For a field, $\otimes_\Lambda$ and $\operatorname{Hom}_\Lambda$ are exact functors and all Λ-modules are both Λ-projective and Λ-injective and so the theory is a great deal simpler for them than for the integers or for a general principal ideal domain. More general base rings would have no impact on the ideas covered in this book. For that reason, and for the sake of simplicity, we shall generally restrict attention to the case of the integers, with a few special comments in the case of a field. Occasionally, we will state results for a principal ideal domain as base ring, but this is only done in order to give a unified statement covering the case of a field in addition to the case of the integers. When this is done, $\otimes$ and Hom should be interpreted as $\otimes_\Lambda$ and Hom_Λ even if those subscripts are omitted.

7. Universal Coefficient Theorems

In this section we will show that the integral homology groups determine the homology and cohomology groups with arbitrary coefficients. Note that this does not mean, and it is not true, that anything you can do with these more general groups can be done with integral homology. In the first place, field coefficients, by their simplicity, often provide a better platform for computation. Much more significant is the fact that cohomology can be given extra structure which is not present in homology, as we shall see in Chapter VI.

Assume, throughout this section, that C_* is a free chain complex, i.e., that each C_p is free abelian. Let Z_p denote the p-cycles, B_p the p-boundaries, and $H_p = Z_p/B_p$ the pth homology group.

Since B_{p-1} is projective (because $B_{p-1} \subset C_{p-1}$ and C_{p-1} is free), the exact sequence

$$0 \to Z_p \to C_p \to B_{p-1} \to 0$$

splits. Let $\phi\colon C_p \to Z_p$ be a splitting homomorphism.

7.1. Theorem (Universal Coefficient Theorem). *For a free chain complex C_* there is a natural, in C_* and G, exact sequence*

$$0 \to \mathrm{Ext}(H_{n-1}(C_*), G) \to H^n(\mathrm{Hom}(C_*, G)) \xrightarrow{\beta} \mathrm{Hom}(H_n(C_*), G) \to 0$$

which splits (naturally in G but not in C_*) and where $(\beta[\![f]\!])([\![c]\!]) = f(c)$.

PROOF. The proof is just an easy diagram chase in the diagram:

$$\begin{array}{ccccccc}
 & & & & & & 0 \\
 & & & & & & \uparrow \\
0 \to & \mathrm{Hom}(B_p, G) & \to & \mathrm{Hom}(C_{p+1}, G) & & & \mathrm{Ext}(H_{p-1}, G) \\
 & \uparrow & & \uparrow\delta & & & \uparrow \\
0 \leftarrow & \mathrm{Hom}(Z_p, G) & \leftarrow & \mathrm{Hom}(C_p, G) & & \leftarrow & \mathrm{Hom}(B_{p-1}, G) \leftarrow 0 \\
 & \uparrow & \llcorner\!\!\longrightarrow\!\!\lrcorner\uparrow & \uparrow\delta & & & \uparrow \\
 & \mathrm{Hom}(H_p, G) & & \mathrm{Hom}(C_{p-1}, G) & & \to & \mathrm{Hom}(Z_{p-1}, G) \to 0 \\
 & \uparrow & & & & & \\
 & 0. & & & & &
\end{array}$$

The diagram is, for the most part, induced by the exact sequences $0 \to Z_p \to C_p \to B_{p-1} \to 0$ and $0 \to B_p \to Z_p \to H_p \to 0$. The rows and end columns are exact. Starting with an element $f \in \mathrm{Hom}(C_p, G)$ killed by δ, we chase left, up, and right. Since that is $\delta f = 0$, and the right arrow is monomorphic, the element goes to $0 \in \mathrm{Hom}(B_p, G)$. By exactness of the first column the element can be pulled back to the $\mathrm{Hom}(H_p, G)$ term. If f was in the image of δ then the first part of the chase kills it. Thus the chase describes the desired homomorphism β. The curved arrow is $\mathrm{Hom}(\phi, 1)$, arising from the splitting map ϕ, and clearly induces, by an even simpler chase, the desired splitting homomorphism. To see the map from $\mathrm{Ext}(H_{p-1}, G)$, go down and left. To see that this produces a cocycle, note that, starting at $\mathrm{Hom}(B_{p-1}, G)$ and going left then up is the same as going left twice then up then right, but going left twice gives zero, as claimed. To see that it is well defined, note that the indeterminacy comes from the $\mathrm{Hom}(C_{p-1}, G)$ term and that goes into a coboundary in $\mathrm{Hom}(C_p, G)$. That the map from $\mathrm{Ext}(H_{p-1}, G)$ to $\mathrm{Hom}(C_p, G)$ goes into the kernel of the map to $\mathrm{Hom}(H_p, G)$ is clear. To see exactness at that point, note that any cocycle of $\mathrm{Hom}(C_p, G)$ mapping to zero in $\mathrm{Hom}(H_p, G)$ must already map to zero in $\mathrm{Hom}(Z_p, G)$. Thus it comes from $\mathrm{Hom}(B_{p-1}, G)$, but that implies that it comes from $\mathrm{Ext}(H_{p-1}, G)$ as claimed. □

7.2. Corollary (Universal Coefficient Theorem). *For singular homology and cohomology there is the exact sequence*

$$0 \to \mathrm{Ext}(H_{n-1}(X, A), G) \to H^n(X, A; G) \xrightarrow{\beta} \mathrm{Hom}(H_n(X, A), G) \to 0$$

which is natural in G and in (X, A) and which splits (naturally in G but not in (X, A)). □

7.3. Corollary. *If $H_{n-1}(X, A)$ and $H_n(X, A)$ are finitely generated then so is $H^n(X, A; \mathbf{Z})$. Indeed*

$$H^n(X, A; \mathbf{Z}) \approx F_n \oplus T_{n-1}$$

(not natural) where F_i and T_i are the free part and torsion part, respectively, of $H_i(X, A)$. □

Now we take up the dual situation of the tensor product.

7.4. Theorem (Universal Coefficient Theorem). *For a free chain complex C_* there is a natural, in C_* and G, exact sequence*

$$0 \to H_n(C_*) \otimes G \xrightarrow{\alpha} H_n(C_* \otimes G) \to H_{n-1}(C_*) * G \to 0$$

which splits (naturally in G but not in C_*) and where $\alpha(\llbracket c \rrbracket \otimes g) = \llbracket c \otimes g \rrbracket$.

Proof. This is a dual argument to Theorem 7.1 using the analogous diagram:

$$\begin{array}{ccccccccc}
 & & & & & & 0 & & \\
 & & & & & & \downarrow & & \\
0 & \leftarrow & B_p\otimes G & \leftarrow & C_{p+1}\otimes G & & H_{p-1}*G & & \\
 & & \downarrow & & \downarrow\partial & & \downarrow & & \\
0 & \rightarrow & Z_p\otimes G & \rightarrow & C_p\otimes G & \rightarrow & B_{p-1}\otimes G & \rightarrow & 0 \\
 & & \downarrow & & \downarrow\partial & & \downarrow & & \\
 & & H_p\otimes G & & C_{p-1}\otimes G & \leftarrow & Z_{p-1}\otimes G & \leftarrow & 0 \\
 & & \downarrow & & & & & & \\
 & & 0. & & & & & &
\end{array}$$

Algebraically, this diagram is identical to that of Theorem 7.1 after rotating it 180°. (The splitting $C_p\otimes G\rightarrow Z_p\otimes G$ implies a splitting $B_{p-1}\otimes G\rightarrow C_p\otimes G$.) Thus an algebraically identical diagram chase gives the result. □

7.5. Corollary (Universal Coefficient Theorem). *For singular homology there is the exact sequence*

$$0\rightarrow H_n(X,A)\otimes G\rightarrow H_n(X,A;G)\rightarrow H_{n-1}(X,A)*G\rightarrow 0$$

which is natural in G and in (X,A) and which splits (naturally in G but not in (X,A)). □

7.6. Example. For an abelian group G let

$$TG=\{g\in G\mid ng=0 \text{ for some integer } n\geq 1\}$$

which is called the "torsion subgroup" of G. The exact sequence $0\rightarrow TG\rightarrow G\rightarrow G/TG\rightarrow 0$ induces the exact sequence $0\rightarrow TG*B\rightarrow G*B\rightarrow G/TG*B$ which implies that $G*B\approx TG*B$ since G/TG is torsion free. The exact sequence $0\rightarrow\mathbf{Z}\rightarrow\mathbf{Q}\rightarrow\mathbf{Q}/\mathbf{Z}\rightarrow 0$ induces the exact sequence.

$$0\rightarrow TG*\mathbf{Q}/\mathbf{Z}\rightarrow TG\otimes\mathbf{Z}\rightarrow TG\otimes\mathbf{Q}.$$

Now $TG\otimes\mathbf{Q}=0$ and so

$$TG\approx TG\otimes\mathbf{Z}\approx TG*\mathbf{Q}/\mathbf{Z}\approx G*\mathbf{Q}/\mathbf{Z}.$$

Therefore Corollary 7.5 gives the exact sequence

$$\boxed{0\rightarrow H_n(X,A)\otimes\mathbf{Q}/\mathbf{Z}\rightarrow H_n(X,A;\mathbf{Q}/\mathbf{Z})\rightarrow TH_{n-1}(X,A)\rightarrow 0.}$$

7.7. Example. This example shows that the splitting in Corollaries 7.2 and 7.5 cannot be natural. It is simply the map $\phi\colon\mathbf{P}^2\rightarrow\mathbf{S}^2$ which (regarding $\mathbf{P}^2$ as a 2-disk with antipodal points on the boundary identified) collapses the "boundary" circle $\mathbf{P}^1$ to a point. Since $H_1(\mathbf{P}^2)\approx\mathbf{Z}_2$ and $H_2(\mathbf{P}^2)=0$, the map $\phi_*\colon H_i(\mathbf{P}^2)\rightarrow H_i(\mathbf{S}^2)$ is trivial for $i=1,2$. If the splitting

$$H_2(X;\mathbf{Z}_2)\approx H_2(X)\otimes\mathbf{Z}_2\oplus H_1(X)*\mathbf{Z}_2$$

were natural then the homomorphism $\phi_*: H_2(\mathbf{P}^2; \mathbf{Z}_2) \to H_2(\mathbf{S}^2; \mathbf{Z}_2)$ would have to be zero. (Both these groups are $\mathbf{Z}_2$.) We claim that this is not the case, i.e., we claim that ϕ_* is an isomorphism. To see that, note that ϕ is cellular for the usual CW-structures on $\mathbf{P}^2$ and $\mathbf{S}^2$. By Theorem 11.6 of Chapter IV, the chain map $\phi_\Delta: C_2(\mathbf{P}^2; \mathbf{Z}_2) \to C_2(\mathbf{S}^2; \mathbf{Z}_2)$ is given by $\phi_\Delta(\sigma) = \deg(\phi_{\tau,\sigma})\tau$ (mod 2) where σ and τ are the unique 2-cells of $\mathbf{P}^2$ and $\mathbf{S}^2$, respectively. But $\phi_{\tau,\sigma}$ is clearly of degree ± 1 in this example; indeed $\phi_{\tau,\sigma}$ is a homeomorphism. Since σ and τ, being the only 2-cells, must generate their respective homology groups, the contention follows.

The following is a corollary of Theorems 7.1 and 7.4:

7.8. Corollary. *Let $\phi: A_* \to B_*$ be a chain map of free chain complexes inducing isomorphisms $\phi_*: H_p(A_*) \xrightarrow{\approx} H_p(B_*)$ for all p. Then $(\phi \otimes 1)_*: H_p(A_* \otimes G) \xrightarrow{\approx} H_p(B_* \otimes G)$ and $\phi^*: H^p(\mathrm{Hom}(B_*, G)) \xrightarrow{\approx} H^p(\mathrm{Hom}(A_*, G))$ for all p.*

PROOF. The Universal Coefficient Theorem (Theorem 7.1) provides the commutative diagram

$$\begin{array}{ccccccccc} 0 & \longrightarrow & \mathrm{Ext}(H_{p-1}(B_*), G) & \longrightarrow & H^p(\mathrm{Hom}(B_*, G)) & \longrightarrow & \mathrm{Hom}(H_p(B_*), G) & \longrightarrow & 0 \\ & & \downarrow \approx & & \downarrow & & \downarrow \approx & & \\ 0 & \longrightarrow & \mathrm{Ext}(H_{p-1}(A_*), G) & \longrightarrow & H^p(\mathrm{Hom}(A_*, G)) & \longrightarrow & \mathrm{Hom}(H_p(A_*), G) & \longrightarrow & 0 \end{array}$$

and the second of the claimed isomorphisms follows from the 5-lemma. The first claimed isomorphism follows the same way from the analogous diagram from Theorem 7.4. □

7.9. Corollary. *If $\phi: (X, A) \to (Y, B)$ is a map such that $\phi_*: H_p(X, A) \to H_p(Y, B)$ is an isomorphism for all p, then*

$$\phi_*: H_p(X, A; G) \to H_p(Y, B; G)$$

and

$$\phi^*: H^p(Y, B; G) \to H^p(X, A; G)$$

are isomorphisms for all p and all G. □

PROBLEMS

1. Prove the analogue of Example 7.7 for cohomology with coefficients in $\mathbf{Z}_2$. That is, show that $\phi^*: H^2(\mathbf{S}^2; \mathbf{Z}_2) \to H^2(\mathbf{P}^2; \mathbf{Z}_2)$ is an isomorphism.

2. Given that the Klein bottle $\mathbf{K}^2$ has $H_0(\mathbf{K}^2) \approx \mathbf{Z}$, $H_1(\mathbf{K}^2) \approx \mathbf{Z} \oplus \mathbf{Z}_2$, and all other integral homology groups zero, compute the homology and cohomology of $\mathbf{K}^2$ in all dimensions for coefficients in $\mathbf{Z}$, and in $\mathbf{Z}_p$, all primes p.

3. If X arises from attaching a 2-disk to a circle by a map of circles of degree 9, then

compute the homology of X with coefficients in $\mathbf{Z}_3$ and in $\mathbf{Z}_9$ and in $\mathbf{Z}_6$. Do the same in case the map was of degree 3, and of degree 6.

4. Suppose $f, g\colon X \to Y$ are maps such that $f_* = g_*\colon H_*(X; \mathbf{Z}) \to H_*(Y; \mathbf{Z})$. There are cases in the literature of the Universal Coefficient Theorem being cited as implying that then $f_* = g_*\colon H_*(X; G) \to H_*(Y; G)$ for any coefficient group G. Show by example that this is false.

5. If $H_*(X)$ is finitely generated then show that $\chi(X) = \sum(-1)^i \dim H_i(X; \Lambda)$ for any field Λ.

6. Show that $H^1(X)$ is torsion free for all X.

7. ◆ If G is finitely generated show that there is a natural exact sequence

$$0 \to H^n(X, A; \mathbf{Z}) \otimes G \to H^n(X, A; G) \to H^{n+1}(X, A; \mathbf{Z}) * G \to 0$$

which splits naturally in G but not in X, A.

8. Excision and Homotopy

8.1. Theorem (Excision). *If $B \subset A \subset X$ and $\bar{B} \subset \operatorname{int}(A)$ then the inclusion map $(X - B, A - B) \hookrightarrow (X, A)$ induces isomorphisms $H^p(X, A; G) \xrightarrow{\approx} H^p(X - B, A - B; G)$ and $H_p(X - B, A - B; G) \xrightarrow{\approx} H_p(X, A; G)$ for all p and G.*

PROOF. The Excision Theorem for integer homology (Theorem 17.8 of Chapter IV) says that this inclusion induces, through the chain map $\Delta_*(X - B, A - B) \to \Delta_*(X, A)$, an isomorphism in homology. Thus the present result follows immediately from this and Corollary 7.9. □

The following result is a similar consequence of Corollary 7.9.

8.2. Theorem. *If $\mathbf{U}$ is a collection of subsets of X whose interiors cover X, and we put $\Delta^*_{\mathbf{U}}(X; G) = \operatorname{Hom}(\Delta^{\mathbf{U}}_*(X), G)$ then $\Delta^*(X; G) \to \Delta^*_{\mathbf{U}}(X; G)$ induces an isomorphism in cohomology and $\Delta^{\mathbf{U}}_*(X) \otimes G \to \Delta_*(X) \otimes G$ also induces an isomorphism in homology.* □

Consider the case of $X = \operatorname{int}(A) \cup \operatorname{int}(B)$, and put $\mathbf{U} = \{A, B\}$. Then we have the exact sequence

$$0 \to \Delta_p(A \cap B) \xrightarrow{i_A \oplus i_B} \Delta_p(A) \oplus \Delta_p(B) \xrightarrow{j_A - j_B} \Delta^{\mathbf{U}}_p(A \cup B) \to 0$$

of Theorem 18.1 of Chapter IV. Since these groups are all free abelian, applying $\operatorname{Hom}(\cdot, G)$ or $(\cdot) \otimes G$ to this sequence yields *short* exact sequences. These, in turn, induce Mayer–Vietoris long exact sequences, proving, via Theorem 8.2:

8.3. Theorem (Mayer–Vietoris). *If $X = \operatorname{int}(A) \cup \operatorname{int}(B)$ then there is the long* *exact sequence*

$$\cdots \to H^p(A \cup B) \xrightarrow{j_A^* - j_B^*} H^p(A) \oplus H^p(B) \xrightarrow{i_A^* + i_B^*} H^p(A \cap B)$$
$$\to H^{p+1}(A \cup B) \to \cdots$$

for any coefficient group G. Also Theorem 18.1 *of Chapter IV holds for any coefficients.* □

8.4. Theorem. *If* $X = +X_\alpha$ *(topological sum) then the inclusions and projections induce an isomorphism* $H^p(X; G) \approx \times H^p(X_\alpha; G)$.

PROOF. This is true already on the cochain level. That is,

$$\operatorname{Hom}(\bigoplus \Delta_*(X_\alpha), G) \approx \times \operatorname{Hom}(\Delta_*(X_\alpha), G). \qquad \square$$

In the proof of the Homotopy Axiom for singular homology with integer coefficients we constructed (in Section 16 of Chapter IV) a natural chain homotopy $D: \Delta_p(X) \to \Delta_{p+1}(I \times X)$ between η_{0_Δ} and η_{1_Δ}, where η_{i_Δ} is the chain map induced by the inclusion $X \hookrightarrow I \times X$ taking x to (i, x). That is, $\partial D + D\partial = \eta_{1_\Delta} - \eta_{0_\Delta}$. Putting $D' = \operatorname{Hom}(D, 1)$, $\eta_i^\Delta = \operatorname{Hom}(\eta_{i_\Delta}, 1)$, and $\delta = \operatorname{Hom}(\partial, 1)$ etc., this becomes

$$D'\delta + \delta D' = \eta_1^\Delta - \eta_0^\Delta,$$

where $D': \Delta^p(I \times X; G) \to \Delta^{p-1}(X; G)$. Thus if $F: I \times X \to Y$ is a homotopy between $f_0, f_1: X \to Y$ then, with $F^\Delta = \operatorname{Hom}(F_\Delta, 1): \Delta^*(Y; G) \to \Delta^*(I \times X; G)$, we have

$$(D'F^\Delta)\delta + \delta(D'F^\Delta) = \eta_1^\Delta \circ F^\Delta - \eta_0^\Delta \circ F^\Delta = (F \circ \eta_1)^\Delta - (F \circ \eta_0)^\Delta = f_1^\Delta - f_0^\Delta.$$

Consequently, $f_1^* = f_0^*: H^p(Y; G) \to H^p(X; G)$.

The naturality of D implies that D and D' induce similar chain homotopies on pairs (X, A). The reader can fill in these details. We state the final result.

8.5. Theorem (Homotopy). *If* $f_0 \simeq f_1: (X, A) \to (Y, B)$ *then, for any coefficient group G,*

$$f_0^* = f_1^*: H^p(Y, B; G) \to H^p(X, A; G)$$

and

$$f_{0*} = f_{1*}: H_p(X, A; G) \to H_p(Y, B; G). \qquad \square$$

9. de Rham's Theorem

We now return to de Rham cohomology (closed forms modulo exact forms) and its relation to singular theory. As mentioned, the relationship is through singular theory based on *smooth* singular simplices. Thus we must know to what extent the development of singular theory can be followed with the restriction to smooth simplices. Of course, we are interested only in (co)homology of smooth manifolds, and smooth maps and homotopies. There are a few things we will need below for this theory. We will need the long exact sequences, the Mayer–Vietoris sequences (for open subsets), and the triviality of the homology of contractible manifolds (like euclidean space). All of this goes along quite easily for smooth simplices with two exceptions. First, one must talk about chains in the standard simplices, but these spaces are not manifolds. They do, however, carry a functional structure as subspaces of euclidean space, and so smoothness makes sense for them. This is really a nonissue. More serious is the proof of Theorem 15.5 of Chapter IV that a contractible space has trivial homology. It does no good to cite the Homotopy Axiom for this, as it is a crucial ingredient of the proof of that axiom. The problem is the "cone" construction of a $(p+1)$-simplex from a given p-simplex and the contracting homotopy. This construction simply does not produce a smooth simplex. However, if you look at where this result is used in proving the Homotopy Axiom, you see that it is used only in the "acyclic models" argument constructing the cross product of chains. There, one has a cycle in the product of two standard simplices and one wishes to conclude it is a boundary. But it is clear that, by induction, this construction need only be done on affine chains, yielding affine chains. In that case, we are in good shape because that is exactly what the cone construction does, i.e., in that case it is the same as the cone operator discussed in Section 17 of Chapter IV. That takes care of the second problem.

Recall that in Section 5 we defined the "de Rham homomorphism" $\Psi^*: H^p_\Omega(M) \to H^p(M; \mathbf{R})$ (using smooth singular theory) which is induced by $\Psi(\omega)(\sigma) = \int_\sigma \omega$ where ω is a p-form and σ is a (smooth) p-simplex in M. The purpose of this section is to prove the following:

9.1. Theorem (de Rham's Theorem). *The homomorphism*

$$\Psi^*: H^p_\Omega(M) \to H^p(M; \mathbf{R})$$

is an isomorphism for all smooth manifolds M.

To prove this we will need to know a few things about de Rham cohomology analogous to some things we know about singular cohomology. The first of these is an analogue of the Mayer–Vietoris sequence. Let U and V be open subsets of the smooth manifold M and consider the sequence

$$0 \to \Omega^p(U \cup V) \to \Omega^p(U) \oplus \Omega^p(V) \to \Omega^p(U \cap V) \to 0,$$

where the first nontrivial map is the difference of the restrictions, and the second is the sum of the restrictions. We claim this sequence is exact. The only part of this that is nontrivial is the surjectivity on the right. Thus suppose ω is a p-form on $U \cap V$. We must show that, on $U \cap V$, $\omega = \omega_U + \omega_V$ for some p-forms ω_U defined on U and ω_V defined on V.

By taking a smooth partition of unity on $U \cup V$ subordinate to the cover $\{U, V\}$ and passing to the function with support in V we get a smooth function $f: U \cup V \to \mathbf{R}$ which is 0 on a neighborhood of $U - V$ and 1 on a neighborhood of $V - U$. Then $f\omega$ is zero on a neighborhood of $U - V$ and so can be extended by 0 to U. Similarly, $(1-f)\omega$ is zero on a neighborhood of $V - U$ and so extends by 0 to V. So, $\omega = f\omega + (1-f)\omega$ gives us our desired decomposition. Thus this sequence generates a long exact sequence in de Rham cohomology. Moreover, the de Rham map $\Psi: \Omega^p(U \cup V) \to \Delta^p(U \cup V)$ can be composed with the map $\Delta^p(U \cup V) \to \Delta^p_{\mathrm{U}}(U \cup V)$, yielding the commutative diagram:

$$\begin{array}{ccccccccc} 0 & \longrightarrow & \Omega^p(U \cup V) & \longrightarrow & \Omega^p(U) \oplus \Omega^p(V) & \longrightarrow & \Omega^p(U \cap V) & \longrightarrow & 0 \\ & & \downarrow & & \downarrow & & \downarrow & & \\ 0 & \longrightarrow & \Delta^p_{\mathrm{U}}(U \cup V) & \longrightarrow & \Delta^p(U) \oplus \Delta^p(V) & \longrightarrow & \Delta^p(U \cap V) & \longrightarrow & 0 \end{array}$$

with exact rows, the vertical maps being Ψ. Therefore there is the associated "ladder" commutative diagram, with exact rows, in homology:

$$\begin{array}{cccccccccc} \cdots \to & H^p_\Omega(U \cup V) & \to & H^p_\Omega(U) \oplus H^p_\Omega(V) & \to & H^p_\Omega(U \cap V) & \to & H^{p+1}_\Omega(U \cup V) & \to \cdots \\ (*) & \downarrow & & \downarrow & & \downarrow & & \downarrow & \\ \cdots \to & H^p(U \cup V) & \to & H^p(U) \oplus H^p(V) & \to & H^p(U \cap V) & \to & H^{p+1}(U \cup V) & \to \cdots. \end{array}$$

9.2. Lemma (Poincaré Lemma). *The de Rham Theorem (Theorem 9.1) is true for any convex open subset U of $\mathbf{R}^n$.*

Proof. We can assume that U contains the origin. We must show:

(i) that any closed p-form ω, $p \geq 1$, on U is exact; and
(ii) that any smooth function f on U with $df = 0$ is constant (this suffices because the de Rham map takes a constant function with value r to the constant 0-cocycle taking value r on each 0-simplex).

Part (ii) is clear since $df = 0$ means that all partials of f are zero, implying that f is locally constant, hence constant since U is connected.

We prove part (i) on $U \subset \mathbf{R}^{n+1}$, for notational convenience, with coordinates $x_0, \ldots, x_n$. For $p \geq 0$, we define

$$\phi: \Omega^{p+1} \to \Omega^p$$

as follows: If $\omega = f(x_0, \ldots, x_n)\, dx_{j_0} \wedge \cdots \wedge dx_{j_p}$ then put

$$\phi(\omega) = \left(\int_0^1 t^p f(tx)\, dt \right) \eta,$$

where

$$\eta = \sum_{i=0}^{p} (-1)^i x_{j_i}\, dx_{j_0} \wedge \cdots \wedge \widehat{dx}_{j_i} \wedge \cdots \wedge dx_{j_p}.$$

Then, using D_k to denote the partial derivative with respect to the kth variable,

$$d\phi(\omega) = \sum_{k=0}^{n} \left(\int_0^1 t^{p+1} D_k f(tx)\, dt \right) dx_k \wedge \eta + \left(\int_0^1 t^p f(tx)\, dt \right) d\eta$$
$$= S + T,$$

where S is the sum term and T the rest. Also

$$d\omega = \sum_{k=0}^{n} D_k f(x)\, dx_k \wedge dx_{j_0} \wedge \cdots \wedge dx_{j_p},$$

so that

$$\phi(d\omega) = \sum_{k=0}^{n} \left(\int_0^1 t^{p+1} D_k f(tx)\, dt \right) (x_k\, dx_{j_0} \wedge \cdots dx_{j_p} - dx_k \wedge \eta)$$
$$= \sum_{k=0}^{n} x_k \left(\int_0^1 t^{p+1} D_k f(tx)\, dt \right) dx_{j_0} \wedge \cdots \wedge dx_{j_p} - S$$
$$= \left(\int_0^1 t^{p+1} \frac{d}{dt} f(tx)\, dt \right) dx_{j_0} \wedge \cdots \wedge dx_{j_p} - S$$
$$= \left\{ t^{p+1} f(tx)]_0^1 - (p+1) \int_0^1 t^p f(tx)\, dt \right\} dx_{j_0} \wedge \cdots \wedge dx_{j_p} - S$$
$$= \omega - T - S,$$

since $d\eta = (p+1) dx_{j_0} \wedge \cdots \wedge dx_{j_p}$. Then $d\phi(\omega) + \phi(d\omega) = \omega$ for $\omega \in \Omega^p(\mathbf{R}^{n+1})$, $p \geq 1$. Thus, if $d\omega = 0$ then $\omega = d(\phi(\omega))$, as required. □

9.3. Lemma. *If the de Rham map Ψ^* is an isomorphism for open sets U, V in M and for $U \cap V$ then it is an isomorphism for $U \cup V$.*

Proof. This follows from the Mayer–Vietoris "ladder" $(*)$ and the 5-lemma. □

9.4. Lemma. *If the de Rham map Ψ^* is an isomorphism for* disjoint *open sets U_α then it is an isomorphism for the union $\bigcup U_\alpha$.*

Proof. This follows from the Product Theorem (Theorem 8.4) and its obvious analogue in de Rham cohomology, and the naturality of Ψ^*. □

The de Rham Theorem (Theorem 9.1) now follows from the next lemma.

9.5. Lemma. *Let M^n be a smooth n-manifold. Suppose that $P(U)$ is a statement about open subsets of M^n, satisfying the following three properties:*

(1) *$P(U)$ is true for U diffeomorphic to a convex open subset of $\mathbf{R}^n$;*
(2) *$P(U)$, $P(V)$, $P(U \cap V) \Rightarrow P(U \cup V)$; and*

(3) *$\{U_\alpha\}$ disjoint, and $P(U_\alpha)$, all $\alpha \Rightarrow P(\bigcup U_\alpha)$.*

Then $P(M)$ is true.

PROOF. First, we shall prove this in case M^n is diffeomorphic to an open subset of $\mathbf{R}^n$. We may as well, and will, think of M^n as equal to an open set in $\mathbf{R}^n$.

By (1) and (2) and an induction it follows that $P(U)$ is true when U is the union of a finite number of convex open subsets, because of the identity

$$(U_1 \cup \cdots \cup U_n) \cap U_{n+1} = (U_1 \cap U_{n+1}) \cup \cdots \cup (U_n \cap U_{n+1}).$$

Let $f: M \to [0, \infty)$ be a proper map (so that it extends to the one-point compactifications). For example, we can use the fact that the one-point compactification M^+ is metrizable and let, for some such metric, $f(x) = 1/\text{dist}(x, \infty)$. Or, we could take a partition of unity $\{f_i\}$ subordinate to a collection of open sets U_i with $\bar{U}_i$ compact and then take $f(x) = \sum_n nf_n(x)$.

Then define $A_n = f^{-1}[n, n+1]$, which is compact since f is proper. Cover A_n by a finite union U_n of convex open sets contained in $f^{-1}(n - \frac{1}{2}, n + \frac{3}{2})$. Then

$$A_n \subset U_n \subset f^{-1}(n - \tfrac{1}{2}, n + \tfrac{3}{2}).$$

It follows that the sets U_{even} are disjoint. Similarly, the sets U_{odd} are disjoint. See Figure V-2.

Since U_n is a finite union of convex open sets, $P(U_n)$ is true for all n. From (3) we deduce that $P(U)$ and $P(V)$ are true, where $U = \bigcup U_{2n}$ and $V = \bigcup U_{2n+1}$. But $U \cap V = (\bigcup U_{2n}) \cap (\bigcup U_{2n+1}) = \bigcup (U_{2i} \cap U_{2j+1})$ (disjoint) and each $U_{2i} \cap U_{2j+1}$ is a finite union of convex open sets. Therefore $P(U \cap V)$ is also true. By (2) it follows that $P(M) = P(U \cup V)$ is true.

Now we know that $P(U)$ is true whenever U is diffeomorphic to an open subset of $\mathbf{R}^n$. In the general case, substitute in (1) and in the above proof, the words "open subset of $\mathbf{R}^n$" for "convex open subset of $\mathbf{R}^n$." The proof clearly applies and shows that $P(M)$ is true for all M. □

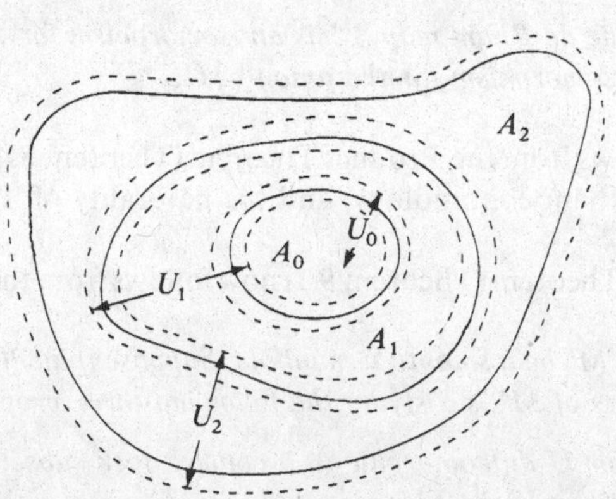

Figure V-2. An "onion" used in the proof of Lemma 9.5.

This completes the proof of the de Rham Theorem. People with knowledge of Riemannian geometry will note that one can prove Lemma 9.5 and hence Theorem 9.1 in "one pass" by substituting "convex" in the sense of Riemannian geometry (unique geodesic segment between two given points) for "convex in $\mathbf{R}^n$" in (1) and in the proof. The given "two pass" proof, however, is a good deal more elementary. The present proof dates to 1962 when the author gave it in a course on Lie Groups.

There is one final wrinkle to this story. de Rham's Theorem as it now stands involves smooth singular cohomology. We would like to replace that by ordinary singular cohomology. There is the inclusion

$$(**) \qquad \Delta_*^{\text{smooth}}(M) \hookrightarrow \Delta_*(M).$$

This induces, via $\text{Hom}(\cdot, G)$, the chain map

$$\Delta^*(M; G) \to \Delta^*_{\text{smooth}}(M; G).$$

Exactly the same proof as the above use of Lemma 9.5 will show that the induced map

$$H^*(M; G) \to H^*_{\text{smooth}}(M; G)$$

is an isomorphism for all smooth manifolds M. (Better would be to use Lemma 9.5 to show that $(**)$ induces a homology isomorphism. Then the result for cohomology would follow from Corollary 7.8.)

PROBLEMS

1. Deduce from de Rham's Theorem that if $U \subset \mathbf{R}^3$ is open and $H_1(U) = 0$ then any vector field $\vec{F}$ on U with curl $\vec{F} = 0$ is a gradient field.

2. Deduce from de Rham's Theorem that if $U \subset \mathbf{R}^3$ is open and $H_2(U) = 0$ then any vector field $\vec{F}$ on U with div $\vec{F} = 0$ has the form $\vec{F} = \text{curl}\,\vec{G}$ for some vector field $\vec{G}$ on U.

3. If $\vec{F}$ is an incompressible (i.e. div $\vec{F} = 0$) vector field on $\mathbf{R}^3 - \{0\}$ then show that there is a real number a and a vector field G on $\mathbf{R}^3 - \{0\}$ such that $\vec{F} = \text{curl}\,\vec{G} + (a/\rho^3)\langle x, y, z\rangle$, where ρ is the distance from the origin.

4. Call a smooth real valued function $f(x, y)$ defined on $\mathbf{R}^2$ "periodic" if $f(x + n, y + m) = f(x, y)$ for all real x, y and integers n, m. Call a pair $\langle f, g\rangle$ of periodic functions, a "periodic pair." Call a periodic pair $\langle f, g\rangle$ "nice" if $f_y \equiv g_x$. If $h(x, y)$ is periodic then $\langle h_x, h_y\rangle$ is a nice periodic pair; call such a nice pair, "excellent."
   (a) Show that there are nice periodic pairs which are not excellent.
   (b) The set of all nice periodic pairs forms an abelian group N under vector addition. The excellent ones form a subgroup E of N. Find the quotient group N/E.
   (c) Describe all nice periodic pairs as explicitly as you can.

5. Suppose that ω is a closed p-form on a smooth manifold M^n such that $\int_c \omega = 0$ for every smooth p-cycle c of M^n. Then show that ω is exact.

10. The de Rham Theory of $\mathbf{CP}^n$ ☼

In this (optional) section we are going to illustrate de Rham's Theorem in a very special case, that of complex projective n-space $\mathbf{CP}^n$, which is a real $2n$-dimensional manifold. Most readers will get a deeper understanding of the theorem from these very explicit manipulations, and discussions, of forms.

Let $(z_0:z_1:z_2)$ be homogeneous coordinates on $\mathbf{CP}^2$. The complement of the complex line $\mathbf{CP}^1 = \{(0:z_1:z_2)\}$ is the affine space $\{(1:u:v)\} \approx \mathbf{C} \times \mathbf{C}$. To describe forms on $\mathbf{CP}^2$ it suffices (by density) to write them down for the affine space. (Of course, it must be checked that they extend smoothly to all of $\mathbf{CP}^2$.)

Let

$$\begin{aligned} u &= re^{2\pi i\theta}, & r, s &\geq 0, \\ v &= se^{2\pi i\phi}, & 0 &\leq \theta, \phi \leq 1. \end{aligned}$$

(Thus r, θ and s, ϕ are *polar* coordinates. Note that r is not smooth, but $r^2 = x^2 + y^2$ is, and hence rdr is smooth, since $2r\,dr = d(r^2)$. Similarly, $2\pi\theta = \arctan(y/x)$ is not even globally defined, but $2\pi r^2\,d\theta = x\,dy - y\,dx$ is a smooth 1-form and $2\pi r dr \wedge d\theta = dx \wedge dy$ is a smooth 2-form. These things should be kept in mind.)

Our strategy is going to be this: We wish to write down a closed 2-form ω on $\mathbf{CP}^2$ which is not exact. But the restriction of that form to affine space will be exact ($\omega = d\eta$ on $\mathbf{C} \times \mathbf{C}$) by de Rham. Thus we shall start with the 1-form η. (Of course, η *cannot* extend to $\mathbf{CP}^2$ or ω will be exact.)

We claim that the 1-form

$$\eta = \frac{r^2\,d\theta + s^2\,d\phi}{1 + r^2 + s^2}$$

works. (The obvious generalization to more variables on $\mathbf{CP}^n$ also works.) Thus define

$$\begin{aligned} (1)\quad \omega &= d\eta \\ &= \frac{(1 + s^2)2r\,dr \wedge d\theta - 2s\,ds \wedge r^2\,d\theta - 2r\,dr \wedge s^2 d\phi + (1 + r^2)2s\,ds \wedge d\phi}{(1 + r^2 + s^2)^2} \end{aligned}$$

where the first equality holds on $\mathbf{C} \times \mathbf{C}$ only.

We must show that ω defines a form on all of $\mathbf{CP}^2$. To do this we must compute the change of variables to the charts $\{(*:1:*)\}$ and $\{(*:*:1)\}$, but, by symmetry, it suffices to consider just the first. It is easier to change variables in the form η and then pass to $d\eta$ and analyze it.

Thus note that

$$\begin{aligned} (1:u:v) &= (u^{-1}:1:u^{-1}v) \qquad (\text{for } u \neq 0) \\ &= ((1/r)e^{2\pi i(-\theta)}:1:(s/r)e^{2\pi i(\phi - \theta)}) \\ &= (r_1 e^{2\pi i\theta_1}:1:s_1 e^{2\pi i\phi_1}). \end{aligned}$$

Hence the change of coordinates is

$$\begin{aligned} r &= 1/r_1, & dr &= -r_1^{-2}\,dr_1, \\ s &= s_1/r_1, & ds &= r_1^{-1}\,ds_1 - s_1 r_1^{-2}\,dr_1, \\ \theta &= -\theta_1, & d\theta &= -d\theta_1, \\ \phi &= \phi_1 - \theta_1, & d\phi &= d\phi_1 - d\theta_1. \end{aligned}$$

Therefore

$$\eta = \frac{(1/r_1^2)(-d\theta_1) + (s_1^2/r_1^2)(d\phi_1 - d\theta_1)}{1 + 1/r_1^2 + s_1^2/r_1^2} = \frac{-(1+s_1^2)\,d\theta_1 + s_1^2\,d\phi_1}{1 + r_1^2 + s_1^2}.$$

(Note that the $s_1^2\,d\phi_1$ part is all right but the part involving $d\theta_1$ is not defined when $r_1 = 0$; which is why η is not a global form on $\mathbf{CP}^2$.) We need only check that

$$d\left(\frac{(1+s_1^2)\,d\theta_1}{1+r_1^2+s_1^2}\right)$$

is a smooth form on this affine space. But this is

$$\frac{-(1+s_1^2)2r_1\,dr_1 \wedge d\theta_1 + [(1+r_1^2+s_1^2)2s_1 - (1+s_1^2)2s_1]\,ds_1 \wedge d\theta_1}{(1+r_1^2+s_1^2)^2}.$$

The term in $r_1\,dr_1 \wedge d\theta_1$ is smooth as noted above. The expression in the square brackets is $2s_1 r_1^2\,ds \wedge d\theta_1 = (2s_1\,ds_1) \wedge (r_1^2\,d\theta_1)$ and so that is smooth also. Hence ω is a 2-form on all of $\mathbf{CP}^2$. Since, on $\mathbf{C} \times \mathbf{C}$, $d\omega = dd\eta = 0$, we have that $d\omega = 0$ on $\mathbf{CP}^2$, by continuity; that is, ω is closed.

We also wish to show that ω is *not* exact. But it *suffices* to show that $\omega \wedge \omega$ is not exact and we wish to compute that anyway. Using the explicit formula (1) for ω, we compute

$$\begin{aligned} \omega \wedge \omega &= \frac{2[4rs(1+s^2)(1+r^2) - 4r^3s^3]}{(1+r^2+s^2)^4}\,dr \wedge d\theta \wedge ds \wedge d\phi \\ &= \frac{8rs}{(1+r^2+s^2)^3}\,dr \wedge d\theta \wedge ds \wedge d\phi. \end{aligned}$$

Using $dr \wedge d\theta \wedge ds \wedge d\phi$ to orient $\mathbf{CP}^2$, we can now compute

$$\begin{aligned} \int_{\mathbf{CP}^2} \omega \wedge \omega &= \int_0^1 \int_0^\infty \int_0^1 \int_0^\infty \frac{8rs}{(1+r^2+s^2)^3}\,dr\,d\theta\,ds\,d\phi \\ &= 8 \int_0^\infty \int_0^\infty \frac{rs}{(1+r^2+s^2)^3}\,dr\,ds = 1. \end{aligned}$$

(The complement $\mathbf{CP}^1$ of the affine space is "small" and will not affect the integral, so we can, and did, ignore it in the computation.) In particular, $\omega \wedge \omega$ is *not* exact, by Stokes' Theorem.

[*Remark.* Similarly, you can integrate ω over $\mathbf{CP}^1 = \{(1:u:0)\}$ and get 1. This implies that ω is an "integral class," that is, it is in

$$H^2(\mathbf{CP}^2;\mathbf{Z}) \subset H^2(\mathbf{CP}^2;\mathbf{R}) \approx H^2_\Omega(\mathbf{CP}^2).$$

This uses the fact that $\mathbf{CP}^1$ "represents" the generator of $H_2(\mathbf{CP}^2;\mathbf{Z})$; i.e., the map $\mathbf{Z} \approx H_2(\mathbf{CP}^1) \to H_2(\mathbf{CP}^2) \approx \mathbf{Z}$ induced by inclusion, is an isomorphism. (The reader might want to try to prove that.) This form ω on $\mathbf{CP}^n$ (in general) is, up to sign, what is called the "first Chern form" for the "canonical line bundle" over $\mathbf{CP}^n$, and its cohomology class, up to sign, is called the "first universal Chern class c_1."]

We treated only $\mathbf{CP}^2$ above merely to simplify notation. It is easy to generalize η and $\omega = d\eta$ to $\mathbf{CP}^n$. Thus we would get a closed 2-form ω on $\mathbf{CP}^n$ such that

$$\int_{\mathbf{CP}^n} \omega \wedge \cdots \wedge \omega \neq 0 \qquad (\text{indeed} = \pm 1),$$

where the wedge is of n copies of ω. Thus none of $\omega, \omega^2, \ldots, \omega^n$ are exact and they must, therefore, give generators of

$$H^2_\Omega(\mathbf{CP}^n), H^4_\Omega(\mathbf{CP}^n), \ldots, H^{2n}_\Omega(\mathbf{CP}^n),$$

which we already know to be $\mathbf{R}$ by the de Rham Theorem and the previous computation of the homology of $\mathbf{CP}^n$ (Problem 7 of Section 10 of Chapter IV).

For fun, we will apply the above computations to some topological facts about $\mathbf{CP}^2$. These things will be done in more generality later, but must await the further development of cohomology theory.

10.1. Theorem. *Let X be any space homotopy equivalent to* $\mathbf{CP}^2$. *Then any map $f: X \to X$ has Lefschetz number $L(f) \neq 0$. (Hence if X is a compact* ENR, *then f must have a fixed point.)*

Proof. It follows easily from the Universal Coefficient Theorem that $L(f)$ can be computed, in the obvious way, using real *co*homology, so we stick to cohomology.

Let $\phi: \mathbf{CP}^2 \to X$ be a homotopy equivalence with homotopy inverse $\psi: X \to \mathbf{CP}^2$. Consider the map $\psi \circ f \circ \phi: \mathbf{CP}^2 \to \mathbf{CP}^2$. We have the commutative diagram

$$\begin{array}{ccc} H^*(\mathbf{CP}^2) & \xrightarrow{\phi^* f^* \psi^* = (\psi \circ f \circ \phi)^*} & H^*(\mathbf{CP}^2) \\ \Big\downarrow{\scriptstyle \psi^* = (\phi^*)^{-1}} & & \Big\uparrow{\scriptstyle \phi^*} \\ H^*(X) & \xrightarrow{\quad f^* \quad} & H^*(X) \end{array}$$

and it follows that $L(f) = L(\psi \circ f \circ \phi)$. (This results from the algebraic formula $\operatorname{trace}(ABA^{-1}) = \operatorname{trace}(B)$.) Thus the space X is immaterial. It suffices to compute on $\mathbf{CP}^2$ itself.

On $\mathbf{CP}^2$ we have that f is homotopic to a smooth map by the Smooth Approximation Theorem (Theorem 11.8 of Chapter II), and hence we can take f to be smooth as far as computing $L(f)$ is concerned.

Then it makes sense to talk about $[\![f^*\omega]\!] = f^*[\![\omega]\!] \in H^2_\Omega(\mathbf{CP}^2)$. Since $[\![\omega]\!]$ generates $H^2_\Omega(\mathbf{CP}^2)$ it follows that $f^*[\![\omega]\!] = r[\![\omega]\!]$, for some $r \in \mathbf{R}$. (That is, $f^*\omega = r\omega + d\mu$ for some μ.) Then $f^*[\![\omega \wedge \omega]\!] = [\![f^*\omega \wedge f^*\omega]\!] = [\![r\omega \wedge r\omega + \text{exact}]\!] = [\![r\omega \wedge r\omega]\!] = r^2[\![\omega \wedge \omega]\!]$. (All of this is, of course, just a special case of the fact that $\wedge$ induces a *natural* product structure on H^*_Ω.)

Thus, $L(f) = \text{trace } f^*|H^0_\Omega + \text{trace} f^*|H^2_\Omega + \text{trace} f^*|H^4_\Omega = 1 + r + r^2$.

Since $1 + r + r^2 \neq 0$ for all real r, we are done. (Actually it follows from the naturality of the de Rham isomorphism and of the Universal Coefficient isomorphism $H^2(\mathbf{CP}^2; \mathbf{R}) \xrightarrow{\approx} \text{Hom}(H_2(\mathbf{CP}^2), \mathbf{R})$, that r must be an integer.) □

Recall that $\mathbf{CP}^2$ can be represented as a CW-complex by $\mathbf{CP}^2 = \mathbf{S}^2 \cup_h \mathbf{D}^4$ where the attaching map $h: \mathbf{S}^3 \to \mathbf{S}^2$ is called the "Hopf map." Explicitly, one can show that h is $(u, v) \mapsto uv^{-1}$ of $\mathbf{S}^3 = \{(u, v) \in \mathbf{C} \times \mathbf{C} \mid |u|^2 + |v|^2 = 1\} \to \mathbf{C} \cup \{\infty\} = \mathbf{S}^2$.

10.2. Corollary. *The Hopf map* $\mathbf{S}^3 \to \mathbf{S}^2$ *is not homotopic to a constant map.*

PROOF. If $h \simeq$ constant then $\mathbf{CP}^2 \simeq \mathbf{S}^2 \cup_{\text{constant}} \mathbf{D}^4 = \mathbf{S}^2 \vee \mathbf{S}^4$ has a map f without fixed points (project to $\mathbf{S}^2$ and follow by the antipodal map on $\mathbf{S}^2$), and hence with $L(f) = 0$, contrary to the theorem. □

10.3. Corollary. $\pi_3(\mathbf{S}^2) \neq 0$. □

It is, in fact, known that $\pi_3(\mathbf{S}^2) \approx \mathbf{Z}$, generated by the class of the Hopf map; see Section 8 of Chapter VII.

Another consequence of the de Rham Theorem is in showing that $\mathbf{CP}^3$ does not have the homotopy type of $\mathbf{S}^2 \times \mathbf{S}^4$ even though these spaces have the same homology groups. One can see this by showing, similarly to one of the previous arguments that any self-map of $\mathbf{CP}^3$ (and thus of any space of its homotopy class) for which the map on $H^2(\mathbf{CP}^3) \approx \mathbf{Z}$ is multiplication by k must be multiplication by k^2 on H^4 and multiplication by k^3 on H^6. On the other hand, the map f on $\mathbf{S}^2 \times \mathbf{S}^4$ which is the projection to $\mathbf{S}^2$ is the identity on H^2 but kills H^4 and H^6.

It is instructive to compute the pull back of ω to $\mathbf{C}^3 - \{0\}$ (or to the unit 5-sphere in this). Thus let $\pi: \mathbf{C}^3 - \{0\} \to \mathbf{CP}^2$ be the map from nonhomogeneous coordinates to homogeneous coordinates; i.e., the map

$$(ae^{2\pi i\alpha}, be^{2\pi i\beta}, ce^{2\pi i\gamma}) \xrightarrow{\pi} \left(1 : \frac{b}{a} e^{2\pi i(\beta - \alpha)} : \frac{c}{a} e^{2\pi i(\gamma - \alpha)}\right).$$

Thus $r = b/a$, $\theta = \beta - \alpha$ and $s = c/a$, $\phi = \gamma - \alpha$ in the notation above. Again it is easier to pull back η first:

$$\pi^*(\eta) = \pi^*\left(\frac{r^2\,d\theta + s^2\,d\phi}{1 + r^2 + s^2}\right) = \frac{b^2(d\beta - d\alpha) + c^2(d\gamma - d\alpha)}{a^2 + b^2 + c^2}$$

$$= \frac{b^2\,d\beta + c^2\,d\gamma - (b^2 + c^2)d\alpha}{a^2 + b^2 + c^2} = \frac{a^2\,d\alpha + b^2\,d\beta + c^2\,d\gamma}{a^2 + b^2 + c^2} - d\alpha.$$

Thus $\pi^*\omega = \pi^*(d\eta) = d\pi^*\eta = d\nu$ (since $dd\alpha = 0$), where

$$\nu = \frac{a^2 d\alpha + b^2 d\beta + c^2 d\gamma}{a^2 + b^2 + c^2}.$$

(Note that ν is *everywhere defined* on $\mathbf{C}^3 - \{0\}$ whereas the "$-d\alpha$" prevents $\pi^*\eta$ from being defined on the set $a = 0$.)

Let us now see how ν restricts to a fiber $\mathbf{S}^1$ (of $\mathbf{S}^5 \to \mathbf{CP}^2$). The inclusion of a typical fiber can be given as $i: \mathbf{S}^1 \to \mathbf{S}^5 \subset \mathbf{C}^3 - \{0\}$ where

$$i(e^{2\pi i\lambda}) = (a_0 e^{2\pi i(\alpha_0 + \lambda)}, b_0 e^{2\pi i(\beta_0 + \lambda)}, c_0 e^{2\pi i(\gamma_0 + \lambda)}),$$

where $a_0^2 + b_0^2 + c_0^2 = 1$. Use λ as the coordinate on $\mathbf{S}^1$. Then $\alpha = \alpha_0 + \lambda$, etc., and

$$i^*\nu = \frac{a^2\,d\lambda + b^2\,d\lambda + c^2\,d\lambda}{a^2 + b^2 + c^2} = d\lambda.$$

(Of course, λ is not globally defined, and so $d\lambda$ is not really exact. Thus $i^*\nu = d\lambda =$ the standard 1-form on $\mathbf{S}^1$. Note that the integral over $\mathbf{S}^1$ of $d\lambda$ is $\int d\lambda = \int_0^1 d\lambda = 1$.)

Passing to cohomology, we claim that we have defined a homomorphism

$$\tau: H^2_\Omega(\mathbf{CP}^2) \to H^1_\Omega(\mathbf{S}^1).$$

In fact, for an arbitrary closed 2-form ω on $\mathbf{CP}^2$ we take $\pi^*\omega$ on $\mathbf{C}^3 - \{0\}$ (or on $\mathbf{S}^5$). Since $H^2_\Omega(\mathbf{C}^3 - \{0\}) = 0$ (by de Rham's Theorem), $\pi^*\omega$ must be exact:

$$\pi^*\omega = d\nu.$$

Then $i^*\nu$ is a 1-form on $\mathbf{S}^1$; automatically closed for dimensional reasons. (It is closed for a better reason also: $d(i^*\nu) = i^*d\nu = i^*\pi^*\omega = (\pi \circ i)^*\omega =$ (constant)$^*\omega = 0$.) Therefore $[\![i^*\nu]\!] \in H^1_\Omega(\mathbf{S}^1)$ is defined, so we put

$$\tau[\![\omega]\!] = [\![i^*\nu]\!] \quad \text{where} \quad \pi^*\omega = d\nu.$$

Of course, we must show that this is well defined: Suppose ν is replaced by $\nu + \nu'$ where $d\nu' = 0$. Then ν' is a closed 1-form on $\mathbf{C}^3 - \{0\}$ and hence it is exact. Thus $\nu' = d\nu''$, and $[\![i^*\nu]\!] = [\![i^*\nu + d(i^*\nu'')]\!] = [\![i^*(\nu + \nu')]\!]$. Similarly, if ω is replaced by $\omega + d\omega'$, then $\pi^*(\omega + d\omega') = \pi^*\omega + d\pi^*\omega' = d(\nu + \pi^*\omega')$. But $i^*(\nu + \pi^*\omega') = i^*\nu + i^*\pi^*\omega' = i^*\nu$, since $i^*\pi^* = (\pi \circ i)^* =$ constant$^* = 0$ on p-forms, $p > 0$.

From our specific calculation, τ is an isomorphism in the present example. Homomorphisms constructed this way are called "transgressions." Note that a similar definition works in singular cohomology with arbitrary coefficients.

11. Hopf's Theorem on Maps to Spheres ☼

In this section we will show that the set of homotopy classes of maps of an n-dimensional CW-complex to an n-sphere can be completely described by cohomology. This is a precursor to what is known as "obstruction theory," and contains, in fact, most of the ideas from that subject.

Let us recall the definition of cellular homology. If σ is an n-cell of a CW-complex K, let $f_\sigma: \mathbf{D}^n \to K$ be the characteristic map and $f_{\partial\sigma}: \mathbf{S}^{n+1} \to K$ its restriction, the "attaching map." We denote by $p_\sigma: K^{(n)} \to \mathbf{S}^n$, the map collapsing everything but the "interior" of σ to the base point. (A choice of homeomorphism of the collapsed space with $\mathbf{S}^n$ is involved here, and it is resolved by assuming that the map

$$p_\sigma \circ f_\sigma: \mathbf{D}^n \to \mathbf{S}^n$$

is a fixed predefined collapsing map γ_n, but this is rather a technicality.) For an n-cell σ and an $(n-1)$-cell τ we defined the "incidence number" to be

$$[\tau:\sigma] = \deg(p_\tau \circ f_{\partial\sigma}).$$

Then $C_n(K)$ was defined to be the free abelian group on the n-cells, and the boundary operator was defined by $\partial\sigma = \sum_\tau [\tau:\sigma]\tau$.

Thus the cellular cochain groups are defined to be

$$C^n(K;G) = \mathrm{Hom}(C_n(K), G).$$

The coboundary operator is, of course, defined by $(\delta c)(\sigma_{n+1}) = c(\partial\sigma)$. Thus

$$(\delta c)(\sigma) = \sum_\tau [\tau:\sigma]c(\tau).$$

The resulting cohomology group is isomorphic to $H^n(K;G)$ by the Universal Coefficient Theorem (Theorem 7.2).

In this section we are going to treat $\mathbf{S}^n$ and $\mathbf{D}^{n+1}$ as explicit CW-complexes, with one cell in dimensions $0, n$ and (for the disk) $n+1$, so here are the boring details of that. The cells will be called $*, e_n$, and e_{n+1}. We let

$$\mathbf{D}^{n+1} = * \cup_{f_{\partial e_n}} \mathbf{D}^n \cup_{f_{\partial e_{n+1}}} \mathbf{D}^{n+1}.$$

We take $f_{e_n}: \mathbf{D}^n \to \mathbf{S}^n$ to be γ_n and $f_{e_{n+1}}: \mathbf{D}^{n+1} \to \mathbf{D}^{n+1}$ to be the identity. Note that $p_{e_n} \circ \gamma_n = p_{e_n} \circ f_{e_n} = \gamma_n$, so that $p_{e_n} = 1$. Thus $[e_n: e_{n+1}] = \deg(p_{e_n} \circ f_{\partial e_{n+1}}) = \deg(1) = 1$. Therefore, the boundary formula is $\partial e_{n+1} = e_n$. (Of course, all that work was just to figure out the sign.)

Now let K be a CW-complex. We wish to study homotopy classes of maps $K \to \mathbf{S}^n$. By the Cellular Approximation Theorem (Theorem 11.4 of

Chapter IV), we know:

(1) Any map $\phi: K \to \mathbf{S}^n$ is homotopic to one that takes $K^{(n-1)}$ to $*$.
(2) If $\phi, \psi: K \to \mathbf{S}^n$ are cellular and $\phi \simeq \psi$ then they are homotopic via a homotopy $K \times I \to \mathbf{S}^n$ which takes $(K \times I)^{(n-1)}$ to the base point $*$. (Note that this skeleton is $K^{(n-1)} \times \partial I \cup K^{(n-2)} \times I$.)

If $\phi: K^{(n)} \to \mathbf{S}^n$ is cellular then recall from Theorem 11.6 of Chapter IV that the induced map $\phi_\Delta: C_n(K) \to C_n(\mathbf{S}^n)$ is given by $\phi_\Delta(\tau) = \deg(\phi_{e,\tau})e$ where $\phi_{e,\tau}: \mathbf{S}^n \to \mathbf{S}^n$ is that map making

$$\begin{array}{ccc} \mathbf{D}^n & \xrightarrow{f_\tau} & K^{(n)} \\ {\scriptstyle p_\tau f_\tau = \gamma_n}\downarrow & & \downarrow{\scriptstyle \phi} \\ \mathbf{S}^n & \xrightarrow{\phi_{e,\tau}} & \mathbf{S}^n \end{array}$$

commutative and where $e = e_n$. Define a cochain $c_\phi \in C^n(K; \mathbf{Z})$ by

$$\boxed{c_\phi(\tau) = \deg \phi_{e,\tau}.}$$

Then

$$\boxed{\phi_\Delta(\tau) = c_\phi(\tau)e.}$$

11.1. Proposition. *If $\phi: K \to \mathbf{S}^n$ then c_ϕ is a cocycle.*

Proof. We have that $(\delta c_\phi)(\sigma) = c_\phi(\partial\sigma) = 0$ since $c_\phi(\partial\sigma)e = \phi_\Delta(\partial\sigma) = \partial\phi_\Delta(\sigma) = 0$ because $\phi_\Delta(\sigma) \in C_{n+1}(\mathbf{S}^n) = 0$. □

The cocycle c_ϕ is called the "obstruction cocycle." The following generalizes Proposition 11.1.

11.2. Proposition. *Let σ be an $(n+1)$-cell of K and suppose that $\phi: K^{(n)} \to \mathbf{S}^n$ is a map defined on the n-skeleton. Then $(\delta c_\phi)(\sigma) = c_\phi(\partial\sigma) = \deg(\phi \circ f_{\partial\sigma})$.*

Proof. Clearly, we may assume that the characteristic map $f_\sigma: \mathbf{D}^{n+1} \to K$ is cellular since we can throw the rest of the $(n+1)$-cells and higher dimensional cells away as far as this computation in concerned. (Note that f_σ is always cellular except on the 0-cell of $\mathbf{D}^{n+1}$, so we are just saying that the 0-cell $*$ can be assumed to go into a 0-cell of K for the purposes of this computation.) Thus f_σ induces a chain map fitting in the diagram:

$$\begin{array}{ccccc} C_{n+1}(\mathbf{D}^{n+1}) & \xrightarrow{(f_\sigma)_\Delta} & C_{n+1}(K) & & \\ \downarrow{\scriptstyle \partial} & & \downarrow{\scriptstyle \partial} & & \\ C_n(\mathbf{S}^n) & \xrightarrow{(f_{\partial\sigma})_\Delta} & C_n(K^{(n)}) & \xrightarrow{\phi_\Delta} & C_n(\mathbf{S}^n). \end{array}$$

Chase this diagram starting with the cell e_{n+1} at the upper left. Going right, down, right gives $\phi_\Delta(\partial\sigma)=c_\phi(\partial\sigma)e$ since $(f_\sigma)_\Delta(e_{n+1})=\sigma$. Going down, right, right gives $\phi_\Delta((f_{\partial\sigma})_\Delta(\partial e_{n+1}))=(\phi\circ f_{\partial\sigma})_\Delta(e_n)=\deg(\phi\circ f_{\partial\sigma})e$. □

Now we will study homotopies of maps of K to $\mathbf{S}^n$. Suppose that $F:(K\times I)^{(n)}\to\mathbf{S}^n$ and let $F(x,0)=\phi_0(x)$ and $F(x,1)=\phi_1(x)$ where $\phi_i: K^{(n)}\to\mathbf{S}^n$. We wish to compare the cochains c_{ϕ_0} and c_{ϕ_1}.

Define $d_F\in C^{n-1}(K;\mathbf{Z})$ by

$$\boxed{d_F(\tau)=c_F(\tau\times I)=\deg F_{e,\tau\times I}.}$$

11.3. Proposition. *In the situation above, and for an n-cell σ of K, we have*

$$(\delta d_F)(\sigma)=\deg(F\circ f_{\partial(\sigma\times I)})+(-1)^{n+1}(c_{\phi_1}-c_{\phi_0})(\sigma).$$

PROOF. We have $(\delta d_F)(\sigma_n)=d_F(\partial\sigma)=c_F(\partial\sigma\times I)=c_F(\partial(\sigma\times I)-(-1)^n\sigma\times\{1\}+(-1)^n\sigma\times\{0\})$. Substitution of $c_F(\partial(\sigma\times I))=\deg F\circ f_{\partial(\sigma\times I)}$, from Proposition 11.2, yields the desired formula. □

Note that the degree term here vanishes if F is defined on $K^{(n)}\times I$. This gives:

11.4. Corollary. *If $\phi_0\simeq\phi_1: K\to\mathbf{S}^n$ then $[\![c_{\phi_0}]\!]=[\![c_{\phi_1}]\!]$ in $H^n(K;\mathbf{Z})$.* □

Thus for $\phi: K\to\mathbf{S}^n$ an arbitrary (not necessarily cellular) map we put

$$\boxed{\xi_\phi=[\![c_{\phi_0}]\!]\in H^n(K),}$$

where ϕ_0 is a cellular approximation to ϕ. Then $\phi\simeq\psi\Rightarrow\xi_\phi=\xi_\psi$.

11.5. Theorem. *If* $\dim(K)=n$, *or if* $n=1$, *and* $\phi_0,\phi_1: K\to\mathbf{S}^n$ *then:*

(1) $\phi_0\simeq\phi_1\Leftrightarrow\xi_{\phi_0}=\xi_{\phi_1}$; *and*
(2) *for any* $\xi\in H^n(K;\mathbf{Z})$, $\exists\phi: K\to\mathbf{S}^n$ *with* $\xi_\phi=\xi$.

PROOF. We will prove (2) first. In the case $\dim(K)=n$ let c be a cocycle representing ξ. Then map $K^{(n)}\to K^{(n)}/K^{(n-1)}\to\vee\{\mathbf{S}^n\}$, one sphere for each n-cell τ, the map to the one-point union of spheres being induced by the projections p_τ. Follow this by putting, on the τth sphere, a map to $\mathbf{S}^n$ of degree $c(\tau)$. If the final map is called ϕ then $c_\phi(\tau)=c(\tau)$ by definition, proving (2).

In the case $n=1$ of (2) we can get the map ϕ on the 1-skeleton as above. Then we try to extend it over a 2-cell σ. Consider the map $\phi\circ f_{\partial\sigma}:\mathbf{S}^1\to\mathbf{S}^1$. By Proposition 11.2, we have $\deg(\phi\circ f_{\partial\sigma})=(\delta c)(\sigma)=0$ since c is a cocycle. Thus this map extends to $\mathbf{D}^2$ and this means, looking at the quotient topology of this cell, that ϕ extends over the cell σ. We can thus extend ϕ over the 2-skeleton. We make the same argument to extend over 3-cells, but, in this case it is the fact that any map $\mathbf{S}^k \to \mathbf{S}^1$ extends to $\mathbf{D}^{k+1}$, for $k \geq 2$, that makes it go through. Thus one gets an extension over the whole complex K.

(Note that this argument works for $n > 1$ to the extent of getting an extension of the map to $K^{(n+1)}$. One cannot go further than that.)

Now we address part (1). We already proved the $\Rightarrow$ half. For the $\Leftarrow$ half, suppose that $\xi_{\phi_0} = \xi_{\phi_1}$. We may assume that the ϕ_i are cellular. Then $(-1)^{n+1}(c_{\phi_0} - c_{\phi_1}) = \delta d$ for some $d \in C^{n-1}(K)$. Define $F: K^{(n)} \times \partial I \cup K^{(n-1)} \times I \to \mathbf{S}^n$ by:

(a) F is cellular (meaning here that it takes the $(n-1)$-skeleton to the base point);
(b) $F = \phi_i$ on $K \times \{i\}$, $i = 0, 1$; and
(c) for τ an $(n-1)$-cell of K, deg $F_{e,\tau \times I} = d(\tau)$.

Then note that $d_F = d$, because, by definition, $d_F(\tau) = c_F(\tau \times I) = \deg F_{e,\tau \times I} = d(\tau)$. Hence, for an n-cell σ, $(\delta d_F)(\sigma) = (-1)^{n+1}(c_{\phi_1} - c_{\phi_0})(\sigma) + \deg F \circ f_{\partial(\sigma \times I)}$ by Proposition 11.3. However, this is equal to $(\delta d)(\sigma) = (-1)^{n+1}(c_{\phi_1} - c_{\phi_0})(\sigma)$, and so deg $F \circ f_{\partial(\sigma \times I)} = 0$. This implies that F extends over the $(n+1)$-cell $\sigma \times I$, by Corollary 16.5 of Chapter II. Therefore it extends to all of $K \times I$ when $\dim(K) = n$. The remainder of the proof in the case $n = 1$ is exactly as for part (2). □

The fact, from Corollary 16.5 of Chapter II, that a map $\mathbf{S}^n \to \mathbf{S}^n$ of degree zero extends over $\mathbf{D}^{n+1}$ (which is equivalent to saying it is homotopic to a constant) is a crucial item in the proof of Theorem 11.5, and in most of algebraic topology. Since some readers have skipped that part of Chapter II, and others may have found the proof there to be difficult to follow, we will give a different proof of this fact at the end of this section in Lemma 11.13.

11.6. Theorem (Hopf Classification Theorem). *Let K be a CW-complex and assume that* $\dim(K) = n$ *or that* $n = 1$. *Then there is a one–one correspondence:*

$$[K; \mathbf{S}^n] \leftrightarrow H^n(K; \mathbf{Z})$$

given by $[\phi] \leftrightarrow \xi_\phi$. □

11.7. Corollary. *The function* degree: $\pi_n(\mathbf{S}^n) \to \mathbf{Z}$ *is a bijection.* □

This was also shown in Corollary 16.4 of Chapter II, using smooth manifold methods. In fact, degree was shown there to be a homomorphism, hence an isomorphism. That is also an easy consequence of Corollary 7.5 of Chapter IV.

11.8. Example. Let $K = \mathbf{P}^2$, the real projective plane. Then $H^2(\mathbf{P}^2; \mathbf{Z}) \approx \mathbf{Z}_2$, as follows from the Universal Coefficient Theorem (Theorem 7.3). Thus there are exactly two homotopy classes of maps from $\mathbf{P}^2$ to $\mathbf{S}^2$, one of which, of course, is the class of the constant map. The other class is represented by the collapse $\mathbf{P}^2 \to \mathbf{P}^2/\mathbf{P}^1 \approx \mathbf{S}^2$, which was shown to be nontrivial in mod 2-homology, and hence not homotopic to a constant, in Example 7.7.

Now we wish to derive an expression for ξ_ϕ for any given map $\phi: K \to \mathbf{S}^n$. Let e be the top cell of $\mathbf{S}^n$ and let $e' \in C^n(\mathbf{S}^n; \mathbf{Z}) = \operatorname{Hom}(C_n(\mathbf{S}^n), \mathbf{Z})$ be defined by $e'(e) = 1$. Define the "orientation class" $\vartheta_n = [\![e']\!] \in H^n(\mathbf{S}^n; \mathbf{Z})$.

11.9. Theorem. *For a map $\phi: K^{(n)} \to \mathbf{S}^n$ we have $\xi_\phi = \phi^*(\vartheta_n)$.*

PROOF. We calculate $\phi^\Delta(e')(\sigma) = e'(\phi_\Delta(\sigma)) = e'(c_\phi(\sigma)\, e) = c_\phi(\sigma)$. Therefore, $c_\phi = \phi^\Delta(e')$. Hence $\xi_\phi = [\![c_\phi]\!] = \phi^*[\![e']\!] = \phi^*(\vartheta_n)$. □

11.10. Corollary. *A map $\phi: K^{(n)} \to \mathbf{S}^n$ is homotopic to a constant if and only if $\phi^*: H^n(\mathbf{S}^n; \mathbf{Z}) \to H^n(K^{(n)}; \mathbf{Z})$ is trivial.* □

11.11. Corollary. *Two maps $K^{(n)} \to \mathbf{S}^n$ are homotopic if and only if they induce the same homomorphism on $H^n(\cdot; \mathbf{Z})$.* □

11.12. Corollary. *For any CW-complex K, $[K; \mathbf{S}^1] \approx H^1(K; \mathbf{Z})$.* □

This is meant to indicate only a one–one correspondence, but the maps of any space into $\mathbf{S}^1$ can be made into a group by using the group structure on $\mathbf{S}^1$, and it can be shown that the correspondence in Corollary 11.12 is a group isomorphism.

As mentioned above, we are going to give another proof of the following important fact.

11.13. Lemma (Hopf). *A map $\mathbf{S}^n \to \mathbf{S}^n$ of degree zero is homotopic to a constant.*

PROOF. For the proof we will regard $\mathbf{S}^n$ as the boundary of the standard $(n+1)$-simplex, so that it is a simplicial complex K. Let $f: |K| \to |K|$ be a map of degree zero. Then, by the Simplicial Approximation Theorem it is homotopic to a simplicial map from some subdivision $K^{[r]}$ to K. We may as well assume f to be this simplicial map. Take one of the vertices v of K as a base point and consider the opposite face τ. In its structure as a sphere, take a disk in that face. Then there is a deformation of the sphere ending with a map $\phi: |K| \to |K|$ taking the interior of that disk diffeomorphically to $|K| - \{v\}$ and taking everything else to v. Let $g = \phi \circ f$. Then $f \simeq g$ and g has a very special nature: On each simplex of $K^{[r]}$, g either maps the entire simplex to the base point v or it is an affine homeomorphism to τ followed by ϕ. Let us call the latter type of simplex "special," and call this type of map on a simplex a "special map." Note that all special maps differ only by an affine homeomorphism of simplices, i.e., by the ordering of the vertices. Each special map collapses the exterior of a disk to the base point and maps the interior diffeomorphically to the complement of the base point. Therefore a special map has a degree which is ± 1. The degree of the original map f is just the sum of these local degrees by Theorem 7.4 of Chapter IV. Thus half of the local degrees are $+1$ and the other half are -1.

Consider two n-simplices of $K^{[r]}$ which have a common $(n-1)$-face. If one of them, σ, is special and the other, μ, is not, we will describe how to perform a homotopy which makes the second one special and the first not, and does not change the map outside these simplices. The homotopy will be constant outside these two simplices and so it suffices to describe it on a space homeomorphic to their union. One can take the union of two regular simplices in $\mathbf{R}^n$ for this purpose. First we "shrink" the special simplex as follows. Take a point inside the special simplex σ and consider a deformation $F: \sigma \times I \to \sigma$ through embeddings starting with the identity and ending with a map shrinking σ into some small neighborhood of the selected point. (By this we mean to describe a homotopy of the map on σ as follows: Embed $\sigma \times I$ in $\sigma \times I$ by $(x, t) \mapsto (F(x, t), t)$ and let θ be the inverse of this embedding (defined, of course, on its image). Then define a homotopy which is $(g \times 1) \circ \theta$ on the image of this embedding and is constant to the base point outside it.) Note that, after this "shrinking," the map is special on the shrunk simplex and is constant to the base point on its complement in σ.

By the same type of construction, we can then "move" the shrunk simplex over to the nonspecial simplex μ. (The space requirements of doing this move dictate how small the previous shrinking had to be.) Now by rotating and stretching we can maneuver the shrunk and moved simplex so that it is a shrunk version of μ. Then we can expand it (the inverse operation to shrinking) until it fits μ exactly. Now the map on μ is special and σ maps to the base point. See Figure V-3. (The figures depict irregular simplices in order to emphasize only what is important.)

Now suppose we have two such adjoining simplices σ and μ which are both special but have opposite local degrees. (Again we are looking at these as adjoining regular simplices in euclidean space.) Then the maps on these

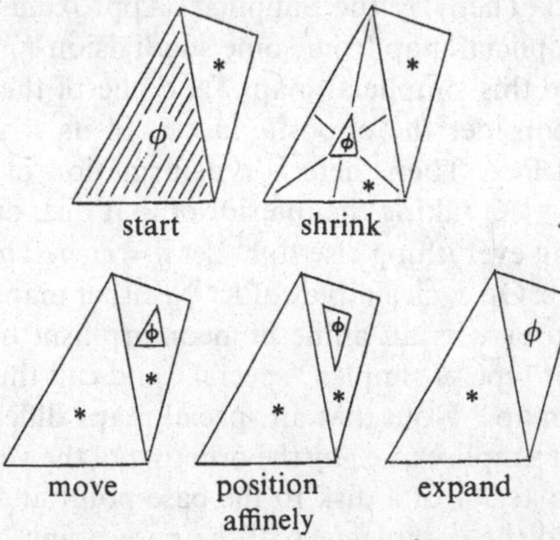

Figure V-3. Moving a special simplex.

simplices are identical except for the ordering of the vertices. By shrinking one of them, rotating, and expanding, we can bring them into a position such that the maps on them are mirror images across the common face, because they have opposite local degrees. Suppose, as we may, the common face to be in the plane $x_1 = 0$ of euclidean space and that the heights of these regular simplices are 1. Let $h: \mathbf{R}^n \to \mathbf{S}^n$ be the given map on these simplices extended to be constant to the base point outside them. Then the fact that the maps on the simplices are mirror images means that $h(x_1, x_2, \ldots, x_n) = h(-x_1, x_2, \ldots, x_n)$. The homotopy $G: \mathbf{R}^n \times I \to \mathbf{R}^n$ given by

$$G(x_1, x_2, \ldots, x_n, t) = \begin{cases} h(x_1, x_2, \ldots, x_n) & \text{for } |x_1| \geq t, \\ h(t, x_2, \ldots, x_n) & \text{for } |x_1| \leq t, \end{cases}$$

starts with h and ends with the constant map to the base point. (Note that this is always constant to the base point outside the two simplices.) Thus, this lets us convert the two special simplices with opposite local degree to two nonspecial simplices. See Figure V-4.

Now let us complete the argument. If there are no special simplices then the map is constant and we are done. If there are special simplices then there must be one of positive degree and one with negative degree. There must be a sequence of n-simplices each having a common $(n-1)$-face with the next, starting with a special simplex with positive degree, ending with one of negative degree with only nonspecial simplices between. (We give a rigorous proof of this in the next paragraph.) Thus, by a sequence of changes of the type we have described, one can cancel out these two special simplices. By an induction on the number of special simplices, the map must be homotopic to a constant, finishing the proof.

Finally, as promised, let us show that there is a sequence of simplices as described above. Let L be the simplicial complex $K^{[r]}$. As a space it is the n-sphere. Then it follows from the Invariance of Domain (Corollary 19.9 of Chapter IV) that no three n-simplices can have a common $(n-1)$-face, since the face plus the interiors of two of them is an n-manifold and so must be open in the sphere. Now starting with a special simplex of local degree $+1$,

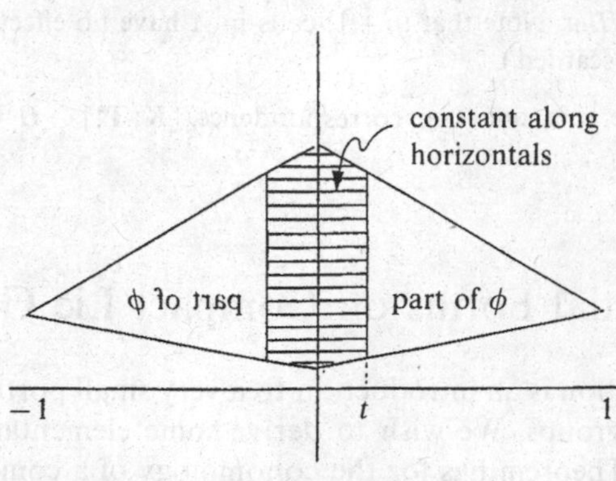

Figure V-4. Cancelling opposite special simplices.

write down a sequence of n-simplices each having a common $(n-1)$-face with the last. If we come to a place where it is impossible to add another n-simplex that is not already in the list, we look for any simplex v in the list that does have an $(n-1)$-face in common with an n-simplex not in the list. If there is one then we retrace our steps (adding repetitions to the list) until we come to v. Then we add that new simplex bordering v to the list. Eventually, we either come to a special simplex with negative local degree, or we cannot continue. We claim the latter possibility cannot happen. If it does happen then the list we have at that point is a set of n-simplices such that any n-simplex with an $(n-1)$-face in common with one of the simplices in the list, is also in the list. We can regard this set of n-simplices, without repetitions, as an n-chain c with mod 2 coefficients (in the simplicial chain complex of L). Since any $(n-1)$-face of any of these simplices is the face of exactly two n-simplices, we see that $\partial c = 0$. But the sum s of all the n-simplices of L is also an n-cycle. Since there are no n-boundaries (there being no $(n+1)$-simplices) the homology group $H_n(L; \mathbf{Z}_2)$ is the same as the group of cycles. Since we have found two different nonzero cycles, we conclude that $H_n(L; \mathbf{Z}_2) \not\approx \mathbf{Z}_2$, contrary to our knowledge of this group. Thus this eventuality cannot happen, and so we have constructed a sequence of n-simplices, possibly with repetitions, from a special simplex of degree $+1$ to one of degree -1. Somewhere within this sequence must be a subsequence from a special simplex of degree $+1$ to one of degree -1 with only nonspecial simplices between, which is what we set out to construct. □

Problems

1. Let $K = \mathbf{S}^2 \cup_f \mathbf{D}^3$ where $f: \mathbf{S}^2 \to \mathbf{S}^2$ is a map of degree p. Find the number of homotopy classes of maps $K \to \mathbf{S}^3$.

2. How many homotopy classes of maps are there from the projective plane to the circle?

3. Show that a map $\mathbf{P}^3 \to \mathbf{S}^3$ extends to $\mathbf{P}^4 \Leftrightarrow$ it is homotopic to a constant map.

4. Let K be a CW-complex of dimension $n+1$ and let $A \subset K$ be a subcomplex. Show that a map $\phi: A \to \mathbf{S}^n$ extends to $K \Leftrightarrow \xi_\phi$ is in the image of the restriction $H^n(K) \to H^n(A)$. (*Hint*: Note that $(n+1)$-cells in A have no effect on this question and so can be discarded.)

5. Show that there is a one–one correspondence $[K; \mathbf{T}^n] \leftrightarrow H^1(K; \mathbf{Z}^n)$ for CW-complexes K.

12. Differential Forms on Compact Lie Groups ☼

This optional section is an introduction to a very small portion of the theory of compact Lie groups. We wish to derive some elementary consequences that de Rham's Theorem has for the cohomology of a compact Lie group.

Let G be a Lie group and let $T_e(G)$ be its tangent space at the identity element $e \in G$. Let $L_\sigma: G \to G$ be left translation: $L_\sigma(\tau) = \sigma\tau$. If $X_e \in T_e(G)$ is a tangent vector at e then $X_\sigma = (L_\sigma)_*(X_e)$ is a tangent vector at $\sigma \in G$. This defines a vector field X on G which is "left-invariant." Such a vector field X is uniquely determined by X_e.

12.1. Proposition. *Any left-invariant vector field is smooth.*

PROOF. For a smooth real valued function f defined near σ, we have

$$X(f)(\sigma) = X_\sigma(f) = (L_\sigma)_*(X_e)(f) = X_e(f \circ L_\sigma).$$

Let $x_1, \ldots, x_n$ be local coordinates at $e \in G$ and put $X_e = \sum a_i \partial/\partial x_i$. Let $y_1, \ldots, y_n$ be local coordinates at σ. Then, for τ near e, we can write

$$y_i(\sigma\tau) = F_i(y_1(\sigma), \ldots, y_n(\sigma), x_1(\tau), \ldots, x_n(\tau)),$$

where the F_i are smooth real valued functions defined near 0 in $\mathbf{R}^n \times \mathbf{R}^n$. Then

$$X_e(f \circ L_\sigma) = \sum_{i=1}^n a_i \frac{\partial}{\partial x_i}(f(F_1(y, x), \ldots, F_n(y, x)))|_{x=0}$$

is smooth in $y_1, \ldots, y_n$ as required. □

Similarly, if ω is a p-form on G then $L_\sigma^*(\omega)$ is another p-form and ω is said to be "left-invariant" if $L_\sigma^*(\omega) = \omega$ for all $\sigma \in G$. As in Proposition 12.1, left-invariant forms can be seen to be smooth and hence are in one–one correspondence with the p-forms on the vector space $T_e(G)$.

If X and Y are left-invariant vector fields then the Lie bracket $[X, Y] = XY - YX$ is also left-invariant. This defines $[X, Y]$ on $T_e(G)$. We define L_G to be $T_e(G)$ with this product $[X, Y]$. Then L_G is called the "Lie algebra" of G.

We shall assume the formula for $d\omega$ given above Definition 2.4. If ω is a left-invariant p-form and $X_0, \ldots, X_p$ are left-invariant vector fields, then $\omega(X_0, \ldots, \hat{X}_i, \ldots, X_p)$ is a constant function and so $X_i(\omega(X_0, \ldots, \hat{X}_i, \ldots, X_p)) = 0$. Therefore, in this case, the formula for $d\omega$ simplifies to

$$\boxed{d\omega(X_0, \ldots, X_p) = \sum_{i<j} (-1)^{i+j} \omega([X_i, X_j], X_0, \ldots, \hat{X}_i, \ldots, \hat{X}_j, \ldots, X_p).}$$

Let $X_1, \ldots, X_n$ be a basis of L_G and let $\omega_1, \ldots, \omega_n$ be the dual 1-forms, i.e., $\omega_i(X_j) = \delta_{i,j}$. The ω_i can be considered as left-invariant 1-forms on G and so $\omega = \omega_1 \wedge \cdots \wedge \omega_n$ is a left-invariant n-form. Let this n-form be fixed once and for all. If $f: G \to \mathbf{R}$ is a smooth function with compact support then we can define

$$\boxed{\int_G f\, d\sigma = \int_G f\omega.}$$

Assuming now that G is compact, we can normalize this integral so that

$$\int_G 1\,d\sigma = 1.$$

By the left-invariance of ω, we have

$$\int_G f\circ L_\tau\,d\sigma = \int_G f\,d\sigma$$

for all $\tau\in G$, which we can write as

$$\int_G f(\tau\sigma)\,d\sigma = \int_G f(\sigma)d\sigma.$$

This left-invariant integral on G is called the "Haar integral."

Now a left-invariant n-form, where $n = \dim G$, is unique up to a constant multiple. Let $R_\tau: G\to G$ be right translation: $\sigma\mapsto\sigma\tau^{-1}$. Then $L_\sigma\circ R_\tau = R_\tau\circ L_\sigma$ and it follows that $R_\tau^*(\omega) = c(\tau)\omega$ where $c(\tau)\in\mathbf{R}$. Now $R_{\sigma\tau}^* = R_\tau^*\circ R_\sigma^*$ and it follows that $c(\sigma\tau) = c(\sigma)c(\tau)$. Therefore $c: G\to\mathbf{R}$ is a homomorphism. Since G is compact and the only compact multiplicative subgroup of $\mathbf{R}-\{0\}$ is $\{\pm 1\}$, we must have that $c(\sigma) = \pm 1$ for all σ, and clearly $c(\sigma) = -1 \Leftrightarrow R_\sigma$ reverses orientation. (For G disconnected, we orient the components of G so that left-translation always preserves orientation.) Hence

$$\int f(\sigma\tau^{-1})d\sigma = \int_G f\circ R_\tau\omega = c(\tau)^{-1}\int_G (f\circ R_\tau)R_\tau^*(\omega)$$

$$= c(\tau)^{-1}\int_G R_\tau^*(f\omega) = \int_G f\omega = \int f\,d\sigma.$$

That is, the Haar integral is automatically right-invariant for G compact.

Suppose now that G acts smoothly on the smooth manifold M; i.e., that there is given a smooth map $G\times M\to M$ taking $(\sigma, x)\mapsto t_\sigma(x)$, such that $t_{\sigma\tau} = t_\sigma\circ t_\tau$ and $t_e = 1_M$. Then a p-form ω on M is said to be "invariant" if $t_\sigma^*(\omega) = \omega$ for all $\sigma\in G$. Let $\Omega^G(M)$ be the set of invariant forms on M. Define

$$I: \Omega(M)\to\Omega^G(M)$$

by

$$I(\omega)(X_1,\dots,X_p) = \int t_\sigma^*\,\omega(X_1,\dots,X_p)d\sigma = \int \omega((t_\sigma)_*X_1,\dots,(t_\sigma)_*X_p)d\sigma.$$

Then

$$t_\tau^*(I(\omega))(X_1,\dots,X_p) = I(\omega)((t_\tau)_*X_1,\dots,(t_\tau)_*X_p)$$

$$= \int \omega((t_\sigma)_*(t_\tau)_*X_1,\dots,(t_\sigma)_*(t_\tau)_*X_p)d\sigma$$

$$= \int \omega((t_{\sigma\tau})_* X_1, \ldots, (t_{\sigma\tau})_* X_p) d\sigma$$

$$= \int \omega((t_\sigma)_* X_1, \ldots, (t_\sigma)_* X_p) d\sigma = I(\omega)$$

by the right-invariance of the integral. Thus $I(\omega) \in \Omega^G(M)$ for all $\omega \in \Omega(M)$.

Suppose that $\omega \in \Omega^G(M)$. Then

$$I(\omega)(X_1, \ldots, X_p) = \int t_\sigma^*(\omega)(X_1, \ldots, X_p) d\sigma = \int \omega(X_1, \ldots, X_p) d\sigma = \omega(X_1, \ldots, X_p)$$

by the normalization of the integral (assuming, as we do, that G is compact).

Let $J : \Omega^G(M) \hookrightarrow \Omega(M)$ be the inclusion. Then we have just shown that

$$IJ = 1 : \Omega^G(M) \to \Omega^G(M).$$

12.2. Lemma. *$I : \Omega(M) \to \Omega^G(M)$ is a chain map; i.e., $dI = Id$.*

PROOF. Using the shorthand notation $X^\sigma = (t_\sigma)_* X$ we have

$$\begin{aligned} d(I(\omega))(X_0, \ldots, X_p) &= \sum_{i=0}^p (-1)^i X_i(I(\omega)(X_0, \ldots, \hat{X}_i, \ldots, X_p)) \\ &\quad + \sum_{i<j} (-1)^{i+j} I(\omega)([X_i, X_j], X_0, \ldots, \hat{X}_i, \ldots, \hat{X}_j, \ldots, X_p) \\ &= \sum_{i=0}^p (-1)^i X_i \int \omega(X_0^\sigma, \ldots, \hat{X}_i^\sigma \ldots, X_p^\sigma) d\sigma \\ &\quad + \sum_{i<j} (-1)^{i+j} \int \omega([X_i^\sigma, X_j^\sigma], X_0^\sigma, \ldots, \hat{X}_i^\sigma, \ldots, \hat{X}_j^\sigma, \ldots, X_p^\sigma) d\sigma \\ &= \int d\omega(X_0^\sigma, \ldots, X_p^\sigma) d\sigma \\ &= I(d\omega)(X_0, \ldots, X_p). \end{aligned}$$

□

Now G acts as a group of automorphisms on $H^p(M; \mathbf{R})$. Let $H^p(M; \mathbf{R})^G$ denote the fixed point set of this action. This is all of $H^p(M; \mathbf{R})$ if G is connected, since then $t_\sigma \simeq 1_M$ for each $\sigma \in G$.

12.3. Theorem. *The inclusion $J : \Omega^G(M) \hookrightarrow \Omega(M)$ induces an isomorphism*

$$J^* : H^p(\Omega^G(M)) \xrightarrow{\approx} H^p(M; \mathbf{R})^G.$$

PROOF. We have $I^*J^* = 1$ since $I \circ J = 1$. Therefore, I^* is onto and J^* is an injection. We must show that the image of J^* is all of $H^*(M;\mathbf{R})^G$.

If $\alpha = [\![\omega]\!] \in H^*(M;\mathbf{R})^G$ then we claim that $J^*I^*(\alpha) = \alpha$, which suffices. That is, we claim that ω and $I(\omega)$ represent the same class in $H^*(M;\mathbf{R})$.

Let $\sigma \in G$. Then $\omega - t_\sigma^*(\omega) = d\eta$ for some $(p-1)$-form η depending on σ, since $[\![\omega]\!]$ is invariant under G. Therefore, for a smooth p-cycle $c \in \Delta_p(M)$, we have

$$\int_c \omega - \int_c t_\sigma^* \omega = \int_c d\eta = \int_{\partial c} \eta = 0.$$

Thus

$$\int_c I(\omega) = \int_c \left(\int_G t_\sigma^* \omega \, d\sigma \right) = \int_G \left(\int_c t_\sigma^* \omega \right) d\sigma$$

$$= \int_G \left(\int_c \omega \right) d\sigma = \left(\int_c \omega \right) \left(\int_G 1 \, d\sigma \right) = \int_c \omega.$$

Hence $\int_c (I(\omega) - \omega) = 0$ for every p-cycle c. That is, the de Rham isomorphism $H^p(\Omega^*(M)) \to H^p(M;\mathbf{R}) \approx \operatorname{Hom}(H_p(M),\mathbf{R})$ kills $[\![I(\omega) - \omega]\!]$. Therefore $[\![I(\omega)]\!] = [\![\omega]\!]$ in $H^*(M;\mathbf{R})$. □

12.4. Corollary. *Let G be a compact connected Lie group. Then $H^p(G;\mathbf{R})$ is isomorphic to $H^p(L_G^*)$, where L_G^* is the chain complex of alternating forms ω on L_G with differential given by*

$$d\omega(X_0,\dots,X_p) = \sum_{i<j} (-1)^{i+j} \omega([X_i, X_j], X_0, \dots, \hat{X}_i, \dots, \hat{X}_j, \dots, X_p). \quad \square$$

We can do better than this by applying Theorem 12.3 to the action of $G \times G$ on G given by $(\sigma \times \tau)(g) = \sigma g \tau^{-1}$. Then the real cohomology of G (compact connected) is given by the "invariant forms," meaning now two-sided invariance. But this is the same as the space of left-invariant forms which are also invariant under conjugation. Let $c_\sigma: G \to G$ be $c_\sigma(\tau) = \sigma\tau\sigma^{-1}$. Then the differential of c_σ defines an automorphism of the vector space $T_e(G) = L_G$ and the invariant forms on G are equivalent to forms on L_G invariant under each $(c_\sigma)_*$. This linear transformation $(c_\sigma)_*$ of (the underlying vector space of) L_G is called $Ad(\sigma) \in \mathbf{Gl}(L_G)$. Then $\sigma \mapsto Ad(\sigma)$ is a homomorphism $Ad: G \to \mathbf{Gl}(L_G)$ called the "adjoint representation" of G. This induces an action of G on forms ω by putting

$$\sigma(\omega)(X_1,\dots,X_p) = \omega(Ad(\sigma^{-1})X_1,\dots,Ad(\sigma^{-1})X_p).$$

12.5. Theorem. *Let G be a compact connected Lie group. Let ω be a multilinear p-form (not necessarily alternating) on L_G. Then the adjoint action of G leaves*

ω invariant $\Leftrightarrow$ the following identity holds:

$$\sum_{i=1}^{p} \omega(X_1, \ldots, X_{i-1}, [Y, X_i], X_{i+1}, \ldots, X_p) = 0$$

for all $Y, X_1, \ldots, X_p \in L_G$.

The proof of Theorem 12.5 will be deferred until the end of the section because it involves many details and would disrupt the flow of ideas.

By an "invariant" form on L_G we mean one invariant under the adjoint action.

12.6. Proposition. *Every invariant alternating form on L_G is closed.*

PROOF. Putting $\epsilon_{i,i} = 0$, $\epsilon_{i,j} = (-1)^j$ if $i < j$ and $\epsilon_{i,j} = (-1)^{j+1}$ if $i > j$, we have

$$d\omega(X_0, \ldots, X_p) = \frac{1}{2} \sum_{i \neq j} (-1)^i \epsilon_{i,j} \omega([X_i, X_j], X_0, \ldots, \hat{X}_i, \ldots, \hat{X}_j, \ldots, X_p)$$

$$= \frac{1}{2} \sum_i (-1)^i \sum_j \epsilon_{i,j} \omega([X_i, X_j], X_0, \ldots, \hat{X}_i, \ldots, \hat{X}_j, \ldots, X_p)$$

$$= 0,$$

since the inner sum is zero by the invariance formula of Theorem 12.5. □

12.7. Corollary. *For a compact connected Lie group G, $H^p(G; \mathbf{R})$ is isomorphic to the vector space of invariant alternating p-forms on L_G.* □

For the remainder of this section, we assume that G is a compact connected Lie group. Let $[L_G, L_G]$ denote the span of the elements of L_G of the form $[X, Y]$ for $X, Y \in L_G$.

12.8. Corollary.

$$[L_G, L_G] = L_G \quad \Leftrightarrow \quad H^1(G; \mathbf{R}) = 0.$$

PROOF. If $[L_G, L_G] \neq L_G$ then there is a nonzero 1-form ω vanishing on $[L_G, L_G]$, and conversely. Such a form is invariant since that just means that $\omega([X, Y]) = 0$ for all $X, Y \in L_G$. □

12.9. Corollary.

$$H^1(G; \mathbf{R}) = 0 \quad \Rightarrow \quad H^2(G; \mathbf{R}) = 0.$$

PROOF. Let ω be an invariant alternating 2-form on L_G. Then

$$\begin{aligned} 0 = d\omega(X, Y, Z) &= -\omega([X, Y], Z) + \omega([X, Z], Y) - \omega([Y, Z], X) \\ &= -\omega([X, Y], Z) - \{\omega([Z, X], Y) + \omega(X, [Z, Y])\} \\ &= -\omega([X, Y], Z) \end{aligned}$$

by invariance. Since $[L_G, L_G] = L_G$ by Corollary 12.8, $\omega \equiv 0$. □

12.10. Theorem. *If* $H^1(G;\mathbf{R})=0$ *then the assignment* $\eta\mapsto\omega$, *where* $\omega(X,Y,Z)=\eta([X,Y],Z)$ *is a one–one correspondence from the space of invariant symmetric 2-forms* η *on* L_G *to that of invariant alternating 3-forms* ω *on* L_G.

PROOF. Given an alternating 3-form ω, define an alternating 2-form ω_Z on L_G by $\omega_Z(X,Y)=\omega(X,Y,Z)$. We claim that ω_Z is closed. We compute

$$\begin{aligned} 0 &= d\omega(X_0,X_1,X_2,Z)\\ &= -\omega([X_0,X_1],X_2,Z)+\omega([X_0,X_2],X_1,Z)-\omega([X_1,X_2],X_0,Z)\\ &\quad -\omega([X_0,Z],X_1,X_2)+\omega([X_1,Z],X_0,X_2)-\omega([X_2,Z],X_0,X_1)\\ &= d\omega_Z(X_0,X_1,X_2), \end{aligned}$$

since the last three terms cancel by the invariance of ω. Since $H^2(G;\mathbf{R})=0$ by Corollary 12.9, we conclude that $\omega_Z=d\zeta_Z$ for some 1-form ζ_Z. That is,

$$\omega(X,Y,Z)=\omega_Z(X,Y)=d\zeta_Z(X,Y)=\zeta_Z([X,Y]).$$

Put $\eta(S,T)=\zeta_T(S)$. This is clearly linear in S. Since $[L_G,L_G]=L_G, \eta(S,T)$ is also linear in T. For $\sigma\in G$ and $X\in L_G$, denote $(c_\sigma)_*(X)$ by X^σ where c_σ is the inner automorphism of G by σ. Then

$$\begin{aligned} \eta([X,Y],Z) &= \omega(X,Y,Z)=\omega(X^\sigma,Y^\sigma,Z^\sigma)\\ &= \eta([X^\sigma,Y^\sigma],Z^\sigma)=\eta([X,Y]^\sigma,Z^\sigma). \end{aligned}$$

Since $[L_G,L_G]=L_G$ it follows that η is invariant.

Now η decomposes uniquely as $\eta=\eta_{\text{sym}}+\eta_{\text{skew}}$ and both terms must be invariant by the uniqueness of the decomposition. Since $H^2(G;\mathbf{R})=0$, we have that $\eta_{\text{skew}}=0$. Thus η is symmetric.

Conversely, if η is given and ω is defined by $\omega(X,Y,Z)=\eta([X,Y],Z)$ then ω is invariant by the same argument. By invariance of η we have $\eta([X,Y],Z)=-\eta(Y,[X,Z])$. Thus an interchange of X and Y or of X and Z changes the sign of ω. It follows that ω is alternating. □

There always exists a nontrivial invariant symmetric 2-form on L_G, for G compact, since if $\langle\cdot,\cdot\rangle$ is a positive definite inner product on L_G then

$$\eta(X,Y)=\int\langle X^\sigma,Y^\sigma\rangle\,d\sigma$$

is a positive definite invariant inner product. Consequently, we have the following result:

12.11. Corollary. *If* G *is nontrivial and* $H^1(G;\mathbf{R})=0$ *then* $H^3(G;\mathbf{R})\neq 0$. □

12.12. Corollary. *The only spheres which are Lie groups are* $\mathbf{S}^0$, $\mathbf{S}^1$, *and* $\mathbf{S}^3$. □

Now let us introduce the notation

$$L_Z = \{X \in L_G \mid [X, Y] = 0 \text{ for all } Y \in L_G\}.$$

12.13. Proposition. *Given an invariant positive definite inner product* $\langle \cdot, \cdot \rangle$ *on* L_G, *we have* $[L_G, L_G] = L_Z^{\perp}$.

PROOF. If $X \in L_Z$ then $\langle X, [Y, Z] \rangle = \langle [X, Y], Z \rangle = 0$ by the invariance formula of Theorem 12.5. Thus $L_Z \subset [L_G, L_G]^{\perp}$. Conversely, if $X \in [L_G, L_G]^{\perp}$ then $\langle [X, Y], Z \rangle = \langle X, [Y, Z] \rangle = 0$ for all Y and Z and so $[X, Y] = 0$ for all Y, whence $X \in L_Z$ by definition. □

12.14. Corollory. $H^1(G; \mathbf{R}) \approx L_Z^*$, *the dual space to* L_Z.

PROOF. The invariant 1-forms on L_G are just the 1-forms vanishing on $[L_G, L_G]$. Thus they are the 1-forms vanishing on $L_Z^{\perp}$, and those can be identified with all 1-forms on L_Z; i.e., with the dual space L_Z^*. □

12.15. Corollary. $\dim H^1(G; \mathbf{R}) = \dim L_Z$. □

From some basic Lie group theory which we cannot derive here, it can be seen that L_Z is just the tangent space at e to the center of G. Thus, the condition $H^1(G; \mathbf{R}) = 0$ is equivalent to the center being finite. A compact connected Lie group with finite center is said to be "semisimple." It is said to be "simple" if it is nontrivial and has no connected normal subgroups except for $\{e\}$ and G itself. Another basic fact, which we must leave unproved, is that any invariant subspace of L_G, for G semisimple, is the tangent space of a closed normal subgroup of G. Moreover, a compact semisimple group is, locally at e, a product of its simple normal subgroups. (More precisely, the universal covering group of G is compact and is the product of simple normal subgroups.) Granting this gives the following corollaries.

12.16. Corollary. *If* G *is simple then* $H^3(G; \mathbf{R}) \approx \mathbf{R}$.

PROOF. Let $\langle X, Y \rangle$ be a positive definite invariant inner product, and let η be an invariant symmetric 2-form on L_G. Put

$$k = \min\{\eta(X, X) \mid \langle X, X \rangle = 1\}$$

and let $\zeta(X, Y) = \eta(X, Y) - k\langle X, Y \rangle$. Then $\zeta(X, X) \geq 0$ for all X and equals 0 for some $X \neq 0$. If $\zeta(X, X) = 0 = \zeta(Y, Y)$ then

$$2\zeta(X, Y) = \zeta(X + Y, X + Y) \geq 0,$$
$$-2\zeta(X, Y) = \zeta(X - Y, X - Y) \geq 0,$$

and so $\zeta(X, Y) = 0$ and $\zeta(X + Y, X + Y) = 0$. Therefore

$$S = \{X \mid \zeta(X, X) = 0\}$$

is a vector subspace of L_G. But ζ is invariant and it follows that S is invariant under Ad. Since G is simple, ζ must be identically zero. Therefore $\eta(\cdot,\cdot) = k\langle\cdot,\cdot\rangle$. By Theorem 12.10, there is then exactly one nontrivial invariant alternating 3-form on L_G up to a constant multiple and so $\dim H^3(G;\mathbf{R}) = 1$.

□

12.17. Corollary. $H^1(G;\mathbf{R}) = 0 \Rightarrow \dim H^3(G;\mathbf{R})$ *equals the number of simple normal factors of G.* □

Proof of Theorem 12.5. For a finite-dimensional vector space V, suppose we are given a linear action of G on V; i.e., a homomorphism $\theta: G \to \mathbf{Gl}(V)$. (Think of V as $\mathbf{R}^k$ if you wish.) Then the differential of θ at e is a linear map $\theta_*: L_G = T_e(G) \to T_I(\mathbf{Gl}(V)) = \mathbf{Endo}(V)$, the space of endomorphisms of V (i.e., the $k \times k$ matrices if V is regarded as $\mathbf{R}^k$). For $v \in V$, we claim that $\theta(\sigma)(v) = v$ for all $\sigma \in G \Leftrightarrow \theta_*(X)(v) = 0$ for all $X \in L_G$. Let $\phi^X: \mathbf{R} \to G$ be the one-parameter subgroup of G with tangent vector $X = \phi^X_*(d/dt)$ at e. (Recall from Theorem 12.10 of Chapter II that this exists for each $X \in L_G$ and that they fill out a neighborhood of e in G and hence generate G.) Then $\theta\phi^X(t)$ is a one-parameter group in $\mathbf{Gl}(V)$ and therefore has the form $\theta\phi^X(t) = e^{tA}$ for some $A \in \mathbf{Endo}(V)$. By definition $\theta_*(X) = \theta_*(\phi^X_*(d/dt)) = (d/dt)e^{tA}|_{t=0} = A$. Thus $\theta\phi^X(t)(v) = e^{tA}v = v + tAv + \frac{1}{2}t^2A^2v + \cdots$. If v is fixed by the action, then this is v and so $Av = (d/dt)(\theta\phi^X(t)(v))|_{t=0} = (d/dt)(v)|_{t=0} = 0$. Conversely, if $Av = 0$ then $e^{tA}v = v + tAv + \frac{1}{2}t^2A^2v + \cdots = v$ and so v is fixed under G since the one-parameter subgroups generate G.

For the adjoint representation $Ad: G \to \mathbf{Gl}(L_G)$, the differential is called $ad = Ad_*: L_G \to \mathbf{Endo}(L_G)$. By definition, if $\phi^Y: \mathbf{R} \to G$ is the one-parameter subgroup with tangent Y, then

$$ad\, Y = Ad_*(Y) = Ad_*\phi^Y_*\left(\frac{d}{dt}\bigg|_0\right) = \frac{d}{dt}\left((Ad \circ \phi^Y)(t)\right)\bigg|_{t=0}$$

and so

$$ad\, Y(X) = \frac{d}{dt}(Ad(\phi^Y(t)))|_{t=0}(X) = \frac{d}{dt}Ad\,(\phi^Y(t)(X))|_{t=0}.$$

Let V be the space of p-forms (not necessarily alternating) on L_G with action given by

$$\theta(\sigma)(\omega)(X_1, \ldots, X_p) = \omega(Ad(\sigma^{-1})X_1, \ldots, Ad(\sigma^{-1})X_p).$$

Then $\theta(\phi^{-Y}(t))(\omega)(X_1, \ldots, X_p) = \omega(Ad(\phi^Y(t))X_1, \ldots, Ad(\phi^Y(t))X_p)$ and so, putting $B_t = Ad(\phi^Y(t))$, we have

$$-\theta_*(Y)(\omega)(X_1, \ldots, X_p) = \theta_*\phi^{-Y}_*\left(\frac{d}{dt}\bigg|_0\right)(\omega)(X_1, \ldots, X_p)$$

$$= \frac{d}{dt}\theta(\phi^{-Y}(t))(\omega)(X_1, \ldots, X_p)|_{t=0}$$

$$= \frac{d}{dt}(\omega(B_t X_1, \ldots, B_t X_p))|_{t=0}$$

$$= \lim_{t\to 0} \frac{1}{t}\{\omega(B_t X_1, \ldots, B_t X_p) - \omega(X_1, \ldots, X_p)\}$$

$$= \lim_{t\to 0} \frac{1}{t}\{(\omega(B_t X_1 - X_1, B_t X_2, \ldots, B_t X_p)$$

$$+ \omega(X_1, B_t X_2 - X_2, B_t X_3, \ldots) + \cdots$$

$$+ \omega(X_1, X_2, \ldots, X_{p-1}, B_t X_p - X_p)\}$$

$$= \omega\left(\lim_{t\to 0} \frac{1}{t}(B_t X_1 - X_1), \lim_{t\to 0} B_t X_2, \ldots\right)$$

$$+ \omega\left(X_1, \lim_{t\to 0} \frac{1}{t}(B_t X_2 - X_2), \lim_{t\to 0} B_t X_3, \ldots\right)$$

$$+ \cdots + \omega\left(X_1, \ldots, X_{p-1}, \lim_{t\to 0} \frac{1}{t}(B_t X_p - X_p)\right)$$

$$= \omega(ad\, Y(X_1), X_2, \ldots, X_p)$$

$$+ \omega(X_1, ad\, Y(X_2), X_3, \ldots, X_p) + \cdots$$

$$+ \omega(X_1, \ldots, X_{p-1}, ad\, Y(X_p))$$

since

$$\frac{d}{dt} B_t X_i|_{t=0} = \frac{d}{dt} Ad\, \phi^Y(t)|_{t=0}(X_i) = ad\, Y(X_i).$$

By the general remarks at the beginning of the proof, $\omega \in V$ is invariant (i.e., $\theta(\sigma)\omega = \omega$ for all $\sigma \in G) \Leftrightarrow \theta_*(Y)(\omega) = 0$ for all $Y \in L_G$.

Therefore Theorem 12.5 will be proved once we show that $ad\, Y(X) = [Y, X]$. For that, consider the left-invariant vector field Y. Left-invariance means that $Y_\sigma = (L_\sigma)_* Y_e$ for all $\sigma \in G$. Thus, for a smooth function f, we have

$$Y_\sigma(f) = (L_\sigma)_*(Y_e)(f) = Y_e(f \circ L_\sigma)$$

$$= \phi^Y_*\left(\frac{d}{dt}\bigg|_0\right)(f \circ L_\sigma) = \frac{d}{dt}\bigg|_0 (f \circ L_\sigma \circ \phi^Y)$$

$$= \frac{d}{dt}(f(\sigma\phi^Y(t)))|_{t=0}.$$

Consider the function $g(t, s, u) = f(\phi^X(t)\phi^Y(s)\phi^X(u))$. We have

$$\frac{\partial}{\partial t}\frac{\partial}{\partial s} g(0, 0, 0) = \frac{\partial}{\partial t}\frac{\partial}{\partial s} f(\phi^X(t)\phi^Y(s))|_{s=t=0}$$

$$= \frac{\partial}{\partial t} Yf(\phi^X(t))|_{t=0}$$

$$= XY(f) \quad \text{at } e.$$

Similarly

$$\frac{\partial}{\partial u}\frac{\partial}{\partial s} g(0,0,0) = \frac{\partial}{\partial u}\frac{\partial}{\partial s} f(\phi^Y(s)\phi^X(u))|_{s=u=0}$$

$$= \frac{\partial}{\partial s}\frac{\partial}{\partial u} f(\phi^Y(s)\phi^X(u))|_{s=u=0}$$

$$= \frac{\partial}{\partial s} Xf(\phi^Y(s))|_{s=0}$$

$$= YX(f) \quad \text{at } e.$$

If $h(s,t) = g(t,s,-t)$ then

$$\frac{\partial}{\partial t}\frac{\partial}{\partial s} h(0,0) = \frac{\partial}{\partial t}\frac{\partial}{\partial s} g(0,0,0) - \frac{\partial}{\partial u}\frac{\partial}{\partial s} g(0,0,0) = XY(f) - YX(f)$$

$$= [X,Y](f) \quad \text{at } e.$$

That is,

$$\frac{\partial}{\partial t}\frac{\partial}{\partial s} f(\phi^X(t)\phi^Y(s)\phi^{-X}(t))|_{s=t=0} = [X,Y]_e(f).$$

On the other hand, $s \mapsto \phi^X(t)\phi^Y(s)\phi^{-X}(t)$ is a one-parameter group whose tangent vector at e is $Ad(\phi^X(t))(Y)$ by definition of Ad. Thus

$$\phi^X(t)\phi^Y(s)\phi^{-X}(t) = \phi^{Ad(\phi^X(t))(Y)}(s)$$

and so

$$\frac{\partial}{\partial s} f(\phi^X(t)\phi^Y(s)\phi^{-X}(t))|_{s=0} = Ad(\phi^X(t))(Y_e)(f),$$

since $(d/ds)f(\phi^Z(s))|_{s=0} = Z_e(f)$ as seen before. Therefore

$$\frac{\partial}{\partial t}\frac{\partial}{\partial s} f(\phi^X(t)\phi^Y(s)\phi^{-X}(t))|_{s=t=0} = \frac{\partial}{\partial t} Ad(\phi^X(t))(Y_e)(f)|_{t=0} = ad\,X(Y_e)(f).$$

Comparing formulas gives the desired result $ad\,X(Y) = [X,Y]$. □

We remark that the common notation for the one-parameter group $\phi^X(t)$ with tangent vector X at e is $\exp(tX)$. That is, $\exp: L_G \to G$ is defined by $\exp(X) = \phi^X(1)$, and so $\exp(tX) = \phi^{tX}(1) = \phi^X(t)$. This is compatible with the previously defined exponential map for the classical matrix groups.

CHAPTER VI

Products and Duality

Algebra is generous; she often gives more than is asked of her.

D'ALEMBERT

1. The Cross Product and the Künneth Theorem

Now we begin discussion of several varieties of products that can be introduced into homology and, especially, cohomology. First, we need to discuss a sign convention.

Suppose that A_*, B_*, C_* and D_* are graded groups. A map $f: A_* \to B_*$ is said to be of degree d if it takes A_i to B_{i+d} for all i. For the most part d has the values 0 or ± 1. We define the tensor product of two graded groups A_* and B_* to be the graded group $A_* \otimes B_*$ where

$$(A_* \otimes B_*)_n = \bigoplus_{i+j=n} A_i \otimes B_j.$$

If $f: A_* \to C_*$ and $g: B_* \to D_*$ then we define $f \otimes g: A_* \otimes B_* \to C_* \otimes D_*$ by

$$(f \otimes g)(a \otimes b) = (-1)^{\deg(a)\deg(g)} f(a) \otimes g(b).$$

Note that the exponent in the sign is the product of the degrees of the items whose order is interchanged. For compositions of such maps, it is easy to check that

$$(f \otimes g) \circ (h \otimes k) = (-1)^{\deg(g)\deg(h)} (f \circ h) \otimes (g \circ k).$$

In particular, for the chain complexes $\Delta_*(X)$ and $\Delta_*(Y)$ we have the chain complex

$$(\Delta_*(X) \otimes \Delta_*(Y))_n = \bigoplus_{i+j=n} \Delta_i(X) \otimes \Delta_j(Y),$$

with boundary operator $\partial_\otimes = \partial \otimes 1 + 1 \otimes \partial$. The sign convention holds here, so this means

$$\partial_{\otimes}(a_p \otimes b_q) = (\partial \otimes 1 + 1 \otimes \partial)(a \otimes b) = \partial a \otimes b + (-1)^p a \otimes \partial b.$$

The subscript on $\partial_{\otimes}$ is used only for stress. It will be dropped in many cases.

If x and y are points then $(\Delta_*(\{x\}) \otimes \Delta_*(\{y\}))_0 = \Delta_0(\{x\}) \otimes \Delta_0(\{y\}) \approx \mathbf{Z}$ generated by $x \otimes y$. Also $\Delta_0(\{x\} \times \{y\}) \approx \mathbf{Z}$ generated by (x, y). Recall that in Section 16 of Chapter IV we had a "cross product" $\times : \Delta_*(X) \times \Delta_*(Y) \to \Delta_*(X \times Y)$. Because this is bilinear, it induces a homomorphism, still called a cross product,

$$\times : \Delta_*(X) \otimes \Delta_*(Y) \to \Delta_*(X \times Y),$$

which, by definition, takes $a \otimes b$ to $a \times b$. The product is natural in X and Y, and is the canonical map $(x \times y = (x, y))$ when X and Y are points. There was the boundary formula $\partial(a \times b) = \partial a \times b + (-1)^{\deg(a)} a \times \partial b$. Thus

$$\begin{aligned}\partial(\times(a \otimes b)) = \partial(a \times b) &= \partial a \times b + (-1)^{\deg(a)} a \times \partial b \\ &= \times(\partial a \otimes b + (-1)^{\deg(a)} a \otimes \partial b) \\ &= \times(\partial_{\otimes}(a \otimes b)).\end{aligned}$$

That is, $\times$ is a chain map.

1.1. Lemma. *If X and Y are contractible then there is a chain contraction of $\Delta_*(X) \otimes \Delta_*(Y)$. Consequently, $H_n(\Delta_*(X) \otimes \Delta_*(Y)) = 0$ for $n > 0$ and is $\mathbf{Z}$, generated by $[\![x_0 \otimes y_0]\!]$, for $n = 0$.*

Proof. In Theorem 15.5 of Chapter IV we constructed such a chain contraction for X. That is, for the chain map $\epsilon: \Delta_p(X) \to \Delta_p(X)$ which is 0 for $p > 0$ and $\epsilon(\sum n_x x) = \sum n_x x_0$, we constructed a map $D: \Delta_p(X) \to \Delta_{p+1}(X)$ such that $\partial D + D\partial = 1 - \epsilon$. Put $E = D \otimes 1 + \epsilon \otimes D$ on $\Delta_*(X) \otimes \Delta_*(Y)$. Then

$$\begin{aligned}E\partial_{\otimes} + \partial_{\otimes} E &= (D \otimes 1 + \epsilon \otimes D)(\partial \otimes 1 + 1 \otimes \partial) + (\partial \otimes 1 + 1 \otimes \partial)(D \otimes 1 + \epsilon \otimes D) \\ &= D\partial \otimes 1 + D \otimes \partial - \epsilon\partial \otimes D + \epsilon \otimes D\partial \\ &\quad + \partial D \otimes 1 + \partial\epsilon \otimes D - D \otimes \partial + \epsilon \otimes \partial D \\ &= D\partial \otimes 1 + \epsilon \otimes D\partial + \partial D \otimes 1 + \epsilon \otimes \partial D \\ &= (D\partial + \partial D) \otimes 1 + \epsilon \otimes (D\partial + \partial D) \\ &= (1 - \epsilon) \otimes 1 + \epsilon \otimes (1 - \epsilon) \\ &= 1 \otimes 1 - \epsilon \otimes 1 + \epsilon \otimes 1 - \epsilon \otimes \epsilon \\ &= 1 \otimes 1 - \epsilon \otimes \epsilon = \text{identity} - \text{augmentation}.\end{aligned}$$

□

1.2. Theorem. *There exists a natural (in X and Y) chain map*

$$\theta: \Delta_*(X \times Y) \to \Delta_*(X) \otimes \Delta_*(Y)$$

which is the canonical map $(x, y) \mapsto x \otimes y$ in degree 0.

Proof. Note that naturality means that for maps $f: X \to X'$ and $g: Y \to Y'$ we have $\theta \circ (f \times g)_\Delta = (f_\Delta \otimes g_\Delta) \circ \theta$. The proof is by the method of acyclic models. Suppose θ is defined for degree less than k, such that $\partial\theta = \theta\partial$ in those degrees. This holds trivially for $k = 1$.

Consider the case $X = \Delta_k = Y$ and let $d_k: \Delta_k \to \Delta_k \times \Delta_k$ be the diagonal map. Then $d_k \in \Delta_k(\Delta_k \times \Delta_k)$. The chain $\theta(\partial d_k) \in (\Delta_*(\Delta_k) \otimes \Delta_*(\Delta_k))_{k-1}$ is defined and $\partial\theta(\partial d_k) = \theta(\partial\partial d_k) = 0$. Thus $\theta(\partial d_k)$ is a cycle in $\Delta_*(\Delta_k) \otimes \Delta_*(\Delta_k)$ (and if $k = 1$ then the augmentation takes this to 0). Therefore $\theta(\partial d_k)$ is a boundary by Lemma 1.1. Let $\theta(d_k)$ be some chain whose boundary is $\theta(\partial d_k)$.

Next, for general X and Y, let $\pi_X: X \times Y \to X$ and $\pi_Y: X \times Y \to Y$ be the projections. Let $\sigma: \Delta_k \to X \times Y$ be any singular k-simplex of $X \times Y$. Then we have the product map $\pi_X\sigma \times \pi_Y\sigma: \Delta_k \times \Delta_k \to X \times Y$. We can express σ as the composition $\sigma = (\pi_X\sigma \times \pi_Y\sigma) \circ d_k: \Delta_k \to X \times Y$. It follows that $\sigma = (\pi_X\sigma \times \pi_Y\sigma)_\Delta(d_k)$ in $\Delta_k(X \times Y)$ where $(\pi_X\sigma \times \pi_Y\sigma)_\Delta$ is the induced chain map $\Delta_*(\Delta_k \times \Delta_k) \to \Delta_*(X \times Y)$. Thus, for θ to be natural, we *must* define

$$\theta(\sigma) = \theta((\pi_X\sigma \times \pi_Y\sigma)_\Delta(d_k)) = ((\pi_X\sigma)_\Delta \otimes (\pi_Y\sigma)_\Delta)(\theta(d_k)).$$

We do so. Then naturality of this definition is clear. To prove the boundary formula, we compute

$$\begin{aligned}
\partial_\otimes \theta(\sigma) &= \partial_\otimes(((\pi_X\sigma)_\Delta \otimes (\pi_Y\sigma)_\Delta \theta(d_k)) \\
&= ((\pi_X\sigma)_\Delta \otimes (\pi_Y\sigma)_\Delta)(\partial_\otimes \theta d_k) \\
&= ((\pi_X\sigma)_\Delta \otimes (\pi_Y\sigma)_\Delta)(\theta\partial d_k) \\
&= \theta(\pi_X\sigma \times \pi_Y\sigma)_\Delta(\partial d_k) \\
&= \theta\partial(\pi_X\sigma \times \pi_Y\sigma)_\Delta(d_k) \\
&= \theta\partial\sigma.
\end{aligned}$$

□

1.3. Theorem. *Any two natural chain maps on $\Delta_*(X \times Y)$ to itself or on $\Delta_*(X) \otimes \Delta_*(Y)$ to itself or on one of these to the other which are the canonical isomorphisms in degree zero, with X and Y points, are naturally chain homotopic.*

Proof. The proofs of all four cases are similar. One uses "models" $\iota_p \otimes \iota_q \in \Delta_*(\Delta_p) \otimes \Delta_*(\Delta_q)$ for the complex $\Delta_*(X) \otimes \Delta_*(Y)$, and $d_p \in \Delta_*(\Delta_p \times \Delta_p)$ for the complex $\Delta_*(X \times Y)$, and then tries to construct a chain homotopy D inductively so that $D\partial + \partial D = \phi - \psi$ where ϕ and ψ are the two chain maps. One does this by computing, inductively on a model, the boundary of $\phi - \psi - D\partial$, showing it is zero and using a chain of which it is a boundary as D of the model. We will do one case in detail and the reader can furnish the details of the other three.

We will take the case of $\phi, \psi: \Delta_*(X \times Y) \to \Delta_*(X) \otimes \Delta_*(Y)$. Take D to be zero on 0-chains. Suppose that we have defined D on chains of degree less than k, for some $k > 0$. We compute

$$\begin{aligned}
\partial_\otimes(\phi - \psi - D\partial)(d_k) &= \partial_\otimes \phi(d_k) - \partial_\otimes \psi(d_k) - \partial_\otimes D\partial d_k \\
&= \phi(\partial d_k) - \psi(\partial d_k) - (\phi - \psi - D\partial)(\partial d_k) = 0.
\end{aligned}$$

Since $\Delta_*(\Delta_k) \otimes \Delta_*(\Delta_k)$ is acyclic by Lemma 1.1, there exists a $(k+1)$-chain whose boundary is $(\phi - \psi - D\partial)(d_k)$. We let Dd_k be one such chain. If σ is any singular simplex of $X \times Y$ then we have $\sigma = (\pi_X\sigma \times \pi_Y\sigma)_\Delta(d_k)$ as seen before. Thus we define $D\sigma = ((\pi_X\sigma)_\Delta \otimes (\pi_Y\sigma)_\Delta)(Dd_k)$. Extending by linearity, this defines D on $\Delta_k(X \times Y)$ for all X and Y, such that $D\partial + \partial D = \phi - \psi$ on chains of degree k. This completes the induction. □

1.4. Corollary (The Eilenberg–Zilber Theorem). *The chain maps*

$$\theta: \Delta_*(X \times Y) \to \Delta_*(X) \otimes \Delta_*(Y)$$

and

$$\times: \Delta_*(X) \otimes \Delta_*(Y) \to \Delta_*(X \times Y)$$

are natural homotopy equivalences which are naturally homotopy inverses of one another. □

The implication of this result is that one can compute the homology and cohomology of a product space $X \times Y$ from the chain complex $\Delta_*(X) \otimes \Delta_*(Y)$. That is,

$$H_p(X \times Y; G) \approx H_p(\Delta_*(X) \otimes \Delta_*(Y) \otimes G),$$

and

$$H^p(X \times Y; G) \approx H^p(\mathrm{Hom}(\Delta_*(X) \otimes \Delta_*(Y), G)).$$

1.5. Theorem (Algebraic Künneth Theorem). *Let K_* and L_* be free chain complexes. Then there is a natural exact sequence*

$$0 \to (H_*(K_*) \otimes H_*(L_*))_n \xrightarrow{\times} H_n(K_* \otimes L_*) \to (H_*(K_*) * H_*(L_*))_{n-1} \to 0$$

which splits (not naturally).

Proof. Recall that $(H_*(A_*) \otimes H_*(B_*))_n$ means $\bigoplus\{H_p(A_*) \otimes H_q(B_*) | p + q = n\}$, and similarly for the torsion product. Let Z_* denote the group of cycles of K_*, and B_* the group of boundaries. There is the exact sequence

$$0 \to Z_n \xrightarrow{i} K_n \xrightarrow{\partial} B_{n-1} \to 0$$

which splits, since B_{n-1} is projective. Tensoring this with L_* gives a short exact sequence

$$0 \to Z_* \otimes L_* \to K_* \otimes L_* \to B_* \otimes L_* \to 0.$$

This induces the long exact sequence

$$(*) \qquad \cdots \xrightarrow{\Delta_*} H_n(Z_* \otimes L_*) \xrightarrow{(i \otimes 1)_*} H_n(K_* \otimes L_*)$$

$$\xrightarrow{(\partial \otimes 1)_*} H_{n-1}(B_* \otimes L_*) \xrightarrow{\Delta_*} H_{n-1}(Z_* \otimes L_*) \to \cdots.$$

Now B_* has trivial differential and $B_* \otimes (\cdot)$ is exact. It follows that a cycle of $B_* \otimes L_*$ is just an element of $B_* \otimes Z_*(L_*)$. This group is generated by elements of the form $\partial k \otimes l = (\partial \otimes 1)(k \otimes l)$ where $\partial l = 0$. Therefore $\Delta_* [\![\partial k \otimes l]\!]$ is represented by $(i \otimes 1)^{-1} \partial_\otimes (k \otimes l) = (i \otimes 1)^{-1}(\partial k \otimes l) = (j \otimes 1)(\partial k \otimes l)$ where $j: B_* \hookrightarrow Z_*$ is the inclusion. Because $B_* \otimes (\cdot)$ and $Z_* \otimes (\cdot)$ are exact, we have the canonical isomorphisms

$$H_*(B_* \otimes L_*) \approx B_* \otimes H_*(L_*) \qquad \text{and} \qquad H_*(Z_* \otimes L_*) \approx Z_* \otimes H_*(L_*).$$

It follows that, under these isomorphisms, Δ_* becomes $j \otimes 1: B_* \otimes H_*(L_*) \to Z_* \otimes H_*(L_*)$. This is part of the exact sequence

$$\begin{aligned} 0 \to H_*(K_*) * H_*(L_*) \to B_* \otimes H_*(L_*) \xrightarrow{j \otimes 1} Z_* \otimes H_*(L_*) \\ \to H_*(K_*) \otimes H_*(L_*) \to 0. \end{aligned}$$

From $(*)$ one derives the exact sequence

$$0 \to (\operatorname{coker} \Delta_*)_n \to H_n(K_* \otimes L_*) \to (\ker \Delta_*)_{n-1} \to 0.$$

But $\operatorname{coker} \Delta_* \approx \operatorname{coker}(j \otimes 1) \approx H_*(K_*) \otimes H_*(L_*)$ and $\ker \Delta_* \approx \ker(j \otimes 1) \approx H_*(K_*) * H_*(L_*)$, and the substitution of these gives the desired Künneth sequence.

If $k \in K_*$ and $l \in L_*$ are cycles, then it is easy to trace through the correspondences to see that the first map in the Künneth sequence takes $[\![k]\!] \otimes [\![l]\!] \in H_*(K_*) \otimes H_*(L_*)$ to $[\![k \otimes l]\!] \in H_*(K_* \otimes L_*)$, so this map is, indeed, the cross product.

We may regard the graded groups $H_*(K_*)$ and $H_*(L_*)$ as chain complexes with zero differentials.

If $\phi: K_* \to Z_*(K_*)$ and $\psi: L_* \to Z_*(L_*)$ are splittings then $\Phi: k \mapsto [\![\phi(k)]\!]$ and $\Psi: l \mapsto [\![\psi(l)]\!]$ are chain maps $\Phi: K_* \to H_*(K_*)$ and $\Psi: L_* \to H_*(L_*)$ since $[\![\phi(\partial k)]\!] = [\![\partial k]\!] = 0 = \partial [\![\phi(k)]\!]$ (the latter by definition). Hence $\Phi \otimes \Psi: K_* \otimes L_* \to H_*(K_*) \otimes H_*(L_*)$ is a chain map inducing

$$(\Phi \otimes \Psi)_*: H_n(K_* \otimes L_*) \to (H_*(K_*) \otimes H_*(L_*))_n,$$

where we identify "chains" with homology in the right-hand side since the differential is zero there. If $\partial k = 0 = \partial l$ then

$$\begin{aligned} (\Phi \otimes \Psi)_*(\times([\![k]\!] \otimes [\![l]\!])) &= (\Phi \otimes \Psi)_*([\![k]\!] \times [\![l]\!]) = (\Phi \otimes \Psi)_*([\![k \otimes l]\!]) \\ &= [\![\Phi(k) \otimes \Psi(l)]\!] = \Phi(k) \otimes \Psi(l) \\ &= [\![\phi(k)]\!] \otimes [\![\psi(l)]\!] = [\![k]\!] \otimes [\![l]\!], \end{aligned}$$

and so this gives the desired splitting. Since it is defined through the splittings ϕ and ψ which are not natural, this splitting may not be natural. Indeed, examples, such as that given for the Universal Coefficient Theorem, show that there does not exist a natural splitting. □

It can be seen that the freeness assumption in Theorem 1.5 can be replaced by the hypothesis that $H_*(K_* * L_*) = 0$, but that would entail extending the proof and we have no need for that generality. (See Dold [1] or Spanier [1] for a proof.) The result can also be proved, in exactly the same way, for free chain complexes over a principal ideal domain instead of **Z**. In particular, it holds over a field, in which case the torsion term does not appear. But over a field $\otimes$ is exact and the result is quite trivial in that case.

1.6. Theorem (Geometric Künneth Theorem). *There is a natural exact sequence*

$$0 \to (H_*(X) \otimes H_*(Y))_n \xrightarrow{\times} H_n(X \times Y) \to (H_*(X) * H_*(Y))_{n-1} \to 0$$

which splits (not naturally). □

For example, we compute

$$\begin{aligned} H_2(\mathbf{P}^2 \times \mathbf{P}^2) &\approx H_0(\mathbf{P}^2) \otimes H_2(\mathbf{P}^2) \oplus H_1(\mathbf{P}^2) \otimes H_1(\mathbf{P}^2) \oplus H_2(\mathbf{P}^2) \otimes H_0(\mathbf{P}^2) \\ &\quad \oplus H_0(\mathbf{P}^2) * H_1(\mathbf{P}^2) \oplus H_1(\mathbf{P}^2) * H_0(\mathbf{P}^2) \\ &\approx \mathbf{Z} \otimes 0 \oplus \mathbf{Z}_2 \otimes \mathbf{Z}_2 \oplus 0 \otimes \mathbf{Z} \oplus \mathbf{Z} * \mathbf{Z}_2 \oplus \mathbf{Z}_2 * \mathbf{Z} \\ &\approx \mathbf{Z}_2 \otimes \mathbf{Z}_2 \approx \mathbf{Z}_2. \end{aligned}$$

For another example, let $☼ \in H_0(\mathbf{S}^n)$ be the canonical generator (class of a point), and $[\mathbf{S}^n] \in H_n(\mathbf{S}^n)$ a generator, the "orientation class." Then $☼ \times ☼$ generates $H_0(\mathbf{S}^n \times \mathbf{S}^m)$, $☼ \times [\mathbf{S}^m]$ generates $H_m(\mathbf{S}^n \times \mathbf{S}^m)$, $[\mathbf{S}^n] \times ☼$ generates $H_n(\mathbf{S}^n \times \mathbf{S}^m)$ and $[\mathbf{S}^n] \times [\mathbf{S}^m]$ generates $H_{n+m}(\mathbf{S}^n \times \mathbf{S}^m)$. (In case $n = m$ then $☼ \times [\mathbf{S}^n]$ and $[\mathbf{S}^m] \times ☼$ are a free basis of $H_n(\mathbf{S}^n \times \mathbf{S}^n)$.)

For a subspace $A \subset X$ the cross product $\Delta_*(X) \otimes \Delta_*(Y) \to \Delta_*(X \times Y)$ carries $\Delta_*(A) \otimes \Delta_*(Y)$ into $\Delta_*(A \times Y)$ and hence induces a chain map $\Delta_*(X, A) \otimes \Delta_*(Y) \to \Delta_*((X, A) \times Y)$. This induces an isomorphism in homology by the 5-lemma. Thus one obtains a natural Künneth exact sequence

$$\boxed{\begin{aligned} 0 &\to (H_*(X, A) \otimes H_*(Y))_n \to H_n((X, A) \times Y) \\ &\to (H_*(X, A) * H_*(Y))_{n-1} \to 0, \end{aligned}}$$

which splits nonnaturally. In general, there is no such result when Y is *also* replaced by a pair (Y, B); see Bredon [2]. There is one, however, if A and B are both open. For open sets $U, V \subset X$ the map $\Delta_*(X \times Y)/(\Delta_*(U \times Y) + \Delta_*(X \times V)) \to \Delta_*((X, U) \times (Y, V))$ induces an isomorphism in homology. It follows that the map $\Delta_*(X, U) \otimes \Delta_*(Y, V) \to \Delta_*((X, U) \times (Y, V))$ induces an isomorphism in homology, and hence that there is a natural Künneth exact sequence

$$\boxed{\begin{aligned} 0 &\to (H_*(X, U) \otimes H_*(Y, V))_n \to H_n((X, U) \times (Y, V)) \\ &\to (H_*(X, U) * H_*(Y, V))_{n-1} \to 0. \end{aligned}}$$

PROBLEMS

1. Verify the example on $\mathbf{P}^2 \times \mathbf{P}^2$ by using a CW-structure to compute its homology directly.
2. Let X_p be the space resulting from attaching an n-cell to $\mathbf{S}^{n-1}$ by a map of degree p. Use the Künneth Theorem to compute the homology of $X_p \times X_q$ for any p, q.
3. For spaces X, Y of bounded finite type, show that $\chi(X \times Y) = \chi(X)\chi(Y)$.
4. Complete the proof of Theorem 1.3.

2. A Sign Convention

Let us ask whether the coboundary operator should have a sign attached to it instead of being given by the formula $\delta f = f \circ \partial$. A singular p-cochain is a map $\Delta_p(X) \to G$ and G can be thought of as a graded group B_* with $B_0 = G$ and $B_i = 0$ for $i \neq 0$. Then f is a map of graded groups of degree $-p$, i.e., it lowers degree by p.

For any two graded groups A_*, B_*, we let $\mathrm{Hom}(A_*, B_*)_{-p} = \mathrm{Hom}(A_*, B_*)^p$ be the set of homomorphisms taking $A_i \to B_{i-p}$. Then $f \in \mathrm{Hom}(A_*, B_*)^p$ and to be consistent with the way signs work in most places in homological algebra we should have the formula

$$\partial(f(a)) = (\delta f)(a) + (-1)^p f(\partial a)$$

in our case. The left-hand side is 0 in our case of a complex B_* vanishing outside degree 0. Thus we should define

$$\boxed{\delta f = (-1)^{\deg(f)+1} f \circ \partial,}$$

and, in fact, we will use this convention in the remainder of the book. Note that this change does not affect which cochains are cocycles and coboundaries and so it does not affect cohomology groups. It does affect some things in minor ways. (For example, the de Rham map $\Omega^*(M) \to \Delta^*(M; \mathbf{R}) = \mathrm{Hom}(\Delta_*, \mathbf{R})$ is no longer a chain map strictly speaking.) For the most part, however, it only affects signs in various formulas that we will use in subsequent parts of the book. This sign convention will have the effect of making a number of formulas appear much more logical than they would otherwise (see Mac Lane [1]).

3. The Cohomology Cross Product

In this section we will consider cohomology with coefficients in a commutative ring Λ with unity. (The work can be generalized a bit to the case of two groups G_1 and G_2 of coefficients and a pairing $G_1 \otimes G_2 \to G$ to a third.)

Recall that we had a natural chain equivalence $\theta: \Delta_*(X \times Y) \to \Delta_*(X) \otimes \Delta_*(Y)$. Let $f \in \Delta^p(X; \Lambda)$ and $g \in \Delta^q(Y; \Lambda)$. Thus $f: \Delta_p(X) \to \Lambda$ and $g: \Delta_q(Y) \to \Lambda$. Then

$$f \otimes g: \Delta_p(X) \otimes \Delta_q(Y) \to \Lambda \otimes \Lambda \to \Lambda$$

is defined using the ring structure of Λ. We define the cohomology cross product

$$f \times g \in \Delta^{p+q}(X \times Y; \Lambda)$$

by

$$\boxed{f \times g = (f \otimes g) \circ \theta.}$$

That is, if $\theta(c) = \sum_{p+q=n} (\sum_i a_i^p \otimes b_i^q)$, then $(f \times g)(c) = (f \otimes g)\theta(c) = \sum_i (-1)^{pq} \cdot f(a_i^p) g(b_i^q)$. (One regards f as zero on $\Delta_i(\cdot)$ for $i \neq p$, etc.) We compute

$$\begin{aligned}
\delta(f \times g) &= (-1)^{p+q+1}(f \times g) \circ \partial \\
&= (-1)^{p+q+1}(f \otimes g \circ \theta \circ \partial) \\
&= (-1)^{p+q+1}(f \otimes g \circ \partial_\otimes \circ \theta) \quad (\text{where } \partial_\otimes = 1 \otimes \partial + \partial \otimes 1) \\
&= (-1)^{p+q+1}[f \otimes g\partial + (-1)^q f\partial \otimes g] \circ \theta \\
&= (-1)^{p+q+1}[(-1)^{q+1} f \otimes \delta g + (-1)^{q+p+1} \delta f \otimes g] \circ \theta \\
&= (\delta f \otimes g) \circ \theta + (-1)^p (f \otimes \delta g) \circ \theta \\
&= \delta f \times g + (-1)^p f \times \delta g,
\end{aligned}$$

giving the coboundary formula

$$\boxed{\delta(f \times g) = \delta f \times g + (-1)^{\deg(f)} f \times \delta g.}$$

It follows that $\times$ induces a product

$$\times: H^p(X; \Lambda) \otimes H^q(Y; \Lambda) \to H^{p+q}(X \times Y; \Lambda).$$

Note that the fact that θ is unique up to chain homotopy implies that this product does not depend on the choice of θ. (The reader should detail this.)

There is another, simpler product called the "Kronecker product" that takes

$$\boxed{H^p(X; G) \otimes H_p(X) \to G.}$$

If $\alpha = [\![f]\!] \in H^p(X; G)$ and $\gamma = [\![c]\!] \in H_p(X)$ then this product is defined by

$$\boxed{\langle \alpha, \gamma \rangle = f(c) \in G.}$$

This is also denoted by $\alpha(\gamma)$. It is clear that $\langle \alpha, \gamma \rangle = \beta(\alpha)(\gamma)$ where $\beta: H^p(X; G) \to \operatorname{Hom}(H_p(X), G)$ is the map appearing in the Universal Coefficient Theorem. The Kronecker product is sometimes referred to as evaluation of a cohomology class on a homology class. It is a special case of the "cap product" which we shall study later. We will need the following formula.

3.1. Proposition. *For cocycles f, g and cycles a, b of X, Y, respectively, we have*

$$(f \times g)(a \times b) = (-1)^{\deg(g)\deg(a)} f(a)g(b).$$

PROOF. $(f \times g)(a \times b) = (f \otimes g)\theta(a \times b) = (f \otimes g)\theta \times (a \otimes b)$. Using that $\theta \circ \times = 1 + D\partial_\otimes + \partial D$ this is $(f \otimes g)(a \otimes b) + (f \otimes g)(D\partial_\otimes + \partial D)(a \otimes b) = (f \otimes g)(a \otimes b) \pm \delta(f \otimes g)(D(a \otimes b)) = (f \otimes g)(a \otimes b) = (-1)^{\deg(g)\deg(a)} f(a)g(b)$ by our sign conventions. Note that $f(a) = 0$ if $\deg(f) \neq \deg(a)$. □

For any space X, we let $1 \in H^0(X)$ denote the class of the augmentation cocycle $\epsilon: \Delta_0(X) \to \mathbf{Z}$ taking each 0-simplex to 1. The reason for this notation will become apparent momentarily. Clearly $f^*(1) = 1$ for any map $f: X \to Y$.

For a one-point space P, easy acyclic model arguments show that the composition

$$\Delta_*(X \times P) \xrightarrow{\theta} \Delta_*(X) \otimes \Delta_*(P) \xrightarrow{1 \otimes \epsilon} \Delta_*(X) \otimes \mathbf{Z} \xrightarrow{\approx} \Delta_*(X)$$

is naturally chain homotopic to the chain map $\Delta_*(X \times P) \xrightarrow{\approx} \Delta_*(X)$ induced by the projection $p_X: X \times P \xrightarrow{\approx} X$. The first of these induces $\alpha \mapsto \alpha \times 1$ of $H^k(X) \to H^k(X \times P)$. By naturality of the cross product applied to the map $X \times Y \to X \times P$, it follows that

$$\boxed{\alpha \times 1 = p_X^*(\alpha) \in H^*(X \times Y)}$$

for any $\alpha \in H^*(X)$. Similarly,

$$\boxed{1 \times \beta = p_Y^*(\beta) \in H^*(X \times Y)}$$

for any $\beta \in H^*(Y)$.

We will now discuss the cross product for relative cohomology. As above, coefficients for cohomology will be in the ring Λ throughout, and will be suppressed. Suppose that $A \subset X$ and let $f \in \Delta^p(X, A) = \ker(\Delta^p(X) \to \Delta^p(A))$. That is, $f \in \Delta^p(X)$ and is zero on A. The following diagram commutes:

$$\begin{array}{ccc} \Delta_*(A \times Y) & \xrightarrow{\theta} & \Delta_*(A) \otimes \Delta_*(Y) \\ \downarrow & & \downarrow \\ \Delta_*(X \times Y) & \xrightarrow{\theta} & \Delta_*(X) \otimes \Delta_*(Y), \end{array}$$

and the vertical maps are monomorphisms. If $c \in \Delta_*(A \times Y)$ then $(f \times g)(c) = (f \otimes g)\theta(c) = 0$ since f is zero on A, and so $f \times g \in \Delta^*(X \times Y, A \times Y) =$

$\Delta^*((X, A) \times Y)$. Thus we get a relative product and commutative diagram:

$$\begin{array}{ccc} H^p(X,A) \otimes H^q(Y) & \xrightarrow{\times} & H^{p+q}((X,A) \times Y) \\ \downarrow & & \downarrow \\ H^p(X) \otimes H^q(Y) & \xrightarrow{\times} & H^{p+q}(X \times Y). \end{array}$$

Consider the diagram:

$$\begin{array}{ccc} H^p(A) \otimes H^q(Y) & \xrightarrow{\times} & H^{p+q}(A \times Y) \\ \downarrow{\scriptstyle \delta^* \otimes 1} & & \downarrow{\scriptstyle \delta^*} \\ H^{p+1}(X,A) \otimes H^q(Y) & \xrightarrow{\times} & H^{p+q+1}((X,A) \times Y). \end{array}$$

We claim that this commutes. We will chase the diagram with representatives $f \in \Delta^p(A)$, $g \in \Delta^q(Y)$ with $\delta f = 0 = \delta g$. Extend f to a cochain $f' \in \Delta^p(X)$. Then $f' \times g$ extends $f \times g$. Since $\delta f'$ is zero on A, $\delta f' \in \Delta^{p+1}(X, A)$. Going down and then right in the diagram yields a class represented by $(\delta f') \times g$. Going right then down, gives a class represented by $\delta(f' \times g)$. Since $\delta g = 0$, these are equal, as was to be shown.

For the case of a relative group in Y we get the diagram

$$\begin{array}{ccc} H^p(X) \otimes H^q(B) & \xrightarrow{\times} & H^{p+q}(X \times B) \\ \downarrow{\scriptstyle 1 \otimes \delta^*} & & \downarrow{\scriptstyle \delta^*} \\ H^p(X) \otimes H^{q+1}(Y,B) & \xrightarrow{\times} & H^{p+q+1}(X \times (Y,B)) \end{array}$$

which does not quite commute. Instead the two directions of travel differ by the sign $(-1)^p$.

The formulas $\alpha \times 1 = p_X^*(\alpha)$ and $1 \times \beta = p_Y^*(\beta)$ also hold for relative classes α and β.

Now let us consider the question of commutativity of the cross product. That is, the relation between $\alpha \times \beta$ and $\beta \times \alpha$.

Let $T: X \times Y \to Y \times X$ be $(x, y) \mapsto (y, x)$ and consider the (noncommutative) diagram

$$\begin{array}{ccc} \Delta_*(X \times Y) & \xrightarrow{\theta_{X,Y}} & \Delta_*(X) \otimes \Delta_*(Y) \\ {\scriptstyle T_\Delta}\downarrow{\scriptstyle \approx} & & \uparrow{\scriptstyle \tau} \\ \Delta_*(Y \times X) & \xrightarrow{\theta_{Y,X}} & \Delta_*(Y) \otimes \Delta_*(X) \end{array}$$

where $\tau(b^q \otimes a^p) = (-1)^{pq} a^p \otimes b^q$. Note that τ is a chain map. By Theorem 1.3, there is a chain homotopy $\tau \circ \theta_{Y,X} \circ T_\Delta \simeq \theta_{X,Y}$. That is, there is a D so that,

$$\tau \theta_{Y,X} T_\Delta - \theta_{X,Y} = D\partial + \partial_\otimes D.$$

Let $\delta f^p = 0 = \delta g^q$. That is, $f \circ \partial = 0 = g \circ \partial$. Note that $(g \otimes f)\circ(D\partial + \partial_\otimes D) = (g \otimes f)\circ D\partial = \pm\,\delta((g \otimes f)\circ D)$ is a coboundary. Now we compute

$$\begin{aligned} T^*([\![g]\!] \times [\![f]\!]) &= T^*([\![g \times f]\!]) = T^*([\![(g \otimes f)\circ \theta_{Y,X}]\!]) \\ &= [\![(g \otimes f)\circ \theta_{Y,X} \circ T_\Delta]\!] \\ &= (-1)^{pq}[\![(f \otimes g)\tau \theta_{Y,X} T_\Delta]\!] \quad \text{(using commutativity of } \Lambda) \\ &= (-1)^{pq}[\![(f \otimes g)\theta_{X,Y}]\!] = (-1)^{pq}[\![f \times g]\!] \\ &= (-1)^{pq}[\![f]\!] \times [\![g]\!]. \end{aligned}$$

Thus we conclude that for cohomology classes $\alpha \in H^p(X)$ and $\beta \in H^q(Y)$, we have

$$\boxed{\alpha \times \beta = (-1)^{pq} T^*(\beta \times \alpha).}$$

In general there is no Künneth Theorem for the cohomology cross product although one can prove such a result under conditions of finite type for the spaces involved. We need only the case of a field of coefficients (for Section 12), and that is easy:

3.2. Theorem. *Let Λ be a field. Then the cross product*

$$(H_*(X;\Lambda) \otimes_\Lambda H_*(Y;\Lambda))_n \to H_n(X \times Y;\Lambda)$$

is an isomorphism. If one of $H_*(X;\Lambda)$ or $H_*(Y;\Lambda)$ is of finite type then the cross product

$$(H^*(X;\Lambda) \otimes_\Lambda H^*(Y;\Lambda))^n \to H^n(X \times Y;\Lambda)$$

is an isomorphism.

Proof. The case of homology can be proved by a simpler version of the proof of Theorem 1.5 and that will be left to the reader. For the case of cohomology, there is the diagram (coefficients in Λ throughout)

$$\begin{array}{ccc} H^*(X) \otimes H^*(Y) & \xrightarrow{\quad\times\quad} & H^*(X \times Y) \\ \downarrow{\scriptstyle \beta \otimes \beta} & & \downarrow{\scriptstyle \beta} \\ \mathrm{Hom}(H_*(X),\Lambda) \otimes \mathrm{Hom}(H_*(Y),\Lambda) & & \mathrm{Hom}(H_*(X \times Y),\Lambda) \\ \downarrow{\scriptstyle \gamma} & \swarrow{\scriptstyle \mathrm{Hom}(\times,1)} & \\ \mathrm{Hom}(H_*(X) \otimes H_*(Y),\Lambda) & & \end{array}$$

which commutes by Proposition 3.1. The maps β and $\beta \otimes \beta$ are isomorphisms since $\mathrm{Hom}(\cdot,\Lambda)$ is exact over a field Λ. The map γ is an isomorphism by the assumption of finite type on one of the factors. $\mathrm{Hom}(\times,1)$ is an isomorphism by the case of homology. Thus the map on top, the cohomology cross product, is an isomorphism. □

It can be seen that γ, and hence $\times$, is injective for all fields Λ in all cases. It can also be seen that it is surjective *only* under the hypothesis of Theorem 3.2.

Problems

1. Show that $\alpha \times (\beta \times \gamma) = (\alpha \times \beta) \times \gamma$ for cohomology classes α, β, γ.
2. Fill in the details of the proof of the formula $\alpha \times 1 = p_X^*(\alpha)$.

4. The Cup Product

The most important product is the "cup product" with which we deal in this section.

4.1. Definition. Let $d: X \to X \times X$ be the diagonal map $d(x) = (x, x)$. Then the *cup product* is the homomorphism

$$\cup : H^p(X) \otimes H^q(X) \to H^{p+q}(X)$$

defined by $\alpha \cup \beta = d^*(\alpha \times \beta)$. (Coefficients are in any commutative ring with unity.)

An immediate consequence of the rules for the cross product is that

$$\alpha \cup \beta = (-1)^{pq} \beta \cup \alpha,$$

where p and q are the degrees of α and β. Also the cup product is natural in X.

The cup product is often denoted by juxtaposition, i.e., $\alpha\beta = \alpha \cup \beta$.

If $p_1: X \times X \to X$ is the projection to the first factor then

$$\alpha \cup 1 = d^*(\alpha \times 1) = d^* p_1^*(\alpha) = (p_1 d)^*(\alpha) = 1^*(\alpha) = \alpha.$$

Similarly, $1 \cup \alpha = \alpha$ so that $1 \in H^0(X)$ is a two sided unity element.

One can define the cup product on the cochain level by $f \cup g = d^\Delta(f \times g)$, i.e.,

$$(f \cup g)(c) = f \otimes g(\theta(d_\Delta(c)))$$

where $c \in \Delta_{p+q}(X)$, $f \in \Delta^p(X)$, $g \in \Delta^q(X)$ and where θ is the Eilenberg–Zilber map of Theorem 1.2. We will give a more explicit formula for this later.

The coboundary formula for the cross product immediately yields one for the cup product:

$$\boxed{\delta(f \cup g) = \delta f \cup g + (-1)^{\deg(f)} f \cup \delta g.}$$

By naturality of the cochain formula, if the cocycle f vanishes on A (meaning on singular simplices entirely in A), then so does $f \cup g$. Thus, if f vanishes on A and g vanishes on B then $f \cup g$ vanishes on A and on B, but not generally on $A \cup B$. But we know, by the discussion of subdivision, that the inclusion $\Delta_*(A) + \Delta_*(B) \hookrightarrow \Delta_*(A \cup B)$ induces an isomorphism in homology, and hence in cohomology, if A and B are open. In this case, the complex $\{f \in \Delta^*(X) \mid f(\sigma) = 0$ if σ is a simplex of A or a simplex of $B\}$ can be used to compute $H^*(X, A \cup B)$. Thus, in this case there is a cup product

$$\cup: H^p(X, A; \Lambda) \otimes H^q(X, B; \Lambda) \to H^{p+q}(X, A \cup B; \Lambda).$$

In particular, this holds if A and B are both open, or if one of them is contained in the other, e.g., if one of them is empty. (The latter case is immediate from the definition.)

Suppose given $\alpha_1, \alpha_2 \in H^*(X)$ and $\beta_1, \beta_2 \in H^*(Y)$. Let $d_X: X \to X \times X$, $d_Y: Y \to Y \times Y$ and $d_{X \times Y}: X \times Y \to X \times Y \times X \times Y$ be the diagonal maps. Let $T: X \times X \times Y \times Y \to X \times Y \times X \times Y$ be given by $T(x_1, x_2, y_1, y_2) = (x_1, y_1, x_2, y_2)$. Then $d_{X \times Y} = T \circ (d_X \times d_Y)$ giving

$$\begin{aligned}
(\alpha_1 \times \beta_1) \cup (\alpha_2 \times \beta_2) &= d^*_{X \times Y}(\alpha_1 \times \beta_1 \times \alpha_2 \times \beta_2) \\
&= (d_X \times d_Y)^* T^*(\alpha_1 \times \beta_1 \times \alpha_2 \times \beta_2) \\
&= (-1)^{\deg(\alpha_2)\deg(\beta_1)}(d_X \times d_Y)^*(\alpha_1 \times \alpha_2 \times \beta_1 \times \beta_2) \\
&= (-1)^{\deg(\alpha_2)\deg(\beta_1)} d^*_X(\alpha_1 \times \alpha_2) \times d^*_Y(\beta_1 \times \beta_2) \\
&= (-1)^{\deg(\alpha_2)\deg(\beta_1)}(\alpha_1 \cup \alpha_2) \times (\beta_1 \cup \beta_2)
\end{aligned}$$

and so we have the formula

$$\boxed{(\alpha_1 \times \beta_1) \cup (\alpha_2 \times \beta_2) = (-1)^{\deg(\alpha_2)\deg(\beta_1)}(\alpha_1 \cup \alpha_2) \times (\beta_1 \cup \beta_2).}$$

In particular, for the projections $p_X: X \times Y \to X$ and $p_Y: X \times Y \to Y$, we have

$$p^*_X(\alpha) \cup p^*_Y(\beta) = (\alpha \times 1) \cup (1 \times \beta) = (\alpha \cup 1) \times (1 \cup \beta) = \alpha \times \beta,$$

so that the cross product can be recovered from the cup product.

We will now discuss more explicit cochain formulas for the cup product. The composition

$$\Delta = \theta \circ d_\Delta: \Delta_*(X) \to \Delta_*(X \times X) \to \Delta_*(X) \otimes \Delta_*(X)$$

is called a "diagonal approximation." Note that on 0-simplices $\Delta(x) = \theta(x, x) = x \otimes x$. Also note that $f \cup g = (f \otimes g) \circ \theta \circ d_\Delta = (f \otimes g) \circ \Delta$.

4.2. Definition. A *diagonal approximation* is a natural chain map

$$\Delta: \Delta_*(X) \to \Delta_*(X) \otimes \Delta_*(X),$$

such that $\Delta(x) = x \otimes x$ on 0-simplices x.

4.3. Theorem. *Any two diagonal approximations are naturally chain homotopic.*

PROOF. This is an easy application of the method of acyclic models which will be left to the reader as Problem 5. □

Thus, for computing the cup product of cohomology classes, any diagonal approximation will do. We will now describe a particular such approximation that makes computations on the cochain level relatively simple.

For a singular simplex $\sigma: \Delta_n \to X$ and for $p+q=n$, $0 \le p, q \le n$, we denote by

$$\sigma\rfloor_p: \Delta_p \to \Delta_n \to X$$

the "front p-face" of σ, which is the composition of σ with the inclusion $e_i \mapsto e_i$ of Δ_p in Δ_n. Similarly, we denote by

$$_q\lfloor\sigma: \Delta_q \to \Delta_n \to X$$

the "back q-face" of σ, which is the composition of σ with the inclusion $e_i \mapsto e_{n-q+i}$ of Δ_q in Δ_n.

4.4. Definition. The *Alexander–Whitney diagonal approximation* is generated by

$$\Delta\sigma = \sum_{p+q=n} \sigma\rfloor_p \otimes {_q\lfloor\sigma} \in \Delta_*(X) \otimes \Delta_*(X),$$

where $\sigma: \Delta_n \to X$.

This is clearly natural and has the right value on 0-simplices. We must check that it is a chain map. By naturality it suffices to do that on $\iota_n = [e_0, \ldots, e_n]$. The computation is straightforward but lengthy and will be omitted.

This diagonal approximation gives us the cup product formula

$$(f^p \cup g^q)(\sigma) = (f \otimes g)\Delta\sigma = (f \otimes g)\sum \sigma\rfloor_p \otimes {_q\lfloor\sigma} = (-1)^{pq} f(\sigma\rfloor_p)\cdot g({_q\lfloor\sigma}).$$

Note that $f \cup g \neq \pm g \cup f$ on the cochain level. In fact, the failure of commutativity on the cochain level turns out to yield some useful "operations" on cohomology classes. We will discuss that later in this chapter (Section 16).

For simplicial chains on an *ordered* simplicial complex and for $p+q=n$, the formula

$$\boxed{(f^p \cup g^q)(\sigma) = (-1)^{pq} f(\sigma\rfloor_p) g({_q\lfloor\sigma})}$$

becomes

$$\boxed{(f^p \cup g^q)\langle v_0, \ldots, v_n\rangle = (-1)^{pq} f\langle v_0, \ldots, v_p\rangle g\langle v_p, \ldots, v_n\rangle.}$$

We now list the major properties of the cup product on $H^*(\cdot)$.

4.5. Theorem. *The following are properties of the cup product on cohomology classes which hold whenever they make sense:*

(1) $\cup$ *is natural; i.e., if* $\lambda: X \to Y$ *then* $\lambda^*(\alpha \cup \beta) = \lambda^*(\alpha) \cup \lambda^*(\beta)$;

(2) $\alpha \cup 1 = \alpha = 1 \cup \alpha$;
(3) $\alpha \cup (\beta \cup \gamma) = (\alpha \cup \beta) \cup \gamma$;
(4) $\alpha \cup \beta = (-1)^{\deg(\alpha)\deg(\beta)} \beta \cup \alpha$; *and*
(5) $\delta^*(\alpha \cup i^*(\beta)) = \delta^*(\alpha) \cup \beta$ *where* $\alpha \in H^*(A)$, $\beta \in H^*(X)$, *and* $i^*: H^*(X) \to H^*(A)$ *and* $\delta^*: H^p(A) \to H^{p+1}(X, A)$ *are the homomorphisms in the exact sequence for* (X, A).

PROOF. We already have (1), (2) and (4). Item (3) follows easily from the Alexander–Whitney diagonal approximation or from the associativity formula for the cross product.

To prove (5), let $f \in \Delta^p(X)$ be such that $[\![f|_A]\!] = \alpha$ and let $g \in \Delta^q(X)$ represent β. Since g is a cocycle, $\delta g = 0$. Now $\alpha \cup i^*\beta$ is represented by $f|_A \cup g|_A = (f \cup g)|_A$. It follows that $\delta^*(\alpha \cup i^*\beta)$ is represented by $\delta(f \cup g) = (\delta f) \cup g$, but this also represents $\delta^*(\alpha) \cup \beta$. □

The graded group $H^*(X)$ together with the cup product on it is called the "cohomology ring" of X or the "cohomology algebra" of X.

Note that Theorem 4.5(5) means that $\delta^*: H^p(A) \to H^{p+1}(X, A)$ is an $H^*(X)$-module homomorphism, as are the other maps in the cohomology sequence of (X, A) by naturality of the cup product.

We will now do three explicit examples of computations with the cup product. The computations will be made much easier by using general results developed later on in the book, but the explicit arguments here should make the nature of the cup product more accessible. We will do some examples using simplicial homology, but we have not fully justified this usage, although that can be done. This is of no importance, the examples being just to make the ideas about cup products clearer. Computations are virtually never done this way in actual practice.

4.6. Example. This is an example of a computation made in simplicial homology. For the cup product to make sense on the cochain level in simplicial homology one must fix an ordering of the vertices of each simplex and apply the Alexander–Whitney formula only to chains written in terms of simplices in that order. Otherwise, it contains inconsistencies. The easiest way to do this is to order all the vertices. For example, we take the projective plane triangulated as in Figure VI-1.

We take the base ring $\Lambda = \mathbf{Z}_2$, so all cohomology will be in $H^*(\mathbf{P}^2; \mathbf{Z}_2)$. Let f be the 1-cochain marked with the 1's in the figure. That is, f takes the value 1 on the marked 1-simplices and the value 0 otherwise. Since every 2-simplex has either 0 or 2 of the 1's, and coefficients are mod 2, it follows that $\delta f = 0$. (For example, we compute $(\delta f)\langle 2,3,4\rangle = f(\partial\langle 2,3,4\rangle) = f(\langle 3,4\rangle - \langle 2,4\rangle + \langle 2,3\rangle) = 0 - 1 + 1 = 0$.)

Consider $f \cup f$. We compute $(f \cup f)\langle 2,4,6\rangle = f\langle 2,4\rangle \cdot f\langle 4,6\rangle = 1 \cdot 1 = 1$. Similarly, $(f \cup f)\langle 1,3,6\rangle = f\langle 1,3\rangle \cdot f\langle 3,6\rangle = 1 \cdot 0 = 0$. In fact, $f \cup f$ can be seen to vanish on all simplices except $\langle 2,4,6\rangle$. Now $f \cup f$ is a cocycle, as are all 2-cochains. However, it is not a coboundary since the coboundary

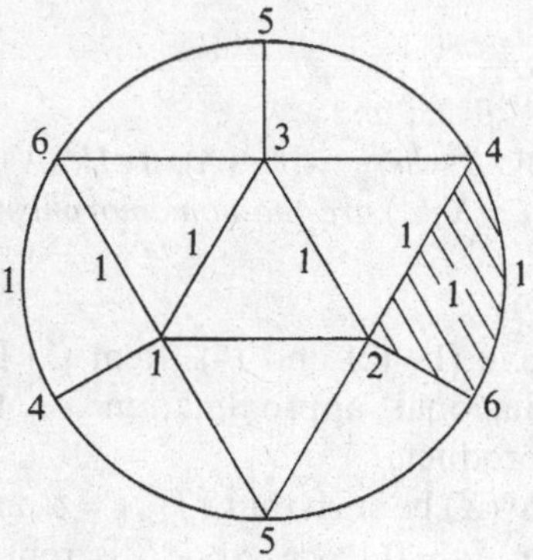

Figure VI-1. Triangulation and cochains on the projective plane.

of a 1-simplex takes value one on exactly two 2-simplices, and it follows that any coboundary takes value 1 on an even number of 2-simplices. Thus $0 \neq [\![f \cup f]\!] \in H^2(\mathbf{P}^2; \mathbf{Z}_2) \approx \mathbf{Z}_2$.

If we let $\alpha = [\![f]\!]$ then we get $\alpha^2 \neq 0$, so that $1, \alpha$, and α^2 generate the cohomology of $\mathbf{P}^2$. Thus, as a ring, $H^*(\mathbf{P}^2; \mathbf{Z}_2) \approx \mathbf{Z}_2[\alpha]/(\alpha^3)$, a "truncated polynomial ring" over $\mathbf{Z}_2$.

4.7. Example. Here we give an alternative derivation of the cohomology ring of $\mathbf{P}^2$ using singular, rather than simplicial, theory. Let f be a singular 1-cocycle representing a generator of the nonzero class $\alpha = [\![f]\!] \in H^1(\mathbf{P}^2; \mathbf{Z}_2) \approx \mathbf{Z}_2$. Note that the map $H^1(\mathbf{P}^2; \mathbf{Z}_2) \to \operatorname{Hom}(H_1(\mathbf{P}^2), \mathbf{Z}_2)$, of the Universal Coefficient Theorem, is an isomorphism since it is onto and both groups are $\mathbf{Z}_2$. Let λ be a loop representing the generator of $H_1(\mathbf{P}^2)$, which exists by the Hurewicz Theorem (Theorem 3.4 of Chapter IV). Then $f(\lambda) = 1$. From covering space theory, for example, we can take this loop to be the "semicircle" in Figure VI-2.

Let const_1 and const_2 denote a constant 1-simplex and 2-simplex, respectively, at the base point of λ. Then $f(\text{const}_1) = f(\partial\, \text{const}_2) = (\delta f)(\text{const}_2) = 0$.

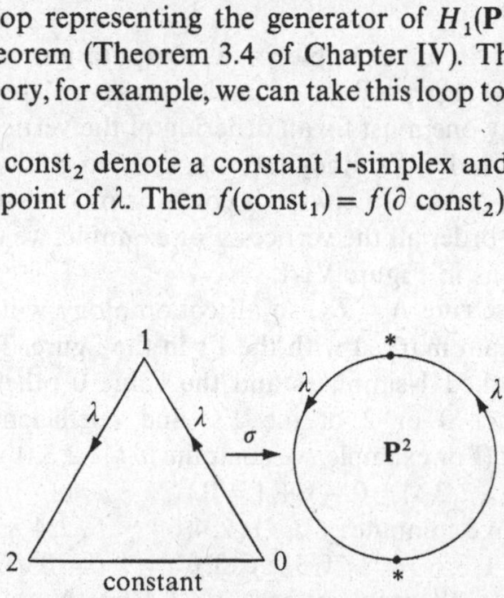

Figure VI-2. A singular 2-simplex in the projective plane.

Consider the 2-simplex σ illustrated in Figure VI-2. (It maps edges (0, 1) and (1, 2) along λ and (0, 2) by a constant map, and takes the interior of the standard 2-simplex homeomorphically onto $\mathbf{P}^2 - \mathbf{P}^1$.) Then $\partial\sigma = 2\lambda - \text{const}_1$. We compute

$$(f \cup f)(\sigma) = -f(\sigma|_{[0,1]}) \cdot f(\sigma|_{[1,2]}) = f(\lambda) \cdot f(\lambda) = 1.$$

Also

$$(f \cup f)(\text{const}_2) = f(\text{const}_1) \cdot f(\text{const}_1) = 0 \cdot 0 = 0.$$

We claim that $f \cup f$ is not a coboundary. If $f \cup f = \delta g$ then

$$\begin{aligned} 1 = (f \cup f)(\sigma) &= (\delta g)(\sigma) = g(\partial\sigma) = g(2\lambda - \text{const}_1) \\ &= 2g(\lambda) - g(\text{const}_1) = g(\text{const}_1) = g(\partial\, \text{const}_2) \\ &= (\delta g)(\text{const}_2) = (f \cup f)(\text{const}_2) = 0. \end{aligned}$$

This contradiction shows that $\alpha^2 = [f \cup f] \neq 0$, as we wished to prove.

4.8. Example. This is one more calculation using simplicial homology. The space will be the torus $\mathbf{T}^2$, and the coefficients for cohomology will be the integers $\mathbf{Z}$. Order the vertices of the torus as shown in Figure VI-3.

Let f be the 1-cochain corresponding to the 1's along the horizontal in Figure VI-3, i.e., it is 1 on any edge mapping to the middle edge by the horizontal projection. Similarly, let g be the 1-cochain along the middle vertical. It is easily checked that these are both cocycles and so they represent classes $\alpha = [\![f]\!]$ and $\beta = [\![g]\!]$ in $H^1(\mathbf{T}^2)$. Computations as in Example 4.6 show that $f \cup g\langle 5, 8, 9\rangle = 1$ and it is zero on all other 2-simplices. It is easy to see that the cohomology class of such a cocycle is a generator of $H^2(\mathbf{T}^2)$. Thus $\alpha \cup \beta$ is a generator. It is not hard to show directly that α and β form a free basis of $H^1(\mathbf{T}^2)$, but we will give an indirect argument for this using only the

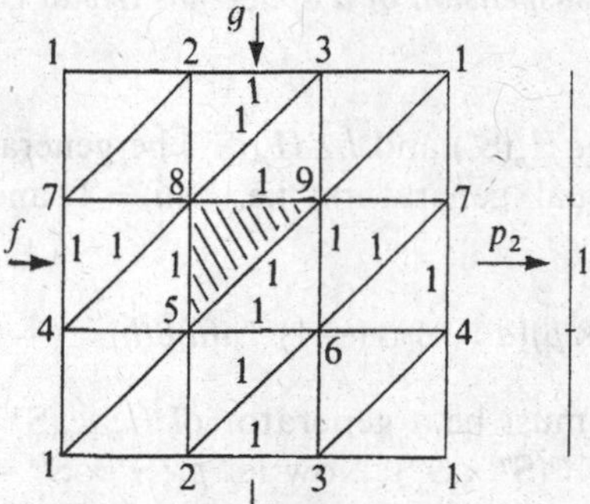

Figure VI-3. Triangulation and cochains on the torus.

fact that $\alpha \cup \beta$ is a generator. Note first that the signed commutative law for the cup product implies that for a cohomology class γ of *odd* degree, $\gamma^2 = -\gamma^2$. Since the cohomology of the torus is free abelian, this means that the square of any one-dimensional class is 0. (In particular, α and β cannot be equal since $\alpha\beta \neq 0$.) Suppose that u and v form a free basis of $H^1(\mathbf{T}^2)$. Then we can write

$$\alpha = au + bv,$$
$$\beta = cu + dv,$$

for integers a, b, c, d. We compute $\alpha\beta = (au + bv)(cu + dv) = acuu + aduv + bcvu + bdvv = (ad - bc)uv$. But, for this to be a generator of $H^2(\mathbf{T}^2) \approx \mathbf{Z}$, we must have that $ad - bc = \pm 1$. But that implies that α and β also form a basis.

It is worthwhile pointing out that $f = p_2^\Delta(c)$ and $g = p_1^\Delta(c)$ where c is a cocycle of $\mathbf{S}^1$ (1 on the "middle" 1-simplex) and the p_i are the projections of the torus to the circle. If $\gamma = [\![c]\!]$ then we deduce that $\alpha \cup \beta = p_2^*(\gamma) \cup p_1^*(\gamma) = (1 \times \gamma) \cup (\gamma \times 1) = -\gamma \times \gamma$.

We will illustrate a similar example using singular theory in Example 4.12.

4.9. Theorem. *Suppose that $X = U \cup V$ where U and V are open, acyclic sets. Then $\alpha \cup \beta = 0$ for all cohomology classes $\alpha, \beta \in H^*(X)$ of positive degree.*

PROOF. Consider the exact sequence $H^p(X, U) \to H^p(X) \to H^p(U)$ for the pair (X, U). If $\alpha \in H^p(X)$ then it maps to 0 in $H^p(U) = 0$, and so it comes from a class $\bar{\alpha} \in H^p(X, U)$. Similarly, $\beta \in H^q(X)$ maps to 0 in $H^q(V) = 0$, and so comes from a class $\bar{\beta} \in H^q(X, V)$. Then $\bar{\alpha} \cup \bar{\beta} \in H^{p+q}(X, U \cup V) = H^{p+q}(X, X) = 0$, and it maps to $\alpha \cup \beta$ via the homomorphism $H^{p+q}(X, U \cup V) \to H^{p+q}(X)$. □

4.10. Corollary. *The projective plane and the 2-torus cannot be written as the union of two open, acyclic sets.* □

4.11. Corollary. *The suspension of a space has trivial cup products in positive degrees.* □

4.12. Example. Let $a \in H_n(\mathbf{S}^n)$ and $b \in H_m(\mathbf{S}^m)$ be generators and let $\alpha \in H^n(\mathbf{S}^n)$ and $\beta \in H^m(\mathbf{S}^m)$ be dual generators, i.e., $\alpha(a) = 1$ and $\beta(b) = 1$. Then, by Proposition 3.1,

$$(\alpha \times \beta)(a \times b) = (-1)^{nm}\alpha(a)\beta(b) = (-1)^{nm}.$$

It follows that $a \times b$ must be a generator of $H_{n+m}(\mathbf{S}^n \times \mathbf{S}^m)$ and $\alpha \times \beta$ must be a generator of $H^{n+m}(\mathbf{S}^n \times \mathbf{S}^m)$. Now let $p_1: \mathbf{S}^n \times \mathbf{S}^m \to \mathbf{S}^n$ and $p_2: \mathbf{S}^n \times \mathbf{S}^m \to \mathbf{S}^m$ be the projections and put $u = p_1^*(\alpha)$ and $v = p_2^*(\beta)$. Then $uv = u \cup v = \alpha \times \beta$ generates $H^{n+m}(\mathbf{S}^n \times \mathbf{S}^m)$.

Now consider the case $n = m$. Here we can take $b = a$ and $\beta = \alpha$. The Künneth Theorem implies that $a \times ☼$ and $☼ \times a$ form a basis of $H_n(\mathbf{S}^n \times \mathbf{S}^n)$.

By Proposition 3.1 we have $(\alpha \times 1)(a \times ☼) = 1$, $(\alpha \times 1)(☼ \times a) = 0$ (because the degrees don't match), $(1 \times a)(a \times ☼) = 0$, and $(1 \times \alpha)(☼ \times a) = 1$. Then it follows from the Universal Coefficient Theorem (Theorem 7.2 of Chapter V) that $u = \alpha \times 1$ and $v = 1 \times \alpha$ form a basis of $H^n(\mathbf{S}^n \times \mathbf{S}^n)$.

4.13. Theorem. *With the notation of Example 4.12 and with $n = m$, let $f\colon \mathbf{S}^n \times \mathbf{S}^n \to \mathbf{S}^n \times \mathbf{S}^n$ be a map of degree ± 1, i.e., that induces an automorphism of $H^{2n}(\mathbf{S}^n \times \mathbf{S}^n)$. (For example, any homeomorphism works.) Let the integers a, b, c, d be defined by*

$$f^*(u) = au + bv,$$
$$f^*(v) = cu + dv.$$

If n is even then the coefficient matrix must be

$$\begin{pmatrix} a & b \\ c & d \end{pmatrix} = \begin{pmatrix} \pm 1 & 0 \\ 0 & \pm 1 \end{pmatrix} \quad or \quad \begin{pmatrix} 0 & \pm 1 \\ \pm 1 & 0 \end{pmatrix}$$

and all these can be realized by obvious maps.

PROOF. Since $\alpha^2 = 0$ we also have $u^2 = 0$ and similarly $v^2 = 0$. But then

$$0 = f^*(u^2) = (f^*(u))^2 = (au + bv)^2 = a^2u^2 + 2abuv + b^2v^2 = 2abuv.$$

Since $uv \neq 0$ we conclude that $ab = 0$. Similarly, $0 = f^*(v^2)$ implies that $cd = 0$. Also $\pm uv = f^*(uv) = (au + bv)(cu + dv) = (ad + bc)uv$, which implies that $ad + bc = \pm 1$. In case $a = 0$, we get $bc = \pm 1$, then $d = 0$ and $b, c = \pm 1$. Similarly, if $b = 0$ then $ad = \pm 1$, then $c = 0$ and $a, d = \pm 1$. Maps realizing these cases are reflections in the factors, reversal of the factors and compositions of these. □

We remark that if, in Theorem 4.13, n is 1, 3, or 7 then any matrix of determinant one can be achieved but for other odd n there are further restrictions; see Corollary 15.14 and Proposition 15.15.

4.14. Theorem. *For n even, $\mathbf{S}^n$ is not an H-space, i.e., there is no map $\mu\colon \mathbf{S}^n \times \mathbf{S}^n \to \mathbf{S}^n$ such that $\mu \circ i_1 \simeq 1$ and $\mu \circ i_2 \simeq 1$, where i_1 and i_2 are the inclusions $i_1(x) = (x, *)$ and $i_2(x) = (*, x)$.*

PROOF. Let $w \in H^n(\mathbf{S}^n)$ be a generator and let $u = p_1^*(w)$ and $v = p_2^*(w)$ where the p_i are the projections. Then $1 = (p_1 \circ i_1)^* = i_1^* p_1^*$ and it follows that $i_1^*(u) = w$ and that $i_2^*(v) = w$. Similarly, $i_1^*(v) = 0 = i_2^*(u)$.

Put $\mu^*(w) = au + bv$. Then $w = 1^*w = (\mu i_1)^*(w) = i_1^*\mu^*(w) = i_1^*(au + bv) = aw$. Thus $a = 1$. Similarly, we can show $b = 1$. Therefore, $\mu^*(w) = u + v$. However, $0 = \mu^*(w^2) = (u + v)^2 = 2uv \neq 0$, a contradiction finishing the proof. □

The cup product can be defined for more general coefficient groups. For example, the Alexander–Whitney diagonal approximation gives a cup product

$$H^p(X; G_1) \otimes H^q(X; G_2) \to H^{p+q}(X; G_1 \otimes G_2)$$

by putting $(f \cup g)(\sigma) = (-1)^{pq} f(\sigma\rfloor_p) \otimes g({}_q\lfloor\sigma)$. In particular, since $\mathbf{Z} \otimes \mathbf{Z}_k \approx \mathbf{Z}_k$, there is the cup product $H^p(X; \mathbf{Z}) \otimes H^q(X; \mathbf{Z}_k) \to H^{p+q}(X; \mathbf{Z}_k)$ and it coincides with the product obtained by first reducing mod k and then taking the cup product over the ring $\mathbf{Z}_k$.

PROBLEMS

1. Suppose that a space X can be covered by n acyclic open sets. Then show that the cup product of any n cohomology classes of positive degree is zero.
2. In the proof of Theorem 4.14, determine the crucial place(s) where the assumption that n is even is used.
3. Show that any map $\mathbf{S}^4 \to \mathbf{S}^2 \times \mathbf{S}^2$ must induce the zero homomorphism on $H_4(\cdot)$.
4. Find maps $\mathbf{S}^1 \times \mathbf{S}^1 \to \mathbf{S}^1 \times \mathbf{S}^1$ of degree one realizing all possible matrices of determinant one, as in Theorem 4.13.
5. Prove Theorem 4.3.
6. Verify that the Alexander–Whitney diagonal approximation Δ of Definition 4.4 is a chain map.
7. Let $\beta_0: H^i(X; \mathbf{Z}_p) \to H^{i+1}(X; \mathbf{Z})$ be the Bockstein homomorphism associated with the exact sequence $0 \to \mathbf{Z} \to \mathbf{Z} \to \mathbf{Z}_p \to 0$ where p is prime. For $a, b \in H^*(X; \mathbf{Z}_p)$, show that $\rho_p\beta_0(ab) = \beta_0(a)b + (-1)^{\deg(a)} a\beta_0(b)$ where ρ_p is reduction mod p. (*Hint*: If $f: \Delta_*(X) \to \mathbf{Z}$ represents a then $\delta f = pg$ for some cocycle $g: \Delta_*(X) \to \mathbf{Z}$ which then represents $\beta_0(a)$.)
8. Let $\beta: H^i(X; \mathbf{Z}_p) \to H^{i+1}(X; \mathbf{Z}_p)$ be the Bockstein homomorphism associated with the exact sequence $0 \to \mathbf{Z}_p \to \mathbf{Z}_{p^2} \to \mathbf{Z}_p \to 0$ where p is prime. For $a, b \in H^*(X; \mathbf{Z}_p)$, show that $\beta(ab) = \beta(a)b + (1)^{\deg(a)} a\beta(b)$. (*Hint*: Show that $\beta = \rho_p \circ \beta_0$ and apply Problem 7.)

5. The Cap Product

Throughout most of this section we take homology and cohomology with coefficients in a commutative ring Λ with unity, but we will suppress this in the notation. As usual we regard a p-cochain f to be defined but zero on i-simplices when $i \neq p$. Define the "cap product" on the chain–cochain level

$$\cap: \Delta^p(X) \otimes \Delta_n(X) \to \Delta_{n-p}(X)$$

by

$$f \cap c = (1 \otimes f)\Delta c,$$

where Δ is some diagonal approximation. If we use the Alexander–Whitney diagonal approximation, then this is $f^p \cap \sigma_{p+q} = (1 \otimes f)(\sum \sigma\rfloor_q \otimes {}_p\lfloor\sigma)$, i.e.,

$$\boxed{f^p \cap \sigma_{p+q} = (-1)^{pq} f({}_p\lfloor\sigma)\sigma\rfloor_q.}$$

5.1. Proposition. *Using the Alexander–Whitney diagonal approximation and with $\epsilon: \Delta_0(X) \to \Lambda$ being the augmentation (taking all 0-simplices to 1), the cap product has the following properties:*

(i) $\epsilon \cap c = c$;
(ii) $f \in \Delta^p(X)$, $c \in \Delta_p(X) \Rightarrow \epsilon(f \cap c) = f(c)$;
(iii) $(f \cup g) \cap c = f \cap (g \cap c)$;
(iv) $\lambda: X \to Y$, $f \in \Delta^p(Y)$, $c \in \Delta_n(X) \Rightarrow \lambda_\Delta(\lambda^\Delta(f) \cap c) = f \cap \lambda_\Delta c$; *and*
(v) $f \in \Delta^p(X) \Rightarrow \partial(f \cap c) = \delta f \cap c + (-1)^p f \cap \partial c$.

Proof. Parts (i) and (ii) are elementary. For part (iii) let $f \in \Delta^p(X)$, $g \in \Delta^q(X)$, and $\sigma \in \Delta_n(X)$. Let $n = p + q + r$. Then the left-hand side of (iii) is

$$\begin{aligned} \text{lhs} &= (-1)^{(p+q)r}(f \cup g)({}_{p+q}\lfloor\sigma)\sigma\rfloor_r \\ &= (-1)^{(p+q)r+pq} f(({}_{p+q}\lfloor\sigma)\rfloor_p) g({}_q\lfloor\sigma)\sigma\rfloor_r. \end{aligned}$$

The right-hand side of (iii) is

$$\begin{aligned} \text{rhs} &= f \cap (-1)^{q(p+r)} g({}_q\lfloor\sigma)\sigma\rfloor_{p+r} \\ &= (-1)^{q(p+r)+pr} f({}_p\lfloor(\sigma\rfloor_{p+r})) g({}_q\lfloor\sigma)\sigma\rfloor_r. \end{aligned}$$

Since ${}_p\lfloor(\sigma\rfloor_{p+r}) = ({}_{p+q}\lfloor\sigma)\rfloor_p$, this is the same as the left-hand side.

Part (iv) is easy and is left to the reader. For part (v), consider the diagram

$$\begin{array}{ccc} \Delta_n(X) \otimes \Delta_p(X) & \xrightarrow{1 \otimes f} & \Delta_n(X) \otimes \Lambda \approx \Delta_n(X) \\ \downarrow \partial \otimes 1 & & \downarrow \partial \\ \Delta_{n-1}(X) \otimes \Delta_p(X) & \xrightarrow{1 \otimes f} & \Delta_{n-1}(X) \otimes \Lambda \approx \Delta_{n-1}(X). \end{array}$$

Starting at the upper left with $c_n \otimes c_p$ and going to the right, we get $(-1)^{pn} c_n \otimes f(c_p)$, which is identified with $(-1)^{pn} f(c_p) c_n$. Going down with this gives $(-1)^{pn} f(c_p)\partial c_n$. Again from the upper left, but going down gives $(\partial c_n) \otimes c_p$. Taking this to the right gives $(-1)^{pn-p} f(c_p)\partial c_n$. Thus the diagram commutes up to the sign $(-1)^p$. Then we calculate

$$\begin{aligned} \partial(f \cap c) &= \partial((1 \otimes f)\Delta c) \\ &= (-1)^p (1 \otimes f)(\partial \otimes 1)\Delta c \\ &= (-1)^p (1 \otimes f)(\partial_\otimes - (1 \otimes \partial))\Delta c \\ &= (-1)^p \{(1 \otimes f)(\partial_\otimes \Delta c) - (1 \otimes f)(1 \otimes \partial)\Delta c\} \\ &= (-1)^p \{(1 \otimes f)\Delta(\partial c) + (-1)^p (1 \otimes \delta f)\Delta c\} \\ &= (\delta f) \cap c + (-1)^p f \cap (\partial c). \end{aligned}$$

□

The boundary formula (v) shows that this product induces the cap product in homology:

$$(*)\qquad \boxed{\cap : H^p(X) \otimes H_n(X) \to H_{n-p}(X).}$$

This is easily seen to be independent of the diagonal approximation Δ used to define it. Properties (i)–(iv) translated to (co)homology classes give:

5.2. Theorem. *The cap product* $(*)$ *satisfies the following properties:*

(1) $1 \cap \gamma = \gamma$;
(2) $\deg(\alpha) = \deg(\gamma) \Rightarrow \epsilon_*(\alpha \cap \gamma) = \alpha(\gamma) = \langle \alpha, \gamma \rangle$;
(3) $(\alpha \cup \beta) \cap \gamma = \alpha \cap (\beta \cap \gamma)$; *and*
(4) $\lambda : X \to Y$, $\alpha \in H^*(Y)$, $\beta \in H_*(X) \Rightarrow \lambda_*(\lambda^*(\alpha) \cap \beta) = \alpha \cap \lambda_*(\beta)$. □

Let us now discuss the cap product for relative (co)homology classes. If $A \subset X$ and c is a chain in A then $f \cap c$ is a chain in A. Thus there are induced cap products

$$\cap : \Delta^p(X) \otimes \Delta_n(X, A) \to \Delta_{n-p}(X, A)$$

and

$$\boxed{\cap : H^p(X) \otimes H_n(X, A) \to H_{n-p}(X, A).}$$

If $f \in \Delta^p(X, A)$, i.e., $f \in \Delta^p(X)$ and is zero on A, then for a chain c in X, $f \cap c$ is a chain in X which is unaffected by modification of c by a chain in A. Thus the cap product induces maps

$$\cap : \Delta^p(X, A) \otimes \Delta_n(X, A) \to \Delta_{n-p}(X),$$

and

$$\boxed{\cap : H^p(X, A) \otimes H_n(X, A) \to H_{n-p}(X).}$$

More generally, if A, B are subspaces such that the inclusion $\Delta_*(A) + \Delta_*(B) \hookrightarrow \Delta_*(A \cup B)$ induces an isomorphism in homology then there is a cap product

$$\boxed{\cap : H^p(X, A) \otimes H_n(X, A \cup B) \to H_{n-p}(X, B).}$$

In particular, this holds when either A or B is empty, or when they are both open.

The formulas of Theorem 5.2 remain valid for relative classes where they make sense.

5.3. Corollary. *The Kronecker product* $\langle \cdot, \cdot \rangle$ *of* $H^p(X) \otimes H_p(X) \to \Lambda$ *satisfies:*

(1) $\langle \alpha \cup \beta, \gamma \rangle = \langle \alpha, \beta \cap \gamma \rangle$; *and*
(2) $\langle f^*(\alpha), \gamma \rangle = \langle \alpha, f_*(\gamma) \rangle$.

Proof. These result from Theorem 5.2(3) and (4) upon applying ϵ_* and using 5.2(2). □

5.4. Theorem. *For $\alpha \in H^*(X)$, $\beta \in H^*(Y)$, $a \in H_*(X)$ and $b \in H_*(Y)$, we have*

$$(\alpha \times \beta) \cap (a \times b) = (-1)^{\deg(\beta)\deg(a)}(\alpha \cap a) \times (\beta \cap b).$$

Proof. Consider the following diagram, where τ is the signed interchange of the two middle factors and f, g are cocycles representing α, β:

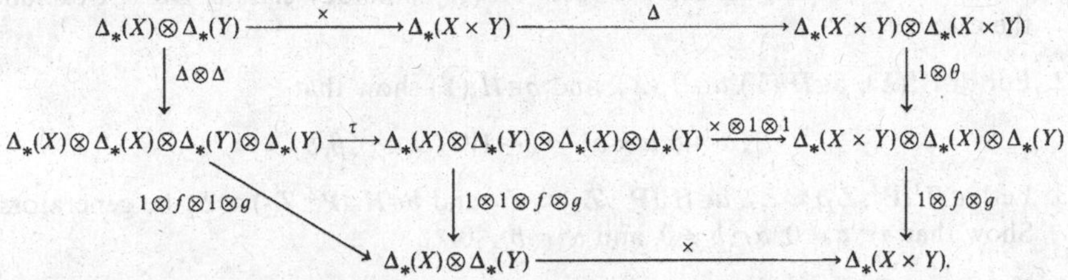

The triangle and square on the bottom commute. The top rectangle commutes up to chain homotopy by an acyclic model argument. The vertical map $1 \otimes f \otimes g$ on the bottom right is a chain map (up to the sign $(-1)^{\deg(f)+\deg(g)}$) when f and g are cocycles, as the reader can verify. Thus the composition from the upper left going down and then right to $\Delta_*(X \times Y)$ is chain homotopic to the composition going right and then down to $\Delta_*(X \times Y)$. Starting with the chain $a \otimes b$ at the upper left, the first of these compositions is

$$\begin{aligned} a \otimes b &\mapsto \Delta(a) \otimes \Delta(b) \mapsto (-1)^{\deg(a)\deg(g)}(f \cap a) \otimes (g \cap b) \\ &\mapsto (-1)^{\deg(a)\deg(g)}(f \cap a) \times (g \cap b). \end{aligned}$$

The other composition is

$$\begin{aligned} a \otimes b \mapsto a \times b \mapsto \Delta(a \times b) \mapsto (1 \otimes f \otimes g)(1 \otimes \theta)\Delta(a \times b) &= (1 \otimes (f \times g))\Delta(a \times b) \\ &= (f \times g) \cap (a \times b) \end{aligned}$$

and so, on the level of homology with a and b cycles, these two compositions induce the same homology class. □

It is easy to generalize the cap product to arbitrary coefficients

$$\cap : \Delta^p(X; G) \otimes \Delta_n(X) \to \Delta_{n-p}(X) \otimes G$$

via the definition

$$f^p \cap \sigma_{p+q} = (-1)^{pq} \sigma \rfloor_q \otimes f({}_p\lfloor \sigma)$$

giving the product

$$\cap : H^p(X; G) \otimes H_n(X) \to H_{n-p}(X; G)$$

and similarly with the various relative cap products. Even more generally we can define

$$\cap : \Delta^p(X; G_1) \otimes \Delta_n(X) \otimes G_2 \to \Delta_{n-p}(X) \otimes G_1 \otimes G_2$$

by

$$f^p \cap (\sigma_{p+q} \otimes g) = (-1)^{pq} \sigma\rfloor_q \otimes f({}_p\lfloor\sigma) \otimes g,$$

yielding the product

$$\cap : H^p(X; G_1) \otimes H_n(X; G_2) \to H_{n-p}(X; G_1 \otimes G_2).$$

Problems

1. For $\mathbf{S}^n \times \mathbf{S}^m$ compute all cap products on (co)homology classes. Do not exclude the case $n = m$.

2. For $\alpha \in H^p(X)$, $\beta \in H^q(Y)$, $a \in H_p(X)$, and $b \in H_q(Y)$ show that
$$\langle \alpha \times \beta, a \times b \rangle = (-1)^{pq} \langle \alpha, a \rangle \langle \beta, b \rangle.$$

3. Let $\alpha \in H^1(\mathbf{P}^2; \mathbf{Z}_2) \approx \mathbf{Z}_2$, $a \in H_1(\mathbf{P}^2; \mathbf{Z}_2) \approx \mathbf{Z}_2$ and $b \in H_2(\mathbf{P}^2; \mathbf{Z}_2) \approx \mathbf{Z}_2$ be generators. Show that $\alpha \cap a \neq 0$, $\alpha \cap b \neq 0$, and $\alpha^2 \cap b \neq 0$.

4. If β: $H^i(X; \mathbf{Z}_p) \to H^{i+1}(X; \mathbf{Z}_p)$ and β: $H_n(X; \mathbf{Z}_p) \to H_{n-1}(X; \mathbf{Z}_p)$ are the Bocksteins associated with the coefficient sequence $0 \to \mathbf{Z}_p \to \mathbf{Z}_{p^2} \to \mathbf{Z}_p \to 0$ and if $a \in H^i(X; \mathbf{Z}_p)$ and $c \in H_n(X; \mathbf{Z}_p)$ then show that $\beta(a \cap c) = \beta(a) \cap c + (-1)^{\deg(a)} a \cap \beta(c)$.

5. If $\beta_0 : H^i(X; \mathbf{Z}_p) \to H^{i+1}(X; \mathbf{Z})$ and β_0: $H_n(X; \mathbf{Z}_p) \to H_{n-1}(X; \mathbf{Z})$ are the Bocksteins associated with the coefficient sequence $0 \to \mathbf{Z} \to \mathbf{Z} \to \mathbf{Z}_p \to 0$ and if $a \in H^i(X; \mathbf{Z}_p)$ and $c \in H_n(X; \mathbf{Z})$ then show that $\beta_0(a \cap c) = \beta_0(a) \cap c$.

6. Classical Outlook on Duality ☼

In most of the remainder of this chapter we shall be concerned with the homological properties of manifolds. In this optional section we will describe the classical viewpoint on this matter. The modern approach, given in subsequent sections, is very powerful and provides easier proofs than does the classical approach, but this comes at a cost of forgoing much of the intuitive content of the latter. It is the intent of this section to provide that intuitive content, without going into details of either theorems or proofs. This section is not used elsewhere in this book and may be skipped with impunity. Details can be found in the classic textbook of Seifert and Threlfall [1].

Let M^n be a connected, compact, triangulated n-manifold. For an n-simplex σ_1 of M, any $(n-1)$-face τ of σ_1 is the face of exactly one other n-simplex σ_2. Then the boundary $\partial(\sigma_1 + \sigma_2)$ contains τ with coefficient ± 2 or 0. If σ_1 and σ_2 are oriented coherently then this coefficient is zero. (This can be taken as a definition.) If all the n-simplices of M can be oriented coherently then the sum c of all those simplices is an n-cycle and the n-cycles are precisely the multiples of c. If a coherent orientation is not possible then there can be no nonzero n-cycles, and M is called "nonorientable." Thus $H_n(M; \mathbf{Z}) \approx \mathbf{Z}$ if M is orientable and is zero otherwise. With $\mathbf{Z}_2$ coefficients, c is always an n-cycle and so $H_n(M; \mathbf{Z}_2) \approx \mathbf{Z}_2$. In case M has a boundary then c is an n-cycle

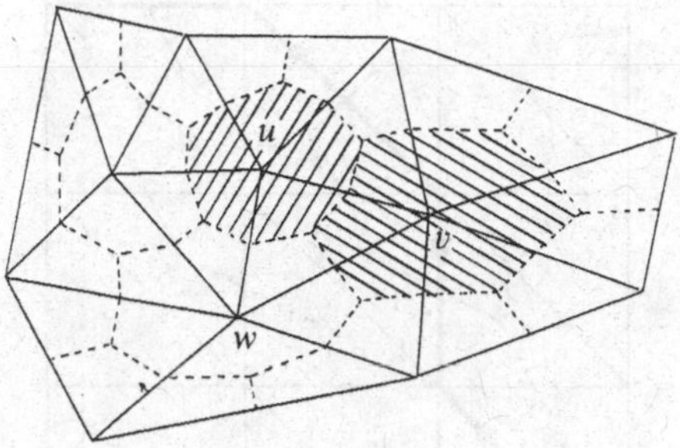

Figure VI-4. Dual cell structure on a manifold.

modulo ∂M; i.e., $H_n(M, \partial M) \approx \mathbf{Z}$ if M is orientable. The class $[\![c]\!] \in H_n(M, \partial M)$ is called the "fundamental homology class" of M, or the "orientation class" of M.

For simplicity in the rest of the discussion we will take $\mathbf{Z}_2$ coefficients, and so questions of orientability and sign are mute.

Consider the barycentric subdivision of M. The closed stars (in the subdivision) of the vertices of the *original* triangulation form a collection of n-cells. See Figure VI-4 where the irrelevant parts of the barycentric subdivision are suppressed and the boundaries of the dual cells are drawn in broken lines.

Two of these n-cells (stars) are hatched in the figure and they correspond to the two vertices u and v of M. The intersection of these two cells is an $(n-1)$-cell (a 1-cell in the figure, where it is made up of two edges of the barycentric subdivision). Consider the correspondence of vertices u, v of M with the n-cells $D(u)$, $D(v)$ of this "dual cell structure," the hatched 2-cells in the figure. To the 1-simplex $\langle u, v\rangle$ of M, we make correspond the $(n-1)$-cell $D(u) \cap D(v)$. Cell this $(n-1)$-cell $D(\langle u, v\rangle)$. Similarly, for a third vertex, w in the figure, we make the 2-simplex $\langle u, v, w\rangle$ correspond to $D(\langle u, v, w\rangle) = D(u) \cap D(v) \cap D(w)$, which is an $(n-2)$-cell (which is a single point, a vertex, of the dual cell structure in the figure).

Now consider, for example, the boundary $\partial\langle u, v\rangle = \langle v\rangle - \langle u\rangle$ of $\langle u, v\rangle$ as a 1-chain. How does this relate to $D(\langle u, v\rangle)$? The answer is that, if we consider $D(\langle u, v\rangle)$ not as a chain but as the cochain taking value 1 on the physical cell $D(\langle u, v\rangle)$, then its coboundary $\delta D(\langle u, v\rangle)$ takes value 1 on the n-cells $D(u)$ and $D(v)$. This can be written $\delta D(\langle u, v\rangle) = D(v) - D(u)$, which we cannot resist rewriting as $D(\partial\langle u, v\rangle)$. This is a general phenomenon and so it can be seen that we can assign, to each (simplicial) p-chain c of M, an $(n-p)$-cochain $D(c)$ (of the dual cell structure) such that $\delta D(c) = D(\partial c)$, and vice versa. But then D clearly induces an isomorphism

$$H_p(M) \approx H^{n-p}(M)$$

called "Poincaré duality."

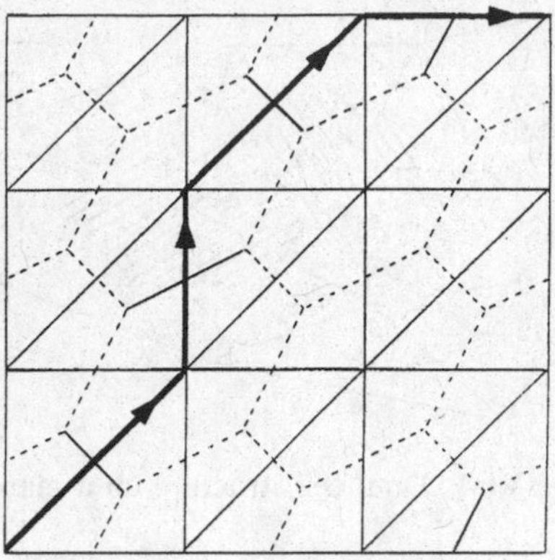

Figure VI-5. Dual cell structure on $\mathbf{T}^2$.

(Actually, cohomology was unknown in the classical period, and so this duality was expressed via homology alone using "intersection numbers." Thus, we are injecting a small bit of modernity into this discussion.)

One can elaborate on this. For example, for orientable manifolds M^n with boundary, one gets isomorphisms

$$H_p(M, \partial M) \approx H^{n-p}(M)$$

and

$$H_p(M) \approx H^{n-p}(M, \partial M).$$

Again in the absolute case, note that since a single vertex v generates $H_0(M)$, the dual generator $D(v)$ of $H^n(M)$ is a single n-cell; i.e., the cocycle taking value 1 on $D(v)$ and value 0 elsewhere.

Figure VI-5 illustrates a triangulation of the torus $\mathbf{T}^2$ and its dual cell subdivision. A 1-cycle, going from the bottom left to the top right, is shown in heavy ruling with arrows, and its dual 1-cocycle is indicated in solid ruling to distinguish it from the other 1-cells of the dual cell structure.

There is not much difficulty in providing the details of this discussion. What is hard is to show that the results are independent of the triangulation and to derive some of the more sophisticated applications of duality. For that, the modern approach is far preferable. Triangulations will not enter the picture in any way. Indeed, we will prove duality for topological manifolds which may not even be triangulable.

7. The Orientation Bundle

In this section we begin studying the special properties that (co)homology has for manifolds. The study starts by examining the nth homology group of an n-manifold and culminates, in Section 8, with the Poincaré–Alexander–Lefschetz Duality Theorem. Our treatment owes much to that of Dold [1]. The entire remainder of this chapter is devoted to applications of duality.

In this section M^n will denote a topological n-manifold. We make no assumptions of smoothness, connectedness, paracompactness, or anything else. (A manifold is, however, Hausdorff by definition.)

We start by defining and studying a notion of orientation for such spaces. For smooth manifolds we already have a notion of orientation, and the two will be related at the end of this section.

The notion of orientation is based on the fact that $H_n(M^n, M^n - \{x\}) \approx \mathbf{Z}$ for any point $x \in M^n$, as we see below.

Let $A \subset M^n$ be a closed set and let $x \in A$. Let G be any coefficient group and denote by

$$j_{x,A}: H_n(M, M - A; G) \to H_n(M, M - \{x\}; G),$$

the map induced from the inclusion.

7.1. Proposition. *If A is a compact, convex subset of $\mathbf{R}^n \subset M$ then $j_{x,A}$ is an isomorphism and both groups are isomorphic to G.*

PROOF. The set A is contained in the interior of some closed n-disk $D \subset \mathbf{R}^n \subset M$. Thus there is the commutative diagram (coefficients in G)

$$\begin{array}{ccc} H_n(M, M - A) & \longrightarrow & H_n(M, M - \{x\}) \\ \uparrow \approx & & \uparrow \approx \quad \text{(excision)} \\ H_n(\mathbf{R}^n, \mathbf{R}^n - A) & \longrightarrow & H_n(\mathbf{R}^n, \mathbf{R}^n - \{x\}) \\ \uparrow \approx & & \uparrow \approx \quad \text{(homotopy)} \\ H_n(D, \partial D) & \xrightarrow{=} & H_n(D, \partial D). \end{array}$$

The group on the bottom is $\mathbf{Z} \otimes G \approx G$ by Theorem 6.6 of Chapter IV. □

7.2. Definition. Let $\Theta_x \otimes G = H_n(M, M - \{x\}; G) \approx G$. For $G = \mathbf{Z}$ we just use Θ_x. Also let $\Theta \otimes G = \bigcup\{\Theta_x \otimes G \mid x \in M\}$ (disjoint) and let p: $\Theta \otimes G \to M$ be the function taking $\Theta_x \otimes G$ to x. Give $\Theta \otimes G$ the following topology: Let $U \subset M$ be open and $\alpha \in H_n(M, M - \bar{U}; G)$. Then for $x \in U, j_{x,\bar{U}}(\alpha) \in \Theta_x \otimes G$. Let $U_\alpha = \{j_{x,\bar{U}}(\alpha) \mid x \in U\} \subset p^{-1}(U) \subset \Theta \otimes G$. Take the U_α as a basis for the topology on $\Theta \otimes G$.

7.3. Proposition. *The sets U_α defined in Definition* 7.2 *are the basis of a topology. With this topology, p: $\Theta \otimes G \to M$, restricted to any component, is a covering map and fiberwise addition is continuous.*

PROOF. A point $\tilde{x} \in \Theta_x \otimes G = H_n(M, M - \{x\}; G)$ satisfies $\tilde{x} = j_{x,A}(\alpha)$ for some $A = \bar{U}$ which is convex in a euclidean neighborhood, by Proposition 7.1. Thus any point is in one of the prospective basis sets.

If $\tilde{x} \in U_\alpha \cap V_\beta$ then $\tilde{x} = j_{x,\bar{U}}(\alpha) = j_{x,\bar{V}}(\beta)$. Take $x \in W \subset U \cap V$, W a convex open set in a euclidean neighborhood of x. Then $j_{x,\bar{W}}$ is an isomorphism by Proposition 7.1. Let $\gamma \in H_n(M, M - \bar{W}; G)$ be such that $j_{x,\bar{W}}(\gamma) = \tilde{x}$. Then the homomorphism $H_n(M, M - \bar{U}; G) \to H_n(M, M - \bar{W}; G)$ must take α to γ since γ is the unique element going to $\tilde{x}$. Similarly, β must go to γ. This shows that $\tilde{x} \in W_\gamma \subset U_\alpha \cap V_\beta$ and hence that we do have a basis for a topology.

To show it is a covering when restricted to components, note that p is open and continuous by definition. Consider an open set U with $\bar{U}$ convex and compact in some euclidean subset of M. Consider the commutative diagram

$$\begin{array}{ccc} U \times H_n(M, M - \bar{U}; G) & \xrightarrow{\varphi} & p^{-1}(U) \\ & \underset{\text{proj}}{\searrow} \quad \underset{p}{\swarrow} & \\ & U & \end{array}$$

where $\varphi(x, \alpha) = j_{x,\bar{U}}(\alpha)$ and $H_n(M, M - \bar{U}; G)$ has the discrete topology. Then φ is open since, for $V \subset U$ open, φ takes $V \times \{\alpha\}$ onto V_α. If $\varphi(x, \alpha) = \varphi(y, \beta)$ then $x = y$ since $p(\varphi(x, \alpha)) = x$ and $p(\varphi(y, \beta)) = y$. Also $j_{x,\bar{U}}(\alpha) = j_{x,\bar{U}}(\beta)$ which implies that $\alpha = \beta$ because $j_{x,\bar{U}}$ is an isomorphism. Hence φ is one–one into.

Next, φ is onto since, for any $y \in U$, the map $j_{y,\bar{U}}: H_n(M, M - \bar{U}; G) \to \Theta_y \otimes G$ is onto. Therefore, for $V \subset U$ open, $\varphi^{-1}(V_\alpha) = V \times \{\alpha\}$. It follows that φ is continuous. Consequently, φ is a homeomorphism.

On a fiber (i.e., inverse image of a point in U) φ is just $j_{x,\bar{U}}$ which is an isomorphism. It follows that the fiberwise group operations correspond to the operations on the second factor of $U \times H_n(M, M - \bar{U}; G)$, and so they are continuous. □

7.4. Definition. For $A \subset M$ closed, the group of *sections over* A of $\Theta \otimes G$ is

$$\Gamma(A, \Theta \otimes G) = \{s: A \to \Theta \otimes G, \text{ continuous} \mid p \circ s = 1\}.$$

This is an abelian group under the operation $(s + s')(x) = s(x) + s'(x)$. Also let $\Gamma_c(A, \Theta \otimes G)$ be the subgroup consisting of sections with compact support, i.e., those sections with value 0 outside some compact set.

7.5. Definition. The topological n-manifold M is said to be *orientable along* A, where $A \subset M$ is closed, if there exists a section $\vartheta_A \in \Gamma(A, \Theta)$ which is a generator of each Θ_x for $x \in A$. The manifold M is said to be *orientable* if it is orientable along M. An *orientation along* A is such a section ϑ_A.

7.6. Proposition. *If M is an n-manifold, then the following conditions are equivalent:*

(1) *M is orientable.*
(2) *M is orientable along all compact subsets.*
(3) *The units in each fiber Θ_x of Θ form a trivial double cover.*
(4) *$\Theta \approx M \times \mathbf{Z}$ via a homeomorphism commuting with the projection to the base.*

Proof. For convenience we use the word "cover" in Proposition 7.6 to mean everything in the definition of a covering space except for the connectivity requirements.

Clearly (1)$\Rightarrow$(2). For (2)$\Rightarrow$(3) it is enough to treat the case in which M is connected. If the subspace of Θ formed by the units is not connected then it is clear that the units form two components each mapping homeomorphically to M. In the only other case, they form a connected double cover. This implies that there is a path λ in Θ from one unit in a fiber to the other unit in that fiber. The image of λ in M is compact and there is no section over it.

The implications (3)$\Rightarrow$(4) and (4)$\Rightarrow$(1) are obvious. $\square$

For $A \subset M$ closed, define the homomorphism

$$J_A: H_n(M, M-A; G) \to \Gamma_c(A, \Theta \otimes G)$$

by

$$J_A(\alpha)(x) = j_{x,A}(\alpha).$$

We must, of course, show that the section $J_A(\alpha)$ has compact support. To prove this, let $c \in \Delta_n(M)$ be a representative of α. Then c is a chain in some compact subset B of M. We claim that $J_A(\alpha)(x) = 0$ for $x \notin B$. To see this, note that c, as a chain in $\Delta_n(M, M-A)$, goes to zero in $\Delta_n(M, M-\{x\})$ since $B \subset M - \{x\}$. Thus α maps to zero in $H_n(M, M-\{x\})$ as claimed.

We must also show that $J_A(\alpha)$ is continuous. For this, again represent α by a chain c. Then ∂c is a chain in $M - A$ (since $\alpha \in H_n(M, M-A)$). Thus ∂c is a chain in some compact subset of $M - A$. Hence, for $x \in A$ there is an open neighborhood U of x such that ∂c is a chain in $M - \bar{U}$. Then c represents a class $\beta \in H_n(M, M-\bar{U})$. But then the section $\varphi(U \times \{\beta\})$ coincides with $J_A(\alpha)$ over $U \cap A$, where φ is as in the proof of Proposition 7.3, establishing continuity.

7.7. Proposition.

(1) *For $A \supset B$ both closed, the following diagram commutes*:

$$\begin{array}{ccc} H_n(M, M-A; G) & \longrightarrow & H_n(M, M-B; G) \\ \downarrow{\scriptstyle J_A} & & \downarrow{\scriptstyle J_B} \\ \Gamma_c(A, \Theta\otimes G) & \longrightarrow & \Gamma_c(B, \Theta\otimes G). \end{array}$$

(2) *For $A, B \subset M$ both closed, the sequence*

$$0 \to \Gamma_c(A \cup B, \Theta\otimes G) \xrightarrow{h} \Gamma_c(A, \Theta\otimes G) \oplus \Gamma_c(B, \Theta\otimes G) \xrightarrow{k} \Gamma_c(A \cap B, \Theta\otimes G)$$

is exact, where h is the sum of restrictions and k is the difference of restrictions.

(3) *If $A_1 \supset A_2 \supset \cdots$ are all compact and $A = \bigcap A_i$, then the restriction homomorphisms $\Gamma(A_i, \Theta\otimes G) \to \Gamma(A, \Theta\otimes G)$ induce an isomorphism*

$$\varinjlim \Gamma(A_i, \Theta\otimes G) \xrightarrow{\approx} \Gamma(A, \Theta\otimes G).$$

(*Note that $\Gamma_c = \Gamma$ in (3) since all the A_i are compact.*)

PROOF. Parts (1) and (2) are elementary. For part (3) suppose that s, s' are sections over some A_i which restrict to the same section over A. Since sections that coincide at a point do so on an open set about the point, there is an open set $U \supset A$ such that $s = s'$ on $U \cap A_i$. But, by compactness, there is a $j > i$ such that $A_j \subset U$. Then s and s' coincide on A_j, which means they give the same element in the direct limit. Thus the map in (3) is one–one.

To show that the map in (3) is onto, we must show that a section s over A extends to a neighborhood of A. For $x \in A$, there is a section $s_x \in \Gamma(U_x, \Theta \otimes G)$, where U_x is some neighborhood of x, and where $s = s_x$ on $U_x \cap A$. Cover A by a finite number of these sets, say $A \subset U_{x_1} \cup \cdots \cup U_{x_k}$. Let $U = \{y \mid \text{all } s_{x_i}(y) \text{ are equal for } y \in U_{x_i}\}$. This is open (because of the finite number of conditions) and contains A. □

7.8. Theorem. *Let M^n be a topological n-manifold and let $A \subset M^n$ be closed. Then:*

(a) $H_i(M, M - A; G) = 0$ *for* $i > n$; *and*
(b) $J_A: H_n(M, M - A; G) \to \Gamma_c(A, \Theta \otimes G)$ *is an isomorphism.*

PROOF. For a given manifold M^n, then for each closed subset A of M^n, let $P_M(A)$ be the statement that the conclusion of the theorem holds for A. Then P_M satisfies the following five properties (i) through (v):

(i) A compact and convex in some euclidean open set in $M^n \Rightarrow P_M(A)$.

This is immediate from Proposition 7.1 and its proof.

(ii) $P_M(A), P_M(B), P_M(A \cap B) \Rightarrow P_M(A \cup B)$.

This follows from the 5-lemma applied to the commutative diagram

$$\begin{array}{ccccccc}
\cdots \to H_{n+1}(M, M-(A\cap B)) & \to & H_n(M, M-(A\cup B)) & \to & H_n(M, M-A) \oplus H_n(M, M-B) & \to & H_n(M, M-(A\cap B)) \\
\| & & \downarrow J_{A\cup B} & & \downarrow J_A \oplus J_B \approx & & \downarrow J_{A\cap B} \approx \\
0 & \to & \Gamma_c(A\cup B, \Theta\otimes G) & \to & \Gamma_c(A, \Theta\otimes G) \oplus \Gamma_c(B, \Theta\otimes G) & \to & \Gamma_c(A\cap B, \Theta\otimes G),
\end{array}$$

in which the first row is a Mayer–Vietoris sequence from Theorem 18.2 of Chapter IV.

(iii) $A_1 \supset A_2 \supset \ldots$, all compact, $P_M(A_i)$ for all $i \Rightarrow P_M(\bigcap A_i)$.

Putting $A = \bigcap A_i$, this follows from the commutative diagram

$$\begin{array}{ccc}
\varinjlim H_p(M, M - A_i) & \xrightarrow{\approx} & H_p(M, M - A) \\
\downarrow \approx & & \downarrow \\
\varinjlim \Gamma_c(A_i, \Theta \otimes G) & \xrightarrow{\approx} & \Gamma_c(A, \Theta \otimes G)
\end{array}$$

where the top isomorphism comes from the fact that any chain is contained in a compact subset. (The isomorphism already holds on the chain level.) (For $p \neq n$, disregard the bottom of the diagram.)

(iv)

A_i compact with disjoint neighborhoods $N_i, P_M(A_i)$ all $i \Rightarrow P_M(\bigcup A_i)$.

This item follows from $H_p(M, M - \bigcup A_i) \approx H_p(\bigcup N_i, \bigcup N_i - \bigcup A_i) \approx \bigoplus H_p(N_i, N_i - A_i) \approx \bigoplus H_p(M, M - A_i)$ and the similar thing for the Γ_c. Let us call such unions with disjoint neighborhoods "separated unions."

(v)	$P_W(A \cap W)$ for all open, relatively compact $W \subset M \Rightarrow P_M(A)$.

This follows from the commutative diagram (disregard the bottom if $p \neq n$)

$$\begin{array}{ccc} \underset{\rightarrow}{\lim}_W H_p(W, W-(A\cap W)) & \xrightarrow{\approx} & H_p(M, M-A) \\ \downarrow \approx & & \downarrow \\ \underset{\rightarrow}{\lim}_W \Gamma_c(A\cap W, \Theta\otimes G) & \xrightarrow{\approx} & \Gamma_c(A, \Theta\otimes G). \end{array}$$

The theorem now follows from the next lemma. This lemma will be used to prove several other important results in subsequent sections.

7.9. Lemma (The Bootstrap Lemma). *Let $P_M(A)$ be a statement about compact sets A in a given n-manifold M^n. If* (i), (ii), *and* (iii) *hold, then $P_M(A)$ is true for all compact A in M^n.*

If M^n is separable metric, and $P_M(A)$ is defined for all closed sets A, and if (i), (ii), (iii), *and* (iv) *hold, then $P_M(A)$ is true for all closed sets A in M^n.*

For general M^n, if $P_M(A)$ is defined for all closed sets A in M, for all M^n, and if all five statements (i)–(v) *hold for all M^n, then $P_M(A)$ is true for all closed $A \subset M$ and all M^n.*

PROOF. In the first case, we get $P_M(A)$ for all finite unions of compact convex sets A_i in a given euclidean open set U by an inductive argument using (i), (ii) and the identity $A \cap (B_1 \cup \cdots \cup B_k) = (A \cap B_1) \cup \cdots \cup (A \cap B_k)$. Then (iii) implies it for all compact sets inside U. Repeating this argument without convexity, one gets $P_M(A)$ for all finite unions of "small" compact sets, "small" meaning "contained in some euclidean neighborhood." Finally, (iii) implies it for all compact sets, since any compact set is the intersection of a sequence of such finite unions. (Here we are using that a compact set in a manifold is separable metrizable.)

For the second case note that the one point compactification M_+ of M can be given a separable metric with distances bounded by 1. Let $f(x) = 1/\text{dist}(x, \infty)$ for $x \in M$. If $C \subset M$ is closed, let

$$A_i = C \cap f^{-1}[2i-2, 2i-1],$$

and

$$B_i = C \cap f^{-1}[2i-1, 2i].$$

Set $A = \bigcup A_i$ and $B = \bigcup B_i$, and note that these are separated unions. Then A_i and B_i are compact, so $P_M(A_i)$ and $P_M(B_i)$ are true. From (iv) it follows that $P_M(A)$ and $P_M(B)$ are true. But $A \cap B$ is also the separated union of the compact sets $A_i \cap B_j$. Therefore $P_M(A \cap B)$ is true. Now (ii) implies that $P_M(C) = P_M(A \cup B)$ is true.

For the last statement, note that an open, relatively compact set $W \subset M$ is separable metrizable and so $P_W(A \cap W)$ is true for all closed A and all such W by the second case of the lemma. Condition (v) then implies $P_M(A)$ for all closed $A \subset M$. □

Note that, for a given abelian group G and a given element $g \in G$, the following maps are natural in A (closed in a given M):

$$H_n(M, M-A) \approx H_n(M, M-A) \otimes \mathbf{Z} \to H_n(M, M-A) \otimes G \to H_n(M, M-A; G)$$

where the middle map is induced by the homomorphism $\mathbf{Z} \to G$ taking 1 to g. It follows that they induce a map $\Theta \to \Theta \otimes G$ commuting with the projection to M. On the fibers this is $\mathbf{Z} \to G$ taking 1 to g. (Note that we only know the additive structure on the fibers, so we cannot distinguish 1 from -1 in a fiber, or g from $-g$.) The (possibly disconnected) cover of a closed connected set A in M provided by the units ± 1 in the fibers of Θ maps to a similar set in $\Theta \otimes G$. Clearly, this map is a homeomorphism unless $g = -g$, in which case the two sheets of the first covering go to a single sheet of the second, which is a section of $\Theta \otimes G$ over A. This proves the following:

7.10. Proposition. *If $A \subset M^n$ is closed and connected then*

$$\Gamma(A, \Theta \otimes G) \approx \begin{cases} G & \text{if } M \text{ is orientable along } A, \\ {}_2G & \text{if } M \text{ is not orientable along } A, \end{cases}$$

where ${}_2G = \{g \in G \mid 2g = 0\}$. (Note that ${}_2G \approx G * \mathbf{Z}_2$.) □

7.11. Corollary. *If A is closed and connected in M^n then*

$$H_n(M, M-A; G) \approx \begin{cases} G & \text{if } A \text{ is compact and } M \text{ is orientable along } A, \\ {}_2G & \text{if } A \text{ is compact and } M \text{ is not orientable along } A, \\ 0 & \text{if } A \text{ is not compact.} \end{cases}$$ □

7.12. Corollary. *If M^n is connected then $H_i(M; G) = 0$ for $i > n$ and*

$$H_n(M; G) \approx \begin{cases} G & \text{if } M \text{ is compact and orientable,} \\ {}_2G & \text{if } M \text{ is compact and not orientable,} \\ 0 & \text{if } M \text{ is not compact.} \end{cases}$$ □

7.13. Corollary. *If M^n is connected then the torsion subgroup $TH_{n-1}(M)$ is $\mathbf{Z}_2$ if M is compact and not orientable and is 0 if M is noncompact or orientable.*

PROOF. By Example 7.6 of Chapter V we have the exact sequence

$$0 \to H_n(M) \otimes \mathbf{Q}/\mathbf{Z} \to H_n(M; \mathbf{Q}/\mathbf{Z}) \to TH_{n-1}(M) \to 0$$

where TA denotes the torsion subgroup of A. If M is noncompact then this sequence looks like $0 \to 0 \to 0 \to ? \to 0$, which takes care of that case. If M is compact and orientable the sequence looks like $0 \to \mathbf{Q}/\mathbf{Z} \to \mathbf{Q}/\mathbf{Z} \to ? \to 0$ and it is easy to show that any monomorphism $\mathbf{Q}/\mathbf{Z} \to \mathbf{Q}/\mathbf{Z}$ is onto, so the "?" is zero. If M is compact but not orientable then the sequence is $0 \to 0 \to {}_2(\mathbf{Q}/\mathbf{Z}) \to ? \to 0$. Clearly ${}_2(\mathbf{Q}/\mathbf{Z}) \approx \mathbf{Z}_2$. □

If M is a compact manifold then it can be shown that $H_*(M)$, and hence $H^*(M)$, is finitely generated. (See Appendix E.) For smooth manifolds, the simple discussion in Corollary E.5 suffices. The following result uses this.

7.14. Corollary. *If M^n is a compact connected n-manifold then*

$$H^n(M; G) \approx \begin{cases} G & \text{if } M \text{ is orientable,} \\ G/2G & \text{if } M \text{ is not orientable.} \end{cases}$$

PROOF. From the Universal Coefficient Theorem we have that

$$H^n(M; G) \approx \mathrm{Hom}(H_n(M), G) \oplus \mathrm{Ext}(H_{n-1}(M), G).$$

If M is orientable then $\mathrm{Hom}(H_n(M), G) \approx \mathrm{Hom}(\mathbf{Z}, G) \approx G$, and $H_{n-1}(M)$ is free abelian by Corollary 7.13, and hence $\mathrm{Ext}(H_{n-1}(M), G) = 0$ by Proposition 6.6 of Chapter V. If M is nonorientable then $H_n(M) = 0$ and $H_{n-1}(M)$ is the direct sum of a free abelian group and $\mathbf{Z}_2$, by Corollary 7.13. Thus $\mathrm{Hom}(H_n(M), G) = 0$ and $\mathrm{Ext}(H_{n-1}(M), G) \approx \mathrm{Ext}(\mathbf{Z}_2, G) \approx G/2G$. □

7.15. Theorem. *If M is a smooth manifold then it is orientable in the sense in which that was defined in terms of charts if and only it is orientable in the sense of this section.*

PROOF. The first sense of orientability is that one can choose an atlas so that all the transition functions have positive Jacobian determinants. Suppose that is the case. At a point $x \in M$ take the inverse ϕ of such a chart, i.e., a diffeomorphism $\phi: \mathbf{R}^n \to U$. We can assume that $\phi(0) = x$. Let $\vartheta \in H_n(\mathbf{R}^n, \mathbf{R}^n - \{0\}) \approx \mathbf{Z}$ be a generator, chosen once and for all. Then $\phi_*(\vartheta) \in H_n(U, U - \{x\}) = \Theta_x$ is a generator. Two such charts can be compared to a third one whose domain is contained in those of both charts. Thus suppose $\psi: \mathbf{R}^n \to V \subset U$ is another such chart with $\psi(0) = x$. Then $\phi^{-1}\psi$ is a diffeomorphism of $\mathbf{R}^n$ into itself, fixing the origin, and with positive Jacobian. We have seen (essentially in Lemma 16.3 of Chapter II) that such a map is isotopic to the identity. (It is isotopic to a linear map which is isotopic to the identity ⇔ its determinant is positive.) Such an isotopy gives a homotopy, to the identity, of the pair $(\mathbf{R}^n, \mathbf{R}^n - \{0\})$ mapping to itself. Thus the induced map on homology is the identity. Therefore the element $s_x = \phi_*(\vartheta) \in \Theta_x$ is independent of the choice of the chart ϕ. The section $s \in \Gamma(\Theta)$ is easily seen to be continuous by comparison to the map $\phi_*: H_n(\mathbf{R}^n, \mathbf{R}^n - \mathbf{D}^n) \to H_n(U, U - D)$ where $D \subset \phi(\mathbf{D}^n)$ is a small disk neighborhood of x.

On the other hand, if one has an orientation in the present sense, i.e., a section s of Θ, then at any point if one looks at any chart, either the element $\phi_*(\vartheta) \in \Theta_x$ given as above by the chart coincides with s_x or not. If it does not, then the chart which is just the composition with a reflection through a hyperplane does give a conforming element. Taking just those charts that do give the orientation section provides an atlas. By the previous remarks, the Jacobian of any change of variables within that atlas must have positive determinant and so this is an orientation of the old sort. □

PROBLEMS

1. If M^m and N^n are manifolds then show that $M \times N$ is orientable $\Leftrightarrow$ both M and N are orientable.
2. If M is a connected manifold such that $\pi_1(M)$ has no subgroups of index 2 then show that M is orientable.
3. If $T: \mathbf{R}^n \to \mathbf{R}^n$ is a map with $T^2 = 1$ then show that T has a fixed point. (*Hint*: Use the method of Section 20 of Chapter IV.)
4. ◆ Repeat Problem 3 replacing $T^2 = 1$ by $T^p = 1$ for some prime p.
5. For a connected nonorientable manifold M^n show that there exists a unique orientable double covering space of M^n.

8. Duality Theorems

In this section we prove one of the most important results in this book. It is a "duality theorem," generally going under the name "Poincaré Duality," which relates a homology group of a compact oriented manifold with the cohomology group in the complementary dimension. The main Duality Theorem (Theorem 8.3) contains generalizations of this due to Lefschetz and Alexander. An easy corollary will be a far-reaching generalization (Corollary 8.8) of the Jordan Curve Theorem. This section will be central to the remainder of this chapter.

Let M^n be an orientable n-manifold and let $\vartheta_M \in \Gamma(M, \Theta)$ be an orientation. For $K \subset M$ compact, ϑ_M restricts to $\vartheta_K \in \Gamma(K, \Theta) = \Gamma_c(K, \Theta) \approx H_n(M, M - K)$ and we will regard ϑ_K as lying in $H_n(M, M - K)$. We let $\vartheta = \{\vartheta_K\}$ be the collection of all these, and will call ϑ an orientation.

For closed sets $L \subset K \subset M$ define

$$\check{H}^p(K, L; G) = \varinjlim \{H^p(U, V; G) \mid (U, V) \supset (K, L),\ U,\ V \text{ open}\}.$$

This group appears to depend on the embedding of K in M and on the particular manifold M, but actually it doesn't. It is naturally isomorphic to what is called the "Čech cohomology group." If K and L are reasonably nice spaces such as euclidean neighborhood retracts (e.g., CW-complexes or topological manifolds) then it is known (see Corollary E.6) that this is naturally isomorphic to singular cohomology. In general, however, it is not. (See Dold [1] for details on these matters, which we shall not need.)

Suppose that $(K, L) \subset (U, V)$ as above. Then there is a cap product

$$\Delta^p(U, V; G) \otimes \left[\frac{\Delta_n(V) + \Delta_n(U-L)}{\Delta_n(U-K)}\right] \xrightarrow{\ \cap\ } \Delta_{n-p}(U-L, U-K; G)$$

given by $f \cap (b+c) = f \cap b + f \cap c = f \cap c$. Note that in homology $H_{n-p}(U-L,\ U-K) \approx H_{n-p}(M-L, M-K)$. Also $H_*((\Delta_*(V) + \Delta_*(U-L))/\Delta_*(U-K)) \approx H_*(U, U-K) \approx H_*(M, M-K)$ by excision, and since $\{V, U-L\}$ is an open cover of U. Thus, in (co)homology we get a cap product

$$H^p(U, V; G) \otimes H_n(M, M-K) \to H_{n-p}(M-L, M-K; G).$$

Also, this is natural in (K, L) (fulfilling the restrictions it must). In particular, one can cap with a class in $H_n(M, M-A)$ for some very large compact set A. On the (co)chain level this is given by the following: Let $f \in \Delta^p(U, V; G)$ and $\gamma \in H_n(M, M-A)$, and let γ be represented by the chain $b + c + d \in \Delta_n(V) + \Delta_n(U-L) + \Delta_n(M-K)$. Then

$$[\![f]\!] \cap \gamma = [\![f \cap (b+c+d)]\!] = [\![f \cap c]\!] \in H_{n-p}(M-L, M-K; G),$$

since $f \cap b = 0$ while $f \cap d$ is a chain in $M-K$.

Thus by capping with ϑ_A for A "very large," we get the homomorphism

$$\cap \vartheta : H^p(U, V; G) \to H_{n-p}(M-L, M-K; G),$$

compatible with changing A, and inclusion maps for (U, V). Thus in the direct limit we get a natural map

$$\boxed{\cap \vartheta : \check{H}^p(K, L; G) \to H_{n-p}(M-L, M-K; G).}$$

Note that, although we assumed M to be orientable, it is really only necessary for it to be orientable along K. Also, if we work over $\mathbf{Z}_2$ as the base ring, then there need be no orientability requirement.

In order to prove some things about this map, we give its (co)chain description. Let $\alpha \in \check{H}^p(K, L; G)$ be represented by the p-cocycle f of (U, V), where (U, V) is some open neighborhood of (K, L). Thus $f = 0$ on V and $\partial f = 0$ on U. Extend f to a cochain on all of M. Represent the orientation ϑ by a chain $a = b + c + d$ where b is a chain in V; c, in $U - L$; d, in $M - K$. Then $f \cap (b+c+d) = f \cap b + f \cap c + f \cap d$. But $f \cap b = 0$, and $f \cap d$ is a chain in $M - K$ so it doesn't matter. Thus $\alpha \cap \vartheta$ is represented by $f \cap c$. Note the special case for which $L = \emptyset$. In that case, we can take $V = \emptyset$ as well, and the chain b does not even enter into consideration.

8.1. Lemma. *The following diagram* (*arbitrary coefficients*) *has exact rows and* *commutes:*

$$\begin{array}{ccccccccc} \cdots \longrightarrow & \check{H}^p(K,L) & \longrightarrow & \check{H}^p(K) & \longrightarrow & \check{H}^p(L) & \longrightarrow & \check{H}^{p+1}(K,L) & \longrightarrow \cdots \\ & \downarrow & & \downarrow & & \downarrow & & \downarrow & \\ \cdots \to & H_{n-p}(M-L, M-K) & \to & H_{n-p}(M, M-K) & \to & H_{n-p}(M, M-L) & \to & H_{n-p-1}(M-L, M-K) & \to \cdots \end{array}$$

where all vertical maps are the cap products with the orientation ϑ.

Proof. The exactness of the top row follows from that of the direct limit functor; see Theorem D.4.

The only thing really in question is the last square (the connecting homomorphisms). Choose $f \in \Delta^p(M; G)$ such that $f|_V \in \Delta^p(V; G)$ represents $\alpha \in H^p(V; G)$ mapping to the class in $\check{H}^p(L; G)$ we wish to chase. Thus $\delta f = 0$ on V.

Represent ϑ by $a = b + c + d \in \Delta_n(V) + \Delta_n(U - L) + \Delta_n(M - K)$. This is the decomposition of a appropriate to the (K, L) pair. But note that $a = 0 + b + (c + d) \in \Delta_n(\varnothing) + \Delta_n(V - \varnothing) + \Delta_n(M - L)$ is the decomposition appropriate to the pair $(L, \varnothing)$. This shows that the same chain a can be used in the definition of both cap products under consideration. Note that, since ϑ is a class of $(M, M - K)$, ∂a must be a chain in $M - K$.

Now we do the chase starting with f. Going right gives δf. Then going down gives $\delta f \cap a$. On the other hand, taking f down first, gives $f \cap a$ and then going right gives $\partial(f \cap a) = (\delta f) \cap a \pm f \cap \partial a$. But $f \cap \partial a$ is a chain in $M - K$ and so it vanishes on passage to homology. □

8.2. Lemma. *Let K and L be compact subsets of the n-manifold M with the orientation ϑ. Then the diagram (arbitrary coefficients)*

$$\begin{array}{ccccccccc} \cdots \longrightarrow & \check{H}^p(K \cup L) & \longrightarrow & \check{H}^p(K) \oplus \check{H}^p(L) & \longrightarrow & \check{H}^p(K \cap L) & \xrightarrow{\delta^*} & \check{H}^{p+1}(K \cup L) & \to \cdots \\ & \downarrow & & \downarrow & & \downarrow & & \downarrow & \\ \cdots \to & H_{n-p}(M, M-(K \cup L)) & \to & H_{n-p}(M, M-K) \oplus H_{n-p}(M, M-L) & \to & H_{n-p}(M, M-(K \cap L)) & \xrightarrow{\partial_*} & H_{n-p-1}(M, M-(K \cup L)) & \to \cdots \end{array}$$

where the vertical maps are the cap products with ϑ, *commutes and has exact rows.*

Proof. Only the square involving the connecting homomorphisms is at issue. The top sequence is induced by the exact cochain sequences

$$\begin{aligned} 0 \to \operatorname{Hom}(\Delta_*(U) + \Delta_*(V), G) &\to \operatorname{Hom}(\Delta_*(U), G) \oplus \operatorname{Hom}(\Delta_*(V), G) \\ &\to \operatorname{Hom}(\Delta_*(U \cap V), G) \to 0 \end{aligned}$$

for open neighborhoods $U \supset K$ and $V \supset L$. Let $\alpha \in \check{H}^p(K \cap L; G)$ be represented by $f \in \Delta^p(M; G)$ where $\delta f = 0$ on $U \cap V$ for some such U, V. Then $\delta^*(\alpha) \in \check{H}^{p+1}(K \cup L; G)$ is obtained as follows. Pull f back to $\langle f, 0 \rangle \in \operatorname{Hom}(\Delta_*(U), G) \oplus \operatorname{Hom}(\Delta_*(V), G)$ and then take its coboundary $\langle \delta f, 0 \rangle$. Then the element $h \in \operatorname{Hom}(\Delta_*(U) + \Delta_*(V), G)$ defined by $h(u + v) = (\delta f)(u)$ represents $\delta^*(\alpha)$. We can extend h arbitrarily to $h \in \operatorname{Hom}(\Delta_*(M), G)$.

Now let ϑ be represented by the chain $a = b + c + d + e$ in $\Delta_*(U \cap V) + \Delta_*(U - L) + \Delta_*(V - K) + \Delta_*(M - (K \cup L))$. The last term e can be disregarded since it will produce chains in $M-(K\cup L)$ having no effect on the final result of the diagram chase. Then $\delta^*(\alpha)\cap\vartheta$ is represented by $h\cap(b+c+d)=h\cap(c+(b+d))=\delta f\cap c$.

On the other hand, the bottom sequence is induced by the exact chain sequence

$$0\to\frac{\Delta_*(M)}{\Delta_*(M-(K\cup L))}\to\frac{\Delta_*(M)}{\Delta_*(M-K)}\oplus\frac{\Delta_*(M)}{\Delta_*(M-L)}\to\frac{\Delta_*(M)}{\Delta_*(M-K)+\Delta_*(M-L)}$$
$$\to 0.$$

Thus $\alpha\cap\vartheta\in H_{n-p}(M,M-(K\cap L);G)$ is represented by $f\cap a$ modulo $\Delta_*(M-K;G)+\Delta_*(M-L;G)$. This pulls back to $\langle f\cap a,0\rangle$ whose boundary is $\langle\partial(f\cap a),0\rangle=\langle\delta f\cap a\pm f\cap\partial a,0\rangle=\langle\delta f\cap c+\delta f\cap d\pm f\cap\partial a,0\rangle$ which is equivalent to $\langle\delta f\cap c,0\rangle$ since $\delta f\cap d\pm f\cap\partial a\in\Delta_*(M-K)$. Therefore $\delta f\cap c$, modulo $\Delta_*(M-(K\cup L))$, represents $\partial_*(\alpha\cap\vartheta)$, proving the commutativity. □

8.3. Theorem (Poincaré–Alexander–Lefschetz Duality). *Let M^n be an n-manifold oriented by ϑ, and let $K\supset L$ be compact subsets of M. Then the cap product*

$$\cap\vartheta\colon \check{H}^p(K,L;G)\to H_{n-p}(M-L,M-K;G)$$

with the orientation class, is an isomorphism.

PROOF. By the 5-lemma and Lemma 8.1 it is enough to prove the theorem in the case $L=\varnothing$. That is, it suffices to show that

$$(*)\qquad \cap\vartheta\colon \check{H}^p(K)\to H_{n-p}(M,M-K)$$

is an isomorphism. (The coefficient group is immaterial and will be suppressed.)

The homomorphism (∗) is an isomorphism when K is a point, since for $p\neq 0$ both groups are zero, and for $p=0$, and for U a neighborhood of x, (∗) is induced by the map $H^0(U)\to H_n(M,M-\{x\})$ taking $1\mapsto 1\cap\vartheta_{\{x\}}=\vartheta_{\{x\}}$.

For compact sets $K\subset M$, let $P_M(K)$ be the statement that, "The homomorphism (∗) is an isomorphism for K for all p." Then it suffices to show that P satisfies the conditions (i), (ii), and (iii) of the Bootstrap Lemma (Lemma 7.9).

For property (i), let K be a convex compact subset of a euclidean open set and let $x\in K$. Then (i) follows from the diagram

$$\begin{array}{ccc}\check{H}^p(K) & \xrightarrow{\cap\vartheta} & H_{n-p}(M,M-K)\\ \downarrow\approx & & \downarrow\approx\\ \check{H}^p(\{x\}) & \xrightarrow[\approx]{\cap\vartheta} & H_{n-p}(M,M-\{x\}).\end{array}$$

Property (ii) follows from Lemma 8.2 and the 5-lemma.

For property (iii), Let $K_1 \supset K_2 \supset \cdots$ be a decreasing sequence of compact sets, all of which satisfy $P_M(\cdot)$, and let $K = \bigcap K_i$. For each i let $U_{i,j}$ be a fundamental system of open neighborhoods of K_i. Because K_i has a metrizable neighborhood, we can use a sequence here, say the $1/j$-neighborhood, but this is only for notational convenience and is not important.) Again, for notational convenience, we can redefine these sets so that for any j, $U_{1,j} \supset U_{2,j} \supset \cdots$, either by using a metric as above or merely noting that we can inductively intersect the original sets with the previous items in this list. Then the $U_{i,j}$ form a fundamental set of neighborhoods of K, so

$$\varinjlim_i \check{H}^p(K_i) = \varinjlim_i \lim_j H^p(U_{i,j}) \xrightarrow{\approx} \varinjlim_{i,j} H^p(U_{i,j}) = \check{H}^p(K)$$

by Theorem D.5. The commutative diagram

$$\begin{array}{ccc} \varinjlim_i \check{H}^p(K_i) & \xrightarrow[\approx]{} & \varinjlim_i H_{n-p}(M, M - K_i) \\ \downarrow\approx & & \downarrow\approx \\ \check{H}^p(K) & \longrightarrow & H_{n-p}(M, M - K) \end{array}$$

then proves property (iii). □

Taking $K = M$, compact, in the diagram of Lemma 8.1, gives the following:

8.4. Corollary (Poincaré–Lefschetz Duality). *If M^n is a compact orientable n-manifold and $L \subset M^n$ is closed, then we have the following diagram with exact rows and all verticals (cap products with the orientation class) being isomorphisms:*

$$\begin{array}{ccccccccc} \cdots \longrightarrow & \check{H}^p(M, L) & \longrightarrow & \check{H}^p(M) & \longrightarrow & \check{H}^p(L) & \longrightarrow & \check{H}^{p+1}(M, L) & \longrightarrow \cdots \\ & \downarrow\approx & & \downarrow\approx & & \downarrow\approx & & \downarrow\approx & \\ \cdots \to & H_{n-p}(M - L) & \to & H_{n-p}(M) & \to & H_{n-p}(M, M - L) & \to & H_{n-p-1}(M - L) & \to \cdots. \end{array}$$

(The isomorphism involving M alone is called "Poincaré Duality.") This holds with arbitrary coefficients, and M^n need not be orientable for $\mathbf{Z}_2$ as base ring.

□

8.5. Corollary. *If L is a proper compact subset of an orientable connected n-manifold M^n then $\check{H}^n(L; G) = 0$ for any coefficient group G.*

PROOF. It is isomorphic to $H_0(M, M - L; G) = 0$, since $M - L \neq \emptyset$. □

8.6. Corollary (Alexander Duality). *If A is a compact subset of $\mathbf{R}^n$ then*

$$\tilde{H}_q(\mathbf{R}^n - A; G) \approx \check{H}^{n-q-1}(A; G).$$

PROOF. $\check{H}^{n-q-1}(A; G) \approx H_{q+1}(\mathbf{R}^n, \mathbf{R}^n - A; G) \approx \tilde{H}_q(\mathbf{R}^n - A; G)$ by the reduced homology sequence of $(\mathbf{R}^n, \mathbf{R}^n - A)$. □

8.7. Corollary (Alexander Duality). *If $A \neq \varnothing$ is a closed subspace of $\mathbf{S}^n$ then*

$$\tilde{H}_q(\mathbf{S}^n - A; G) \approx \tilde{\check{H}}^{n-q-1}(A; G).$$

PROOF. We have $\check{H}^{n-q-1}(A; G) \approx H_{q+1}(\mathbf{S}^n, \mathbf{S}^n - A; G)$. Also $H_{q+1}(\mathbf{S}^n, \mathbf{S}^n - A; G) \approx \tilde{H}_q(\mathbf{S}^n - A; G)$ except when $q + 1 = n$. In that case we have the commutative diagram with exact rows

$$\begin{array}{ccccccc}
 & & H^0(\mathbf{S}^n) & \longrightarrow & \check{H}^0(A) & \longrightarrow & \tilde{\check{H}}^0(A) \longrightarrow 0 \\
 & & \downarrow \approx & & \downarrow \approx & & \\
H_n(\mathbf{S}^n - A) & \xrightarrow{0} & H_n(\mathbf{S}^n) & \rightarrow & H_n(\mathbf{S}^n, \mathbf{S}^n - A) & \rightarrow & H_{n-1}(\mathbf{S}^n - A) \xrightarrow{0}
\end{array}$$

which makes the result clear. (The 0's on the bottom row are because the inclusion map $\mathbf{S}^n - A \hookrightarrow \mathbf{S}^n$ factors through a contractible space $\mathbf{S}^n -$ point.) $\square$

8.8. Corollary (Generalized Jordan Curve Theorem). *Let M^n be a connected, orientable, compact n-manifold with $H_1(M^n; \Lambda) = 0$ over some ring Λ with unity. Let A be a proper closed subset of M^n. Then $\check{H}^{n-1}(A; \Lambda)$ is a free Λ-module whose rank is one less than the number of components of $M^n - A$.*

PROOF. The number of components of $M - A$ is $\operatorname{rank}(H_0(M - A)) = 1 + \operatorname{rank}(\tilde{H}_0(M - A))$. Since $H_1(M) = 0$ and $\tilde{H}_0(M) = 0$, the exact sequence of $(M, M - A)$ gives $\tilde{H}_0(M - A) \approx H_1(M, M - A)$ and the latter is isomorphic to $\check{H}^{n-1}(A)$ by duality. $\square$

8.9. Corollary. *Let M^n be a connected, orientable, and compact n-manifold with $H_1(M^n; \mathbf{Z}) = 0$. Then no nonorientable compact $(n-1)$-manifold N^{n-1} can be embedded in M^n.*

PROOF. This is because $H^{n-1}(N; \mathbf{Z}) \approx \mathbf{Z}_2$ is not free. $\square$

For example, this last corollary implies that real projective $2n$-space cannot be embedded in $\mathbf{S}^{2n+1}$. Of course, in the proof of Corollary 8.9, we are using the fact from Appendix E that $\check{H}^*(N) \approx H^*(N)$. Note that in the case for which M^n and N^{n-1} are smooth and N^{n-1} is smoothly embedded, this follows immediately from the Tubular Neighborhood Theorem (Theorem 11.14 of Chapter II).

8.10. Theorem (Poincaré). *There is a compact 3-manifold having the homology groups of $\mathbf{S}^3$ but which is not simply connected.*

PROOF. Consider the group I of rotational symmetries of a regular icosahedron, the "icosahedral group." We have $I \subset \mathbf{SO}(3)$ and it is well known that I is isomorphic to the alternating group A_5 on five letters. (This can be

Figure VI-6. Shows that $I = A_5$.

seen geometrically by considering the five tetrahedra inscribed in a dodecahedron (which is dual to the icosahedron) and the permutations of them induced by the action of I. See Figure VI-6.) Also well known is the fact that this group is simple. Consider the homomorphism $\mathbf{S}^3 \to \mathbf{SO}(3)$, where $\mathbf{S}^3$ is the group of unit quaternions. The inverse image of I in $\mathbf{S}^3$ is a group I', of which I is the quotient by the subgroup $\{\pm 1\} \subset I'$. The dodecahedron has an inscribed cube, so that I contains the rotation group of a cube. Assuming the cube to be aligned with the coordinate axes, this implies that the quaternions i, j, k are in I'. Thus $iji^{-1}j^{-1} = ijij = k^2 = -1$ is in the commutator subgroup $[I', I']$. The image of $[I', I']$ in I is $[I, I] = I$, and it follows that $[I', I'] = I'$. The space in question is $\Sigma^3 = \mathbf{S}^3/I'$. From covering space theory, we have $\pi_1(\Sigma^3) \approx I'$ and so $H_1(\Sigma^3) \approx \pi_1(\Sigma^3)/[\pi_1, \pi_1] = 0$. By the Universal Coefficient Theorem, $H^1(\Sigma^3) \approx \mathrm{Hom}(H_1(\Sigma^3), \mathbf{Z}) = 0$. By Poincaré duality, $H_2(\Sigma^3) \approx H^1(\Sigma^3) = 0$. □

This example occupies an interesting niche in the history of topology. Poincaré originally conjectured that a manifold which is a homology sphere is homeomorphic to a sphere. When the above counterexample, called the "Poincaré dodecahedral space," and others came to light, the conjecture was modified to include the hypothesis of simple connectivity. Today, for smooth manifolds, that conjecture is known to be true with the single exception of dimension three, where it remains an open and very important conjecture called, of course, the "Poincaré Conjecture."

Another interesting fact about this space $\Sigma^3 = \mathbf{SO}(3)/I = \mathbf{S}^3/I'$ is that it is the unique example of an "exotic homogeneous homology sphere." That is, if Σ^n is a closed n-manifold with the homology groups of $\mathbf{S}^n$ and if G is a compact Lie group acting transitively on Σ^n, with isotropy group H, then, with the single exception of this example, Σ^n is diffeomorphic to $\mathbf{S}^n$ and G acts on it linearly, i.e., as a subgroup of $\mathbf{O}(n+1)$. (See Bredon [1].) For another context in which this space occurs, see Section 18.

PROBLEMS

1. ◇ If M^n is a connected, orientable, and compact n-manifold with $H_1(M^n; \mathbf{Z}) = 0$ and if $N^{n-1} \subset M^n$ is a compact connected $(n-1)$-manifold, then show that $M^n - N^{n-1}$ has exactly two components with N^{n-1} as the topological boundary of each.
2. Give a counterexample to Problem 1 if the condition $H_1(M^n; \mathbf{Z}) = 0$ is dropped.
3. Show, by example, that Corollary 8.8 would be false if $\check{H}$ were replaced by H.
4. For a locally compact space X, define $H_c^p(X) = \varinjlim H^p(X, X-K)$ where K ranges over the compact subsets of X. (This is called "cohomology with compact supports.") For an oriented n-manifold M^n, define a cap product $\cap \vartheta: H_c^p(M^n) \to H_{n-p}(M^n)$ and show that it is an isomorphism. (*Hint*: For $U \subset X$ open with $\bar{U}$ compact, $\varinjlim \check{H}^p(X, X-U) = \varinjlim H^p(X, X-\bar{U})$.)
5. Using Problem 4, show that, for a connected n-manifold M^n, $H_c^n(M^n) \approx \mathbf{Z}$ for M^n orientable and $H_c^n(M^n) \approx \mathbf{Z}_2$ for M^n nonorientable.
6. If M^{2n+1} is a compact connected $(2n+1)$-manifold, possibly nonorientable, show that the Euler characteristic of M^{2n+1} is zero. (Assume the fact that $H_*(M)$ is finitely generated.)
7. If M^3 is a compact, connected, and nonorientable 3-manifold, show that $H_1(M)$ is infinite. (*Hint*: Use Problem 6.)
8. If $U \subset \mathbf{R}^3$ is open, show that $H_1(U)$ is torsion free. (*Hint*: This would be false for $U \subset \mathbf{R}^n, n > 3$.)
9. Show that Corollary 8.9 remains true if the hypothesis that $H_1(M; \mathbf{Z}) = 0$ is weakened to $H_1(M; \mathbf{Z}_2) = 0$.
10. Rework Problems 6–9 of Section 19 of Chapter IV in light of the results of the present section.

9. Duality on Compact Manifolds with Boundary

We remark that, in general, if M^n is compact then the orientation ϑ is simply an element of $H_n(M^n)$ which is a generator on each component. In this case, we usually denote it by $[M] \in H_n(M)$. This class $[M]$ is called the "orientation class" or "fundamental class" of M.

Let M^n be a compact n-manifold with boundary ∂M. We shall assume that there is a neighborhood of ∂M in M^n which is a product $\partial M \times [0, 2)$, with ∂M corresponding to $\partial M \times \{0\}$. This is clearly the case for smooth manifolds and it is also known to always be the case for paracompact topological manifolds, by a theorem of M. Brown [2]. Also, one can avoid such an assumption merely by adding an external collar. For simplicity of notation, we will treat $\partial M \times [0, 2)$ as a subspace of M.

Assume that M^n is connected and orientable, by which we mean that its interior $M - \partial M$ is orientable. Then we have the following isomorphisms:

$$\begin{aligned} H_n(M, \partial M) &\approx H_n(M, \partial M \times [0,1)) && \text{(homotopy)} \\ &\approx H_n(\text{int}(M), \partial M \times (0,1)) && \text{(excision)} \\ &\approx H^0(M - \partial M \times [0,1)) && \text{(duality)} \\ &\approx H^0(M) && \text{(homotopy)} \\ &\approx \mathbf{Z}. \end{aligned}$$

The orientation class $\vartheta \in H_n(\text{int}(M), \partial M \times (0,1))$ corresponds to a class $[M] \in H_n(M, \partial M)$. At the other end of this sequence of isomorphisms, the orientation class corresponds to $1 \in H^0(M)$, the class of the augmentation cocycle taking all 0-simplices to 1.

Consider the following sequence of isomorphisms:

$$\begin{aligned} H^p(M; G) &\approx H^p(M - \partial M \times [0,1); G) && \text{(homotopy)} \\ &\approx H_{n-p}(\text{int}(M), \partial M \times (0,1); G) && \text{(duality, cap with } \vartheta) \\ &\approx H_{n-p}(M, \partial M \times [0,1); G) && \text{(excision)} \\ &\approx H_{n-p}(M, \partial M; G) && \text{(homotopy).} \end{aligned}$$

By naturality of the cap product, the resulting isomorphism $H^p(M; G) \approx H_{n-p}(M, \partial M; G)$ is the cap product with the orientation class $[M] \in H^n(M, \partial M)$.

9.1. Lemma. *If M^n is compact and orientable then ∂M is orientable and $[\partial M] = \partial_*[M]$ is an orientation class, where ∂_* is the connecting homomorphism of the exact sequence of the pair $(M, \partial M)$.*

PROOF. Let A be a component of ∂M, and put $B = \partial M - A$ (possibly empty). Consider the exact homology sequence of the triple $(M, A \cup B, B)$. Part of it is the homomorphism $\partial_*: H_n(M, A \cup B) \to H_{n-1}(A \cup B, B)$. The first group is $H_n(M, \partial M)$ and the second is isomorphic, by excision, to $H_{n-1}(A)$. If c is a chain representing $[M] \in H_n(M, \partial M)$ then $[M] = [\![c]\!]$ goes to $[\![$part of ∂c in $A]\!]$ in $H_{n-1}(A)$. Thus we are to show that the part of ∂c in A is an orientation, i.e., that it gives a generator of $H_{n-1}(A)$.

For any coefficient group G, we have

$$\begin{aligned} H_n(M, B; G) &\approx H_n(M, B \times [0,1); G) \\ &\approx H_n(\text{int}(M), B \times (0,1); G) \\ &\approx \Gamma_c(\text{int}(M) - B \times (0,1), \Theta \otimes G) \\ &= 0, \end{aligned}$$

since $\text{int}(M) - B \times (0,1)$ is connected and *non*-compact. By the Universal Coefficient Theorem,

$$0 = H_n(M, B; \mathbf{Q}/\mathbf{Z}) \approx H_n(M, B) \otimes \mathbf{Q}/\mathbf{Z} \oplus TH_{n-1}(M, B),$$

see Example 7.6 of Chapter V. Hence, $H_{n-1}(M, B)$ is torsion free and the exact sequence of the triple $(M, A \cup B, B)$ has the segment

$$0 \to H_n(M, \partial M) \xrightarrow{\partial_*} H_{n-1}(A) \to (\text{torsion free}).$$

But $H_n(M, \partial M) \approx \mathbf{Z}$, and $H_{n-1}(A)$ is either $\mathbf{Z}$ (if orientable) or 0 (if not). Thus $H_{n-1}(A)$ must be $\mathbf{Z}$ and ∂_* must be onto to make the cokernel torsion free. □

9.2. Theorem. *If M^n is an oriented, compact, connected n-manifold with boundary, then the diagram (arbitrary coefficients)*

$$\begin{array}{ccccccccc}
\cdots \longrightarrow & H^p(M) & \xrightarrow{i^*} & H^p(\partial M) & \xrightarrow{\delta^*} & H^{p+1}(M, \partial M) & \xrightarrow{j^*} & H^{p+1}(M) & \longrightarrow \cdots \\
 & \approx \downarrow \cap[M] & (-1)^p & \approx \downarrow \cap[\partial M] & (-1)^{p+1} & \approx \downarrow \cap[M] & 1 & \approx \downarrow \cap[M] & \\
\cdots \longrightarrow & H_{n-p}(M, \partial M) & \xrightarrow[\partial_*]{} & H_{n-p-1}(\partial M) & \xrightarrow[i_*]{} & H_{n-p-1}(M) & \xrightarrow[j_*]{} & H_{n-p-1}(M, \partial M) & \longrightarrow \cdots
\end{array}$$

with exact rows, commutes up to the indicated signs. This also holds, without the orientability restriction, over the base ring $\mathbf{Z}_2$.

PROOF. All the vertical isomorphisms, except the third, result from previous theorems or remarks. The third one will follow from the 5-lemma as soon as we have proved the commutativity.

Let $c \in \Delta_n(M)$ represent the orientation class $[M] \in H_n(M, \partial M)$. Then ∂c is a chain in ∂M.

For the first square, let $f \in \Delta^p(M)$ be a cocycle. Then going right and then down gives a class represented by $f|_{\partial M} \cap \partial c = f \cap \partial c = (-1)^p \partial(f \cap c)$. Going down then right gives $\partial(f \cap c)$.

For the second square, let $f \in \Delta^p(M)$ with $\delta f = 0$ on ∂M. Then going right then down gives $(\delta f) \cap c = \partial(f \cap c) + (-1)^{p+1} f \cap \partial c$ which is homologous to $(-1)^{p+1} f \cap \partial c = (-1)^{p+1} f|_{\partial M} \cap \partial c$. Going down then right gives $f|_{\partial M} \cap \partial c$.

Commutativity of the third square is obvious. □

9.3. Corollary. $\cap[M]$: *$H^p(M, \partial M; G) \to H_{n-p}(M; G)$ is an isomorphism.* □

It is often desirable to have a version of duality entirely in terms of cohomology and the cup product. To this end, let Λ be a principal ideal domain and, with the notation of Example 7.6 of Chapter V, put

$$\bar{H}^p(\cdot) = H^p(\cdot)/TH^p(\cdot),$$

the "torsion free part" of the pth cohomology group. Note that if Λ is a field then $\bar{H} = H$. We shall assume the fact, proved in Appendix E, that $H_*(M; \Lambda)$ is finitely generated. Then it follows that $\operatorname{Ext}(H_*(M), \Lambda)$ is all torsion so that the Universal Coefficient Theorem gives the isomorphism $\bar{H}^p(M; \Lambda) \xrightarrow{\approx} \operatorname{Hom}(\bar{H}_p(M), \Lambda)$.

9.4. Theorem. *Let M^n be a compact, connected, oriented (over Λ) n-manifold with boundary. Then the cup product pairing*

$$\bar{H}^p(M; \Lambda) \otimes_\Lambda \bar{H}^{n-p}(M, \partial M; \Lambda) \to H^n(M, \partial M; \Lambda) \approx H_0(M; \Lambda) \approx \Lambda$$

taking $(\alpha\otimes\beta)\mapsto\alpha\cup\beta\mapsto\langle\alpha\cup\beta,[M]\rangle\in\Lambda$, *is a duality pairing. That is, the map*

$$\bar{H}^p(M;\Lambda)\to\operatorname{Hom}_\Lambda(\bar{H}^{n-p}(M,\partial M;\Lambda),\Lambda),$$

taking $\alpha\mapsto\bar{\alpha}$ *where* $\bar{\alpha}(\beta)=\langle\alpha\cup\beta,[M]\rangle$, *is an isomorphism.*

Proof. We have the isomorphism $\bar{H}^p(M;\Lambda)\xrightarrow{\approx}\operatorname{Hom}(\bar{H}_p(M),\Lambda)\xrightarrow{\approx}\operatorname{Hom}(\bar{H}^{n-p}(M,\partial M),\Lambda)$ (given by the Universal Coefficient Theorem and cap with $[M]$, respectively), taking, say, α to α^* and then to α°, where $\alpha^*(\gamma)=\langle\alpha,\gamma\rangle$. We claim that $\alpha^\circ=\bar{\alpha}$, which would prove the desired isomorphism. We compute $\alpha^\circ(\beta)=\alpha^*(\beta\cap[M])=\langle\alpha,\beta\cap[M]\rangle=\langle\alpha\cup\beta,[M]\rangle=\bar{\alpha}(\beta)$. □

Problems

1. If M^n and N^n are compact connected oriented n-manifolds, one defines their "connected sum" $M\#N$ as follows: Take a nicely embedded n-disk in each, remove its interior, and paste the remainders together via an orientation *reversing* homeomorphism on the boundary spheres of these disks. Show that the cohomology ring of $M\#N$ is isomorphic to the ring resulting from the direct product of the rings for M and N with the unity elements (in dimension 0) identified and the orientation classes identified. Similarly, the multiples of these identifications must also be made. (The orientation cohomology class of M is that class $\vartheta\in H^n(M)$ which is Kronecker dual to $[M]$, i.e., such that $\vartheta[M]=1$. It can also be described as the class that is Poincaré dual to the standard generator in $H_0(M)$, the class represented by any 0-simplex.) In particular, cup products of positive dimensional classes, one from each of the two original manifolds, are zero.

2. Suppose that N^n is a compact, orientable, smooth n-manifold embedded smoothly in the compact, orientable m-manifold M^m. Let W be a closed tubular neighborhood of N in M. Show that there exists an isomorphism $H_p(N)\approx H_{m-n+p}(W,\partial W)$.

3. ⌖ Let M^n be a compact manifold with boundary $\partial M=A\cup B$ where A and B are $(n-1)$-manifolds with common boundary $A\cap B$. Since $A\cap B$ is a neighborhood retract in both A and B (see Appendix E) the inclusion $\Delta_*(A)+\Delta_*(B)\hookrightarrow\Delta_*(A\cup B)$ induces an isomorphism in homology, and so there is a cap product

$$\cap\colon H^p(M,A)\otimes H_n(M,A\cup B)\to H_{n-p}(M,B).$$

Take the orientation class $[A]$ to come from $[\partial M]=\partial_*[M]$ via $H_{n-1}(\partial M)=H_{n-1}(A\cup B)\to H_{n-1}(A\cup B,B)\approx H_{n-1}(A,A\cap B)$ (by excision and homotopy). Show that the diagram

$$\begin{array}{ccccccccc}
\cdots\to & H^p(M,A) & \to & H^p(M) & \to & H^p(A) & \to & H^{p+1}(M,A) & \to\cdots\\
& \downarrow{\cap[M]} & & \approx\downarrow{\cap[M]} & & \approx\downarrow{\cap[A]} & & \downarrow{\cap[M]} & \\
& & & & & H_{n-p-1}(A,\partial A) & & & \\
& & & & & \downarrow\approx & & & \\
\cdots\to & H_{n-p}(M,B) & \to & H_{n-p}(M,\partial M) & \to & H_{n-p-1}(A\cup B,B) & \to & H_{n-p-1}(M,B) & \to\cdots
\end{array}$$

commutes up to sign. Deduce that there is the duality isomorphism

$$\boxed{\cap[M]\colon H^p(M,A)\xrightarrow{\approx}H_{n-p}(M,B).}$$

4. Verify, by direct computation, the isomorphism $H^p(M, A) \approx H_{3-p}(M, B)$ for $M^3 = \mathbf{S}^1 \times \mathbf{D}^2$ and where A is a nice 2-disk in ∂M and B is the closure of the complement of A in ∂M.

5. If M^m and N^n are compact orientable manifolds with boundary, show that $H_{m-p}((M, \partial M) \times N) \approx H^{n+p}(M \times (N, \partial N))$.

10. Applications of Duality

In this section we will give several applications of duality to problems about manifolds. It is standard terminology to refer to compact manifolds without boundary as "closed" manifolds. We shall occasionally use the fact, from Appendix E, that such manifolds have finitely generated homology.

10.1. Proposition. *Let M^n be a closed, connected, orientable manifold and let $f: \mathbf{S}^n \to M$ be a map of nonzero degree. Then $H_*(M^n; \mathbf{Q}) \approx H_*(\mathbf{S}^n; \mathbf{Q})$. If, moreover, $\deg(f) = \pm 1$, then $H_*(M^n; \mathbf{Z}) \approx H_*(\mathbf{S}^n; \mathbf{Z})$.*

PROOF. For the last part, suppose $H_q(M; \mathbf{Z}) \neq 0$ for some $q \neq 0, n$. Then it can easily be seen from the Universal Coefficient Theorem that there is a field Λ such that $H^q(M; \Lambda) \neq 0$. For the first part, take $\Lambda = \mathbf{Q}$.

If $0 \neq \alpha \in H^q(M; \Lambda)$ then there is a $\beta \in H^{n-q}(M; \Lambda)$ with $\alpha \cup \beta \neq 0$. Thus $\alpha\beta = k \cdot \gamma$, where γ is a generator of $H^n(M; \Lambda)$ and $0 \neq k \in \Lambda$. Therefore $0 = 0 \cdot 0 = f^*(\alpha) f^*(\beta) = f^*(\alpha\beta) = f^*(k\gamma) = k \cdot \deg(f) \cdot \text{generator} \neq 0$. □

10.2. Proposition. *The cohomology rings of the real, complex, and quaternionic projective spaces are:*

$$H^*(\mathbf{RP}^n; \mathbf{Z}_2) \approx \mathbf{Z}_2[\alpha]/\alpha^{n+1} \quad \text{where} \quad \deg(\alpha) = 1,$$
$$H^*(\mathbf{CP}^n; \mathbf{Z}) \approx \mathbf{Z}[\alpha]/\alpha^{n+1} \quad \text{where} \quad \deg(\alpha) = 2,$$
$$H^*(\mathbf{QP}^n; \mathbf{Z}) \approx \mathbf{Z}[\alpha]/\alpha^{n+1} \quad \text{where} \quad \deg(\alpha) = 4.$$

PROOF. We already know the additive (co)homology groups of these spaces. The arguments for all of these are essentially the same so we will give it only for the case of complex projective space. The proof is by induction on n. Suppose it holds for $n-1$, i.e., that there is an element $\alpha \in H^2(\mathbf{CP}^{n-1})$ such that $1, \alpha, \alpha^2, \ldots, \alpha^{n-1}$ generate the homology groups in those dimensions. Now $\mathbf{CP}^n$ is obtained from $\mathbf{CP}^{n-1}$ by attaching a $2n$-cell. It follows from the exact sequence of the pair $(\mathbf{CP}^n, \mathbf{CP}^{n-1})$ that $H^i(\mathbf{CP}^n) \to H^i(\mathbf{CP}^{n-1})$ is an isomorphism for $i \leq 2n-2$. Thus it makes sense to identify α and its powers up to α^{n-1} with their preimages in $H^i(\mathbf{CP}^n)$ in this range. (This is just a notational convenience.) Also, of course, the case $n = 1$ is trivial, so we can assume $n \geq 2$. Thus we have the classes $\alpha \in H^2(\mathbf{CP}^n)$ and $\alpha^{n-1} \in H^{2n-2}(\mathbf{CP}^n)$. By Theorem 9.4, the product $\alpha^n = \alpha \cup \alpha^{n-1}$ must be a generator of $H^{2n}(\mathbf{CP}^n)$. □

10.3. Corollary. *Any homotopy equivalence* $\mathbf{CP}^{2n} \to \mathbf{CP}^{2n}$ *preserves orientation for* $n \geq 1$.

Proof. Such a map f must be an isomorphism on $H^2(\mathbf{CP}^{2n}) \approx \mathbf{Z}$ and so, for a generator α we must have $f^*(\alpha) = \pm\alpha$. Therefore $f^*(\alpha^{2n}) = (f^*(\alpha))^{2n} = (\pm\alpha)^{2n} = \alpha^{2n}$. The contention follows since this is a top dimensional generator. □

We will now study to a small extent the cohomology of manifolds that are boundaries of other manifolds.

10.4. Theorem. *Let* Λ *be a field (coefficients for all homology and cohomology). Let* V^{2n+1} *be an oriented (unless* $\Lambda = \mathbf{Z}_2$*) compact manifold with* $\partial V = M^{2n}$ *connected. Then* $\dim H^n(M^{2n})$ *is even and*

$$\dim[\ker(i_*: H_n(M) \to H_n(V))] = \dim[\operatorname{im}(i^*: H^n(V) \to H^n(M))] = \tfrac{1}{2}\dim H^n(M).$$

Moreover, $\operatorname{im}(i^*) \subset H^n(M)$ *is self-annihilating, i.e., the cup product of any two classes in it is zero.*

Proof. Consider this portion of the Poincaré–Lefschetz diagram:

$$\begin{array}{ccccc} H^n(V) & \xrightarrow{i^*} & H^n(M) & \xrightarrow{\delta^*} & H^{n+1}(V,M) \\ & & {\scriptstyle\cap[M]}\big\downarrow{\scriptstyle\approx} & & {\scriptstyle\approx}\big\downarrow{\scriptstyle\cap[V]} \\ & & H_n(M) & \xrightarrow{i_*} & H_n(V). \end{array}$$

From the diagram we see that $\{\operatorname{im}(i^*)\} \cap [M] = \{\ker(\delta^*)\} \cap [M] = \ker(i_*)$. Thus $\operatorname{rank}(i^*) = \dim \operatorname{im}(i^*) = \dim \ker(i_*) = \dim H_n(M) - \operatorname{rank}(i_*) = \dim H^n(M) - \operatorname{rank}(i^*)$, since i^* and i_* are Kronecker duals of one another (this is the fact that the rank of a transposed matrix equals the rank of the original). Therefore, $\dim H^n(M) = 2\cdot\operatorname{rank}(i^*) = 2\cdot\dim(\ker(i_*))$.

Now if $\alpha, \beta \in H^n(V)$ then $\delta^*(i^*(\alpha) \cup i^*(\beta)) = (\delta^* i^*)(\alpha \cup \beta) = 0$ since $\delta^* i^* = 0$ by exactness. But $\delta^*: H^{2n}(M) \to H^{2n+1}(V, M)$ is a monomorphism since it is dual to $i_*: H_0(M) \to H_0(V)$. Thus $i^*(\alpha) \cup i^*(\beta) = 0$ as claimed. □

10.5. Corollary. *If* $M^m = \partial V$ *is connected with* V *compact, then the Euler characteristic* $\chi(M)$ *is even; also see Problem* 1.

Proof. If $\dim(M)$ is odd then Poincaré duality on M pairs odd and even dimensions and so $\chi(M) = 0$ for all closed M. For M of dimension $2n$, we have that $\chi(M) \equiv \dim H^n(M; \mathbf{Z}_2)$ modulo 2. For $M = \partial V$, the latter is 0 (mod 2) by Theorem 10.4. □

10.6. Corollary. $\mathbf{RP}^{2n}, \mathbf{CP}^{2n}$, *and* $\mathbf{QP}^{2n}$ *are not boundaries of compact manifolds.* □

We remark that all *orientable* two- and three-dimensional closed manifolds are boundaries. The Klein bottle is a nonorientable 2-manifold which is a boundary.

10.7. Definition (H. Weyl). Let M be a closed oriented manifold. The *signature* of M is defined to be 0 if $\dim(M)$ is not divisible by 4. If $\dim(M) = 4n$, then signature(M) is defined to be the signature of the quadratic form $\langle \alpha, \beta \rangle = (\alpha \cup \beta)[M]$ on $H^{2n}(M; \mathbf{R})$.

Recall that a quadratic form over the reals is the sum and difference of squares. Its "signature" is the sum of the signs on those squares. Another term used for this is "index."

10.8. Corollary (Thom). *If* $M^{4n} = \partial V^{4n+1}$ *is connected with* V *compact and orientable then* signature(M) = 0.

PROOF. Let $W = H^{2n}(M; \mathbf{R})$ and let $\dim(W) = 2k$. The quadratic form (over $\mathbf{R}$) of Definition 10.7 is equivalent to the sum of, say, r positive squares and, thus, $2k - r$ negative ones. That is, there is a subspace W^+ on which the form is positive definite and another subspace W^- on which it is negative definite with $\dim W^+ = r$ and $\dim W^- = 2k - r$. By Theorem 10.4, there is a subspace $U \subset W$ of dimension k such that $\langle \alpha, \beta \rangle = 0$ on U. Clearly $U \cup W^+ = \{0\}$ and so the sum $r + k$ of their dimensions cannot be greater than the dimension $2k$ of W. That is, $r + k \leq 2k$, so that $r \leq k$.

Similarly $U \cap W^- = \{0\}$, so that $(2k - r) + k \leq 2k$, i.e., $k \leq r$.

Thus $r = k$ and the signature is zero. □

10.9. Example. The connected sum (see Problem 1 in Section 9) $M^4 = \mathbf{CP}^2 \# \mathbf{CP}^2$ is not the boundary of an orientable 5-manifold. To see this, note that the ring of M^4 is generated by classes $\alpha, \beta \in H^2(M)$, with $\alpha\beta = 0$ and $\alpha^2 = \beta^2$, so that its quadratic form is the identity 2×2 matrix whose signature is 2 (or -2 for the other orientation).

Of course, a more general argument shows that the signature is additive with respect to the connected sum operation on oriented manifolds.

However, $\mathbf{CP}^2 \# -\mathbf{CP}^2$ is the boundary of the orientable 5-manifold $V^5 = (\mathbf{CP}^2 - U) \times I$, where U is an open 4-disk in $\mathbf{CP}^2$. ($-\mathbf{CP}^2$ stands for $\mathbf{CP}^2$ with the opposite orientation.) The only difference in the cohomology ring is that $\beta^2 = -\alpha^2$, but that is enough, of course, to make the signature zero. Naturally, this is a general fact having nothing to do with $\mathbf{CP}^2$ specifically.

Also we claim that $\mathbf{CP}^2 \# \mathbf{CP}^2$ is the boundary of a nonorientable 5-manifold. To see this consider $(\mathbf{CP}^2 \times I) \# (\mathbf{RP}^2 \times \mathbf{S}^3)$, where the sum is done away from the two boundary components. Now run an arc from one of the boundary components through an orientation reversing loop in $\mathbf{RP}^2 \times \mathbf{S}^3$ and then to the other boundary component. Done nicely this arc has a product neighborhood, and we can remove that. This leaves $\mathbf{CP}^2 \# \mathbf{CP}^2$ as the boundary of the resulting, nonorientable, 5-manifold. Again, this is a general construction and has nothing to do with $\mathbf{CP}^2$ specifically.

In Section 20 of Chapter IV, we proved the Borsuk–Ulam Theorem by a somewhat special argument. Now we shall reprove it using a fairly direct argument with the ring structure of the cohomology of real projective space.

10.10. Theorem. *If $m > n$ then for any map $\phi: \mathbf{P}^m \to \mathbf{P}^n$, the induced homomorphism $\phi_{\#}: \pi_1(\mathbf{P}^m) \to \pi_1(\mathbf{P}^n)$ is trivial.*

PROOF. This is clear for $n = 1$ since the homomorphism in question is $\mathbf{Z}_2 \to \mathbf{Z}$. Thus take $m > n > 1$. Since $\pi_1(\mathbf{P}^n)$ is abelian, it is naturally isomorphic to $H_1(\mathbf{P}^n)$ by the Hurewicz Theorem (Theorem 3.4 of Chapter IV). Therefore it suffices to show that $\phi_*: H_1(\mathbf{P}^m) \to H_1(\mathbf{P}^n)$ is trivial.

Similarly, the Universal Coefficient Theorem gives a natural epimorphism

$$H^1(\mathbf{P}^n; \mathbf{Z}_2) \to \operatorname{Hom}(H_1(\mathbf{P}^n), \mathbf{Z}_2)$$

which is an isomorphism since both groups are isomorphic to $\mathbf{Z}_2$. Thus it suffices to show that $\phi^*: H^1(\mathbf{P}^n; \mathbf{Z}_2) \to H^1(\mathbf{P}^m; \mathbf{Z}_2)$ is trivial. But, if it is not trivial then $\phi^*(\alpha) = \beta \neq 0$ for some $\alpha \in H^1(\mathbf{P}^n; \mathbf{Z}_2)$. Then $0 = \phi^*(0) = \phi^*(\alpha^m) = (\phi^*(\alpha))^m = \beta^m \neq 0$. □

Now we will use Theorem 10.10 to reprove Theorem 20.1 of Chapter IV, and hence the rest of the results in that section.

10.11. Corollary. *If $\phi: \mathbf{S}^m \to \mathbf{S}^n$ is any map with $\phi(-x) = -\phi(x)$ for all x, then $m \leq n$.*

PROOF. ϕ induces a map $\mathbf{P}^m \to \mathbf{P}^n$. It suffices, by Theorem 10.10, to show that this cannot be trivial on π_1. Consider any path γ in $\mathbf{S}^m$ from some point x to $-x$. Then $\phi \circ \gamma$ is a path in $\mathbf{S}^n$ from $\phi(x)$ to $\phi(-x) = -\phi(x)$. In $\mathbf{P}^m$, γ projects to a loop α, and in $\mathbf{P}^n$, $\phi \circ \gamma$ also projects to a loop β. These are not homotopically trivial loops since they lift to nonloops. Therefore $\phi_{\#}([\alpha]) = [\beta]$ is nontrivial. □

10.12. Theorem. *For $n \geq 2$, $\mathbf{P}^n$ cannot be embedded in $\mathbf{S}^{n+1}$.*

PROOF. For n even this already follows from Corollary 8.9. For n odd, Corollary 8.8 implies that if $\mathbf{P}^n \subset \mathbf{S}^{n+1}$ then $\mathbf{S}^{n+1} = A \cup B$ with $A \cap B = \mathbf{P}^n$. (See Problem 1 of Section 8.) We will show that this is incompatible with what we know about the cohomology of $\mathbf{P}^n$ with $\mathbf{Z}_2$ coefficients. For the remainder of the proof we assume coefficients in $\mathbf{Z}_2$, which will be suppressed from the notation. We are also going to assume that $\check{H}^*(\mathbf{P}^n) \approx H^*(\mathbf{P}^n)$. This is always true, and is proved in Corollary E.6. One could just assume that $\mathbf{P}^n$ is embedded smoothly, from which this follows immediately from the Tubular Neighborhood Theorem (Theorem 11.4 of Chapter II).

By naturality of the cup product, there is a cup product on $\check{H}^*$ induced by the direct limit. The exactness of the direct limit functor and the Mayer–Vietoris sequence for open sets implies that there is an exact Mayer–Vietoris sequence of the form

$$\cdots \to H^i(\mathbf{S}^{n+1}) \to \check{H}^i(A) \oplus \check{H}^i(B) \to H^i(\mathbf{P}^n) \to H^{i+1}(\mathbf{S}^{n+1}) \to \cdots.$$

It follows that $\check{H}^i(A) \oplus \check{H}^i(B) \to H^i(\mathbf{P}^n)$ is an isomorphism for $0 < i < n$. Also the latter group is $\mathbf{Z}_2$. Hence we may assume that $\check{H}^1(A) \approx \mathbf{Z}_2$ and $\check{H}^1(B) = 0$. There is a class $\alpha \in \check{H}^1(A)$ mapping to $0 \neq \beta \in H^1(\mathbf{P}^n) \approx \mathbf{Z}_2$. Thus α^n maps to $\beta^n \neq 0$, and so the map $\check{H}^n(A) \to H^n(\mathbf{P}^n)$ is onto.

Also, since A and B are proper closed subsets of $\mathbf{S}^{n+1}$, we have, from Corollary 8.5, that $\check{H}^{n+1}(A) = 0 = \check{H}^{n+1}(B)$.

These conclusions are contrary to the exactness of

$$\check{H}^n(A) \oplus \check{H}^n(B) \to H^n(\mathbf{P}^n) \to H^{n+1}(\mathbf{S}^{n+1}) \to 0$$

(part of the Mayer–Vietoris sequence above). □

Our last application is to lens spaces, which have been mentioned before. Consider the sphere $\mathbf{S}^3$ as $\{(u, v) \in \mathbf{C}^2 \mid |u|^2 + |v|^2 = 1\}$. Let $\omega = e^{2\pi i/p}$, a primitive pth root of unity. For q relatively prime to p, consider the map $T_q: \mathbf{S}^3 \to \mathbf{S}^3$ given by $T_q(u, v) = (\omega u, \omega^q v)$. This has period p and none of the iterates $T_q, T_q^2, \ldots, T_q^{p-1}$ has fixed points. That is, T_q generates a free action of the cyclic group of order p on $\mathbf{S}^3$. The orbit space of this action is called $L(p, q)$, and is known as a "lens space." Then $L(p, q)$ is an orientable 3-manifold and has $\mathbf{S}^3$ as its universal covering space, with p sheets.

For given p, we aim to attach a homotopy invariant to $L(p, q)$ that depends on q. For this, we need to recall the "Bockstein homomorphism." The short exact sequence $0 \to \mathbf{Z} \xrightarrow{p} \mathbf{Z} \to \mathbf{Z}_p \to 0$ of coefficient groups induces the long exact sequence

$$\cdots \to H^n(X; \mathbf{Z}) \xrightarrow{p} H^n(X; \mathbf{Z}) \xrightarrow{\rho} H^n(X; \mathbf{Z}_p) \xrightarrow{\beta_0} H^{n+1}(X; \mathbf{Z}) \to \cdots.$$

Let $\beta = \rho \circ \beta_0: H^n(X; \mathbf{Z}_p) \to H^{n+1}(X; \mathbf{Z}_p)$. This, and β_0 itself, is called the Bockstein homomorphism. It is natural in X, so that it is what is called a "cohomology operation."

For the case of $L = L(p, q)$, we have that $\pi_1(L(p, q)) \approx \mathbf{Z}_p$ by covering space theory. Using the Hurewicz Theorem, the Universal Coefficient Theorem, and Poincaré duality, we can calculate that

$$\begin{array}{lll} H_1(L; \mathbf{Z}) \approx \mathbf{Z}_p, & H^1(L; \mathbf{Z}) \approx 0, & H_1(L; \mathbf{Z}_p) \approx \mathbf{Z}_p \approx H^1(L; \mathbf{Z}_p), \\ H_2(L; \mathbf{Z}) \approx 0, & H^2(L; \mathbf{Z}) \approx \mathbf{Z}_p, & H_2(L; \mathbf{Z}_p) \approx \mathbf{Z}_p \approx H^2(L; \mathbf{Z}_p), \\ H_3(L; \mathbf{Z}) \approx \mathbf{Z}, & H^3(L; \mathbf{Z}) \approx \mathbf{Z}, & H_3(L; \mathbf{Z}_p) \approx \mathbf{Z}_p \approx H^3(L; \mathbf{Z}_p). \end{array}$$

The exact sequence

$$H^1(L; \mathbf{Z}) \to H^1(L; \mathbf{Z}_p) \xrightarrow{\beta_0} H^2(L; \mathbf{Z}) \xrightarrow{0} H^2(L; \mathbf{Z}) \xrightarrow{\rho} H^2(L; \mathbf{Z}_p)$$

shows that β_0 and ρ are isomorphisms here, and so

$$\beta: H^1(L; \mathbf{Z}_p) \to H^2(L; \mathbf{Z}_p)$$

is an isomorphism.

Now let $a \in H^1(L(p, q); \mathbf{Z}_p)$ be a generator. Then $\beta(a)$ is a generator and $a\beta(a) \in H^3(L(p, q); \mathbf{Z}_p)$ is a generator by Poincaré duality (Theorem 9.4). Let $[L(p, q)] \in H_3(L(p, q); \mathbf{Z}_p)$ be the mod p reduction of a generator of $H_3(L(p, q); \mathbf{Z}) \approx \mathbf{Z}$. This is unique up to sign. For any generator $a \in H^1(L(p, q); \mathbf{Z}_p)$ we have that $\langle a\beta(a), [L(p, q)] \rangle$ is a generator of $\mathbf{Z}_p$. For any other generator $b = na$ (n prime to p), we have $\langle b\beta(b), [L(p, q)] \rangle = \langle na\beta(na), [L(p, q)] \rangle = n^2 \langle a\beta(a), [L(p, q)] \rangle$. Thus, the generator $\langle a\beta(a), [L(p, q)] \rangle$ is determined, independent of the choice of a and of the orientation, up to $\pm$ a square factor, prime to p, in $\mathbf{Z}_p$. Let $i \cong j$ be the equivalence relation in $\mathbf{Z}_p$ meaning that $i \equiv \pm jn^2 (\text{mod } p)$ for some $n \in \mathbf{Z}$, prime to p. Then we define an invariant $t_q \in \mathbf{Z}_p$, modulo the equivalence relation $\cong$, by

$$\boxed{t_q \cong \langle a\beta(a), [L(p, q)] \rangle}$$

where $a \in H^1(L(p, q); \mathbf{Z}_p)$ is a generator.

10.13. Lemma. *The invariants t_q satisfy $t_q \cong qt_1$.*

Proof. First we remark that detailed calculations can show that $t_1 \cong 1$, but we do not need that. If $v = re^{i\theta} \in \mathbf{C}$ and $k \in \mathbf{Z}$ then we will use $v^{(k)}$ to denote $re^{ki\theta}$. Consider the map $\phi: \mathbf{S}^3 \to \mathbf{S}^3$ where $\phi(u, v) = (u, v^{(q)})$. Note that $\phi T_1(u, v) = \phi(\omega u, \omega v) = (\omega u, \omega^q v^{(q)})$, and $T_q\phi(u, v) = T_q(u, v^{(q)}) = (\omega u, \omega^q v^{(q)})$. Thus $\phi T_1 = T_q\phi$, which means that ϕ carries the action by T_1 on $\mathbf{S}^3$ to that by T_q on $\mathbf{S}^3$. (That is, ϕ is equivariant with respect to these actions.) Consequently, ϕ induces a map $\psi: L(p, 1) \to L(p, q)$. From covering space theory, $\pi_1(L(p, 1))$ is generated by the loop which is the image of any path in $\mathbf{S}^3$ from the base point $*$ to $T_1(*)$, and similarly for $L(p, q)$ and T_q. Since ϕ carries the path for T_1 to that for T_q, we conclude that

$$\psi_\#: \pi_1(L(p, 1)) \to \pi_1(L(p, q))$$

is an isomorphism. By the naturality of the Hurewicz homomorphism, it follows that $\psi_*: H_1(L(p, 1)) \to H_1(L(p, q))$ is an isomorphism and consequently $\psi^*: H^1(L(p, q); \mathbf{Z}_p) \to H^1(L(p, 1); \mathbf{Z}_p)$ is an isomorphism.

Now ϕ is easily seen to have degree q. (For example, it is the double suspension of the map $v \mapsto v^q$ on $\mathbf{S}^1$.) The diagram

$$\begin{array}{ccc} H_3(\mathbf{S}^3; \mathbf{Z}) & \xrightarrow{\phi_*} & H_3(\mathbf{S}^3; \mathbf{Z}) \\ \downarrow & & \downarrow \\ H_3(L(p, 1); \mathbf{Z}) & \xrightarrow{\psi_*} & H_3(L(p, q); \mathbf{Z}) \end{array}$$

shows that ψ has degree q, since the verticals both have degree p because they are p-fold coverings. Therefore $\psi_*[L(p,1)] = \pm q[L(p,q)]$, the $\pm$ allowing for a difference in choice of orientations.

Let $a \in H^1(L(p,q); \mathbf{Z}_p)$ be a generator and put $b = \psi^*(a)$. Then $\beta(b) = \psi^*(\beta(a))$ and $b\beta(b) = \psi^*(a\beta(a))$ and we have

$$\begin{aligned} t_1 &\cong \langle b\beta(b), [L(p,1)]\rangle \\ &= \langle \psi^*(a\beta(a)), [L(p,1)]\rangle \\ &= \langle a\beta(a), \psi_*[L(p,1)]\rangle \\ &= \langle a\beta(a), \pm q[L(p,q)]\rangle \\ &= \pm q\langle a\beta(a), [L(p,q)]\rangle \\ &\cong qt_q. \end{aligned}$$

Thus $t_q \cong q^2 t_q \cong qt_1$, as claimed. □

10.14. Theorem. *If $L(p,q) \simeq L(p,q')$ then $qq' \equiv \pm n^2$* (mod p) *for some integer n. (That is, either qq' or $-qq'$ is a quadratic residue* mod p.)

Proof. If $h: L(p,q) \to L(p,q')$ is a homotopy equivalence then it has degree ± 1. Therefore $h_*[L(p,q)] = \pm [L(p,q')]$. Then, if $b \in H^1(L(p,q'); \mathbf{Z}_p)$ is a generator and $a = h^*(b)$, we have

$$\begin{aligned} t_q &\cong \langle a\beta(a), [L(p,q)]\rangle \\ &= \langle h^*(b\beta(b)), [L(p,q)]\rangle \\ &= \langle b\beta(b), h_*[L(p,q)]\rangle \\ &= \pm\langle b\beta(b), [L(p,q')]\rangle \\ &\cong t_{q'}. \end{aligned}$$

Hence $q't_1 \cong t_{q'} \cong t_q \cong qt_1$, and so $qq' \cong qq \cong 1$. □

10.15. Theorem. *There exist two orientable closed 3-manifolds with the same fundamental groups and homology groups, but which are not homotopy equivalent.*

Proof. The lens space $L(5,1)$ is not homotopy equivalent to $L(5,2)$ since neither ± 2 is a quadratic residue mod 5. □

Much more is known about these interesting spaces. The converse of Theorem 10.14 is true, due to J.H.C. Whitehead [1]; see Example 11.17 of Chapter VII. It is not hard to see that $L(p,q) \approx L(p,q')$ if $q' \equiv \pm q^{\pm 1}$ (mod p) and it is known (hard) that the converse is also true; see Brody [1]. This implies, for example, that $L(7,1)$ and $L(7,2)$ are not homeomorphic, even though $L(7,1) \simeq L(7,2)$ since $1 \cdot 2 \equiv 3^2$ (mod 7). Also see Milnor [4]. Another curious fact is that the connected sum of three projective planes embeds in $L(14,3)$ but not in $L(14,1)$ even though $L(14,1) \simeq L(14,3)$ since $1 \cdot 3 \equiv -5^2$ (mod 14); see Bredon and Wood [1].

PROBLEMS

1. If V^{2n+1} is a compact manifold with boundary M^{2n} then show that $\chi(M^{2n}) = 2\chi(V^{2n+1})$.

2. If M^{4n+2} is a closed orientable manifold, show that $\chi(M)$ is even.

3. ◆ The generalized lens space $L^{2n-1}(p; q_1, \ldots, q_n)$ is defined to be the orbit space of the free action of $\mathbf{Z}_p$ on $\mathbf{S}^{2n-1}$, the unit sphere in $\mathbf{C}^n$, generated by $(u_1, \ldots, u_n) \mapsto (\omega^{q_1} u_1, \ldots, \omega^{q_n} u_n)$ where the q_i are relatively prime to p. If $L^{2n-1}(p; q_1, \ldots, q_n) \simeq L^{2n-1}(p; q_1', \ldots, q_n')$, show that $q_1' \cdots q_n' \equiv \pm q_1 \cdots q_n k^n \pmod{p}$ for some integer k. (*Hint*: First show, as in Proposition 10.2, that $H^i(L^{2n-1}(p; q_1, \ldots, q_n); \mathbf{Z}_p) \approx \mathbf{Z}_p$ for $0 \le i \le 2n-1$ and that there is a one-dimensional generator a such that, if $b = \beta(a)$, then b^k generates H^{2k} and ab^k generates H^{2k+1}; i.e., $H^*(L^{2n-1}(p; q_1, \ldots, q_n); Z_p) \approx \wedge(a) \otimes \mathbf{Z}_p[b]/(b^n)$, for p odd.) The converse is also true.

4. If the lens space $L(p, q)$ admits a homotopy equivalence $L(p, q) \to L(p, q)$ which reverses orientation then show that -1 is a quadratic residue mod p.

5. Show that the connected sum $L(3, 1) \# L(3, 1)$ is not homotopy equivalent to $L(3, 1) \# -L(3, 1)$.

6. Reprove Theorem 20.6 of Chapter IV from the point of view of the present section.

7. Show that there can be no 5-manifold M^5 with $H_0(M) \approx \mathbf{Z}$, $H_2(M) \approx \mathbf{Z}_3$, $H_5(M) \approx \mathbf{Z}$ and all other homology groups zero. (*Hint*: Use Problem 8 of Section 4.)

8. ◆ Let M^3 be a closed oriented 3-manifold. Consider the exact sequence

$$0 \to H_2(M) \otimes \mathbf{Q}/\mathbf{Z} \to H_2(M; \mathbf{Q}/\mathbf{Z}) \xrightarrow{\gamma} TH_1(M) \to 0$$

of Example 7.6 of Chapter V. Let $D: H_2(M; \mathbf{Q}/\mathbf{Z}) \xrightarrow{\approx} H^1(M; \mathbf{Q}/\mathbf{Z})$ be the inverse of the Poincaré duality isomorphism $\cap[M]: H^1(M; \mathbf{Q}/\mathbf{Z}) \xrightarrow{\approx} H_2(M; \mathbf{Q}/\mathbf{Z})$. For $a, b \in TH_1(M)$ put $lk(a, b) = \langle D\gamma^{-1}(a), b \rangle \in \mathbf{Q}/\mathbf{Z}$. Show that this is a well-defined pairing $TH_1(M) \otimes TH_1(M) \to \mathbf{Q}/\mathbf{Z}$. (It is called the "linking pairing.") Also show that $lk(a, b) = lk(b, a)$. If $H_1(M) \approx \mathbf{Z}_p$ and $0 \ne a \in H_1(M)$ then show that $lk(a, a) \ne 0$.

11. Intersection Theory ☼

In this section we show how to define the "intersection" of homology classes in a compact oriented manifold. If the classes are fundamental classes of submanifolds in "general position" then we relate their intersection product to the physical intersection of the submanifolds. Also, the intersection product is the Poincaré dual of the cup product. This provides a nice geometric interpretation of the cup product which is intuitively useful. Through the correspondence, algebraic information about the ring structure in the cohomology of the manifold yields geometric information about the manifold, and conversely.

Historically, the idea of the intersection product preceded that of cohomology and hence of the cup product. Today, it is of less importance since, unlike the cup product, it is limited to homology classes in a manifold. It is due to Lefschetz.

Let M^n be a compact, oriented, connected manifold, possibly with boundary, although we are concerned mainly with the closed case. Let $D: H_i(M, \partial M) \to H^{n-i}(M)$ or $D: H_i(M) \to H^{n-i}(M, \partial M)$ be the inverse of the Poincaré duality isomorphism. That is,

$$\boxed{D(a) \cap [M] = a.}$$

If the manifold M must be specified, we use the notation D_M for D.

Define the "intersection product"

$\bullet$: $H_i(M) \otimes H_j(M) \to H_{i+j-n}(M)$, or
$\bullet$: $H_i(M, \partial M) \otimes H_j(M) \to H_{i+j-n}(M)$, or
$\bullet$: $H_i(M, \partial M) \otimes H_j(M, \partial M) \to H_{i+j-n}(M, \partial M)$, by

$a \bullet b = D^{-1}(D(b) \cup D(a)) = (D(b) \cup D(a)) \cap [M] = D(b) \cap (D(a) \cap [M]) = D(b) \cap a$, that is,

$$\boxed{D(a \bullet b) = D(b) \cup D(a).}$$

(Note the reversal in order.) We have

$$\boxed{a \bullet b = (-1)^{(n-\deg(a))(n-\deg(b))} b \bullet a.}$$

Also, $D(a \bullet (b \bullet c)) = D(b \bullet c) \cup D(a) = (D(c) \cup D(b)) \cup D(a) = \cdots = D((a \bullet b) \bullet c)$, so that

$$\boxed{a \bullet (b \bullet c) = (a \bullet b) \bullet c.}$$

Our aim is to give a way to see the intersection product geometrically. This can be used, for example, to compute the dual cup product from obvious geometric information. First we must define and study the "Thom class" of a disk bundle and the "Thom isomorphism."

Let N^n be a connected, oriented, closed n-manifold and let $\pi: W^{n+k} \to N^n$ be a k-disk bundle over N. For those unfamiliar with the concept of a bundle, this means the following: Each point of N has a neighborhood U such that $\pi^{-1}(U)$ has the structure of a product

$$\phi: \pi^{-1}(U) \approx U \times \mathbf{D}^k$$

such that the projection π to U corresponds to the projection $U \times \mathbf{D}^k \to U$. (For a smooth bundle, which we do not require here, ϕ must be smooth.) Over two such sets U and V with homeomorphisms ϕ and ψ, the composition

$\psi \circ \phi^{-1}$, defined on $U \cap V$, induces homeomorphisms $\theta_x: \mathbf{D}^k \to \mathbf{D}^k$ for $x \in U \cap V$. We require these maps θ_x to be linear.

Thus W^{n+k} is an $(n+k)$-manifold with boundary ∂W being a $(k-1)$-sphere bundle over N^n.

Let us also assume that W as well as N is oriented. (In this case the fibers are oriented and the linear maps θ_x preserve orientation. Presently, we will give the orientation of the fibers $\mathbf{D}^k$ specifically.) Alternatively to having W^{n+k} and N^n oriented, we could use $\mathbf{Z}_2$ coefficients in all the (co)homology in this section.

The origin in each fiber $\mathbf{D}^k$ provides an inclusion (a "section") of N in W which we will call $i: N \to W$. For simplicity of notation we will regard N^n as a subspace of W^{n+k}. Note that $i^*: H^*(W) \to H^*(N)$ is an isomorphism inverse to $\pi^*: H^*(N) \to H^*(W)$.

11.1. Definition. In the above situation, the *Thom class* of the disk bundle π is the class $\tau \in H^k(W, \partial W)$ given by

$$\tau = D_W(i_*[N]).$$

Equivalently,

$$\tau \cap [W] = i_*[N].$$

Sometimes we will regard $\tau \in H^k(W, W - N) \approx H^k(W, \partial W)$.

11.2. Definition (Hopf–Freudenthal). If $f: N^n \to M^m$ is a map from a compact, oriented n-manifold N to a compact, oriented m-manifold M, taking ∂N into ∂M, then

$$f^!: H^{n-p}(N) \to H^{m-p}(M) \qquad \text{and} \qquad f^!: H^{n-p}(N, \partial N) \to H^{m-p}(M, \partial M),$$

are both defined by

$$f^! = D_M f_* D_N^{-1}.$$

Also

$$f_!: H_{m-p}(M) \to H_{n-p}(N) \qquad \text{and} \qquad f_!: H_{m-p}(M, \partial M) \to H_{n-p}(N, \partial N),$$

are both defined by

$$f_! = D_N^{-1} f^* D_M.$$

These are called "transfer" maps or "shriek" maps. (Also see Section 14.)

11.3. Theorem (Thom Isomorphism Theorem). *If $\pi: W \to N$ is a k-disk bundle over the connected oriented closed n-manifold N^n, then there is the "Thom Isomorphism"*

$$H^p(N) \xrightarrow[\approx]{\pi^*} H^p(W) \xrightarrow[\approx]{\cup \tau} H^{p+k}(W, \partial W)$$

which coincides with $i^!$. Similarly, $i_! = \pm\pi_*(\tau\cap(\cdot)): H_{p+k}(W,\partial W)\to H_p(N)$ is an isomorphism.

PROOF. Note that $i^!$ is an isomorphism since i_* is, and similarly for $i_!$. Thus it suffices to prove that the two indicated maps coincide. Let $\beta = i^*(\alpha)$, so that $\alpha = \pi^*(\beta)$. Then we compute

$$\begin{aligned} i^!(\beta) &= D_W i_* D_N^{-1}(\beta) \\ &= D_W i_*(i^*(\alpha)\cap[N]) \\ &= D_W(\alpha\cap i_*[N]) \quad \text{(by Theorem 5.2)} \\ &= D_W(\alpha\cap(\tau\cap[W])) \quad \text{(by definition of } \tau) \\ &= D_W((\alpha\cup\tau)\cap[W]) \quad \text{(by Theorem 5.2)} \\ &= \alpha\cup\tau = \pi^*(\beta)\cup\tau. \end{aligned}$$

Similar computations yield the homology isomorphism, which we will not need. □

Now we wish to prove a similar Thom Isomorphism for disk bundles over base spaces that are not necessarily manifolds, but which are embedded in manifolds.

Assume that, in the above setup, $A\subset N$ is a closed subset. Put $\tilde{A} = \pi^{-1}(A)\subset W$ and $\partial\tilde{A} = \tilde{A}\cap\partial W$. (By this notation we do not imply that $\partial\tilde{A}$ is a manifold or the boundary of anything.)

11.4. Lemma. *If $A\subset N$ is closed, then $\check{H}^i(\tilde{A},\partial\tilde{A}) = 0$ for $i<k$.*

PROOF. Let $P_N(A)$ be the statement that the conclusion of this lemma is true for the closed set A. Then $P_N(A)$ is certainly true if A is a compact convex subset of a euclidean set in N. The Mayer–Vietoris sequence shows that $P_N(A), P_N(B), P_N(A\cap B)\Rightarrow P_N(A\cup B)$. The fact that Čech cohomology commutes with direct limits (Theorem 8.3, proof of (iii)), shows that if $\{A_i\}$ is a decreasing sequence of closed sets, then $P_N(A_i)$, all i, $\Rightarrow P_N(\bigcap A_i)$. Thus the present lemma follows from the Bootstrap Lemma (Lemma 7.9). □

11.5. Lemma. *The restriction $\tau_x\in\check{H}^k(\tilde{A},\partial\tilde{A})$ of τ, when $A=\{x\}$, is a generator.*

PROOF. Note that here, $(\tilde{A},\partial\tilde{A})\approx(\mathbf{D}^k,\mathbf{S}^{k-1})$. Suppose, first, that $\tau_x = 0$ for some x. Then consideration of a neighborhood (as A) of x in N shows that $\tau_y = 0$ for all y near x. Since N is connected this implies that $\tau_y = 0$ for all $y\in N$.

For closed sets $A\subset N$, let $P_N(A)$ be the statement that $\tau_A = 0$, where τ_A is the restriction of τ to $(\tilde{A},\partial\tilde{A})$. Then $P_N(A)$ is true when A is a convex set in some euclidean open set in N, since the restriction to a point in the set is an isomorphism on the cohomology involved and $\tau_x = 0$.

If $P_N(A)$ and $P_N(B)$ hold, then $P_N(A \cup B)$ follows from the diagram

$$\begin{array}{ccccc} 0 & \longrightarrow & H^k(W \cup W, \partial(W \cup W)) & \longrightarrow & H^k(W, \partial W) \oplus H^k(W, \partial W) \\ & & \downarrow & & \downarrow \\ 0 & \longrightarrow & \check{H}^k(\tilde{A} \cup \tilde{B}, \partial(\tilde{A} \cup \tilde{B})) & \longrightarrow & \check{H}^k(\tilde{A}, \partial\tilde{A}) \oplus \check{H}^k(\tilde{B}, \partial\tilde{B}) \end{array}$$

which is a portion of a Mayer–Vietoris diagram, and where the zeros on the left are from Lemma 11.4.

If $P_N(A_i)$ holds for each set in a decreasing sequence of closed sets then P_N also holds for the intersection, by the fact that the direct limits commute with Čech cohomology, as in the proof of Lemma 11.4. Thus, by the Bootstrap Lemma (Lemma 7.9), we have that $P_N(N)$ holds. But this means that $\tau = 0$, and it is not.

We conclude that the hypothesis, that any $\tau_x = 0$, is untenable.

Suppose then that τ_x is not a generator for some x. Say $\tau_x \in H^k(\mathbf{D}^k, \mathbf{S}^{k-1}) \approx \mathbf{Z}$ is $\pm m$, and let p be a prime dividing m. Then if we pass to $\mathbf{Z}_p$ coefficients we deduce that τ_x becomes 0. This still provides a contradiction, as above, and proves the lemma. □

11.6. Corollary. *If N is connected then the class $\tau \in H^k(W, \partial W)$ is (up to sign) the unique class whose restriction to the fiber over each point is a generator. Thus we can think of τ as defining the orientation in the fibers.*

Proof. The cup product with τ gives the Thom Isomorphism $H^0(N) \approx H^0(W) \to H^k(W, \partial W)$. Thus τ generates $H^k(W, \partial W) \approx \mathbf{Z}$, which makes the claim obvious. □

11.7. Theorem (Thom Isomorphism Theorem). *For any compact $A \subset N$, the map*

$$\pi(\cdot) \cup \tau_A : \check{H}^i(A) \to \check{H}^{i+k}(\tilde{A}, \partial\tilde{A})$$

is an isomorphism.

Proof. This is another argument using the Bootstrap Lemma (Lemma 7.9), essentially the same as its use in Lemma 11.5. The details should be supplied with ease by the reader. □

Remark. One can get the Thom Isomorphism Theorem, from the above, for any oriented disk bundle over a compact space embeddable in euclidean space, since all such bundles arise as above. We indicate the proof of this: Embed the base space A in some sphere $\mathbf{S}^n$. The bundle then extends over a neighborhood of A in $\mathbf{S}^n$. (One must know a bit more about bundles than we are assuming to justify this statement.) One can then take a neighborhood in the form of a manifold with boundary. By doubling this (joining two copies along the common boundary) one gets a manifold N and a disk bundle over it whose restriction to the subspace A is the original bundle. Consequently, our assumptions are not as restrictive as they appear.

Now we continue with the program of showing how one can interpret the intersection product geometrically. Below, all manifolds will be assumed compact and oriented, and possibly with boundary. We also assume them to be smooth manifolds.

Let $i_N^W: N^n \to W^w$ be a smooth embedding of smooth manifolds with boundary, and assume that N meets ∂W transversely in ∂N.

11.8. Definition. In the above situation, denote $i_{N*}^W[N] \in H_n(W, \partial W)$ by $[N]_W$. Also define $\tau_N^W = D_W([N]_W) \in H^{w-n}(W)$, the *Thom class*. Here $D_W: H_n(W, \partial W) \xrightarrow{\approx} H^{w-n}(W)$.

Note that the Thom class τ_N^W is the *image* of the Thom class of the normal $(w-n)$-disk bundle ν_N^W of N in W via

$$H_{w-n}(\text{tube}, \partial\,\text{tube}) \approx H^{w-n}(W, W\text{-tube}) \to H^{w-n}(W),$$

where the first map is the inverse of the excision isomorphism. One can see this by "doubling" W to get rid of the boundary.

By dualizing the definition $\tau_N^W = D_W[N]_W$, we get

$$\text{(1)} \qquad \boxed{\tau_N^W \cap [W] = i_{N*}^W[N] = [N]_W.}$$

Now suppose that K^k and N^n are two such submanifolds of W^w. Then note that

$$\text{(2)} \qquad \boxed{[K]_W \bullet [N]_W = (\tau_K^W \cap [W]) \bullet (\tau_N^W \cap [W]) = (\tau_N^W \cup \tau_K^W) \cap [W].}$$

Again, our intention is to interpret this intersection product geometrically.

Now assume that K meets N transversely in W. (In symbols, $K \pitchfork N$.) Then ν_K^W restricts to $\nu_{K\cap N}^N$. (See Figure VI-7.) That is, $\nu_{K\cap N}^N = \nu_K^W|_{K\cap N}$.

This implies that the Thom class of K in W restricts to the Thom class of $K \cap N$ in N, since its restriction to any point gives the generator in the cohomology of the fiber modulo its boundary, and the Thom class is characterized by this. Of course, actually, this only characterizes the Thom class

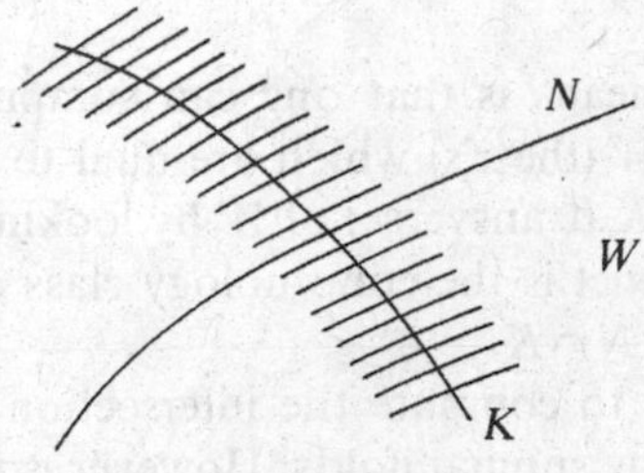

Figure VI-7. Transverse intersection and normal bundle.

up to sign, so demanding that the resulting class be the Thom class on the nose is really an orientation convention: it specifies the orientation of $K \cap N$ when that of N, K, and W are given. Note that it depends on the order in which we take K and N. In actual examples it is difficult to determine the correct orientation on $K \cap N$ from this discussion, but it usually does not matter much. We will indicate a practical way of doing that later. Thus we have the formula

$$\boxed{\tau^W_{K \cap N} = (i^W_N)^*(\tau^W_K).} \tag{3}$$

Note that changing the orientation of W has the effect of changing the sign of both Thom classes, and also changes the sign of the duality converting the cup product of the Thom classes to the intersection class. Thus, in all, reversing the orientation of W reverses the signs of intersection products.

Now we are finally able to give the geometric interpretation of the intersection product.

11.9. Theorem. *With the assumptions above, including that $K \pitchfork N$ in W, we have*

$$\tau^W_{K \cap N} = \tau^W_K \cup \tau^W_N$$

and, equivalently,

$$[K \cap N]_W = [N]_W \bullet [K]_W.$$

Proof. We compute:

$$\begin{aligned}
[K \cap N]_W &= (i^W_{K \cap N})_*[K \cap N] \\
&= (i^W_N)_*(i^N_{K \cap N})_*[K \cap N] \\
&= (i^W_N)_*(\tau^N_{K \cap N} \cap [N]) && \text{(by (1))} \\
&= (i^W_N)_*((i^W_N)^*(\tau^W_N) \cap [N]) && \text{(by (3))} \\
&= \tau^W_K \cap (i^W_N)_*[N] && \text{(by Theorem 5.2(4))} \\
&= \tau^W_K \cap (\tau^W_N \cap [W]) && \text{(by (1))} \\
&= (\tau^W_K \cup \tau^W_N) \cap [W] && \text{(by Theorem 5.2(3))} \\
&= [N]_W \bullet [K]_W && \text{(by (2))},
\end{aligned}$$

as claimed. □

What the theorem means is that one can compute the cup product of cohomology classes of W (the τ's) which are dual to the orientation classes of submanifolds N and K (transverse) of W by looking at the intersection of N and K. The cup product is the cohomology class dual to the orientation class of the intersection $N \cap K$.

We have shown how to compute the intersection product geometrically only for classes carried by submanifolds. However, some conclusions can be derived in more generality. For example:

11.10. Theorem. *Let M^n be an orientable closed manifold and let $A, B \subset M$. Let $\alpha \in H_p(A)$ and $\beta \in H_q(B)$ and let α_M and β_M denote their images in $H_*(M)$. If $\alpha_M \bullet \beta_M \neq 0$ then $A \cap B \neq \emptyset$.*

PROOF. Suppose A and B are open. Let D_M^A be the duality isomorphism $D_M^A: H_p(A) \to \check{H}^{n-p}(M, M-A)$ and similarly with B. Then $D_M^B(\beta) \cup D_M^A(\alpha) \in \check{H}^{2n-p-q}(M, M-(A \cap B))$ and so we could define

$$\alpha \bullet \beta = (D_M^B(\beta) \cup D_M^A(\alpha)) \cap [M] \in H_{p+q-n}(A \cap B)$$

and it is easy to see that the map $H_{p+q-n}(A \cap B) \to H_{p+q-n}(M)$, induced by inclusion, takes $\alpha \bullet \beta$ into $\alpha_M \bullet \beta_M$, so that the latter would have to be zero if $A \cap B = \emptyset$. This proves the open case. Unfortunately, one cannot define such intersection products for arbitrary (or even closed) subsets, in general, in singular theory. (That can be done in more advanced theories such as sheaf cohomology and Borel–Moore homology.)

However, if A and B are arbitrary and $A \cap B = \emptyset$ then, since every singular chain is carried by a compact set, α and β are images of classes in some compact subsets $K \subset A$ and $L \subset B$. Then K and L have open neighborhoods U and V with $U \cap V = \emptyset$ since M is normal. Passage to the induced classes $\alpha_U \in H_p(U)$ and $\beta_V \in H_q(V)$ then provides the desired conclusion. □

Implicit in the proof of Theorem 11.10, of course, is the statement that if $\alpha, \beta \in H_*(M)$ are carried by subsets A and B of M then $\alpha \bullet \beta$ is represented by a cycle carried by any given neighborhood of $A \cap B$. Figure VI-8 illustrates a typical example showing that "any given neighborhood" cannot be deleted from the previous sentence. The manifold M is the 3-torus depicted as a cube of side 2 about the origin with opposite faces identified. To explain the example: one starts with sets A_0 and B, both tori in M. The torus A_0 is given by $z = 0$. The torus B is the cylinder along the x-axis based on the path from $(y, z) = (-1, -1)$ to $(-\frac{1}{2}, 0)$ to $(\frac{1}{2}, 0)$ to $(1, 1)$ in the y–z-plane. The torus B can be deformed to the torus B_0 which is the cylinder based on the curve $y = z$. Now $A_0 \cap B_0$ is a circle which carries a nonzero 1-homology class which is $[A_0] \bullet [B_0] = [A_0] \bullet [B]$. (Note that $A_0 \pitchfork B_0$.) The right side of the

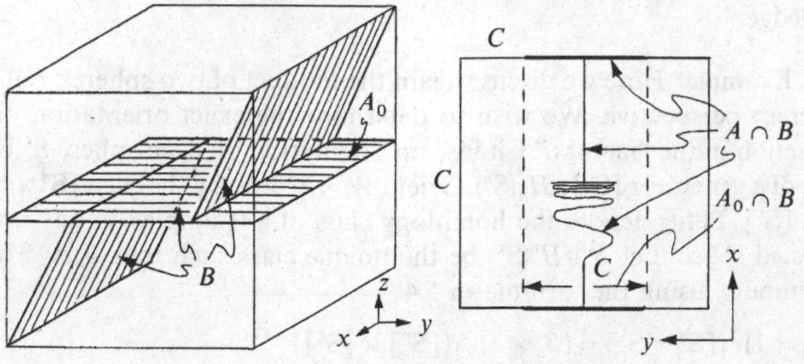

Figure VI-8. Wild intersection in 3-torus.

diagram represents the torus A_0; i.e., the square in the x–y-plane. Let C in this square be the union of the boundary of the square with a $\sin(1/x)$ type curve running from $(x, y) = (-1, 0)$ to $(1, 0)$ and contained in $|y| \le \frac{1}{2}$. The set C divides the square into two pieces that we will call the left and right sides. Define a function $f(x, y)$ on the square by

$$f(x,y) = \begin{cases} \operatorname{dist}((x,y),C) & \text{on the left,} \\ -\operatorname{dist}((x,y),C) & \text{on the right.} \end{cases}$$

Let $A = \operatorname{graph}(f)$ as a subset of M. Then A is a torus that deforms in M to A_0. Thus $[A]\bullet[B] = [A_0]\bullet[B_0] \neq 0$. However $A \cap B$ is the $\sin(1/x)$ type curve together with the segment $\{\pm 1\} \times [-\frac{1}{2}, \frac{1}{2}] \times \{0\}$ with the ± 1 being identified in M. Hence $A \cap B$ is simply connected and so carries no nonzero 1-class, finishing the example.

We now turn to some applications of these results on intersections.

11.11. Example. Perhaps the simplest example is the product of spheres $W = \mathbf{S}^n \times \mathbf{S}^m$. Let $K = \mathbf{S}^n \times \{y_0\}$ and $N = \{x_0\} \times \mathbf{S}^m$. Then $[K]_W$ and $[N]_W$ generate $H_n(W)$ and $H_m(W)$. Let $\alpha = \tau_K^W$ and $\beta = \tau_N^W$ be the dual classes. Then, dropping the subscript "W,"

$$[N]\bullet[K] = [K \cap N] = \pm[(x_0, y_0)],$$

which is a generator of $H_0(W)$. Thus $\alpha \cup \beta$ is dual to $\pm[(x_0, y_0)]$, and so it is a generator of $H^{n+m}(\mathbf{S}^n \times \mathbf{S}^m)$. (Note that we don't even have to know that α and β are generators of their respective cohomology classes, since, if they were not, then their cup product could not possibly be a generator of its.)

Similarly, $[K]\bullet[K] = [K]\bullet[K'] = [\varnothing] = 0$, where K' is a shifted copy of K, and this implies that $\alpha^2 = 0$. Similarly, $\beta^2 = 0$. Thus, this obvious geometrical information about K and N completely determines the cohomology ring.

Still in the example of a product of spheres, but looking at it from the other way around, the fact that we know there are the cohomology classes α and β whose cup product is nonzero (say, from Poincaré duality, cup product version) implies that the dual homology classes carried by submanifolds *must* intersect. This gives a geometric consequence of algebraic knowledge.

11.12. Example. Here we discuss again the product of two spheres, but from a different perspective. We wish to determine the exact orientation on the intersection of the "factors." Choose an orientation on each sphere $\mathbf{S}^n$ by the choice of a generator $[\mathbf{S}^n] \in H_n(\mathbf{S}^n)$. Orient $W = \mathbf{S}^n \times \mathbf{S}^k$ by the class $[\mathbf{S}^n \times \mathbf{S}^k] = [\mathbf{S}^n] \times [\mathbf{S}^k]$. Let $\star$ denote the homology class of a 0-simplex in any arcwise connected space. Let $\vartheta_n \in H^n(\mathbf{S}^n)$ be the unique class such that $\vartheta_n \cap [\mathbf{S}^n] = \star$. We compute, using the formula in 5.4:

$$\begin{aligned} (\vartheta_n \times 1) \cap [\mathbf{S}^n \times \mathbf{S}^k] &= (\vartheta_n \times 1) \cap ([\mathbf{S}^n] \times [\mathbf{S}^k]) \\ &= (\vartheta_n \cap [\mathbf{S}^n]) \times (1 \cap [\mathbf{S}^k]) = \star \times [\mathbf{S}^k], \end{aligned}$$

$$(1 \times \vartheta_k) \cap [\mathbf{S}^n \times \mathbf{S}^k] = (-1)^{nk}(1 \cap [\mathbf{S}^n]) \times (\vartheta_k \cap [\mathbf{S}^k]) = (-1)^{nk}[\mathbf{S}^n] \times \star.$$

It follows that $D([\mathbf{S}^n] \times ☼) = (-1)^{nk}(1 \times \vartheta_k)$ and $D(☼ \times [\mathbf{S}^k]) = \vartheta_n \times 1$. Thus

$$\begin{aligned}[\mathbf{S}^n] \bullet [\mathbf{S}^k] &= \{(\vartheta_n \times 1) \cup (-1)^{nk}(1 \times \vartheta_k)\} \cap [\mathbf{S}^n] \times [\mathbf{S}^k] \\ &= (-1)^{nk}(\vartheta_n \times \vartheta_k) \cap ([\mathbf{S}^n] \times [\mathbf{S}^k]) \\ &= (-1)^{nk}(-1)^{nk}(\vartheta_n \cap [\mathbf{S}^n]) \times (\vartheta_k \cap [\mathbf{S}^k]) \\ &= ☼ \times ☼ = ☼.\end{aligned}$$

The implication of this is as follows. If N^n and K^k are smooth, compact, connected manifolds meeting transversely in M^{n+k} then at each point x in the finite set $N \cap K$, the sign of $[x]$ in the class $[N] \bullet [K]$ is determined as follows (note that the order of writing N and K is important). Choose a frame at x in N, consistent with the orientation of N, follow it by such a frame in K and compare it with such a frame in the total space W. If it matches, in the sense of defining the same orientation, then assign $[x]$ a plus sign, else give $[x]$ a minus sign. (In a coordinate system about x, this can be done by writing down tangent vectors to N as column vectors, then those of K. Give $[x]$ the sign of the resulting determinant.) Of course, we have not proved this contention other than for the product of spheres example, but it is strongly intuitively plausible. A proof would require extending the range of definition of the intersection product to relative classes, and discussion of naturality. Since we don't really need this, and since it is not very important, we omit the details. The idea, however, is contained in the proof of Theorem 11.10.

In Problem 1, the more general case of the product of three spheres is considered. That allows a general description of how the intersection is oriented. Here is the description of that (without proof): If N^n and K^k are compact, oriented manifolds in the compact oriented manifold W^w, then the intersection product $[N] \bullet [K] = [K \cap N]$, where the component of $K \cap N$ containing a point x is oriented as follows. Take a frame at x in $K \cap N$, then fill it out to be a frame of N, consistent with the orientation of N, then fill it out with vectors tangent to K so that the frame of $K \cap N$ and these last vectors make a frame of K consistent with its orientation. By transversality, these give a full frame of W. If this frame is consistent with the orientation of W, then give the component of $K \cap N$ containing x the orientation defined by the frame chosen in this procedure. If not, give it the opposite orientation. (It should be noted that this works out this nice way because of the reversal of order in the definition of the intersection product $[N] \bullet [K] = (v_K \cup v_N) \cap [W]$. Since some other authors do not reverse the order, there is a sign difference between our intersection product and theirs. Our intersection product does agree with that of most *classical* authors, e.g., Seifert and Threlfall [1].)

11.13. Example. Consider $W = \mathbf{CP}^2$. The two copies of $\mathbf{CP}^1$ given in homogeneous coordinates by $\{(z_0 : z_1 : 0)\}$ and $\{(0 : z_1 : z_2)\}$ intersect in a single point $\mathbf{CP}^0$. If we call one of these N and the other K, we deduce that $[K] \bullet [N] = [N \cap K]$, which is a generator of $H_0(\mathbf{CP}^2)$. It follows, as if we did not already know it, that both of these are generators of $H_2(\mathbf{CP}^2)$, so $[N] = \pm [K]$ with sign depending on how one chooses the orientations, and not mattering much. If α is the dual class to one of these, then we conclude that α^2 generates $H^4(\mathbf{CP}^2)$. This completely characterizes the ring structure of $H^*(\mathbf{CP}^2)$.

11.14. Example. Consider $W = \mathbf{P}^2$, the real projective plane. This is not orientable, but the results of this section still apply as long as we use (co)homology with coefficients in $\mathbf{Z}_2$ exclusively. In the common picture of the projective plane as a disk with antipodal points on the boundary identified, we know that a curve going from a point on the boundary, through the middle and to the antipodal point on the boundary represents the nonzero element of $H_1(\mathbf{P}^2; \mathbf{Z}_2)$, since it represents the generator of $\pi_1(\mathbf{P}^2)$. One can take two such curves that intersect transversely in one point. It follows that the dual cohomology class $\alpha \in H^1(\mathbf{P}^2; \mathbf{Z}_2)$ must have $\alpha^2 \neq 0$, an algebraic conclusion from a simple geometric fact.

On the other hand, the fact that $\alpha^2 \neq 0$ implies that *any* two curves joining antipodal points on the boundary of the disk *must* intersect, a geometric conclusion from an algebraic fact.

11.15. Example. Suppose that N and K are submanifolds of W and have one, or any odd number of transverse intersection points. Then $[K] \in H_*(W; \mathbf{Z}_2)$ cannot be zero. For example, suppose that we have an embedding $N^n \subset W^{n+1}$ that is one-sided. That is, one can follow a normal vector around some curve in N and come back to the original point with the normal reversed. Then the tip of the normal describes a path in the complement of N and we can join its ends by an arc cutting through N at the single, original, point. This gives a circle that intersects N transversely in a single point. Thus, the circle defines a nonzero class in $H_1(W; \mathbf{Z}_2)$. (Also $[N]$ gives a nonzero class in $H_n(W; \mathbf{Z}_2)$.) Thus, we have shown that if $H_1(W; \mathbf{Z}_2) = 0$, then one cannot embed a codimension 1 surface in W so as to be one-sided. The condition $H_1(W; \mathbf{Z}_2) = 0$ implies that W is orientable, and in that case "one-sided" is the same as N being nonorientable. So one could rephrase the conclusion to say that such a W cannot contain a codimension 1, nonorientable manifold, a result we had previously by another method.

The question arises as to which homology classes can be realized by embedded submanifolds. A general answer is that not all of them can be so realized but enough to them can be realized so as to be useful. We conclude this section by showing that all codimension 1 and 2 classes of a smooth manifold can be so realized as the fundamental classes of embedded smooth submanifolds. The proof requires Hopf's Theorem from Section 11 of Chapter V. The proof for codimension 2 requires a result from Chapter VII and also Section 15 of Chapter II. However, this material will not be used elsewhere. In the case of codimension 1, the submanifold must be allowed to be disconnected as the example of $\mathbf{S}^1$ shows. (Also see Problem 6.) For higher codimension, one can always connect components of the submanifold by running tubes from one to another. (This does not change the represented homology class since the difference is a boundary, the boundary of the solid tubes.) For a much deeper exploration of this matter, see Thom [1].

11.16. Theorem (Thom). *If M^n is a smooth, orientable, closed manifold then any homology class in $H_{n-1}(M)$ or in $H_{n-2}(M)$ is represented by the fundamental class of a smooth submanifold, possibly disconnected in the case of codimension* 1.

PROOF. First consider the case of a class $a \in H_{n-1}(M^n) \approx H^1(M^n)$. By Hopf's Theorem 11.6 of Chapter V and Theorem 11.9 of Chapter V there is a map $f: M \to \mathbf{S}^1$ such that $D_M(a) = f^*(u)$ where $u \in H^1(\mathbf{S}^1)$ is a generator. We can assume that f is smooth. Let $x \in \mathbf{S}^1$ be a regular value. Then $u = D_{\mathbf{S}^1}[x]$, the Thom class of the normal bundle of $\{x\}$ in $\mathbf{S}^1$. We claim that $N^{n-1} = f^{-1}(x)$ is the required manifold. Note that f gives a bundle map from the normal bundle of N to that of $\{x\}$. It follows from this and Corollary 11.6 that $f^*(u)$ is the Thom class of the normal bundle of N. That is, $f^*(u) = D_M[N]_M$, and so $D_M(a) = f^*(u) = D_M[N]_M$. Consequently, $a = [N]_M$ as claimed.

The case of codimension 2 is similar but uses the fact from Section 12 of Chapter VII that $H^2(M^n) \approx [M; \mathbf{CP}^k]$ for k large; $k > n$ is more than large enough. Again the correspondence is $f^*(\alpha) \leftrightarrow [f]$ where $f: M \to \mathbf{CP}^k$ and $\alpha \in H^2(\mathbf{CP}^k)$ is a generator. Note that α is dual to $[\mathbf{CP}^{k-1}] \in H_{2k-2}(\mathbf{CP}^k)$; see Problem 3. Instead of taking a regular value, one modifies f by a homotopy to be a map transverse to $\mathbf{CP}^{k-1}$ using Corollary 15.6 of Chapter II. (Actually the simpler Corollary 15.4 of Chapter II suffices, by moving $\mathbf{CP}^{k-1}$ instead.) Assuming, then, that $f \pitchfork \mathbf{CP}^{k-1}$ we put $N^{n-2} = f^{-1}(\mathbf{CP}^{k-1})$. Again f gives a bundle map from the normal bundle of N in M to that of $\mathbf{CP}^{k-1}$ in $\mathbf{CP}^k$. The remainder of the argument is identical to that of codimension 1, and shows that $[N]_M$ is dual to $f^*(\alpha)$, which is an arbitrary class in $H^2(M) \approx H_{n-2}(M)$. □

PROBLEMS

1. Consider the product $\mathbf{S}^n \times \mathbf{S}^k \times \mathbf{S}^r$. Orient such that $[\mathbf{S}^n \times \mathbf{S}^k] = [\mathbf{S}^n] \times [\mathbf{S}^k]$, and so on, keeping the ordering of the three spheres as given. Also, for notational purposes, identify $\mathbf{S}^n \times \mathbf{S}^k$ with $\mathbf{S}^n \times \mathbf{S}^k \times ☼$, etc. Then show that:

$$[\mathbf{S}^n \times \mathbf{S}^k] \bullet [\mathbf{S}^n \times \mathbf{S}^r] = [\mathbf{S}^n] \quad (\text{meaning } [\mathbf{S}^n] \times ☼ \times ☼),$$
$$[\mathbf{S}^n \times \mathbf{S}^k] \bullet [\mathbf{S}^k \times \mathbf{S}^r] = [\mathbf{S}^k],$$
$$[\mathbf{S}^n \times \mathbf{S}^r] \bullet [\mathbf{S}^k \times \mathbf{S}^r] = [\mathbf{S}^r].$$

2. Discuss the intersection theory of $\mathbf{CP}^n \# \mathbf{CP}^n$.

3. ✧ Use intersection theory to show that the fundamental class of $\mathbf{CP}^i \subset \mathbf{CP}^n$ generates $H^{2i}(\mathbf{CP}^n)$ for all $0 \le i \le n$.

4. For homology classes a, b of M^n, show that $a \bullet b = (-1)^{n(n-\deg a)} d_!(b \times a)$, where $d: M \to M \times M$ is the diagonal map.

5. Prove the omitted homology case of Theorem 11.3 determining the appropriate sign.

6. If M^n is a smooth closed orientable n-manifold and $N^{n-1} \subset M^n$ is a smoothly embedded closed connected oriented $(n-1)$-manifold with $0 \neq [N]_M \in H_{n-1}(M)$ then show that $[N]_M$ is indivisible; i.e., $[N]_M \neq ka$ for any integer $k > 1$ and class $a \in H_{n-1}(M)$. (This can be proved without smoothness, but is easier with it.)
7. For $a, b \in H_*(M^m)$ and $c, d \in H_*(N^n)$ show that
$$(a \bullet b) \times (c \bullet d) = (-1)^{(m-\deg a)(n-\deg d)}(a \times c) \bullet (b \times d)$$
and use this to rework Problem 1.
8. Use intersection theory to determine the cohomology ring of a 2-sphere with k handles attached (i.e., the connected sum of k tori).

12. The Euler Class, Lefschetz Numbers, and Vector Fields ☼

In this section we introduce the "Euler class" of a vector bundle over a manifold M, and, in particular, of the tangent bundle, or of the normal bundle of M when M is embedded in another manifold. This is closely related to the Thom class and also to the Euler characteristic, from which it derives its name. We use this to give a geometric version of the Lefschetz Fixed Point Theorem (in Theorem 12.6) which was the original viewpoint of Lefschetz. This is then applied to the study of vector fields, culminating in the classical theorem of Poincaré and Hopf concerning the "indices" of vector fields.

Since a vector bundle over a manifold M is the normal bundle of M in the total space of the bundle, it suffices to treat normal bundles.

12.1. Definition. Let N^n be a smoothly embedded submanifold of W^w, both oriented, and with N meeting ∂W transversely in ∂N, if there are boundaries. Then the *Euler class* of the normal bundle to N in W is
$$\chi_N^W = (i_N^W)^*(\tau_N^W) \in H^{w-n}(N).$$

Since $\tau_N^W \in H^{w-n}(W)$, the Euler class must vanish if this group is zero. This is a triviality, but it is useful enough to write down:

12.2. Proposition. *In the above situation, if $H^{w-n}(W) = 0$, then $\chi_N^W = 0$.* □

Let T be a closed tubular neighborhood of N in W. Then there is the commutative diagram
$$\begin{array}{ccccc} H^{w-n}(W, W-T) & \longrightarrow & H^{w-n}(W) & & \\ \downarrow\approx & & \downarrow & & \\ H^{w-n}(T, \partial T) & \longrightarrow & H^{w-n}(T) & \xrightarrow{\approx} & H^{w-m}(N). \end{array}$$
The Thom class τ is in $H^{w-n}(T, \partial T)$ and maps into the class $\tau_N^W \in H^{w-n}(W)$. Both map into the Euler class $\chi_N^W \in H^{w-n}(N)$.

Suppose, in this situation, that there is a nonzero section of the normal bundle, i.e., a map $N \to \partial T$ such that the composition $N \to \partial T \to N$ with the projection is the identity. Then the second map in the exact sequence $H^*(T, \partial T) \to H^*(T) \to H^*(\partial T)$ is a monomorphism, and so the first map is zero. But, by the above diagram, that implies that the Euler class is zero. This proves:

12.3. Theorem. *If there is a nonzero cross section of the normal bundle of N in W (both oriented) then the Euler class* χ_N^W *of the normal bundle of N in W is zero.* □

Thus if the Euler class is not zero then there is no nonzero normal section. One speaks of the Euler class as an "obstruction" to having a nonzero normal section.

We wish to study the Euler class of the tangent bundle. To do that we note that the tangent bundle is realized as the normal bundle of the diagonal Δ in $N \times N$. To see this, note that the tangent bundle of $N \times N$ is just the product of the tangent bundle of N with itself. The tangent bundle of Δ sits in this as the diagonal. Either one of the factors in the tangent bundle of $N \times N$, restricted to Δ, meets the tangent bundle of Δ trivially, and thus projects isomorphically onto the quotient when we divide out by the tangent bundle of Δ. But that quotient is just the normal bundle of Δ in $N \times N$ by definition, showing that the tangent bundle of N is isomorphic to this normal bundle.

For the next few results, we are going to assume that N^n is a closed, oriented n-manifold. At the end of the section we shall show how to generalize to the bounded case. Put $W = N \times N$ and let $d: N \to \Delta \subset W$ be the diagonal map. Then put

$$\boxed{\tau = \tau_\Delta^{N \times N} \in H^n(N \times N),}$$

which is the Thom class of the tangent bundle of N. Also put

$$\boxed{\chi = d^*(\tau) = d^*(\tau|_\Delta) = d^*(\chi_\Delta^{N \times N}) \in H^n(N),}$$

the "Euler class of the tangent bundle" of N.

We wish to "compute" this class. Recall the properties of the Kronecker product given in Corollary 5.3. They will be used extensively in the remainder of this section.

12.4. Theorem. *Take coefficients of (co)homology in some field. Let* $B = \{\alpha\}$ *be a basis of* $H^*(N)$. *Let* $\{\alpha^\circ\}$ *be the dual basis of* $H^*(N)$. *That is,*

$$\langle \alpha^\circ \cup \beta, [N] \rangle = \delta_{\alpha,\beta}.$$

(Note that $\deg(\alpha^\circ) = n - \deg(\alpha)$.*) Then*

$$\tau = \sum_{\alpha \in B} (-1)^{\deg(\alpha)} \alpha^\circ \times \alpha \in H^n(N \times N).$$

Therefore, also

$$\chi = d^*\tau = \sum_{\alpha \in B} (-1)^{\deg(\alpha)} \alpha^\circ \cup \alpha.$$

PROOF. By the cohomology version of the Künneth Theorem (Theorem 3.2), we can write

$$\tau = \sum A_{\alpha',\beta'} \alpha'^\circ \times \beta'.$$

Note that $\deg(\alpha') = \deg(\beta')$. For the following computation, choose basis elements α, β of degree p. Then we shall compute $\langle (\alpha \times \beta^\circ) \cup \tau, [W] \rangle$ in two ways:

$$\begin{aligned}
\langle (\alpha \times \beta^\circ) \cup \tau, [W] \rangle &= \langle \alpha \times \beta^\circ, \tau \cap [W] \rangle \\
&= \langle \alpha \times \beta^\circ, d_*[N] \rangle \\
&= \langle d^*(\alpha \times \beta^\circ), [N] \rangle \\
&= \langle \alpha \cup \beta^\circ, [N] \rangle \\
&= (-1)^{p(n-p)} \langle \beta^\circ \cup \alpha, [N] \rangle \\
&= (-1)^{p(n-p)} \delta_{\alpha,\beta}.
\end{aligned}$$

On the other hand

$$\begin{aligned}
\langle (\alpha \times \beta^\circ) \cup \tau, [W] \rangle &= \langle (\alpha \times \beta^\circ) \cup \textstyle\sum A_{\alpha',\beta'} (\alpha'^\circ \times \beta'), [W] \rangle \\
&= (-1)^{n-p} A_{\alpha,\beta} \langle (\alpha \cup \alpha^\circ) \times (\beta^\circ \cup \beta), [N] \times [N] \rangle \\
&\qquad \text{(since one gets zero for } \alpha', \beta' \neq \alpha, \beta, \text{ all of degree } p) \\
&= (-1)^{n-p+p(n-p)+n} A_{\alpha,\beta} \langle \alpha^\circ \cup \alpha, [N] \rangle \langle \beta^\circ \cup \beta, [N] \rangle \\
&= (-1)^{p(n-p)-p} A_{\alpha,\beta}.
\end{aligned}$$

Thus we conclude that $A_{\alpha,\beta} = (-1)^p \delta_{\alpha,\beta}$. □

12.5. Corollary. *The evaluation of the Euler class on the orientation class is the Euler characteristic. That is,* $\langle \chi, [N] \rangle = \chi(N)$.

PROOF. We can use the rational field for calculation. Then we have

$$\langle \chi, [N] \rangle = \sum_\alpha (-1)^{\deg(\alpha)} \langle \alpha^\circ \cup \alpha, [N] \rangle = \sum_\alpha (-1)^{\deg(\alpha)} = \chi(N),$$

since the sums are over a basis of cohomology. □

Suppose now that $f: N \to N$ is a map of a closed, oriented, connected manifold into itself. Let $(1 \times f): N \to N \times N$ denote the composition $(1 \times f) \circ d$. Then $(1 \times f)(N) = \{(x, f(x)) \in N \times N\} = \Gamma$ is the graph of f. Orient Γ by $[\Gamma] = (1 \times f)_* [N]$ and $W = N \times N$ by $[W] = [N] \times [N]$. Define the "intersection number" of $[\Gamma]$ and $[\Delta]$ as

$$\boxed{[\Gamma] \cdot [\Delta] = \epsilon_*([\Gamma] \bullet [\Delta]).}$$

12.6. Theorem. *The Lefschetz number $L(f) = [\Gamma]\cdot[\Delta]$, the intersection number.*

PROOF. Let $\gamma \in H^n(W)$ be the Poincaré dual to Γ, i.e., $\gamma \cap [W] = [\Gamma]$. Let $f^*(\alpha) = \sum_\beta f_{\beta,\alpha}\beta$, where α, β run over a basis of $H^*(N;\mathbf{Q})$. As above, let τ be the Thom class of the tangent bundle, i.e., the dual of the diagonal $[\Delta]$. We compute

$$\begin{aligned}
[\Gamma]\cdot[\Delta] &= \epsilon_*([\Gamma]\bullet[\Delta]) = \epsilon_*((\tau\cup\gamma)\cap[W])\\
&= \langle \tau\cup\gamma, [W]\rangle = \langle \tau, \gamma\cap[W]\rangle\\
&= \langle \tau, [\Gamma]\rangle = \langle \tau, (1\times f)_*[N]\rangle\\
&= \langle (1\times f)^*(\tau), [N]\rangle\\
&= \sum_\alpha (-1)^{\deg(\alpha)}\langle (1\times f)^*(\alpha^\circ \times \alpha), [N]\rangle\\
&= \sum_\alpha (-1)^{\deg(\alpha)}\langle \alpha^\circ \cup f^*(\alpha), [N]\rangle\\
&= \sum_\alpha (-1)^{\deg(\alpha)}\left\langle \alpha^\circ \cup \sum_\beta f_{\beta,\alpha}\beta, [N]\right\rangle\\
&= \sum_\alpha (-1)^{\deg(\alpha)} f_{\alpha,\alpha}\\
&= L(f). \qquad \square
\end{aligned}$$

From this, but just for manifolds, we again deduce that if f has no fixed points then $L(f) = 0$, since $\Gamma \cap \Delta = \varnothing$ in that case. This holds even when f is not smooth and N is only a topological manifold since the dual classes γ, τ live in $H^*(W, W - \Gamma)$ and $H^*(W, W - \Delta)$, respectively, so that $\gamma \cup \tau \in H^*(W, W) = 0$ when $\Gamma \cap \Delta = \varnothing$.

Note that Theorem 12.6 implies that $L(f)$ can be computed from local data about the fixed points if f is in "general position," i.e., $\Gamma \pitchfork \Delta$. In fact:

12.7. Corollary. *If N is a smooth, closed, orientable manifold and $f: N \to N$ is smooth and such that the differential f_{*_x} does not have 1 as an eigenvalue at any fixed point x of f, then*

$$L(f) = \sum_x \operatorname{sign}(\det(I - f_{*_x}))$$

where the sum is over all fixed points x.

PROOF. The hypothesis can easily be seen to be equivalent to $\Gamma \pitchfork \Delta$. Thus, we need only determine the "orientation" (sign) to be attached to each point of $\Gamma \cap \Delta$ to give the proper intersection product $[\Gamma]\bullet[\Delta]$. According to Example 11.12, we just have to write down tangent vectors to Γ and then to Δ and compute the sign of the resulting determinant. This gives the determinant:

$$\begin{vmatrix} I & I \\ f_* & I \end{vmatrix} = \begin{vmatrix} I - f_* & 0 \\ f_* & I \end{vmatrix} = |I - f_*|$$

as claimed. Even if one does not want to take advantage of Example 11.12, it is clear that the sign that should go with x is the one claimed up to a sign factor depending only on the dimension n of N. Thus that sign can be determined just by working out one example in each dimension. The map of $\mathbf{S}^n$ to itself which is the reflection in the line through the poles, is an example that works to give the desired result. The reader can fill in the details of this. □

We remark that the hypothesis that f be smooth is stronger than necessary. It need only be smooth in a neighborhood of each fixed point since it can be approximated by a homotopic smooth map with no additional fixed points and identical to f in the neighborhood of each fixed point.

Now consider the case of a k-disk bundle W^{n+k} over N^n. As usual, its Thom class τ is defined by $\tau \cap [W] = i^W_{N*}[N]$. Then the "self-intersection class" of N in W is defined to be $[N]_W \bullet [N]_W \in H_{n-k}(W)$. We compute:

$$\begin{aligned}[N]_W \bullet [N]_W &= (\tau \cup \tau) \cap [W] = \tau \cap (\tau \cap [W]) = \tau \cap i_*[N] \\ &= i_*(i^*(\tau) \cap [N]) = i_*(\chi \cap [N]),\end{aligned}$$

where χ is the Euler class of the bundle. Since $i_*: H_{n-k}(N) \to H_{n-k}(W)$ is an isomorphism, by the homotopy axiom, we may as well think of this as $\chi \cap [N] \in H_{n-k}(N)$. Thus we have:

12.8. Proposition. *If W^{n+k} is an oriented k-disk bundle over the orientable closed manifold N^n, then the self-intersection class of N in W is the Poincaré dual, in N, of the Euler class of the bundle.* □

In case $k = n$ then the augmentation of the self-intersection class is the "self-intersection number" $[N]_W \cdot [N]_W = \epsilon_*([N]_W \bullet [N]_W) = \langle \chi, [N] \rangle$, also called the "Euler number" of the bundle. Thus if $\chi = 0$ then the self-intersection number is zero. This can be realized geometrically by moving the zero section of the bundle so it is transverse to the zero section and then counting the finite number of intersections with signs as discussed previously for the Lefschetz number. It is possible to show that a positive intersection point can be cancelled against a negative one. (See Figure VI-9 for the suggestion

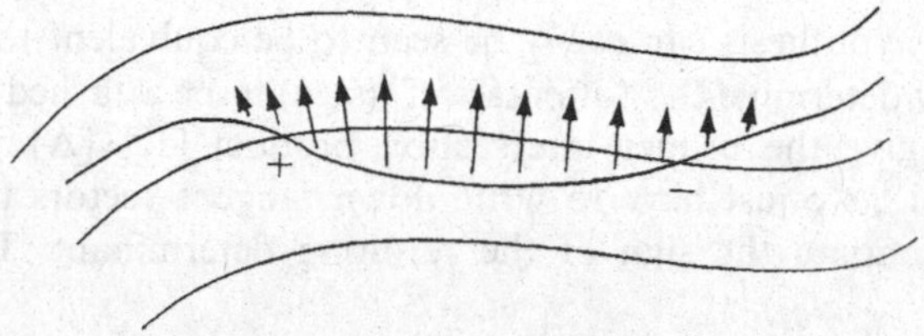

Figure VI-9. Cancellation of opposite intersection points.

of this). Thus, if the Euler number is zero, one can get rid of intersections altogether. That is, one can prove that if the Euler number is zero then the bundle has a nonzero section. In particular, when the bundle is the tangent disk bundle, then the Euler number is the Euler characteristic, and one has the result that a closed orientable manifold has a nonzero tangent vector field if and only if its Euler characteristic is zero. (A proof, but from a different perspective, is given in Corollary 14.5 of Chapter VII.)

In Section 18 we give an important application of the use of the intersection product to a very interesting construction of differential topology. That section can be read at this point if so desired.

We devote the remainder of this section to showing how to generalize Theorem 12.6 and Corollary 12.7 to the case of manifolds with boundary, and to some classical consequences of that generalization. To do this, we must generalize some of the material in the last section slightly.

We are going to be concerned with $W = N \times N$ where N has a boundary. The "corner" on ∂W will not cause any trouble. In this case, $\partial W = N \times \partial N \cup \partial N \times N = A \cup B$ naturally breaks up into these two pieces, and that will be crucial to the discussion.

Let us first look at the general case of a manifold W^w with boundary $A \cup B$ where A and B are $(w-1)$-manifolds with common boundary $A \cap B$. By Section 9, Problem 3, there is the duality isomorphism

$$\cap[W]\colon H^p(W, A) \xrightarrow{\approx} H_{w-p}(W, B),$$

whose inverse will be denoted by D_W. We depend on the context to distinguish this from its analogue with A and B reversed. Let N^n and K^k be manifolds with boundary such that $(N, \partial N) \subset (W, A)$ and $(K, \partial K) \subset (W, B)$. Also assume that N and K meet transversely in $N \cap K \subset W - \partial W$. Then, as in Definition 11.8, there are the classes

$$[N]_W \in H_n(W, A), \qquad [K]_W \in H_k(W, B),$$

which dualize to the Thom classes

$$\tau_N^W = D_W[N]_W \in H^{w-n}(W, B), \qquad \tau_K^W = D_W[K]_W \in H^{w-k}(W, A).$$

We still have the intersection class of these:

$$[N]_W \bullet [K]_W = (\tau_K^W \cup \tau_N^W) \cap [W] \in H_{n+k-w}(W).$$

Since $N \cap K \subset W - \partial W$, we can take $[K \cap N]_W \in H_{n+k-w}(W)$ with Thom class

$$\tau_{K \cap N}^W = D_W[K \cap N]_W \in H^{2w-n-k}(W, \partial W).$$

The proof of Theorem 11.9 still applies to give $[K \cap N]_W = [N]_W \bullet [K]_W$ and $\tau_{K \cap N}^W = \tau_K^W \cup \tau_N^W$.

Now we specialize to $W = N^n \times N^n$ as in the discussion below Theorem 12.3, but now where N has a boundary. We regard $(\Delta, \partial\Delta) \subset N \times (N, \partial N)$ and, consequently, $\tau = \tau_\Delta^{N \times N} \in H^n((N, \partial N) \times N)$.

In Theorem 12.4 take α to range over a basis B of $H^*(N)$ and $\alpha^\circ \in H^*(N, \partial N)$ then denotes the corresponding element of the dual basis. There is no difficulty in adapting the proof of Theorem 12.4 to show that

$$\tau = \sum_{\alpha \in B} (-1)^{\deg(\alpha)} \alpha^\circ \times \alpha \in H^n((N, \partial N) \times N).$$

Now we consider the situation of Theorem 12.6. Thus let $f: N \to N$ be a map. Here we assume, however, that f has no fixed points on the boundary ∂N. As before, put

$$1 \mathbin{\underline{\times}} f = (1 \times f) \circ d: (N, \partial N) \to (N, \partial N) \times N$$

and let $\Gamma = \{(x, f(x)) \in N \times N\}$ be the graph of f. Note that $(\Gamma, \partial\Gamma) \subset (N, \partial N) \times N$ and so $[\Gamma] \in H_n((N, \partial N) \times N)$, $[\Gamma] \bullet [\Delta] \in H_0(N \times N)$ and $[\Gamma] \cdot [\Delta] = \epsilon_*([\Gamma] \bullet [\Delta])$ are defined by the discussion above, since we took $[\Delta] \in H_n(N \times (N, \partial N))$. Then the proofs of Theorem 12.6 and Corollary 12.7 go through verbatim to give, finally, the following result.

12.9. Theorem (Lefschetz). *Let $f: N \to N$ be a map on the compact, orientable manifold N with boundary. Assume that f has no fixed points in ∂N. Then, with the above notation,*

$$L(f) = [\Gamma] \cdot [\Delta].$$

If, moreover, f is smooth and f_{*x} does not have eigenvalue 1 *at any fixed point x of f, then*

$$L(f) = \sum_x \operatorname{sign} \det(I - f_{*x}),$$

the sum ranging over the fixed points of f. □

12.10. Definition. Let $N^n \subset \mathbf{R}^n$ be a compact smooth domain. For $x \in \partial N$, let ξ_x be the outward unit normal to ∂N at x. Then $x \mapsto \xi_x$ is a map $G: \partial N \to \mathbf{S}^{n-1}$ called the *Gauss map*.

12.11. Theorem (Hopf). *In the situation of Definition* 12.10, *the degree of the Gauss map G equals the Euler characteristic $\chi(N)$.*

Proof. Let T be a closed tubular neighborhood of ∂N and define a map $f: T \cap N \to (\partial T) \cap N$ projecting inward normal vectors to their heads. Note that on ∂N we have $f(x) = x - \xi_x$. We can extend f by the identity to the remainder of N, so that $f: N \to N$ and is homotopic to the identity.

It follows that $L(f) = L(1) = \chi(N)$ and it remains to show that $L(f) = \deg G$.

We can approximate f by a smooth map ϕ which coincides with f on ∂N and is a close enough approximation so that ϕ maps N into N, and $\|x - \phi(x)\| \leq 1$ for all $x \in N$.

Put $\psi(x) = x - \phi(x): N \to \mathbf{D}^n$ and note that $\psi|_{\partial N} = G$. From the commutative diagram

$$\begin{array}{ccc} H_{n-1}(\partial N) & \xrightarrow{G_*} & H_{n-1}(\mathbf{S}^{n-1}) \\ \approx \uparrow \partial_* & & \partial_* \uparrow \approx \\ H_n(N, \partial N) & \xrightarrow{\psi_*} & H_n(\mathbf{D}^n, \mathbf{S}^{n-1}) \end{array}$$

we have $\deg G = \deg \psi$.

Now let $y \in \operatorname{int} \mathbf{D}^n$ be a regular value of ψ. Then $\psi^{-1}(y) = \{x_1, \ldots, x_k\}$ is finite and there are disks D_i about the x_i mapping diffeomorphically to a disk D about y with $N - \bigcup D_i$ going to $\mathbf{D}^n - D$. Each map $H_n(N, \partial N) \to H_n(N, N - \{x_i\}) \approx H_n(D_i, D_i - \{x_i\})$ is an isomorphism, providing an orientation of D_i. The commutative diagram

$$\begin{array}{ccc} H_n(N, \partial N) & \xrightarrow{\psi_*} & H_n(\mathbf{D}^n, \mathbf{S}^{n-1}) \\ \downarrow & & \\ H_n(N, N - \{x_1, \ldots, x_k\}) & & \\ \uparrow \approx & & \downarrow \approx \\ H_n(\bigcup D_i, \bigcup (D_i - \{x_i\})) & & \\ \downarrow \approx & & \\ \bigoplus H_n(D_i, D_i - \{x_i\}) & \longrightarrow & H_n(\mathbf{D}^n, \mathbf{D}^n - \{y\}) \\ \uparrow \approx & & \uparrow \approx \\ \bigoplus H_n(D_i, \partial D_i) & \xrightarrow{\psi_*} & H_n(D, \partial D) \end{array}$$

shows that $\psi_*[N] = \sum_i \psi_*[D_i]$. This means that $\deg \psi$ is the sum of the local degrees at the x_i, and we know the latter are the signs of the Jacobians of ψ at x_i. But $\psi = I - \phi$, so that the differentials satisfy

$$\psi_* = I - \phi_*.$$

We conclude that $\operatorname{sign}(\det(\psi_{*x})) = \operatorname{sign}(\det(I - \phi_{*x}))$ at each $x = x_i$, so that

$$\deg G = \deg \psi = \sum_x \operatorname{sign} \det(I - \phi_{*x}) = L(\phi) = L(1) = \chi(N)$$

by Theorem 12.9 and the interim results. □

The map ψ in the proof of Theorem 12.11 can be thought of as a vector field on N and the latter part of the proof can be generalized a bit to the case of vector fields with isolated zeros, as follows. If ξ is a vector field on N and $x \in \operatorname{int} N$ is an isolated zero then take a disk D about x containing no other zeros. Then the map

$$\partial D \to \mathbf{S}^{n+1}$$

given by $y \mapsto \xi(x)/\|\xi(x)\|$ is defined. Its degree is called the "index" of ξ at x. See Figure VI-10. If one removes disks D_i about each zero of ξ then the above map extends to $N - \bigcup D_i \to \mathbf{S}^{n-1}$ and essentially the same proof shows that the sum of the indices equals the degree of this map on $\partial N \to \mathbf{S}^{n-1}$. If ξ points outward from N on ∂N then this map is homotopic to the Gauss map and so its degree is $\chi(N)$. Therefore, we have:

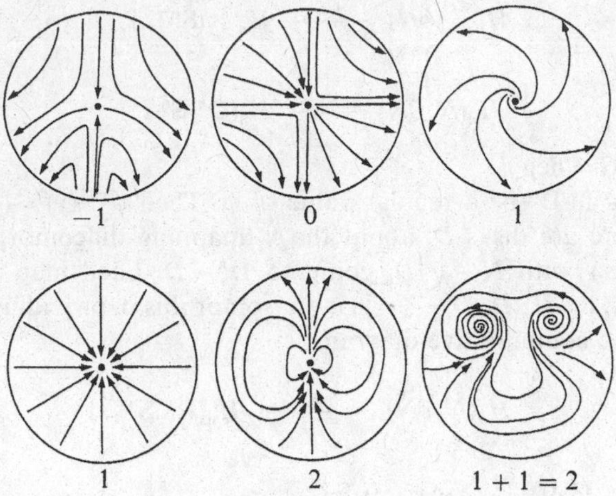

Figure VI-10. Indices of vector fields and index sum.

12.12. Proposition. *If $N^n \subset \mathbf{R}^n$ is a compact smooth domain and ξ is a vector field on N having only isolated zeros, all in* int N, *then the sum of the indices of ξ at its zeros is equal to the degree of $x \mapsto \xi(x)/\|\xi(x)\|$ on $\partial N \to \mathbf{S}^{n+1}$. In particular, if ξ always points outward on ∂N, then this is the Euler characteristic $\chi(N)$.* □

One can extend this result to vector fields on arbitrary smooth manifolds N, possibly with boundary, as follows:

Let ξ be a vector field on N which points out of N on ∂N (if any), and has only isolated zeros. Then $-\xi$ induces a local flow $\{f_t\}$ on N defined only for $t \in [0, \epsilon)$, some ϵ. (Actually it is defined on all of $\mathbf{R}_+$, because of the compactness of N and the fact that $-\xi$ points inward on ∂N. However, we do not need that.) The program is to put $f = f_t$ for some sufficiently small t, relate the indices of ξ to local data for f near its fixed points and to apply Theorem 12.9. There are a few technical difficulties:

(a) the index of ξ at a zero must be defined in the general case;
(b) f may have more fixed points than the zeros of ξ;
(c) in a coordinate patch around a zero of ξ, f is not quite the same as the displacement map $x \mapsto x - \xi(x)$; and
(d) f may not be transverse to the diagonal; i.e., it may have eigenvalues 1.

Using local coordinates $U \xrightarrow{\approx} \mathbf{R}^n$ at a zero p of ξ, one can define the index of ξ at p as the index of the induced field on $\mathbf{R}^n$. It is not very hard to see this is independent of the choice of local coordinates. The idea of one method of doing this is in Lemma 16.3 of Chapter II. We shall not detail that.

Next, let $p_1, \ldots, p_k \in N$ be the zeros of ξ, fix local coordinates at each p_i and take disjoint disks D_i about p_i as origin, in terms of the local coordinates there. Since $-\xi$ has no zeros on $N - \bigcup D_i$, there is an $\epsilon > 0$ so that f_t has no fixed points outside $\bigcup \operatorname{int} D_i$ for any $0 < t < \epsilon$. (For, if not, there is a sequence of points x_n such that $f(x_n, t) = f_t(x_n) = x_n$ for some $0 < t < 1/n$. By passing to a subsequence we can assume that $x_n \to x$ with $\xi(x) \neq 0$. Then $f(x_n, t) = x_n$ for t in a $1/n$-dense set in $[0, \epsilon)$ and continuity then implies that $f(x, t) = x$ for all t, whence $\xi(x) = 0$, a contradiction.) Also, in the given local coordinates about p_i, we have

$$\lim_{t \to 0} \frac{x - f_t(x)}{t} = \xi(x),$$

and so we can also take ϵ so small that $x - f_t(x)$ and $\xi(x)$ are never antipodal on ∂D_i, for all i and $0 < t < \epsilon$. Now put $f = f_t$ for some such $0 < t < \epsilon$. Note that, although we have guaranteed that f has no fixed points outside the D_i, it may well have more than one fixed point in a given D_i.

Since, in the coordinates about p_i, $x - f(x)$ is never antipodal to $\xi(x)$ on ∂D_i, it follows that $x \mapsto x - f(x) = \eta_i(x)$ has the same degree as does $x \mapsto \xi(x)$ as maps $\partial D_i \to \mathbf{R}^n - \{0\}$. Therefore the index I_i of ξ at p_i equals the degree of $\eta_i|_{\partial D_i}$. Of course, η_i is defined only on the coordinate patch.

Let $F_i \subset E_i \subset D_i$ be concentric disks about p_i such that int F_i contains all the zeros on η_i and let $\lambda: D_i \to [0, 1]$ be smooth, equal to 0 outside E_i and to 1 on F_i. Let $q \in F_i \subset \mathbf{R}^n$ be a regular value of η_i and consider

$$\eta_i'(x) = \eta_i(x) - \lambda(x)q.$$

On F_i, this has only "nondegenerate" zeros (meaning the differential is nonsingular at the zero). Also, q can be taken so close to the origin that η' has no zeros on $D_i - F_i$. Also, on ∂D_i, η_i' equals η_i. Thus the index I_i of ξ at p_i, which equals the degree of $\eta_i|_{\partial D_i}$, also equals the index sum of η_i' on D_i by Proposition 12.12.

Put $f'(x) = x - \eta_i'(x)$ on D_i and $f'(x) = f(x)$ on $N - \bigcup D_i$. Note that $f' \simeq f \simeq 1$. Now f' does not have eigenvalue 1 at any fixed point, and so the sum of its local indices at fixed points in D_i equals I_i. By Theorem 12.9, we can now conclude that the index sum $\sum I_i$ of ξ equals the index sum (the right side of the second displayed equation in Theorem 12.9) of f' which equals $\chi(N)$. Because of the use of Theorem 12.9 we have implicitly assumed that N^n is orientable. This assumption can be avoided by the passage to the orientable double covering of N^n. This doubles both the index sum and the Euler characteristic. (The doubling of the Euler characteristic can be proved via a triangulation of N or by appeal to the Smith exact sequence in Section 20 of Chapter IV.) Summarizing, we have:

12.13. Theorem (Poincaré–Hopf). *If ξ is a tangent vector field on the compact manifold N^n which has only isolated zeros and points outward along the boundary ∂N, if any, then the sum of the indices of ξ at its zeros equals the Euler* *characteristic of N:*

$$\sum I_i = \chi(N).$$ □

For another approach to this topic, see Milnor [3].

12.14. Corollary (Fundamental Theorem of Algebra). *If $p(z)$ is a complex polynomial of positive degree then $p(z)$ has a zero.*

PROOF. Let $p(z) = z^n + a_1 z^{n-1} + \cdots + a_n, n > 0$. For $z \in \mathbf{D}^2$ define

$$\xi(z) = (1 - |z|^2)^n p(z/(1 - |z|^2)) = z^n + a_1(1 - |z|^2)z^{n-1} + \cdots + a_n(1 - |z|^2)^n.$$

The last expression shows that ξ is defined and smooth on all of $\mathbf{D}^2$ and equals $z \mapsto z^n$ on $\partial\mathbf{D}^2$. Therefore ξ gives a vector field on $\mathbf{D}^2$ which has degree n on $\partial\mathbf{D}^2$. If p has no zeros then neither does ξ. By Proposition 12.12, the degree $n > 0$ of ξ on $\partial\mathbf{D}^2$ must equal the index sum of ξ on $\mathbf{D}^2$. Therefore ξ must have a zero. □

As a more serious application we shall briefly discuss "maximal tori" in a compact connected Lie group G. We shall assume as known that a compact connected abelian Lie group is a torus; i.e., a product of circle groups. The proof of this is not hard but would lead us too far afield. We shall also assume as known that if a compact Lie group acts smoothly on a manifold then there are coordinates about any fixed point in which the group acts by orthogonal transformations. The proof of this is easy if one knows some Riemannian geometry. In our case, the action considered is by conjugation and this special case can be handled by the discussion at the end of Section 12 of Chapter V.

There is clearly a maximal compact connected abelian subgroup, hence a "maximal torus," T of G. An automorphism of T is given by a unimodular matrix with integer entries. It follows that the normalizer N of T in G has N/T finite, since otherwise there would be a one-parameter group in N, but not in T, which commutes with T, and the closure of this, together with T, would give a larger compact connected abelian subgroup of G.

12.15. Corollary. *Let T be a maximal torus of the compact connected Lie group G and let N be its normalizer in G. Then:*

(a) $\chi(G/T) =$ *order* N/T;
(b) $\chi(G/N) = 1$; *and*
(c) *any other maximal torus of G is conjugate to T.*

PROOF. It is easy to see that there is a one-parameter subgroup H of T whose closure is T. Consider the action of H on G/T by left translation. We have that $HgT = gT \Leftrightarrow g^{-1}HgT = T \Leftrightarrow g^{-1}Tg = T \Leftrightarrow g \in N$ and so the fixed point set of H is the finite set N/T. Thus H generates a vector field on G/T with exactly these zeros. By the previous remarks, the vector field is tangent to a small sphere (in appropriate coordinates) about any fixed point. A tangent field on such a sphere is homotopic to the outward normal field and so has index 1. By Theorem 12.13, the number of zeros is $\chi(G/T)$, proving (a). A similar argument on G/N yields (b). For (c) let T' be some other maximal torus of G and consider the action of H on G/T'. By a similar argument, the number of fixed points is $\chi(G/T') > 0$ and so a fixed point exists. That is, there is a $g \in G$ such that $HgT' = gT'$. This implies that $g^{-1}Tg = T'$. □

The finite group N/T is called the "Weyl group" of G. It, and its action on T, is known to characterize G up to "local isomorphism."

Problems

1. Consider $\mathbf{S}^3$ as the unit quaternions. Let $f: \mathbf{S}^3 \to \mathbf{S}^3$ be $f(q) = q^2$. Use the methods of the present section to find the Lefschetz number of f.
2. Consider $\mathbf{S}^3$ as the unit quaternions. Let $f: \mathbf{S}^3 \to \mathbf{S}^3$ be $f(q) = q^3$. Use the methods of the present section to find the Lefschetz number of f.
3. Consider $\mathbf{S}^3$ as the unit quaternions. Let $f: \mathbf{S}^3 \times \mathbf{S}^3 \to \mathbf{S}^3 \times \mathbf{S}^3$ be $f(p, q) = (pq, pq^2)$. Use the methods of the present section to find the Lefschetz number of f. Also check your answer by computing it from its definition.
4. Let A be a 2×2 matrix with integer entries. As a linear transformation of the plane, this induces a map f on the torus $\mathbf{R} \times \mathbf{R}/\mathbf{Z} \times \mathbf{Z}$ to itself. Show that the number of fixed points of f is $|1 - \text{trace}(A) + \det(A)|$ unless the latter is zero. (*Hint*: Compute $L(f)$ two different ways.) Also verify this result on at least two nontrivial examples.
5. Let $p \in \mathbf{S}^n$ be fixed. For $q \in \mathbf{S}^n$, map q to the point on the great circle from p to q of twice the distance q is from p along the circle. (For example, $-p$ maps to p and the equator having p as a pole maps to $-p$.) Use Corollary 12.7 to find the degree of this map $\mathbf{S}^n \to \mathbf{S}^n$.
6. Let $f: M^2 \to M^2$ be a smooth self-map on the closed connected orientable 2-manifold M^2. Call a fixed point x of f "elliptic" if the differential f_* at x satisfies the inequality $1 + \det f_* > \text{trace } f_*$. If f has only isolated fixed points, all elliptic, and has at least one of them, then show that $M^2 = \mathbf{S}^2$.
7. Let M^n be a closed oriented manifold. Let ϵ be the trivial line bundle over M. Then M is said to be "stably parallelizable" if $\tau \oplus \epsilon$ is trivial where τ is the tangent bundle of M. Show how to define a "Gauss map" on such a manifold that generalizes the Gauss map $M^n \to \mathbf{S}^n$ on the boundary of a smooth domain in $\mathbf{R}^{n+1}$, and which has degree zero if M is parallelizable. (The map will depend on some arbitrary choices.)
8. If K is a finite simplicial complex then it is known that K can be embedded in the interior of some compact smooth manifold M^n with boundary, with K as a deformation retract of M^n. Assuming this, use Theorem 12.9 to rederive Hopf's version, Theorem 23.4 of Chapter IV, of the Lefschetz Fixed Point Theorem.
9. Let T be a toral subgroup of the compact connected Lie group G and assume that T is not a maximal torus. Then show that $\chi(G/T) = 0$.

13. The Gysin Sequence ☼

In this section we discuss a useful exact sequence that relates the cohomology of the base space of a sphere bundle to that of the total space. This section assumes knowledge of the Thom class and Thom isomorphism from Section 11 and the definition of the Euler class from Section 12.

Let A be a compact space which is a neighborhood retract in some euclidean space. (These restrictions are stronger than necessary, but our applications are all compact smooth manifolds anyway.) Let $\tilde{\pi}: \tilde{A} \to A$ be a k-disk bundle. Then we have the Thom class $\tau \in H^k(\tilde{A}, \partial\tilde{A})$. We will use the notation $j^*: H^*(\tilde{A}, \partial\tilde{A}) \to H^*(\tilde{A})$ and $i^*: H^*(\tilde{A}) \to H^*(A)$.

We define the Euler class $\chi = i^*j^*(\tau) \in H^k(A)$. This is a generalization of the notion of Euler class in the previous section.

13.1. Lemma. *There is the following commutative diagram of $H^*(A)$-module homomorphisms:*

$$\begin{array}{ccc} H^p(A) & \xrightarrow{\cup\chi} & H^{p+k}(A) \\ {\scriptstyle\approx}\downarrow{\scriptstyle\tilde{\pi}^*(\cdot)\cup\tau} & & {\scriptstyle i^*}\uparrow{\scriptstyle\approx}\ \downarrow{\scriptstyle\tilde{\pi}^*} \\ H^{p+k}(\tilde{A}, \partial\tilde{A}) & \xrightarrow{j^*} & H^{p+k}(\tilde{A}). \end{array}$$

PROOF. The module structure is via the cup product. For $\tilde{A}$ it is $\tilde{\pi}^*$ followed by the cup product. The commutativity is by

$$i^*j^*(\tilde{\pi}^*(\beta) \cup \tau) = i^*(\tilde{\pi}^*(\beta) \cup j^*(\tau)) = i^*\tilde{\pi}^*(\beta) \cup i^*j^*(\tau) = \beta \cup \chi. \qquad \square$$

13.2. Theorem. *For an oriented $(k-1)$-sphere bundle $\pi: X \to A$ there is the exact "Gysin sequence"*

$$\cdots \to H^p(A) \xrightarrow{\cup\chi} H^{p+k}(A) \xrightarrow{\pi^*} H^{p+k}(X) \xrightarrow{\sigma^*} H^{p+1}(A) \to \cdots,$$

where all maps are $H^*(A)$-module homomorphisms.

PROOF. This comes from the associated k-disk bundle $\tilde{\pi}: \tilde{A} \to A$ and the cohomology sequence of the pair $(\tilde{A}, \partial\tilde{A}) = (\tilde{A}, X)$, by replacing the terms involving $\tilde{A}$ using Lemma 13.1. Thus

$$\sigma^* = (\tilde{\pi}^*(\cdot) \cup \tau)^{-1} \circ \delta^* = h \circ \delta^*,$$

where $h = (\tilde{\pi}^*(\cdot) \cup \tau)^{-1}: H^{p+k+1}(\tilde{A}, X) \xrightarrow{\approx} H^{p+1}(A)$. That σ^* is an $H^*(A)$-homomorphism means that for $\alpha \in H^i(A)$ and $\beta \in H^j(X)$ we have $\sigma^*(\pi^*(\alpha) \cup \beta) = (-1)^i \alpha \cup \sigma^*(\beta)$. To see this let $k: X \hookrightarrow \tilde{A}$ and compute:

$$\begin{aligned} \sigma^*(\pi^*(\alpha) \cup \beta) &= h\delta^*(\pi^*(\alpha) \cup \beta) \\ &= (-1)^{ij} h\delta^*(\beta \cup \pi^*(\alpha)) \\ &= (-1)^{ij} h\delta^*(\beta \cup k^*\tilde{\pi}^*(\alpha)) \end{aligned}$$

$$\begin{aligned}
&= (-1)^{ij} h(\delta^*(\beta) \cup \bar{\pi}^*(\alpha)) \qquad \text{(by Theorem 4.5)}\\
&= (-1)^{ij+i(j+1)} h(\bar{\pi}^*(\alpha) \cup \delta^*\beta)\\
&= (-1)^{i} h(\bar{\pi}^*(\alpha) \cup \delta^*\beta)\\
&= (-1)^{i} \alpha \cup h\delta^*\beta\\
&= (-1)^{i} \alpha \cup \sigma^*\beta,
\end{aligned}$$

where the penultimate equation is from the fact that h is an $H^*(A)$-module homomorphism. □

13.3. Corollary. *If a sphere bundle has a section then its Euler class $\chi = 0$.*

PROOF. We had this before. Here it follows from the fact that π^* is a monomorphism when there is a section s, since $s^* \circ \pi^* = (\pi \circ s)^* = 1^* = 1$. Then apply the sequence, as displayed, with $p = 0$. □

13.4. Corollary. *Suppose $\mathbf{S}^{n+k-1} \to M^n$ is a bundle with $\mathbf{S}^{k-1}$ as fiber, $k > 1$. Then $n = kr$ for some integer r and M has cohomology ring*

$$H^*(M^n) \approx \mathbf{Z}[\chi]/(\chi^{r+1}),$$

where $\chi \in H^*(M)$ is the Euler class of the bundle.

PROOF. A portion of the Gysin sequence is

$$H^{i+k-1}(\mathbf{S}^{n+k-1}) \to H^i(M) \xrightarrow{\cup\chi} H^{i+k}(M) \to H^{i+k}(\mathbf{S}^{n+k-1}).$$

Suppose that $0 \neq \alpha \in H^{i+k}(M)$ for some $i + k > 0$. We have $0 < i + k \leq n < n + k - 1$. Thus, the right end of the displayed sequence is zero. Hence $\alpha = \beta \cup \chi$ for some $\beta \in H^i(M)$. This cannot be repeated indefinitely and can only end when i becomes zero. This implies that $1, \chi, \chi^2, \ldots, \chi^r$, for some r, is an additive basis for $H^*(M)$, which must be free abelian. □

It is known that the situation of Corollary 13.4 can happen only for $k = 1, 2, 4$, or 8; see Section 15.

13.5. Example. Let T^{2n+1} be the unit tangent bundle to $\mathbf{S}^n$. This is also called the real Stiefel manifold $V_{n+1,2}$, the space of 2-frames in $\mathbf{R}^{n+1}$ (the first vector of the frame is the base position on the sphere, the second is the tangent vector). This is an $(n-1)$-sphere bundle over the n-sphere $\mathbf{S}^{n-1} \to T^{2n-1} \to \mathbf{S}^n$. The Euler class $\chi \in H^n(\mathbf{S}^n)$ is zero for n odd, and twice a generator for n even, because the Euler characteristic of $\mathbf{S}^n$ is zero or two, respectively. The critical parts of the Gysin sequence are

$$0 \to H^{n+1}(T) \to H^0(\mathbf{S}^n) \xrightarrow{\cup\chi} H^n(\mathbf{S}^n) \to H^n(T) \to 0$$

and

$$0 \to H^{2n-1}(T) \to H^n(\mathbf{S}^n) \to 0.$$

It follows that

$$H^i(T) = \begin{cases} \mathbf{Z} & \text{for } i = 0, 2n-1, \\ \mathbf{Z} & \text{for } n \text{ odd and } i = n, n-1, \\ \mathbf{Z}_2 & \text{for } n \text{ even and } i = n, \\ 0 & \text{otherwise.} \end{cases}$$

13.6. Example. Consider an $\mathbf{S}^1$-bundle M^3 over $\mathbf{S}^2$. The Gysin sequence is

$$0 \to H^1(M) \to H^0(\mathbf{S}^2) \xrightarrow{\cup\chi} H^2(\mathbf{S}^2) \to H^2(M) \to 0.$$

It follows that if $\chi = 0$ then M has the cohomology ring of $\mathbf{S}^1 \times \mathbf{S}^2$. If χ is a generator of $H^2(\mathbf{S}^2)$ then M has the homology of $\mathbf{S}^3$. If χ is k times a generator, then

$$H^i(M) = \begin{cases} \mathbf{Z} & \text{for } i = 0, 3, \\ \mathbf{Z}_k & \text{for } i = 2, (k \neq 0), \\ 0 & \text{otherwise.} \end{cases}$$

Consider $M^3 = \mathbf{D}^2 \times \mathbf{S}^1 \cup_f \mathbf{D}^2 \times \mathbf{S}^1$ where $f: \mathbf{S}^1 \times \mathbf{S}^1 \to \mathbf{S}^1 \times \mathbf{S}^1$ is $f(z, w) = (z, z^k w)$. Since f is a diffeomorphism, projection to the first coordinate makes M^3 an $\mathbf{S}^1$-bundle over $\mathbf{S}^2$. Using the Seifert–Van Kampen Theorem, the fundamental group of M is

$$\pi_1(M) \approx \{x, y \mid x = y, 1 = y^k\} \approx \mathbf{Z}_k.$$

(For $k = 0$, the fundamental group is $\mathbf{Z}$ and for $k = \pm 1$, it is trivial.) Thus, for $|k| > 1, H_1(M) \approx \mathbf{Z}_k$. By Poincaré duality, $H_2(M) = 0$. By the Universal Coefficient Theorem, $H^2(M) \approx \mathbf{Z}_k$. Therefore, these examples realize all the possible Euler classes. Incidentally, these manifolds M^3 are our old friends, the lens spaces $L(k, 1)$.

13.7. Example. We will calculate the cohomology ring of the complex Stiefel manifold $V_{n,n-k}$, the space of complex $(n-k)$-frames in $\mathbf{C}^n$. Selecting the last $n - k - 1$ vectors in an $(n-k)$-frame gives a map $V_{n,n-k} \to V_{n,n-k-1}$ which can be seen to be an $\mathbf{S}^{2k+1}$-bundle (see Section 8 of Chapter VII). We aim to show that

$$\boxed{H^*(V_{n,n-k}) \approx \wedge(x_{2k+1}, x_{2k+3}, \ldots, x_{2n-1}),}$$

the exterior algebra on the x_i where $\deg(x_i) = i$. That is, it is the same cohomology ring as for the product $\mathbf{S}^{2k+1} \times \mathbf{S}^{2k+3} \times \ldots \times \mathbf{S}^{2n-1}$ of spheres.

The proof will be by induction on k. The Euler class $\chi \in H^{2k+2}(V_{n,n-k-1}) = 0$, by the inductive assumption. Thus, the Gysin sequence degenerates into short exact sequences such as

$$0 \to H^j(V_{n,n-k-1}) \xrightarrow{\pi^*} H^j(V_{n,n-k}) \xrightarrow{\sigma^*} H^{j-2k-1}(V_{n,n-k-1}) \to 0.$$

By induction, the group on the right is free abelian, and hence the sequence splits, and the middle group is free abelian.

Let $x = x_{2k+1} \in H^{2k+1}(V_{n,n-k})$ be such that $\sigma^*(x) = 1 \in H^0(V_{n,n-k-1})$. Since x is an odd-dimensional class, $2x^2 = 0$. Since it is in a torsion free group, $x^2 = 0$.

Recall that $\sigma^*(\pi^*(y) \cup x) = (-1)^{\deg(y)} y \cup \sigma^*(x) = (-1)^{\deg(y)} y$. Define the map

$$\Phi\colon H^*(V_{n,n-k-1}) \otimes \Lambda(x) \to H^*(V_{n,n-k})$$

by $\Phi(y \otimes 1) = \pi^*(y)$, and $\Phi(y \otimes x) = \pi^*(y) \cup x$. This is an algebra homomorphism since $x^2 = 0$. The map on $y \otimes x$ splits $\pm\sigma^*$, and so Φ is an additive isomorphism. Hence it is an algebra isomorphism, as claimed.

Note that $V_{n,n} = \mathbf{U}(n)$, the unitary group in n variables. Therefore, we have

$$\boxed{H^*(\mathbf{U}(n)) \approx \Lambda(x_1, x_3, \ldots, x_{2n-1}),}$$

meaning it has the same cohomology ring as $\mathbf{S}^1 \times \mathbf{S}^3 \times \cdots \times \mathbf{S}^{2n-1}$. The group $\mathbf{U}(1)$ is isomorphic to $\mathbf{S}^1$. Also, $\mathbf{U}(2) \approx \mathbf{S}^1 \times \mathbf{S}^3$, but not as a group. Although $\mathbf{U}(3)$ has the cohomology ring of $\mathbf{S}^1 \times \mathbf{S}^3 \times \mathbf{S}^5$ it is not even of the same homotopy type. They can be distinguished, as we will see latter, by their homotopy groups and also by the action of certain "cohomology operations" on them; see Section 8 of Chapter VII, Problems 1 and 3.

14. Lefschetz Coincidence Theory ☼

If $f, g\colon X \to Y$ are two maps, then a "coincidence" of f and g is a point $x \in X$ such that $f(x) = g(x)$. This is a generalization of the notion of a fixed point since the two ideas coincide when $X = Y$ and $g = 1_X$. In this section we describe the Lefschetz theory of coincidences for maps $f, g\colon N^n \to M^n$ of manifolds of equal dimension. We could have treated this case in the discussion of fixed point theory in Section 12, deriving the results there as corollaries. We did not do that for two reasons. First, the coincidence theory is somewhat more complicated and the Lefschetz coincidence number is much more difficult to compute than the fixed point number. Second, and most pertinent, the two most interesting classes of examples (1) coincidences of f, g where one of f and g is a homeomorphism and (2) maps to spheres, reduce to studying fixed points or other methods. (If g is a homeomorphism, coincidences of f and g equal fixed points of $g^{-1}f$. Two maps $f, g\colon N^n \to \mathbf{S}^n$ are coincidence free up to homotopy $\Leftrightarrow f \simeq -g \Leftrightarrow \deg(f) = (-1)^{n+1}\deg(g)$.) However, the theory is quite interesting and clearly worthy of inclusion in a book like this.

Let N^n and M^n be closed orientable manifolds and let $f, g\colon N \to M$. Then $f\colon N \to M$ has graph $\Gamma_f = \{(x, f(x)) \in N \times M\}$ and $g\colon N \to M$ has graph Γ_g. Both graphs are submanifolds of $N \times M$. We define the Lefschetz coincidence number of f and g to be

$$\boxed{L(f,g) = [\Gamma_f]\cdot[\Gamma_g] = \epsilon_*([\Gamma_f]\bullet[\Gamma_g]) \in \mathbf{Z},}$$

where $[\Gamma_f] = (1 \times f)_*[N]$, $[\Gamma_g] = (1 \times g)_*[N]$, and $\epsilon_*: H_0(N \times M) \to \mathbf{Z}$ is the augmentation. Note that this does not depend on the orientations chosen for N and M as long as we orient $N \times M$ by the product orientation. Also note that $L(f,g) \in \mathbf{Z}_2$ can be defined if N or M is nonorientable. By Theorem 12.6 we have $L(f) = L(f, 1)$.

It is immediate from the definition that $L(f,g) \neq 0$ implies the existence of a coincidence; i.e., a point of intersection of Γ_f and Γ_g. Our task is to give a formula for this number.

First we must digress to establish some further formulas for the shriek (transfer) maps $f_!$ and $f^!$ of Definition 11.2 and for the intersection product.

14.1. Proposition. *For $f: N^n \to M^m$ a map of oriented manifolds, we have:*

(1) $f_*(b \cap [N]) = f^!(b) \cap [M]$ *and* $f_!(a \cap [M]) = f^*(a) \cap [N]$;
(2) $f_![M] = [N]$;
(3) $f_!(a \cap b) = f^*(a) \cap f_!(b)$;
(4) $f_*(a \cap f_!(b)) = (-1)^{(m-\deg(b))(m-n)} f^!(a) \cap b$;
(5) $f^!(f^*(a) \cup b) = a \cup f^!(b)$;
(6) $n = m \Rightarrow f_* f_! = \deg(f) = f^! f^*$;
(7) $n = m \Rightarrow f_! f_* = \deg(f)$ *on* $\operatorname{im}(f_!)$ *and* $f^* f^! = \deg(f)$ *on* $\operatorname{im}(f^*)$;
(8) $(fg)_! = g_! f_!$ *and* $(fg)^! = f^! g^!$.

PROOF. Formulas (1) follow directly from the definitions. The case $a = 1$ yields (2). For (3),

$$\begin{aligned} f^*(a) \cap f_!(b) &= f^*(a) \cap D_N^{-1} f^* D_M(b) \\ &= f^*(a) \cap (f^*(D_M(b)) \cap [N]) \\ &= f^*(a \cup D_M(b)) \cap [N] \\ &= f_!((a \cup D_M(b)) \cap [M]) \\ &= f_!(a \cap (D_M(b) \cap [M])) \\ &= f_!(a \cap b). \end{aligned}$$

The reader can prove (4) and (5), which we will not be using. For (6),

$$\begin{aligned} f_* f_!(a) &= f_* D_N^{-1} f^* D_M(a) \\ &= f_*(f^* D_M(a) \cap [N]) \\ &= D_M(a) \cap f_*[N] \\ &= D_M(a) \cap \deg(f)[M] \\ &= \deg(f) a. \end{aligned}$$

Also,

$$\begin{aligned} f^! f^*(a) &= D_M f_* D_N^{-1} f^*(a) \\ &= D_M f_*(f^*(a) \cap [N]) \\ &= D_M(a \cap f_*[N]) \\ &= D_M(a \cap [M]) \deg(f) \\ &= a \deg(f). \end{aligned}$$

For (7), $f_! f_* f_!(a) = f_!(\deg(f)\, a) = \deg(f) f_!(a)$. The reader can prove the other half of (7). Formula (8) follows immediately from the definition of shrieking. □

14.2. Proposition. *If $f: N^n \to M^m$ is a map of oriented manifolds then*

$$f_!(a \bullet b) = f_!(a) \bullet f_!(b).$$

If $n = m$ then

$$f_*(a) \bullet f_*(b) = \deg(f) f_*(a \bullet b) \quad \text{on } \operatorname{im}(f_!).$$

PROOF. We compute

$$\begin{aligned} f_!(a) \bullet f_!(b) &= D_N^{-1} f^* D_M(a) \bullet D_N^{-1} f^* D_M(b) \\ &= D_N^{-1}(f^* D_M(b) \cup f^* D_M(a)) \\ &= D_N^{-1} f^*(D_M(b) \cup D_M(a)) \\ &= D_N^{-1} f^* D_M(a \bullet b) \\ &= f_!(a \bullet b). \end{aligned}$$

The second formula is left to the reader. □

14.3. Proposition. *For $f: K^k \to N^n$, and $g: L^l \to M^m$ maps of oriented manifolds, we have*

$$(f \times g)^!(a \times b) = (-1)^{(n+k)\deg(b) + n(m-l)} f^!(a) \times g^!(b)$$

and

$$(f \times g)_!(a \times b) = (-1)^{(n+k)(m-\deg(b))} f_!(a) \times g_!(b).$$

PROOF. We compute

$$\begin{aligned} (f \times g)^!(a \times b) &= D_{N \times M}(f \times g)_* D_{K \times L}^{-1}(a \times b) \\ &= D_{N \times M}(f \times g)_*((a \times b) \cap [K] \times [L]) \\ &= (-1)^{k \deg(b)} D_{N \times M}(f \times g)_*((a \cap [K]) \times (b \cap [L])) \\ &= (-1)^{k \deg(b)} D_{N \times M}\{f_*(a \cap [K]) \times g_*(b \cap [L])\} \\ &= (-1)^{k \deg(b)} D_{N \times M}\{(f^!(a) \cap [N]) \times (g^!(b) \cap [M])\} \\ &= (-1)^{k \deg(b) + n(\deg(b) + m - l)} D_{N \times M}\{(f^!(a) \times g^!(b)) \cap ([N] \times [M])\} \\ &= (-1)^{(n+k)\deg(b) + n(m-l)} f^!(a) \times g^!(b). \end{aligned}$$

The second formula will not be used and is left for the reader to derive. □

Now, finally, we return to the program of providing computational formulas for the coincidence number $L(f, g)$.

14.4. Theorem. *For $f, g: N^n \to M^n$ maps of closed oriented manifolds, the Lefschetz coincidence number is given by each of the following formulas. (Traces are computed with coefficients in the rationals, or in $\mathbf{Z}_p$ where $L(f, g)$ must be reduced mod p. The subscript i on* tr *indicates the trace on the* ith *degree (co)homology.)*

(1) $L(f, g) = \sum_i (-1)^i \operatorname{tr}_i(f^* g^!)$;
(2) $L(f, g) = \sum_i (-1)^i \operatorname{tr}_i(g^! f^*)$;
(3) $L(f, g) = \sum_i (-1)^i \operatorname{tr}_i(f_* g_!)$;
(4) $L(f, g) = \sum_i (-1)^i \operatorname{tr}_i(g_! f_*)$; *and*
(5) $L(f, g) = \epsilon_*(g \mathbin{\underline{\times}} f)_![\Delta_M]$ *where* $g \mathbin{\underline{\times}} f: N \to M \times M$ *is* $(g \times f) \circ d$.
Also
(6) $L(f, g) = (-1)^n L(g, f)$.

Proof. Let $\gamma_f = D_{N \times M}[\Gamma_f]$ and $\gamma_g = D_{N \times M}[\Gamma_g]$. Consider $1 \mathbin{\underline{\times}} f: N \to N \times M$ and $1 \times g: N \times N \to N \times M$. Then, with notation from Theorem 12.4, we compute

$$\begin{aligned}
L(f, g) &= \epsilon_*([\Gamma_f] \bullet [\Gamma_g]) = \epsilon_*((\gamma_g \cup \gamma_f) \cap [N \times M]) \\
&= \epsilon_*(\gamma_g \cap (\gamma_f \cap [N \times M])) = \epsilon_*(\gamma_g \cap [\Gamma_f]) \\
&= \langle \gamma_g, [\Gamma_f] \rangle = \langle D_{N \times M}[\Gamma_g], [\Gamma_f] \rangle \\
&= \langle D_{N \times M}(1 \times g)_*[\Delta], (1 \mathbin{\underline{\times}} f)_*[N] \rangle \\
&= \langle (1 \mathbin{\underline{\times}} f)^* D_{N \times M}(1 \times g)_*(\tau \cap [N \times N]), [N] \rangle \\
&= \langle (1 \mathbin{\underline{\times}} f)^*(1 \times g)^!(\tau), [N] \rangle \\
&= \sum_\alpha (-1)^{\deg \alpha} \langle (1 \mathbin{\underline{\times}} f)^*(1 \times g)^!(\alpha^\circ \times \alpha), [N] \rangle \\
&= \sum_\alpha (-1)^{\deg \alpha} \langle (1 \mathbin{\underline{\times}} f)^*(\alpha^\circ \times g^!(\alpha)), [N] \rangle \qquad \text{(by 14.3)} \\
&= \sum_\alpha (-1)^{\deg \alpha} \langle \alpha^\circ \cup f^* g^!(\alpha), [N] \rangle \\
&= \sum_i (-1)^i \operatorname{tr}_i(f^* g^!)
\end{aligned}$$

because $\langle \alpha^\circ \cup \beta, [N] \rangle = \delta_{\alpha,\beta}$. This proves (1).

Formula (2) is from the algebraic fact that $\operatorname{tr}(AB) = \operatorname{tr}(BA)$. Formula (6) is immediate from the definition. The commutative diagram

$$\begin{array}{ccccc}
H^{n-p}(N) & \xrightarrow{f^!} & H^{n-p}(M) & \xrightarrow{g^*} & H^{n-p}(N) \\
\downarrow{\scriptstyle \cap[N]} & & \downarrow{\scriptstyle \cap[M]} & & \downarrow{\scriptstyle \cap[N]} \\
H_p(N) & \xrightarrow{f_*} & H_p(M) & \xrightarrow{g_!} & H_p(N)
\end{array}$$

implies that $\operatorname{tr} g^* f^! = \operatorname{tr} g_! f_*$ and this, together with (6), gives (3). (The sign $(-1)^n$ disappears because of the dimension shift between the rows of the above diagram.) Formula (3) is equivalent to (4) by $\operatorname{tr}(AB) = \operatorname{tr}(BA)$.

It remains to prove formula (5). We will not use this formula but include it because of its intrinsic interest. To prove it, we compute

$$\begin{aligned}
\epsilon_*(g \times f)_![\Delta_M] &= \epsilon_*(D_N^{-1}(g \times f)^* D_{M\times M}[\Delta_M]) \\
&= \epsilon_*((g \times f)^*(\tau) \cap [N]) \\
&= \langle (g \times f)^*(\tau), [N] \rangle \\
&= \sum_\alpha (-1)^{\deg \alpha} \langle g^*(\alpha^\circ) \cup f^*(\alpha), [N] \rangle \\
&= \sum_\alpha (-1)^{\deg \alpha} \langle g^*(\alpha^\circ), f^*(\alpha) \cap [N] \rangle \\
&= \sum_\alpha (-1)^{\deg \alpha} \langle \alpha^\circ, g_*(f^*(\alpha) \cap [N]) \rangle \\
&= \sum_\alpha (-1)^{\deg \alpha} \langle \alpha^\circ, g^! f^*(\alpha) \cap [M] \rangle \\
&= \sum_\alpha (-1)^{\deg \alpha} \langle \alpha^\circ \cup g^! f^*(\alpha), [M] \rangle \\
&= \sum_i (-1)^i \operatorname{tr}_i(g^! f^*) \\
&= L(f, g)
\end{aligned}$$

by (2). Note that the α in the proof of (5) are for M and differ from the α in the proof of (1) which are for N. Also see Problem (10). □

14.5. Theorem. *If N^n and M^n are smooth closed oriented n-manifolds and $f, g: N \to M$ are smooth and such that the difference of differentials $g_* - f_*$ is nonsingular at each coincidence point of f and g then*

$$L(f, g) = \sum_x \operatorname{sign} \det(g_* - f_*)_x,$$

where the sum is over all coincidences $x \in N^n$ of f and g.

PROOF. That $g_* - f_*$ be nonsingular is precisely the condition that Γ_f and Γ_g be transverse at a point of their intersection. The orientation ± 1 attached to an intersection point x is either $\operatorname{sign} \det(g_* - f_*)_x$ or its negative depending only on the dimension n. This must be consistent with the case $g = 1$ of Corollary 12.7, and so the indicated sign is correct. □

Now we will prove some immediate corollaries and compute some examples.

14.6. Corollary. *For $f, g: N^n \to M^n$ and $h: K^n \to N^n$ then we have*

$$L(fh, gh) = \deg(h) L(f, g).$$

Proof. This follows from $\operatorname{tr}((fh)_*(gh)_!) = \operatorname{tr}(f_* h_* h_! g_!) = \deg(h)\operatorname{tr}(f_* g_!)$ by Proposition 14.1(6). □

14.7. Corollary. *For $f: N^n \to M^n$ we have $L(f, f) = \deg(f)\chi(M)$.*

Proof. We have $L(f, f) = L(1_M f, 1_M f) = \deg(f)L(1_M, 1_M) = \deg(f)\chi(M)$. □

14.8. Corollary. *If g is homotopic to a homeomorphism with degree ± 1 then $L(f, g) = \pm L(fg^{-1})$.* □

The following extraordinary result shows that Corollary 14.8 holds, in a fashion, even for many nonhomeomorphisms. If h^* (resp., h_*) is an endomorphism of (co)homology groups, *not necessarily induced by a map h*, we shall still use the notation $L(h^*) = \sum_i (-1)^i \operatorname{tr}(h^i)$ and similarly for h_*.

14.9. Theorem. *If $f, g: N^n \to M^n$ are maps between closed oriented manifolds with the same Betti numbers in each dimension and with $\deg(g) \neq 0$ then g^* is nonsingular and*

$$L(f, g) = \deg(g)L(f^*(g^*)^{-1}) = \deg(g)L(f_* g_*^{-1}).$$

Proof. By Proposition 14.1(6) we have $g_* g_! = \deg(g)$, so that $g_! = \deg(g)g_*^{-1}$ (over a field of coefficients). Hence $L(f, g) = \sum_i (-1)^i \operatorname{tr}_i(f_* g_!) = \deg(g)\sum_i (-1)^i \operatorname{tr}_i(f_* g_*^{-1}) = \deg(g)L(f_* g_*^{-1})$ and similarly for cohomology. □

What this means is that if g (or f) has nonzero degree then the algebra of computing $L(f, g)$ is essentially the same as that for computing the *fixed point number*.

For doing direct computations with the formulas of Theorem 14.4 one needs a description of a convenient way to calculate the matrix of $g^!$ (or $g_!$). For this, start with a basis $u, v, \ldots$ of cohomology (for both N and M if they differ). If one then takes the Poincaré dual basis $x, y, \ldots$ for homology then the duality isomorphisms D have the identity matrix and so the matrix for $g^!$ is identical to that for g_* in this basis. The basis of cohomology which is Kronecker product dual to $x, y, \ldots$ is just the basis $u^\circ, v^\circ, \ldots$ which is Poincaré cup product dual to $u, v, \ldots$. Since g^* is Kronecker dual to g_* it follows that the matrix for g_* is the transpose of that for g^* in the $u^\circ, v^\circ, \ldots$ basis. Thus if F is the matrix of f^* in the $u, v, \ldots$ basis and G_0 is the matrix for g^* in the $u^\circ, v^\circ, \ldots$ basis then

$$\boxed{\operatorname{tr}(f^* g^!) = \operatorname{tr}(F G_0^t).}$$

In the examples, we will use G for the matrix of g^* in the original $u, v, \ldots$ basis and will use $G^!$ for the matrix of $g^!$. Thus $G^! = G_0^t$ in the notation above.

In all our computations, we will take $1 \in H^0(N)$ as the basis element and $\vartheta = 1^\circ \in H^n(N)$ as the basis element. Then $\langle \vartheta, [N] \rangle = 1$.

In homology, if one prefers to compute there, one could use a basis $x, y, \ldots$ for H_* and its intersection product dual $x^\circ, y^\circ, \ldots$ in the same way as indicated in the above remarks about cohomology.

14.10. Example. Let $f, g: \mathbf{S}^2 \times \mathbf{S}^2 \to \mathbf{S}^2 \times \mathbf{S}^2$. First we will compute $L(f,g)$ using Theorem 14.9 assuming, then, that $\deg(g) \neq 0$. We have

$$L(f,g) = \deg(g) L(f^*(g^*)^{-1}) = \deg(g)\left(1 + \operatorname{tr}(FG^{-1}) + \frac{\deg(f)}{\deg(g)}\right)$$

where F and G are the matrices of f and g on $H^2(\mathbf{S}^2 \times \mathbf{S}^2)$. Thus

$$L(f,g) = \deg(f) + \deg(g) + \frac{\deg(g)}{|G|}\operatorname{tr}(F \operatorname{adj}(G)).$$

Let us use the matrix notation

$$F = \begin{pmatrix} a & b \\ c & d \end{pmatrix}, \qquad G = \begin{pmatrix} A & B \\ C & D \end{pmatrix},$$

in the basis $u = \vartheta \times 1, v = 1 \times \vartheta$ of $H^1(\mathbf{S}^2 \times \mathbf{S}^2)$, where $\langle \vartheta, [\mathbf{S}^2] \rangle = 1$. Then $f^*(uv) = (au + cv)(bu + dv) = (ad + bc)uv$, so that

$$\deg(f) = ad + bc.$$

Similarly,

$$\deg(g) = AD + BC.$$

Thus

$$L(f,g) = ad + bc + AD + BC + \frac{AD + BC}{AD - BC}\operatorname{tr}\left[\begin{pmatrix} a & b \\ c & d \end{pmatrix}\begin{pmatrix} D & -B \\ -C & A \end{pmatrix}\right]$$

$$= ad + bc + AD + BC + \frac{AD + BC}{AD - BC}(aD - bC - cB + dA).$$

Although easy to compute, this looks very strange. It doesn't even look like an integer, which it must be. However, maps $\mathbf{S}^2 \times \mathbf{S}^2 \to \mathbf{S}^2 \times \mathbf{S}^2$ are very restricted. From the proof of Theorem 4.13 we must always have $AC = 0$ and $BD = 0$. For $AD + BC = \deg(g) \neq 0$ we conclude that either $A = 0 = D$ or $B = 0 = C$. In these cases the expression for $L(f,g)$ simplifies to

$$A = 0 = D \quad \Rightarrow \quad L(f,g) = ad + bc + BC - (-bC - cB) = ad + (b + B)(c + C),$$

$$B = 0 = C \quad \Rightarrow \quad L(f,g) = ad + bc + AD + (aD + dA) = (a + A)(d + D) + bc.$$

Note that both cases are covered by the nice formula

$$L(f,g) = (a + A)(d + D) + (b + B)(c + C) = \begin{vmatrix} a + A & -(b + B) \\ c + C & d + D \end{vmatrix}.$$

From such an attractive formula, it is reasonable to expect that the formula also holds in the general case in which $\deg(g)$ may be zero. We shall show that this is, indeed, the case by computing $L(f,g)$ from the formulas of Theorem 14.4, which we wish to illustrate anyway.

In general, $\mathrm{tr}_0(f^*g^!) = \deg(g)$ since $f^* = 1$ in the standard basis $\{1\}$ for H^0 and $g^! = Dg_*D^{-1}$ is multiplication by $\deg(g)$ since g_* is operating in the top dimension. Similarly, in the top dimension, $f^*g^! = \deg(f)\cdot 1$ and so its trace is $\deg(f)$. It remains to compute $\mathrm{tr}_2(f^*g^!)$ and, for this, we must lay down an explicit basis for H^2 and its cup product dual basis.

Start with the choice above of $u = \vartheta \times 1$ and $v = 1 \times \vartheta$. Then $u^\circ = v$ and $v^\circ = u$. (Note that for odd-dimensional spheres we would have $u^\circ = -v$ and $v^\circ = u$ in order for $u^\circ u = \vartheta = v^\circ v$.)

Using, as above, the matrix G for g^*, we have

$$g^*(u) = Au + Cv,$$
$$g^*(v) = Bu + Dv.$$

Then we compute

$$g^*(u^\circ) = g^*(v) = Bu + Dv = Du^\circ + Bv^\circ,$$
$$g^*(v^\circ) = g^*(u) = Au + Cv = Cu^\circ + Av^\circ,$$

so that the matrix for g_* and hence for $g^!$ is

$$G^! = \begin{pmatrix} D & B \\ C & A \end{pmatrix}.$$

We conclude that

$$\begin{aligned} L(f,g) &= \deg(g) + \mathrm{tr}(FG^!) + \deg(f) \\ &= AD + BC + (aD + bC + cB + dA) + ad + bc \\ &= (a + A)(d + D) + (b + B)(c + C) \end{aligned}$$

as claimed.

14.11. Example. We will study the n-torus $\mathbf{T}^n = \mathbf{S}^1 \times \cdots \times \mathbf{S}^1$ here. The coincidence number will be computed in two completely different ways, first algebraically based on Theorem 14.9, and then geometrically.

The cohomology of $\mathbf{T}^n$ is an exterior algebra $\wedge(x_1, \ldots, x_n)$ on n generators of degree 1 and f^* is determined by its action on the x_i. The Lefschetz fixed point number makes sense for any endomorphism A on $\mathbf{C}^n$: $L(A) = \sum_i (-1)^i \mathrm{tr}(\wedge^i A\colon \wedge^i(\mathbf{C}^n))$ and not only for those induced by maps of the torus. Assume, to begin, that A is diagonalizable over $\mathbf{C}$. Then there are n independent eigenvectors $v_1, \ldots, v_n$ with eigenvalues $\lambda_1, \ldots, \lambda_n$. Then $\{v_{s_1} \wedge \cdots \wedge v_{s_k} \mid s_1 < \cdots < s_k\}$ is a basis of $\wedge^k \mathbf{C}^n$ and each is an eigenvector of $\wedge^k A$ since $Av_{s_1} \wedge \cdots \wedge Av_{s_k} = \lambda_{s_1} \cdots \lambda_{s_k} v_{s_1} \wedge \cdots \wedge v_{s_k}$. Therefore, $\mathrm{tr} \wedge^k A = \sum\{\lambda_{s_1} \cdots \lambda_{s_k} \mid s_1 < \cdots < s_k\}$, so that

$$L(A) = (1 - \lambda_1)(1 - \lambda_2) \cdots (1 - \lambda_n) = |I - A|.$$

Since $L(A)$ and $|I - A|$ are continuous functions of A and the diagonalizable matrices are dense in the space of all complex $n \times n$ matrices, this formula holds in general.

Now for maps $f, g: \mathbf{T}^n \to \mathbf{T}^n$ with $\deg(g) \neq 0$, and letting F and G be the matrices for f^*, g^* on $H^1(\mathbf{T}^n)$, we have

$$L(f,g) = \deg(g) L(FG^{-1}) = |G| \cdot |I - FG^{-1}| = |G - F|.$$

Since both sides of this equation are continuous functions in F and G, it follows that the formula

$$L(f,g) = |G - F|$$

is valid for all g, not just those of nonzero degree $|G|$. (Although we have not derived a formula for $g^!$, it is clear that there is one which is a polynomial function of the entries of G.)

Now we will show that the same formula can be derived from a purely geometric discussion. For any maps $f, g: \mathbf{T}^n \to \mathbf{T}^n$, the induced endomorphisms F, G on $H^1(\mathbf{T}^n; \mathbf{R})$ are identical to those induced on $H^1(\mathbf{T}^n; \mathbf{R}) \approx \mathbf{R}^n$ by the maps $\bar{F}, \bar{G}: \mathbf{T}^n \to \mathbf{T}^n$ induced from F, G. In fact, $\bar{F} \simeq f$ and $\bar{G} \simeq g$, but we don't need that. This implies that $L(f,g) = L(\bar{F}, \bar{G})$ and we can concentrate on the latter. Assume, for the moment, that $G - F$ is nonsingular.

By Theorem 14.5 the local coincidence number of $\bar{F}, \bar{G}$ at any coincidence is $\operatorname{sign}|G - F|$ and it remains to find the number of coincidences. But coincidences of $\bar{F}$ and $\bar{G}$ are identical to the zeros of $\bar{H} = \overline{G - F}$; i.e., the points in a fundamental domain which map into lattice points by $H = G - F: \mathbf{R}^n \to \mathbf{R}^n$. But $\bar{H}: \mathbf{T}^n \to \mathbf{T}^n$ is a covering map and so its number of zeros equals its number of sheets. This, in turn, is the factor by which $\bar{H}$ increases volume, and that equals $\operatorname{abs}|H|$. Therefore, by Theorem 14.5, $L(f,g) = L(\bar{F}, \bar{G}) = \operatorname{sign}|H| \operatorname{abs}|H| = |H| = |G - F|$. If $H = G - F$ is singular, then $\bar{H}$ is onto a proper subtorus of $\mathbf{T}^n$. If $A \in \mathbf{R}^n$ is a sufficiently small vector not in range(H) then $F'(x) = F(x) + A$ induces a map $\bar{F}': \mathbf{T}^n \to \mathbf{T}^n$ which is homotopic to $\bar{F}$ and which has no coincidences with $\bar{G}$ since the similarly defined $\bar{H}'$ has an image which is disjoint from that of $\bar{H}$. Hence $L(f,g) = L(\bar{F}', \bar{G}) = 0 = |G - F|$ in this case also.

14.12. Example. In order to illustrate the case in which $N \neq M$, let us consider maps $f, g: \mathbf{S}^2 \times \mathbf{S}^2 \to \mathbf{CP}^2$. Here Theorem 14.9 is not available because the restriction on Betti numbers is not satisfied, and so we must use a formula from Theorem 14.4. Since we know the contribution in degrees 0 and 4 from the discussion in Example 14.10, it suffices to compute $\operatorname{tr} f^* g^!$ in degree 2. We will use the basis $u, v \in H^2(\mathbf{S}^2 \times \mathbf{S}^2)$ as constructed in Example 14.10. Therefore $u^\circ = v$ and $v^\circ = u$. Let $t \in H^2(\mathbf{CP}^2)$ be a generator with $\mathbf{Z}$ coefficients and orient $\mathbf{CP}^2$ so that $\langle t^2, [\mathbf{CP}^2] \rangle = 1$. Then $t = t^\circ$. Put

$$f^*(t) = au + bv, \qquad g^*(t) = Au + Bv.$$

Then $f^*(t^2) = 2abuv$, so that $\deg(f) = 2ab$. Similarly, $\deg(g) = 2AB$. Now

$$g^*(t^\circ) = Au + Bv = Bu^\circ + Av^\circ$$

so that $g^!$ has matrix (B, A). Also, f^* has matrix $(a, b)^t$ and so

$$\begin{aligned} L(f, g) &= \deg(g) + \operatorname{tr}(a, b)^t(B, A) + \deg(f) \\ &= 2AB + aB + bA + 2ab, \end{aligned}$$

finishing this example.

Now let us briefly discuss the case of bounded manifolds. If one of the maps, say g, takes ∂N into ∂M then the proofs can be carried through as in the discussion above Theorem 12.9 where Γ_g takes the place of Δ. There are no difficulties with concluding the following generalization of Theorems 14.4 and 14.5.

14.13. Theorem. *Let $f: N^n \to M^n$ and $g: (N, \partial N) \to (M, \partial M)$ be maps. Assume that f and g have no coincidences on ∂N. Then with $[\Gamma_f] \in H_n((N, \partial N) \times M)$ and $[\Gamma_g] \in H_n(N \times (M, \partial M))$*

$$L(f, g) = [\Gamma_f] \cdot [\Gamma_g]$$

can be computed by

$$L(f, g) = \sum_i (-1)^i \operatorname{tr}_i(f^* g^!) \quad \text{on } H^*(N)$$

or by

$$L(f, g) = \sum_i (-1)^i \operatorname{tr}_i(f_* g_!) \quad \text{on } H_*(M).$$

Moreover, if f and g are also smooth and if the differential $g_* - f_*$ is nonsingular at each coincidence, then

$$L(f, g) = \sum_x \operatorname{sign} \det(g_* - f_*)_x$$

where the sum ranges over all coincidences $x \in N^n$ of f and g. □

For example, we have the following generalization of the Brouwer Fixed Point Theorem:

14.14. Corollary. *If $g: (\mathbf{D}^n, \partial \mathbf{D}^n) \to (\mathbf{D}^n, \partial \mathbf{D}^n)$ has nonzero degree then any map $f: \mathbf{D}^n \to \mathbf{D}^n$ has a coincidence with g.*

PROOF. The only nontrivial dimension for $f_* g_!$ is dimension 0. There we have that $g_! = D^{-1} g^* D$ is multiplication by $\deg(g)$. Hence $L(f, g) = \deg(g)$. □

In a sense, the bounded case reduces to the closed case by doubling both manifolds. Let N_1, N_2 be the two copies of N in the doubled manifold and similarly for M_1, M_2. Let $f_1, g_1: N_1 \to M_1$ be copies of f, g. On N_2 let $f_2: N_2 \to M_1$ also be a copy of f and $g_2: N_2 \to M_2$ a copy of g. (See Figure IV-11.) Then the only coincidence of $f_1 \cup f_2$ with $g_1 \cup g_2$ is between f_1 and g_1.

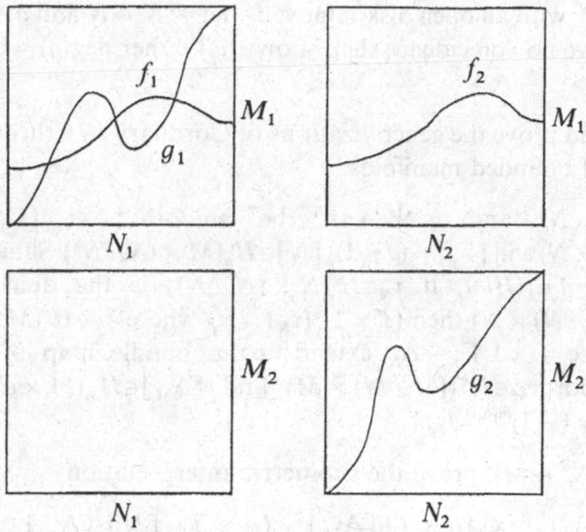

Figure VI-11. Coincidence doubling.

Consequently

$$L(f, g) = L(f_1 \cup f_2, g_1 \cup g_2).$$

Of course, the doubled manifolds have more homology, so the computation of the coincidence numbers is potentially more difficult, although it can be shown that the situation in homology is analogous to Figure VI-11, so that the traces are all the same as in the bounded situation.

PROBLEMS

1. For maps $f, g: M^n \to \mathbf{S}^n$, compute $L(f, g)$.
2. If $f, g: N^n \to M^m$ and f is constant, show that $L(f, g) = \deg(g)$.
3. Let $f, g: \mathbf{S}^n \times \mathbf{S}^m \to \mathbf{S}^n \times \mathbf{S}^m$ with $n \neq m$.
   (a) Compute $L(f, g)$.
   (b) If $n \neq m$ are both odd and $f_* \neq g_*$ on both H_n and H_m then show that f, g have a coincidence.
   (c) If $n \neq m$ are both even and $f_* \neq -g_*$ on both H_n and H_m then show that f, g have a coincidence.
   (d) Show, by examples, that the conditions $f_* \neq g_*$ in (b) and $f_* \neq -g_*$ in (c) are necessary.
4. For $f, g: \mathbf{CP}^n \to \mathbf{CP}^n$, find $L(f, g)$. Use this to show that, for n even, any two maps $f, g: \mathbf{CP}^n \to \mathbf{CP}^n$ of nonzero degree must have a coincidence.
5. Using mod 2 homology, show that any two maps $f, g: \mathbf{P}^2 \to \mathbf{P}^2$ of nonzero degree have a coincidence. Give two proofs, one by coincidence theory mod 2 and one by other methods.
6. Rederive Corollary 14.14 using the technique of doubling.

7. Let N^4 be $\mathbf{CP}^2$ with an open disk removed. Let $f: N \to N$ and $g: (N, \partial N) \to (N, \partial N)$. If f and g have no coincidence then show that either $\deg(g) = 0$ or $f_* = -g_*$ on $H_2(N)$.

8. Formulate and prove the generalizations of Corollary 14.6 through Theorem 14.9 in the case of bounded manifolds.

9. Let $f: M^m \to N^n$ and $g: N^n \to M^m$ be smooth. Let $[\Gamma_f] = (1 \times f)_*[M] \in H_m((M, \partial M) \times N)$ and $[\Gamma_g] = (g \times 1)_*[N] \in H_n(M \times (N, \partial N))$. Show that $[\Gamma_f] \cdot [\Gamma_g] = L(f \circ g) = L(g \circ f)$. (*Hint*: If $\tau_N \in H^n(N \times (N, \partial N))$ is the dual of the diagonal $[\Delta_N] \in H_n((N, \partial N) \times N)$ then $(f \times 1)^*(\tau_N) = \gamma_f$ where $\gamma_f \in H^n(M \times (N, \partial N))$ is dual to $[\Gamma_f]$ since $f \times 1: \Gamma_f \to \Delta_N$ extends to a bundle map of normal bundles. Similarly, with $\tau_M \in H^m((M, \partial M) \times M)$ and $[\Delta_M] \in H_m(M \times (M, \partial M))$, one has $(1 \times g)^*(\tau_M) = (-1)^{m(n+1)}\gamma_g$.)

10. ◆ For $f, g: N^n \to M^n$, prove the geometric interpretation

$$\epsilon_*(g \times f)_![\Delta_M] = (g \times f)_*[N] \cdot [\Delta_M]$$

of formula (5) of Theorem 14.4.

15. Steenrod Operations ☼

In this section we describe certain "cohomology operations" discovered by Steenrod, related to squaring $a \mapsto a^2$. The definitive reference is Steenrod and Epstein [1] (also see Mosher and Tangora [1]) and we shall follow that work by laying down axioms for the operations and discussing the immediate applications. The proof of the existence of the operations, i.e., their construction, is left to the following section. We shall not consider the more difficult matter of uniqueness.

The ith "Steenrod square," $i \geq 0$, is a cohomology operation

$$\boxed{\mathrm{Sq}^i: H^n(X, A; \mathbf{Z}_2) \to H^{n+i}(X, A; \mathbf{Z}_2)}$$

(meaning that it is a natural transformation of functors of (X, A)) which is a homomorphism satisfying the following axioms:

Axiom (1) $\mathrm{Sq}^0 = 1$.
Axiom (2) $\deg(x) = i \Rightarrow \mathrm{Sq}^i(x) = x^2$.
Axiom (3) $i > \deg(x) \Rightarrow \mathrm{Sq}^i(x) = 0$.
Axiom (4) (Cartan formula.) $\mathrm{Sq}^k(xy) = \sum_{i=0}^k \mathrm{Sq}^i(x) \cup \mathrm{Sq}^{k-i}(y)$.

We shall also assume the following two properties which can be shown to follow from the axioms:

Property (5) Sq^1 is the Bockstein (connecting) homomorphism for the coefficient sequence

$$0 \to \mathbf{Z}_2 \to \mathbf{Z}_4 \to \mathbf{Z}_2 \to 0.$$

Property (6) (Adem relations.) If $0 < a < 2b$ then

$$\boxed{\operatorname{Sq}^a \operatorname{Sq}^b = \sum_{j=0}^{[a/2]} \binom{b-1-j}{a-2j} \operatorname{Sq}^{a+b-j} \operatorname{Sq}^j}$$

where the binomial coefficient is taken mod 2.

15.1. Proposition. *For* $x \in H^*(X; \mathbf{Z}_2)$, $y \in H^*(Y; \mathbf{Z}_2)$, *and* $x \times y \in H^*(X \times Y; \mathbf{Z}_2)$ *we have*

$$\operatorname{Sq}^n(x \times y) = \sum_{i=0}^{n} \operatorname{Sq}^i(x) \times \operatorname{Sq}^{n-i}(y).$$

PROOF. From Axiom 4 we have $\operatorname{Sq}^n(x \times y) = \operatorname{Sq}^n((x \times 1) \cup (1 \times y)) = \sum_i \operatorname{Sq}^i(x \times 1) \operatorname{Sq}^{n-i}(1 \times y)$. But, if $p_X : X \times Y \to X$ is the projection then $\operatorname{Sq}^i(x \times 1) = \operatorname{Sq}^i(p_X^*(x)) = p_X^* \operatorname{Sq}^i(x) = \operatorname{Sq}^i(x) \times 1$. Consequently $\operatorname{Sq}^n(x \times y) = \sum_i \operatorname{Sq}^i(x \times 1) \operatorname{Sq}^{n-i}(1 \times y) = \sum_i (\operatorname{Sq}^i(x) \times 1) \cup (1 \times \operatorname{Sq}^{n-i}(y)) = \sum_i \operatorname{Sq}^i(x) \times \operatorname{Sq}^{n-i}(y)$. □

15.2. Proposition. Sq^i *commutes with* δ^*. *That is, the following diagram commutes:*

$$\begin{array}{ccc} H^n(A; \mathbf{Z}_2) & \xrightarrow{\delta^*} & H^{n+1}(X, A; \mathbf{Z}_2) \\ \Big\downarrow \operatorname{Sq}^i & & \Big\downarrow \operatorname{Sq}^i \\ H^{n+i}(A; \mathbf{Z}_2) & \xrightarrow{\delta^*} & H^{n+i+1}(X, A; \mathbf{Z}_2). \end{array}$$

PROOF. Let $Y = X \times \{0\} \cup A \times I$. Then, by naturality, it suffices to prove the result for the pair $(Y, A \times I)$. Another naturality argument implies that it suffices to prove it for the pair $(Y, A \times \{1\})$. Put $C = A \times \{1\}$ and $B = A \times [0, \frac{1}{2}] \cup X \times \{0\} \subset Y$. In the diagram

$$\begin{array}{ccc} H^n(B \cup C; \mathbf{Z}_2) & \longrightarrow & H^n(C; \mathbf{Z}_2) \\ \Big\downarrow \delta^* & & \Big\downarrow \delta^* \\ H^{n+1}(Y, B \cup C; \mathbf{Z}_2) & \longrightarrow & H^{n+1}(Y, C; \mathbf{Z}_2) \end{array}$$

the map on top is onto and it follows that it suffices to prove the result for the pair $(Y, B \cup C)$. By the excision (and homotopy) isomorphisms, it suffices to prove the result for the pair $(A \times [\frac{1}{2}, 1], A \times \{\frac{1}{2}\} \cup A \times \{1\})$, which is equivalent to the pair $(A \times I, A \times \partial I)$. Now an element of $H^*(A \times \partial I; \mathbf{Z}_2)$ has the form $x \times y$ for $x \in H^*(A; \mathbf{Z}_2)$ and $y \in H^0(\partial I)$. Then $\delta^*(x \times y) = x \times \delta^* y$. (There is a sign $(-1)^{\deg(x)}$, which can be dropped since coefficients are in $\mathbf{Z}_2$.) Therefore

$$\begin{aligned} \operatorname{Sq}^i(\delta^*(x \times y)) &= \operatorname{Sq}^i(x \times \delta^* y) \\ &= \operatorname{Sq}^i(x) \times \operatorname{Sq}^0(\delta^*(y)) \quad \text{(by Proposition 15.1)} \\ &= \operatorname{Sq}^i(x) \times \delta^*(y) \quad \text{(by Axiom (1))} \end{aligned}$$

$$
\begin{aligned}
&= \delta^*(\mathrm{Sq}^i(x) \times y) \\
&= \delta^*(\mathrm{Sq}^i(x \times y)) \qquad \text{(by Proposition 15.1, Axioms (1) and (3))}
\end{aligned}
$$

finishing the proof. □

15.3. Proposition. Sq^i *commutes with (unreduced) suspension. That is, the following diagram commutes:*

$$
\begin{array}{ccc}
\tilde{H}^n(X; \mathbf{Z}_2) & \xrightarrow{\Sigma} & \tilde{H}^{n+1}(\Sigma X; \mathbf{Z}_2) \\
\downarrow \mathrm{Sq}^i & & \downarrow \mathrm{Sq}^i \\
\tilde{H}^{n+i}(X; \mathbf{Z}_2) & \xrightarrow{\Sigma} & \tilde{H}^{n+i+1}(\Sigma X; \mathbf{Z}_2).
\end{array}
$$

PROOF. The suspension can be defined as the composition

$$\tilde{H}^n(X) \xrightarrow{\delta^*} H^{n+1}(CX, X) \xleftarrow{\approx} H^{n+1}(\Sigma X)$$

and so the result follows from Proposition 15.2 and the naturality of Sq^i. □

For example, consider $\mathbf{CP}^2$. We know that $H^*(\mathbf{CP}^2; \mathbf{Z}_2)$ has (only) the nonzero elements $1 \in H^0(\mathbf{CP}^2; \mathbf{Z}_2)$, $x \in H^2(\mathbf{CP}^2; \mathbf{Z}_2)$ and $x^2 \in H^4(\mathbf{CP}^2; \mathbf{Z}_2)$. We have $\mathrm{Sq}^2(x) = x^2 \neq 0$. Recall that

$$\mathbf{CP}^2 \approx \mathbf{S}^2 \cup_h \mathbf{D}^4$$

where $h: \mathbf{S}^3 \to \mathbf{S}^2$ is the Hopf map. This implies that

$$\Sigma\mathbf{CP}^2 \approx \mathbf{S}^3 \cup_{\Sigma h} \mathbf{D}^5,$$
$$\Sigma^2\mathbf{CP}^2 \approx \mathbf{S}^4 \cup_{\Sigma^2 h} \mathbf{D}^6,$$

and so on. It follows from Proposition 15.3 that $\mathrm{Sq}^2: H^3(\Sigma\mathbf{CP}^2; \mathbf{Z}_2) \to H^5(\Sigma\mathbf{CP}^2; \mathbf{Z}_2)$ is nonzero.

This implies that Σh is not homotopically trivial, for, if it were, then $\Sigma\mathbf{CP}^2 \simeq \mathbf{S}^3 \vee \mathbf{S}^5$. But the commutative diagram

$$
\begin{array}{ccc}
H^3(\mathbf{S}^3) & \xrightarrow{\mathrm{Sq}^2} & H^5(\mathbf{S}^3) = 0 \\
\approx \downarrow \mathrm{proj}^* & & \downarrow \\
H^3(\mathbf{S}^3 \vee \mathbf{S}^5) & \xrightarrow{\mathrm{Sq}^2} & H^5(\mathbf{S}^3 \vee \mathbf{S}^5)
\end{array}
$$

shows that $\mathrm{Sq}^2 = 0$ on $H^3(\mathbf{S}^3 \vee \mathbf{S}^5; \mathbf{Z}_2)$ contrary to its being nonzero on $H^3(\Sigma\mathbf{CP}^2; \mathbf{Z}_2)$. Similarly, none of the suspensions $\Sigma^n h$ of h are homotopically trivial. Thus we have:

15.4. Corollary. $\pi_{n+1}(\mathbf{S}^n) \neq 0$ *for all* $n \geq 2$. □

Similar considerations apply to the other Hopf maps.

It is convenient to put $H(X) = \bigoplus_i H^i(X; \mathbf{Z}_2)$ and to define

$$\mathrm{Sq}\colon H(X) \to H(X)$$

to be $\mathrm{Sq} = \mathrm{Sq}^0 + \mathrm{Sq}^1 + \mathrm{Sq}^2 + \cdots$. This makes sense by Axiom (3). Then the Cartan formula becomes

$$\mathrm{Sq}(xy) = \mathrm{Sq}(x)\mathrm{Sq}(y).$$

In particular

$$\mathrm{Sq}(x^k) = (\mathrm{Sq}(x))^k.$$

If $\deg(x) = 1$ then we compute $\mathrm{Sq}(x^k) = (\mathrm{Sq}(x))^k = (x + x^2)^k = x^k(1 + x)^k = \sum_i \binom{k}{i} x^{k+i}$. Thus we have:

15.5. Proposition. *For* $\deg(x) = 1$ *we have* $\mathrm{Sq}^i(x^k) = \binom{k}{i} x^{k+i}$. □

Because of this result and the Adem relations, it is helpful to have a convenient way to compute the mod 2 binomial coefficients.

15.6. Proposition. *If* $a = \sum_j a_j 2^j$ *and* $b = \sum_j b_j 2^j$ *then*

$$\binom{a}{b} \equiv \times \binom{a_j}{b_j} \pmod 2.$$

Proof. Consider the polynomial ring $\mathbf{Z}_2[x]$. We compute

$$\begin{aligned}
(1 + x)^a &= (1 + x)^{\Sigma a_j 2^j} \\
&= \times\ (1 + x)^{a_j 2^j} \\
&= \times\ (1 + x^{2^j})^{a_j} \quad (\text{since } (1 + x)^2 = 1 + 2x + x^2 \equiv 1 + x^2) \\
&= \times \sum_k \binom{a_j}{k} x^{k 2^j}.
\end{aligned}$$

But the coefficient of x^b in $(1 + x)^a$ is $\binom{a}{b}$, while its coefficient in the last expression is $\times\ \binom{a_j}{b_j}$. □

Note that $\binom{0}{1} = 0$ while $\binom{1}{0} = \binom{1}{1} = \binom{0}{0} = 1$. Thus $\binom{a}{b} = 1 \Leftrightarrow (b_j = 1 \Rightarrow a_j = 1)$.

15.7. Corollary. *If* $\deg(x) = 1$ *then*

$$\mathrm{Sq}^i(x^{2^k}) = \begin{cases} x^{2^k} & \text{for } i = 0, \\ x^{2^{k+1}} & \text{for } i = 2^k, \\ 0 & \text{otherwise.} \end{cases}$$ □

15.8. Theorem. *If i is not a power of 2 then* Sq^i *is decomposable, meaning that it is a sum of compositions of Steenrod squaring operations of smaller degree than i.*

PROOF. The Adem relations can be rewritten in the form

$$\binom{b-1}{a}\mathrm{Sq}^{a+b} = \mathrm{Sq}^a\mathrm{Sq}^b + \sum_{j=1}^{[a/2]}\binom{b-1-j}{a-2j}\mathrm{Sq}^{a+b-j}\mathrm{Sq}^j$$

where $0 < a < 2b$. Thus if $\binom{b-1}{a} \equiv 1 \pmod 2$ then Sq^{a+b} is decomposable. But if i is not a power of 2 we can write $i = a + b$ where $b = 2^k$ and $0 < a < 2^k$. Since $b - 1 = 1 + 2 + 2^2 + \cdots + 2^{k-1}$ and $a \le b - 1$ we see from Proposition 15.6 that $\binom{b-1}{a} \equiv 1 \pmod 2$. □

For example, there are the relations

$$\mathrm{Sq}^3 = \mathrm{Sq}^1\mathrm{Sq}^2,$$
$$\mathrm{Sq}^5 = \mathrm{Sq}^1\mathrm{Sq}^4,$$
$$\mathrm{Sq}^6 = \mathrm{Sq}^2\mathrm{Sq}^4 + \mathrm{Sq}^5\mathrm{Sq}^1.$$

Also note the relations

$$\mathrm{Sq}^1\mathrm{Sq}^{2n} = \mathrm{Sq}^{2n+1},$$
$$\mathrm{Sq}^1\mathrm{Sq}^{2n+1} = 0,$$
$$\mathrm{Sq}^{2n-1}\mathrm{Sq}^n = 0,$$
$$\mathrm{Sq}^2\mathrm{Sq}^2 = \mathrm{Sq}^3\mathrm{Sq}^1.$$

15.9. Corollary. *If i is not a power of 2 and if X is a space such that $H^k(X; \mathbf{Z}_2) = 0$ for all $n < k < n + i$ then $\mathrm{Sq}^i: H^n(X; \mathbf{Z}_2) \to H^{n+i}(X; \mathbf{Z}_2)$ is zero.* □

15.10. Corollary. *If $x \in H^n(X; \mathbf{Z}_2)$ and $x^2 \neq 0$ then $\mathrm{Sq}^{2^i}(x) \neq 0$ for some i with $0 < 2^i \le n$.* □

15.11. Corollary. *If $H^*(X; \mathbf{Z}_2) = \mathbf{Z}_2[x]$ or $\mathbf{Z}_2[x]/(x^q)$ for some $q > 2$ then $n = \deg(x)$ is a power of 2.* □

It has been shown by Adams, using a much deeper study of Steenrod squares, that $n = 1, 2, 4$, or 8 are the only possibilities in the situation of Corollary 15.11. See Atiyah [1].

15.12. Corollary. *If M^{2n} is a closed $2n$-manifold with $H_i(M; \mathbf{Z}_2) = 0$ for $0 < i < n$ and with $H_n(M; \mathbf{Z}_2) \approx \mathbf{Z}_2$ then n is a power of 2 (in fact, $n = 1, 2, 4$, or 8 by Adams).* □

15.13. Corollary. *If $\mathbf{S}^{2n-1}$ is a fiber bundle over $\mathbf{S}^n$ with fiber $\mathbf{S}^{n-1}$ then n is a power of 2 ($n = 1, 2, 4$, or 8 by Adams).*

PROOF. If $f: \mathbf{S}^{2n-1} \to \mathbf{S}^n$ is a bundle projection with fiber $\mathbf{S}^{n-1}$ then M_f is a $2n$-manifold with boundary $\mathbf{S}^{2n-1}$ and so C_f is a closed $2n$-manifold and has homology as in Corollary 15.12. □

Let $f: \mathbf{S}^{2n-1} \to \mathbf{S}^n$ be any map, $n \geq 2$. Put $X = C_f = \mathbf{S}^n \cup_f \mathbf{D}^{2n}$. Then X is a CW-complex with three cells, in dimensions 0, n, and $2n$. The inclusion $\mathbf{S}^n \hookrightarrow X$ induces an isomorphism $H^n(X) \xrightarrow{\approx} H^n(\mathbf{S}^n)$, so take $0 \neq x \in H^n(X; \mathbf{Z})$ to correspond to the orientation class of $\mathbf{S}^n$. Similarly, the collapse $X \to \mathbf{S}^{2n}$ induces an isomorphism $H^{2n}(\mathbf{S}^{2n}) \xrightarrow{\approx} H^{2n}(X)$, so take $0 \neq y \in H^{2n}(X; \mathbf{Z})$ to correspond to the orientation class of $\mathbf{S}^{2n}$. Then

$$\boxed{x^2 = h_f y}$$

for some integer h_f. This integer h_f is called the "Hopf invariant" of f. For the Hopf maps $\mathbf{S}^3 \to \mathbf{S}^2$, $\mathbf{S}^7 \to \mathbf{S}^4$, and $\mathbf{S}^{15} \to \mathbf{S}^8$, C_f is a manifold and so $h_f = \pm 1$ in those cases by Poincaré duality.

15.14. Corollary. *Let $f: \mathbf{S}^{2n-1} \to \mathbf{S}^n$. If the Hopf invariant h_f is odd then n is a power of 2 ($n = 1, 2, 4$, or 8 by Adams).* □

Now suppose we have a map $f: \mathbf{S}^p \times \mathbf{S}^q \to \mathbf{S}^r$. Then f induces a map $\mathbf{D}^{p+1} \times \mathbf{S}^q \to \mathbf{D}_+^{r+1}$ by coning off. Similarly, it induces $\mathbf{S}^p \times \mathbf{D}^{q+1} \to \mathbf{D}_-^{r+1}$ and so there is an induced map

$$\bar{f}: \mathbf{S}^{p+q+1} \approx \mathbf{D}^{p+1} \times \mathbf{S}^q \cup \mathbf{S}^p \times \mathbf{D}^{q+1} \to \mathbf{D}_+^{r+1} \cup \mathbf{D}_-^{r+1} \approx \mathbf{S}^{r+1}.$$

(Compare Definition 8.6 of Chapter VII.) Specialize to the case $f: \mathbf{S}^{n-1} \times \mathbf{S}^{n-1} \to \mathbf{S}^{n-1}$. If the restriction of f to $\mathbf{S}^{n-1} \times \{*\} \to \mathbf{S}^{n-1}$ has degree p and the restriction to $\{*\} \times \mathbf{S}^{n-1} \to \mathbf{S}^{n-1}$ has degree q then we say that f has "bidegree" (p, q).

15.15. Proposition. *If $f: \mathbf{S}^{n-1} \times \mathbf{S}^{n-1} \to \mathbf{S}^{n-1}$ has bidegree (p, q) then the induced map $\bar{f}: \mathbf{S}^{2n-1} \to \mathbf{S}^n$ has Hopf invariant $\pm pq$.*

PROOF. We shall use integer coefficients for cohomology throughout this proof, which is based on that in Steenrod–Epstein [1]. The mapping cone C_f can be regarded as the space

$$C = (\mathbf{D}_+^n \cup \mathbf{D}_-^n) \cup_{\bar{f}} (\mathbf{D}^n \times \mathbf{D}^n)$$

since $\partial(\mathbf{D}^n \times \mathbf{D}^n) = \mathbf{D}^n \times \mathbf{S}^{n-1} \cup \mathbf{S}^{n-1} \times \mathbf{D}^n$. Thus we have a map

$$g: (\mathbf{D}^n \times \mathbf{D}^n, \mathbf{D}^n \times \mathbf{S}^{n-1}, \mathbf{S}^{n-1} \times \mathbf{D}^n) \to (C, \mathbf{D}_+^n, \mathbf{D}_-^n).$$

There is the commutative diagram

$$\begin{array}{ccc} H^n(C, \mathbf{D}_+^n) \times H^n(C, \mathbf{D}_-^n) & \xrightarrow{\ \cup\ } & H^{2n}(C, \mathbf{S}^{n-1}) \\ \downarrow \approx & & \downarrow \approx \\ H^n(C) \times H^n(C) & \xrightarrow{\ \cup\ } & H^{2n}(C) \end{array}$$

and so if we let $x_- \in H^n(C, \mathbf{D}_+^n)$ and $x_+ \in H^n(C, \mathbf{D}_-^n)$ correspond to the generator

$x \in H^n(C)$ then $x_+ \cup x_-$ corresponds to x^2. The commutative diagram

$$\begin{array}{ccccccc}
H^n(C) & & \xleftarrow{\quad\approx\quad} & & H^n(C, \mathbf{D}^n_-) & \xrightarrow{g^*} & H^n(\mathbf{D}^n \times \mathbf{D}^n, \mathbf{S}^{n-1} \times \mathbf{D}^n) \\
\downarrow\approx & & & & \downarrow\approx & & \downarrow\approx \\
H^n(\mathbf{S}^n) & \xleftarrow{\approx} & H^n(\mathbf{S}^n, \mathbf{D}^n_-) & \xrightarrow{\approx} & H^n(\mathbf{D}^n_+, \mathbf{S}^{n-1}) & \xrightarrow{g^*} & H^n(\mathbf{D}^n \times \{*\}, \mathbf{S}^{n-1} \times \{*\}) \\
 & & & & \approx\uparrow\delta^* & & \approx\uparrow\delta^* \\
 & & & & H^{n-1}(\mathbf{S}^{n-1}) & \underset{\times p}{\xrightarrow{g^*}} & H^{n-1}(\mathbf{S}^{n-1} \times \{*\})
\end{array}$$

shows that $g^*(x_+) = \pm pw \times 1$ where $w \in H^n(\mathbf{D}^n, \mathbf{S}^{n-1})$ is a generator and hence $w \times 1 \in H^n(\mathbf{D}^n \times \mathbf{D}^n, \mathbf{S}^{n-1} \times \mathbf{D}^n)$ is a generator. Similarly, we have that $g^*(x_-) = \pm q1 \times w$. But $g^*: H^{2n}(C, \mathbf{S}^n) \to H^{2n}(\mathbf{D}^n \times \mathbf{D}^n, \partial(\mathbf{D}^n \times \mathbf{D}^n))$ is an isomorphism and carries $x_+ \cup x_-$ to $\pm pq(w \times 1) \cup (1 \times w) = \pm pq(w \times w)$. Thus $x_+ \cup x_- = pq(\text{generator})$. Since x_+ and x_- each map to $x \in H^n(C)$ we conclude that $x^2 = pqy$ for some generator $y \in H^{2n}(C)$. □

15.16. Corollary. *If* $\mathbf{S}^{n-1}$ *is parallelizable then n is a power of* 2 ($n = 1, 2, 4,$ *or* 8 *by Adams*).

Proof. If $\mathbf{S}^{n-1}$ is parallelizable then there is a map $\phi: \mathbf{S}^{n-1} \to \mathbf{SO}(n)$ assigning to $x \in \mathbf{S}^{n-1}$ a matrix with first column x and the rest making an $(n-1)$-frame orthogonal to x. Then $\phi(x)e_1 = x$ if e_1 is the first standard basis vector. Define $f: \mathbf{S}^{n-1} \times \mathbf{S}^{n-1} \to \mathbf{S}^{n-1}$ by $f(x, y) = \phi(x) \cdot y$. Then $f(e_1, y) = \phi(e_1) \cdot y$ has degree 1 as a function of y. Also $f(x, e_1) = \phi(x) \cdot e_1 = x$ has degree 1. Therefore $\bar{f}: \mathbf{S}^{2n-1} \to \mathbf{S}^n$ has Hopf invariant 1. □

We conclude this section by briefly introducing the "Steenrod cyclic reduced powers" which are the analogues of the squares for odd primes p.

Let $\beta: H^i(X; \mathbf{Z}_p) \to H^{i+1}(X; \mathbf{Z}_p)$ be the Bockstein associated with the coefficient sequence $0 \to \mathbf{Z}_p \to \mathbf{Z}_{p^2} \to \mathbf{Z}_p \to 0$.

For an odd prime p, the Steenrod cyclic reduced power operation $\mathcal{P}^k$, $k \geq 0$, is a natural transformation

$$\boxed{\mathcal{P}^k: H^q(X, A; \mathbf{Z}_p) \to H^{q+2k(p-1)}(X, A; \mathbf{Z}_p)}$$

which is a homomorphism satisfying the following axioms:

Axiom (1_p) $\mathcal{P}^0 = 1$.
Axiom (2_p) $\deg(x) = 2k \Rightarrow \mathcal{P}^k(x) = x^p$.
Axiom (3_p) $2k > \deg(x) \Rightarrow \mathcal{P}^k(x) = 0$.
Axiom (4_p) (Cartan formula.) $\mathcal{P}^k(x \cup y) = \sum_{i=0}^{k} \mathcal{P}^i(x) \cup \mathcal{P}^{k-i}(y)$.

In addition, we will assume the following fact that can be shown to follow from the axioms:

Property (5_p) (Adem relations.)

(a) If $0 < a < pb$ then

$$\mathscr{P}^a\mathscr{P}^b = \sum_{i=0}^{[a/p]} (-1)^{a+i}\binom{(p-1)(b-i)-1}{a-pi}\mathscr{P}^{a+b-i}\mathscr{P}^i.$$

(b) If $0 < a \leq pb$ then

$$\mathscr{P}^a\beta\mathscr{P}^b = \sum_{i=0}^{[a/p]} (-1)^{a+i}\binom{(p-1)(b-i)}{a-pi}\beta\mathscr{P}^{a+b-i}\mathscr{P}^i + \sum_{i=0}^{[(a-1)/p]} (-1)^{a+i-1}\binom{(p-1)(b-i)-1}{a-pi-1}\mathscr{P}^{a+b-i}\beta\mathscr{P}^i.$$

Our first application uses the simplest Adem relation $\mathscr{P}^1\mathscr{P}^1 = 2\mathscr{P}^2$ (since $-\binom{(p-1)-1}{1} = -(p-2) \equiv 2 \pmod p$).

15.17. Proposition. *Let* $\alpha \in H^4(\mathbf{QP}^n; \mathbf{Z}_p)$. *Then* $\mathscr{P}^1(\alpha) = \pm 2\alpha^{(p+1)/2}$.

Proof. Note that this says nothing unless $2(p+1) \leq 4n$ since otherwise $\mathscr{P}^1(\alpha)$ is in a trivial group. By naturality, it suffices to prove the formula for $n \geq p$, and for a generator α. In that case $\mathscr{P}^2(\alpha) = \alpha^p \neq 0$. Now $\mathscr{P}^1(\alpha) \in H^{2(p+1)}(\mathbf{QP}^n; \mathbf{Z}_p)$ and so $\mathscr{P}^1(\alpha) = k\alpha^{(p+1)/2}$ for some $k \in \mathbf{Z}_p$. Then

$$\begin{aligned}
2\alpha^p = 2\mathscr{P}^2(\alpha) &= \mathscr{P}^1\mathscr{P}^1(\alpha) = \mathscr{P}^1(k\alpha^{(p+1)/2}) \\
&= k\mathscr{P}^1(\alpha\cdot\alpha\cdot\cdots\cdot\alpha) \quad ((p+1)/2 \text{ times}) \\
&= k[\mathscr{P}^1(\alpha)\cdot\alpha\cdot\cdots\cdot\alpha + \alpha\cdot\mathscr{P}^1(\alpha)\cdot\cdots\cdot\alpha + \cdots + \alpha\cdot\cdots\cdot\alpha\cdot\mathscr{P}^1(\alpha)] \\
&= k((p+1)/2)(k\alpha^{(p+1)/2})\alpha^{(p-1)/2} \\
&= k^2((p+1)/2)\alpha^p.
\end{aligned}$$

Therefore $k^2(p+1)/2 \equiv 2 \pmod p$ so that $k^2 \equiv k^2(p+1) \equiv 4 \pmod p$. Consequently, $k \equiv \pm 2 \pmod p$. □

15.18. Corollary. *If* $n \geq 2$ *and* $f: \mathbf{QP}^n \to \mathbf{QP}^n$ *then* $f^*(\alpha)$ *is either* 0 *or* α *for* $\alpha \in H^4(\mathbf{QP}^n; \mathbf{Z}_3)$.

Proof. Of course, the only other possibility is that $f^*(\alpha) = -\alpha$ and so what we are claiming is that that is impossible. Suppose that $f^*(\alpha) = -\alpha$. By Proposition 15.17, $\mathscr{P}^1(\alpha) = k\alpha^2 \neq 0$, where $k = \pm 2$. Then

$$\mathscr{P}^1 f^*(\alpha) = \mathscr{P}^1(-\alpha) = -\mathscr{P}^1(\alpha) = -k\alpha^2.$$

But this equals

$$f^*\mathscr{P}^1(\alpha) = f^*(k\alpha^2) = kf^*(\alpha)^2 = k\alpha^2,$$

a contradiction since $-1 \not\equiv 1 \pmod 3$. □

15.19. Corollary. $\mathbf{QP}^n$ *has the fixed point property for* $n \geq 2$.

PROOF. If $\alpha \in H^4(\mathbf{QP}^n; \mathbf{Z})$ is a generator and $f: \mathbf{QP}^n \to \mathbf{QP}^n$ is a map, then $f^*(\alpha) \neq -\alpha$ by Corollary 15.18. It follows as in Section 23 of Chapter IV that the Lefschetz number $L(f) \neq 0$. □

15.20. Theorem. *There is no "Cayley projective 3-space," i.e., there is no space* X *with* $H^*(X; \mathbf{Z}_3) \approx \mathbf{Z}_3[a]/(a^4)$ *where* $\deg(a) = 8$.

PROOF. For $p = 3$, we use the Adem relation $\mathcal{P}^1\mathcal{P}^3 = \mathcal{P}^4$. If X exists then this shows that $\mathcal{P}^4(a) = 0$. But $\mathcal{P}^4(a) = a^3 \neq 0$ by Axiom (2_3). □

15.21. Corollary. *The 7-sphere does not carry the structure of a topological group.* □

PROBLEMS

1. For $x \in \mathbf{S}^{n-1}$ let $T_x: \mathbf{R}^n \to \mathbf{R}^n$ be the reflection through the line $\mathbf{R}x$; i.e., $T_x(y) = 2\langle x, y\rangle x - y$. For n even, show that the map $f: \mathbf{S}^{n-1} \times \mathbf{S}^{n-1} \to \mathbf{S}^{n-1}$ given by $f(x, y) = T_x(y)$ has bidegree $(2, -1)$. Conclude that for n even, there exists a map $\mathbf{S}^{2n-1} \to \mathbf{S}^n$ of Hopf invariant -2.

2. For maps $\mathbf{S}^{2n-1} \xrightarrow{f} \mathbf{S}^n \xrightarrow{g} \mathbf{S}^n$ show that $h_{g\circ f} = \deg(g)^2 h_f$.

3. For maps $\mathbf{S}^{2n-1} \xrightarrow{f} \mathbf{S}^{2n-1} \xrightarrow{g} \mathbf{S}^n$ show that $h_{g\circ f} = \deg(f) h_g$.

4. (a) Show that $\mathbf{CP}^{2n+1}$ is an $\mathbf{S}^2$-bundle over $\mathbf{QP}^n$.
   (b) Use (a) to give another proof of Proposition 15.17 showing that the sign there is $+$ if α is the reduction of a generator of $H^4(\mathbf{QP}^n; \mathbf{Z})$ mapping to the square of a generator of $H^2(\mathbf{CP}^{2n+1}; \mathbf{Z})$.

5. Read Definition 3.1 of Chapter VII for the definition of an "H-space."
   (a) If $\mathbf{S}^{n-1}$ is an H-space, show that n is a power of 2 ($n = 1, 2, 4, 8$ by Adams).
   (b) If X is an H-space with unity e and $A \subset X$ is a retract of X with $e \in A$ then show that A is an H-space.
   (c) If $X \times Y$ is an H-space then show that both X and Y are H-spaces.
   (d) Show that $\mathbf{S}^1 \times \mathbf{S}^3 \times \mathbf{S}^5$ is not homotopy equivalent to $\mathbf{U}(3)$ even though they have the same cohomology rings by Example 13.7.

16. Construction of the Steenrod Squares ☼

In this section we shall construct the Steenrod squaring operations. We shall not construct the cyclic reduced powers, but that can be done in essentially the same way. We shall prove the axioms for the squares but the proof of the Adem relations is beyond our present capabilities and so the applications given in the previous section are not completely proved in this book. For proofs of the Adem relations, see Bullett and MacDonald [1], Steenrod and Epstein [1], or Mosher and Tangora [1].

We will carry the development as far as conveniently possible using coefficients in an arbitrary commutative ring with unity. We could simplify the formulas by taking $\mathbf{Z}_2$ coefficients throughout, but the more general outlook indicates much better how to convert the arguments to the case of the cyclic reduced powers for odd primes. (In the latter case, keep our present $T-1$ but the substitute $T^{p-1}+\cdots+T^2+T+1$ for our present $T+1$.)

As we hinted at previously, the squares owe their existence to the fact that the cup product is not (signed) commutative at the cochain level.

Consider any diagonal approximation

$$\Delta_0 : \Delta_*(X) \to \Delta_*(X) \otimes \Delta_*(X).$$

Define the chain map

$$T : \Delta_*(X) \otimes \Delta_*(X) \to \Delta_*(X) \otimes \Delta_*(X)$$

by $T(\sigma_p \otimes \sigma_q) = (-1)^{pq}\sigma_q \otimes \sigma_p$, i.e., the signed interchange of factors. Then $T \circ \Delta_0$ is another diagonal approximation. Since any two diagonal approximations are naturally chain homotopic there is a natural chain homotopy

$$\Delta_1 : \Delta_*(X) \to \Delta_*(X) \otimes \Delta_*(X)$$

which is a map of degree $+1$ such that

$$(T-1)\Delta_0 = T\Delta_0 - \Delta_0 = \partial\Delta_1 + \Delta_1\partial.$$

If Δ_0 could be taken to be signed commutative, then we could take $\Delta_1 = 0$ and the subsequent development would be trivial and would imply that $\mathrm{Sq}^i(a) = 0$ unless $i = \deg(a)$. Since that is not the case, Δ_0 cannot be signed commutative. Note that

$$(T+1)(T-1) = T^2 - 1 = 0,$$

so that

$$\partial[(T+1)\Delta_1] + [(T+1)\Delta_1]\partial = (T+1)[\partial\Delta_1 + \Delta_1\partial] = 0.$$

This means that $(T+1)\Delta_1$ (with the usual sign conventions) is a natural chain map of degree 1. The zero mapping $0 : \Delta_n(X) \to (\Delta_*(X) \otimes \Delta_*(X))_{n+1}$ is another such map and the method of acyclic models shows easily that there is a natural chain homotopy Δ_2 between them. That is

$$(T+1)\Delta_1 = \partial\Delta_2 - \Delta_2\partial.$$

Applying the operator $(T-1)$ to this gives

$$0 = (T-1)(T+1)\Delta_1 = \partial(T-1)\Delta_2 - (T-1)\Delta_2\partial$$

which means that $(T-1)\Delta_2$ is a chain map of degree $+2$. Again 0 is also such a natural chain map and the method of acyclic models provides a chain homotopy Δ_3 between them. One can continue this ad infinitum, constructing a sequence of natural chain homotopies Δ_n of degree n such that

$$\boxed{(T+(-1)^{n+1})\Delta_n = \partial\Delta_{n+1} + (-1)^n\Delta_{n+1}\partial.}$$

Now we pass to the cochain complexes. For any commutative ring Λ with unity, the maps Δ_k induce maps

$$\mathrm{Hom}(\Delta_*(X),\Lambda)\otimes\mathrm{Hom}(\Delta_*(X),\Lambda)\to\mathrm{Hom}(\Delta_*(X)\otimes\Delta_*(X),\Lambda)$$
$$\xrightarrow{\mathrm{Hom}(\Delta_k,1)}\mathrm{Hom}(\Delta_*(X),\Lambda),$$

that is,

$$\boxed{\Delta^*(X;\Lambda)\otimes\Delta^*(X;\Lambda)\xrightarrow{h_k}\Delta^*(X;\Lambda).}$$

The usual sign convention is in use, giving $h_k(c^n) = (-1)^{kn}c^n\circ\Delta_k$. Then h_k is of degree $-k$; i.e., $h_k(f^p\otimes g^q)$ has degree $p+q-k$. Define the "cup-i" product of cochains f and g by

$$f\cup_i g = h_i(f\otimes g).$$

Then $\cup_0$ is the usual cup product $\cup$. We shall use the same letter T to denote $\mathrm{Hom}(T,1)$. Then note that $h_k Tc^n = (-1)^{kn}c^n T\Delta_k$. Then we have the dual formulas

$$\boxed{h_n(1+(-1)^{n+1}T) = h_{n+1}\delta + (-1)^n\delta h_{n+1}.}$$

16.1. Proposition. *If* $\deg(a) = q$ *and if* $n-q$ *is odd, or if* $2\Lambda = 0$, *then*

$$h_{n+1}(\delta a\otimes\delta a) = (-1)^{n+1}\delta h_{n+1}(a\otimes\delta a) - \delta h_n(a\otimes a).$$

PROOF. We compute

$$\begin{aligned}
h_{n+1}(\delta a\otimes\delta a) &= h_{n+1}\delta(a\otimes\delta a)\\
&= (-1)^{n+1}\delta h_{n+1}(a\otimes\delta a) + h_n(a\otimes\delta a + (-1)^{n+1}\delta a\otimes a)\\
&= (-1)^{n+1}\delta h_{n+1}(a\otimes\delta a) + h_n(a\otimes\delta a + (-1)^q\delta a\otimes a)\\
&= (-1)^{n+1}\delta h_{n+1}(a\otimes\delta a) + h_n(-1)^q\delta(a\otimes a)\\
&= (-1)^{n+1}\delta h_{n+1}(a\otimes\delta a) + (-1)^{n+1}h_n\delta(a\otimes a)\\
&= (-1)^{n+1}\delta h_{n+1}(a\otimes\delta a) - (-1)^{n+1}(-1)^{n-1}\delta h_n(a\otimes a) + 0\\
&= (-1)^{n+1}\delta h_{n+1}(a\otimes\delta a) - \delta h_n(a\otimes a).
\end{aligned}$$ □

16.2. Proposition. *Let* $\deg(a) = q = \deg(b)$. *Assume that* $n-q$ *is odd or that* $2\Lambda = 0$. *Then*

$$h_n((a+b)\otimes(a+b)) = h_n(a\otimes a) + h_n(b\otimes b) + h_{n+1}\delta(a\otimes b) + (-1)^n\delta h_{n+1}(a\otimes b).$$

PROOF. This follows from the identity (true under the stated conditions)

$$h_n(a\otimes b+b\otimes a)=h_n(1+(-1)^qT)(a\otimes b)=h_{n+1}\delta(a\otimes b)+(-1)^n\,\delta h_{n+1}(a\otimes b).$$

□

16.3. Theorem. *If $q-n$ is odd or if $2\Lambda=0$ then $a\mapsto h_n(a\otimes a)$ induces a natural homomorphism*

$$\mathrm{Sq}_n\colon H^q(X;\Lambda)\to H^{2q-n}(X;\Lambda).$$

PROOF. If a is a cocycle then $h_n(a\otimes a)$ is a cocycle by Proposition 16.1. By Proposition 16.2, $h_n((a+\delta b)\otimes(a+\delta b))\sim h_n(a\otimes a)+h_n(\delta b\otimes\delta b)\sim h_n(a\otimes a)$, the latter homology by Proposition 16.1. Thus Sq_n is defined. The formula of Proposition 16.2 shows that Sq_n is additive. □

Now put $j=q-n$ and define $\mathrm{Sq}^j=\mathrm{Sq}_n=\mathrm{Sq}_{q-j}$. Then we have a natural homomorphism

$$\mathrm{Sq}^j\colon H^q(X;\Lambda)\to H^{q+j}(X;\Lambda)$$

defined when j is odd or when $2\Lambda=0$.

16.4. Theorem. *If $a\in H^q(X;\Lambda)$ then $\mathrm{Sq}^q(a)=a^2$ when it is defined.*

PROOF. On $H^q(X;\Lambda)$, $\mathrm{Sq}^q=\mathrm{Sq}_0$ which is induced by $h_0=\smile$. □

16.5. Theorem. *The operations Sq^j are independent of the choice of the Δ_i.*

PROOF. Suppose we are given another natural sequence Δ_i' satisfying

$$(T+(-1)^{k+1})\Delta_k'=\partial\Delta_{k+1}'+(-1)^k\Delta_{k+1}'\partial.$$

Then $\Delta_0-\Delta_0'\colon\Delta_*(X)\to\Delta_*(X)\otimes\Delta_*(X)$ is a natural chain map inducing zero in homology. The method of acyclic models then provides a chain homotopy D_1 such that

$$\Delta_0-\Delta_0'=\partial D_1+D_1\partial.$$

Multiplying this by $(T-1)$ gives

$$\partial(T-1)D_1+(T-1)D_1\partial=(T-1)\Delta_0-(T-1)\Delta_0'=(\partial\Delta_1+\Delta_1\partial)-(\partial\Delta_1'+\Delta_1'\partial).$$

This is the same as

$$(\Delta_1-\Delta_1'-(T-1)D_1)\partial+\partial(\Delta_1-\Delta_1'-(T-1)D_1)=0$$

which means that $\Delta_1-\Delta_1'-(T-1)D_1$ is a natural chain map of degree $+1$. Therefore there exists a map D_2 with

$$\Delta_1-\Delta_1'-(T-1)D_1=D_2\partial-\partial D_2.$$

Multiplying this by $(T+1)$ and going through the analogous argument shows that there is a D_3 with

$$\Delta_2 - \Delta_2' - (T+1)D_2 = D_3\partial + \partial D_3.$$

Continuing with this and dualizing Δ_i and Δ_i' to h_i and h_i' and the D_i to E_i gives the formulas

$$h_i - h_i' - E_i(T + (-1)^i) = E_{i+1}\delta + (-1)^i \delta E_{i+1}.$$

For a cocycle a, and for the conditions which $\mathrm{Sq}_i(a)$ is defined, we see that

$$h_i(a \otimes a) - h_i'(a \otimes a) = \pm \delta E_{i+1}(a \otimes a). \qquad \square$$

16.6. Theorem. *For all a and j we have $2\,\mathrm{Sq}^j(a) = 0$ when it is defined.*

PROOF. If $2\Lambda = 0$ then this is clear. If $j = n - q$ is odd and $\delta a = 0$ then

$$\begin{aligned} 2h_n(a \otimes a) &= h_n(a \otimes a + a \otimes a) \\ &= h_n(1 + (-1)^q T)(a \otimes a) \\ &= h_n(1 + (-1)^{n+1} T)(a \otimes a) \\ &= h_{n+1}\delta(a \otimes a) + (-1)^n \delta h_{n+1}(a \otimes a) \\ &= (-1)^n \delta h_{n+1}(a \otimes a). \end{aligned} \qquad \square$$

Now if $A \subset X$ and a is a cocycle of X which vanishes on A then by naturality of h_n we have that $h_n(a)$ vanishes on A. Therefore the operations h_n are defined on $\Delta^*(X, A)$ and satisfy all the formulas given for them. Consequently, the squaring operation Sq^j is defined on $H^q(X, A; \Lambda)$ when j is odd or when j is arbitrary and $2\Lambda = 0$.

16.7. Theorem. *If $A \subset X$ and j is odd or $2\Delta = 0$ then the following diagram commutes:*

$$\begin{array}{ccc} H^q(A;\Lambda) & \xrightarrow{\delta^*} & H^{q+1}(X, A; \Lambda) \\ \downarrow{\scriptstyle \mathrm{Sq}^j} & & \downarrow{\scriptstyle \mathrm{Sq}^j} \\ H^{q+j}(A;\Lambda) & \xrightarrow{\delta^*} & H^{q+j+1}(X, A; \Lambda). \end{array}$$

PROOF. Assume either that $2\Lambda = 0$ or that $n - q$ is odd. Let $a \in \Delta^q(A)$ with $\delta a = 0$. Let $a = i^\Delta(b)$ for a cochain b on X and let $\delta b = j^\Delta(c)$ so that $c \in \Delta^{q+1}(X, A; \Lambda)$ represents $\delta^*[\![a]\!]$. Then $h_{n+1}(c \otimes c)$ represents $\mathrm{Sq}^{q-n}[\![c]\!] = \mathrm{Sq}^{q-n}\delta^*[\![a]\!]$. Now

$$\begin{aligned} j^\Delta h_{n+1}(c \otimes c) &= h_{n+1}(j^\Delta c \otimes j^\Delta c) = h_{n+1}(\delta b \otimes \delta b) \\ &= (-1)^{n+1}\delta h_{n+1}(b \otimes \delta b) - \delta h_n(b \otimes b) \\ &= \delta\{(-1)^{n+1} h_{n+1}(b \otimes \delta b) - h_n(b \otimes b)\} \end{aligned}$$

and

$$\begin{aligned} i^{\Delta}\{(-1)^{n+1}h_{n+1}(b\otimes\delta b)-h_n(b\otimes b)\} &= (-1)^{n+1}h_{n+1}(i^{\Delta}b\otimes\delta i^{\Delta}b)-h_n(i^{\Delta}b\otimes i^{\Delta}b)\\ &= (-1)^{n+1}h_{n+1}(a\otimes\delta a)-h_n(a\otimes a)\\ &= -h_n(a\otimes a)\sim h_n(a\otimes a)\end{aligned}$$

(by Theorem 16.6) and so $h_{n+1}(c\otimes c)$ represents $\delta^*\mathrm{Sq}^{q-n}[\![a]\!]$. □

16.8. Corollary. Sq^j *commutes with suspension.*

PROOF. The same argument as in Proposition 15.3 applies. □

We shall now specialize to the cases $\Lambda=\mathbf{Z}$ or $\Lambda=\mathbf{Z}_2$. Then Sq^j is defined on $H^q(X;\mathbf{Z})$ for j odd, and on $H^q(X;\mathbf{Z}_2)$ for all j. Let

$$\boxed{\rho\colon H^*(X;\mathbf{Z})\to H^*(X;\mathbf{Z}_2)}$$

be reduction mod 2; i.e., the map induced by the epimorphism $\mathbf{Z}\to\mathbf{Z}_2$. By naturality of the construction of the h_n, ρ commutes with the h_n.

16.9. Lemma. *If $q-n$ is even and if $a\in\Delta^q(X)$ then*

$$\begin{aligned} h_{n+1}(\delta a\otimes\delta a) &= (-1)^{n+1}\delta h_{n+1}(a\otimes\delta a)+2h_n(a\otimes\delta a)\\ &\quad -\delta h_n(a\otimes a)+(-1)^{n+1}2h_{n-1}(a\otimes a).\end{aligned}$$

PROOF. We compute

$$\begin{aligned} h_{n+1}(\delta a\otimes\delta a) &= (-1)^{n+1}\delta h_{n+1}(a\otimes\delta a)+h_n(a\otimes\delta a+(-1)^{n+1}\delta a\otimes a)\\ &= (-1)^{n+1}\delta h_{n+1}(a\otimes\delta a)+h_n((-1)^{n+1}\delta(a\otimes a)+2a\otimes\delta a)\\ &= (-1)^{n+1}\delta h_{n+1}(a\otimes\delta a)+2h_n(a\otimes\delta a)-\delta h_n(a\otimes a)\\ &\quad +(-1)^{n+1}h_{n-1}(a\otimes a+(-1)^{n+q}a\otimes a)\\ &= (-1)^{n+1}\delta h_{n+1}(a\otimes\delta a)+2h_n(a\otimes\delta a)-\delta h_n(a\otimes a)\\ &\quad +(-1)^{n+1}2h_{n-1}(a\otimes a). \end{aligned}$$ □

Putting $a=b$ in Lemma 16.9 and solving for $\delta h_n(b\otimes b)$ gives:

16.10. Corollary. *If $b\in\Delta^q(X)$, $q-n$ is even, and $\delta b=2c$, then*

$$\begin{aligned}\delta h_n(b\otimes b) &= 2[(-1)^{n+1}h_{n-1}(b\otimes b)\pm\delta h_{n+1}(b\otimes c)\\ &\quad +2\{h_n(b\otimes c)-h_{n+1}(c\otimes c)\}].\end{aligned}$$ □

Now let $a\in\Delta^q(X;\mathbf{Z}_2)$ with $\delta a=0$ and let $a=\rho(b)$. Then $\delta b=2c$ for some integral cochain c and $[\![c]\!]=\beta[\![a]\!]$ where $\beta\colon H^q(X;\mathbf{Z}_2)\to H^{q+1}(X;\mathbf{Z})$ is the Bockstein. (This is the definition of β.)

Then $\rho h_n(b\otimes b)=h_n(a\otimes a)$ represents $\mathrm{Sq}_n[\![a]\!]$. By Corollary 16.10, $\delta h_n(b\otimes b)=2\{\pm h_{n-1}(b\otimes b)\pm\delta(?)+2(?)\}$, and so $\pm h_{n-1}(b\otimes b)+2(?)$ represents $\beta\mathrm{Sq}_n[\![a]\!]$. Hence $\rho(h_{n-1}(b\otimes b))=h_{n-1}(a\otimes a)$ represents $\rho\beta\mathrm{Sq}_n[\![a]\!]=\beta_2\mathrm{Sq}_n[\![a]\!]$ where $\beta_2=\rho\circ\beta$. But $h_{n-1}(a\otimes a)$ also represents $\mathrm{Sq}_{n-1}[\![a]\!]\in H^q(X;\mathbf{Z}_2)$.

Put $q-n=2i$ so that $\mathrm{Sq}_n=\mathrm{Sq}^{2i}$ and $\mathrm{Sq}_{n-1}=\mathrm{Sq}^{2i+1}$. Then for $\alpha\in H^q(X;\mathbf{Z}_2)$ we have just shown that $\mathrm{Sq}^{2i+1}(\alpha)=\beta_2\mathrm{Sq}^{2i}(\alpha)$.

The coefficient diagram

$$\begin{array}{ccccccccc} 0 & \longrightarrow & \mathbf{Z} & \xrightarrow{2} & \mathbf{Z} & \longrightarrow & \mathbf{Z}_2 & \longrightarrow & 0 \\ & & \downarrow & & \downarrow & & \downarrow & & \\ 0 & \longrightarrow & \mathbf{Z}_2 & \longrightarrow & \mathbf{Z}_4 & \longrightarrow & \mathbf{Z}_2 & \longrightarrow & 0 \end{array}$$

shows that β_2 is the Bockstein for the lower sequence.

Also, if $\delta b=0$ then $h_{n-1}(b\otimes b)$ represents $\mathrm{Sq}_{n-1}[\![b]\!]$ and also represents $\beta\,\mathrm{Sq}_n(\rho[\![b]\!])$ by Corollary 16.10. That is, on $H^*(X;\mathbf{Z})$ we have that $\mathrm{Sq}^{2i+1}=\beta\circ\mathrm{Sq}^{2i}\circ\rho$. Thus we have shown that the following diagram commutes:

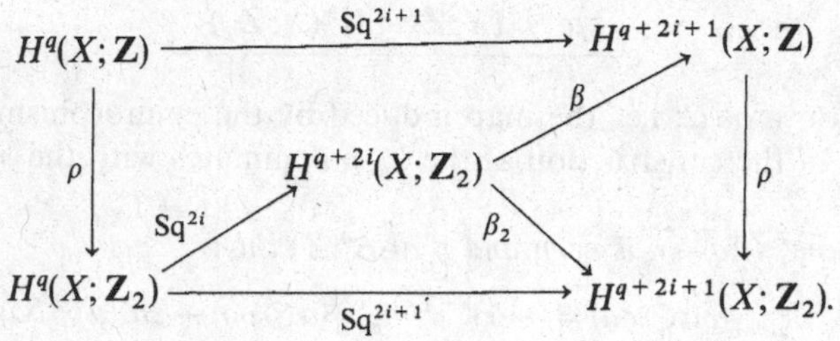

16.11. Theorem. (1) $j<0\Rightarrow\mathrm{Sq}^j=0$. (2) $\mathrm{Sq}^0=1$.

Proof. Both of these are true for a point, (1) trivially, and (2) since $\mathrm{Sq}^0(x)=x^2=x$ for a zero-dimensional class x by Theorem 16.4. It follows that they hold, in general, on $H^0(X;\mathbf{Z}_2)$ for any space X. By naturality, they hold on $\tilde{H}^0(\mathbf{S}^0)$. By Corollary 16.8 they hold on $H^n(\mathbf{S}^n;\mathbf{Z}_2)$. Let K be a CW-complex and let $\alpha\in H^n(K^{(n)};\mathbf{Z}_2)$. Since $H^n(K^{(n)};\mathbf{Z})\to H^n(K^{(n)};\mathbf{Z}_2)$ is onto, it follows from Theorems 11.6 and 11.9 of Chapter V that there is a map $\phi\colon K^{(n)}\to\mathbf{S}^n$ and an element $\vartheta\in H^n(\mathbf{S}^n;\mathbf{Z}_2)$ such that $\phi^*(\vartheta)=\alpha$. Thus $\mathrm{Sq}^0(\alpha)=\mathrm{Sq}^0(\phi^*(\vartheta))=\phi^*(\mathrm{Sq}^0(\vartheta))=\phi^*(\vartheta)=\alpha$, and hence the result is true on $H^n(K^{(n)};\mathbf{Z}_2)$. Since $H^n(K;\mathbf{Z}_2)\to H^n(K^{(n)};\mathbf{Z}_2)$ is monomorphic, it follows that the result is true on $H^n(K;\mathbf{Z}_2)$ for any CW-complex K and all n.

To extend the proof to arbitrary spaces requires some material from the next chapter. For any space X one can find a CW-complex K and a map $K\to X$ which induces an isomorphism on homotopy groups. (One does this by induction: killing homotopy classes in the kernel and adding cells to make the mapping onto in homotopy.) Whitehead's Theorem (Theorem 11.2 of Chapter VII) implies that the map induces an isomorphism on homology and hence on cohomology. The present result for X then follows from that for K by naturality. □

Since $\mathrm{Sq}^1 = \beta_2 \mathrm{Sq}^0$ we have $\mathrm{Sq}^1 = \beta_2$ which is Property (5) of Section 15.

We have now proved all the axioms (at least on CW-complexes) except for the Cartan formula. We will now restrict attention to coefficients in $\mathbf{Z}_2$. In particular, the formulas for the Δ_i simplify to $(1+T)\Delta_i = \Delta_{i+1}\delta + \delta\Delta_{i+1}$.

For spaces X and Y define the "shuffle" map

$$\lambda\colon \Delta_*(X)\otimes\Delta_*(X)\otimes\Delta_*(Y)\otimes\Delta_*(Y)\to\Delta_*(X)\otimes\Delta_*(Y)\otimes\Delta_*(X)\otimes\Delta_*(Y)$$

by $\lambda(a\otimes b\otimes c\otimes d) = a\otimes c\otimes b\otimes d$, where we do not need a sign because of the mod 2 coefficients which are understood. Define

$$D_n\colon \Delta_*(X)\otimes\Delta_*(Y)\to\Delta_*(X)\otimes\Delta_*(X)\otimes\Delta_*(Y)\otimes\Delta_*(Y)$$

by $D_n = \sum_i(\Delta_{2i}\otimes\Delta_{n-2i} + \Delta_{2i-1}\otimes T\Delta_{n-2i+1}) = \Delta_{\text{even}}\otimes\Delta + \Delta_{\text{odd}}\otimes T\Delta$. We claim that

$$\partial D_{n+1} + D_{n+1}\partial = (1\otimes 1 + T\otimes T)D_n.$$

We shall compute both sides and compare:

$$\begin{aligned}
\text{lhs} &= (\partial\Delta_{\text{even}})\otimes\Delta + \Delta_{\text{even}}\otimes\partial\Delta + (\partial\Delta_{\text{odd}})\otimes T\Delta + \Delta_{\text{odd}}\otimes T\partial\Delta\\
&\quad + (\Delta_{\text{even}}\partial)\otimes\Delta + \Delta_{\text{even}}\otimes\Delta\partial + (\Delta_{\text{odd}}\partial)\otimes T\Delta + \Delta_{\text{odd}}\otimes T\Delta\partial\\
&= (\partial\Delta_{\text{even}} + \Delta_{\text{even}}\partial)\otimes\Delta + \Delta_{\text{even}}\otimes(\partial\Delta + \Delta\partial)\\
&\quad + (\partial\Delta_{\text{odd}} + \Delta_{\text{odd}}\partial)\otimes T\Delta + \Delta_{\text{odd}}\otimes T(\partial\Delta + \Delta\partial)\\
&= (1+T)\Delta_{\text{odd}}\otimes\Delta + \Delta_{\text{even}}\otimes(1+T)\Delta\\
&\quad + (1+T)\Delta_{\text{even}}\otimes T\Delta + \Delta_{\text{odd}}\otimes T(1+T)\Delta.
\end{aligned}$$

Note that $T(1+T) = 1+T$. Now for the right-hand side:

$$\begin{aligned}
\text{rhs} &= (1\otimes 1 + T\otimes T)D_n\\
&= [(1+T)\otimes T + 1\otimes(1+T)](\Delta_{\text{even}}\otimes\Delta + \Delta_{\text{odd}}\otimes T\Delta)\\
&= (1+T)\Delta_{\text{even}}\otimes T\Delta + \Delta_{\text{even}}\otimes(1+T)\Delta\\
&\quad + (1+T)\Delta_{\text{odd}}\otimes TT\Delta + \Delta_{\text{odd}}\otimes(1+T)T\Delta
\end{aligned}$$

and so the contention follows since $TT = 1$ and $(1+T)T = (1+T)$.

By the Eilenberg–Zilber Theorem and Theorem 16.5 we can replace $\Delta_*(X\times Y)$ by $\Delta_*(X)\otimes\Delta_*(Y)$ and $\Delta^*(X\times Y)$ by $\mathrm{Hom}(\Delta_*(X)\otimes\Delta_*(Y),\mathbf{Z}_2) = (\Delta(X)\otimes\Delta(Y))^*$, so we can define the squares on $X\times Y$ by means of the redefined maps

$$\begin{aligned}
\Delta_n = \lambda\circ D_n\colon \Delta_*(X)\otimes\Delta_*(Y) &\to \Delta_*(X)\otimes\Delta_*(X)\otimes\Delta_*(Y)\otimes\Delta_*(Y)\\
&\to(\Delta_*(X)\otimes\Delta_*(Y))\otimes(\Delta_*(X)\otimes\Delta(Y))
\end{aligned}$$

which dualize to

$$h_n\colon (\Delta(X)\otimes\Delta(Y))^*\otimes(\Delta(X)\otimes\Delta(Y))^*\to(\Delta(X)\otimes\Delta(Y))^*$$

by $h_n(\alpha\otimes\beta)(c) = (\alpha\times\beta)(\Delta_n(c)) = (\alpha\times\beta)\lambda D_n(c)$ since we are using mod 2 coefficients

We compute:

$$
\begin{aligned}
h_n&((\alpha\times\beta)\otimes(\gamma\times\omega))\\
&=((\alpha\times\beta)\times(\gamma\times\omega))\circ\lambda\circ D_n\\
&=(\alpha\times\gamma\times\beta\times\omega)\sum(\Delta_{2i}\otimes\Delta_{n-2i}+\Delta_{2i-1}\otimes T\Delta_{n-2i+1})\\
&=\sum_i[(\alpha\times\gamma)\Delta_{2i}\times(\beta\times\omega)\Delta_{n-2i}+(\alpha\times\gamma)\Delta_{2i-1}\times(\beta\times\omega)T\Delta_{n-2i+1}]\\
&=\sum_i[h_{2i}(\alpha\otimes\gamma)\times h_{n-2i}(\beta\otimes\omega)+h_{2i-1}(\alpha\otimes\gamma)\times h_{n-2i+1}(\omega\otimes\beta)].
\end{aligned}
$$

Putting $\gamma=\alpha$ and $\omega=\beta$ we deduce that

$$
\begin{aligned}
h_n((\alpha\times\beta)\otimes(\alpha\times\beta))&=\sum_i(h_{2i}(\alpha\otimes\alpha)\times h_{n-2i}(\beta\otimes\beta)\\
&\qquad+h_{2i-1}(\alpha\otimes\alpha)\times h_{n-2i+1}(\beta\otimes\beta))\\
&=\sum_i h_i(\alpha\otimes\alpha)\times h_{n-i}(\beta\otimes\beta)
\end{aligned}
$$

and so $\mathrm{Sq}_n(\alpha\times\beta)=\sum_i\mathrm{Sq}_i(\alpha)\times\mathrm{Sq}_{n-i}(\beta)$. This translates to $\mathrm{Sq}^{p+q-n}(\alpha\times\beta)=\sum_i\mathrm{Sq}^{p-i}(\alpha)\times\mathrm{Sq}^{q-n+i}(\beta)$. By reindexing, this becomes

$$\mathrm{Sq}^n(\alpha\times\beta)=\sum_{i=0}^n\mathrm{Sq}^i(\alpha)\times\mathrm{Sq}^{n-i}(\beta).$$

Letting $d\colon X\to X\times X$ be the diagonal map, we get

$$
\begin{aligned}
\mathrm{Sq}^n(\alpha\cup\beta)&=\mathrm{Sq}^n d^*(\alpha\times\beta)=d^*\,\mathrm{Sq}^n(\alpha\times\beta)=\sum_{i=0}^n d^*(\mathrm{Sq}^i(\alpha)\times\mathrm{Sq}^{n-i}(\beta))\\
&=\sum_{i=0}^n\mathrm{Sq}^i(\alpha)\cup\mathrm{Sq}^{n-i}(\beta)
\end{aligned}
$$

which is the Cartan formula.

17. Stiefel–Whitney Classes ☼

In this section we apply Steenrod squares and the Thom isomorphism to introduce certain important "characteristic classes" which are invariants of vector bundles and of manifolds (via the tangent bundle). These classes will, in turn, be applied to derive some interesting nonembedding results for manifolds.

All cohomology in this section is taken with $\mathbf{Z}_2$ coefficients. In particular, there are no orientation requirements.

Suppose that $\pi\colon W^{n+k}\to N^n$ is a k-disk bundle and let $i\colon N\to W$ be the zero section. As in Section 11, let $\tau\in H^k(W,\partial W)$ be the Thom class. By Theorem 11.3 the Thom isomorphism

$$\Phi\colon H^p(N)\xrightarrow{\approx}H^{p+k}(W,\partial W)$$

is given by $\Phi(u)=\pi^*(u)\cup\tau$.

We can also treat general vector bundles W over N either by passing to an associated disk bundle or, directly, by taking $\tau \in H^k(W, W-N)$ and $\Phi: H^p(N) \to H^{p+k}(W, W-N)$. It will be convenient to use that setting in this section.

17.1. Definition. For a k-plane bundle ξ, as above, the qth *Stiefel–Whitney class* $w_q \in H^q(N; \mathbf{Z}_2)$ is defined by

$$w_q = \Phi^{-1} \operatorname{Sq}^q(\tau).$$

It is sometimes useful to deal with the "total" Stiefel–Whitney class

$$w = w_0 + w_1 + w_2 + \cdots$$

and so

$$\boxed{w = \Phi^{-1} \operatorname{Sq}(\tau).}$$

Note that $w_0 = 1$ since $\tau = \Phi(1)$. Also note that $w_q = 0$ for $q > k$ since $\deg(\tau) = k$ and by Section 15, Axiom 3.

The classes w_1 and w_k have immediate interpretations:

17.2. Proposition. *For a k-plane bundle,*

(1) $w_1 = 0 \Leftrightarrow$ *the vector bundle is orientable; and*
(2) w_k *is the mod 2 reduction of the Euler class* χ.

PROOF. We have $w_1 = \Phi^{-1} \operatorname{Sq}^1(\tau) = \Phi^{-1}\beta(\tau)$ where β is the Bockstein in the exact sequence

$$H^k(W, W-N; \mathbf{Z}_4) \to H^k(W, W-N; \mathbf{Z}_2) \xrightarrow{\beta} H^{k+1}(W, W-N; \mathbf{Z}_2)$$

and so $\beta(\tau) = 0$ if and only if τ is the mod 2 reduction of a $\mathbf{Z}_4$ class. But the existence of a mod 4 Thom class is equivalent to having an orientation on the bundle.

For part (2) we note that $\Phi(w_k) = \operatorname{Sq}^k(\tau) = \tau^2 = \pi^* i^*(\tau) \cup \tau$, but it is also $\pi^*(w_k) \cup \tau$ and so $w_k = i^*(\tau) = \chi$ (mod 2) by definition. □

The following fact is immediate from the naturality of the definition of the w_i:

17.3. Proposition. *If* ξ *and* η *are vector bundles and* $f: \xi \to \eta$ *is a bundle map then* $w_i(\xi) = f^*(w_i(\eta))$. □

It follows that $w(\epsilon) = 1$ for any trivial vector bundle ϵ, since ϵ is induced from the trivial bundle over a point.

If ξ and η are vector bundles over B_ξ and B_η, respectively, then one can form the product bundle $\xi \times \eta$ over $B_\xi \times B_\eta$. It is clear that the Thom class

τ for the product bundle is $\tau_\xi \times \tau_\eta$. Thus

$$\mathrm{Sq}(\tau) = \mathrm{Sq}(\tau_\xi) \times \mathrm{Sq}(\tau_\eta).$$

Also

$$(w(\xi) \times w(\eta)) \cup (\tau_\xi \times \tau_\eta) = (w(\xi) \cup \tau_\xi) \times (w(\eta) \cup \tau_\eta) = \mathrm{Sq}(\tau_\xi) \times \mathrm{Sq}(\tau_\eta)$$

and it follows that

$$w(\xi \times \eta) = w(\xi) \times w(\eta).$$

If $B_\xi = B_\eta = B$ and we apply the diagonal map $d: B \to B \times B$ to this, we get:

17.4. Theorem (Whitney Duality).

$$w(\xi \oplus \eta) = w(\xi)w(\eta). \qquad \square$$

In particular, $w(\xi \oplus \epsilon) = w(\xi)$ for a trivial bundle ϵ.

Now let N^n be a closed manifold. We define the Stiefel–Whitney class $w_q(N)$ to be the Stiefel–Whitney class w_q of the tangent bundle of N. Recall that the tangent bundle of N is the pullback, via the diagonal map $d: N \to N \times N$, of the normal bundle η of the diagonal $\Delta \subset N \times N$. Hence $w_q(N) = d^*w_q(\eta)$. The Thom class of η maps to $\tau \in H^n(N \times N)$, the dual of $[\Delta]$, as in Section 11. We have

$$w_q(\eta) \cup \tau = \mathrm{Sq}^q(\tau)$$

by definition of the w_q and the naturality of this equation. Dualizing this gives

$$\begin{aligned} \mathrm{Sq}^q(\tau) \cap [N \times N] &= (w_q(\eta) \cup \tau) \cap [N \times N] \\ &= w_q(\eta) \cap (\tau \cap [N \times N]) \\ &= w_q(\eta) \cap d_*[N] \\ &= d_*(d^*(w_q(\eta)) \cap [N]) \\ &= d_*(w_q(N) \cap [N]). \end{aligned}$$

Therefore we have the formula

$$\boxed{d_*(w_q(N) \cap [N]) = \mathrm{Sq}^q(\tau) \cap [N \times N]}$$

which characterizes the $w_q(N)$. This is equivalent to the nice formula

$$\boxed{d^! w_q(N) = \mathrm{Sq}^q(\tau).}$$

We would like to convert this formula into one more easily computable. To this aim, consider the homomorphism $u \mapsto \langle \mathrm{Sq}^i(u), [N] \rangle$ of $H^{n-i}(N; \mathbf{Z}_2) \to \mathbf{Z}_2$. By cup product duality (Theorem 9.4), the map

$$H^i(N; \mathbf{Z}_2) \to \mathrm{Hom}(H^{n-i}(N; \mathbf{Z}_2), \mathbf{Z}_2)$$

given by $v \mapsto \langle u \cup v, [N] \rangle$, is an isomorphism. Therefore there is an element $v_i \in H^i(N; \mathbf{Z}_2)$ taken by this into the homomorphism $u \mapsto \langle \mathrm{Sq}^i(u), [N] \rangle$. That is, there exists a unique class $v_i \in H^i(N; \mathbf{Z}_2)$ such that

$$\langle \mathrm{Sq}^i(u), [N] \rangle = \langle v_i \cup u, [N] \rangle$$

for all $u \in H^{n-i}(N; \mathbf{Z}_2)$. These classes v_i are called Wu classes and we can form the total Wu class

$$v = 1 + v_1 + v_2 + \cdots$$

(note that $v_0 = 1$).

17.5. Theorem (Wu). *The total Stiefel–Whitney class w and total Wu class v of a manifold N^n satisfy*

$$w = \mathrm{Sq}(v).$$

That is,

$$w_q = \sum_j \mathrm{Sq}^{q-j}(v_j).$$

Proof. Let $p: N \times N \to N$ be the projection to the second factor. With the notation from Theorem 12.4, we compute

$$\begin{aligned}
w_q \cap [N] &= p_* d_*(w_q \cap [N]) \\
&= p_*(\mathrm{Sq}^q(\tau) \cap [N \times N]) \\
&= p_*\left(\mathrm{Sq}^q\left(\sum_\alpha \alpha^\circ \times \alpha \right) \cap [N] \times [N] \right) \\
&= p_* \sum_\alpha \sum_j (\mathrm{Sq}^j(\alpha^\circ) \times \mathrm{Sq}^{q-j}(\alpha)) \cap ([N] \times [N]) \\
&= \sum_\alpha \sum_j p_*((\mathrm{Sq}^j(\alpha^\circ) \cap [N]) \times (\mathrm{Sq}^{q-j}(\alpha) \cap [N])) \\
&= \sum_\alpha \sum_j \langle \mathrm{Sq}^j(\alpha^\circ), [N] \rangle \mathrm{Sq}^{q-j}(\alpha) \cap [N].
\end{aligned}$$

Therefore, letting $n_{j,\alpha}$ be the αth component of v_j in the basis of $H^*(N; \mathbf{Z}_2)$ formed by the α's,

$$\begin{aligned}
w_q &= \sum_\alpha \sum_j \langle \mathrm{Sq}^j(\alpha^\circ), [N] \rangle \mathrm{Sq}^{q-j}(\alpha) \\
&= \sum_\alpha \sum_j \langle \alpha^\circ \cup v_j, [N] \rangle \mathrm{Sq}^{q-j}(\alpha) \\
&= \sum_\alpha \sum_j n_{j,\alpha} \langle \alpha^\circ \cup \alpha, [N] \rangle \mathrm{Sq}^{q-j}(\alpha) \\
&= \sum_\alpha \sum_j n_{j,\alpha} \mathrm{Sq}^{q-j}(\alpha) \\
&= \sum_j \mathrm{Sq}^{q-j}\left(\sum_\alpha n_{j,\alpha} \alpha \right) \\
&= \sum_j \mathrm{Sq}^{q-j}(v_j).
\end{aligned}$$

□

In principle, this provides a method of computing the w_i by knowledge of the action of the Sq^j on $H^*(N; \mathbf{Z}_2)$ since the v_i can be computed from that information.

For example, on $\mathbf{CP}^2$, with α being the generator of $H^2(\mathbf{CP}^2; \mathbf{Z}_2)$, we have $\mathrm{Sq}^2(\alpha) = \alpha^2$ and so $v_2 = \alpha$ and $v = 1 + \alpha$. Then $w = \mathrm{Sq}(v) = 1 + \alpha + \alpha^2$ and so $w_2 = \alpha$ and $w_4 = \alpha^2$.

Now we wish to compute the Stiefel–Whitney classes of real projective space $\mathbf{P}^n$. This could, in principle, be done using Theorem 17.5 but there is a better way as follows. We will prove, by induction, that $w(\mathbf{P}^n) = (1+\alpha)^{n+1}$ where $\alpha \in H^1(\mathbf{P}^n; \mathbf{Z}_2)$ is the generator, If τ_n is the tangent bundle of $\mathbf{P}^n$ then $\tau_n|_{\mathbf{P}^{n-1}} \approx \tau_{n-1} \oplus \gamma$ where γ is the normal line bundle to $\mathbf{P}^{n-1}$ in $\mathbf{P}^n$. Now γ is not orientable since exactly one of $\mathbf{P}^n$ and $\mathbf{P}^{n-1}$ is orientable. Therefore, by Proposition 17.2, $w(\gamma) = 1 + \alpha$. By induction, $w(\tau_n)|_{\mathbf{P}^{n-1}} = w(\tau_{n-1})w(\gamma) = (1+\alpha)^n(1+\alpha) = (1+\alpha)^{n+1}$. Since $H^i(\mathbf{P}^n; \mathbf{Z}_2) \to H^i(\mathbf{P}^{n-1}; \mathbf{Z}_2)$ is an isomorphism for $i \neq n$, this shows that $(1+\alpha)^{n+1}$ is correct for $w(\mathbf{P}^n)$ except possibly for $w_n(\mathbf{P}^n)$. By Proposition 17.2, $w_n(\mathbf{P}^n) = \chi(\mathbf{P}^n)\alpha^n = (n+1)\alpha^n \pmod 2$, completing the induction. Using the identity $(1+\alpha)^2 = 1 + \alpha^2$ over $\mathbf{Z}_2$, and hence $(1+\alpha)^{2^i} = 1 + \alpha^{2^i}$, we have:

17.6. Theorem. *Let $n + 1 = \sum n_i 2^i$ be the binary representation of $n + 1$. Then*

$$w(\mathbf{P}^n) = (1+\alpha)^{n+1} = \mathop{\times}_{n_i = 1} (1 + \alpha^{2^i})$$

where α is the nonzero class in $H^1(\mathbf{P}^n; \mathbf{Z}_2)$. □

For example, since $11 = 1 + 2 + 8$ we have

$$w(\mathbf{P}^{10}) = (1+\alpha)(1+\alpha^2)(1+\alpha^8) = 1 + \alpha + \alpha^2 + \alpha^3 + \alpha^8 + \alpha^9 + \alpha^{10}$$

since $\alpha^{11} = 0$.

Now suppose that N^n is embedded, or just immersed, in some $\mathbf{R}^k$. Let τ denote its tangent bundle (not to be confused with the Thom class, which will not be needed below) and v its normal bundle. Then $\tau \oplus v = \epsilon^k$. Let $w_i = w_i(\tau) = w_i(N)$ and $\bar{w}_i = w_i(v)$, and put $\bar{w} = 1 + \bar{w}_1 + \bar{w}_2 + \cdots$. Then, by Whitney duality,

$$w\bar{w} = 1.$$

It is not hard to see from this that $\bar{w}$ can be computed from w. In particular, it does not depend on the particular immersion.

For $\mathbf{P}^n$ we have

$$\bar{w}(\mathbf{P}^n) = (1+\alpha)^{-n-1} = (1+\alpha)^{2^s - n - 1},$$

where 2^s is the smallest power of 2 for which $2^s \geq n + 1$ (actually, for *any* $2^s \geq n + 1$). For example, $\bar{w}(\mathbf{P}^{10}) = (1+\alpha)^{16-11} = (1+\alpha)^5 = (1+\alpha)(1+\alpha^4) = 1 + \alpha + \alpha^4 + \alpha^5$. It follows that $\dim(v) \geq 5$, so that $\mathbf{P}^{10}$ cannot be immersed in $\mathbf{R}^{14}$.

For $\mathbf{P}^{2^k}$ we have $\bar{w}(\mathbf{P}^{2^k}) = (1+\alpha)^{2^{k+1} - 2^k - 1} = (1+\alpha)^{2^k - 1} = 1 + \alpha + \alpha^2 + \cdots + \alpha^{2^k - 1}$. Consequently:

17.7. Theorem. *For $n = 2^k$, $\mathbf{P}^n$ cannot be immersed in $\mathbf{R}^{2n-2}$.* □

It is a theorem of Whitney that any closed n-manifold, $n > 1$, admits an immersion into $\mathbf{R}^{2n-1}$ and so Theorem 17.7 is best possible.

One can generalize Theorem 17.6 to complex and quaternionic projective spaces:

17.8. Theorem. *Let N be a closed manifold with $H^*(N; \mathbf{Z}_2) \approx \mathbf{Z}_2[\alpha]/(\alpha^{m+1})$ for some $\alpha \in H^r(N; \mathbf{Z}_2)$. Then $w(N) = (1+\alpha)^{m+1}$ and $\bar{w}(N) = (1+\alpha)^{2^s - m - 1}$ for any $2^s \geq m + 1$.*

Proof. Our proof follows Milnor–Stasheff [1], an excellent reference for continuing study in the direction of this section. We have $\mathrm{Sq}(\alpha) = \alpha + \alpha^2$ and so, from the Cartan formula,

$$\mathrm{Sq}(\alpha^k) = (\mathrm{Sq}(\alpha))^k = \alpha^k(1+\alpha)^k.$$

Therefore, for the Wu class v_{ri} we have that

$$\langle \alpha^{m-i} \cup v_{ri}, [N] \rangle = \langle \mathrm{Sq}^{ri}(\alpha^{m-i}), [N] \rangle$$

is the coefficient of α^m in $\mathrm{Sq}(\alpha^{m-i}) = \alpha^{m-i}(1+\alpha)^{m-i}$, which is $\binom{m-i}{i}$. Therefore $v_{ri} = \binom{m-i}{i}\alpha^i$ and so

$$\begin{aligned} w = \mathrm{Sq}(v) &= \sum_i \binom{m-i}{i} \mathrm{Sq}(\alpha^i) \\ &= \sum_i \binom{m-i}{i} \alpha^i(1+\alpha)^i \\ &= \sum_i \sum_j \binom{m-i}{i}\binom{i}{j} \alpha^{i+j}. \end{aligned}$$

In principle this can be computed. But the computation is independent of $r = \deg(\alpha)$ and so it would come out the same as the computation for $\mathbf{P}^m$. For $\mathbf{P}^m$ we know this gives

$$(1+\alpha)^{m+1} = \sum_k \binom{m+1}{k} \alpha^k.$$

Consequently, this formula persists in the general case. □

For example, for the Cayley projective plane K we have $w(K) = 1 + \alpha + \alpha^2$ and $\bar{w}(K) = 1 + \alpha$ where $\deg(\alpha) = 8$. In particular, K does not immerse in $\mathbf{R}^{16+7} = \mathbf{R}^{23}$.

Finally we state, without proof, the following deep and fundamental result of Thom. (The "only if" part is due to Pontryagin and is not hard; the reader may wish to try proving it.)

17.9. Theorem (Thom). *A smooth closed connected manifold M^n is the boundary of a smooth compact $(n+1)$-manifold if and only if its fundamental class $\vartheta \in H^n(M^n; \mathbf{Z}_2)$ is not a product of Stiefel–Whitney classes of M^n.* □

PROBLEMS

1. For the Klein bottle $\mathbf{K}^2$ compute $w(\mathbf{K}^2)$ and $\bar{w}(\mathbf{K}^2)$.
2. For an orientable 3-manifold M^3 show that $\mathrm{Sq} = 1$ on M^3 and deduce that $w(M) = 1$.
3. For $M^3 = \mathbf{P}^2 \times \mathbf{S}^1$ show that $w(M) = 1 + a + a^2$ and $\bar{w}(M) = 1 + a$ for some $0 \neq a \in H^1(M; \mathbf{Z}_2)$.
4. If M^3 is the nonorientable $\mathbf{S}^2$-bundle over $\mathbf{S}^1$ show that $w(M) = 1 + a$ and $\bar{w}(M) = 1 + a$ for some $0 \neq a \in H^1(M; \mathbf{Z}_2)$.
5. ◆ Recall from Problem 4 of Section 15 that $\mathbf{CP}^{2n+1}$ is fibered by 2-spheres with base space $\mathbf{QP}^n$. The tangent vectors to the fibers give a 2-plane bundle ξ over $\mathbf{CP}^{2n+1}$. Show that $w(\xi) = 1$.
6. ◆ For the canonical line bundle ξ over $\mathbf{RP}^\infty$, show that $w(\xi) = 1 + a$ where $0 \neq a \in H^1(\mathbf{RP}^\infty; \mathbf{Z}_2)$. Use this to show that there is no vector bundle η over $\mathbf{RP}^\infty$ such that $\xi \oplus \eta$ is trivial.
7. ◆ Let N^n be a given manifold with $w(N) \neq 1$. If $i > 0$ is minimal such that $w_i(N) \neq 0$ then show that i is a power of two.

18. Plumbing ☼

The purpose of this section is to illustrate the method of intersection theory in doing a homology calculation in an important and nontrivial situation, and to discuss some interesting consequences for differential topology. This section can be read after the material on intersection theory up to and including Proposition 12.8.

Suppose that ξ and η are two smooth n-disk bundles over smooth n-manifolds M^n and N^n. Around any given point of M there is a neighborhood $A \approx \mathbf{D}^n$ and a trivialization

$$\phi: E(\xi|_A) \xrightarrow{\approx} \mathbf{D}^n \times \mathbf{D}^n,$$

i.e., a diffeomorphism commuting with the projections $p_\xi: E(\xi|_A) \to A$ and $p_1: \mathbf{D}^n \times \mathbf{D}^n \to \mathbf{D}^n$, where $p_1(x, y) = x$. Similarly, let $B \approx \mathbf{D}^n$ be a neighborhood of a point in N and take a trivialization

$$\psi: E(\eta|_B) \xrightarrow{\approx} \mathbf{D}^n \times \mathbf{D}^n.$$

Let $\theta: E(\eta|_B) \xrightarrow{\approx} E(\xi|_A)$ be $\theta = \phi^{-1}\chi\psi$ where $\chi: \mathbf{D}^n \times \mathbf{D}^n \to \mathbf{D}^n \times \mathbf{D}^n$ is the exchange of factors $\chi(x, y) = (y, x)$. Then define

$$P^{2n} = E(\xi) \cup_\theta E(\eta)$$

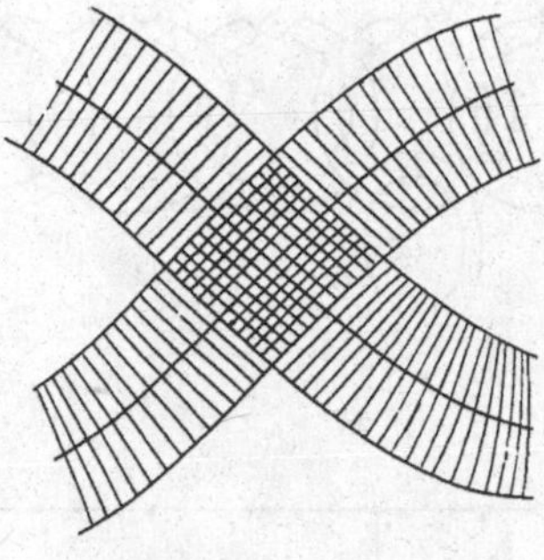

Figure VI-12. Simple plumbing.

called the "plumbing" of $E(\xi)$ and $E(\eta)$. See Figure VI-12. Note that the identification θ matches the base of one bundle with the fiber of the other.

The space P^{2n} is a topological $2n$-manifold with boundary and is close to being a smooth manifold, but it has "corners." There is a canonical way to "straighten" these corners and so to produce P^{2n} as a smooth manifold, or one can modify the construction to do that. We will not detail that, but we will wish to discuss P^{2n} as a smooth manifold later in this section.

There are obvious generalizations of this construction. For instance, one can plumb several disk bundles to a given one using disjoint coordinate patches to carry the identifications.

We will now restrict attention to the case in which the manifolds M^n, N^n are both $\mathbf{S}^n$ and the disk bundles are each the tangent disk bundle of $\mathbf{S}^n$. If one is given a finite tree (or, more generally, a connected graph) T, let $P^{2n}(T)$ be the $2n$-manifold with boundary obtained by taking a copy of $\mathbf{S}^n$ for each vertex of T and plumbing the tangent disk bundles of two of these if there is an edge in T joining the corresponding vertices. This is illustrated in Figure VI-13 for an important tree named E_8.

We wish to compute the homology of $Q = Q^{2n-1}(T) = \partial P^{2n}(T)$. Note that $P^{2n} = P^{2n}(T)$ is homotopy equivalent to the one-point union of copies of $\mathbf{S}^n$, one copy for each vertex of T. (These spheres are the "cores" of the various disk bundles, and two cores meet in exactly one point at any plumbing.) Thus $H_i(P^{2n})$ is nonzero in exactly one nonzero dimension, $i = n$. In that dimension, $H_n(P^{2n})$ is free abelian on k generators where k is the number of vertices of the tree T. We have the duality diagram

$$\begin{array}{ccccccccc}
0 \longrightarrow & H^{n-1}(Q) & \longrightarrow & H^n(P,Q) & \xrightarrow{j^*} & H^n(P) & \longrightarrow & H^n(Q) & \longrightarrow 0 \\
& \approx\downarrow & & \approx\downarrow\cap[P] & & \approx\downarrow\cap[P] & & \approx\downarrow & \\
0 \longrightarrow & H_n(Q) & \longrightarrow & H_n(P) & \xrightarrow{j_*} & H_n(P,Q) & \longrightarrow & H_{n-1}(Q) & \longrightarrow 0.
\end{array}$$

Other parts of this diagram show that Q can have no homology in dimensions other than $0, n-1, n$ and $2n-1$. To compute $H_*(Q)$ it suffices to know the map $j_*: H_n(P) \to H_n(P,Q)$. Equivalently, it suffices to know the composition

E_8

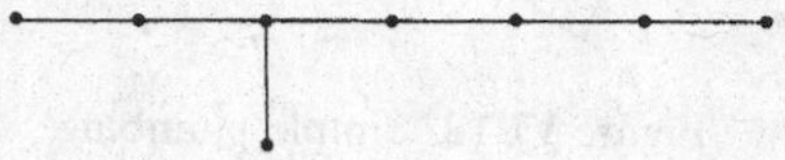

Figure VI-13. The Milnor plumbing.

of j_* with the isomorphisms $D: H_n(P, Q) \xrightarrow{\approx} H^n(P)$ (the inverse of $(\cdot) \cap [P]$) and $\beta: H^n(P) \xrightarrow{\approx} \text{Hom}(H_n(P), \mathbf{Z})$ (the evaluation, which is isomorphic since $H_*(P)$ is free abelian). This composition takes $a \in H_n(P)$ to the homomorphism $\beta D j_*(a)$ where, for $b \in H_n(P)$, we have

$$\begin{aligned} \{\beta D j_*(a)\}(b) &= \langle D j_*(a), b \rangle \\ &= \langle j^* D(a), b \rangle \qquad \text{(by the diagram)} \\ &= \langle D(a), j_*(b) \rangle \\ &= \langle D(a), j^* D(b) \cap [P] \rangle \qquad \text{(by the diagram)} \\ &= \langle D(a) \cup j^* D(b), [P] \rangle \\ &= \langle D(a) \cup D(b), [P] \rangle \\ &= \epsilon_*(b \bullet a) \\ &= b \cdot a, \end{aligned}$$

the "intersection number" of b and a. Therefore, the matrix of j_* is equivalent to the intersection matrix $I(T)$ on $H_n(P)$. Note that $I(T)$ is symmetric if n is even and skew symmetric if n is odd.

Let $a_1, \ldots, a_k$ be the basis of $H_n(P)$ represented by the k core n-spheres. Since the disk bundles are the tangent bundles of $\mathbf{S}^n$ we have $a_i \cdot a_i = \chi(\mathbf{S}^n) = 1 + (-1)^n$ by Proposition 12.8. Also, two core spheres meet transversely in exactly one point if they correspond to an edge of T. For n even, $a_i \cdot a_j = 1$ if (i, j) is an edge of T. Otherwise $a_i \cdot a_j = 0$. For n odd, $a_i \cdot a_j = \pm 1$ and $a_j \cdot a_i = \mp 1$ for an edge (i, j). The sign depends on how we orient things and is not important. (A change of basis $a_i \mapsto -a_i$ just changes the sign in the ith row and column and, since T is a tree, it is easily seen that all arrangements of signs are possible.)

We are particularly interested in cases where Q is a homology sphere. (Also note that Q is simply connected for $n \geq 3$, an easy consequence of the Seifert–Van Kampen Theorem.) By the discussion above, Q is a homology sphere if and only if the intersection matrix $I(T)$ is unimodular, i.e., has determinant ± 1.

For n odd, say $n = 2m + 1$, the case of the connected graph A_2 with two vertices (i.e., a single plumbing) yields the matrix

$$\begin{pmatrix} 0 & 1 \\ -1 & 0 \end{pmatrix}$$

which is unimodular. Hence $\partial P^{4m+2}(A_2)$ is a homology sphere. In fact, it follows from the Generalized Poincaré Conjecture, proved by Smale [1], that this is homeomorphic to $\mathbf{S}^{4m+1}$ for $m \geq 1$. It was proved by Kervaire (see Kervaire and Milnor [1]) that this is not diffeomorphic to $\mathbf{S}^{4m+1}$ for suitable values of m; e.g., for $m = 2$. This "exotic sphere," a smooth manifold which is a topological sphere but is not diffeomorphic to the standard sphere, is known as the "Kervaire sphere."

For n even, say $n = 2m$, and for the tree E_8 of Figure VI-13, the intersection matrix (numbering the "bottom" vertex last) is

$$\begin{bmatrix} 2 & 1 & 0 & 0 & 0 & 0 & 0 & 0 \\ 1 & 2 & 1 & 0 & 0 & 0 & 0 & 0 \\ 0 & 1 & 2 & 1 & 0 & 0 & 0 & 1 \\ 0 & 0 & 1 & 2 & 1 & 0 & 0 & 0 \\ 0 & 0 & 0 & 1 & 2 & 1 & 0 & 0 \\ 0 & 0 & 0 & 0 & 1 & 2 & 1 & 0 \\ 0 & 0 & 0 & 0 & 0 & 1 & 2 & 0 \\ 0 & 0 & 1 & 0 & 0 & 0 & 0 & 2 \end{bmatrix}$$

and this turns out to be unimodular. Hence $\Sigma^{4m-1} = \partial P^{4m}(E_8)$ is a homology sphere for all $m \geq 1$. For $m = 1$ it is not simply connected and, in fact, can be seen to be the Poincaré dodecahedral space of Theorem 8.10. For $m > 1$, Σ^{4m-1} is homeomorphic to $\mathbf{S}^{4m-1}$. Milnor has shown that Σ^{4m-1} is not diffeomorphic to $\mathbf{S}^{4m-1}$ for $m \geq 2$. This manifold is called the "Milnor sphere." See Kervaire and Milnor [1], Hirzebruch and Mayer [1], and Kosinski [1] for much more on this topic.

Addition of a cone over the boundary sphere of $P^{4m}(E_8)$, $m > 1$, provides a topological manifold which cannot be smoothed.

Problems

1. Compute $H_*(\partial P^{2n}(A_k))$ where $A_k =$  (k vertices).

2. Compute $H_*(\partial P^{2n}(D_k))$ where $D_k =$ 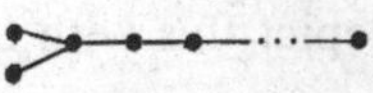 (k vertices).

CHAPTER VII
Homotopy Theory

I believe that we lack another analysis properly geometric or linear which expresses location directly as algebra expresses magnitude.

G.W. Leibniz
(letter to Huygens, 1679)

1. Cofibrations

One of the fundamental questions in topology is the "extension problem." This asks for criteria for being able to extend a map $g: A \to Y$ defined on a subspace A of X to all of X. Of course, this cannot always be done as is shown by the case $A = Y = \mathbf{S}^n$, $X = \mathbf{D}^{n+1}$.

It is natural to ask whether or not this is a homotopy-theoretic problem. That is, does the answer depend only on the homotopy class of g? The answer to this is "not generally" as is shown by the space $X = [0, 1]$, $A = \{0\} \cup \{1/n | n = 1, 2, \ldots\}$, and $Y = CA$, the cone on A. The map g which is the canonical inclusion of A in Y cannot be extended to X, since the extension would have to be discontinuous at $\{0\}$. However, $g \simeq g'$, where g' is the constant map of A to the vertex of the cone, and g' obviously extends to X.

However, it turns out that some very mild conditions on the spaces will ensure that this problem is homotopy theoretic, as we now discuss.

1.1. Definition. Let (X, A) and Y be given spaces. Then (X, A) is said to have the *homotopy extension property* with respect to Y if the following diagram can always be completed to be commutative:

$$\begin{array}{ccc} A \times I \cup X \times \{0\} & \longrightarrow & Y \\ \cap\downarrow & \nearrow & \\ X \times I. & & \end{array}$$

Note that one can also depict this with the following type of diagram:

$$\begin{array}{ccccc} A \times \{0\} & & \longrightarrow & & A \times I \\ \cap\downarrow & & \swarrow & & \downarrow\cap \\ & \nearrow & Y & \nwarrow & \\ X \times \{0\} & & \longrightarrow & & X \times I. \end{array}$$

If (X, A) has the homotopy extension property with respect to Y then extensibility of maps $g: A \to Y$ clearly depends only on the homotopy class of g.

1.2. Definition. Let $f: A \to X$ be a map. Then f is called a *cofibration* if one can always fill in the following commutative diagram:

$$
\begin{array}{ccc}
A \times \{0\} & \longrightarrow & A \times I \\
{\scriptstyle f \times 1}\downarrow & \searrow \; Y \; \nwarrow & \downarrow {\scriptstyle f \times 1} \\
X \times \{0\} & \longrightarrow & X \times I
\end{array}
$$

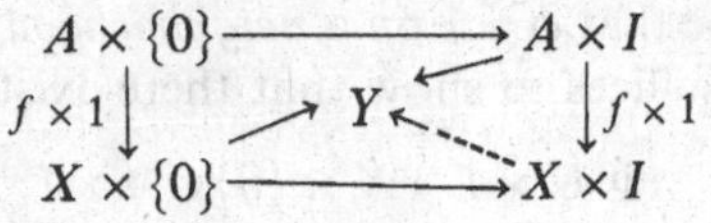

for *any* space Y.

Note that if f is an inclusion then this is the same as the homotopy extension property for all Y. That attribute is sometimes referred to as the "absolute homotopy extension property."

1.3. Theorem. *For an inclusion $A \subset X$ the following are equivalent:*

(1) *The inclusion map $A \hookrightarrow X$ is a cofibration.*
(2) $A \times I \cup X \times \{0\}$ *is a retract of $X \times I$.*

PROOF. For (1) $\Rightarrow$ (2), consider the diagram of Definition 1.2 with $Y = A \times I \cup X \times \{0\}$. The filled-in map is the desired retraction.

For (2) $\Rightarrow$ (1), composing the retraction of (2) with a map $A \times I \cup X \times \{0\} \to Y$ gives the homotopy extension property for all Y, which, as mentioned, is equivalent to (1). $\qquad\qquad\qquad\qquad\qquad\qquad\qquad\qquad\qquad\qquad\qquad\qquad\quad\square$

1.4. Corollary. *If A is a subcomplex of a CW-complex X, then the inclusion $A \hookrightarrow X$ is a cofibration.*

PROOF. One constructs a retraction $((A \cup X^{(r)}) \times I) \cup (X \times \{0\}) \to (A \times I) \cup (X \times \{0\})$ by induction on r. If it has been defined for the $(r-1)$-skeleton then extending it over an r-cell is simply a matter of extending a map on $S^{r-1} \times I \cup D^r \times \{0\}$ over $D^r \times I$, which can always be done because the pair $(D^r \times I, S^{r-1} \times I \cup D^r \times \{0\})$ is homeomorphic to $(D^r \times I, D^r \times \{0\})$, see Figure VII-6 on p. 451.

These maps for each cell fit together to give a map on the r-skeleton because of the weak topology on $X \times I$. The union of these maps for all r gives a map on $X \times I$, again because of the weak topology of $X \times I$. $\quad\square$

The main technical result for proving that particular inclusions are cofibrations is the following. Note that conditions (1) and (2) always hold if X is metric.

1.5. Theorem. *Assume that $A \subset X$ is closed and that there exists a neighborhood U of A and a map $\phi: X \to I$, such that:*

(1) $A = \phi^{-1}(0)$;
(2) $\phi(X - U) = \{1\}$; *and*
(3) U *deforms to* A *through* X *with* A *fixed. That is, there is a map* $H: U \times I \to X$ *such that* $H(a, t) = a$ *for all* $a \in A$, $H(u, 0) = u$, *and* $H(u, 1) \in A$ *for all* $u \in U$.

Then the inclusion $A \hookrightarrow X$ *is a cofibration. The converse also holds.*

PROOF. We can assume that $\phi = 1$ on a *neighborhood* of $X - U$, by replacing ϕ with $\min(2\phi, 1)$. It suffices to show that there exists a map

$$\Phi: U \times I \to X \times \{0\} \cup A \times I$$

such that $\Phi(x, 0) = (x, 0)$ for $x \in U$ and $\Phi(a, t) = (a, t)$ for $a \in A$ and all t, since then the map $r(x, t) = \Phi(x, t(1 - \phi(x)))$ for $x \in U$ and $r(x, t) = (x, 0)$ for $x \notin U$ gives the desired retraction $X \times I \to A \times I \cup X \times \{0\}$.

We define Φ by

$$\Phi(u, t) = \begin{cases} H(u, t/\phi(u)) \times \{0\} & \text{for } \phi(u) > t, \\ H(u, 1) \times \{t - \phi(u)\} & \text{for } \phi(u) \leq t. \end{cases}$$

We need only show that Φ is continuous at those points $(u, 0)$ such that $\phi(u) = 0$, i.e., at points $(a, 0)$ for $a \in A$.

Note that $H(a, t) = a$ for all $t \in I$. Thus, for W a neighborhood of a, there is a neighborhood $V \subset W$ of a such that $H(V \times I) \subset W$. Therefore, $t < \epsilon$ and $u \in V$ imply that $\Phi(u, t) \in W \times [0, \epsilon]$, and hence that Φ is continuous.

We will now prove the converse.

Let $r: X \times I \to A \times I \cup X \times \{0\}$ be a retraction, let $s(x) = r(x, 1)$ and put $U = s^{-1}(A \times (0, 1])$. Let p_X, p_I be the projections of $X \times I$ to its factors. Then put $H = p_X \circ r: U \times I \to X$. This satisfies (3). For (1) and (2), put $\phi(x) = \max_{t \in I} |t - p_I r(x, t)|$ which makes sense since I is compact. That this satisfies (1) and (2) is clear and it remains to show that ϕ is continuous. Let $f(x, t) = |t - p_I r(x, t)|$ and $f_t(x) = f(x, t)$, all of which are continuous. Then

$$\phi^{-1}((-\infty, b]) = \{x \mid f(x, t) \leq b \text{ for all } t\} = \bigcap_{t \in I} f_t^{-1}((-\infty, b])$$

is an intersection of closed sets and so is closed. Similarly

$$\phi^{-1}([a, \infty)) = \{x \mid f(x, t) \geq a \text{ for some } t\} = p_X(f^{-1}([a, \infty)))$$

is closed since p_X is closed by Proposition 8.2 of Chapter I. Since the complements of the intervals of the form $[a, \infty)$ and $(-\infty, b]$ give a subbase for the topology of **R**, the contention follows. □

It can be shown that, in the situation of Theorem 1.5, $X \times \{0\} \cup A \times I$ is a *deformation* retract of $X \times I$. See Dugundji [1], pp. 327–328.

Suppose that $f: X \to Y$ is any map. Recall that the "mapping cylinder" M_f of f is defined to be the quotient space

$$M_f = ((X \times I) + Y)/((x, 0) \sim f(x)).$$

The inclusion $i: X \hookrightarrow M_f$ clearly satisfies Theorem 1.5 and hence is a cofibration. Also, the retraction $r: M_f \to Y$ is a homotopy equivalence with homotopy inverse being the inclusion $Y \hookrightarrow M_f$. The diagram

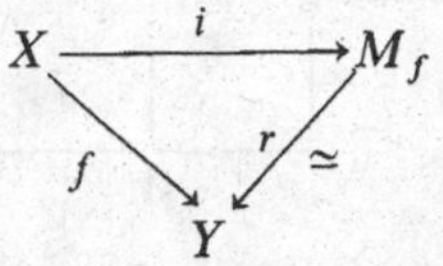

commutes. This shows that *any* map f is a cofibration, up to a homotopy equivalence of spaces.

Also recall the definition of the "mapping cone" of $f: X \to Y$ as the quotient space

$$C_f = M_f/X \times \{1\} \approx M_f \cup CX.$$

In the case of an inclusion $i: A \hookrightarrow X$, we have $C_i = X \cup CA$. There is the map

$$C_i \xrightarrow{h} X/A,$$

defined as the quotient map $X \cup CA \to X \cup CA/CA$ composed with the inverse of the homeomorphism $X/A \to X \cup CA/CA$. It is natural to ask whether h is a homotopy equivalence. This is not always the case, but the following gives a sufficient condition for it to be so.

1.6. Theorem. *If $A \subset X$ is closed and the inclusion $i: A \hookrightarrow X$ is a cofibration then $h: C_i \to X/A$ is a homotopy equivalence. In fact, it is a homotopy equivalence of pairs*

$$(X/A, *) \simeq (C_i, CA) \simeq (C_i, v),$$

where v is the vertex of the cone.

PROOF. The mapping cone $C_i = X \cup CA$ consists of three different types of points, the vertex $v = \{A \times \{1\}\}$, the rest of the cone $\{(a,t) | 0 \leq t < 1\}$ where $(a, 0) = a \in A \subset X$, and points in X itself, which we identify with $X \times \{0\}$ to simplify definitions of maps.

Define $f: A \times I \cup X \times \{0\} \to C_i$ as the collapsing map and extend f to $\bar{f}: X \times I \to C_i$ by the definition of cofibration. Then $\bar{f}(a, 1) = v$, $\bar{f}(a, t) = (a, t)$ and $\bar{f}(x, 0) = x$.

Put $\bar{f}_t = \bar{f}|_{X \times \{t\}}$. Since $\bar{f}_1(A) = \{v\}$, there is the factorization $\bar{f}_1 = g \circ j$, where $j: X \to X/A$ is the quotient map and $g: X/A \to C_i$. (g is continuous by definition of the quotient topology.)

We claim that g is a homotopy equivalence and a homotopy inverse to h.

First we will prove that $hg \simeq 1$. There is the homotopy $h\bar{f}_t: X \to X/A$. For all t, this takes A into the point $\{A\}$. Thus it factors to give the homotopy

$$hg \simeq \{h\bar{f}_1\} \simeq \{h\bar{f}_0\} = \{j\} = 1.$$

Next we will show that $gh \simeq 1$. For this, consider $W = (X \times I)/(A \times \{1\})$ and the maps illustrated in Figure VII-1. The map $\bar{f}'$ is induced by $\bar{f}$. The

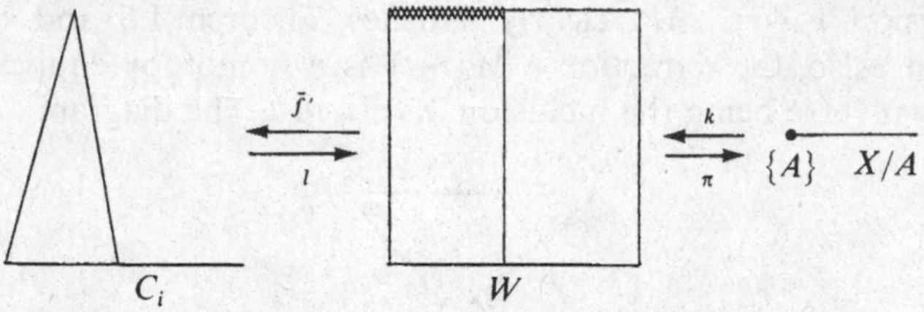

Figure VII-1. A homotopy equivalence and homotopy inverse.

map k is the "top face" map. We see that

$$\begin{aligned} \bar{f}'\circ l &= 1, \\ \pi\circ k &= 1 \quad \text{(which we don't need)}, \\ k\circ\pi &\simeq 1, \\ \bar{f}'\circ k &= g \quad \text{(definition of } g), \\ \pi\circ l &= h. \end{aligned}$$

Hence $g\circ h = \bar{f}'\circ(k\circ\pi)\circ l \simeq \bar{f}'\circ l = 1$, as claimed. □

A nonexample of Theorem 1.6 is $A = \{0\}\cup\{1/n \mid n = 1, 2, \ldots\}$, and $X = [0, 1]$. Here C_i is not homotopy equivalent to X/A, which is a one-point union of an infinite sequence of circles with radii going to zero. (C_i has homeomorphs of circles joined along edges, but the circles do not tend to a point and so any prospective homotopy equivalence $X/A \to C_i$ would be discontinuous at the image of $\{0\}$ in X/A.)

1.7. Corollary. *If $A \subset X$ is closed and the inclusion $A \hookrightarrow X$ is a cofibration then the map $j: (X, A) \to (X/A, *)$ induces isomorphisms*

$$H_*(X, A) \xrightarrow{\approx} H_*(X/A, *) \approx \tilde{H}_*(X/A),$$

and

$$\tilde{H}^*(X/A) \approx H^*(X/A, *) \xrightarrow{\approx} H^*(X, A).$$

PROOF. $H_*(X/A, *) \approx H_*(C_i, CA) \approx H_*(X \cup A \times [0, \frac{1}{2}], A \times [0, \frac{1}{2}]) \approx H_*(X, A)$. □

A nonexample is $X = \mathbf{S}^2$ with $A \subset X$ the "$\sin(1/x)$" subspace pictured in Figure VII-2. Here $X/A \approx \mathbf{S}^2 \vee \mathbf{S}^2$, so that $\tilde{H}_2(X/A) \approx \mathbf{Z}\oplus\mathbf{Z}$. But $H_1(A) = 0 = H_2(A)$, so that $H_2(X, A) \approx H_2(X) \approx \mathbf{Z}$. It follows that the inclusion $A \to \mathbf{S}^2$ is not a cofibration.

Let us recall the notion of the pointed category and some notational items. The pointed category has, as objects, spaces with a base point $*$, and,

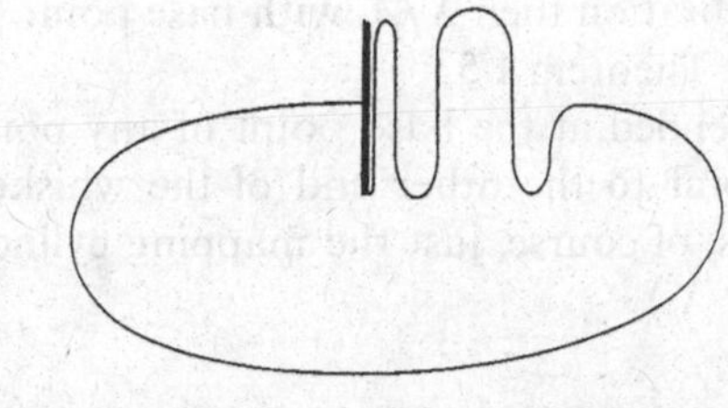

Figure VII-2. A pseudo-circle.

as maps, those maps of spaces preserving the base point. There is also the category of pairs of pointed spaces. There is also the notion of homotopies in this category, those homotopies which preserve the base point.

If $f: X \to Y$ is a pointed map then the reduced mapping cylinder of f is the quotient space M_f of $(X \times I) \cup Y$ modulo the relations identifying $(x, 0)$ with $f(x)$ and identifying the set $\{*\} \times I$ to the base point of M_f.

The reduced mapping cone is the quotient of the reduced mapping cylinder M_f gotten by identifying the image of $X \times \{1\}$ to a point, the base point.

The one-point union of pointed spaces X and Y is the quotient $X \vee Y$ of the disjoint union $X + Y$ obtained by identifying the two base points.

The wedge, or smash, product is the pointed space $X \wedge Y = X \times Y / X \vee Y$.

The circle $\mathbf{S}^1$ is defined as $I/\partial I$ with base point $\{\partial I\}$.

The reduced suspension of a pointed space X is $SX = X \wedge \mathbf{S}^1$. It can also be considered as the quotient space $X \times I/(X \times \partial I \cup \{*\} \times I)$.

As remarked before, $\mathbf{S}^n \wedge \mathbf{S}^m$ is the one-point compactification of $\mathbf{R}^n \times \mathbf{R}^m$ and hence is homeomorphic to $\mathbf{S}^{n+m}$. Thus we can, and will in this chapter, redefine $\mathbf{S}^n$ inductively by letting $\mathbf{S}^{n+1} = S\mathbf{S}^n$. Also note that

$$S(SX) = (SX) \wedge \mathbf{S}^1 = (X \wedge \mathbf{S}^1) \wedge \mathbf{S}^1 = X \wedge \mathbf{S}^2, \quad \text{etc.}$$

The preceding results of this section can all be rephrased in terms of the pointed category. Extending the proofs is elementary, mostly a matter of seeing that the unreduced versions become the reduced versions by taking the quotient of spaces by sets involving the base point. For example, Theorem 1.6 would say that if A is a closed, pointed, subspace of the pointed space X and if the inclusion $i: A \to X$ is a cofibration (same definition since the base point is automatically taken care of) then $X/A \simeq C_i$, where the latter is now the reduced mapping cone, and the homotopies involved must preserve the base points.

1.8. Definition. A base point $x_0 \in X$ is said to be *nondegenerate* if the inclusion $\{x_0\} \hookrightarrow X$ is a cofibration. A pointed Hausdorff space X with nondegenerate base point is said to be *well-pointed.*

Any pointed manifold or CW-complex is clearly well-pointed. A pointed space that is not well-pointed is $\{0\} \cup \{1/n \mid n \geq 1\}$ with 0 as base point. The reduced suspensions of this also fail to be well-pointed.

If $A \hookrightarrow X$ is a cofibration then X/A, with base point $\{A\}$, is well-pointed as follows easily from Theorem 1.5.

If a whisker is appended at the base point of any pointed space X, then changing the base point to the other end of the whisker provides a well-pointed space. (This is, of course, just the mapping cylinder of the inclusion of the base point into X.)

1.9. Theorem. *If X is well-pointed then so are the reduced cone CX and the reduced suspension SX. Moreover, the collapsing map $\Sigma X \to SX$, of the unreduced suspension to the reduced suspension, is a homotopy equivalence.*

Proof. Denote the base point of X by $*$. Consider a homeomorphism

$$h: (I \times I, I \times \{0\} \cup \partial I \times I) \xrightarrow{\approx} (I \times I, I \times \{0\})$$

which clearly exists. Then the induced homeomorphism

$$1 \times h: X \times I \times I \xrightarrow{\approx} X \times I \times I$$

carries $X \times I \times \{0\} \cup X \times \partial I \times I$ to $X \times I \times \{0\}$. Hence it takes $A = X \times I \times \{0\} \cup X \times \partial I \times I \cup \{*\} \times I \times I$ to $X \times I \times \{0\} \cup \{*\} \times I \times I$. Therefore, the pair $(X \times I \times I, A)$ is homeomorphic to the pair $I \times (X \times I, X \times \{0\} \cup \{*\} \times I)$. Since $X \times \{0\} \cup \{*\} \times I$ is a retract of $X \times I$ by the definition of "well-pointed," it follows that A is a retract of $X \times I \times I$. This implies that the inclusion $X \times \partial I \cup (*) \times I \hookrightarrow X \times I$ is a cofibration. Therefore, $SX = X \times I/(X \times \partial I \cup \{*\} \times I)$ is well-pointed. A similar argument using a homeomorphism $(I \times I, I \times \{0\} \cup \{1\} \times I) \xrightarrow{\approx} (I \times I, I \times \{0\})$ shows that the inclusion $X \times \{1\} \cup \{*\} \times I \hookrightarrow X \times I$ is a cofibration and so $CX = X \times I/(X \times \{1\} \cup \{*\} \times I)$ is well-pointed.

The fact that $X \times \partial I \cup \{*\} \times I \hookrightarrow X \times I$ is a cofibration implies that the induced inclusion $I \approx \{*\} \times I \hookrightarrow X \times I/\{X \times \{0\}, X \times \{1\}\} = \Sigma X$ is a cofibration by an easy application of Theorem 1.5. By Theorem 1.6, $\Sigma X \simeq \Sigma X \cup CI \simeq \Sigma X/I = SX$ via the collapsing map. □

Problems

1. Find $H_*(\mathbf{P}^2, \mathbf{P}^1)$ using methods or results from this section. (If $\mathbf{P}^2$ is thought of as the unit disk in the plane with antipodal points on the boundary identified, then $\mathbf{P}^1$ corresponds to the "boundary" circle.)

2. Find $H_*(\mathbf{T}^2, \{*\} \times \mathbf{S}^1 \cup \mathbf{S}^1 \times \{*\})$ using methods or results from this section.

3. For a space X consider the pair (CX, X). What do the results of this section tell you about the homology of these, and related, spaces?

4. If $f: A \to X$ is a cofibration then show that f is an embedding. If X is also Hausdorff then show that $f(A)$ is closed in X. (*Hint*: Consider M_f.)

5. If $A \subset X$ is closed and $i: A \hookrightarrow X$ is a cofibration and A is contractible, show that the collapse $X \to X/A$ is a homotopy equivalence.

2. The Compact-Open Topology

Let X be a locally compact Hausdorff space, and Y any Hausdorff space. By Y^X we mean the set of *continuous* functions $X \to Y$.

2.1. Definition. The *compact-open topology* on Y^X is the topology generated by the sets $M(K, U) = \{f \in Y^X \mid f(K) \subset U\}$, where $K \subset X$ is compact and $U \subset Y$ is open.

Recall that "generated" here means that these sets form a subbasis for the open sets. In the remainder of this book, unless otherwise noted, Y^X will always be given the compact-open topology.

2.2. Lemma. *Let* $\mathbf{K}$ *be a collection of compact subsets of* X *containing a neighborhood base at each point of* X. *Let* $\mathbf{B}$ *be a subbasis for the open sets of* Y. *Then the sets* $M(K, U)$, *for* $K \in \mathbf{K}$ *and* $U \in \mathbf{B}$, *form a subbasis for the compact-open topology.*

PROOF. Note that $M(K, U) \cap M(K, V) = M(K, U \cap V)$, which implies that it suffices to consider the case in which $\mathbf{B}$ is a basis. We need to show that the indicated sets form a neighborhood basis at each point $f \in Y^X$. Thus it suffices to show that if $K \subset X$ is compact and $U \subset Y$ is open, and $f \in M(K, U)$, then there exist $K_1, \ldots, K_n \in \mathbf{K}$ and $U_1, \ldots, U_n \in \mathbf{B}$ such that $f \in \bigcap M(K_i, U_i) \subset M(K, U)$.

For each $x \in K$, there is an open set $U_x \in \mathbf{B}$ with $f(x) \in U_x \subset U$, and there exists a $K_x \in \mathbf{K}$ which is a neighborhood of x such that $f(K_x) \subset U_x$. Thus $f \in M(K_x, U_x)$.

By the compactness of K there exist points $x_1, \ldots, x_n$ such that $K \subset K_{x_1} \cup \cdots \cup K_{x_n}$. Then $f \in \bigcap M(K_{x_i}, U_{x_i}) \subset M(K, U)$. □

2.3. Proposition. *For* X *locally compact Hausdorff, the "evaluation map"* $e: Y^X \times X \to Y$, *defined by* $e(f, x) = f(x)$, *is continuous.*

PROOF. If f and x are given, let U be an open neighborhood of $f(x)$. Since f is continuous, there is a compact neighborhood K of x such that $f(K) \subset U$. Thus $f \in M(K, U)$ and $M(K, U) \times K$ is taken into U by the evaluation e. Since $M(K, U) \times K$ is a neighborhood of (f, x) in $Y^X \times X$, we are done. □

2.4. Theorem. *Let* X *be locally compact Hausdorff and* Y *and* T *arbitrary Hausdorff spaces. Given a function* $f: X \times T \to Y$, *define, for each* $t \in T$, *the function* $f_t: X \to Y$ *by* $f_t(x) = f(x, t)$. *Then* f *is continuous* $\Leftrightarrow$ *both of the following conditions hold:*

(a) *each* f_t *is continuous; and*
(b) *the function* $T \to Y^X$ *taking* t *to* f_t *is continuous.*

PROOF. The implication $\Leftarrow$ follows from the fact that f is the composition of the map $X \times T \to Y^X \times X$ taking (x, t) to (f_t, x), with the evaluation $Y^X \times X \to Y$.

For the implication $\Rightarrow$, (a) follows from the fact that f_t is the composition $X \to X \times T \to Y$ of the inclusion $x \mapsto (x, t)$ with f. To prove (b), let $t \in T$ be given and let $f_t \in M(K, U)$. It suffices to show that there exists a neighborhood W of t in T such that $t' \in W \Rightarrow f_{t'} \in M(K, U)$. (That is, it suffices to prove the conditions for continuity for a subbasis only.)

For $x \in K$, there are open neighborhoods $V_x \subset X$ of x and $W_x \subset T$ of t such that $f(V_x \times W_x) \subset U$. By compactness, $K \subset V_{x_1} \cup \cdots \cup V_{x_n} = V$ say. Put $W = W_{x_1} \cap \cdots \cap W_{x_n}$. Then $f(K \times W) \subset f(V \times W) \subset U$, so that $t' \in W \Rightarrow f_{t'} \in M(K, U)$ as claimed. □

This theorem implies that a homotopy $X \times I \to Y$, with X locally compact, is the same thing as a path $I \to Y^X$ in Y^X.

An often used consequence of Theorem 2.4 is that in order to show a function $T \to Y^X$ to be continuous, it suffices to show that the associated function $X \times T \to Y$ is continuous.

2.5. Theorem (The Exponential Law). *Let X and T be locally compact Hausdorff spaces and let Y be an arbitrary Hausdorff space. Then there is the homeomorphism*

$$Y^{X \times T} \xrightarrow{\approx} (Y^X)^T$$

taking f to f^*, where $f^*(t)(x) = f(x, t) = f_t(x)$.

PROOF. Theorem 2.4 says that the assignment $f \mapsto f^*$ is a bijection. We must show it and its inverse to be continuous. Let $U \subset Y$ be open, and $K \subset X$, $K' \subset T$ compact. Then

$$\begin{aligned} f \in M(K \times K', U) &\Leftrightarrow (t \in K', x \in K \Rightarrow f_t(x) = f(x, t) \in U) \\ &\Leftrightarrow (t \in K' \Rightarrow f_t \in M(K, U)) \\ &\Leftrightarrow f^* \in M(K', M(K, U)). \end{aligned}$$

Now the $K \times K'$ give a neighborhood basis for $X \times T$. Therefore the $M(K \times K', U)$ form a subbasis for the topology of $Y^{X \times T}$.

Also, the $M(K, U)$ give a subbasis for Y^X and therefore the $M(K', M(K, U))$ give a subbasis for the topology of $(Y^X)^T$.

Since these subbases correspond to one another under the exponential correspondence, the theorem is proved. □

2.6. Proposition. *If X is locally compact Hausdorff and Y and W are Hausdorff then there is the homeomorphism*

$$Y^X \times W^X \xrightarrow{\approx} (Y \times W)^X$$

given by $(f, g) \mapsto f \times g$.

PROOF. This is clearly a bijection. If $K, K' \subset X$ are compact, and $U \subset Y$ and

$V \subset W$ are open then we have

$$
\begin{aligned}
(f,g) \in M(K,U) \times M(K',V) \quad &\Leftrightarrow \quad (x \in K \Rightarrow f(x) \in U) \quad \text{and} \quad (x \in K' \Rightarrow g(x) \in V)\\
&\Leftrightarrow \quad (x \in K \Rightarrow (f \times g)(x) \in U \times W) \quad \text{and}\\
&\qquad\ (x \in K' \Rightarrow (f \times g)(x) \in Y \times V)\\
&\Leftrightarrow \quad (f \times g) \in M(K, U \times W) \cap M(K', Y \times V).
\end{aligned}
$$

Thus $(f,g) \mapsto f \times g$ is open.

Also, $(f,g) \in M(K,U) \times M(K,V) \Leftrightarrow (f \times g) \in M(K, U \times V)$, which implies that the function in question is continuous. □

2.7. Proposition. *If X and T are locally compact Hausdorff spaces and Y is an arbitrary Hausdorff space then there is the homeomorphism*

$$Y^{X+T} \xrightarrow{\approx} Y^X \times Y^T$$

taking f to $(f \circ i_X, f \circ i_Y)$.

PROOF. This is an easy exercise left to the reader. □

2.8. Theorem. *For X locally compact and both X and Y Hausdorff, Y^X is a covariant functor of Y and a contravariant functor of X.*

PROOF. A map $\phi\colon Y \to Z$ induces $\phi^X\colon Y^X \to Z^X$, by $\phi^X(f) = \phi \circ f$. We must show that ϕ^X is continuous. By Theorem 2.4 it suffices to show that $Y^X \times X \to Z$, taking (f,x) to $\phi(f(x))$, is continuous. But this is the composition $\phi \circ e$ of ϕ with the evaluation, which is continuous.

Next, for $\psi\colon X \to T$, both spaces locally compact, we must show that $Y^\psi\colon Y^T \to Y^X$, taking f to $f \circ \psi$, is continuous. It suffices, by Theorem 2.4, to show that $Y^T \times X \to Y$, taking (f,x) to $f(\psi(x))$, is continuous. But this is just the composition $e \circ (1 \times \psi)$, which is continuous. □

2.9. Corollary. *For $A \subset X$ both locally compact and X, Y Hausdorff, the restriction $Y^X \to Y^A$ is continuous.* □

2.10. Theorem. *For X, Y locally compact, and X, Y, Z Hausdorff, the function*

$$Z^Y \times Y^X \to Z^X$$

taking (f,g) to $f \circ g$, is continuous.

PROOF. It suffices, by Theorem 2.4, to show that the function $Z^Y \times Y^X \times X \to Z$, taking (f,g,x) to $(f \circ g)(x)$, is continuous. But this is the composition $e \circ (1 \times e)$. □

All of these things, and the ones following, have versions in the pointed category, the verification of which is trivial.

We finish this section by showing that, for Y metric, the compact-open topology is identical to a more familiar concept.

2.11. Lemma. *Let Y be a metric space, let C be a compact subset of Y, and let $U \supset C$ be open. Then there is an $\epsilon > 0$ such that $B_\epsilon(C) \subset U$.*

PROOF. Cover C by a finite number of balls of the form $B_{\epsilon(x_i)}(x_i)$ such that $B_{2\epsilon(x_i)}(x_i) \subset U$. Put $\epsilon = \min(\epsilon(x_i))$. Suppose $x \in B_\epsilon(C)$. Then there is a $c \in C$ with $\operatorname{dist}(x, c) < \epsilon$ and an i such that $\operatorname{dist}(c, x_i) < \epsilon(x_i)$. Thus $x \in B_{2\epsilon(x_i)}(x_i) \subset U$. □

2.12. Theorem. *If X is compact Hausdorff and Y is metric then the compact-open topology is induced by the uniform metric on Y^X, i.e., the metric given by* $\operatorname{dist}(f, g) = \sup\{\operatorname{dist}(f(x), g(x)) \mid x \in X\}$.

PROOF. For $f \in Y^X$, it suffices to show that a basic neighborhood of f in each of these topologies contains a neighborhood of f in the other topology.

Let $\epsilon > 0$ be given. Let $N = B_\epsilon(f) = \{g \in Y^X \mid \operatorname{dist}(f(x), g(x)) < \epsilon \text{ for all } x \in X\}$. Given x, there is a compact neighborhood N_x of x such that $p \in N_x \Rightarrow f(p) \in B_{\epsilon/2}(f(x))$. Cover X by $N_{x_1} \cup \cdots \cup N_{x_k}$. We claim that

$$V = M(N_{x_1}, B_{\epsilon/2}(f(x_1))) \cap \cdots \cap M(N_{x_k}, B_{\epsilon/2}(f(x_k))) \subset N.$$

To see this, let $g \in V$, i.e., $x \in N_{x_i} \Rightarrow g(x) \in B_{\epsilon/2}(f(x_i))$. But $f(x) \in B_{\epsilon/2}(f(x_i))$ and so it follows that $g \in V \Rightarrow \operatorname{dist}(f(x), g(x)) < \epsilon$ for all x. That is, $V \subset N$.

Conversely, suppose that $f \in M(K_1, U_1) \cap \cdots \cap M(K_r, U_r)$, i.e., $f(K_i) \subset U_i$ for $i = 1, \ldots, r$. By Lemma 2.11, there is an $\epsilon > 0$ such that $B_\epsilon(f(K_i)) \subset U_i$ for all $i = 1, \ldots, r$. If $x \in K_i$ then $B_\epsilon(f(x)) \subset B_\epsilon(f(K_i)) \subset U_i$. Therefore, if $g \in B_\epsilon(f)$ and $x \in K_i$ then $g(x) \in B_\epsilon(f(x)) \subset U_i$. Thus $g \in M(K_i, U_i)$ for all i and so $B_\epsilon(f) \subset \bigcap M(K_i, U_i)$. □

2.13. Corollary. *If X is locally compact Hausdorff and Y is metric then the compact-open topology on Y^X is the topology of uniform convergence on compact sets. That is, a net $f_\alpha \in Y^X$ converges to $f \in Y^X$ in the compact-open topology $\Leftrightarrow f_\alpha|_K$ converges uniformly to $f|_K$ for each compact set $K \subset X$.*

PROOF. For $\Rightarrow$ recall from Corollary 2.9 that $Y^X \to Y^K$ is continuous. Thus $f_\alpha|_K \to f|_K$ in the compact-open topology. But Y^K has the topology of the uniform metric and so $f_\alpha|_K$ converges to $f|_K$ uniformly.

For $\Leftarrow$, suppose that $f_\alpha|_K$ converges uniformly to $f|_K$ for each compact $K \subset X$. Let $f \in M(K, U)$. Then there exists an $\epsilon > 0$ so that $B_\epsilon(f(K)) \subset U$. There is an α such that $\beta > \alpha \Rightarrow \operatorname{dist}(f_\beta(x), f(x)) < \epsilon$ for all $x \in K$. That is, $f_\beta(x) \in B_\epsilon(f(K)) \subset U$. Thus $\beta > \alpha \Rightarrow f_\beta \in M(K, U)$. This implies that f_α converges to f in the compact-open topology. □

PROBLEMS

1. Consider the Tychonoff topology on the set Y^X of continuous functions from X to Y, i.e., the subspace topology as a subspace of the space of all functions $X \to Y$

with the product topology. Which parts, if any, of Theorem 2.4 hold in this topology? Give either proofs or counterexamples.

2. Describe the Tychonoff topology on Y^X in a manner similar to the description in Corollary 2.13 of the compact-open topology.

3. Prove Proposition 2.7.

4. Show that I^I is not compact in the compact-open topology.

3. H-Spaces, H-Groups, and H-Cogroups

An H-space or H-group is a space with a product that satisfies some of the laws of a group but only *up to homotopy*. An H-cogroup is a dual notion. The "H" stands for "Hopf" or for "Homotopy."

3.1. Definition. An *H-space* is a pointed space X with base point e, together with a map

$$\bullet: X \times X \to X$$

sending (x, y) to $x \bullet y$, such that $e \bullet e = e$, and the maps $X \to X$ taking x to $x \bullet e$ and x to $e \bullet x$ are each homotopic rel$\{e\}$ to the identity.

It is *homotopy associative* if the maps $X \times X \times X \to X$ taking (x, y, z) to $(x \bullet y) \bullet z$ and to $x \bullet (y \bullet z)$ are homotopic rel$\{(e, e, e)\}$.

It has a *homotopy inverse* $\hat{}: X \to X$ if $\hat{e} = e$ and the maps $X \to X$ taking x to $x \bullet \hat{x}$ and to $\hat{x} \bullet x$ are each homotopic rel$\{e\}$ to the constant map to $\{e\}$.

An *H-group* is a homotopy associative *H*-space with a given homotopy inverse.

There are two main classes of examples. The first is the class of topological groups. The second is the class of "loop spaces." The loop space on a space X is the space

$$\boxed{\Omega X = (X, *)^{(\mathbf{S}^1, *)},}$$

i.e., $X^{\mathbf{S}^1}$ in the pointed category. The product is concatenation of loops, and the homotopy inverse is loop reversal. ΩX is a pointed space with base point being the constant loop at $*$.

If $f: X \to Z$ and $g: Y \to W$ are maps then let $f \vee g: X \vee Y \to Z \vee W$ be the induced map on the one-point union. Also let $\nabla: Z \vee Z \to Z$ be the codiagonal; i.e., the identity on both factors. If $f: X \to Z$ and $g: Y \to Z$ then let $f \underline{\vee} g: X \vee Y \to Z$ be the composition $f \underline{\vee} g = \nabla \circ (f \vee g)$; i.e., the map which is f on X and g on Y.

3.2. Definition. An *H-cogroup* is a pointed space Y and a map $\gamma: Y \to Y \vee Y$ such that the following three conditions are satisfied:

(1) The constant map $*: Y \to Y$ to the base point is a homotopy identity. That is, the compositions $(* \veebar 1)\circ\gamma$ and $(1 \veebar *)\circ\gamma$ of $Y \xrightarrow{\gamma} Y \vee Y \to Y$ are both homotopic to the identity rel base point.
(2) It is homotopy associative. That is, the compositions $(\gamma \vee 1)\circ\gamma$ and $(1 \vee \gamma)\circ\gamma$ of $Y \xrightarrow{\gamma} Y \vee Y \to Y \vee Y \vee Y$ are homotopic to one another rel base point.
(3) There is a homotopy inverse $i: Y \to Y$. That is, $(1 \veebar i)\circ\gamma$ and $(i \veebar 1)\circ\gamma$ of $Y \xrightarrow{\gamma} Y \vee Y \to Y$ are both homotopic to the constant map to the base point rel base point.

There is one important class of examples, the reduced suspensions. The "coproduct" $\gamma: SX \to SX \vee SX$ is given by

$$\gamma(t, x) = \begin{cases} (2t, x)_1 & \text{if } t \le \frac{1}{2}, \\ (2t-1, x)_2 & \text{if } t \ge \frac{1}{2}, \end{cases}$$

where the subscripts indicate in which copy of SX in the one-point union the indicated point lies. The homotopy inverse is just reversal of the t parameter.

3.3. Theorem. *In the pointed category:*

(1) *Y an H-group $\Rightarrow [X; Y]$ is a group with multiplication induced by $(f \bullet g)(x) = f(x) \bullet g(x)$;*
(2) *X an H-cogroup $\Rightarrow [X; Y]$ is a group with multiplication induced by $f * g = (f \veebar g)\circ\gamma$; and*
(3) *X an H-cogroup and Y an H-space $\Rightarrow$ the two multiplications above on $[X; Y]$ coincide and are abelian.*

PROOF. Part (1) is obvious. Part (2) is nearly as obvious. For example, to show associativity in part (2) note that $(f * g) * h = [((f \veebar g)\circ\gamma) \veebar h]\circ\gamma$ which equals the composition

$$X \xrightarrow{\gamma} X \vee X \xrightarrow{\gamma \vee 1} (X \vee X) \vee X \xrightarrow{(f \veebar g) \veebar h} Y.$$

The first composition is homotopic to $(1 \vee \gamma)\circ\gamma$, and the last map is equal to $f \veebar (g \veebar h)$. This provides the homotopy to $f * (g * h)$. The other parts of (2) are no harder.

For (3) we need the following lemma:

3.4. Lemma. *In the situation of Theorem* 3.3(3), *and for $f, g: X \to Y$, we have*

$$(f \bullet g) * (h \bullet k) = (f * h) \bullet (g * k).$$

PROOF. For a particular point $x \in X$ suppose that $\gamma(x) = (w, *) \in X \vee X$. Then

$$(f \bullet g) * (h \bullet k)(x) = ((f \bullet g) \veebar (h \bullet k))(w, *) = (f \bullet g)(w) = f(w) \bullet g(w).$$

Also

$$(f*h)\bullet(g*k)(x)=(f*h)(x)\bullet(g*k)(x)=f(w)\bullet g(w).$$

The case $\gamma(x)=(*, w')$ is similar and will be omitted. □

Returning to the proof of (3) in Theorem 3.3, note that for both products, the identity 1 is given by the constant map to the base point. Operating in $[X; Y]$, we have

$$(\alpha\bullet\beta)*(\gamma\bullet\delta)=(\alpha*\gamma)\bullet(\beta*\delta).$$

Thus

$$\alpha*\beta=(1\bullet\alpha)*(\beta\bullet 1)=(1*\beta)\bullet(\alpha*1)=\beta\bullet\alpha,$$

and

$$\alpha*\beta=(\alpha\bullet 1)*(1\bullet\beta)=(\alpha*1)\bullet(1*\beta)=\alpha\bullet\beta.$$

Therefore, $\alpha\bullet\beta=\beta\bullet\alpha=\alpha*\beta$. □

4. Homotopy Groups

For now on, unless otherwise indicated, we regard the n-sphere $\mathbf{S}^n$ as having the cogroup structure as the reduced suspension $\mathbf{S}^n=S\mathbf{S}^{n-1}=\mathbf{S}^{n-1}\wedge\mathbf{S}^1$. The 0-sphere $\mathbf{S}^0$ is $\{0,1\}$ with base point $\{0\}$.

Then, for a space X with base point x_0, we define the nth homotopy group

$$\boxed{\pi_n(X,x_0)=[\mathbf{S}^n,*;X,x_0].}$$

This is a group for $n\geq 1$ with the product defined by Theorem 3.3(2).

Note that $\mathbf{S}^n=\mathbf{S}^{n-1}\wedge\mathbf{S}^1=\mathbf{S}^1\wedge\cdots\wedge\mathbf{S}^1=I^n/\partial I^n$. Thus we can regard

$$\pi_n(X,x_0)=[I^n,\partial I^n;X,x_0].$$

In this context, if f, $g\colon I^n\to X$ are maps taking ∂I^n to x_0, then the group structure is induced by

$$(f*g)(t_1,\ldots,t_n)=\begin{cases}f(t_1,\ldots,2t_n) & \text{for } t_n\leq\frac{1}{2},\\ g(t_1,\ldots,2t_n-1) & \text{for } t_n\geq\frac{1}{2}.\end{cases}$$

4.1. Theorem. *If X is an H-space then the multiplication in $\pi_n(X,x_0)$ is induced by the H-space multiplication and is abelian for $n\geq 1$.*

PROOF. This is a direct corollary of Theorem 3.3. □

4.2. Lemma. *In the pointed category $[SX;Y]\approx[X;\Omega Y]$ as groups.*

Proof. Recall the correspondence

$$f: X \times \mathbf{S}^1 \to Y \quad \leftrightarrow \quad f': X \to Y^{\mathbf{S}^1},$$

given by $f'(x)(t) = f(x,t)$. The map $g: X \wedge \mathbf{S}^1 \to Y$, using the composition f of g with the quotient map $X \times \mathbf{S}^1 \to X \wedge \mathbf{S}^1$, corresponds to $f': X \to Y^{\mathbf{S}^1}$ where $f'(x)(*) = f(x,*) = *$ and $f'(*)(t) = *$, i.e., $f'(*) = *$, the base point of ΩY. Thus the correspondence induces a one–one correspondence between *pointed* maps $SX \to Y$ and pointed maps $X \to \Omega Y$. Also pointed homotopies correspond. For f, $g: SX \to Y$ the product in $[SX; Y]$ is induced by

$$(f * g)(x,t) = \begin{cases} f(x,2t) & \text{for } t \le \frac{1}{2}, \\ g(x,2t-1) & \text{for } t \ge \frac{1}{2}. \end{cases}$$

This is equal to $(f*g)'(x)(t)$. The multiplication in $[X;\Omega Y]$ is $(f' \bullet g')(x) = f'(x)*g'(x)$, where $*$ is loop concatenation. At t this is $f'(x)(2t)$ for $t \le \frac{1}{2}$ and is $g'(x)(2t-1)$ for $t \ge \frac{1}{2}$. Thus $(f*g)' = f' \bullet g'$. □

Remark. For X compact we have $Y^{SX} \approx (\Omega Y)^X$ by the exponential law. The proof of Lemma 4.2 shows that the H-space operations on these spaces correspond. Therefore there are the group isomorphisms $[SX; Y] = \pi_0(Y^{SX}) \approx \pi_0((\Omega Y)^X) = [X; \Omega Y]$.

All we have done goes over immediately to the case of pointed pairs (X, A), with base point in A. For example, $[SX, SA; Y, B]$ is a group and is canonically isomorphic to $[X, A; \Omega Y, \Omega B]$.

In particular, consider $\mathbf{D}^n = \mathbf{D}^1 \wedge \mathbf{S}^{n-1} \supset \mathbf{S}^0 \wedge \mathbf{S}^{n-1} = \mathbf{S}^{n-1}$. Then $(\mathbf{D}^n, \mathbf{S}^{n-1}) = S^{n-1}(\mathbf{D}^1, \mathbf{S}^0)$, the $(n-1)$-fold reduced suspension. Hence we define the relative homotopy group by

$$\boxed{\pi_n(Y, B, *) = [\mathbf{D}^n, \mathbf{S}^{n-1}; Y, B] = [\mathbf{S}^{n-1}(\mathbf{D}^1, \mathbf{S}^0); Y, B].}$$

This is a group for $n \ge 2$. Under composition with the projection $I^n \to \mathbf{D}^1 \wedge \mathbf{S}^1 \wedge \cdots \wedge \mathbf{S}^1 = \mathbf{D}^n$, ∂I^n corresponds to (is the inverse image of) $\mathbf{S}^{n-1}$ and the inverse image of the base point is the set $J^{n-1} = (I \times \partial I^{n-1}) \cup (\{0\} \times I^{n-1})$. Thus

$$\pi_n(Y, B, *) \approx [I^n, \partial I^n, J^{n-1}; Y, B, *].$$

4.3. Corollary. *$\pi_n(Y,*)$ is abelian for $n \ge 2$ and $\pi_n(Y, B, *)$ is abelian for $n \ge 3$. Moreover, the group structure is independent of the suspension coordinate used to define it.*

Proof. Consider the correspondence of Lemma 4.2: $[\mathbf{S}^1 \wedge \cdots \wedge \mathbf{S}^1; Y] \approx [\mathbf{S}^{n-1}; \Omega Y]$. The loop structure corresponds, by definition, to the suspension in the last coordinate. These yield identical group operations by Lemma 4.2. The product from the loop space in $[\mathbf{S}^{n-1}; \Omega Y]$ is the same as that from any of the suspension coordinates by Theorem 3.3(3). But the latter clearly is identical to the product on $[\mathbf{S}^1 \wedge \cdots \wedge \mathbf{S}^1; Y]$ using the same factor $\mathbf{S}^1$ as the suspension coordinate used for $[\mathbf{S}^{n-1};\Omega Y]$, and this is arbitrary. The product in $[\mathbf{S}^{n-1};\Omega Y]$ is abelian for $n-1\geq 1$ by Theorem 4.1. The relative case is similar. □

4.4. Corollary. $\pi_n(Y,*)\approx\pi_{n-1}(\Omega Y,*)\approx\cdots\approx\pi_1(\Omega^{n-1}Y,*)\approx\pi_0(\Omega^n Y,*)$ *and similarly in the relative case.* □

4.5. Theorem. *Let A be a closed subspace of X containing the base point $*$. Suppose that $F: X\times I\to X$ is a deformation of X contracting A to $*$; i.e.,*

$$F(A\times I)\subset A,$$
$$F(x,0)=x,$$
$$F(A\times\{1\})=*,$$
$$F(\{*\}\times I)=*.$$

Then the quotient map $\phi: X\to X/A$ is a homotopy equivalence. Similarly for pairs (X,X') with $A\subset X'$.

PROOF. Let $\psi: X/A\to X$ be induced by $F|_{X\times\{1\}}$. Then $\psi\circ\phi\simeq 1_X$ via F. The homotopy F induces $F': (X/A)\times I\to X/A$ and this is a homotopy between $\phi\circ\psi$ and $1_{X/A}$. □

5. The Homotopy Sequence of a Pair

In this section we develop an exact sequence of homotopy groups analogous to the exact homology sequence of a pair. It is, of course, an indispensable tool in the study of homotopy groups. Everything in this section is in the pointed category. In particular, mapping cones and suspensions are reduced.

5.1. Definition. A sequence

$$A\xrightarrow{f}B\xrightarrow{g}C$$

of pointed spaces (or pointed pairs) is called *coexact* if, for each pointed space (or pair) Y, the sequence of sets (pointed homotopy classes)

$$[C;Y]\xrightarrow{g^\#}[B;Y]\xrightarrow{f^\#}[A;Y]$$

is exact, i.e., $\operatorname{im}(g^\#)=(f^\#)^{-1}(*)$.

5.2. Theorem. *For any map $f: A\to X$ and for the inclusion $i: X\hookrightarrow C_f$ the sequence*

$$A\xrightarrow{f}X\xrightarrow{i}C_f$$

is coexact.

Proof. Clearly $i \circ f \simeq *$, the constant map to the base point, so the sequence is of "order two." Suppose given $\phi: X \to Y$ with $\phi \circ f \simeq *$ via the homotopy F. Then F on $A \times I$ and ϕ on X fit together to give a map $C_f \to Y$ extending ϕ. □

5.3. Corollary. *If $f: A \hookrightarrow X$ is a cofibration, where $A \subset X$ is closed, then*

$$A \to X \to X/A$$

is coexact.

Proof. This follows from Theorem 5.2 and the fact that we can replace C_f by its homotopy equivalent space X/A (by Theorem 1.6). □

5.4. Corollary. *Let $f: A \to X$ be any map. Then the sequence*

$$A \xrightarrow{f} X \xrightarrow{i} C_f \xrightarrow{j} C_i \xrightarrow{k} C_j$$

is coexact, where j and k are the obvious inclusions. □

Now we will replace, in this sequence, C_i and C_j by simpler things. Note that $C_i = C_f \cup_X CX$. (See Figure VII-3.)

By pulling the cone CX towards its vertex and stretching the mapping cylinder of f to accommodate it, we see that C_i has a deformation carrying CX to the base point through itself. Thus, Theorem 4.5 implies that the collapsing map $C_i \to C_i/CX = C_f/X = SA$ is a homotopy equivalence. Similarly, $C_j \simeq SX$. Under these homotopy equivalences we wish to show that k becomes Sf. Consider Figure VII-4.

The map l stretches the top cone of SA to the cylinder part of $C_f \subset C_i$ and is Cf on the bottom cone. The map $coll_1$ is the collapse of the bottom of the picture and gives the homotopy equivalence $C_i \simeq SA$ obtained above. The map $coll_2$ is the collapse of the top of C_j in the picture (the dashed lines) and is the homotopy equivalence $C_j \simeq SX$.

Clearly, $coll_1 \circ l \simeq 1$, so that l is a homotopy inverse of $coll_1$, i.e., $l \circ coll_1 \simeq 1$ as well. Also, $coll_2 \circ k \circ l = Sf \circ g \simeq Sf$, where g is the collapse of the top cone of SA. Composing this with $coll_1$ on the right gives $coll_2 \circ k \simeq coll_2 \circ k \circ l \circ coll_1 \simeq$

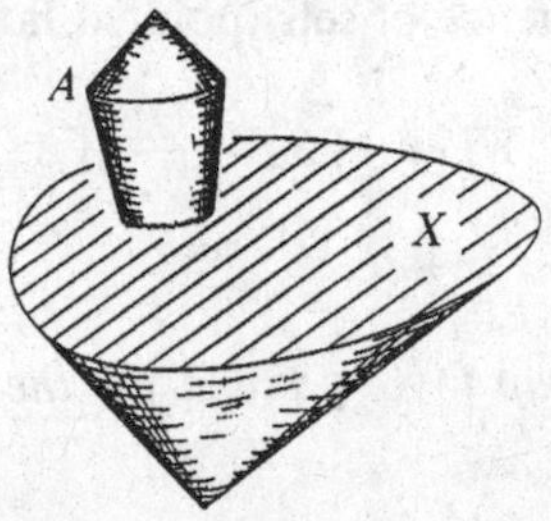

Figure VII-3. The mapping cone C_i.

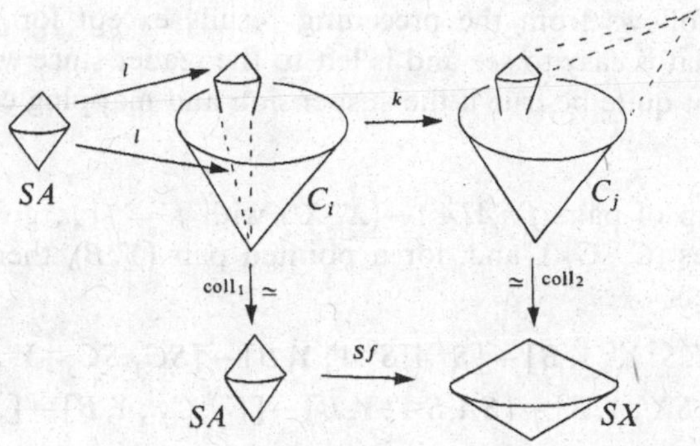

Figure VII-4. Homotopy equivalences of mapping cones.

$Sf \circ coll_1$. This shows that the diagram

$$\begin{array}{ccc} C_i & \xrightarrow{k} & C_j \\ {\scriptstyle coll_1}\downarrow & & \downarrow{\scriptstyle coll_2} \\ SA & \xrightarrow{Sf} & SX \end{array}$$

is homotopy commutative, as we wished to show. We have shown:

5.5. Corollary. *Given any map $f: A \to X$ of pointed spaces, the sequence*

$$A \xrightarrow{f} X \xrightarrow{i} C_f \xrightarrow{g} SA \xrightarrow{Sf} SX$$

is coexact, where $g: C_f \to SA$ is the composition of the collapse $C_f \to C_f/X$ with the homotopy equivalence $SA \simeq C_f/X$ induced by the inclusion of $A \times I$ in $(A \times I) + X$ followed by the quotient map to C_f and then the collapsing of the subspace X of C_f. □

5.6. Lemma. *Coexactness is preserved by suspension.*

PROOF. Suppose $A \to B \to C$ is coexact. Then the sequence

$$[SC; Y] \to [SB; Y] \to [SA; Y]$$

is equivalent to the sequence

$$[C; \Omega Y] \to [B; \Omega Y] \to [A; \Omega Y]$$

which is exact. □

5.7. Corollary (Barratt–Puppe). *If $f: A \to X$ is any map then the sequence*

$$A \xrightarrow{f} X \xrightarrow{i} C_f \xrightarrow{g} SA \xrightarrow{Sf} SX \xrightarrow{Si} SC_f \xrightarrow{Sg} S^2 A \xrightarrow{S^2 f} \cdots$$

is coexact. Also $SC_f \approx C_{Sf}$, etc. Similarly for maps of pairs of pointed spaces.

Proof. This follows from the preceding results except for the statement $SC_f \approx C_{Sf}$. That is easy to see and is left to the reader since we do not need it. It would not quite be true if the suspension and mapping cones were not reduced. □

Thus a map of pairs $f:(A, A') \to (X, X')$ with $f' = f|_{A'}$, gives the pair of mapping cones $(C_f, C_{f'})$, and, for a pointed pair (Y, B), there is the exact sequence of sets

$$\begin{aligned}\cdots &\to [S^2X, S^2X'; Y, B] \to [S^2A, S^2A'; Y, B] \to [SC_f, SC_{f'}; Y, B]\\ &\to [SX, SX'; Y, B] \to [SA, SA'; Y, B] \to [C_f, C_{f'}; Y, B] \to [X, X'; Y, B]\\ &\to [A, A'; Y, B],\end{aligned}$$

where the terms involving suspensions consist of groups and homomorphisms. The rest contains only pointed sets and maps.

Consider the special case of the inclusion $f:(\mathbf{S}^0, \mathbf{S}^0) \hookrightarrow (\mathbf{D}^1, \mathbf{S}^0)$. The pair of reduced mapping cones C_f is a triangle with one side collapsed (because of the reduction) with the subspace consisting of two sides of the triangle with one of those collapsed. Clearly this is homotopy equivalent to the pair $(\mathbf{S}^1, *)$. Thus we have the coexact sequence

$$(\mathbf{S}^0, \mathbf{S}^0) \to (\mathbf{D}^1, \mathbf{S}^0) \to (\mathbf{S}^1, *) \to (\mathbf{S}^1, \mathbf{S}^1) \to (\mathbf{D}^2, \mathbf{S}^1)$$

where the second map is the result of collapsing $\mathbf{S}^0$ to the base point.

By suspending this $r-1$ times, we get the coexact sequence (pointed)

$$(\mathbf{S}^{r-1}, \mathbf{S}^{r-1}) \to (\mathbf{D}^r, \mathbf{S}^{r-1}) \to (\mathbf{S}^r, *) \to (\mathbf{S}^r, \mathbf{S}^r) \to (\mathbf{D}^{r+1}, \mathbf{S}^r).$$

All these fit together to give a long coexact sequence. Now

$$\begin{aligned}[\mathbf{S}^r, \mathbf{S}^r; Y, B] &= [\mathbf{S}^r; B] = \pi_r(B),\\ [\mathbf{S}^r, *; Y, B] &= [\mathbf{S}^r; Y] = \pi_r(Y),\\ [\mathbf{D}^r, \mathbf{S}^{r-1}; Y, B] &= \pi_r(Y, B).\end{aligned}$$

Thus we obtain the "exact homotopy sequence" of the pair (Y, B):

$$\begin{aligned}\cdots &\longrightarrow \pi_{r+1}(Y, B) \longrightarrow \pi_r(B) \xrightarrow{i_\#} \pi_r(Y) \xrightarrow{j_\#} \pi_r(Y, B) \xrightarrow{\partial_\#} \pi_{r-1}(B) \longrightarrow \cdots\\ \cdots &\longrightarrow \pi_1(Y, B) \longrightarrow \pi_0(B) \longrightarrow \pi_0(Y),\end{aligned}$$

where all are groups and homomorphisms until the last three, which are only pointed sets and maps. Tracing through the definitions shows easily that $i_\#$ is induced by the inclusion $B \hookrightarrow Y, j_\#$ is induced by the inclusion $(Y, *) \hookrightarrow (Y, B)$, and $\partial_\#$ is induced by the restriction to $\mathbf{S}^{r-1} \subset \mathbf{D}^r$.

Given a map $f:(\mathbf{D}^n, \mathbf{S}^{n-1}, *) \to (Y, B, *)$, it is important to know when $[f] = 0$ in $\pi_n(Y, B, *)$. The following gives one such criterion.

5.8. Theorem. *For a map $f:(\mathbf{D}^n, \mathbf{S}^{n-1}, *) \to (Y, B, *)$, $[f] = 0$ in $\pi_n(Y, B, *) \Leftrightarrow$ f is homotopic,* rel $\mathbf{S}^{n-1}$, *to a map into B.*

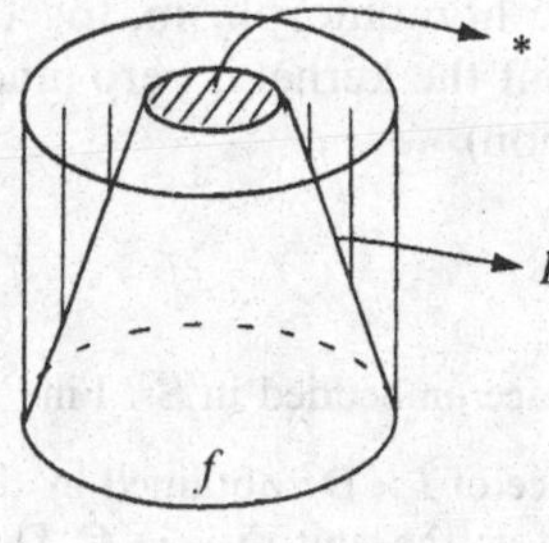

Figure VII-5. A relative homotopy.

PROOF. For $\Leftarrow$, let $f \simeq g$ rel $\mathbf{S}^{n-1}$, where $g: \mathbf{D}^n \to B$. Then g is homotopic within B to the constant map to the base point since $\mathbf{D}^n$ is contractible.

For $\Rightarrow$, suppose we have a homotopy $F: \mathbf{D}^n \times I \to Y$, rel base point, keeping $\mathbf{S}^{n-1}$ going into B and ending with the constant map to the base point. Think of $\mathbf{D}^n \times I$, not as a cylinder but as a frustum of a cone as in Figure VII-5. Consider a cylinder circumscribed about the cone as in the figure. Extend the map F to be constant along verticals on the outside of the cone. This gives the desired homotopy rel $\mathbf{S}^{n-1}$ to a map into B. □

5.9. Definition. A pair (X, A) is said to be *n-connected* if, for each $0 \le r \le n$, every map $(\mathbf{D}^r, \mathbf{S}^{r-1}) \to (X, A)$ is homotopic rel $\mathbf{S}^{r-1}$ to a map into A. (Here $\mathbf{S}^{0-1} = \emptyset$.) That is, each path component of X touches A and $\pi_r(X, A, a) = 0$ for all $a \in A$, and all $1 \le r \le n$.

5.10. Proposition. $\pi_r(\mathbf{S}^n) = 0$ *for all* $r < n$.

PROOF. Of course, we have already proved this in a previous chapter. One uses either smooth or simplicial approximation to change a map $\mathbf{S}^r \to \mathbf{S}^n$ to a homotopic map which misses a point and then uses that the complement of a point in $\mathbf{S}^n$ is contractible. □

5.11. Proposition. *The pair* $(\mathbf{D}^{n+1}, \mathbf{S}^n)$ *is n-connected.*

PROOF. This follows from the exact sequence $\pi_r(\mathbf{D}^{n+1}) \to \pi_r(\mathbf{D}^{n+1}, \mathbf{S}^n) \to \pi_{r-1}(\mathbf{S}^n)$ for $r \le n$. □

Recall that for a map $f: \mathbf{S}^n \to \mathbf{S}^n$, $\deg(f)$ is defined to be that integer such that $f_*: H_n(\mathbf{S}^n) \to H_n(\mathbf{S}^n) \approx \mathbf{Z}$ is multiplication by $\deg(f)$.

5.12. Theorem. *The function* $\deg: \pi_n(\mathbf{S}^n) \to \mathbf{Z}$ *is an isomorphism.*

PROOF. This was proved by Thom–Pontryagin Theory in Corollary 16.4 of Chapter II. An alternative proof of this is given by Theorem 7.4 of Chapter IV which implies that deg is a homomorphism, together with Lemma 11.13 of Chapter V which shows that the kernel is zero (and which is independent of the remainder of that section). □

PROBLEMS

1. Let X be the "figure 8" space embedded in $\mathbf{S}^2$. Find $\pi_2(\mathbf{S}^2, X)$.

2. Let M be the quotient space of $I \times \mathbf{D}^2$ obtained by the identification $(0, z) = (1, \bar{z})$, where the disk is regarded as the unit disk in $\mathbf{C}$. Denote the boundary (a Klein bottle) of this 3-manifold by K. Find $\pi_2(M, K)$.

3. If $A \subset X$ and $\pi_1(A) \to \pi_1(X)$ is a monomorphism, show that $\pi_2(X, A)$ is abelian.

6. Fiber Spaces

The notion of a fibration is a generalization of that of a fiber bundle and is of central importance in homotopy theory. We study it here. One of the major results is an exact sequence (Theorem 6.7) of homotopy groups which is crucial for many calculations.

6.1. Definition. A map $p\colon Y \to B$ has the *homotopy lifting property* with respect to the space X if the following commutative diagram can always be completed as indicated:

$$\begin{array}{ccc} X \times \{0\} & \longrightarrow & Y \\ \downarrow & \nearrow & \downarrow p \\ X \times I & \longrightarrow & B. \end{array}$$

6.2. Definition. A map $p\colon Y \to B$ is a *Hurewicz fiber space* if it has the homotopy lifting property with respect to all spaces X. It is a *Serre fibration*, or simply *fibration*, if it has the homotopy lifting property with respect to $X = \mathbf{D}^n$, for all n.

The simplest example of a fibration is the projection $Y = B \times F \to B$, since, given the two maps $f\colon X \times I \to B$ and $g\colon X \times \{0\} \to B \times F$ such that $f(x,0) = pg(x,0)$, the diagram can be completed by the lift $f'(x,t) = f(x,t) \times qg(x,0)$ of f, where $q\colon B \times F \to F$ is the projection.

A common notation for a fibration p with "fiber" $F = p^{-1}(*)$, is $F \to Y \xrightarrow{p} B$.

6.3. Proposition. *If $p\colon Y \to B$ is a fibration and (K, L) is a CW pair (a CW-complex K and subcomplex L), then the following commutative diagram can be completed as indicated:*

$$\begin{array}{ccc} K \times \{0\} \cup L \times I & \longrightarrow & Y \\ \big\downarrow & \nearrow & \big\downarrow p \\ K \times I & \longrightarrow & B. \end{array}$$

PROOF. By induction on the skeletons of $K - L$ we need only prove this for the case

$$\begin{array}{ccc} \mathbf{D}^n \times \{0\} \cup \mathbf{S}^{n-1} \times I & \longrightarrow & Y \\ \big\downarrow & \nearrow & \big\downarrow p \\ \mathbf{D}^n \times I & \longrightarrow & B. \end{array}$$

But the left-hand side of this diagram, as a pair, is homeomorphic to the pair $(\mathbf{D}^n \times I, \mathbf{D}^n \times \{0\})$, so the completion of the diagram is direct from Definition 6.2. (See Figure VII-6.) □

6.4. Theorem. *Let $p: Y \to B$ be a fibration and (K, L) a CW pair such that L is a strong deformation retract of K. Then the following commutative diagram can be completed as indicated:*

$$\begin{array}{ccc} L & \xrightarrow{g} & Y \\ \big\downarrow & \nearrow & \big\downarrow p \\ K & \xrightarrow[f]{} & B. \end{array}$$

PROOF. Consider the commutative diagram

$$\begin{array}{ccccccc} L \times \{0\} & \longrightarrow & L \times I \cup K \times \{1\} & \xrightarrow{\text{proj} \cup r} & L & \xrightarrow{g} & Y \\ \big\downarrow & & \big\downarrow & \overset{\phi}{\nearrow} & \big\downarrow & & \big\downarrow p \\ K \times \{0\} & \longrightarrow & K \times I & \xrightarrow[F]{} & K & \xrightarrow[f]{} & B \end{array}$$

where F is the hypothesized deformation. Thus $F(k, 0) = k$, $F(k, 1) \in L$ and $F(l, t) = l$, for all $k \in K$, $l \in L$ and $t \in I$. The map ϕ exists by Proposition 6.3. We

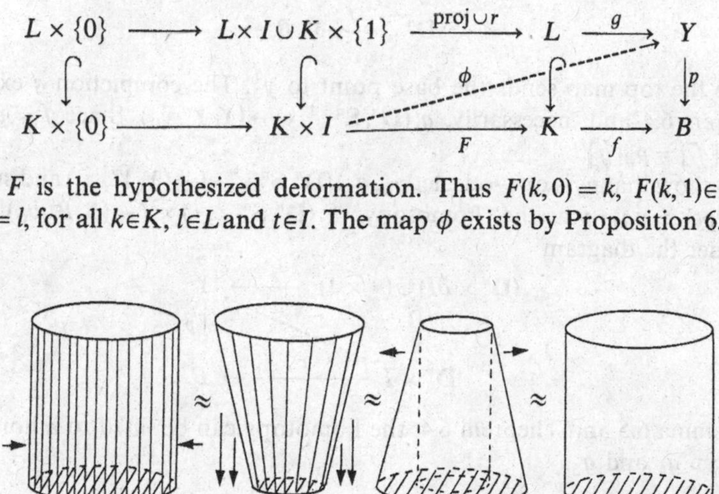

Figure VII-6. A homeomorphism of pairs.

have

$$p\circ\phi(k,0)=f(F(k,0))=f(k)$$

and

$$\phi(l,0)=(g\circ\text{proj})(l,0)=g(l).$$

Hence the restriction of ϕ to $K\times\{0\}$ gives the desired completion. □

For example, K could be $\mathbf{D}^n$ and L a single point in $\mathbf{D}^n$, and then the theorem implies that a map of a disk to the base space can always be lifted with the lift already specified at one point of the disk.

6.5. Lemma. *If $p\in\mathbf{S}^{n-1}$ then $\mathbf{D}^n\times\partial I\cup\{p\}\times I$ is a strong deformation retract of $\mathbf{D}^n\times I$.*

PROOF. This follows from Problem 3 of Section 11 of Chapter IV. □

6.6. Theorem. *Let $p\colon Y\to B$ be a fibration, and suppose $p(y_0)=b_0$. Suppose that $b_0\in B'\subset B$ and put $Y'=p^{-1}(B')$. Then*

$$p_\#\colon\pi_n(Y,Y',y_0)\to\pi_n(B,B',b_0)$$

is an isomorphism for all n.

PROOF. To show that $p_\#$ is onto let $f\colon(\mathbf{D}^n,\mathbf{S}^{n-1},*)\to(B,B',b_0)$. Apply Theorem 6.4 to the diagram

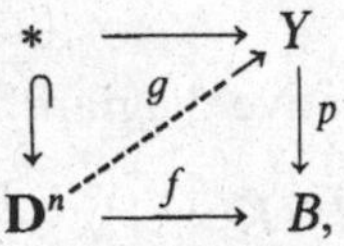

where the top map sends the base point to y_0. The completion g exists by Theorem 6.4 and, necessarily, $g\colon(\mathbf{D}^n,\mathbf{S}^{n-1},*)\to(Y,Y',y_0)$. Since $f=p\circ g$ we have $[f]=p_\#[g]$.

To show that $p_\#$ is one–one, let $g_0,g_1\colon(\mathbf{D}^n,\mathbf{S}^{n-1},*)\to(Y,Y',y_0)$ and suppose that $pg_0\simeq pg_1$ via the homotopy $F\colon(\mathbf{D}^n,\mathbf{S}^{n-1},*)\times I\to(B,B',b_0)$. Then consider the diagram

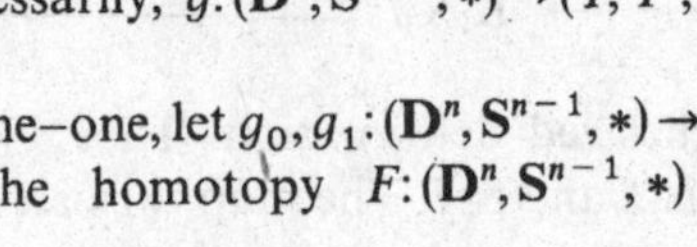

By Lemma 6.5 and Theorem 6.4, the homotopy can be lifted to a homotopy between g_1 and g_2. □

Now consider the homotopy sequence of the pair (Y,F) where $p\colon Y\to B$ is a fibration and $F=p^{-1}(*)$ is the fiber. The term $\pi_n(Y,F)$ may be replaced, by Theorem 6.6, with $\pi_n(B)$. The resulting homomorphism $\pi_n(Y)\to\pi_n(B)$ is clearly induced by p and so will continue to be denoted by $p_\#$. The resulting homomorphism

$$\partial_\# \circ p_\#^{-1}: \pi_n(B) \to \pi_{n-1}(F)$$

can be described as follows: Let $f: \mathbf{S}^n \to B$ represent an element of $\pi_n(B)$. Compose it with the collapsing map $\mathbf{D}^n \to \mathbf{S}^n$ to give a map $\mathbf{D}^n \to B$, then lift it to a map into Y, with base points corresponding. This represents $p_\#^{-1}([f]) \in \pi_n(Y, F)$. Restricting this map to $\mathbf{S}^{n-1}$ gives a map $\mathbf{S}^{n-1} \to F$ representing an element of $\pi_{n-1}(F)$. It is reasonable to still call this homomorphism $\partial_\#: \pi_n(B) \to \pi_{n-1}(F)$. Thus we have:

6.7. Theorem. *If $p: Y \to B$ is a fibration and if $y_0 \in Y$, $b_0 = p(y_0)$, and $F = p^{-1}(b_0)$, then taking y_0 as the base point of Y and of F and b_0 as the base point of B, we have the exact sequence:*

$$\cdots \longrightarrow \pi_n(F) \xrightarrow{i_\#} \pi_n(Y) \xrightarrow{p_\#} \pi_n(B) \xrightarrow{\partial_\#} \pi_{n-1}(F) \longrightarrow \cdots$$

$$\cdots \longrightarrow \pi_1(Y) \longrightarrow \pi_1(B) \longrightarrow \pi_0(F) \longrightarrow \pi_0(Y) \longrightarrow \pi_0(B). \qquad \square$$

For example, a covering map is clearly a fibration. In that case the fiber F is discrete and so $\pi_n(F) = 0$ for $n \geq 1$. Thus the exact sequence implies that $p_\#: \pi_n(Y) \to \pi_n(B)$ is an isomorphism for $n \geq 2$ and a monomorphism for $n = 1$. Of course, it is easy to deduce this directly. In particular $\pi_n(\mathbf{S}^1) \approx \pi_n(\mathbf{R}^1) = 0$ for $n \geq 2$.

Later we will show that the Hopf map $\mathbf{S}^3 \to \mathbf{S}^2$ is a fibration with fiber $\mathbf{S}^1$. For this example the exact sequence looks like

$$\pi_n(\mathbf{S}^1) \to \pi_n(\mathbf{S}^3) \to \pi_n(\mathbf{S}^2) \to \pi_{n-1}(\mathbf{S}^1).$$

The groups on the ends are trivial for $n \geq 3$. Therefore $\pi_n(\mathbf{S}^3) \approx \pi_n(\mathbf{S}^2)$ for $n \geq 3$. In particular, we deduce the interesting fact that $\pi_3(\mathbf{S}^2) \approx \mathbf{Z}$. In fact, by looking closely at the homomorphism we see that the Hopf map itself represents the generator of $\pi_3(\mathbf{S}^2)$.

6.8. Theorem. *If $p: Y \to B$ is a fibration with fiber F and if there is a homotopy $H: F \times I \to Y$ between the inclusion and a constant map, then the sequence*

$$0 \to \pi_n(Y) \to \pi_n(B) \to \pi_{n-1}(F) \to 0$$

is split exact for $n \geq 2$. Thus $\pi_n(B) \approx \pi_n(Y) \times \pi_{n-1}(F)$. In particular, $\pi_1(F)$ is abelian.

PROOF. For any map $f: \mathbf{S}^{n-1} \to F \subset Y$, the homotopy gives a construction of an extension $f': \mathbf{D}^n \to Y$. The composition $p \circ f': (\mathbf{D}^n, \mathbf{S}^{n-1}) \to (B, *)$ represents an element $\lambda([f]) \in \pi_n(B)$. It is clear that $\lambda: \pi_{n-1}(F) \to \pi_n(B)$ is a homomorphism and that $\partial_\# \circ \lambda = 1$. $\square$

We will show later (in Section 8) that, besides the Hopf fibration $\mathbf{S}^1 \to \mathbf{S}^3 \to \mathbf{S}^2$, there are Hopf fibrations $\mathbf{S}^3 \to \mathbf{S}^7 \to \mathbf{S}^4$ and $\mathbf{S}^7 \to \mathbf{S}^{15} \to \mathbf{S}^8$. In all these, the fiber is homotopically trivial in the total space. Thus Theorem 6.8 provides the isomorphisms

$$\begin{aligned}\pi_n(\mathbf{S}^2) &\approx \pi_n(\mathbf{S}^3) \oplus \pi_{n-1}(\mathbf{S}^1),\\ \pi_n(\mathbf{S}^4) &\approx \pi_n(\mathbf{S}^7) \oplus \pi_{n-1}(\mathbf{S}^3),\\ \pi_n(\mathbf{S}^8) &\approx \pi_n(\mathbf{S}^{15}) \oplus \pi_{n-1}(\mathbf{S}^7),\end{aligned}$$

for $n \geq 2$. (These are written additively since these groups are all abelian.)

6.9. Theorem. *If $p\colon Y \to B$ is a fibration with fiber F and if F is a retract of Y then the sequence*

$$0 \to \pi_n(F) \to \pi_n(Y) \to \pi_n(B) \to 0$$

is split exact for $n \geq 1$. Thus $\pi_n(Y) \approx \pi_n(B) \times \pi_n(F)$.

Proof. The retraction $r\colon Y \to F$ gives a homomorphism $r_\#\colon \pi_n(Y) \to \pi_n(F)$ with $r_\# \circ i_\# = 1$. □

6.10. Corollary. $\pi_n(X \times Y) \approx \pi_n(X) \times \pi_n(Y)$ *via the projections.* □

6.11. Theorem. *For a map $p\colon Y \to B$, the property of being a fibration is a local property in B.*

Proof. The theorem means that p is a fibration if each point $b \in B$ has a neighborhood U such that $p\colon p^{-1}(U) \to U$ is a fibration.

Suppose we are given the commutative diagram

$$\begin{array}{ccc} \mathbf{D}^n \times \{0\} & \xrightarrow{\;f\;} & Y \\ \big\downarrow & \nearrow & \big\downarrow p \\ \mathbf{D}^n \times I & \xrightarrow[F]{} & B. \end{array}$$

Take a triangulation of $\mathbf{D}^n$, e.g., regard $\mathbf{D}^n$ as $|\Delta_n|$. Pull back the covering of B by the open sets U to a covering of $\mathbf{D}^n$. We can then subdivide $\mathbf{D}^n$ sufficiently finely and take an integer k sufficiently large so that for all $0 \leq i < k$, and all simplices σ in the subdivision of $\mathbf{D}^n$, we have that $F(\sigma \times [i/k, (i+1)/k])$ is contained in some open set U over which p is a fibration.

Let K be the subdivided $\mathbf{D}^n$. Suppose we have lifted the homotopy F to F', keeping the diagram commutative, over the subspace $K^{(j)} \times [0, (i+1)/k] \cup K \times [0, i/k]$ for some j and i. If σ is a $(j+1)$-cell of K and f_σ is its characteristic map, then we get, by composition, the commutative diagram

$$\begin{array}{ccc} (\mathbf{S}^j \times [i/k, (i+1)/k]) \cup (\mathbf{D}^{j+1} \times \{i/k\}) & \xrightarrow{\;F' \circ f_\sigma\;} & p^{-1}(U) \\ \big\downarrow & \nearrow & \big\downarrow p \\ \mathbf{D}^{j+1} \times [i/k, (i+1)/k] & \xrightarrow[F \circ f_\sigma]{} & U \end{array}$$

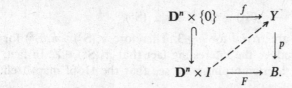

where U is an open set containing σ and over which p is a fibration. This diagram can be completed by Proposition 6.3. These lifts fit together to give a lift over $K^{(j+1)} \times [0, (i+1)/k] \cup (K \times [0, i/k])$. By induction on j, and because of the weak topology on $K \times I$ we finally get a lift over $K \times [0, (i+1)/k]$. This is just our assumption with -1 replacing j and $i+1$ replacing i. An induction on i finishes the proof. $\qquad\square$

6.12. Corollary. *A bundle projection is a fibration.* $\qquad\square$

We finish this section by giving some constructions of important fibrations.

6.13. Theorem. *Let X be locally compact and $A \subset X$ closed. If the inclusion of A in X is a cofibration then the restriction map $Y^X \to Y^A$ is a Hurewicz fibration for all Y.*

PROOF. The commutative diagram

$$
\begin{array}{ccc}
W \times \{0\} & \longrightarrow & Y^X \\
\cap \big\downarrow & \nearrow{}^{\phi} & \big\downarrow{}^{p} \\
W \times I & \longrightarrow & Y^A
\end{array}
$$

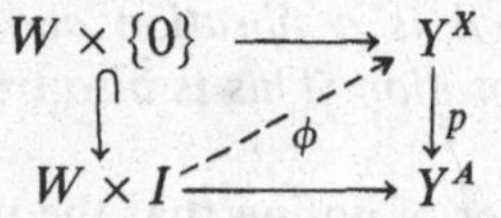

corresponds, via the exponential law, to the commutative diagram

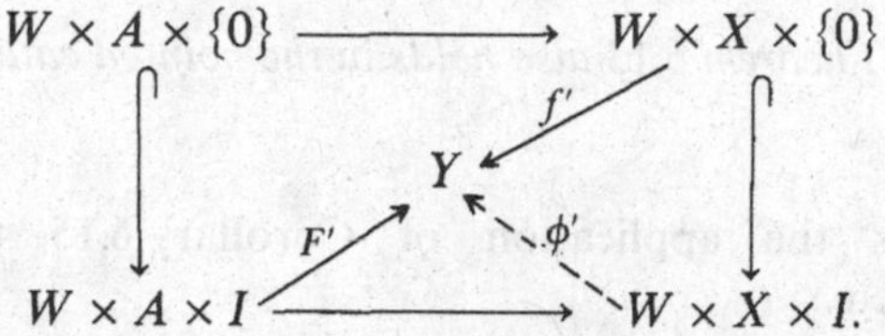

Thus it suffices to show that the inclusion $W \times A \hookrightarrow W \times X$ is a cofibration. By Theorem 1.3 this is equivalent to showing that $(W \times A \times I) \cup (W \times X \times \{0\})$ is a retract of $W \times X \times I$. But the product of the identity on W and a retraction of $X \times I$ to $A \times I \cup X \times \{0\}$ does this. $\qquad\square$

Our next result in this direction concerns "pullbacks." If $p: Y \to B$ and $f: X \to B$ are maps then the "pullback" p' of p via f is the projection $p': f^*Y \to X$ where

$$\boxed{f^*Y = \{(x, y) \in X \times Y \mid f(x) = p(y)\}.}$$

It is easy to see that the pullback satisfies the universal property that, given any commutative diagram

$$
\begin{array}{ccc}
W & \longrightarrow & Y \\
\big\downarrow & & \big\downarrow{}^{p} \\
X & \xrightarrow{f} & B
\end{array}
$$

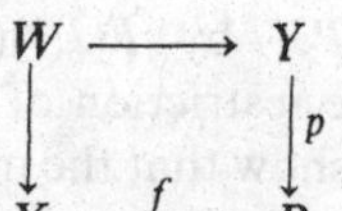

there exists a unique map $W \to f^*Y$ compatible with the maps to X and Y.

6.14. Theorem. *If $p: Y \to B$ is a fibration and $f: X \to B$ is any map then $p': f^*Y \to X$ is a fibration. This is also true for Hurewicz fibrations.*

PROOF. Consider the commutative diagram (with $W = \mathbf{D}^n$ in the case of Serre fibrations):

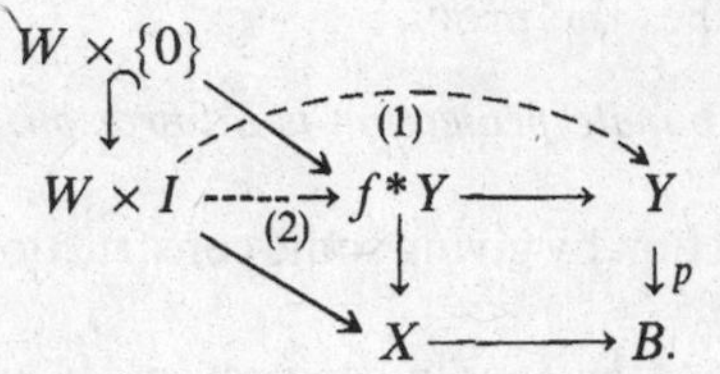

The prospective map marked (1) exists, maintaining commutativity, since p is a fibration. Then the map marked (2) exists by the universal property of pullbacks. □

6.15. Corollary. *If $p: Y \to B$ is a fibration and $B' \subset B$ then the restriction $p': p^{-1}(B') \to B'$ of p is a fibration. This is also true for Hurewicz fibrations.*

PROOF. This is just a matter of noting that the pullback to a subspace is the same as the restriction to it. □

6.16. Corollary. *Theorem* 6.13 *also holds in the pointed category and for pairs, etc.*

PROOF. This is the application of Corollary 6.15 to the inclusion $(Y, *)^{(A,*)} \subset Y^A$, etc. □

For a pointed space X, we define the "path space" of X to be $PX = (X, *)^{(I,0)}$. Similarly, we can regard the loop space ΩX as $(X, *)^{(I,\{0,1\})} \approx (X, *)^{(\mathbf{S}^1,*)}$.

6.17. Corollary. *For a pointed space X the map $PX \to X$ taking λ to $\lambda(1)$ is a fibration with fiber ΩX. This is called the "path-loop fibration of X."*

PROOF. In the pointed category this is the restriction map $X^I \to X^{\partial I}$, which is a fibration by Theorem 6.13, and the latter space, the base, is homeomorphic to X via the evaluation at 1. □

6.18. Proposition. *In the pointed category, the path space PX is contractible.*

PROOF. Define $F: PX \times I \to PX$ by $F(\lambda, t)(s) = \lambda(ts)$. To see that this is continuous, note that it is the restriction of the same thing on $X^I \times I \to X^I$. By Theorem 2.4, it suffices to show that the map $X^I \times I \times I \to X$ taking (λ, t, s) to $\lambda(ts)$ is continuous, but that is clear. □

6.19. Corollary. $\partial_{\#}: \pi_i(X) \xrightarrow{\approx} \pi_{i-1}(\Omega X)$.

Proof. Of course, we already had this. Here, it is a consequence of the exact sequence of the fibration $\Omega X \to PX \to X$ and the fact that the homotopy groups of PX are trivial since it is contractible. □

Problems

1. Let M^3 be the 3-manifold defined as the quotient space of $I \times \mathbf{S}^2$ by the identification $\{0\} \times \{x\} \sim \{1\} \times \{Tx\}$, where $T: \mathbf{S}^2 \to \mathbf{S}^2$ is a reflection through a plane in $\mathbf{R}^3$. Find $\pi_1(M)$ and $\pi_2(M)$.

2. ⌖ Let $f: X \to Y$ be a map and M_f its mapping cylinder. Let $k: X \times I \to M_f$ be the canonical map. Put
$$P_f = (M_f, Y, X)^{(I, \{0\}, \{1\})}.$$
Define maps $s: X \to P_f$, $p: P_f \to X$, and $\pi: P_f \to Y$ by
$$s(x)(t) = k(x, t),$$
$$p(\lambda) = \lambda(1),$$
$$\pi(\lambda) = \lambda(0).$$
Show that π and p are fibrations. Also show that $p \circ s = 1_X$ and $s \circ p \simeq 1_{P_f}$ via a homotopy that preserves each fiber of p (called a "fiber homotopy"). Also show that $f = \pi \circ s$. Thus any map $f: X \to Y$ is homotopy equivalent to a fibration $\pi: P_f \to Y$.

3. Find $\pi_i(\mathbf{CP}^n)$ for $1 \le i \le 2n + 1$.

4. Let $p: X \to Y$ be a fibration and suppose that $s: X \to W$ and $t: W \to Y$ are maps with $p = t \circ s$. If t is injective, show that s is a fibration. If s is onto, show that t is a fibration.

5. Consider the path-loop fibration $p: PX \to X$. Let W be the set of homotopy classes rel ∂I of paths $I \to X$. Give W the quotient topology from the canonical map $PX \to W$. Rederive Theorem 8.4 of Chapter III (existence of universal covering spaces) in this context.

6. Let X be the "figure 8" space embedded in $\mathbf{S}^2$. Find $\pi_2(\mathbf{S}^2/X)$. Compare Problem 1 of Section 5.

7. If X is well-pointed, show that PX and ΩX are well-pointed.

8. Show that $\pi_2(\mathbf{S}^1 \vee \mathbf{S}^2)$ is not finitely generated.

7. Free Homotopy

In this section we will study "free homotopies" $\mathbf{S}^n \times I \to X$, i.e., homotopies not respecting the base point. We will also derive the effect that changing the base point has on homotopy groups.

Let $*$ be a base point in $\mathbf{S}^n$ and consider the space $X^{\mathbf{S}^n}$ of *nonpointed* maps from $\mathbf{S}^n$ to X. The map $p: X^{\mathbf{S}^n} \to X$, defined by $p(f) = f(*)$, is a fibration since it is just restriction to $\{*\}$.

7.1. Definition. $f \in X^{\mathbf{S}^n}$ is *freely homotopic* to $g \in X^{\mathbf{S}^n}$ along $\sigma \in X^I$, $(f \simeq_\sigma g)$ if there exists a path $\bar{\sigma}: I \to X^{\mathbf{S}^n}$ such that $p \circ \bar{\sigma} = \sigma$ and $\bar{\sigma}(0) = f$ and $\bar{\sigma}(1) = g$; that is, if there is a homotopy $I \times \mathbf{S}^n \to X$ between f and g such that on $I \times \{*\} \to X$ it is σ.

We note some, mostly elementary, properties of this definition:

(1) $\sigma: I \to X,\ g \in X^{\mathbf{S}^n} \ni g(*) = \sigma(1) \Rightarrow \exists f \ni f \simeq_\sigma g.$

To prove this, just lift the path σ to $\bar{\sigma}: I \to X^{\mathbf{S}^n}$ so that $\bar{\sigma}(1) = g$, and let $f = \bar{\sigma}(0)$.

(2) $f \simeq_\sigma g \simeq_\tau h \Rightarrow f \simeq_{\sigma * \tau} h.$
(3) $f \simeq_\sigma g \Rightarrow g \simeq_{\sigma^{-1}} f.$
(4) $f \simeq g$ rel $* \Rightarrow f \simeq_e g$ where e is the constant path at $p(f)$.
(5) $f \simeq_\sigma g, \sigma \simeq \tau$ rel $\partial I \Rightarrow f \simeq_\tau g.$

To see this, use Theorem 6.4 to complete the following diagram:

$$\begin{array}{ccc} I \times \{0\} \cup \partial I \times I & \xrightarrow{\bar{\sigma} \times \{0\} \cup \{0\} \times f \cup \{1\} \times g} & X^{\mathbf{S}^n} \\ \big\downarrow & \nearrow & \big\downarrow \\ I \times I & \xrightarrow{F} & X. \end{array}$$

(6) $f \simeq_\sigma g, f' \simeq_\sigma g \Rightarrow f \simeq f'$ rel $*$.

This follows from (2) through (5).

(7) $f \simeq_\sigma g, f' \simeq f$ rel $*, g' \simeq g$ rel $*, \sigma' \simeq \sigma$ rel $\partial I \Rightarrow f' \simeq_{\sigma'} g'.$

This follows from several of the preceding facts.

(8) $\phi: X \to Y, f \simeq_\sigma g \Rightarrow \phi \circ f \simeq_{\phi \circ \sigma} \phi \circ g.$
(9) $f \simeq_\sigma f', g \simeq_\sigma g' \Rightarrow f * g \simeq_\sigma f' * g'.$

To see this, consider the diagram induced by the comultiplication $\gamma: \mathbf{S}^n \to \mathbf{S}^n \vee \mathbf{S}^n$:

$$\begin{array}{ccccc} X^{\mathbf{S}^n} \times X^{\mathbf{S}^n} & \supset & X^{\mathbf{S}^n \vee \mathbf{S}^n} & \xrightarrow{\mu} & X^{\mathbf{S}^n} \\ \big\downarrow & & \big\downarrow & & \big\downarrow \\ X \times X & \supset & X & \xrightarrow{=} & X. \end{array}$$

The map on top takes (f, g) to $f * g$. The vertical maps are evaluation at the base point. The inclusion on the lower left is as the diagonal. The free homotopy from f to f' is just a path σ' from f to f' over σ. The other free homotopy is a path σ'' from g to g' over σ. This pair (σ', σ'') of paths gives a path in $X^{\mathbf{S}^n \vee \mathbf{S}^n}$ over (σ, σ). Thus $\mu(\sigma', \sigma'')$ is a path in $X^{\mathbf{S}^n}$ from $f * g$ to $f' * g'$, proving (9).

These facts imply the following:

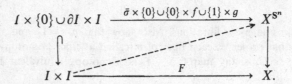

7.2. Theorem. *For each path $\sigma: I \to X$, there is an isomorphism*

$$\sigma_n: \pi_n(X, \sigma(1)) \xrightarrow{\approx} \pi_n(X, \sigma(0))$$

such that

(a) $[f] = \sigma_n[g] \Leftrightarrow f \simeq_\sigma g$;
(b) $\sigma \simeq \tau \text{ rel } \partial I \Rightarrow \sigma_n = \tau_n$;
(c) $\sigma(1) = \tau(0) \Rightarrow (\sigma * \tau)_n = \sigma_n \circ \tau_n$;
(d) σ *constant* $\Rightarrow \sigma_n = 1$; *and*
(e) (*naturality*) $\phi: X \to Y$ *and* $\tau = \phi \circ \sigma \Rightarrow$ *the following diagram commutes:*

$$\begin{array}{ccc} \pi_n(X, \sigma(1)) & \xrightarrow{\sigma_n} & \pi_n(X, \sigma(0)) \\ \downarrow \phi_\# & & \downarrow \phi_\# \\ \pi_n(Y, \tau(1)) & \xrightarrow{\tau_n} & \pi_n(Y, \tau(0)). \end{array}$$ □

7.3. Corollary. *If $f: (\mathbf{S}^n, *) \to (X, x_0)$, then $[f] = 0$ in $\pi_n(X, x_0) \Leftrightarrow f$ is freely homotopic to a constant map.*

PROOF. If $f \simeq_\sigma e_{x_0}$ then $[f] = \sigma_n([e_{x_0}]) = \sigma_n(0) = 0$. □

7.4. Corollary. *The group $\pi_1(X, x_0)$ acts as a group of automorphisms on $\pi_n(X, x_0)$. Moreover, for $n = 1$, the automorphisms are inner. More precisely, for $\alpha \in \pi_1(X)$, $\beta \in \pi_n(X)$, let $\alpha(\beta) = \sigma_n(\beta)$ where σ is a loop representing α. Then:*

$$\alpha(\beta \pm \gamma) = \alpha(\beta) \pm \alpha(\gamma),$$
$$(\alpha\alpha')(\beta) = \alpha(\alpha'(\beta)),$$
$$1(\beta) = \beta.$$

Also, for $n = 1$, $\alpha(\beta) = \alpha\beta\alpha^{-1}$. □

7.5. Definition. A space X is called *n-simple* if $\pi_1(X, x_0)$ acts trivially on $\pi_n(X, x_0)$ for all $x_0 \in X$. It is called *simple* if it is n-simple for all $n \geq 1$.

For X arcwise connected, one needs the condition of Definition 7.5 only for one x_0. Evidently, X being 1-simple is the same as $\pi_1(X, x_0)$ being abelian for all x_0.

7.6. Example. Let $n \geq 2$ and let X be the nonorientable $\mathbf{S}^n$-bundle over $\mathbf{S}^1$. That is,

$$X = (\mathbf{S}^n \times I)/\sim \; \to I/\partial I = \mathbf{S}^1,$$

where $\sim$ is the equivalence relation $(x, 0) \sim (Tx, 1)$ for all $x \in \mathbf{S}^n$, where T is the reflection of $\mathbf{S}^n$ through $\mathbf{S}^{n-1}$ (or any homeomorphism of degree -1). The exact sequence for this fibration yields the exact sequences $0 \to \pi_1(X) \to \pi_1(\mathbf{S}^1) \to 0$ and $0 \to \pi_n(\mathbf{S}^n) \to \pi_n(X) \to 0$, since $n \geq 2$. Thus $\pi_1(X) \approx \mathbf{Z}$

and $\pi_n(X) \approx \mathbf{Z}$. The image of $\{*\} \times I$ gives a loop in X representing the generator $\alpha \in \pi_1(X)$. The sphere $\mathbf{S}^n$ (the fiber) represents the generator $\beta \in \pi_n(X)$, and it "travels" along this loop and comes back as the map $T: \mathbf{S}^n \to \mathbf{S}^n$. This shows that $\alpha(\beta) = -\beta$, so this space X is not n-simple.

Now we shall briefly discuss the case of the relative homotopy groups. Consider the map

$$p: (X, A)^{(\mathbf{D}^n, \partial \mathbf{D}^n)} \to A$$

which is the evaluation at the base point $* \in \partial \mathbf{D}^n$. To see that this is a fibration note that it is the composition $\pi \circ \eta'$ in the diagram

$$\begin{array}{ccc} (X, A)^{(\mathbf{D}^n, \partial \mathbf{D}^n)} & \longrightarrow & X^{\mathbf{D}^n} \\ \downarrow{\scriptstyle \eta'} & & \downarrow{\scriptstyle \eta} \\ A^{\partial \mathbf{D}^n} & \longrightarrow & X^{\partial \mathbf{D}^n} \\ \downarrow{\scriptstyle \pi} & & \\ A^{(*)} = A. & & \end{array}$$

The map η is the restriction and so is a fibration. The map η' is a pullback of η and so is a fibration. The map π is a restriction and so is a fibration. Finally, the composition $p = \pi \circ \eta'$ of fibrations is a fibration.

Thus, just as in the absolute case, paths in A operate on $\pi_n(X, A)$ with base points running along the paths. The following consequence is clear.

7.7. Theorem. *$\pi_1(A, *)$ acts as a group of automorphisms on $\pi_n(X, A, *)$, $\pi_n(A, *)$ and $\pi_n(X, *)$, the latter via $i_\#: \pi_1(A, *) \to \pi_1(X, *)$. These actions commute with $i_\#$, $j_\#$, and $\partial_\#$.* □

There is more to say in the case $n = 2$:

7.8. Theorem. *Consider $\partial_\#: \pi_2(X, A, *) \to \pi_1(A, *)$. For $\alpha, \beta \in \pi_2(X, A, *)$ we have*

$$(\partial_\#(\alpha))(\beta) = \alpha\beta\alpha^{-1}.$$

Proof. A representative $\mathbf{D}^2 \to X$ of β can be taken so that everything maps into the base point except for a small disk, and this disk can be placed anywhere. Thus, we see that a representative of $\alpha\beta\alpha^{-1}$ can be taken as in the first part of Figure VII-7.

In the figure, σ represents $\partial_\#(\alpha)$. In the first part, the "pie slice" represents β. The parts marked α and α^{-1} are mirror images. The second part of the figure is the same map, and only the division lines have changed. The third part represents a stage of a homotopy; the map changes only inside the portion with the vertical rulings, and that part is constant along verticals. The fourth part is the final result of that homotopy, and the map on the left portion of the disk is constant along verticals. The fifth part of the figure represents a free homotopy along σ. The map changes only in the extreme

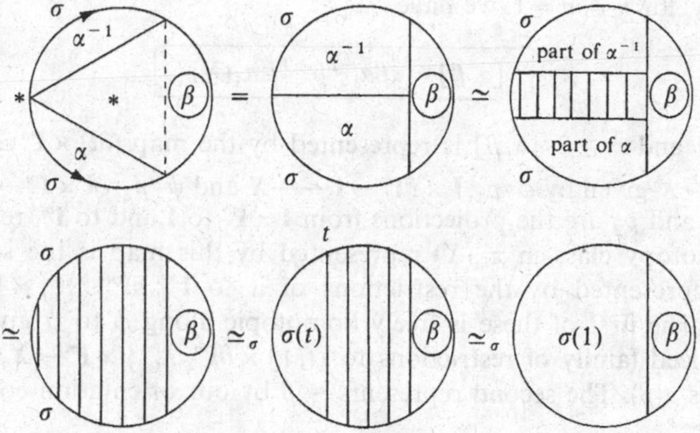

Figure VII-7. Relative action of the fundamental group.

left-hand portion of the disk, and there it is constant equal to $\sigma(t)$, where t is the parameter running from 0 at the leftmost point of the disk and 1 at the vertical separating the β part. The sixth and last part of the figure represents the final result of this free homotopy. It clearly represents β. Thus, the figure gives a free homotopy $\alpha\beta\alpha^{-1} \simeq_\sigma \beta$. Since σ represents $\partial_\#(\alpha)$, this shows that $(\partial_\#(\alpha))(\beta) = \alpha\beta\alpha^{-1}$. □

We conclude this section with a brief discussion of the (J.H.C.) "Whitehead product." This material is not used elsewhere in the book. From our conventions in this chapter, $\mathbf{S}^n$ is represented as the quotient space of $\mathbf{I}^n$ collapsing the boundary to a point. The collapsing map $f_{\sigma_n}: \mathbf{I}^n \to \mathbf{S}^n$ provides the characteristic map for a CW-complex structure on $\mathbf{S}^n$ with one n-cell σ_n. Regard $\mathbf{S}^n \times \mathbf{S}^m$ as the product complex as in Section 12 of Chapter IV. Then $f_{\sigma_n \times \sigma_m} = f_{\sigma_n} \times f_{\sigma_m}: \mathbf{I}^n \times \mathbf{I}^m \to \mathbf{S}^n \times \mathbf{S}^m$ is the characteristic map for the $(n+m)$-cell of the product. Let $g = f_{\partial(\sigma_n \times \sigma_m)}: \partial(\mathbf{I}^n \times \mathbf{I}^m) \to \mathbf{S}^n \vee \mathbf{S}^m$ denote the corresponding attaching map. Then g defines an element

$$[g] \in \pi_{n+m-1}(\mathbf{S}^n \vee \mathbf{S}^m).$$

If $\phi: \mathbf{S}^n \to X$ and $\psi: \mathbf{S}^m \to X$ are maps representing $\alpha = [\phi] \in \pi_n(X)$ and $\beta = [\psi] \in \pi_m(X)$, then there is the map

$$\phi \veebar \psi: \mathbf{S}^n \vee \mathbf{S}^m \to X$$

and so we get an element

$$\boxed{[\alpha, \beta] = (\phi \veebar \psi)_\#[g] \in \pi_{n+m-1}(X)}$$

called the "Whitehead product" of α and β.

Clearly, for $n=m=1$, we have

$$[\alpha,\beta]=\alpha\beta\alpha^{-1}\beta^{-1}\in\pi_1(X).$$

For $n=1$ and $m\geq 2$, $[\alpha,\beta]$ is represented by the map $(\partial\mathbf{I})\times\mathbf{I}^m\cup\mathbf{I}\times\partial\mathbf{I}^m=\partial(\mathbf{I}\times\mathbf{I}^m)\to X$ given by $\phi\circ p_1:\mathbf{I}\times\partial\mathbf{I}^m\to\mathbf{I}\xrightarrow{\phi}X$ and $\psi\circ p_2:\partial\mathbf{I}\times\mathbf{I}^m\to\mathbf{I}^m\xrightarrow{\psi}X$, where p_1 and p_2 are the projections from $\mathbf{I}\times\mathbf{I}^m$ to $\mathbf{I}$ and to $\mathbf{I}^m$, respectively. The homotopy class in $\pi_m(X)$ represented by this map is the sum of the classes represented by the restrictions of it to $\mathbf{I}\times\partial\mathbf{I}^m\cup\{1\}\times\mathbf{I}^m$ and to $\{0\}\times\mathbf{I}^m$. The first of these is freely homotopic along ϕ to ψ given by the parametrized family of restrictions to $[t,1]\times\partial\mathbf{I}^m\cup\{1\}\times\mathbf{I}^m\to X$, and so it represents $\alpha(\beta)$. The second represents $-\beta$ by our orientation conventions. Therefore

$$[\alpha,\beta]=\alpha(\beta)-\beta\in\pi_m(X)$$

for $\alpha\in\pi_1(X)$ and $\beta\in\pi_m(X)$, $m\geq 2$. Thus X is m-simple$\Leftrightarrow[\alpha,\beta]=0$ for all $\alpha\in\pi_1(X)$ and $\beta\in\pi_m(X)$.

Clearly

$$[\alpha,\beta]=(-1)^{nm}[\beta,\alpha]$$

when $n+m\geq 3$.

It is not hard to see that

$$[\alpha,\beta_1+\beta_2]=[\alpha,\beta_1]+[\alpha,\beta_2]$$

when $m\geq 2$.

It is also known that the Whitehead product satisfies the following "Jacobi identity" for $\alpha\in\pi_n(X),\beta\in\pi_m(X)$, and $\gamma\in\pi_p(X)$ for $n,m,p\geq 2$:

$$(-1)^{np}[[\alpha,\beta],\gamma]+(-1)^{nm}[[\beta,\gamma],\alpha]+(-1)^{mp}[[\gamma,\alpha],\beta]=0.$$

Problems

1. Let M^3 be the 3-manifold with boundary resulting from $I\times\mathbf{D}^2$ by the identification $\{1\}\times\{z\}=\{0\}\times\{\bar z\}$ and let K be its boundary, a Klein bottle. Completely describe the action of $\pi_1(K)$ on $\pi_2(M,K)$.

2. ◇ If $F\to X\to B$ is a fibration with F, X, and B arcwise connected, define an action of $\pi_1(B)$ on each $\pi_n(F)$ and derive as many properties of it you can.

3. Let $*\in A\subset X$. Define an action of $\pi_1(X)$ on the set $\pi_1(X,A)$, denoted by $(\alpha,\beta)\mapsto\alpha\beta$ of $\pi_1(X)\times\pi_1(X,A)\to\pi_1(X,A)$, such that:
(a) $\alpha(\alpha'\beta)=(\alpha\alpha')\beta$ for all $\alpha,\alpha'\in\pi_1(X),\beta\in\pi_1(X,A)$;
(b) $\partial_\#\beta=\partial_\#\beta'\Leftrightarrow\alpha\beta=\beta'$ for some $\alpha\in\pi_1(X)$;
(c) $\alpha\in\pi_1(X),\beta\in\pi_1(X,A)\Rightarrow(\alpha\beta=\beta\Leftrightarrow\alpha=1)$;

(d) $\alpha \in \pi_1(X) \Rightarrow \alpha 1 = j_{\#}\alpha$; and
(e) $\alpha \in \pi_1(X) \Rightarrow (\alpha 1 = 1 \Leftrightarrow (\alpha = i_{\#}(\gamma)$ for some $\gamma \in \pi_1(A)))$.
(Here we use the notation

$$\pi_1(A) \xrightarrow{i_\#} \pi_1(X) \xrightarrow{j_\#} \pi_1(X, A) \xrightarrow{\partial_\#} \pi_0(A).)$$

4. ✧ If $B \subset A \subset X$, derive the "exact sequence of the triple (X, A, B)":

$$\cdots \to \pi_n(A, B) \to \pi_n(X, B) \to \pi_n(X, A) \to \pi_{n-1}(A, B) \to \cdots \to \pi_1(X, A).$$

5. If X is an H-space then show that all Whitehead products vanish on X.
6. Show that $\mathbf{S}^n$ is an H-space $\Leftrightarrow$ the Whitehead product $[\iota_n, \iota_n] = 0$, where $\iota_n \in \pi_n(\mathbf{S}^n)$ is represented by the identity map.
7. Prove the identity $[\alpha, \beta_1 + \beta_2] = [\alpha, \beta_1] + [\alpha, \beta_2]$ for all $\alpha \in \pi_n(X)$ and $\beta_i \in \pi_m(X)$, $m \geq 2$.

8. Classical Groups and Associated Manifolds

Recall from Section 12 of Chapter II that a "Lie group" is a topological group G which is also a smooth manifold such that the product $G \times G \to G$, $(g, h) \mapsto gh$ and the inverse $G \to G, g \mapsto g^{-1}$ are smooth.

The classical groups are by far the most important examples, and the only ones we will deal with here.

Also recall from Section 12 of Chapter II that if G is a closed subgroup of $\mathbf{Gl}(n, \mathbf{C})$ then it is a Lie group, and, from Section 13 of Chapter II, that if $K \subset H \subset G$ are closed subgroups then the canonical map $G/K \to G/H$ is a fiber bundle map with fiber H/K.

In this section we are going to calculate a number of homotopy groups for spaces related to the classical groups.

8.1. Example. Consider the determinant homomorphism $\det: \mathbf{U}(n) \to \mathbf{S}^1$. Its kernel is $\mathbf{SU}(n)$ and so there is the fibration $\mathbf{SU}(n) \hookrightarrow \mathbf{U}(n) \to \mathbf{S}^1$.

If $A \in \mathbf{U}(n)$ then there is the unique decomposition $A = BC$ where $B \in \mathbf{SU}(n)$ and C is a diagonal matrix of the form

$$\begin{bmatrix} e^{i\theta} & 0 & 0 & \cdots & 0 \\ 0 & 1 & 0 & \cdots & 0 \\ 0 & 0 & 1 & \cdots & 0 \\ \cdots & \cdots & \cdots & \cdots & \cdots \\ 0 & 0 & 0 & \cdots & 1 \end{bmatrix}.$$

Thus $\mathbf{U}(n) \approx \mathbf{SU}(n) \times \mathbf{S}^1$ as a space, but not as a group.

Either from the fibration or the homeomorphism, one concludes that

$$\pi_i(\mathbf{U}(n)) \approx \pi_i(\mathbf{SU}(n)) \quad \text{for} \quad i \geq 2.$$

8.2. Example. Consider the "Grassmann manifold" $G_{n,k}$ of k-planes through the origin in $\mathbf{R}^n$. For $g \in \mathbf{O}(n)$ and $V \in G_{n,k}$, $g(V)$ is another k-plane. Any k-frame is taken into any other by some $g \in \mathbf{O}(n)$, so the same is also true of k-planes. Thus $\mathbf{O}(n)$ acts transitively on $G_{n,k}$. The isotropy group of the standard $\mathbf{R}^k \subset \mathbf{R}^n$ is clearly $\mathbf{O}(k) \times \mathbf{O}(n-k)$. It follows that

$$G_{n,k} \approx \mathbf{O}(n)/(\mathbf{O}(k) \times \mathbf{O}(n-k)).$$

Similar considerations apply to complex and quaternionic k-planes. The Grassmann manifold for $k=1$ is just projective n-space, in all three cases. Therefore, one gets the homeomorphisms

$$\mathbf{O}(n)/(\mathbf{O}(1) \times \mathbf{O}(n-1)) \approx \mathbf{RP}^{n-1},$$
$$\mathbf{U}(n)/(\mathbf{U}(1) \times \mathbf{U}(n-1)) \approx \mathbf{CP}^{n-1},$$
$$\mathbf{Sp}(n)/(\mathbf{Sp}(1) \times \mathbf{Sp}(n-1)) \approx \mathbf{QP}^{n-1}.$$

Recall the similar Stiefel manifold $V_{n,k}$ of k-frames in n-space. In Section 15 of Chapter I we had that $V_{n,k} \approx \mathbf{O}(n)/\mathbf{O}(n-k)$. Thus we have the fibration

$$\mathbf{O}(k) \to V_{n,k} \to G_{n,k}.$$

The case $k=1$ gives the fibrations

$$\mathbf{S}^0 \to \mathbf{S}^{n-1} \to \mathbf{RP}^{n-1},$$
$$\mathbf{S}^1 \to \mathbf{S}^{2n-1} \to \mathbf{CP}^{n-1},$$
$$\mathbf{S}^3 \to \mathbf{S}^{4n-1} \to \mathbf{QP}^{n-1}.$$

The case $n=2$ gives the Hopf fibrations

$$\mathbf{S}^0 \to \mathbf{S}^1 \to \mathbf{S}^1,$$
$$\mathbf{S}^1 \to \mathbf{S}^3 \to \mathbf{S}^2,$$
$$\mathbf{S}^3 \to \mathbf{S}^7 \to \mathbf{S}^4.$$

Since, in the fibration $\mathbf{S}^1 \to \mathbf{S}^{2n+1} \to \mathbf{CP}^n$, the fiber is null homotopic in the total space, we have the isomorphism

$$\pi_r(\mathbf{CP}^n) \approx \pi_r(\mathbf{S}^{2n+1}) \oplus \pi_{r-1}(\mathbf{S}^1).$$

This is $\mathbf{Z}$ for $r=2$ and $\pi_r(\mathbf{S}^{2n+1})$ for $r \neq 2$. In particular, $\pi_r(\mathbf{CP}^\infty) \approx \mathbf{Z}$ for $r=2$ and is trivial otherwise. The case $n=1$ and $r=3$ gives that $\pi_3(\mathbf{S}^2) \approx \mathbf{Z}$.

8.3. Theorem. *For $n \geq 3$, $\pi_{n+1}(\mathbf{S}^n) \approx \mathbf{Z}_2$, generated by a suspension of the Hopf map $\mathbf{S}^3 \to \mathbf{S}^2$.*

Proof. This proof requires two items from the optional Section 16 of Chapter II and Section 15 of Chapter VI. First, note that the Hopf fibration $\mathbf{S}^3 \to \mathbf{S}^2$ is the projection $\{(u,v) \in \mathbf{C} \times \mathbf{C} \mid |u|^2 + |v|^2 = 1\} \to \mathbf{CP}^1 = \{(u:v)\}$ taking nonhomogeneous coordinates to homogeneous ones. But $(u:v) = (uv^{-1}:1)$, so that the map can be considered as $(u,v) \mapsto uv^{-1} \in \mathbf{C}^+ \approx \mathbf{S}^2$, the one-point compactification of $\mathbf{C}$. Let $h(u,v)=uv^{-1}$ denote this (compactified) map. Let $g(u,v)=(\bar{u},\bar{v})$ of $\mathbf{S}^3\to\mathbf{S}^3$ and $r(u)=\bar{u}$ of $\mathbf{S}^2\to\mathbf{S}^2$. Then $h\circ g=r\circ h$. However, g is a reflection in a 2-plane in $\mathbf{C}\times\mathbf{C}$, which is the same as a rotation in the orthogonal plane through an angle π. Therefore $h\simeq h\circ g=r\circ h$. Note that r is a reflection in a plane of $\mathbf{S}^2$ and hence has degree -1.

Now consider the suspensions of h and r. We get $Sh\simeq Sr\circ Sh$. But Sr, having degree -1, is homotopic to any other map of degree -1, and so is homotopic to the reversal T of the suspension parameter. Thus we have that $Sh\simeq T\circ Sh$. But $T\circ Sh=Sh\circ T$, using T for the reversal of the suspension parameter in $S\mathbf{S}^3$ as well as in $S\mathbf{S}^2$. Hence $[Sh]=[T\circ Sh]=[Sh\circ T]=-[Sh]$ in $\pi_4(\mathbf{S}^3)$ by definition of the inverse in homotopy groups. Consequently, for $\alpha=[h]\in\pi_3(\mathbf{S}^2)$, and $S\alpha\in\pi_4(\mathbf{S}^3)$, we have that $2S\alpha=0$. But α generates $\pi_3(\mathbf{S}^2)$ because of the exact sequence

$$0=\pi_3(\mathbf{S}^1)\to\pi_3(\mathbf{S}^3)\xrightarrow{h_\#}\pi_3(\mathbf{S}^2)\to\pi_2(\mathbf{S}^1)=0$$

which shows that $\alpha=h_\#[1_{\mathbf{S}^3}]$. By Theorem 16.7 of Chapter II, $S\colon\pi_3(\mathbf{S}^2)\to\pi_4(\mathbf{S}^3)$ is onto and so $S\alpha$ generates $\pi_4(\mathbf{S}^3)$. Thus $\pi_4(\mathbf{S}^3)$ is either $\mathbf{Z}_2$ generated by $S\alpha$ or it is zero. By Corollary 15.4 of Chapter VI, it is not zero. By Theorem 16.7 of Chapter II, the suspensions give isomorphisms

$$\mathbf{Z}_2\approx\pi_4(\mathbf{S}^3)\xrightarrow{\approx}\pi_5(\mathbf{S}^4)\xrightarrow{\approx}\pi_6(\mathbf{S}^5)\xrightarrow{\approx}\cdots$$

and so we are finished. □

From the Hopf fibration $\mathbf{S}^1\to\mathbf{S}^3\to\mathbf{S}^2$ it follows that $\pi_n(\mathbf{S}^2)\approx\pi_n(\mathbf{S}^3)$ for all $n\geq 3$. Thus we have that $\pi_4(\mathbf{S}^2)\approx\pi_4(\mathbf{S}^3)\approx\mathbf{Z}_2$, and one sees easily that the nonzero element is the class of the composition $h\circ Sh$ of the Hopf map with its suspension. The suspensions of this are also nontrivial as we now show with the aid of the optional Section 15 of Chapter VI.

8.4. Proposition. *The element* $[S^{n-2}(h\circ Sh)]\in\pi_{n+2}(\mathbf{S}^n)$ *is nontrivial for all* $n\geq 2$.

Proof. Let $f=S^{n-1}h\colon\mathbf{S}^{n+2}\to\mathbf{S}^{n+1}$ and $g=S^{n-2}h\colon\mathbf{S}^{n+1}\to\mathbf{S}^n$. Suppose that $g\circ f\simeq *$. Then the map $g\colon\mathbf{S}^{n+1}\to\mathbf{S}^n$ extends to $\phi\colon C_f\to\mathbf{S}^n$. Note that the mapping cone C_ϕ of ϕ is a three cell complex

$$C_\phi=\mathbf{S}^n\cup_g\mathbf{D}^{n+2}\cup\mathbf{D}^{n+4}$$

and that the quotient space

$$C_\phi/\mathbf{S}^n=\mathbf{S}^{n+2}\cup_{Sf}\mathbf{D}^{n+4}$$

is the suspension $SC_f\simeq S^n\mathbf{CP}^2$. Also, the subcomplex $C_g=\mathbf{S}^n\cup_g\mathbf{D}^{n+2}$ of C_ϕ is homotopy equivalent to $S^{n-2}\mathbf{CP}^2$. By the results in Section 15 of Chapter VI we know that $\mathrm{Sq}^2\colon H^n(C_g;\mathbf{Z}_2)\to H^{n+2}(C_g;\mathbf{Z}_2)$ and $\mathrm{Sq}^2\colon H^{n+2}(SC_f;\mathbf{Z}_2)\to$

$H^{n+4}(SC_f; \mathbf{Z}_2)$ are isomorphisms. From the commutative diagram

$$\begin{array}{ccccc}
H^n(C_g;\mathbf{Z}_2) & \xrightarrow[\approx]{\mathrm{Sq}^2} & H^{n+2}(C_g;\mathbf{Z}_2) & & \\
\uparrow\approx & & \uparrow\approx & & \\
H^n(C_\phi;\mathbf{Z}_2) & \xrightarrow{\mathrm{Sq}^2} & H^{n+2}(C_\phi;\mathbf{Z}_2) & \xrightarrow{\mathrm{Sq}^2} & H^{n+4}(C_\phi;\mathbf{Z}_2) \\
 & & \uparrow\approx & & \uparrow\approx \\
 & & H^{n+2}(C_\phi/\mathbf{S}^n;\mathbf{Z}_2) & \xrightarrow[\approx]{\mathrm{Sq}^2} & H^{n+4}(C_\phi/\mathbf{S}^n;\mathbf{Z}_2)
\end{array}$$

it follows that $\mathrm{Sq}^2\mathrm{Sq}^2$ is nontrivial on $H^n(C_\phi; \mathbf{Z}_2)$. But there is the Adem relation $\mathrm{Sq}^2\mathrm{Sq}^2 = \mathrm{Sq}^3\mathrm{Sq}^1$ and the latter is zero on $H^n(C_\phi; \mathbf{Z}_2)$ since $H^{n+1}(C_\phi; \mathbf{Z}_2) = 0$. □

It is known that $\pi_{n+2}(\mathbf{S}^n) \approx \mathbf{Z}_2$ for $n \geq 2$ and so the elements given by Proposition 8.4 are the only nonzero classes in this group.

8.5. Example. In this example we compute the homotopy groups of the classical groups as far as we can do it at this stage of our knowledge. First, consider the fibration

$$\mathbf{O}(n-1) \to \mathbf{O}(n) \to \mathbf{S}^{n-1}$$

The exact sequence of this fibration looks like

$$\pi_{n-1}(\mathbf{S}^{n-1}) \to \pi_{n-2}(\mathbf{O}(n-1)) \to \pi_{n-2}(\mathbf{O}(n)) \to 0 \to \cdots.$$

It follows that

$$\pi_r(\mathbf{O}(n-1)) \to \pi_r(\mathbf{O}(n)) \text{ is an } \begin{cases} \text{isomorphism for } r < n-2, \\ \text{epimorphism for } r = n-2. \end{cases}$$

Similarly, the complex and quaternionic cases yield that

$$\pi_r(\mathbf{U}(n-1)) \to \pi_r(\mathbf{U}(n)) \text{ is an } \begin{cases} \text{isomorphism for } r < 2n-2, \\ \text{epimorphism for } r = 2n-2. \end{cases}$$

$$\pi_r(\mathbf{Sp}(n-1)) \to \pi_r(\mathbf{Sp}(n)) \text{ is an } \begin{cases} \text{isomorphism for } r < 4n-2, \\ \text{epimorphism for } r = 4n-2. \end{cases}$$

Thus, for r fixed and n increasing, these groups stabilize at some point. The stable values are denoted by $\pi_r(\mathbf{O})$, $\pi_r(\mathbf{U})$, and $\pi_r(\mathbf{Sp})$, respectively. (There are spaces **O**, **U**, and **Sp** which have these homotopy groups but this is not of concern to us here.) Therefore, we have

$$\begin{aligned}
\pi_r(\mathbf{O}) &= \pi_r(\mathbf{SO}) = \pi_r(\mathbf{O}(n)) && \text{for} \quad n \geq r+2, \\
\pi_r(\mathbf{U}) &= \pi_r(\mathbf{U}(n)) && \text{for} \quad n \geq [(r+2)/2], \\
\pi_r(\mathbf{Sp}) &= \pi_r(\mathbf{Sp}(n)) && \text{for} \quad n \geq [(r+2)/4].
\end{aligned}$$

One can also use the same method to calculate π_0 and the dimensions of these manifolds. We leave this to the reader, but will use the results.

Now consider the action of $\mathbf{Sp}(1)$, the unit quaternions, on the quaternions by conjugation. This preserves the norm and leaves the real axis fixed. Thus it operates on the orthogonal complement of the real axis. This gives a homomorphism $\phi: \mathbf{Sp}(1) \to \mathbf{SO}(3)$. The kernel of ϕ is the center of $\mathbf{Sp}(1)$ which is easily seen to be $\{1, -1\} \approx \mathbf{Z}_2$. Thus there is the monomorphism $\mathbf{Sp}(1)/\mathbf{Z}_2 \to \mathbf{SO}(3)$. Since these are both connected compact 3-manifolds, Invariance of Domain implies that ϕ is onto. Thus $\mathbf{SO}(3) \approx \mathbf{Sp}(1)/\mathbf{Z}_2 \approx \mathbf{RP}^3$.

This, plus the fibration $\mathbf{S}^1 \to \mathbf{U}(2) \to \mathbf{S}^3$, allows us to make the following computations:

$$\begin{aligned}
\pi_1(\mathbf{SO}) &= \pi_1(\mathbf{SO}(3)) \approx \mathbf{Z}_2, & \pi_1(\mathbf{Sp}) &= \pi_1(\mathbf{Sp}(1)) = 0,\\
\pi_1(\mathbf{U}) &= \pi_1(\mathbf{U}(1)) \approx \mathbf{Z}, & \pi_2(\mathbf{Sp}) &= \pi_2(\mathbf{Sp}(1)) = 0,\\
\pi_2(\mathbf{U}) &= \pi_2(\mathbf{U}(1)) = 0, & \pi_3(\mathbf{Sp}) &= \pi_3(\mathbf{Sp}(1)) \approx \mathbf{Z},\\
\pi_3(\mathbf{U}) &= \pi_3(\mathbf{U}(2)) \approx \mathbf{Z}, & \pi_4(\mathbf{Sp}) &= \pi_4(\mathbf{Sp}(1)) \approx \mathbf{Z}_2.
\end{aligned}$$

Now, using $\mathbf{S}^3$ for the group $\mathbf{Sp}(1)$, consider the homomorphism $\mathbf{S}^3 \times \mathbf{S}^3 \to \mathbf{SO}(4)$ given by the action of this group on the quaternions given by $(q_1, q_2)(q) = q_1 q q_2^{-1}$. The kernel of this is the set of (q_1, q_2) such that $q_1 q q_2^{-1} = q$ for all quaternions q. Looking at this for $q = 1$ gives $q_1 q_2^{-1} = 1$, i.e., $q_1 = q_2$. Then the equation becomes $q_1 q q_1^{-1} = q$ for all q, i.e., that q_1 is in the center. Thus the kernel has only the two elements $(1,1)$ and $(-1, -1)$. Thus there is the monomorphism $\mathbf{S}^3 \times \mathbf{S}^3/\mathbf{Z}_2 \to \mathbf{SO}(4)$. But these are both compact connected 6-manifolds and so this is onto. It follows that $\pi_i(\mathbf{SO}(4)) \approx \pi_i(\mathbf{S}^3) \oplus \pi_i(\mathbf{S}^3)$ for $i \geq 2$. Thus $\pi_2(\mathbf{O}) = \pi_2(\mathbf{SO}) = \pi_2(\mathbf{SO}(4)) = 0$.

The reader might note that all the groups $\pi_2(G)$ we have computed for a Lie group G have been zero. It turns out that this is no accident, but that is all we can say about it at this point of our knowledge.

Finally, consider the action of $\mathbf{Sp}(2)$ by conjugation on 2×2 (quaternionic) Hermitian matrices of trace 0. (We leave it to the reader to show that there is such an action; it is not completely trivial.) This is a real euclidean space of dimension 5. Thus, this gives a homomorphism $\mathbf{Sp}(2) \to \mathbf{SO}(5)$, and the kernel can be seen to be $\mathbf{Z}_2$. It follows that $\pi_3(\mathbf{O}) = \pi_3(\mathbf{SO}) = \pi_3(\mathbf{SO}(5)) \approx \pi_3(\mathbf{Sp}(2)) = \pi_3(\mathbf{Sp}) \approx \mathbf{Z}$.

All of these stable groups are known. A striking result known as "Bott periodicity" says that $\pi_{i+2}(\mathbf{U}) \approx \pi_i(\mathbf{U})$, $i \geq 0$. Also $\pi_{i+4}(\mathbf{O}) \approx \pi_i(\mathbf{Sp})$, and $\pi_{i+4}(\mathbf{Sp}) \approx \pi_i(\mathbf{O})$ for $i \geq 0$. Granting this, and the computations we have mentioned, gives the following table:

$i \pmod 8 =$	0	1	2	3	4	5	6	7,
$\pi_i(\mathbf{U}) =$	0	$\mathbf{Z}$	0	$\mathbf{Z}$	0	$\mathbf{Z}$	0	$\mathbf{Z}$,
$\pi_i(\mathbf{O}) =$	$\mathbf{Z}_2$	$\mathbf{Z}_2$	0	$\mathbf{Z}$	0	0	0	$\mathbf{Z}$,
$\pi_i(\mathbf{Sp}) =$	0	0	0	$\mathbf{Z}$	$\mathbf{Z}_2$	$\mathbf{Z}_2$	0	$\mathbf{Z}$.

Now we wish to construct the last of the "Hopf fibrations." To do this we need the following construction:

8.6. Definition. If X and Y are spaces then the *join* $X * Y$ is the space obtained from $X \times Y \times I$ by identifying $\{x\} \times Y \times \{0\}$ to a point for each $x \in X$, and identifying $X \times \{y\} \times \{1\}$ to a point for each $y \in Y$.

When talking about the join $X * Y$ we can use "coordinates" (x, y, t), but remember that the point represented by these coordinates is independent of y when $t = 0$ and of x when $t = 1$.

The join $X * Y$ can be regarded as a subset of the product $CY \times CX$ of cones. It is the inverse image of the set $\{(t, s) \mid s + t = 1\}$ under the mapping $CY \times CX \to I \times I$, the map to the cone parameter values.

The join $X * Y$ can also be thought of as the union $(X \times CY) \cup (CX \times Y)$, which is the union of the mapping cylinders of the projections $X \leftarrow X \times Y \to Y$.

8.7. Proposition.

$$\mathbf{S}^n * \mathbf{S}^m \approx \mathbf{S}^{n+m+1}.$$

Proof. Consider the map $\phi: \mathbf{S}^n * \mathbf{S}^m \to \mathbf{S}^{n+m+1}$ induced by the map $\psi: \mathbf{S}^n \times \mathbf{S}^m \times I \to \mathbf{S}^{n+m+1}$ given by $\psi(x, y, t) = x \cos(\pi t/2) + y \sin(\pi t/2)$, where $x \in \mathbf{R}^{n+1}$ and $y \in \mathbf{R}^{m+1}$. It is clear that this does induce a one–one onto map on the join. Since the join is compact and the sphere is Hausdorff, ϕ must be a homeomorphism by Theorem 7.8 of Chapter I. □

As usual, we shall use ΣW to denote the unreduced suspension of W. Then a map $\phi: X \times Y \to W$ induces a map $h_\phi: X * Y \to \Sigma W$ by $h_\phi(x, y, t) = (\phi(x, y), t)$.

Suppose we are given a map $\phi: X \times X \to X$, denoted by $(x, y) \mapsto xy$ just as if this were a group. Then the left translation $L_x(y) = xy$ and right translation $R_x(y) = yx$ are defined.

8.8. Proposition. *Let $\phi: X \times X \to X$ be a map such that L_x and R_x are homeomorphisms for each $x \in X$. Then $h_\phi: X * X \to \Sigma X$ is a fiber bundle with fiber X.*

Proof. Let $C = X \times [0, 1)/X \times \{0\} \subset \Sigma X$, the open cone. Consider the map

$$X \times (X \times [0, 1)) \to X \times X \times I$$

given by $(x, y, t) \mapsto (x, L_x^{-1}(y), t)$. This clearly induces a map $\psi: X \times C \to X * X$, given by the same "coordinate formula."

The following diagram

$$\begin{array}{ccccccc} X \times C & \xrightarrow{\psi} & X * X & \qquad & (x, y, t) & \longmapsto & (x, L_x^{-1}(y), t) \\ \Big\downarrow{\scriptstyle \text{proj}} & & \Big\downarrow{\scriptstyle h_\phi} & & \Big\downarrow & & \Big\downarrow \\ C & \longrightarrow & \Sigma X & & (y, t) & \longmapsto & (y, t) = (x \cdot L_x^{-1}(y), t) \end{array}$$

provides a trivialization of h_ϕ over the open set where $t \neq 1$. Similar constructions yield a trivialization over the open set where $t \neq 0$. □

We define the Cayley numbers to be pairs (a, b) of quaternions with the product

$$(a,b)(c,d) = (ac - \bar{d}b, da + b\bar{c}).$$

If $x = (a, b)$ then we put $\bar{x} = (\bar{a}, -b)$. It is easy to compute that $x\bar{x} = |a|^2 + |b|^2 = \bar{x}x$ and that $|xy| = |x|\,|y|$ where $|x| = (x\bar{x})^{1/2}$. (Use that $\text{Re}(pq) = \text{Re}(qp)$ for quaternions p, q.) It is also easy to see that this product is distributive over addition on both sides. (It is *not*, however, associative.)

Thus this product gives a map $\mathbf{S}^7 \times \mathbf{S}^7 \to \mathbf{S}^7$. We claim that the maps L_x and R_x are homeomorphisms. By invariance of domain it suffices to show that they are one–one into.

Suppose that $L_x(y) = L_x(y')$. That is, $xy = xy'$. Then $x(y - y') = 0$ and so $|y - y'| = |x|\,|y - y'| = |xy - xy'| = |0| = 0$. Thus $y = y'$. A similar argument holds for R_x.

Thus we have proved that there is a fiber bundle

$$\mathbf{S}^{15} = \mathbf{S}^7 * \mathbf{S}^7 \to \Sigma\mathbf{S}^7 = \mathbf{S}^8$$

with fiber $\mathbf{S}^7$, as we promised.

PROBLEMS

1. ◆ Show that $\pi_4(\mathbf{U}) \approx \pi_4(\mathbf{U}(3)) = 0$ without using Bott periodicity. (*Hint*: Use Section 13 of Chapter II, Problem 4 and Section 15 of Chapter VI, Problem 5.)
2. Show that $\pi_4(\mathbf{O}) = 0$ without using Bott periodicity. (*Hint*: Consider the fibration $\mathbf{SO}(5) \to \mathbf{SO}(6) \to \mathbf{S}^5$ and use the fact from Corollary 15.16 of Chapter VI that $\mathbf{S}^5$ is not parallelizable.)
3. ◆ Show that $\mathbf{SU}(3)$ has a CW-structure with single cells in dimensions 0, 3, 5, and 8 (only). Show that the attaching map for the 5-cell cannot be homotopically trivial. Deduce that $\text{Sq}^2: H^3(\mathbf{SU}(n); \mathbf{Z}_2) \to H^5(\mathbf{SU}(n); \mathbf{Z}_2)$ is nonzero for all $n \geq 3$. (*Hint*: Consider the fibration $\mathbf{SU}(2) \to \mathbf{SU}(3) \to \mathbf{S}^5$.)
4. Use Bott periodicity to show that $\pi_5(\mathbf{S}^2) \approx \mathbf{Z}_2$.

9. The Homotopy Addition Theorem

In this section we aim to prove a result that says that if B^n is an n-disk and we have a map $f: (B^n, \partial B^n) \to (X, A)$ and if there are disjoint n-disks σ in B^n such that f takes the complement of the small disks to A, then the element of $\pi_n(X, A)$ represented by f is the sum of the elements represented by the restriction of f to the small disks σ.

There are some technicalities to clear up, even with having this make sense. We have to take care of base points and we have to make clear what is meant by the homotopy class represented by such a map. We shall use $*$ to denote base points in cases where this will not cause confusion. It is not meant to denote the same point in different spaces.

This result will be used in the proof of the Hurewicz Theorem in the next section. There it is applied only to a nice situation, the standard simplex and its faces. Thus most of the complications of the proof of the general case could be avoided. We prefer to do it in full generality, however.

9.1. Lemma. *If* $f:(\mathbf{D}^n, \mathbf{S}^{n-1}, *) \to (\mathbf{D}^n, \mathbf{S}^{n-1}, *)$ *induces the identity homomorphism*

$$f_* = 1: H_n(\mathbf{D}^n, \mathbf{S}^{n-1}) \to H_n(\mathbf{D}^n, \mathbf{S}^{n-1})$$

then $f \simeq 1$.

PROOF. There is the following commutative diagram, part of the homology sequence of the pair $(\mathbf{D}^n, \mathbf{S}^{n-1})$:

$$\begin{array}{ccc} H_n(\mathbf{D}^n, \mathbf{S}^{n-1}) & \xrightarrow{\approx} & H_{n-1}(\mathbf{S}^{n-1}) \\ \downarrow{\scriptstyle f_*=1} & & \downarrow{\scriptstyle f'_*} \\ H_n(\mathbf{D}^n, \mathbf{S}^{n-1}) & \xrightarrow{\approx} & H_{n-1}(\mathbf{S}^{n-1}) \end{array}$$

where f' denotes the restriction of f to $\mathbf{S}^{n-1}$. It follows that f'_* is also the identity on $H_{n-1}(\mathbf{S}^{n-1})$. By Theorem 5.12 it follows that f' is homotopic to the identity. Thus $[f'] = 1$ in $\pi_{n-1}(\mathbf{S}^{n-1}, *)$. Since $\partial_\#: \pi_n(\mathbf{D}^n, \mathbf{S}^{n-1}, *) \to \pi_{n-1}(\mathbf{S}^{n-1}, *)$ is an isomorphism taking the class of the identity to the class of the identity, it follows that $[f] = 1$. □

Now let σ be an n-cell (i.e., homeomorphic to $\mathbf{D}^n$) with a base point in its boundary. An "anchoring" of σ is a choice of a pointed homeomorphism $\vartheta_\sigma: \mathbf{D}^n \to \sigma$. An "anchored" n-cell is such a σ together with its anchoring ϑ_σ. (This is temporary terminology that will be discarded after Corollary 9.3.)

Choose, once and for all, a generator $[\mathbf{D}^n] \in H_n(\mathbf{D}^n, \mathbf{S}^{n-1}) \approx \mathbf{Z}$. Let $[\sigma] = \vartheta_\sigma^*([\mathbf{D}^n])$.

For two anchored n-cells σ and τ, a "map of degree one" from σ to τ is a map $g: (\sigma, \partial\sigma, *) \to (\tau, \partial\tau, *)$ such that $g_*[\sigma] = [\tau]$.

9.2. Corollary. *If σ and τ are anchored n-cells then any two maps $f, g: \sigma \to \tau$ of degree one are homotopic* rel $*$ *as maps of pairs* $(\sigma, \partial\sigma) \to (\tau, \partial\tau)$.

PROOF. The map $\vartheta_\tau^{-1} \circ f \circ \vartheta_\sigma: (\mathbf{D}^n, \mathbf{S}^{n-1}, *) \to (\mathbf{D}^n, \mathbf{S}^{n-1}, *)$ induces the identity in homology and hence is homotopic to the identity by Lemma 9.1. The same is true for g replacing f and the result follows. □

Suppose that σ is an anchored n-cell and that $f: (\sigma, \partial\sigma, *) \to (X, A, *)$ is a map. Then $f \circ \vartheta_\sigma$ represents a class $[f \circ \vartheta_\sigma] \in \pi_n(X, A, *)$. We shall denote this class simply by $[f]$. The following corollary shows that there is no danger doing so.

9.3. Corollary. *If σ and τ are anchored n-cells, $g:(\tau, \partial\tau, *) \to (\sigma, \partial\sigma, *)$ is of degree one and $f:(\sigma, \partial\sigma, *) \to (X, A, *)$, then $[f \circ g] = [f]$.*

PROOF. In homology we have $(g \circ \vartheta_\tau)_* = (\vartheta_\sigma)_*$. By Corollary 9.2 it follows that $g \circ \vartheta_\tau \simeq \vartheta_\sigma$. Therefore $[f] = [f \circ \vartheta_\sigma] = [f \circ g \circ \vartheta_\tau] = [f \circ g]$. □

An "orientation" of an n-cell σ is a choice of a generator $[\sigma] \in H_n(\sigma, \partial\sigma)$. An anchoring $\vartheta_\sigma: \mathbf{D}^n \to \sigma$ induces an orientation $[\sigma] = \vartheta_{\sigma*}[\mathbf{D}^n]$ as above. It follows from Corollaries 9.2 and 9.3 that $[\sigma]$ determines ϑ_σ up to homotopy. Therefore Corollaries 9.2 and 9.3 hold with "anchored" replaced by "oriented."

Now suppose that B^n is an oriented n-cell and that $\sigma \subset B^n$ is another oriented n-cell. We do *not* assume that $\partial\sigma \subset \partial B^n$. We say that these cells are "oriented coherently" if, for some point $x \in \text{int}(\sigma)$, the orientations $[\sigma] \in H_n(\sigma, \partial\sigma)$ and $[B^n] \in H_n(B^n, \partial B^n)$ correspond under the isomorphisms

$$H_n(\sigma, \partial\sigma) \approx H_n(\sigma, \sigma - \{x\}) \approx H_n(B^n, B^n - \{x\}) \approx H_n(B^n, \partial B^n).$$

It is easy to see, by replacing the point x by a small disk about it, that this concept does not depend on the choice of the point x.

9.4. Lemma. *For $n \geq 2$, the complement of the interiors of a finite number of disjoint n-disks in $\text{int}(B^n)$ is arcwise connected. For $n \geq 3$ it is simply connected.*

PROOF. Selected a point from the interior of each disk. Then there is a strong deformation retraction of the complement of those points to the complement of the interiors of the disks, so it suffices to treat the complement of a finite number of points. This complement is of the homotopy type of the one-point union of some $(n-1)$-spheres. This is arcwise connected (an invariant of homotopy type). For $n \geq 3$, this is a CW-complex with one 0-cell and some $(n-1)$-cells. Any map from $\mathbf{S}^1$ to it is homotopic to a cellular map, and that must be a constant map to the 0-cell. (An easy proof for the complement of a finite number of points can also be given using smooth approximation and transversality.) □

Now suppose that $n \geq 2$ and that B^n is an oriented n-cell. Let $\sigma_1, \ldots, \sigma_k \subset B^n$ be n-cells with disjoint interiors oriented coherently with B^n.

Let $f:(B^n, B^n - \bigcup \text{int}(\sigma_i), *) \to (X, A, *)$. Then $f|_{\sigma_i}$ represents an element of $\pi_n(X, A, ?)$, where ? is the point of A to which the base point of σ goes. We wish to replace this with an element of $\pi_n(X, A, *)$. To do this, choose a path in $B^n - \bigcup \text{int}(\sigma_j)$ from the base point $* \in \partial B^n$ to the base point of σ_i, which is in $\partial\sigma_i$. Use this path to produce, as in Section 7, an element $\alpha_i \in \pi_n(X, A, *)$.

If $n \geq 3$ then the class α_i is independent of the path used to define it since any two of those paths are homotopic. In the case $n = 2$, α_i is determined only up to operation by an element of

$$f_{\#}(\pi_1(B^n - \bigcup \mathrm{int}(\sigma_j))) = f_{\#}\partial_{\#}\pi_2(B^n, B^n - \bigcup \mathrm{int}(\sigma_j)) \subset \partial_{\#}\pi_2(X, A, *).$$

The operation by the latter group is by conjugacy by Theorem 7.8. Thus the image of α_i is unique in the abelianized group

$$\tilde{\pi}_2(X, A, *) = \pi_2/[\pi_2, \pi_2].$$

For consistency of notation we define, for use in this section, $\tilde{\pi}_n = \pi_n$ for $n \geq 3$. These considerations allow us to state the main result.

9.5. Theorem (The Relative Homotopy Addition Theorem). *Let $n \geq 2$. Let B^n be an oriented n-cell. Let $\sigma_1, \ldots, \sigma_k$ be n-cells in B^n with disjoint interiors, and oriented coherently with B^n. Let*

$$f:(B^n, B^n - \bigcup \mathrm{int}(\sigma_i), *) \to (X, A, *).$$

Let $\alpha \in \tilde{\pi}_n(X, A, *)$ be represented by f, and let $\alpha_i \in \tilde{\pi}_n(X, A, *)$ be represented by $f|_{\sigma_i}$ as above. Then $\alpha = \sum_{i=1}^k \alpha_i$.

Proof. Before starting let us stress that all we know about the disks σ_i is what comes from invariance of domain: their interiors as disks are the same as their interiors as subspaces of B^n. We do not know anything about their shapes. They could look like amoebas, and continuous amoebas instead of smooth ones at that. Indeed, we shall picture them that way.

The proof will be by induction on the number k of the cells σ_i. First, we give the inductive step. Note that we can shrink the disks σ_i using their internal coordinates. By this we mean a homotopy which does nothing outside the disks but changes the map on the disks themselves so that it is concentrated on a small concentric subdisk outside of which the map (any stage of the homotopy) is constant along rays (defined in the σ_i coordinates of course). Clearly the disks can be made so small, in terms of the large disk, that they are contained in disjoint metric (large disk) disks about some points. One can then take the one of the metric disks farthest south and move it downwards (a homotopy) until it is in the bottom half disk of B^n without disturbing the other disks. Then the others can be moved up (by a stretch of B^n, for example) until they are in the upper half disk. Now, by induction, the sum of the α_i in the top half of B^n gives the same element as the entire upper half as an n-cell. The case $k = 1$ implies that the α_i in the bottom half is the same as the element defined by the entire lower half disk.

Thus it remains to prove the case $k = 1$, and the case $k = 2$ where the disk σ_1 is the upper half disk of B^n and σ_2 is the lower half disk.

We take up the latter case, $k = 2$, first. This is done by a sequence of homotopies illustrated in Figure VII-8.

The first step is a combination of a homeomorphism changing the shape of B^n into a cube and a simple homotopy that makes the map take the left-hand face of the cube to the base point. The next homotopy squeezes the cube right leaving a larger part going to the base point. The next stage

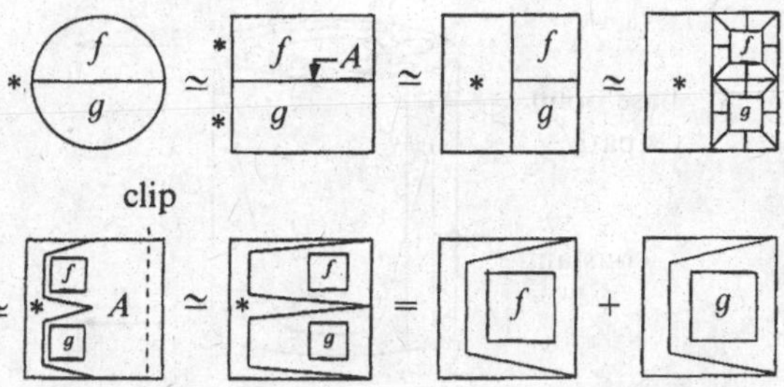

Figure VII-8. Case $k=2$ of the Homotopy Addition Theorem.

shrinks the small cells using the large coordinates. Next a deformation is performed on the cube which moves the small cells left. The vertical dividing line bulges in towards the left. This leaves the entire right half of the cube going to A under the map. The next homotopy just slices more and more from the right half of the map, and it ends with the map that was on the left half. At this time the division between the upper and lower halves of the cube goes entirely into the base point. But then the definition of addition shows that the element represented by the entire cube is the sum of that defined by the top half and that defined by the bottom half. By the case $k=1$, these are just α_1 and α_2.

Thus it remains to prove the case $k=1$. (No, that is not trivial.) For this case we may assume $B^n = \mathbf{D}^n$. By a homeomorphism (of degree one), or a homotopy, we can also assume that the center 0 of $\mathbf{D}^n$ is in $\text{int}(\sigma)$. Let τ be a concentric (with respect to $\mathbf{D}^n$) disk about 0 and contained in $\text{int}(\sigma)$. (See Figure VII-9.) We shall complete the proof in three "stages."

Stage 1. By using the linear structure of σ we can do a shrinking homotopy rel $B^n - \text{int}(\sigma)$ ending with a map taking $B^n - \text{int}(\tau)$ into A. (See Figure VII-10, but note that this is on σ instead of on B^n.) This shows that we may as well

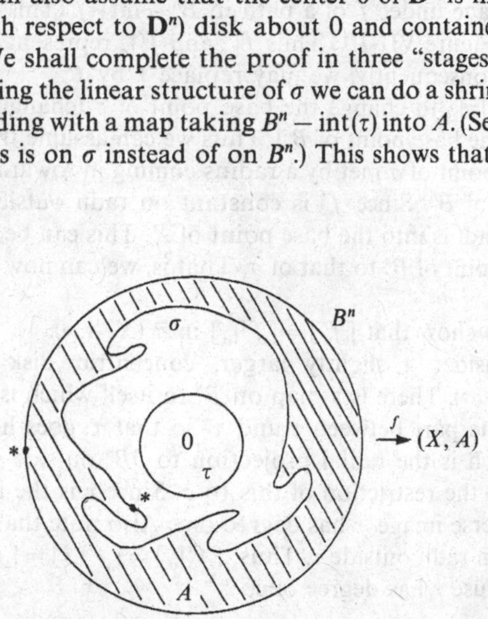

Figure VII-9. Initial state of case $k=1$.

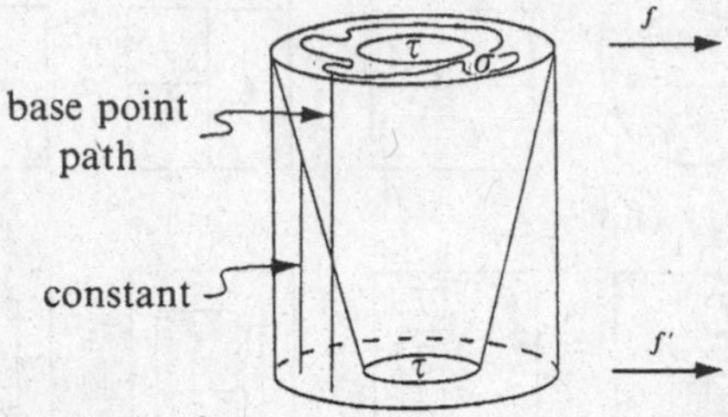

Figure VII-10. Stage 2 homotopy.

assume that

$$f(B^n - \operatorname{int}(\tau)) \subset A.$$

Stage 2. First recall that the element $[f_\sigma] \in \tilde{\pi}_n(X, A, *)$ was defined using the image under f of a path from the base point of B^n to that of σ and going through $B^n - \operatorname{int}(\sigma)$. But note that since the larger set $B^n - \operatorname{int}(\tau)$ is simply connected and is (now) taken by f into A, we can allow images of paths in that set instead.

Now consider a shrinking homotopy using the structure of B^n which ends with B^n shrunk to the size of τ. (See Figure VII-10.) Let f' be the end of the homotopy. Then $f \simeq f'$ rel $*$. Also, throughout the homotopy, $B^n - \operatorname{int}(\tau)$ goes into A. Also note that f' is constant on radii (of B^n) between $\partial\tau$ and ∂B^n.

The restriction of this homotopy to $\sigma \times I$ is a *free* homotopy along a path which is the image under f of a path in $B^n - \operatorname{int}(\tau)$. (This is illustrated as a vertical line in Figure VII-10.) Thus $f|_\sigma$ and $f'|_\sigma$ represent the same element of $\tilde{\pi}_n(X, A, *)$. Consequently, we may replace f by f'.

We are also free to change the base point of σ (changing, along with it, the path from the base point of B^n). Thus we can assume that the base point of σ is the first point of σ met by a radius coming in towards the center from the base point of B^n. Since f' is constant on radii outside τ, it takes this portion of the radius into the base point of X. This can be used for the path from the base point of B^n to that of σ. That is, we can now forget about that path.

It remains to show that $[f'] = [f'|_\sigma]$ in $\tilde{\pi}_n(X, A, *)$.

Stage 3. Consider a slightly larger, concentric, disk τ' about τ still contained in $\operatorname{int}(\sigma)$. There is a map on B^n to itself which is the identity on τ and stretches the part between τ and τ' so that τ' goes homeomorphically to B^n, and which is the radial projection to ∂B^n outside τ'. Let $r:(\sigma, \partial\sigma) \to (B^n, \partial B^n)$ denote the restriction of this to σ. Since r is the identity on τ, and τ is its own inverse image, r has degree one. Also note that $f'|_\sigma = f' \circ r$ since f' is constant on radii outside τ. Thus, $[f'|_\sigma] = [f' \circ r] = [f']$, where the last equality is because r has degree one. □

9.6. Theorem (The Absolute Homotopy Addition Theorem). *Let* $n \geq 1$. *Let*

S^{n+1} be an oriented $(n+1)$-sphere and let $\tau_1, \ldots, \tau_k$ be $(n+1)$-cells in S^{n+1} with disjoint interiors oriented coherently with S^{n+1}. Let $f\colon (S^{n+1} - \bigcup \operatorname{int}(\tau_i)) \to Y$. Then

$$\sum_{i=1}^{k} [f|_{\partial \tau_i}] = 0 \quad \text{in } \tilde{\pi}_n(Y),$$

where $\tilde{\pi}_n(Y) = \pi_n(Y)$ for $n > 1$ and is the abelianized $\pi_1(Y)$ for $n = 1$.

Proof. Of course, one has to make clear the meaning of these homotopy classes and of the orientations. This is done similarly to the relative case and shall be left to the reader. For simplicity of notation, we may as well assume that the sphere is the standard sphere $\mathbf{S}^{n+1}$. Put $X = Y \cup_f \mathbf{S}^{n+1}$. The canonical map from $\mathbf{S}^{n+1}$ into this is an extension of f and we shall retain the notation f for it. Thus we have that

$$f\colon (\mathbf{S}^{n+1}, \mathbf{S}^{n+1} - \bigcup \operatorname{int}(\tau_i)) \to (X, Y).$$

Compose this with the map $c\colon \mathbf{D}^{n+1} \to \mathbf{S}^{n+1}$ collapsing the boundary of the disk to the base point of the sphere. (The base point of the sphere is, of course, assumed to lie outside any of the τ_i.) For simplicity of notation we will use the same notation τ_i for the inverse images of these disks in $\mathbf{D}^{n+1}$ Thus we get a map

$$f\colon (\mathbf{D}^{n+1}, \mathbf{D}^{n+1} - \bigcup \operatorname{int}(\tau_i)) \to (X, Y).$$

On the pair $(\mathbf{D}^{n+1}, \mathbf{S}^n)$, f represents $j_\#[f] \in \pi_{n+1}(X, Y, *)$ where $j_\#\colon \pi_{n+1}(X, *) \to \pi_{n+1}(X, Y, *)$ is from the exact sequence of a pair. The Relative Homotopy Addition Theorem (Theorem 9.5) implies that $j_\#[f] = \sum_i [f|_{\tau_i}]$ in $\tilde{\pi}_{n+1}(X, Y, *)$. Applying $\partial_\#$ we get

$$0 = \partial_\# j_\#[f] = \sum_{i=1}^{k} \partial_\#[f|_{\tau_i}] = \sum_{i=1}^{k} [f|_{\partial \tau_i}]$$

in $\tilde{\pi}_n(Y, *)$. □

10. The Hurewicz Theorem

Previously we proved the Hurewicz Theorem linking $\pi_1(X, *)$ with $H_1(X)$. Here we shall generalize this to higher dimensions. We shall also derive a relative version of the theorem.

To simplify notation in this section we are going to regard the standard n-simplex Δ_n and the disk $\mathbf{D}^n$ as the same space. This should not cause confusion, particularly because of the fact, from the last section, that to specify a homotopy class it suffices to have a map from any pointed n-disk B that is oriented, i.e., has a given generator of $H_n(B, \partial B)$.

Since the identity $\iota = \iota_n$ on Δ_n is a singular n-cycle of $(\Delta_n, \partial\Delta_n)$, its homology class $\vartheta = [\![\iota]\!]$ can be taken as the orientation class. Any map $f\colon (\Delta_n, \partial\Delta_n, *) \to$

$(X, A, *)$ represents a homotopy class $[f] \in \pi_n(X, A, *)$. It is also a cycle and so represents a homology class

$$[\![f]\!] = f_*(\vartheta) \in H_n(X, A).$$

Now, for two such maps f and g,

$$[f] = [g] \quad \Leftrightarrow \quad f \simeq g \quad \Rightarrow \quad f_* = g_* \quad \Rightarrow \quad [\![f]\!] = [\![g]\!].$$

Thus the following definition makes sense (except for the word "homomorphism" which we must justify).

10.1. Definition. The *Hurewicz homomorphism* $h_n: \pi_n(X, A, *) \to H_n(X, A)$ is defined by $h_n[f] = [\![f]\!]$. That is, $h_n[f] = f_*(\vartheta)$.

10.2. Lemma. *The Hurewicz map h_n is a homomorphism.*

PROOF. Consider the comultiplication $\gamma: (\mathbf{D}^n, \mathbf{S}^{n-1}) \to (\mathbf{D}^n, \mathbf{S}^{n-1}) \vee (\mathbf{D}^n, \mathbf{S}^{n-1})$. Let p_i, $i = 1, 2$, be the two projections $(\mathbf{D}^n, \mathbf{S}^{n-1}) \vee (\mathbf{D}^n, \mathbf{S}^{n-1}) \to (\mathbf{D}^n, \mathbf{S}^{n-1})$. Then $p_i \circ \gamma \simeq 1$ for each i. Let i_1, i_2 be the two inclusions $(\mathbf{D}^n, \mathbf{S}^{n-1}) \hookrightarrow (\mathbf{D}^n, \mathbf{S}^{n-1}) \vee (\mathbf{D}^n, \mathbf{S}^{n-1})$. Then $p_i \circ i_j = 1$ if $i = j$ and is the constant map if $i \neq j$. The composition

$$H_n(\mathbf{D}^n, \mathbf{S}^{n-1}) \xrightarrow{\gamma_*} H_n(\mathbf{D}^n \vee \mathbf{D}^n, \mathbf{S}^{n-1} \vee \mathbf{S}^{n-1})$$

$$\xrightarrow{p_{1*} \oplus p_{2*}} H_n(\mathbf{D}^n, \mathbf{S}^{n-1}) \oplus H_n(\mathbf{D}^n, \mathbf{S}^{n-1})$$

takes ϑ to (ϑ, ϑ). But $p_{1*} \oplus p_{2*}$ is an isomorphism with inverse $i_{1*} + i_{2*}$. Thus

$$\gamma_*(\vartheta) = i_{1*}(\vartheta) + i_{2*}(\vartheta).$$

If $f, g: (\mathbf{D}^n, \mathbf{S}^{n-1}, *) \to (X, A, *)$ then $((f \veebar g) \circ \gamma)_* = (f \veebar g)_* \gamma_*$. Therefore

$$\begin{aligned} h_n([f] + [g]) &= h_n([(f \veebar g) \circ \gamma]) \\ &= ((f \veebar g) \circ \gamma)_*(\vartheta) \\ &= (f \veebar g)_*(i_{1*}(\vartheta) + i_{2*}(\vartheta)) \\ &= ((f \veebar g) \circ i_1)_*(\vartheta) + ((f \veebar g) \circ i_2)_*(\vartheta) \\ &= f_*(\vartheta) + g_*(\vartheta) \\ &= h_n[f] + h_n[g]. \end{aligned}$$ □

10.3. Corollary. *For $n > 2$, $\ker(h_n)$ contains the subgroup generated by the elements of the form $\alpha(\beta) - \beta$ for $\alpha \in \pi_1(A, *)$ and $\beta \in \pi_n(X, A, *)$. In case $n = 2$, $\ker(h_2)$ contains the subgroup generated by the $\alpha(\beta)\beta^{-1}$, which contains the commutator subgroup of $\pi_2(X, A, *)$.*

PROOF. Let $[g] = \alpha[f]$ for some $\alpha \in \pi_1(A)$. Then f and g are freely homotopic, which implies that $f_* = g_*$. Thus $[\![f]\!] = [\![g]\!]$. If $\beta = [f]$ then this means that $h_n[\alpha(\beta)] = h_n(\beta)$. Since h_n is a homomorphism, we conclude that $h_n(\alpha(\beta) - \beta) = 0$. The proof needs only minor changes to cover the noncommutative case $n = 2$. □

10.4. Definition. Let $\Delta_k^{(n)}(X, A)$ denote the subgroup of $\Delta_k(X, A)$ generated by those singular simplices $\sigma: \Delta_k \to X$ which carry the n-skeleton of Δ_k into A, modulo $\Delta_k(A)$.

Since $\Delta_k^{(n)}(X, A)$ is a subcomplex of $\Delta_k(X, A)$ we get a homomorphism

$$H_k^{(n)}(X, A) \to H_k(X, A),$$

where the first group denotes, of course, the homology of this subcomplex. If (X, A) is n-connected, we are going to show that this is an isomorphism.

10.5. Theorem. *If (X, A) is n-connected then the inclusion $\Delta_*^{(n)}(X, A) \hookrightarrow \Delta_*(X, A)$ is a chain equivalence.*

PROOF. First we will define, for each singular simplex $\sigma: \Delta_k \to X$, a map $P(\sigma): I \times \Delta_k \to X$, such that the following four conditions are satisfied:

(1) $P(\sigma)(0, y) = \sigma(y)$;
(2) $P(\sigma)(1, \cdot) \in \Delta_k^{(n)}(X)$;
(3) $\sigma \in \Delta_k^{(n)}(X) \Rightarrow P(\sigma)(t, \cdot) = \sigma$ for all t; and
(4) $P(\sigma) \circ (1 \times F_k^i) = P(\sigma^{(i)})$,

where F_k^i is the ith face map and $\sigma^{(i)} = \sigma \circ F_k^i$.

The function P is already defined for $\sigma \in \Delta_k^{(n)}(X)$. We shall define it on other simplices by induction on k. Suppose P is defined for all i-simplices, $i < k$. If σ is a k-simplex, then $P(\sigma^{(i)})$ is defined for all i. These fit together to give a map $(I \times \partial\Delta_k) \cup (\{0\} \times \Delta_k) \to X$. Let us denote this by $P(\partial\sigma)$.

If $k \leq n$ then

$$P(\partial\sigma): (I \times \partial\Delta_k \cup \{0\} \times \Delta_k, \{1\} \times \partial\Delta_k) \to (X, A).$$

Since $\pi_k(X, A) = 0$ by assumption, this map can be filled in taking the top $\{1\} \times \Delta_k$ of the cylinder into A. Define $P(\sigma)$ to be any such map.

If $k > n$ then just use the Homotopy Extension Property on $(\Delta_k, \partial\Delta_k)$ to define $P(\sigma)$. This completes the inductive construction of P.

Now define

$$\phi: \Delta_k(X, A) \to \Delta_k^{(n)}(X, A)$$

by $\phi(\sigma) = P(\sigma)(1, \cdot) = P(\sigma)_\Delta(\iota_1^{(0)} \times \iota_k)$ where $\iota_1^{(0)}: \Delta_0 \to \{1\} \subset I = \Delta_1$ is the 0-face of ι_1.

Define a chain homotopy $D: \Delta_*(X, A) \to \Delta_*(X, A)$ by

$$D(\sigma) = P(\sigma)_\Delta(\iota_1 \times \iota_k).$$

Then

$$\begin{aligned}\partial D(\sigma) &= \partial(P(\sigma)_\Delta(\iota_1 \times \iota_k))\\ &= P(\sigma)_\Delta(\partial\iota_1 \times \iota_k - \iota_1 \times \partial\iota_k)\\ &= P(\sigma)_\Delta[\iota_1^{(0)} \times \iota_k - \iota_1^{(1)} \times \iota_k - \sum_i (-1)^i \iota_1 \times \iota_k^{(i)}].\end{aligned}$$

But

$$\begin{aligned}D(\partial\sigma) &= D\left(\sum_i (-1)^i \sigma^{(i)}\right) = \sum_i (-1)^i D(\sigma^{(i)})\\ &= \sum_i (-1)^i P(\sigma^{(i)})_\Delta(\iota_1 \times \iota_{k-1})\\ &= \sum_i (-1)^i P(\sigma)_\Delta(1 \times F_k^i)_\Delta(\iota_1 \times \iota_{k-1}) \quad \text{(by (4))}\\ &= \sum_i (-1)^i P(\sigma)_\Delta(\iota_1 \times \iota_k^{(i)}).\end{aligned}$$

Thus $\partial D\sigma + D\partial\sigma = P(\sigma)_\Delta(\iota_1^{(0)} \times \iota_k) - P(\sigma)_\Delta(\iota_1^{(1)} \times \iota_k) = \phi(\sigma) - \sigma$, the latter equation by (1). It follows that inclusion$\circ\phi \simeq 1$. By (3), $\phi\circ$inclusion $= 1$. □

10.6. Corollary. *If (X, A) is n-connected then $H_i(X, A) = 0$ for all $i \le n$.* □

10.7. Theorem (The Relative Hurewicz Theorem). *Suppose that $A \subset X$ are both arcwise connected and that (X, A) is $(n-1)$-connected, $n \ge 2$. Then $H_1(X, A) = 0$ for all $i < n$ and*

$$h_n: \pi_n(X, A, *) \to H_n(X, A)$$

is an epimorphism whose kernel is the subgroup generated by the elements $\omega(\beta) - \beta$ for all $\omega \in \pi_1(A, *)$ and $\beta \in \pi_n(X, A, *)$. In particular, if $\pi_1(A) = 1$ then h_n is an isomorphism.

Proof. We know that $\omega(\beta) - \beta \in \ker(h_n)$. Let $\pi_n^*(X, A)$ denote the quotient by the subgroup generated by these elements. (Note that this is independent of the base point and that $\pi_2^*(X, A)$ is abelian.)

If $f: (\Delta_n, \partial\Delta_n) \to (X, A)$, then $f \in \Delta_n^{(n-1)}(X, A)$ and so h_n induces a map (which we call by the same name):

$$h_n: \pi_n^*(X, A) \to H_n^{(n-1)}(X, A) \approx H_n(X, A).$$

For a singular simplex $f: \Delta_n \to X$ which represents a generator of $\Delta_n^{(n-1)}(X, A)$, we have $f: (\Delta_n, \partial\Delta_n) \to (X, A)$ and therefore $[f] \in \pi_n^*(X, A)$ is defined. (If $f: \Delta_n \to A$ then $[f] = 0$.) Thus the assignment $f \mapsto [f] \in \pi_n^*(X, A)$ generates a homomorphism

$$\phi: \Delta_n^{(n-1)}(X, A) \to \pi_n^*(X, A).$$

Also, since $\Delta_{n-1}^{(n-1)}(X, A) = 0$, $\Delta_n^{(n-1)}(X, A)$ is entirely made up of cycles. We claim that the boundaries are in the kernel of ϕ. To see this, suppose that g is a singular simplex in $\Delta_{n+1}^{(n-1)}(X, A)$. Then with $j_{\#}: \pi_n(X) \to \pi_n(X, A)$, we have

$$\begin{aligned}\phi(\partial g) &= \sum_i (-1)^i \phi(g \circ F^i) \\ &= \sum_i (-1)^i [g \circ F^i] \\ &= \pm j_{\#}[g|_{\partial\Delta_{n+1}}]\end{aligned}$$

by the Homotopy Addition Theorem (Theorem 9.5) applied to $\partial\Delta_{n+1}$. But the last term is zero since $[g|_{\partial\Delta_{n+1}}] = 0$ in $\pi_n(X)$ because it extends to Δ_{n+1} as a map to X.

Thus ϕ induces a homomorphism $\bar{\phi}: H_n^{(n-1)}(X, A) \to \pi_n^*(X, A)$ given by $\bar{\phi}[\![f]\!] = [f]$. This is a two sided inverse of h_n by construction. □

By taking $A = \{*\}$, and noting that for $n > 0$, $H_n(X) \approx H_n(X, *)$, we get:

10.8. Corollary (The Absolute Hurewicz Theorem). *If X is $(n-1)$-connected for some $n \geq 2$, then $h_n: \pi_n(X, *) \to H_n(X)$ is an isomorphism.* (*Also see Problem* 4 *of Section* 11.) □

10.9. Corollary. *If $A \subset X$ are both* 1*-connected and $H_i(X, A) = 0$ for $i < n$, then $h_n: \pi_n(X, A) \to H_n(X, A)$ is an isomorphism.*

Proof. The exact sequence $\pi_1(X) \to \pi_1(X, A) \to \pi_0(A)$ of pointed sets has both ends trivial, and so the middle is also trivial. Since X and A are both simply connected, Theorem 10.7 implies that the first nonvanishing terms, $\pi_i(X, A)$ or $H_i(X, A)$, are isomorphic. Thus $\pi_2(X, A) = \cdots = \pi_{n-1}(X, A) = 0$ and the terms in dimension n are isomorphic. □

10.10. Corollary. *If X is* 1*-connected and $H_i(X) = 0$ for $i < n$ then $\pi_i(X, *) = 0$ for $i < n$ and $\pi_n(X, *) \approx H_n(X)$.* □

10.11. Corollary. *A connected CW-complex K is contractible $\Leftrightarrow \pi_1(K) = 1$ and $\tilde{H}_*(K) = 0$.*

Proof. This implies that $\pi_i(K)$ is trivial for all i. Start with the map

$$\phi: K \times \{0\} \cup K \times \{1\} \to K$$

given by $\phi(x, 0) = x$ and $\phi(x, 1) = *$. Extend this to $K \times I \to K$ by induction on the skeletons. Since all homotopy groups of K are trivial, there is nothing to prevent this from being carried out. □

10.12. Example. The dunce cap space X (Figure I-6) is a CW-complex with one 0-cell, one 1-cell and one 2-cell attached by the loop with word a^2a^{-1}. Thus the fundamental group is $\{a \mid a^2a^{-1} = 1\} = 1$. The boundary map

$C_2 \to C_1$ in the cellular chain complex takes a generator x (the two cell) to $2a - a = a \in C_1$. Thus there are no 2-cycles. Consequently, $\tilde{H}_*(X) = 0$. It then follows from Corollary 10.11 that X is contractible.

10.13. Example. Let $X = \Sigma(\mathbf{S}^3/I')$ be the unreduced suspension of the Poincaré dodecahedral space of Theorem 8.10 of Chapter VI. By the Seifert–Van Kampen Theorem (Theorem 9.4 of Chapter III), X is simply connected. By the suspension isomorphism, X is a homology 4-sphere. Thus, by Corollary 10.10, $\pi_i(X) = 0$ for $i < 4$ and $\pi_4(X) \approx \mathbf{Z}$.

To end this section, let us discuss a technical detail that we glossed over in the definition of the Hurewicz homomorphism h_n. We said at the start that we would take Δ_n to equal $\mathbf{D}^n$, but actually one must choose some homeomorphism between them at least up to homotopy. It is clear that a change in this choice can only affect the sign of h_n, i.e., different choices yield homomorphisms that differ only by sign. It is also clear that one can achieve any desired sign by the appropriate choice of homeomorphisms.

PROBLEMS

1. Let K and L be finite CW-complexes and consider the join $K * L$. If K is contractible, show that $K * L$ is contractible. If K and L are simply connected, show that $K * L$ is 4-connected. Generalize.

2. Compute $\pi_i(\mathbf{CP}^2, \mathbf{CP}^1)$ for i as large as you can.

3. Let $f: \mathbf{S}^3 \to \mathbf{S}^3/I'$ be the canonical map where the latter space is the Poincaré dodecahedral space of Theorem 8.10 of Chapter VI. Let $X = (\mathbf{S}^3/I') \cup_f \mathbf{D}^4$. Show that $\pi_2(X)$, $\pi_3(X)$, $H_1(X)$, and $H_2(X)$ are trivial but that $H_3(X) \neq 0$.

11. The Whitehead Theorem

11.1. Proposition. *For $A \subset X$ the following diagram commutes, and is called the "homotopy–homology ladder":*

$$\begin{array}{ccccccccc}
\cdots \longrightarrow & \pi_n(A) & \longrightarrow & \pi_n(X) & \longrightarrow & \pi_n(X, A) & \longrightarrow & \pi_{n-1}(A) & \longrightarrow \cdots \\
& \downarrow{\scriptstyle h_n} & & \downarrow{\scriptstyle h_n} & & \downarrow{\scriptstyle h_n} & & \downarrow{\scriptstyle h_{n-1}} & \\
\cdots \longrightarrow & H_n(A) & \longrightarrow & H_n(X) & \longrightarrow & H_n(X, A) & \longrightarrow & H_{n-1}(A) & \longrightarrow \cdots .
\end{array}$$

PROOF. This is clear from the definitions of the Hurewicz homomorphisms, except for the sign of the square involving the connecting homomorphisms. But, as remarked at the end of Section 10, h_n was not quite pinned down as to sign. One is free to choose the sign so that this diagram does commute, or one can leave it open and treat this ladder as only commuting up to sign in that square. This makes no difference as far as our uses of this result go. □

Suppose that $f: X \to Y$ is a map. Recall that, up to homotopy equivalence, one can regard f as an inclusion by replacing Y by the mapping cylinder M_f. Thus the ladder can be applied to the pair (M_f, X) and then Y can be substituted for M_f in the absolute groups. This yields the following homotopy–homology ladder for a map:

$$\begin{array}{ccccccccc} \cdots \longrightarrow & \pi_i(X) & \xrightarrow{f_\#} & \pi_i(Y) & \longrightarrow & \pi_i(M_f, X) & \longrightarrow & \pi_{i-1}(X) & \longrightarrow \cdots \\ & \downarrow{\scriptstyle h_i} & & \downarrow{\scriptstyle h_i} & & \downarrow{\scriptstyle h_i} & & \downarrow{\scriptstyle h_{i-1}} & \\ \cdots \longrightarrow & H_i(X) & \xrightarrow{f_*} & H_i(Y) & \longrightarrow & H_i(M_f, X) & \longrightarrow & H_{i-1}(X) & \longrightarrow \cdots. \end{array}$$

11.2. Theorem (J.H.C. Whitehead). *Let $n \geq 1$. Given a map $f: X \to Y$ with X and Y arcwise connected, we have:*

I(a). $f_\#$ *is an* $\left\{\begin{matrix} \textit{isomorphism for } i < n \\ \textit{epimorphism for } i = n \end{matrix}\right\} \Rightarrow$ *the same for* f_*,

I(b). $f_\#$ *is an isomorphism for* $i \leq n \Rightarrow$ *the same for* f_*.

II. *If X and Y are simply connected then:*

f_* *is an* $\left\{\begin{matrix} \textit{isomorphism for } i < n \\ \textit{epimorphism for } i = n \end{matrix}\right\} \Rightarrow$ *the same for* $f_\#$.

PROOF. We shall prove part I(b). The exact homotopy sequence of (M_f, X) shows that $\pi_i(M_f, X) = 0$ for $i \leq n$. By the Hurewicz Theorem we get that $H_i(M_f, X) = 0$ for $i \leq n$ *and* that $h_{n+1}: \pi_{n+1}(M_f, X) \to H_{n+1}(M_f, X)$ is an epimorphism. Thus the relevant part of the homotopy–homology ladder is

$$\begin{array}{ccccccc} \pi_{n+1}(M_f, X) & \xrightarrow{0} & \pi_n(X) & \xrightarrow{f_\#} & \pi_n(Y) & \longrightarrow & 0 \\ \downarrow & & \downarrow & & \downarrow & & \\ H_{n+1}(M_f, X) & \longrightarrow & H_n(X) & \xrightarrow{f_*} & H_n(Y) & \longrightarrow & 0 \end{array}$$

and the result follows from a diagram chase.

The proof of I(a) is similar but easier and will be omitted. For part II, the exact homology sequence of the pair (M_f, X) shows that $H_i(M_f, X) = 0$ for $i \leq n$. The exact sequence $\pi_1(Y) \to \pi_1(M_f, X) \to \pi_0(X)$ shows that $\pi_1(M_f, X) = 0$. Since $\pi_1(X) = 0$, the kernel of the first nontrivial relative Hurewicz homomorphism is zero. It follows that $\pi_i(M_f, X) = 0$ for $i \leq n$. The result can now be read off from the exact homotopy sequence of (M_f, X). □

11.3. Example. This example shows that the theorem would be false in absence of the hypothesis that the isomorphism is induced by a *map* of spaces. The product $\mathbf{S}^2 \times \mathbf{S}^4$ has the same homology groups as does $\mathbf{CP}^3$. The homotopy sequence of the fibration $\mathbf{S}^1 \to \mathbf{S}^7 \to \mathbf{CP}^3$ shows that $\pi_3(\mathbf{CP}^3) = 0$, whereas $\pi_3(\mathbf{S}^2 \times \mathbf{S}^4) = \pi_3(\mathbf{S}^2) \oplus \pi_3(\mathbf{S}^4) \approx \mathbf{Z}$.

11.4. Example. This example shows that the analogue of the Whitehead Theorem is false in the relative case. Consider the inclusion map f: $(\mathbf{D}^3_+, \mathbf{S}^2) \hookrightarrow (\mathbf{S}^3, \mathbf{D}^3_-)$. This gives an isomorphism in homology (because it is essentially an excision map). However $\pi_4(\mathbf{D}^3, \mathbf{S}^2) \approx \pi_3(\mathbf{S}^2) \approx \mathbf{Z}$ while $\pi_4(\mathbf{S}^3, \mathbf{D}^3) \approx \pi_4(\mathbf{S}^3, *) \approx \mathbf{Z}_2$ by Theorem 8.3. This example also shows, of course, that homotopy groups do not satisfy an excision property.

11.5. Example. Let $n, m > 1$. The inclusion map $f: \mathbf{S}^n \vee \mathbf{S}^m \hookrightarrow \mathbf{S}^n \times \mathbf{S}^m$ induces an isomorphism

$$H_i(\mathbf{S}^n \vee \mathbf{S}^m) \to H_i(\mathbf{S}^n \times \mathbf{S}^m) \qquad \text{for} \quad i < n + m.$$

By the Whitehead Theorem it follows that

$$f_\#: \pi_i(\mathbf{S}^n \vee \mathbf{S}^m) \approx \pi_i(\mathbf{S}^n \times \mathbf{S}^m) \qquad \text{for} \quad i < n + m - 1.$$

It is not an isomorphism for $i = n + m - 1$, as the reader is asked to show in Problem 1.

Now we shall apply the Whitehead Theorem to the study of the effect on homotopy groups of attaching cells to a space. Assume that X is arcwise connected and let $\{f_\alpha: \mathbf{S}^n \to X\}$ be a family of maps, where $n \geq 1$. Put

$$Y = X \cup_{\{f_\alpha\}} \{\mathbf{D}^{n+1}_\alpha\}.$$

11.6. Proposition. *In the above situation, $\pi_i(X) \to \pi_i(Y)$ is an isomorphism for $i < n$ and is an epimorphism for $i = n$. If X is simply connected then the kernel of $\pi_n(X) \to \pi_n(Y)$ is the subgroup generated by the $[f_\alpha]$.*

PROOF. As in the proof of Lemma 11.2 of Chapter IV, an approximation argument shows that (Y, X) is n-connected. Thus $\pi_i(Y, X) = 0$ for $i \leq n$. From the exact homotopy sequence we get that $\pi_i(X) \to \pi_i(Y)$ is an isomorphism for $i < n$ and an epimorphism for $i = n$. Now assume that $\pi_1(X) = 1$ and consider the commutative diagram

$$\begin{array}{ccccccc}
\pi_{n+1}(Y, X) & \xrightarrow{\partial_\#} & \pi_n(X) & \xrightarrow{i_\#} & \pi_n(Y) & \longrightarrow & 0 \\
\downarrow \approx & & \downarrow & & \downarrow & & \\
H_{n+1}(Y, X) & \longrightarrow & H_n(X) & \longrightarrow & H_n(Y) & \longrightarrow & 0.
\end{array}$$

The isomorphism on the left is by the Relative Hurewicz Theorem, and it implies that $\pi_{n+1}(Y, X)$ is generated by the $[g_\alpha]$ where $g_\alpha: (\mathbf{D}^{n+1}, \mathbf{S}^n) \to (Y, X)$ is the characteristic map for the cell α. (This follows since the homology group is generated by the Hurewicz images of these classes.) Therefore $\ker(i_\#)$ is generated by the $\partial_\#[g_\alpha] = [f_\alpha]$. □

11.7. Theorem. *Let $n \geq 2$ and let X be arcwise connected and semilocally 1-connected (i.e., X has a universal covering space). Let $f_\alpha: \mathbf{S}^n \to X$ and*

$$Y = X \cup_{\{f_\alpha\}} \{\mathbf{D}^{n+1}_\alpha\}$$

as above. Then:

$$\pi_i(X)\to\pi_i(Y) \quad \textit{is an isomorphism for} \quad i<n,$$

and

$$\pi_n(X)\to\pi_n(Y) \quad \textit{is an epimorphism with kernel generated by} \quad \{\omega[f_\alpha] \mid \omega\in\pi_1(X)\}.$$

PROOF. Note that $i_\#(\omega[f_\alpha])=i_\#\omega(0)=0$ in $\pi_n(Y)$, the attached disk killing it. Let X' be the universal covering space of X and choose a base point over the one in X. Let $\Delta\approx\pi_1(X)$ be the group of deck transformations. We can lift each f_α to $f'_\alpha\colon \mathbf{S}^n\to X'$. If $\omega\in\Delta$ then $\omega\circ f'_\alpha$ is another lift of f_α.

Note that $\omega\circ f'_\alpha$ is freely homotopic to a pointed map and this free homotopy covers a free homotopy of f_α in X about a loop in X representing $\omega^{\pm 1}\in\pi_1(X)$. Thus $[\omega\circ f'_\alpha]\in\pi_n(X')$ maps to $\omega^{\pm 1}[f_\alpha]\in\pi_n(X)$.

Put

$$Y'=X'\cup_{\{\omega\circ f_\alpha\}}\{\mathbf{D}_\alpha^{n+1}\}.$$

This is a covering space of Y by Theorem 8.10 of Chapter IV, with deck transformation group Δ. Consider the commutative diagram

$$\begin{array}{ccccc}
\pi_n(X') & \longrightarrow & \pi_n(Y') & \longrightarrow & 0 \\
\big\downarrow\approx & & \big\downarrow\approx & & \\
\pi_n(X) & \longrightarrow & \pi_n(Y). & &
\end{array}$$

By Proposition 11.6 the kernel of the epimorphism on top is the group generated by the $[\omega\circ f'_\alpha]$. These map to $\omega^{\pm 1}[f_\alpha]$ in $\pi_n(X)$ and the result follows. □

Proposition 11.6 can be used to prove the following result about "killing" homotopy groups.

11.8. Theorem. *Let X be arcwise connected and let $n\geq 2$. Then there exists a space $Y\supset X$ obtained by attaching cells to X (called a "relative CW-complex") such that*

$$\pi_i(X)\to\pi_i(Y) \quad \textit{is an isomorphism for} \quad i<n,$$

and $\pi_i(Y)=0$ for $i\geq n$.

PROOF. Choose maps $f_\alpha\colon\mathbf{S}^n\to X$ representing generators of $\pi_n(X)$. Put

$$Y_1=X\cup_{\{f_\alpha\}}\{\mathbf{D}_\alpha^{n+1}\}.$$

Then $\pi_i(X)\to\pi_i(Y_1)$ is an isomorphism for $i<n$ and $\pi_n(Y_1)=0$.

Now do the same thing with generators of $\pi_{n+1}(Y_1)$ producing a space $Y_2\supset Y_1\supset X$ such that $\pi_i(Y_2)\to\pi_i(Y_1)$ is an isomorphism for $i<n+1$ and $\pi_{n+1}(Y_2)=0$, etc. Taking Y to be the union of the Y_j and using the fact that $\pi_i(Y)=\varinjlim_j\pi_i(Y_j)$ finishes the proof. (The fact about direct limits was proved earlier for CW-complexes, and the proof applies to the more general relative CW-complexes.) □

11.9. Corollary. *Let $n \geq 2$ and let π be an abelian group. Then there exists a space $X = K(\pi, n)$ such that $\pi_i(X) = 0$ for $i \neq n$ and $\pi_n(X) \approx \pi$. This also holds for $n = 1$ where π is any group.*

PROOF. The case $n = 1$ will be left to the reader. For $n \geq 2$, let π have a presentation of the form

$$R \xrightarrow{\alpha} F \to \pi \to 0$$

(exact) where F is free abelian. Let $\{f_\alpha\}$ be a basis of F and consider the one-point union

$$W = \vee_{\{f_\alpha\}} \{\mathbf{S}^n_\alpha\}.$$

Then $\pi_i(W) = 0$ for $i < n$, and $\pi_n(W) \approx H_n(W) \approx F$. For each $r \in R$ let $g_r: \mathbf{S}^n \to W$ represent $\alpha(r) \in F = \pi_n(W)$. (Just taking r from a set of generators of R suffices.) Put

$$Y_1 = W \cup_{\{\alpha(r)\}} \{\mathbf{D}^{n+1}_\alpha\}.$$

Then $\pi_1(Y_1) = 0$ for $i < n$ and $\pi_n(Y_1) \approx F/\alpha(R) \approx \pi$. Now add higher-dimensional cells to Y_1 to kill all the higher-dimensional homotopy. This yields the desired "$K(\pi, n)$" space. □

11.10. Theorem. *Let X be $(n-1)$-connected, $n \geq 2$, and let $\pi = \pi_n(X)$. Then there is a fibration*

$$\begin{array}{ccc} K(\pi, n-1) & \longrightarrow & E \\ & & \downarrow p \\ & & X & \xrightarrow{\theta} & K(\pi, n) \end{array}$$

where $\theta_\#: \pi_n(X) \to \pi_n(K(\pi, n)) \approx \pi$ is an isomorphism. (This notation means that p is the pullback of the path-loop fibration of $K(\pi, n)$ via θ, and so the top map is the inclusion of the fiber in the total space.) Moreover, E is n-connected and $p_\#: \pi_i(E) \to \pi_i(X)$ is an isomorphism for $i \neq n$.

PROOF. Let Y be the space obtained from X by attaching cells to X to kill the ith homotopy group of X for all $i \geq n+1$. Then Y is a $K(\pi, n)$ and the inclusion $X \hookrightarrow Y$ induces an isomorphism on nth homotopy groups. The fibration in question, then, is that induced by the inclusion $X \hookrightarrow Y$. The fiber F is the loop space of Y and so is a $K(\pi, n-1)$. There is the following commutative diagram, where the top row is the homotopy sequence of this induced fibration over X and the bottom is part of that for the path-loop fibration of Y:

$$\begin{array}{ccccccccc} 0 & \longrightarrow & \pi_n(E) & \longrightarrow & \pi_n(X) & \longrightarrow & \pi_{n-1}(F) & \longrightarrow & \pi_{n-1}(E) & \longrightarrow & 0 \\ & & & & \approx \downarrow & & \downarrow = & & & & \\ & & & & \pi_n(Y) & \xrightarrow{\approx} & \pi_{n-1}(F). & & & & \end{array}$$

It follows that $\pi_n(E)$ and $\pi_{n-1}(E)$ are both trivial. The remainder follows from the rest of the top sequence. □

11.11. Definition. A map $f: X \to Y$ is an *n-equivalence* if $f_\#: \pi_i(X) \to \pi_i(Y)$ is an isomorphism for $i < n$ and an epimorphism for $i = n$. If f is an n-equivalence for all n then it is called a *weak homotopy equivalence* or an *∞-equivalence.*

Note that the condition in Definition 11.11 is equivalent to $\pi_i(M_f, X)$ being 0 for $i \le n$.

11.12. Theorem. *For $n \le \infty$, a map $f: X \to Y$ is an n-equivalence if and only if, for every relative CW-pair (K, L) with* $\dim(K - L) \le n$ *any commutative diagram*

$$\begin{array}{ccc} L & \xrightarrow{h} & X \\ {\scriptstyle i}\downarrow & & \downarrow{\scriptstyle f} \\ K & \xrightarrow{g} & Y \end{array}$$

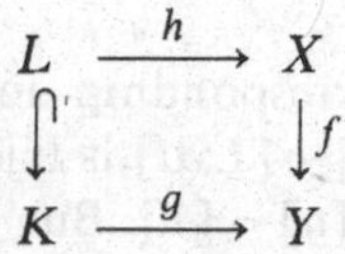

can be completed to

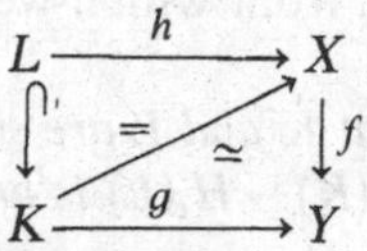

where the top triangle commutes and the bottom triangle commutes up to a homotopy rel L.

PROOF. The implication $\Leftarrow$ is just the definition of $\pi_i(M_f, X) = 0$ using $(K, L) = (\mathbf{D}^i, \mathbf{S}^{i-1})$.

Let $i: X \hookrightarrow M_f$ be the inclusion and $p: M_f \to Y$ the projection. Since $M_f \supset Y$, the map g can be regarded as a map $g: K \to M_f$ and f can be regarded as a map $f: X \to M_f$. Then $g|_L = f \circ h \simeq i \circ h$ since $f \simeq i: X \to M_f$. By the homotopy extension property applied to (K, L), there is a homotopy $F: K \times I \to M_f$ of g to a map $g': K \to M_f$ such that $g'|_L = i \circ h$, and $p \circ F: K \times I \to Y$ is a homotopy rel L. Thus $p \circ g \simeq p \circ g'$ rel L.

Now extend the map $h \times I \cup g' \times \{0\}: L \times I \cup K \times \{0\} \to M_f$ to $G: K \times I \to M_f$ such that $G(K \times \{1\}) \subset X$ by induction over skeletons of (K, L) using that $\pi_i(M_f, X) = 0$ for $i \le n$ and that $\dim(K - L) \le n$. Define $\phi: K \to X$ by $\phi(x) = G(x, 1) \in X$.

Then for $x \in L$, $\phi(x) = G(x, 1) = G(x, 0) = h(x)$, meaning that the top triangle commutes. Also, $i \circ \phi = G(\cdot, 1) \simeq G(\cdot, 0) = g'$ rel L. Thus $f \circ \phi = p \circ i \circ \phi \simeq p \circ g' \simeq p \circ g = g$ rel L. □

11.13. Corollary. *If $f: X \to Y$ is an n-equivalence $(n \le \infty)$ and K is a CW-*

complex, then

$$f_{\#}: [K; X] \to [K; Y]$$

is bijective for $\dim(K) < n$ *and surjective for* $\dim(K) = n$. *This also holds in the pointed category.*

PROOF. The onto part is by application of Theorem 11.12 to $(K, \varnothing)$. The one–one part is by application of Theorem 11.12 to $(K \times I, K \times \partial I)$. In the pointed category use the base point instead of $\varnothing$. □

11.14. Corollary. *Let* $f: K \to L$ *be a map between connected CW-complexes. Then* f *is a homotopy equivalence if and only if* $f_{\#}: \pi_i(K) \to \pi_i(L)$ *is an isomorphism for all* i.

PROOF. Select base points corresponding under f and restrict attention to pointed maps. Then $f_{\#}: [L; K] \to [L; L]$ is bijective by Corollary 11.13. Thus there is a $[g] \in [L; K]$ with $f_{\#}[g] = [1]$. But $f_{\#}[g] = [f \circ g]$, so $f \circ g \simeq 1$.

On homotopy groups we have $1_{\#} = (f \circ g)_{\#} = f_{\#} \circ g_{\#}$. But $f_{\#}$ is an isomorphism so it follows that $g_{\#}$ is also an isomorphism in all dimensions. Then by the same argument used for f applied to g, there is a map $h: K \to L$ such that $g \circ h \simeq 1$. Thus $f \simeq f \circ g \circ h \simeq h$, from which we get $1 \simeq g \circ h \simeq g \circ f$. □

11.15. Corollary. *Suppose that* K *and* L *are simply connected CW-complexes. If* $f: K \to L$ *is such that* $f_*: H_i(K) \to H_i(L)$ *is an isomorphism for all* i, *then* f *is a homotopy equivalence.* □

11.16. Example. Consider the suspension $\Sigma(\mathbf{S}^n \times \mathbf{S}^m)$ of the product of two spheres, $n, m > 0$. We have the composition

$$\Sigma(\mathbf{S}^n \times \mathbf{S}^m) \to \Sigma(\mathbf{S}^n \times \mathbf{S}^m) \vee \Sigma(\mathbf{S}^n \times \mathbf{S}^m) \vee \Sigma(\mathbf{S}^n \times \mathbf{S}^m) \to \mathbf{S}^{n+1} \vee \mathbf{S}^{m+1} \vee \mathbf{S}^{n+m+1},$$

where the first map is from the coproduct and the second is the one-point union of the maps $\Sigma\pi_1, \Sigma\pi_2$, and $\Sigma\eta$, where π_1 and π_2 are the projections to the factors of the product and $\eta: \mathbf{S}^n \times \mathbf{S}^m \to \mathbf{S}^n \wedge \mathbf{S}^m \approx \mathbf{S}^{n+m}$. It is easily seen that this composition is an isomorphism in homology. Thus it is a homotopy equivalence by Corollary 11.15.

11.17. Example. We shall prove the converse of Theorem 10.14 of Chapter VI, thereby giving a complete homotopy classification of the lens spaces $L(p, q)$. We must show that if $\pm qq'$ is a quadratic residue mod p then $L(p, q) \simeq L(p, q')$. The condition is equivalent to the existence of integers k, n, and m, prime to p, such that $n^2kq' + mp = \pm 1$ and $kq \equiv 1 \pmod{p}$. With the notation from the proof of Lemma 10.13 of Chapter VI consider the map $\theta: \mathbf{S}^3 \to \mathbf{S}^3$ given by $\theta(u, v) = (u^{(n)}, v^{(kq'n)})$. Then it can be checked immediately that $\theta T_q = T^n_{q'}\theta$; i.e., θ carries the $\mathbf{Z}_p$-action generated by T_q to that generated by $T^n_{q'}$. Now consider p disjoint disks in $\mathbf{S}^3$ permuted by T_q. By pinching the boundaries of these disks to points, we get a space $W = S_0^3 \cup S_1^3 \cup \cdots \cup S_p^3$ (one point unions but at different points) and an equivariant map $\mathbf{S}^3 \to W$ where $\mathbf{S}^3$ and S_0^3 have the T_q-action and the other S_i^3 are permuted by T_q. Map $W \to \mathbf{S}^3$ by putting θ on S_0^3, and a map of degree m on S_1^3 propagated to maps of degree m on the other S_i^3 by equivariance. Then the composition $\Phi: \mathbf{S}^3 \to W \to \mathbf{S}^3$ has degree $\deg(\theta) + mp = n^2kq' + mp = \pm 1$ and carries the T_q action to the $T_{q'}^n$ action. Since Φ has degree ± 1 it induces isomorphisms $\Phi: \pi_i(\mathbf{S}^3) \to \pi_i(\mathbf{S}^3)$ for all i. The induced map $\Psi: L(p, q) \to L(p, q')$ on the orbit spaces then gives isomorphisms $\Psi_\#: \pi_i(L(p, q)) \to \pi_i(L(p, q'))$ for all i; (see the proof of Lemma 10.13 of Chapter VI). Thus Ψ is a homotopy equivalence by Corollary 11.14 as desired. This discussion generalizes easily to prove the converse of Problem 3 of Section 10 of Chapter VI; i.e., the higher-dimensional analog of the present example.

We finish this section with a brief discussion of the classification problem in topology. This is the problem of finding a way to tell whether or not two spaces are homeomorphic. This is too ambitious, so let us modify it so as to be less demanding. Let us ask for a decision procedure to determine whether or not two finite polyhedra are homotopy equivalent. Perhaps this does not sound too ambitious, but, in fact, it is, as we now explain. Suppose we are given a group G in terms of a finite number of generators and relations. Then we can construct a finite simplicial complex having G as its fundamental group by taking a one-point union of circles, one for each generator, and then attaching 2-cells (which can be done simplicially) to kill the relations. (Perhaps such a construction should be called "fratricide.") If we had such a decision procedure, then that procedure could be used to decide whether or not G is the trivial group (i.e., whether or not the space is simply connected). The problem of finding a decision procedure for determining whether or not a group G, defined by generators and relations, is trivial, is essentially what is known as the "word problem" in group theory. The word problem is known to be unsolvable (proved in 1955 by Novikov [1]), i.e., it is known that there exists no such decision procedure. Thus we have the following fact.

11.18. Theorem. *There does not exist any decision procedure for determining whether or not a given two-dimensional finite polyhedron is simply connected.*

□

Also, it follows from Section 9 of Chapter III, Problem 13 that there is no decision procedure for deciding whether or not a given 4-manifold is simply connected.

This should not be taken as discouraging. After all, the simply connected spaces make up a large segment of interest in topology. Moreover, the result can be viewed as proof that topologists will never find themselves out of work.

Problems

1. Show that $\pi_{n+m-1}(\mathbf{S}^n \vee \mathbf{S}^m) \to \pi_{n+m-1}(\mathbf{S}^n \times \mathbf{S}^m)$ is not an isomorphism.

2. Finish Example 11.16 by showing that the indicated map is an isomorphism in homology. (*Hint*: Show the second map is onto in homology.)

3. If K is a simply connected CW-complex with $H_n(K) \approx \mathbf{Z}$ and $\tilde{H}_i(K) = 0$ for $i \neq n$, then show that $K \simeq \mathbf{S}^n$.

4. Prove this amendment to the Absolute Hurewicz Theorem: Suppose that X is $(n-1)$-connected, $n \geq 2$. Then the Hurewicz homomorphism $h_i: \pi_i(X) \to H_i(X)$ is an isomorphism for $i \leq n$ and an epimorphism for $i = n+1$. (*Hint*: Consider the pair (Y, X) where Y is a space obtained from X by attaching n-cells to kill $\pi_n(X)$.)

5. Consider $\alpha \in \pi_1(\mathbf{S}^1 \vee \mathbf{S}^2)$ and $\beta \in \pi_2(\mathbf{S}^1 \vee \mathbf{S}^2)$ given by the inclusions of the factors. Let $f: \mathbf{S}^2 \to \mathbf{S}^1 \vee \mathbf{S}^2$ represent $2\beta - \alpha(\beta) \in \pi_2(\mathbf{S}^1 \vee \mathbf{S}^2)$ and put $X = (\mathbf{S}^1 \vee \mathbf{S}^2) \cup_f \mathbf{D}^3$. Show that the inclusion $\mathbf{S}^1 \hookrightarrow X$ induces an isomorphism on π_1 and on H_* but is not a homotopy equivalence.

6. A "graph" is a CW-complex of dimension 1. A "tree" is a connected graph with no cycles in the sense of graph theory; i.e., having no simple closed curves.
   (a) Show that a tree is contractible; i.e., prove the infinite case of Lemma 7.7 of Chapter III.
   (b) Show that a connected graph has the homotopy type of the one-point union of circles (possibly infinite in number); i.e., of a graph with a single vertex (the infinite case of Lemma 7.13 of Chapter III).
   (c) Show that the fundamental group of any connected graph is free; i.e., prove the infinite case Theorem 7.14 of Chapter III.
   (d) Show that a subgroup of a free group is free; i.e., prove the infinite case of Corollary 8.2 of Chapter III.

7. For any space X construct a CW-complex K and a map $f: K \to X$ which is a weak homotopy equivalence. (This is called a "CW-approximation" to X.) Use this to remove the hypothesis in Theorem 11.7 that X is semilocally 1-connected.

12. Eilenberg–Mac Lane Spaces

An arcwise connected space Y is called an "Eilenberg–Mac Lane space of type (π, n)" if $\pi_n(Y) \approx \pi$ and $\pi_i(Y) = 0$ for $i \neq n$. We have already met these spaces in the last section where Corollary 11.9 proved their existence as CW-complexes, where, of course, π must be abelian for $n > 1$. In this section we shall also require π to be abelian for $n = 1$. Such a space is also called simply a "space of type (π, n)" or a "$K(\pi, n)$."

The purpose of this section is to show that there exists a natural equivalence of functors

$$[K; K(\pi, n)] \approx H^n(K; \pi),$$

on the category of CW-complexes K and maps. (Compare Hopf's Theorem 11.6 of Chapter V.)

Note that if Y is of type $(\pi, n+1)$ then the loop space ΩY is of type (π, n), as follows from the exact homotopy sequence of the path-loop fibration of Y.

Also $[K; \Omega Y] \approx [SK; Y]$ by Lemma 4.2. It is also clear that any map $K^{(n+1)} \to \Omega Y$ extends to K since $\pi_i(\Omega Y) = 0$ for $i > n$, and any partial homotopy $K^{(n+1)} \times I \cup K \times \partial I \to \Omega Y$ extends to $K \times I$ for the same reason. Therefore $[K; \Omega Y] \approx [K^{(n+1)}; \Omega Y]$.

The sequence

$$K^{(n)} \to K^{(n+1)} \to K^{(n+1)}/K^{(n)} \to SK^{(n)} \to SK^{(n+1)} \to \cdots$$

is coexact by Corollaries 1.4, 5.3, and 5.5. Thus, for Y of type $(\pi, n+1)$, there is the diagram

$$(*)\qquad \begin{array}{ccccccc}
 & & [S^2K^{(n-1)}/S^2K^{(n-2)}; Y] & \longrightarrow & [S^2K^{(n-1)}; Y] & \longrightarrow & 0 \\
 & & & \searrow^{\delta_{n-1}} & \downarrow & & \\
 & & & & [SK^{(n)}/SK^{(n-1)}; Y] & & \\
 & & & & \downarrow & \searrow^{\delta_n} & \\
0 & \longrightarrow & [SK^{(n+1)}; Y] & \longrightarrow & [SK^{(n)}; Y] & \longrightarrow & [K^{(n+1)}/K^{(n)}; Y] \\
 & & & & \downarrow & & \\
 & & & & 0 & &
\end{array}$$

in which the long rows and column are exact and the diagonal maps marked δ are defined by commutativity. The 0 at the left end of the third tier is by $[SK^{(n+1)}/SK^{(n)}; Y] = 0$ since $SK^{(n+1)}/SK^{(n)}$ is a bouquet of $(n+2)$-spheres and $\pi_{n+2}(Y) = 0$. Similarly, the 0 at the end of the exact column is by $[SK^{(n-1)}; Y] = 0$, by Corollary 11.13, since $\dim(SK^{(n-1)}) \le n$ and $\pi_i(Y) = 0$ for $i \le n$. The 0 on the right of the first row is for the same type of reason.

An easy diagram chase gives

$$[K; \Omega Y] \approx [SK; Y] \approx [SK^{(n+1)}; Y] \approx \ker(\delta_n)/\operatorname{im}(\delta_{n-1}).$$

It remains to identify the maps δ_n and δ_{n-1}. They differ only by a change of the index n, so it suffices to look at δ_n. This is induced by the composition

$$K^{(n+1)}/K^{(n)} \xleftarrow{\approx} K^{(n+1)} \cup CK^{(n)} \to SK^{(n)} \to SK^{(n)}/SK^{(n-1)}.$$

Recall that the first map is the homotopy equivalence given by collapsing the cone to a point. The second map is the collapse of $K^{(n+1)}$, and the last is the collapse of $SK^{(n-1)}$.

Now $K^{(n+1)}/K^{(n)}$ is a bouquet of $(n+1)$-spheres, one for each $(n+1)$-cell σ of K. Similarly, $SK^{(n)}/SK^{(n-1)} = S(K^{(n)}/K^{(n-1)})$ is a bouquet of $(n+1)$-spheres, one for each n-cell τ of K.

For an $(n+1)$-cell σ, consider the characteristic map

$$f_\sigma : \mathbf{D}^{n+1} \to K^{(n+1)}.$$

This extends to

$$f_\sigma \cup Cf_{\partial\sigma} \colon \mathbf{D}^{n+1} \cup C\mathbf{S}^n \to K^{(n+1)} \cup CK^{(n)}.$$

Letting $\bar{f}_\sigma \colon \mathbf{S}^{n+1} \to K^{(n+1)}/K^{(n)}$ be the (inclusion) map induced by f_σ, we have the commutative diagram

$$\begin{array}{ccccccc}
\mathbf{S}^{n+1} & \xleftarrow{\approx} & \mathbf{D}^{n+1} \cup C\mathbf{S}^n & \xrightarrow{\approx} & S\mathbf{S}^n & & \mathbf{S}^{n+1} \\
\downarrow{\scriptstyle \bar{f}_\sigma} & & \downarrow{\scriptstyle f_\sigma \cup Cf_{\partial\sigma}} & & \downarrow{\scriptstyle Sf_{\partial\sigma}} & & \uparrow{\scriptstyle S\bar{p}_\tau} \\
K^{(n+1)}/K^{(n)} & \xleftarrow{\approx} & K^{(n+1)} \cup CK^{(n)} & \longrightarrow & SK^{(n)} & \longrightarrow & SK^{(n)}/SK^{(n-1)} = \bigvee_\tau \mathbf{S}^{n+1}
\end{array}$$

where $\bar{p}_\tau$ is the projection of $K^{(n)}/K^{(n-1)}$ to the τth sphere in the bouquet. It follows that δ_n takes the σth sphere to the τth sphere by the map $S(p_\tau f_{\partial\sigma})$; i.e., a map of degree $\deg(p_\tau f_{\partial\sigma})$.

Now an element of $[\bigvee_\sigma \mathbf{S}^{n+1}; Y]$ can be regarded as a function that assigns to each $(n+1)$-cell σ of K, an element of $[\mathbf{S}^{n+1}; Y] = \pi_{n+1}(Y) = \pi$. That is, it is a cellular cochain in $C^{n+1}(K; \pi) = \operatorname{Hom}(C_{n+1}(K), \pi)$. Similarly, an element of $[\bigvee_\tau \mathbf{S}^{n+1}; Y]$ is a function assigning to each n-cell τ of K, an element of π. That is, it is a cochain in $C^n(K; \pi)$. We have shown that the map $[SK^{(n)}/SK^{(n-1)}; Y] \to [K^{(n+1)}/K^{(n)}; Y]$ corresponds to the homomorphism

$$\delta \colon C^n(K; \pi) \to C^{n+1}(K; \pi)$$

given by $\delta f(\sigma) = \sum_\tau \deg(p_\tau f_{\partial\sigma}) f(\tau)$, where σ is an $(n+1)$-cell and τ ranges over the n-cells. But the right-hand side is $f(\sum_\tau \deg(p_\tau f_{\partial\tau})\tau) = f(\partial\sigma)$. Therefore, δ is precisely the cellular coboundary up to sign, justifying our use of that symbol.

We have constructed the isomorphism

$$[K; \Omega Y] \approx H^n(K; \pi),$$

which is natural in K.

We can replace ΩY by a CW-complex L since the construction of a CW-complex L of type (π, n) in Corollary 11.9 makes it clear how to also define a weak homotopy equivalence $L \to \Omega Y$ (or into any $K(\pi, n)$). This is actually a homotopy equivalence because Milnor [1] has shown that ΩY has the homotopy type of a CW-complex when Y has, but we neither need nor will prove this fact. By Corollary 11.13, $[K; L] \approx [K; \Omega Y]$ for all CW-complexes K. Replacing $[SK; Y]$ by $[K; \Omega Y]$ and then by $[K; L]$, the important part of diagram $(*)$ becomes

$$\begin{array}{ccccl}
 & & & & [K^{(n)}/K^{(n-1)}; L] \approx C^n(K; \pi) = \operatorname{Hom}(C_n(K), \pi) \\
 & & & & \downarrow \\
0 & \longrightarrow & [K; L] & \longrightarrow & [K^{(n)}; L] \\
 & & & & \downarrow \\
 & & & & 0.
\end{array}$$

Starting with a map $\phi: K \to L$ representing $[\phi] \in [K; L]$, chasing it to $C^n(K; \pi)$ is given by first restricting it to $K^{(n)}$ then (or prior on K) passing to a homotopic map that takes $K^{(n-1)}$ to the base point of L, and then passing to the induced map $\phi': K^{(n)}/K^{(n-1)} \to L$. Finally, this gives a cochain $c_{\phi'}$ on K by $c_{\phi'}(\tau) = [\phi' \circ \bar{f}_\tau] \in \pi_n(L) = \pi$, where $\bar{f}_\tau: \mathbf{S}^n \to K^{(n)}/K^{(n-1)} = \vee_\tau \mathbf{S}^n$ is the inclusion of the τth sphere induced by the characteristic map $f_\tau: \mathbf{D}^n \to K^{(n)}$. As shown, $c_{\phi'}$ is a cocycle when it comes from $\phi: K \to L$ this way. (One can also see that directly.) The fact that $[\phi] \mapsto [\![c_{\phi'}]\!]$ is a bijection means that the class $[\![c_{\phi'}]\!]$ depends only on $[\phi]$ and this means that the cocycle $c_{\phi'}$ depends on the choice of ϕ', given ϕ, only up to a coboundary. (One can also see this directly, but we do not need that.)

Describing the correspondence the other direction is as easy: Starting with a class $\xi \in H^n(K; \pi)$, represent it by a cocycle $c: C_n(K) \to \pi$ and construct a map

$$f_c: K^{(n)}/K^{(n-1)} = \bigvee_\tau \mathbf{S}^n \to L$$

by putting a representative $\mathbf{S}^n \to L$ of $c(\tau) \in \pi = \pi_n(L)$ on the τth sphere. This, then, induces a map $K^{(n)} \to L$ and it extends to $f: K \to L$ because c is a cocycle and by the main discussion.

If we take the space L of type (π, n) to be as constructed in Corollary 11.9 then $L^{(n-1)} = \{*\}$ and so $L^{(n)} = \vee_\tau \mathbf{S}^n$ where the n-cells τ correspond to given generators of π. Then it is clear that $1 \in [L; L]$ corresponds to the class $u \in H^n(L; \pi)$ represented by the cocycle c taking each n-cell τ to the corresponding generator of π. Then $c^*: H_n(L) \to \pi$ is an isomorphism. (Also recall that the Hurewicz map $\pi_n(L) \to H_n(L)$ is an isomorphism.) A class $v \in H^n(L; \pi)$ which corresponds to an isomorphism $H_n(L) \to \pi$ is called a "characteristic class." This is defined for *any* space with π as the first nonzero homotopy group.

Let us denote by $T: [K; L] \xrightarrow{\approx} H^n(K; \pi)$ our natural equivalence of functors. Then $T(1) = u$. For a map $f: K \to L$, the commutative diagram

$$\begin{array}{ccc} [L;L] & \xrightarrow{T} & H^n(L;\pi) \\ \downarrow f^{\#} & & \downarrow f^* \\ [K;L] & \xrightarrow{T} & H^n(K;\pi) \end{array}$$

shows that $f^{\#}(1) = [f]$ and $T[f] = f^*(T(1)) = f^*(u)$. More generally, any map $f: K \to K'$ of CW-complexes induces

$$\begin{array}{ccc} [K';L] & \xrightarrow{T} & H^n(K';\pi) \\ \downarrow f^{\#} & & \downarrow f^* \\ [K;L] & \xrightarrow{T} & H^n(K;\pi). \end{array}$$

If $f: L \to L$ is a homotopy equivalence, then f^* is an isomorphism. It follows that $T[f] = f^*(u)$ is characteristic. Conversely, if f is such that $f^*(u)$ is characteristic, then $f^{\#}: [L; L] \to [L; L]$ is a bijection, and so there is a map $g: L \to L$ such that $f^{\#}[g] = 1$. This implies that $g \circ f \simeq 1$ and hence $f^*g^* = 1$, so that g^* is also an isomorphism and $f \circ g \simeq 1$. This essentially means that any characteristic class $u \in H^n(L; \pi)$ is as good as any other.

For any space Y of type (π, n) there is a weak homotopy equivalence $L \to Y$ and this induces $[K; L] \xrightarrow{\approx} [K; Y]$. This allows the results for $[K; L]$ to be transferred to $[K; Y]$. Summarizing, we get:

12.1. Theorem. *Let Y be a space of type $(\pi, n), \pi$ abelian, and let $u \in H^n(Y; \pi)$ be characteristic. Then there is a natural equivalence of functors*

$$T_u: [K; Y] \to H^n(K; \pi)$$

of CW-complexes K, given by $T_u[f] = f^*(u)$. □

Note that if (K, A) is a relative CW-complex then K/A is a CW-complex and so it follows that, in the situation of Theorem 12.1,

$$[K/A; Y] \approx H^n(K/A; \pi) \approx H^n(K, A; \pi).$$

There are three cases of well-known spaces of type (π, n). The most obvious one is $\mathbf{S}^1$ which is a $K(\mathbf{Z}, 1)$. Also $\mathbf{CP}^\infty = \bigcup \mathbf{CP}^n$, with the weak topology, is a $K(\mathbf{Z}, 2)$. This follows from the fibrations $\mathbf{S}^1 \to \mathbf{S}^{2n+1} \to \mathbf{CP}^n$ and the fact that $\pi_i(\mathbf{CP}^\infty) = \varinjlim \pi_i(\mathbf{CP}^n)$. Similarly, $\mathbf{P}^\infty$ is a $K(\mathbf{Z}_2, 1)$, and, more generally, an infinite lens space is a $K(\mathbf{Z}_p, 1)$.

Let us now discuss an application to "cohomology operations."

12.2. Definition. A *cohomology operation θ of type $(n, \pi; k, \omega)$* is a natural transformation

$$\theta: H^n(\cdot\,; \pi) \to H^k(\cdot\,; \omega)$$

of functors of CW-complexes. It need not consist of homomorphisms.

For example, $\alpha \mapsto \alpha^2$, for $\alpha \in H^n(\cdot\,; \mathbf{Z})$ is a cohomology operation of type $(n, \mathbf{Z}; 2n, \mathbf{Z})$, and similarly with the higher powers and other coefficient groups. Another example is the Bockstein $\beta_0: H^n(\cdot\,; \mathbf{Z}_p) \to H^{n+1}(\cdot\,; \mathbf{Z})$, which is of type $(n, \mathbf{Z}_p; n+1, \mathbf{Z})$. Similarly, the Bockstein $\beta: H^n(\cdot\,; \mathbf{Z}_p) \to H^{n+1}(\cdot\,; \mathbf{Z}_p)$ is of type $(n, \mathbf{Z}_p; n+1, \mathbf{Z}_p)$.

12.3. Theorem (Serre). *There is a one–one correspondence between the cohomology operations of type $(n, \pi; k, \omega)$ and the elements of $H^k(K(\pi, n); \omega)$, which is given by $\theta \mapsto \theta(u)$ where $u \in H^n(K(\pi, n); \pi)$ is characteristic.*

PROOF. This is equivalent, via Theorem 12.1, to the statement that operations

$$\psi: [X; K(\pi, n)] \to [X; K(\omega, k)]$$

correspond to elements of $[K(\pi,n);K(\omega,k)]$ via $\psi\mapsto\psi(1)$. To simplify notation, let $K=K(\pi,n)$ and $L=K(\omega,k)$.

Given $f:X\to K$ we have the diagram

$$\begin{array}{ccc} [K;K] & \xrightarrow{\psi} & [K;L] \\ \downarrow{\scriptstyle f^{\#}} & & \downarrow{\scriptstyle f^{\#}} \\ [X;K] & \xrightarrow{\psi} & [X;L], \end{array}$$

which, on elements, is

$$\begin{array}{ccc} [1] & \mapsto & \psi(1) \\ \downarrow & & \downarrow \\ [f] & \mapsto & \psi[f]. \end{array}$$

Thus, $\psi[f]=f^{\#}\psi(1)=[g\circ f]$ where $g:K\to L$ represents $\psi(1)\in[K;L]$. Conversely, $[g]\in[K;L]$ induces the operation ψ, by defining $\psi[f]=[g\circ f]$. □

For example, the fact that $H^{2n}(\mathbf{CP}^\infty;\mathbf{Z})\approx\mathbf{Z}$ implies that all cohomology operations $\theta: H^2(\cdot;\mathbf{Z})\to H^{2n}(\cdot;\mathbf{Z})$ have the form $\theta(\alpha)=k\alpha^n$ for some $k\in\mathbf{Z}$.

On the other hand, the fact that $\alpha\mapsto\alpha^2$ of $H^4(X;\mathbf{Z})\to H^8(X;\mathbf{Z})$ is nontrivial on some space X (e.g., $\mathbf{CP}^\infty$) implies that $H^8(K(\mathbf{Z},4);\mathbf{Z})\neq 0$.

Similarly, the fact that $H^2(\mathbf{P}^\infty;\mathbf{Z})\approx\mathbf{Z}_2$ implies that there is exactly one nontrivial operation $H^1(\cdot;\mathbf{Z}_2)\to H^2(\cdot;\mathbf{Z})$. Since the Bockstein in that case is nontrivial (e.g., on $\mathbf{P}^2$), it is that unique operation.

12.4. Corollary. *No nontrivial cohomology operation lowers dimension.*

Proof. This follows from the fact that $H^k(K(\pi,n);\omega)=0$ for $0<k<n$ by the Hurewicz and Universal Coefficient Theorems, or simply by the construction of $K(\pi,n)$ in Corollary 11.9, which has trivial $(n-1)$-skeleton. □

In the next section we will need some technical items about connections between characteristic elements, and another matter. This will fill out the remainder of this section. It is suggested that a first time reader skip this material and refer back to the statements, which are quite believable, when they are used in the following section.

In the remainder of this section, and in the following sections, we shall make the blanket assumption that all pointed spaces under consideration are *well-pointed.*

Let the "suspension isomorphism" in cohomology be defined as the composition

$$S:\tilde{H}^n(X)\xrightarrow{\delta^*}H^{n+1}(CX,X)\xleftarrow{\approx}\tilde{H}^{n+1}(SX)\approx H^{n+1}(SX)$$

(for $n\geq 0$). Sometimes this is defined with a difference in sign. This would have no effect on our main formulas, just on some details of the derivations.

We also use the analogous definition for the suspension isomorphism in homology and the suspension homomorphism for homotopy groups.

12.5. Lemma. *If $f: X \to Y$ is a map between $(n-1)$-connected spaces which induces an isomorphism on $\pi_n(X) \to \pi_n(Y) \approx \pi$ and if $u \in H^n(Y; \pi)$ is characteristic, then $f^*(u) \in H^n(X; \pi)$ is characteristic.*

PROOF. There is the commutative diagram

$$\begin{array}{ccc} H^n(Y;\pi) & \xrightarrow[\approx]{\beta_Y} & \operatorname{Hom}(H_n(Y),\pi) \\ \downarrow{\scriptstyle f^*} & & \approx\downarrow{\scriptstyle \operatorname{Hom}(f_*,1)} \\ H^n(X;\pi) & \xrightarrow[\approx]{\beta_X} & \operatorname{Hom}(H_n(X),\pi) \end{array}$$

where the β's are the maps in the Universal Coefficient Theorem (Theorem 7.2 of Chapter V). By definition, $u \in H^n(Y; \pi)$ is characteristic $\Leftrightarrow \beta_Y(u): H_n(Y) \to \pi$ is an isomorphism. We have that $\beta_X(f^*(u))(a) = \beta_Y(u)(f_*(a))$ by commutativity. Thus $\beta_X(f^*(u)) = \beta_Y(u) \circ f_*$ is an isomorphism, implying that $f^*(u)$ is characteristic. □

12.6. Lemma. *The class $u \in H^n(Y; \pi)$ is characteristic, where Y is $(n-1)$-connected, $\Leftrightarrow Su \in H^{n+1}(SY; \pi)$ is characteristic.*

PROOF. The Hurewicz Theorem implies that SY is n-connected. It is an immediate consequence of the definition that the following diagram commutes up to sign (which can be seen to be $(-1)^{n+1}$):

$$\begin{array}{ccc} H^n(Y;\pi) & \xrightarrow[\approx]{\beta_Y} & \operatorname{Hom}(H_n(Y),\pi) \\ \approx\downarrow{\scriptstyle S} & & \approx\uparrow{\scriptstyle \operatorname{Hom}(S,1)} \\ H^{n+1}(SY;\pi) & \xrightarrow[\approx]{\beta_{SY}} & \operatorname{Hom}(H_{n+1}(SY),\pi). \end{array}$$

Then $\beta_{SY}(S(u))(Sa) = \pm\beta_Y(u)(a)$ and so $\beta_{SY}(S(u)) = \pm\beta_Y(u) \circ S^{-1}$ is an isomorphism. □

12.7. Lemma. *The diagram*

$$\begin{array}{ccc} \pi_n(X) & \xrightarrow{S} & \pi_{n+1}(SX) \\ \downarrow & & \downarrow \\ H_n(X) & \xrightarrow{S} & H_{n+1}(SX) \end{array}$$

commutes, where the verticals are the Hurewicz maps.

PROOF. The suspension for homotopy is defined as the composition along the top of the commutative diagram

$$\begin{array}{ccccc} \pi_n(X) & \overset{\approx}{\longleftarrow} & \pi_{n+1}(CX,X) & \longrightarrow & \pi_{n+1}(SX) \\ \downarrow & & \downarrow & & \downarrow \\ H_n(X) & \overset{\approx}{\longleftarrow} & H_{n+1}(CX,X) & \overset{\approx}{\longrightarrow} & H_{n+1}(SX,*) \end{array}$$

and the lemma follows. □

For any space K consider the map $\lambda: S\Omega K \to K$ which is the adjoint to $1: \Omega K \to \Omega K$. That is, λ is induced by the evaluation map $K^I \times I \to K$. The class $[\lambda]$ corresponds to $[1]$ under the bijection $[S\Omega K; K] \leftrightarrow [\Omega K; \Omega K]$. The diagram (of sets)

$$\begin{array}{ccc} (K^I)^X & \searrow & \\ \downarrow & & \\ (K^I \times I)^{X \times I} & \longrightarrow & K^{X \times I} \end{array}$$

commutes where the horizontal map is induced by the evaluation, the diagonal one is the exponential correspondence $f'(x,t) = f(x)(t)$, and the vertical map is $f \mapsto f \times 1$ where $(f \times 1)(x,t) = (f(x),t)$. This induces the diagram

$$\begin{array}{ccc} [X;\Omega K] & \overset{\approx}{\searrow} & \\ S\downarrow & & \\ [SX; S\Omega K] & \overset{\lambda_\#}{\longrightarrow} & [SX;K], \end{array}$$

where the diagonal is the adjoint (exponential) correspondence. Thus this diagram commutes.

Now if $K = K(\pi, n+1)$ then we conclude that the diagram

$$\begin{array}{ccccc} H_n(\Omega K) & \approx & [S^n;\Omega K] & \overset{\approx}{\searrow} & \\ S\downarrow\approx & & \downarrow & & \\ H_{n+1}(S\Omega K) & \approx & [S^{n+1}; S\Omega K] & \overset{\lambda_\#}{\longrightarrow} & [S^{n+1};K] \end{array}$$

commutes and it follows that

$$\lambda_\#: \pi_{n+1}(S\Omega K) \overset{\approx}{\longrightarrow} \pi_{n+1}(K)$$

is an isomorphism.

Now choose any characteristic class $u \in H^{n+1}(K;\pi)$. By Lemma 12.5, $\lambda^* u \in H^{n+1}(S\Omega K;\pi)$ is characteristic. By Lemma 12.6, $v = S^{-1}\lambda^* u \in H^n(\Omega K;\pi)$ is characteristic. These remarks imply the following result:

12.8. Proposition. *Let $K = K(\pi, n+1)$ and let $u \in H^{n+1}(K;\pi)$ be characteristic. Then $\lambda^*\mu \in H^{n+1}(S\Omega K;\pi)$ and $v = S^{-1}\lambda^* u \in H^n(\Omega K;\pi)$ are characteristic and* *the following diagram commutes:*

$$
\begin{array}{ccccc}
[A;\Omega K] & \xrightarrow{S} & [SA;S\Omega K] & \xrightarrow{\lambda_{\#}} & [SA;K] \\
\approx\big\downarrow T_v & & \approx\big\downarrow T_{\lambda^* u} & \approx \swarrow T_u & \\
H^n(A;\pi) & \xrightarrow{S} & H^{n+1}(SA;\pi). & &
\end{array}
$$

□

12.9. Proposition. *For a cofibration $A \hookrightarrow X$ let $c: X \cup CA \to SA$ be the collapsing map. Then the composition (for arbitrary coefficients)*

$$H^n(A) \xrightarrow{S} \tilde{H}^{n+1}(SA) \xrightarrow{c^*} \tilde{H}^{n+1}(X \cup CA) \xrightarrow{\approx} H^{n+1}(X, A)$$

is $-\delta^*$, where δ^* is the connecting homomorphism for the exact sequence of (X, A).

Proof. Consider the diagram

$$
\begin{array}{ccccc}
H^n(A) & \xrightarrow{\delta^*} & H^{n+1}(CA, A) & \xleftarrow{\approx} & H^{n+1}(SA, *) \\
\uparrow\approx & & \uparrow\approx & & \| \\
H^n(A \cup *, *) & \xrightarrow{\delta^*} & H^{n+1}(CA, A \cup *) & \xleftarrow{\approx} & H^{n+1}(SA, *) \\
\downarrow\approx & & \downarrow\approx & & \| \\
H^n(A \times \partial I, A \times \{1\}) & \xrightarrow{\delta^*} & H^{n+1}(A \times I, A \times \partial I) & \xleftarrow{\approx} & H^{n+1}(SA, *) \\
\downarrow\approx & & \downarrow -1 & & \downarrow -1 \\
H^n(A \times \partial I, A \times \{0\}) & \xrightarrow{\delta^*} & H^{n+1}(A \times I, A \times \partial I) & \xleftarrow{\approx} & H^{n+1}(SA, *) \\
\uparrow\approx & & \uparrow\approx & & \downarrow\approx \\
H^n(A \times \{1\} \cup X \times \{0\}, X \times \{0\}) & \xrightarrow{\delta^*} & H^{n+1}(A \times I \cup X \times \{0\}, A \times \{1\} \cup X \times \{0\}) & \xleftarrow{\approx} & H^{n+1}(S \cup CA, X) \\
\downarrow\approx & & \downarrow & & \downarrow \\
H^n(A \times \{1\}) & \xrightarrow{\delta^*} & H^{n+1}(A \times I \cup X \times \{0\}, A \times \{1\}) & \xleftarrow{\approx} & H^{n+1}(X \cup CA, *) \\
\uparrow\approx & & \uparrow\approx & & \uparrow\approx \\
H^n(A \times I) & \xrightarrow{\delta^*} & H^{n+1}(A \times I \cup X \times \{0\}, A \times I) & \xleftarrow{\approx} & H^{n+1}(X \cup CA, CA) \\
\downarrow\approx & & \downarrow\approx & & \downarrow\approx \\
H^n(A) & \xrightarrow{\delta^*} & H^{n+1}(X, A) & \xleftarrow{\approx} & H^{n+1}(X, A).
\end{array}
$$

Some of the δ^* maps in the diagram are from exact sequences of triples. The horizontal isomorphisms on the right are induced by obvious maps as are the vertical homomorphisms. The composition along the left is the identity and so, from the upper left, all the way down and then right to $H^{n+1}(X, A)$ is just δ^*. The composition along the top is S, by definition. The composition from the upper right, all the way down and then left to $H^{n+1}(X, A)$ is $-c^*$, the sign caused by the inversion of the parameter $SA \to SA$ midway down. Hence $c^* \circ S = -\delta^*$ as claimed. □

12.10. Lemma. *Let $i_0, i_1: X \to X \times \partial I$ be $i_0(x) = (x, 0)$ and $i_1(x) = (x, 1)$. Then for $\delta^*: H^n(X \times \partial I) \to H^{n+1}(X \times I, X \times \partial I) \approx H^{n+1}(SX)$, with any coefficients, we have $S^{-1}\delta^* = i_0^* - i_1^*$.*

Proof. We know that $(i_0^*, i_1^*): H^n(X \times \partial I) \xrightarrow{\approx} H^n(X) \oplus H^n(X)$. Let j_0, j_1: $H^n(X) \to H^n(X \times \partial I)$ induce the inverse isomorphism, so that $i_0^* j_0 = 1 = i_1^* j_1$

and $i_0^*j_1 = 0 = i_1^*j_0$. Then $j_0 i_0^* + j_1 i_1^* = 1$. Clearly, j_0 is the composition

$$j_0 \colon H^n(X) \xleftarrow{\approx} H^n(X \times \partial I, X \times \{1\}) \xrightarrow{h^*} H^n(X \times \partial I)$$

induced by $x \mapsto (x, 0)$ and the inclusion $h \colon (X \times \partial I, \varnothing) \hookrightarrow (X \times \partial I, X \times \{1\})$. Also $j_1 = \omega^* j_0$ where ω is the reversal of the I parameter. Consider the following commutative diagram, similar to that in the proof of Proposition 12.9:

$$\begin{array}{ccccc}
H^n(X) & \xrightarrow{\delta^*} & H^{n+1}(CX, X) & \xleftarrow{\approx} & H^{n+1}(SX) \\
\uparrow \approx & & \uparrow \approx & & \| \\
H^n(X \cup *, *) & \xrightarrow{\delta^*} & H^{n+1}(CX, X \cup *) & \longleftarrow & H^{n+1}(SX) \\
\downarrow \approx & & \downarrow \approx & & \| \\
H^n(X \times \partial I, X \times \{1\}) & \xrightarrow{\delta^*} & H^{n+1}(X \times I, X \times \partial I) & \longleftarrow & H^{n+1}(SX) \\
\downarrow h^* & & \| & & \\
H^n(X \times \partial I) & \xrightarrow{\delta^*} & H^{n+1}(X \times I, X \times \partial I). & &
\end{array}$$

This shows that $S^{-1}\delta^* j_0 = 1$, since S is the composition from top left to bottom right, going right then down. Then $S^{-1}\delta^* j_1 = S^{-1}\delta^*\omega^* j_0 = S^{-1}\omega^*\delta^* j_0 = -S^{-1}\delta^* j_0 = -1$, since ω induces -1 on $H^*(SX)$. Consequently, $S^{-1}\delta^* = S^{-1}\delta^* \circ 1 = S^{-1}\delta^*(j_0 i_0^* + j_1 i_1^*) = i_0^* - i_1^*$. □

Problems

1. Show that any $K(\mathbf{Z}, n)$ is infinite dimensional for each even $n > 0$.

2. Show that any $K(\mathbf{Z}_2, n)$ is infinite dimensional for each $n > 0$.

3. Show that there are no nontrivial cohomology operations of type $(1, \mathbf{Z}; k, \omega)$ for any $k > 1$ and any ω.

4. ◆ Show that there are no nontrivial cohomology operations of type $(n, \mathbf{Z}; n+1, \omega)$ for any $n > 0$ and any ω.

5. Rederive Hopf's Classification Theorem (Theorem 11.6 of Chapter V) as a corollary of the results of this section. (*Hint*: Use Corollary 11.13 and Theorem 11.8.)

13. Obstruction Theory ☼

In this section and the next we impose the blanket assumption that all pointed spaces under consideration are *well-pointed*. This is not an important restriction and is made merely to avoid having to distinguish between reduced and unreduced suspensions.

We shall now attack the fundamental lifting and extension problems in homotopy theory. Suppose that $F \to Y \to B$ is a fibration. The lifting problem is the question of finding criteria for being able to complete the commutative diagram

$$\begin{array}{ccc} A & \longrightarrow & Y \\ \cap\!\downarrow & \nearrow & \downarrow \\ X & \longrightarrow & B. \end{array}$$

The extension problem is the question of giving criteria for being able to complete the diagram

$$\begin{array}{ccc} A & \longrightarrow & Y \\ \cap\!\downarrow & \nearrow & \\ X. & & \end{array}$$

But the extension problem is simply the special case $B = \{*\}$ of the lifting problem and so it suffices to discuss the latter.

As with any problem as difficult as this, it is desirable, perhaps necessary, to break the problem into a sequence of simpler problems. This is exactly what obstruction theory does. We first take up the case in which the fibration $Y \to B$ is induced from the path-loop fibration over a $K(\pi, n+1)$ via some map $\theta: B \to K(\pi, n+1)$ and then try to fit these together to gain information about the general case.

13.1. Definition. Let $p_0: PB_0 \to B_0$ be the path-loop fibration over some space B_0 and let $\theta: B \to B_0$ be a map. Then the induced fibration $p_\theta: E_\theta \to B$ is called the *principal fibration* induced by θ. That is, $E_\theta \to B$ is the pullback:

$$\begin{array}{ccc} E_\theta = \theta^*(PB_0) & \longrightarrow & PB_0 \\ {\scriptstyle p_\theta}\downarrow & & \downarrow{\scriptstyle p_0} \\ B & \xrightarrow{\ \theta\ } & B_0. \end{array}$$

Suppose given such a principal fibration and consider maps $f: S \to E_\theta$, where S is any space. Then, by the definition of a pullback, such maps f are in one–one correspondence with pairs $f_1: S \to B$, $f_2: S \to PB_0$ of maps such that $\theta f_1 = p_0 f_2$.

In the pointed category, we have $PB_0 = B_0^I$ so that such maps $f_2: S \to B_0^I$ correspond to homotopies $\psi: S \times I \to B_0$ with $\psi(s, 0) = *$ and $\psi(s, 1) = p_0 f_2(s) = \theta f_1(s)$; see Section 2. Therefore, there is a one–one correspondence between liftings $f: S \to E_\theta$ of a given $f_1: S \to B$ and homotopies $\psi: S \times I \to B_0$ such that $\psi(s, 0) = *$ and $\psi(s, 1) = \theta f_1(s)$; i.e., homotopies from $*$ to θf_1.

Next let (X, A) be a relative CW-complex, and specialize to the case $B_0 = K(\pi, n+1)$, so that the fiber is $\Omega B_0 = K(\pi, n)$. Consider the commutative diagram

$$\begin{array}{ccccc} A & \xrightarrow{f_A} & E_\theta & \longrightarrow & PB_0 \\ \downarrow & \overset{g}{\nearrow} & \downarrow & & \downarrow \\ X & \xrightarrow[f_X]{} & B & \xrightarrow[\theta]{} & B_0. \end{array}$$

We shall call this diagram a "lifting problem f of type (π, n)," and the prospective map g, a "solution" to this lifting problem. Then the lifting problem corresponds to a map

$$\psi: A \times I \cup X \times \partial I \to B_0$$

such that $\psi(x, 0) = *$ and $\psi(x, 1) = \theta f_X(x)$. A solution g corresponds to an extension of ψ to $X \times I \to B_0$. Since ψ takes $X \times \{0\}$ to the base point, it defines a map

$$\phi_f: X \cup CA = (A \times I \cup X \times \partial I)/X \times \{1\} \xrightarrow{\approx} (A \times I \cup X \times \partial I)/X \times \{0\} \to B_0,$$

where the middle map is the parameter flip in I, a technicality to allow use of our standard definition of CA, etc. If a solution exists to the lifting problem then ϕ_f extends to $CX = X \times I/X \times \{1\}$ and so $\phi_f \simeq *$. Conversely, if $\phi_f \simeq *$ then ϕ_f extends to CX by the homotopy extension property (since $(CX, X \cup CA)$ is a relative CW-complex). Therefore, the lifting problem f has a solution if and only if the element

$$[\phi_f] \in [X \cup CA; B_0]$$

is trivial. Recall that the collapsing map $X \cup CA \to X/A$ is a homotopy equivalence by Theorem 1.6, and so, using Corollary 1.7,

$$[X \cup CA; B_0] \approx [X/A; B_0] \approx H^{n+1}(X/A; \pi) \approx H^{n+1}(X, A; \pi).$$

Let $c_f^{n+1} \in H^{n+1}(X, A; \pi)$ be the cohomology class corresponding to $[\phi_f]$. That is,

$$c_f^{n+1} = T_u[\phi_f] = \phi_f^*(u)$$

where $u \in H^{n+1}(B_0; \pi)$ is characteristic.

By the naturality of these constructions, the image of c_f^{n+1} in $H^{n+1}(X; \pi)$ is the class corresponding to the same lifting problem after A is forgotten (i.e., A is taken to be empty). Let $j: X \hookrightarrow X \cup CA$, so that $j^*: H^*(X, A) \to H^*(X)$. Note that $\phi_f \circ j = \theta \circ f_X$. By the diagram

$$\begin{array}{ccc} [X \cup CA; B_0] & \xrightarrow{j^\#} & [X; B_0] \\ \downarrow{\scriptstyle T_u} & & \downarrow{\scriptstyle T_u} \\ H^{n+1}(X \cup CA; \pi) & \xrightarrow{j^*} & H^{n+1}(X; \pi) \end{array}$$

we have $j^*(c_f) = j^* T_u[\phi_f] = T_u(j^\#[\phi_f]) = T_u[\theta f_X] = (\theta f_X)^*(u) = f_X^* \theta^*(u)$ where $u \in H^{n+1}(B_0; \pi)$ is characteristic. Thus we have proved:

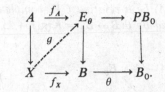

13.2. Theorem. *Let (X, A) be a relative CW-complex and let $E_\theta \to B$ be the principal fibration induced by $\theta: B \to B_0 = K(\pi, n+1)$. Then, given the lifting problem f of type (π, n):*

$$
\begin{array}{ccc}
A & \xrightarrow{\ f_A\ } & E_\theta \\
\Big\downarrow & \nearrow & \Big\downarrow p_\theta \\
X & \xrightarrow[\ f_X\ ]{} & B \xrightarrow[\ \theta\ ]{} K(\pi, n+1)
\end{array}
$$

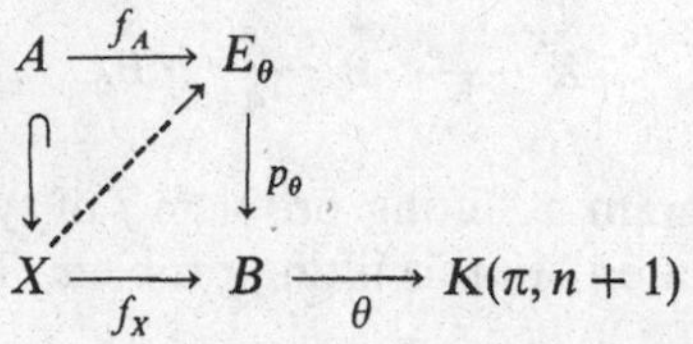

a solution g exists $\Leftrightarrow$ a certain "obstruction class" $c_f^{n+1} \in H^{n+1}(X, A; \pi)$ vanishes. Moreover, c_f^{n+1} goes to $f_X^* \theta^*(u) \in H^{n+1}(X; \pi)$ where u is characteristic. □

Now suppose we have not only one but two solutions g_0, g_1 to the lifting problem f:

$$
\begin{array}{ccc}
A & \xrightarrow{\ f_A\ } & E_\theta \\
\Big\downarrow & {\scriptstyle g_1}\!\!\nearrow\!\!\nearrow\!{\scriptstyle g_0} & \Big\downarrow p_\theta \\
X & \xrightarrow[\ f_X\ ]{} & B \xrightarrow[\ \theta\ ]{} K(\pi, n+1).
\end{array}
$$

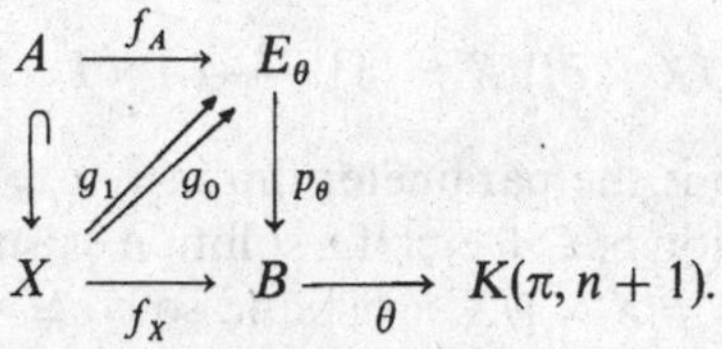

We ask for a similar obstruction to making $g_0 \simeq g_1$ rel A, by a fiber homotopy. But this is just the lifting problem F:

$$
\begin{array}{ccc}
A \times I \cup X \times \{0\} \cup X \times \{1\} & \xrightarrow{\ f_A \circ \mathrm{proj} \cup g_0 \times \{0\} \cup g_1 \times \{1\}\ } & E_\theta \\
\Big\downarrow & {}\!\!\nearrow\!\!\!{}^{\displaystyle G} & \Big\downarrow p_\theta \\
X \times I & \xrightarrow[\ f_X \circ \mathrm{proj}\ ]{} & B \xrightarrow[\ \theta\ ]{} K(\pi, n+1)
\end{array}
$$

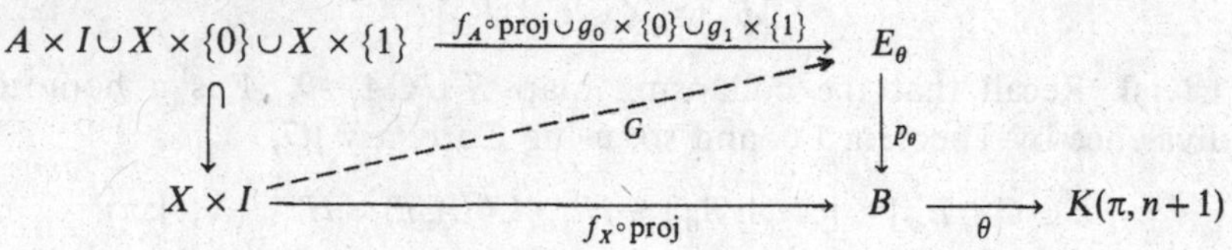

and so there is the obstruction

$$
\begin{aligned}
[\phi_F] \in &[X \times I \cup C(A \times I \cup X \times \partial I); K(\pi, n+1)] \\
\approx &[X \times I/(A \times I \cup X \times \partial I); K(\pi, n+1)] \\
\approx &[S(X/A); K(\pi, n+1)] \\
\approx &[X/A; \Omega K(\pi, n+1)] \\
\approx &[X/A; K(\pi, n)] \\
\approx &H^n(X, A; \pi).
\end{aligned}
$$

The corresponding cohomology class is called the "difference class" of g_0 and g_1 and is denoted by

$$
d^n(g_0, g_1) \in H^n(X, A; \pi).
$$

By definition and Proposition 12.8, $d^n(g_0, g_1) = S^{-1} c_F^{n+1}$ where $c_F^{n+1} \in H^{n+1}((X, A) \times (I, \partial I); \pi)$ and where $S: H^n(X, A; \pi) \xrightarrow{\ \approx\ } H^{n+1}((X, A) \times (I, \partial I); \pi)$ is the suspension isomorphism.

Suppose we have three such liftings g_0, g_1, and g_2. Then there is the lifting problem

$$\begin{array}{ccccc} A \times [0,2] \cup X \times \{0\} \cup X \times \{1\} \cup X \times \{2\} & \longrightarrow & E_\theta & & \\ \cap\downarrow & \nearrow & \downarrow p_\theta & & \\ X \times I & \longrightarrow & B & \longrightarrow & K(\pi, n+1) \end{array}$$

for which the above constructions produce a class in

$$[X \times [0,2]/(A \times [0,2] \cup X \times \{0,1,2\}); K(\pi, n+1)]$$
$$\approx [S(X/A) \vee S(X/A); K(\pi, n+1)].$$

Now, forgetting the middle solution g_1 amounts to composing this with the coproduct $S(X/A) \to S(X/A) \vee S(X/A)$. In terms of the difference classes, this implies that

$$d^n(g_0, g_2) = d^n(g_0, g_1) + d^n(g_1, g_2).$$

(The fact that the coproduct corresponds to addition of cohomology classes results immediately from the explicit correspondence between $[\cdot; K(\pi, n+1)]$ and $H^{n+1}(\cdot; \pi)$ given in Section 12.) Therefore we have:

13.3. Theorem. *For two solutions g_0, g_1 of the lifting f of type (π, n) there is an obstruction $d^n(g_0, g_1) \in H^n(X, A; \pi)$ which vanishes if and only if g_0 is fiber homotopic* rel A *to g_1. These satisfy the relation*

$$d^n(g_0, g_2) = d^n(g_0, g_1) + d^n(g_1, g_2). \qquad \square$$

The next step in studying the general lifting problem for a fibration $p: Y \to B$ is to attempt to decompose p into a sequence $\cdots \to Y_3 \to Y_2 \to Y_1 \to B$ of fibrations, each principal and induced from a path-loop fibration over some $K(\pi, q+1)$. Such a decomposition is called a "Moore–Postnikov decomposition" of p. If B is a point, the decomposition is called a "Postnikov decomposition" of Y.

13.4. Definition. A map $f: Y \to B$ between arcwise connected spaces is called *simple* if $f_\#\pi_1(Y) \supset [\pi_1(B), \pi_1(B)]$ (the commutator subgroup) and the pair (M_f, Y) is simple (meaning that $\pi_1(Y)$ acts trivially on $\pi_n(M_f, Y)$ for all $n \geq 1$).

The following theorem is the main technical result for the construction of Moore–Postnikov decompositions:

13.5. Theorem. *Suppose given a simple map $f_n: Y \to Y_n$ which is an n-equivalence. Then there exists a principal fibration $p: Y_{n+1} \to Y_n$ induced by a map $\theta: Y_n \to K(\pi, n+1)$ and a lifting $f_{n+1}: Y \to Y_{n+1}$ of f_n such that f_{n+1} is an $(n+1)$-equivalence and is simple.*

Proof. We can replace Y_n by M_{f_n} and f_n by the inclusion $Y \hookrightarrow M_{f_n}$ (using the homotopy lifting property). Therefore, upon renaming M_{f_n} as X and Y as A, it suffices to prove the following lemma:

13.6. Lemma. *Let $i: A \hookrightarrow X$ be a simple cofibration with (X, A) n-connected and put $\pi = \pi_{n+1}(X, A) \approx H_{n+1}(X, A) \approx H_{n+1}(X/A)$. Let $v \in H^{n+1}(X/A; \pi)$ be characteristic and let $X \to X/A \xrightarrow{\theta} K = K(\pi, n+1)$ represent v; i.e., $\theta^*(u) = v$ for $u \in H^{n+1}(K; \pi)$ characteristic. Let $p: E \to X$ be the induced principal fibration with fiber $F = \Omega K = K(\pi, n)$. Let $g: A \to E$ be the map $g(a) = (a, c)$ where c is the constant path at the base point. Then g is an $(n+1)$-equivalence and is simple.*

Proof. Note that, in the case $n = 0$, simplicity implies that $\pi_1(X/A) \approx \pi_1(X \cup CA) \approx \pi_1(X)/i_\# \pi_1(A)$ is abelian, and so there is no difficulty with the notion of characteristic elements in this dimension.

Consider the commutative diagram

$$\begin{array}{ccc} & & \pi_q(E) \\ & \overset{g_\#}{\nearrow} & \downarrow{\scriptstyle p_\#} \\ \pi_q(A) & \xrightarrow{i_\#} & \pi_q(X). \end{array}$$

We have that $i_\#$ is isomorphic for $q < n$ and $p_\#$ is isomorphic for $q \neq n, n+1$, and so $g_\#$ is isomorphic for $q < n$. We must show that $g_\#$ is isomorphic for $q = n$ and epimorphic for $q = n+1$. If we have this, then the exact sequences for (M_i, A), (M_g, A), and $E \to X$ show that $\pi_j(M_g, A) \to \pi_j(M_i, A)$ is monomorphic for $j > n+1$ and hence for all j since $\pi_j(M_g, A) = 0$ for $j \leq n+1$. This implies that $\pi_1(A)$ acts trivially on all $\pi_j(M_g, A)$ and hence that g is simple. Therefore it suffices to prove this contention about $g_\#$.

We can extend the composition of $\theta: X/A \to K$ with the collapse $X \to X/A$ to a map $\phi: X \cup CA \to K$ taking CA to the base point and hence define the pullback diagram

$$\begin{array}{ccc} E' & \longrightarrow & PK \\ \downarrow & & \downarrow \\ X \cup CA & \xrightarrow{\phi} & K = K(\pi, n+1). \end{array}$$

Regard F as the fiber over the vertex of CA. Let $\lambda: \pi_{q+1}(X, A) \to \pi_q(F)$ be the composition along the top of the commutative diagram

$$\begin{array}{ccccccc} \lambda: \pi_{q+1}(X, A) & \longrightarrow & \pi_{q+1}(X \cup CA, CA) & \xleftarrow{\approx} & \pi_{q+1}(X \cup CA, *) & \longrightarrow & \pi_q(F) \\ \downarrow & & \downarrow & & & \underset{\phi_\#}{\searrow} \quad \overset{\approx}{\nearrow} & \\ H_{q+1}(X, A) & \xrightarrow{\approx} & H_{q+1}(X \cup CA, CA) & & & \pi_{q+1}(K). & \end{array}$$

The assumption that v is characteristic implies that $\phi_\#$ is an isomorphism for $q = n$, and hence for $q \leq n$ since both groups vanish for $q < n$. The Hurewicz

Theorem implies that the two verticals on the left are isomorphisms for $q \le n$. Consequently, λ is an isomorphism for $q \le n$.

We claim that the following diagram commutes up to sign:

$$\begin{array}{ccccccccc}
\cdots \longrightarrow & \pi_{q+1}(A) & \longrightarrow & \pi_{q+1}(X) & \longrightarrow & \pi_{q+1}(X, A) & \longrightarrow & \pi_q(A) & \longrightarrow \cdots \\
& \downarrow g_\# & & \downarrow = & & \downarrow \lambda & & \downarrow g_\# & \\
\cdots \longrightarrow & \pi_{q+1}(E) & \longrightarrow & \pi_{q+1}(X) & \longrightarrow & \pi_q(F) & \longrightarrow & \pi_q(E) & \longrightarrow \cdots.
\end{array}$$

The desired result will follow from this, the 5-lemma, and the fact that $\pi_q(F) = 0$ for $q \ne n$.

The commutativity of the first two squares is trivial. For the third square, the composition going down then right is illustrated by the top of Figure VII-11. The composition right then down is illustrated by the bottom of the figure.

Most of the top of the figure is the description of λ as follows. We consider an element α of $\pi_{q+1}(X, A)$ as represented by a map on the lower half $(q+1)$-disk as suggested by the figure. Cone off the top to give the extension to the full disk shown in the second part of the figure. Then lift the map to a map into E as indicated in the third part. Restricting the map to the boundary $\mathbf{S}^q$ gives a map $\mathbf{S}^q \to F$ which represents $\lambda(\alpha)$. This is the fourth part of the figure. As a map into E there is a homotopy to the restriction of the third part of the diagram to the map on the top hemisphere and the equator of the disk, illustrated by the fifth part of the figure. This completes the description of going down by λ and then right in the diagram in question.

For the composition right then down, also consider the diagram

$$\begin{array}{ccccc}
\pi_{q+1}(X, A) \longrightarrow & \pi_q(A) & \overset{\approx}{\longleftarrow} & \pi_{q+1}(CA, A) & \\
& \downarrow g_\# & & \downarrow g_\# & \nwarrow \approx \searrow \\
& \pi_q(E) & \longleftarrow & \pi_{q+1}(E', E) & \longleftarrow \quad \pi_{q+1}(E'_{CA}, E_A).
\end{array}$$

Taking $\alpha \in \pi_{q+1}(X, A)$ to $\pi_q(A)$ is illustrated by the first two parts of the bottom of Figure VII-11. The diagram shows that the effect of $g_\#$ can be described by first coning off to give a map into CA represented by the third

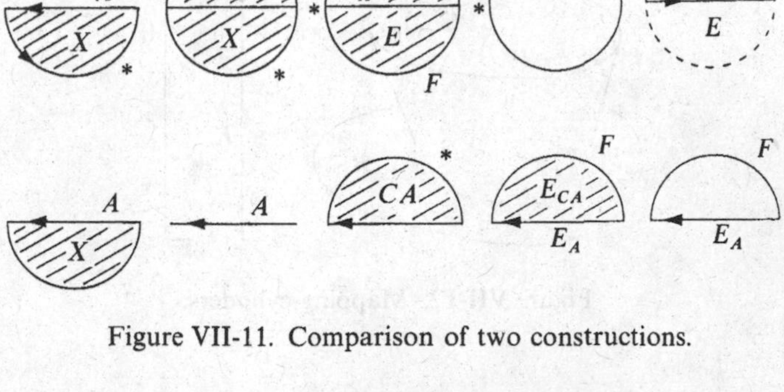

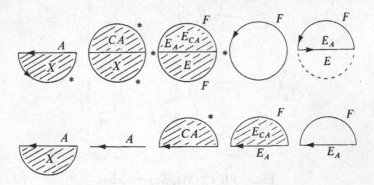

Figure VII-11. Comparison of two constructions.

part of the figure, then lifting to get the fourth part of the figure, then restricting to the boundary to get the last part of the figure. The only difference of this final result from that given by the top of the figure is one of orientation, and so the square commutes up to sign as claimed. □

13.7. Theorem. *Let $g: Y \to B$ be a simple map. Then there exists a sequence*

$$\cdots \to Y_{n+1} \xrightarrow{p_n} Y_n \to \cdots \to Y_0 = B$$

of principal fibrations induced by maps $Y_n \to K(\pi_n, n+1)$, and maps $g_n: Y \to Y_n$ factoring g and such that $p_n \circ g_{n+1} = g_n$ and g_n is an n-equivalence. Moreover, $\pi_n = \pi_{n+1}(M_g, Y)$. (Thus for $B = *$, we have $\pi_n = \pi_{n+1}(CY, Y) \approx \pi_n(Y)$.)

PROOF. The first map $Y_1 \to Y_0$ is essentially just the covering map corresponding to the normal subgroup $g_{\#}\pi_1(Y)$ of $\pi_1(B)$, provided that B is locally arcwise connected and semilocally 1-connected. Most of the theorem follows immediately from Theorem 13.5 and it remains only to identify the groups π_n.

Since $Y_{n+1} \to Y_n$ is a fibration with fiber being a $K(\pi_n, n)$ we have that

$$\pi_q(Y_{n+1}) \to \pi_q(Y_n) \text{ is } \begin{cases} \text{isomorphic for } q \neq n, n+1, \\ \text{epimorphic for } q = n, \\ \text{monomorphic for } q = n+1. \end{cases}$$

It follows that

$$\pi_q(Y_n) \to \pi_q(B) \text{ is } \begin{cases} \text{isomorphic for } q > n, \\ \text{monomorphic for } q = n. \end{cases}$$

Let K be the mapping cylinder of $Y_n \to B$. Then it follows that $\pi_q(K, Y_n) = 0$ for $q \geq n+1$. Let $V = M_{g_n}$, the mapping cylinder of $g_n: Y \to Y_n$ and let W be the union of V and K along Y_n; see Figure VII-12. Then $(W, Y) \simeq (M_g, Y)$, by

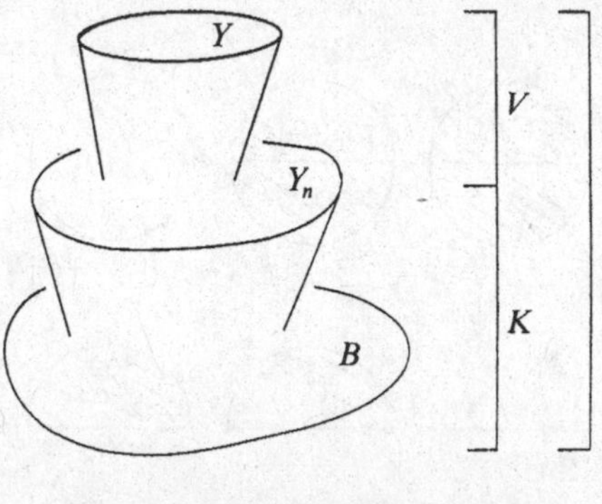

Figure VII-12. Mapping cylinders.

Theorem 14.19 of Chapter I since $K \simeq B$. We have the exact sequence

$$\pi_{n+2}(W, V) \to \pi_{n+1}(V, Y) \to \pi_{n+1}(W, Y) \to \pi_{n+1}(W, V).$$

But (K, Y_n) is a strong deformation retract of (W, V) and so the groups on the ends of this sequence are zero. It follows from the proof of Theorem 13.5, i.e., the proof of Lemma 13.6, that $\pi_n = \pi_{n+1}(V, Y)$ and hence it follows from the preceding remarks that $\pi_n = \pi_{n+1}(V, Y) \approx \pi_{n+1}(W, Y) \approx \pi_{n+1}(M_g, Y)$ as claimed. □

Recall from Problem 2 of Section 6 that any map $g: Y \to B$ is homotopy equivalent to a fibration. The fiber of such a fibration is called the "homotopy fiber" of g.

13.8. Proposition. *If F is the homotopy fiber of $q: Y \to B$ as in Theorem 13.7, then $\pi_n \approx \pi_n(F)$.*

PROOF. It suffices to treat the case of a fibration $g: Y \to B$. Let F, F_n and F'_n be the fibers of the fibrations $g: Y \to B$, $Y_{n+1} \to Y_n$, and $Y_{n+1} \to B$, respectively. The 5-lemma (in the strong form of Problem 2 of Section 5 in Chapter IV) applied to the diagram

$$\begin{array}{ccccccccc}
\pi_{n+1}(Y_{n+1}) & \longrightarrow & \pi_{n+1}(Y_n) & \longrightarrow & \pi_n(F_n) & \longrightarrow & \pi_n(Y_{n+1}) & \longrightarrow & \pi_n(Y_n) \\
\downarrow = & & \downarrow \approx & & \downarrow & & \downarrow = & & \downarrow \text{mono} \\
\pi_{n+1}(Y_{n+1}) & \longrightarrow & \pi_{n+1}(B) & \longrightarrow & \pi_n(F'_n) & \longrightarrow & \pi_n(Y_{n+1}) & \longrightarrow & \pi_n(B) \\
\text{epi} \uparrow & & \uparrow = & & \uparrow & & \uparrow \approx & & \uparrow = \\
\pi_{n+1}(Y) & \longrightarrow & \pi_{n+1}(B) & \longrightarrow & \pi_n(F) & \longrightarrow & \pi_n(Y) & \longrightarrow & \pi_n(B)
\end{array}$$

shows that the middle verticals are isomorphic, and so $\pi_n(F) \approx \pi_n(F_n) = \pi_n$. □

Whenever one has a sequence of maps

$$\cdots \xrightarrow{p_2} Y_2 \xrightarrow{p_1} Y_1 \xrightarrow{p_0} Y_0$$

one can define their "inverse limit" $\varprojlim Y_n = \{y = (y_0, y_1, y_2, \ldots) \in Y_0 \times Y_1 \times \cdots \mid p_n(y_{n+1}) = y_n\}$ with the topology induced from the product topology on $\times\, Y_n$. If one has maps $g_n: Y \to Y_n$ such that $p_n \circ g_{n+1} = g_n$ for all n then there is the induced map $g_\infty: Y \to \varprojlim Y_n$ given by

$$g_\infty(y) = (g_0(y), g_1(y), g_2(y), \ldots).$$

13.9. Proposition. *In the situation of Theorem 13.7, the projection $\varprojlim Y_i \to Y_n$ is a fibration and an n-equivalence.*

PROOF. That this is a fibration is a trivial exercise on the definition of the inverse limit and of a fibration. Suppose that (K, L) is a CW-pair with

$\dim(K - L) \leq n$. Then the lifting problem

$$
\begin{array}{ccc}
L & \longrightarrow & Y_{m+1} \\
\cap & \nearrow & \downarrow \\
K & \longrightarrow & Y_m
\end{array}
$$

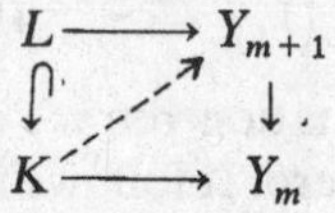

has a solution for $m \geq n$ since $H^{m+1}(K, L; \pi_m) = 0$. It follows that the lifting problem

$$
\begin{array}{ccc}
L & \longrightarrow & \varprojlim Y_i \\
\cap & \nearrow & \downarrow \\
K & \longrightarrow & Y_n
\end{array}
$$

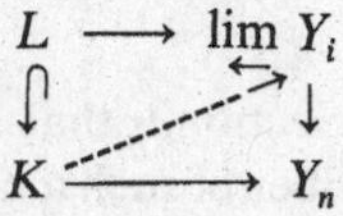

has a solution. Therefore the result follows from Theorem 11.12.   □

13.10. Corollary. *The map* $g_\infty : Y \to \varprojlim Y_n$ *is a weak homotopy equivalence.*

PROOF. In the diagram

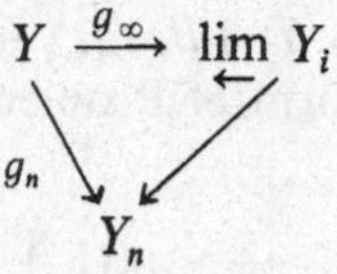

both of the maps to Y_n are n-equivalences and it follows that g_∞ is an $(n - 1)$-equivalence, for all n.   □

Let us now summarize our results to this point. For a map $p : Y \to B$ which is not a fibration, a solution to the "lifting problem" f:

$$
\begin{array}{ccc}
A & \xrightarrow{f_A} & Y \\
\cap & \nearrow & \downarrow p \\
X & \xrightarrow[f_X]{} & B
\end{array}
$$

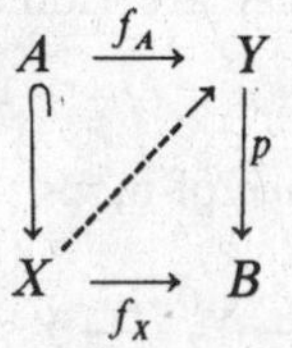

is a completion of the form

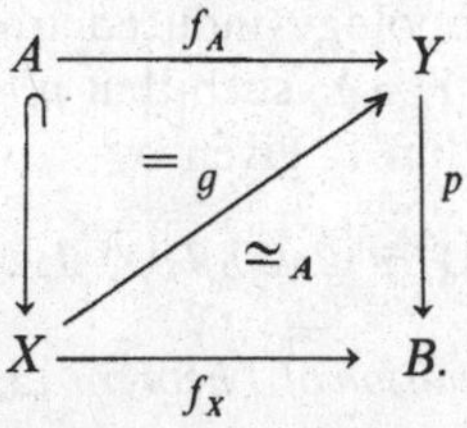

where the lower triangle commutes only up to homotopy rel A. The reader can show that this does correspond to the regular lifting problem when p is replaced by a fibration, i.e., that the lower triangle can be made to commute in that case. Then we have shown:

13.11. Theorem. *Let (X, A) be a relative CW-complex and let $p\colon Y \to B$ be a simple map with homotopy fiber F. Then for the lifting problem f:*

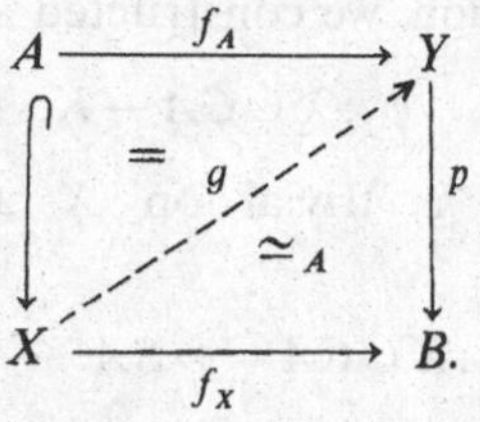

there exists a sequence of obstructions $c_f^{n+1} \in H^{n+1}(X, A; \pi_n(F))$, where all previous obstructions must be zero before the next one is defined, and where different choices of previous liftings may lead to different obstructions, such that there is a complete sequence of obstructions of which all are zero $\Leftrightarrow$ there is a solution to the lifting problem. Also, if g_0 and g_1 are solutions and if $p \circ g_0 \simeq p \circ g_1$ rel *A via the homotopy $G\colon X \times I \to B$, then there exists a sequence of obstructions*

$$d_G^n(g_0, g_1) \in H^n(X, A; \pi_n(F))$$

to lifting the homotopy G. □

Let us now specialize, for the remainder of this section, to the case $B = *$, in which the lifting problem becomes the simpler extension problem:

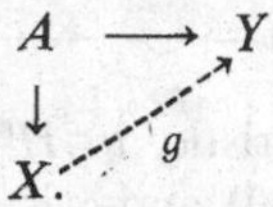

Then the map $Y \to *$ being simple reduces to the space Y being simple and arcwise connected. Then we have a sequence of obstructions

$$c_f^{n+1} \in H^{n+1}(X, A; \pi_n(Y))$$

to the existence of an extension to X. Also, for two extensions $g_0, g_1\colon X \to Y$ we get a sequence of obstructions $d^n(g_0, g_1) \in H^n(X, A; \pi_n(Y))$ to the existence of a homotopy rel A between g_0 and g_1.

Let us specialize further to the case in which Y is $(n-1)$-connected. Then the first nontrivial obstruction to extending $f\colon A \to Y$ to $g\colon X \to Y$ is

$$c_f^{n+1} \in H^{n+1}(X, A; \pi_n(Y)).$$

This is called the "primary obstruction." In this case we wish to identify the obstructions more concretely.

The first nontrivial lifting problem then comes from the diagram

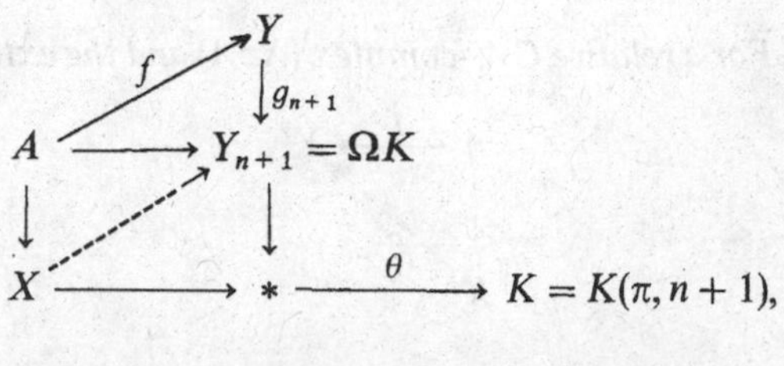

where $\pi = \pi_n(Y)$. For simplicity of notation we shall occasionally use f for the composition $A \to Y_{n+1}$ of f with g_{n+1}.

In defining the obstruction, we constructed a map

$$\phi_f: X \cup CA \to K.$$

In the present case, ϕ_f is trivial on X and so it factors through $(X \cup CA)/X = CA/A \approx SA$:

$$\phi_f: X \cup CA \xrightarrow{c} SA \xrightarrow{f'} K.$$

It is easy to check that $-f'$ is the map corresponding to f via the exponential law $K^{SA} \approx \Omega K^A$ (see Theorem 2.4) where the "$-$" indicates composition with the map $SA \to SA$ reversing the suspension parameter. (The exponential correspondence is defined even when A is not locally compact and it induces the isomorphism $[A; \Omega K] \approx [SA; K]$ of groups.) This setup then induces the diagram

$$\begin{array}{ccccccc}
[A;Y] & & & & & & \\
\downarrow & & & & & & \\
[A;\Omega K] & \xrightarrow{\approx} & [SA;K] & \xrightarrow{c^\#} & [X \cup CA; K] & \xleftarrow{\approx} & [X/A;K] \\
\downarrow T_v & & \downarrow T_u & & & & \downarrow T_u \\
H^n(A;\pi) & \xrightarrow{S} & H^{n+1}(SA;\pi) & & \xrightarrow{c^*} & & H^{n+1}(X,A;\pi)
\end{array}$$

where $v \in H^n(\Omega K; \pi)$ is characteristic, $u \in H^{n+1}(K; \pi)$ is characteristic (recall $K = K(\pi, n+1)$ and $\Omega K = K(\pi, n)$), and S is the suspension isomorphism. We shall take v as in Proposition 12.8 and then the left-hand square commutes by Proposition 12.8. Starting with $[f] \in [A; Y]$, taking it down and then to the extreme right to $[X/A; K]$ and then down to $H^{n+1}(X, A; \pi)$ yields $-c_f^{n+1}$ by definition of the latter. (The "$-$" is because there was a reversal of the suspension coordinate in the definition of c_f.) By commutativity and Proposition 12.9, we have $c_f^{n+1} = -c^*ST_v[f] = -c^*Sf^*(v) = \delta^* f^*(v)$. Similar remarks hold for the first nontrivial obstruction for a homotopy between two extensions g_0 and g_1. This obstruction is called the "primary difference" $d^n(g_0, g_1) = S^{-1}c_G^{n+1} \in H^n(X, A; \pi_n(Y))$, where $G: A \times I \cup X \times \partial I \to Y$ is $f \times 1$ on $A \times I$ and g_i on $X \times \{i\}$. Now $j^*d^n(g_0, g_1) = S^{-1}\delta^* G_X^*(v)$ where G_X is the map $X \times \partial I \to Y$ contained in G. Thus

$$j^*d^n(g_0, g_1) = S^{-1}\delta^* G_X^*(v) = i_0^* G_X^*(v) - i_1^* G_X^*(v) = g_0^*(v) - g_1^*(v)$$

by Lemma 12.10. We conclude:

13.12. Theorem. *For a relative CW-complex (X, A) and the extension problem*

$$\begin{array}{ccc}
A & \xrightarrow{f} & Y \\
\downarrow & \nearrow_{g} & \\
X & &
\end{array}$$

where Y is $(n-1)$-connected (and simple if $n=1$), the primary obstruction to extension to X is

$$c_f^{n+1} = \delta^* f^*(v) \in H^{n+1}(X, A; \pi)$$

for some characteristic class $v \in H^n(Y; \pi_n(Y)) \approx \mathrm{Hom}(H_n(Y), \pi_n(Y))$.

Similarly, if $g_0, g_1 : X \to Y$ *are two extensions of* $f: A \to Y$ *then the primary obstruction to a homotopy* rel *A between them, called the "primary difference," is*

$$d^n(g_0, g_1) = S^{-1}\delta^* G^*(v) \in H^n(X, A; \pi_n(Y)),$$

where $G: A \times I \cup X \times \partial I \to Y$ *is* $f \times 1$ *on* $A \times I$ *and* g_i *on* $X \times \{i\}$. *Also*

$$j^* d^n(g_0, g_1) = g_0^*(v) - g_1^*(v) \in H^n(X; \pi_n(Y)),$$

where $j^*: H^n(X, A; \pi_n(Y)) \to H^n(X; \pi_n(Y))$ *is the canonical map.* □

It is worth noting that there is an easy direct proof that this *is* an obstruction to an extension, since, if f extends to $g: X \to Y$, then the diagram

$$\begin{array}{ccccc} H^n(Y;\pi) & = & H^n(Y;\pi) & & \\ \Big\downarrow{\scriptstyle g^*} & & \Big\downarrow{\scriptstyle f^*} & & \\ H^n(X;\pi) & \xrightarrow{i^*} & H^n(A;\pi) & \xrightarrow{\delta^*} & H^{n+1}(X,A;\pi) \end{array}$$

shows that $\delta^* f^* = 0$ since it equals $\delta^* i^* g^* = 0$ because $\delta^* i^* = 0$.

In order to derive some concrete applications let us further restrict to the case in which the only possible nonzero obstruction is the primary one.

13.13. Corollary. *Suppose that (X, A) is a relative CW-complex and that Y is $(n-1)$-connected. Assume that $H^{i+1}(X, A; \pi_i(Y)) = 0$ for all $i > n$. Then a map $f: A \to Y$ can be extended to $g: X \to Y$ if and only if $\delta^* f^*: H^n(Y; \pi) \to H^{n+1}(X, A; \pi)$ is trivial, where $\pi = \pi_n(Y)$.*

PROOF. If $\delta^* f^* = 0$ then the extension exists since the only obstruction is $c_f^{n+1} = \delta^* f^*(v) = 0$. The converse follows from the preceding remark. □

Similarly, for $A = \emptyset$, we get:

13.14. Corollary. *Let X be a CW-complex. Let Y be $(n-1)$-connected and assume that $H^i(X; \pi_i(Y)) = 0$ for all $i > n$. Let $\pi = \pi_n(Y)$. Then two maps $g_0, g_1: X \to Y$ are homotopic $\Leftrightarrow g_0^* = g_1^*: H^n(Y; \pi) \to H^n(X; \pi)$.* □

If we assume only one possible nonzero obstruction to both the extension and the homotopy problems we get the following generalization of Hopf's Theorem (Theorem 11.6 of Chapter V) on maps to spheres.

13.15. Theorem. *Let (X, A) be a CW-pair. Let Y be $(n-1)$-connected, and* *simple if* $n=1$. *Assume that*

$$H^i(X,A;\pi_i(Y))=0=H^{i+1}(X,A;\pi_i(Y))$$

for all $i>n$. *Let* $f_0\colon X\to Y$ *be given. Using* $[X;Y]_A$ *to denote the homotopy classes rel* A *of maps* $X\to Y$ *which equal* f_0 *on* A, *there is a one–one correspondence*

$$[X;Y]_A\leftrightarrow H^n(X,A;\pi_n(Y))$$

given by $[f_1]\mapsto d^n(f_0,f_1)$.

Proof. We will first show that the indicated correspondence is onto. Let $W=A\times I\cup X\times\partial I$ and $W_0=A\times I\cup X\times\{0\}$. Consider the extension problem (with $\pi=\pi_n(Y)$):

$$\begin{array}{ccc} W_0 & \xrightarrow{f_0\circ \text{proj}} & Y_{n+1}=K(\pi,n) \\ \big\downarrow & \nearrow & \\ W. & & \end{array}$$

Note that, under $[X;Y_{n+1}]\xrightarrow{\text{proj}}[W_0;Y_{n+1}]\xrightarrow{T_v}H^n(W_0;\pi)$, $f_0^\#$ goes to $f_0^*(v)$.

Let $\alpha\in H^n(X,A;\pi)$ so that $S\alpha\in H^{n+1}((X,A)\times(I,\partial I);\pi)=H^{n+1}(X\times I,W;\pi)$. Consider the commutative diagram (coefficients in π):

$$\begin{array}{ccccc} & & H^n(W,W_0) & & \\ & \swarrow & & \searrow & \\ & H^n(W) & \xrightarrow{\delta^*} & H^{n+1}(X\times I,W) & \\ \nearrow & \big\downarrow & & \big\downarrow & \\ H^n(X\times I) & & & & \\ \searrow & H^n(W_0) & \xrightarrow{\delta^*} & H^{n+1}(X\times I,W_0)=0. & \end{array}$$

An easy diagram chase shows that there is an element $\xi\in H^n(W)$ going to $f_0^*(v)$ in $H^n(W_0)$ and to $S\alpha$ in $H^{n+1}(X\times I,W)$. By $H^n(W;\pi)\approx[W;Y_{n+1}]$, this means that there exists a map $F_{n+1}\colon W\to Y_{n+1}$ such that $\delta^*F_{n+1}^*(v)=S\alpha$ and which equals f_0 on $X\times\{0\}$. Let $f_1\colon X\to Y_{n+1}$ correspond to the restriction of F_{n+1} to $X\times\{1\}$.

Now we can extend F_{n+1} to $F\colon W\to Y$ because the obstructions to doing this are in $H^{i+1}(X,A;\pi_i(Y))=0$ for $i>n$. Also $d^n(f_0,f_1)=S^{-1}\delta^*F^*(v)=\alpha$, completing the proof that the correspondence is onto.

To show that the correspondence is one–one, suppose that $d^n(f_0,f_1)=d^n(f_0,f_2)$. By the additivity property (Theorem 13.3) of the difference obstructions, we conclude that $d^n(f_1,f_2)=0$. This is the primary obstruction to making $f_1\simeq f_2$ rel A. The rest of the obstructions are in $H^i(X,A;\pi_i(Y))=0$ for $i>n$, so we are done. □

13.16. Corollary. *Let* Y *be* $(n-1)$*-connected. If* X *is a* CW*-complex such that* $H^{i+1}(X;\pi_i(Y))=0=H^i(X;\pi_i(Y))$ *for all* $i>n$, *then there is a one–one*

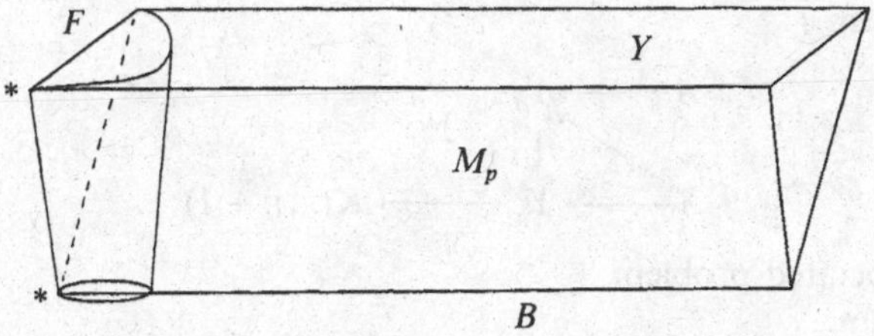

Figure VII-13. Mapping cylinder of a fibration.

correspondence

$$[X;Y] \leftrightarrow H^n(X;\pi_n(Y))$$

given by $f \mapsto f^*(u)$ *where* $u \in H^n(Y;\pi_n(Y))$ *is characteristic.*

PROOF. If $f_0: X \to Y$ is a constant map then $-d^n(f_0, f) = f^*(u) - f_0^*(u) = f^*(u)$ by Theorem 13.12. □

PROBLEMS

1. Let $p: Y \to B$ be a fibration with fiber F. Because of the inclusion $(CF, F) \hookrightarrow (M_p, Y)$ there is the homomorphism
$$\pi_n(F) \xleftarrow{\approx} \pi_{n+1}(CF, F) \longrightarrow \pi_{n+1}(M_p, Y).$$
Show that the diagram
$$\begin{array}{ccccccccc} \cdots \longrightarrow & \pi_{n+1}(B) & \longrightarrow & \pi_n(F) & \longrightarrow & \pi_n(Y) & \longrightarrow & \pi_n(B) & \longrightarrow \cdots \\ & \downarrow \approx & & \downarrow & & \downarrow = & & \downarrow \approx & \\ \cdots \longrightarrow & \pi_{n+1}(M_p) & \longrightarrow & \pi_{n+1}(M_p, Y) & \longrightarrow & \pi_n(Y) & \longrightarrow & \pi_n(M_p) & \longrightarrow \cdots \end{array}$$
commutes, and hence that $\pi_n(F) \xrightarrow{\approx} \pi_{n+1}(M_p, Y)$, giving a more conceptual proof of Proposition 13.8. Figure VII-13 provides a hint.

2. If $p: Y \to B$ is a fibration with arcwise connected fiber F, show that p is simple $\Leftrightarrow \pi_1(B)$ acts trivially on $\pi_n(F)$ for all $n \geq 1$; see Problem 2 of Section 7. (*Hint*: Use Problem 1.)

3. ♦ If $p: Y \to B$ is a fiber-orientable sphere bundle then show that p is simple. (*Hint*: Use Problem 2 and the fact that a map $\mathbf{S}^n \to \mathbf{S}^n$ of degree one is homotopic to the identity.)

14. Obstruction Cochains and Vector Bundles ☼

We again take up the lifting problem. Let (X, A) be a relative CW-complex and let $X^{(k)}$ be the union of A with the k-skeleton of X. Consider the principal lifting problem f:

$$\begin{array}{ccccc} A & \longrightarrow & Y_{n+1} & & \\ \downarrow & \nearrow & \downarrow & & \\ X & \longrightarrow & Y_n & \longrightarrow & K(\pi, n+1) \end{array}$$

and the associated problem

$$\begin{array}{ccc} A & \longrightarrow & Y_{n+1} \\ \downarrow & \nearrow & \downarrow \\ X^{(n)} & \longrightarrow & Y_n. \end{array}$$

The obstruction to the latter is in $H^{n+1}(X^{(n)}, A; \pi) = 0$ and so the lifting exists.

Next consider the lifting problem

$$\begin{array}{ccc} X^{(n)} & \longrightarrow & Y_{n+1} \\ \downarrow & \nearrow & \downarrow \\ X & \longrightarrow & Y_n \end{array}$$

and the lifting problem

$$\begin{array}{ccc} X^{(n)} & \longrightarrow & Y_{n+1} \\ \downarrow & \nearrow & \downarrow \\ X^{(n+1)} & \longrightarrow & Y_n. \end{array}$$

The associated obstructions are related by the maps

$$\begin{array}{l} c_f^{n+1} \in H^{n+1}(X, A; \pi) \\ \qquad\qquad \uparrow \\ \qquad H^{n+1}(X, X^{(n)}; \pi) = Z^{n+1}(X, A; \pi) = \text{cocycles} \\ \qquad\qquad \downarrow \text{mono} \\ \bar{c}_f^{n+1} \in H^{n+1}(X^{(n+1)}, X^{(n)}; \pi) = C^{n+1}(X, A; \pi) \end{array}$$

where we don't name the middle obstruction. As indicated, one can identify the middle group with the cellular cocycles because it is canonically isomorphic to $H^{n+1}(X^{(n+2)}, X^{(n)}; \pi)$ and the sequence

$$0 \to H^{n+1}(X^{(n+2)}, X^{(n)}; \pi) \longrightarrow H^{n+1}(X^{(n+1)}, X^{(n)}; \pi) \xrightarrow{\delta^*} H^{n+2}(X^{(n+2)}, X^{(n+1)}; \pi)$$

is exact.

This relationship between the three obstructions shows that $\bar{c}_f^{n+1} \in C^{n+1}(X, A; \pi)$ is a cocycle and it represents $c_f^{n+1} = [\![\bar{c}_f^{n+1}]\!] \in H^{n+1}(X, A; \pi)$. It is important to realize that it *depends* on the choice of the lifting to $X^{(n)} \to Y_{n+1}$.

Let σ be an $(n+1)$-cell of X and consider the diagram

$$\begin{array}{ccccc} \partial\sigma & \longrightarrow & X^{(n)} & \longrightarrow & Y_{n+1} \\ \downarrow & & \downarrow & \nearrow & \downarrow \\ \sigma & \longrightarrow & X & \longrightarrow & Y_n. \end{array}$$

Then the obstruction $\bar{c}_f^{n+1}$ maps to that for the problem

$$\begin{array}{ccccc} \partial\sigma & \longrightarrow & Y_{n+1} & & \\ \downarrow & \nearrow & \downarrow & & \\ \sigma & \longrightarrow & Y_n & \longrightarrow & K(\pi, n+1) \end{array}$$

which is in $H^{n+1}(\sigma, \partial\sigma; \pi) \approx \pi$. Since $\sigma \to Y_n$ is homotopically trivial, this lifting problem is equivalent to the extension problem

$$\begin{array}{ccc} \partial\sigma & \xrightarrow{h} & K(\pi, n) \\ \downarrow & \nearrow & \downarrow \\ \sigma & \longrightarrow & * \end{array}$$

and the obstruction here is just the cochain $\sigma \mapsto [h] \in \pi_n(K(\pi, n)) \approx \pi$. It is suggestive to use the notation $\partial_\#[f \circ \chi_\sigma] \in \pi_n(F) \approx \pi$ for this class $[h]$, where $\chi_\sigma: \sigma \to X$ is the characteristic map for the cell σ and $F = \Omega K(\pi, n+1) = K(\pi, n)$ is the fiber of $Y_{n+1} \to Y_n$. Thus

$$\bar{c}_f^{n+1}(\sigma) = \partial_\#[f \circ \chi_\sigma].$$

(The actual identification of $\bar{c}_f^{n+1}(\sigma)$ as an element of π depends on the choice of characteristic class for $K(\pi, n+1)$. A different choice acts by an automorphism of π independent of σ. None of this matters from a practical standpoint.)

One can start with this formula for $\bar{c}_f^{n+1}$ as another approach to obstruction theory. That is, in fact, a more traditional method; see G. Whitehead [1].

Now let us pass to the general lifting problem:

$$\begin{array}{ccc} A & \longrightarrow & Y \\ \downarrow & \nearrow & \downarrow p \\ X & \longrightarrow & B \end{array}$$

and let $\cdots \to Y_2 \to Y_1 \to Y_0 = B$ be a Moore–Postnikov decomposition of p.

The following diagram *indicates* some of the relationships among the various associated lifting problems:

$$\begin{array}{ccccc} [X; Y_n]_A & \xleftrightarrow{\ll 3 \gg} & [X^{(n)}; Y_n]_A & \xleftarrow{\ll 1 \gg} & [X^{(n)}; Y]_A \\ \ll 4 \gg \Big\downarrow \Big\uparrow & & \Big\uparrow & \searrow^{\ll 2 \gg} \quad [X^{(n-1)}; Y]_A \quad \swarrow & \Big\uparrow \\ & & & \nwarrow & \\ [X; Y_{n+1}]_A & \xleftrightarrow{\ll 3 \gg} & [X^{(n+1)}; Y_{n+1}]_A & \xleftarrow{\ll 1 \gg} & [X^{(n+1)}; Y]_A. \end{array}$$

Notes on this diagram:

$\ll 1 \gg$ onto since obstructions to lifting are zero;
$\ll 2 \gg$ liftings of $\ll 1 \gg$ are homotopic on $X^{(n-1)}$;

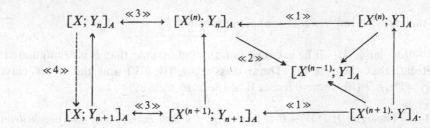

«3» here obstructions to extension and homotopy are all zero;
«4» $c_f^{n+1} \in H^{n+1}(X, A; \pi_n(F))$ is the only obstruction to this lifting.

In addition we have shown:

14.1. Theorem. *Let (X, A) be a relative CW-complex and $p: Y \to B$ a simple fibration with fiber F. Consider the lifting problem:*

$$\begin{array}{ccc} A & \xrightarrow{f_A} & Y \\ \cap\big\downarrow & \nearrow & \big\downarrow p \\ X & \xrightarrow[f_X]{} & B. \end{array}$$

Suppose we are given a lifting $f_n: X^{(n)} \to Y$. Then $\bar{c}_f^{n+1} \in C^{n+1}(X, A; \pi_n(F))$ is defined and is a cocycle representing c_f^{n+1}. It is given by

$$\bar{c}_f^{n+1}(\sigma) = \partial_\#[f \circ \chi_\sigma] \in \pi_n(F)$$

where $\chi_\sigma: (\mathbf{D}^{n+1}, \mathbf{S}^n) \to (X, X^{(n)})$ is the characteristic map for the $(n+1)$-cell σ. Also, f_n extends to a lifting $X^{(n+1)} \to Y \Leftrightarrow \bar{c}_f^{n+1} = 0$. Moreover, the restriction $f_{n-1}: X^{(n-1)} \to Y$ of f_n extends to a lifting $X^{(n+1)} \to Y \Leftrightarrow 0 = c_f^{n+1} = [\![\bar{c}_f^{n+1}]\!] \in H^{n+1}(X, A; \pi_n(F))$. □

Similar considerations apply to obstructions to homotopies.

We shall now apply these remarks to the case of an orientable k-plane (vector) bundle $p: E(\xi) \to B$ where B is a CW-complex of dimension n. The orientability of the bundle ξ implies the simplicity of the associated sphere bundle of p by Problem 3 of Section 13. Then obstructions to a nonzero section of ξ are in $H^{i+1}(B; \pi_i(\mathbf{R}^k - \{0\})) \approx H^{i+1}(B; \pi_i(\mathbf{S}^{k-1}))$ and this is zero for $i \geq n$ and for $i < k-1$. Therefore all obstructions vanish if $k > n$. This proves:

14.2. Corollary. *If ξ^k is an orientable k-plane bundle over a CW-complex B of dimension n and if $k > n$ then there exists an n-plane bundle η^n over B such that*

$$\xi^k \approx \eta^n \oplus \epsilon^{k-n}$$

where ϵ^{k-n} is the trivial $(k-n)$-plane bundle. □

Now let $\xi: W \to B$ be a k-disk bundle and assume that B is triangulated. Recall that there is the Thom class $\tau_\xi \in H^k(W, \partial W)$ and the Euler class $\chi_\xi = i^*\tau_\xi \in H^k(B)$ where $i: B \hookrightarrow W$ is the zero section.

14.3. Theorem. *If $\xi: W \to B$ is an oriented k-disk bundle over the polyhedron B then the primary obstruction to a nonzero section is the Euler class*

$$\chi_\xi = i^*\tau_\xi \in H^k(B).$$

Proof. There are no obstructions to construction of a section $j_0: B^{(k-1)} \to \partial W$

and j_0 extends to a map $j: B \to W$ by local triviality. Then $\chi_\xi = i^*\tau_\xi = j^*\tau_\xi$ since $i \simeq j$. Now $j:(B^{(k)}, B^{(k-1)}) \to (W, \partial W)$ and so

$$j^*: H^k(W, \partial W) \to H^k(B^{(k)}, B^{(k-1)}) = C^k(B)$$

takes τ_ξ to a cochain $c^k = j^*(\tau_\xi)$. For a k-cell σ of B, we have that

$$c^k(\sigma) = j_\sigma^*(\tau_\xi|_\sigma),$$

where j_σ is the restriction of j to $(\sigma, \partial\sigma) \to (W, \partial W)$. Now W is trivial over σ and so this can be thought of as a map $(\sigma, \partial\sigma) \to (\mathbf{D}^k, \mathbf{S}^{k-1})$. Since the Thom class represents a generator of $H^k(\mathbf{D}^k, \mathbf{S}^{k-1})$ in each fiber, it is clear that $j_\sigma^*(\tau_\xi|_\sigma)$ is just the degree of j_σ which is the same as $\partial_\#[j \circ \chi_\sigma] = c_j^k(\sigma)$ by Theorem 14.1, showing that the primary obstruction is

$$c_j^k = [\![c^k]\!] = j^*(\tau_\xi) = \chi_\xi$$

as claimed. □

14.4. Corollary. *If ξ is an orientable n-plane bundle over the n-dimensional complex B then ξ has a nonzero section $\Leftrightarrow 0 = \chi_\xi \in H^n(B)$.* □

14.5. Corollary (Hopf). *A smooth connected orientable closed manifold M^n has a nonzero tangent vector field $\Leftrightarrow \chi(M) = 0$.* □

14.6. Corollary. *Let $M^n \hookrightarrow \mathbf{S}^{n+2}$ be an embedded orientable smooth submanifold. Then the normal bundle is trivial.*

Proof. By Proposition 12.2 of Chapter VI, the Euler class χ of the normal bundle ν^2 to M in $\mathbf{S}^{n+2}$ is zero. The higher obstructions to a section are in $H^{i+1}(M; \pi_i(\mathbf{S}^1)) = 0$ for $i > 1$. Therefore, there is a nonzero section. Thus the normal bundle splits as $\nu^2 = \epsilon^1 \oplus \xi^1$ for some line bundle ξ^1. But ν^2 is orientable, whence ξ^1 is orientable, which means that it is trivial. Therefore ν^2 is trivial. □

14.7. Theorem. *If $M^n \hookrightarrow \mathbf{S}^{n+2}$ is a smooth embedding of the closed orientable manifold M^n then $M^n = \partial V^{n+1}$ for some compact orientable manifold $V^{n+1} \subset \mathbf{S}^{n+2}$.*

Proof. By Corollary 14.6 the normal bundle is trivial and so we can regard a tubular neighborhood of M^n as an embedding $M^n \times \mathbf{D}^2 \subset \mathbf{S}^{n+2}$. Let $K = \mathbf{S}^{n+2} - \operatorname{int}(M^n \times \mathbf{D}^2)$ and consider the exact Mayer–Vietoris sequence

$$H^1(\mathbf{S}^{n+2}) \to H^1(K) \oplus H^1(M \times \mathbf{D}^2) \to H^1(M \times \mathbf{S}^1) \to H^2(\mathbf{S}^{n+2}).$$

It follows that

$$H^1(K) \oplus H^1(M) \approx H^1(M \times \mathbf{S}^1) \approx [M \times \mathbf{S}^1; \mathbf{S}^1].$$

Consider the diagram

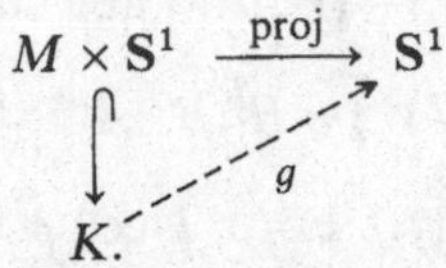

Assume first that $H^1(M)=0$. Then $H^1(K) \xrightarrow{\approx} H^1(M \times \mathbf{S}^1) \approx [M \times \mathbf{S}^1; \mathbf{S}^1]$ and this implies that the prospective map $g: K \to \mathbf{S}^1$ exists. We can assume that g is smooth. Then put $V_0^{n+1} = g^{-1}(x)$ where x is a regular value. Since V_0^{n+1} intersects $M^n \times \mathbf{S}^1$ in $M^n \times \{x\}$, it extends in an obvious manner to the desired V with $\partial V = M$. V is orientable because it has a trivial normal bundle. It can be assumed that g is the projection to $\mathbf{S}^1$ in some neighborhood of $M \times \mathbf{S}^1$ of the form $M \times \mathbf{S}^1 \times I$ and then the resulting V is smooth.

If $H^1(M) \neq 0$ then the element of $H^1(M \times \mathbf{S}^1)$ corresponding to the projection $M \times \mathbf{S}^1 \to \mathbf{S}^1$ goes into the pair $(\alpha, \beta) \in H^1(K) \oplus H^1(M)$ by the isomorphism $H^1(M \times \mathbf{S}^1) \approx H^1(K) \oplus H^1(M)$ above. The element $\beta \in H^1(M)$ corresponds to a map $f: M \to \mathbf{S}^1$. Then the map $M \times \mathbf{S}^1 \to \mathbf{S}^1$ taking $(x, t) \mapsto (f(x)^{-1}t)$ gives a reframing of M. The reader can check that under this reframing, the projection $M \times \mathbf{S}^1 \times \mathbf{S}^1$ now corresponds to $(\alpha, 0) \in H^1(K) \oplus H^1(M)$, and then the argument in the case $H^1(M) = 0$ applies to this case as well. □

A particular case of Theorem 14.7 is a smooth embedding of $\mathbf{S}^1$ in $\mathbf{S}^3$ (or $\mathbf{R}^3$), called a "knot." Thus every knot bounds an orientable surface in $\mathbf{R}^3$. Such a surface is a sphere with handles and with a disk removed. The number of handles is called the "genus" of the surface. For a given knot there is such a surface with minimal genus and then that genus is called the "genus of the knot." Figure VII-14 shows a knot of genus 1 (the cloverleaf) and part of the orientable surface of minimal genus it bounds.

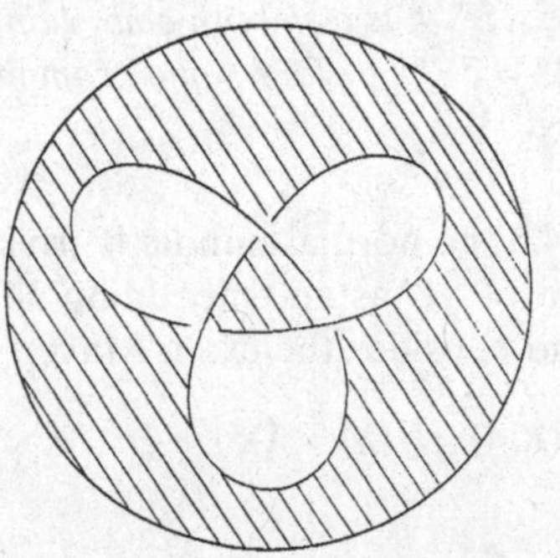

Figure VII-14. Knot of genus 1 and spanning surface.

14.8. Corollary. *There is no smooth embedding of* $\mathbf{CP}^2$ *in* $\mathbf{S}^6$.

PROOF. If there were such an embedding then $\mathbf{CP}^2 = \partial V^5$, with V^5 orientable, contrary to Corollary 10.6 of Chapter VI. □

More generally, by the results in Section 10 of chapter VI, a codimension 2 closed orientable submanifold of $\mathbf{S}^n$ must have even Euler characteristic and zero signature.

14.9. Theorem. *An orientable k-plane bundle* ξ *over an n-dimensional complex* B *is stable for* $k > n$. *That is, if* ξ, ξ' *are two such bundles and if* $\xi \oplus \eta \approx \xi' \oplus \eta$ *for some vector bundle* η *then* $\xi \approx \xi'$, *if* $k > n$.

PROOF. This proof will use some unproved, but elementary, facts about vector bundles. By Theorem 14.2 of Chapter II and the remark below it, there exists a vector bundle ν such that $\nu \oplus \eta$ is trivial. This implies that it suffices to prove the theorem in case $\eta = \epsilon$ is the trivial line bundle. If $\xi \oplus \epsilon \approx \xi' \oplus \epsilon$ then there is a bundle ρ over $B \times I$ which is $\xi \oplus \epsilon$ on one end and $\xi' \oplus \epsilon$ on the other. Then ϵ provides a section s of ρ over $B \times \partial I$. The obstructions to extending s to $B \times I$ are in $H^{i+1}(B \times (I, \partial I); \pi_i(\mathbf{S}^k))$ which is 0 for $i < k$ (hence for $i \leq n$) and for $i + 1 > n + 1$; hence for all i. Therefore s does extend. The complement to s is a bundle over $B \times I$ whose ends are ξ and ξ'. But a bundle over $B \times I$ is isomorphic to a product with I of a bundle over B and so $\xi \approx \xi'$ as claimed. □

14.10. Corollary. *If* M^n *is a smooth closed stably parallelizable submanifold of* $\mathbf{R}^{n+k}$ *and* $k > n$, *then the normal bundle is trivial.*

PROOF. The Whitney sum $\tau \oplus \nu$ is the restriction of the tangent bundle of $\mathbf{R}^{n+k}$ and so is trivial. But $\tau \oplus \epsilon^k$ is also trivial for $k > n$ by the definition of "stably parallelizable" and by Theorem 14.9. It follows from Theorem 14.9 that $\nu \approx \epsilon^k$. □

Appendices

Dare to be naive.

BUCKMINSTER FULLER

A. The Additivity Axiom

In the proof of the uniqueness of homology theories for CW-complexes we needed, in the case of infinite-dimensional complexes, a result that asserted that the map $\varinjlim_j H_p(K^{(j)}) \to H_p(K)$ is an isomorphism; see Section 10 of Chapter IV. This will be established here. (Note that this is trivial for singular theory and the whole point is in proving it from only the axioms.) It is a consequence of the Additivity Axiom, due to Milnor. Also used is the Mayer–Vietoris sequence, which can be proved from the axioms according to Problem 4 of Section 18 of Chapter IV. We also need Corollary 11.14 of Chapter VII.

First, for a CW-complex K, consider the product complexes $K^{(n)} \times [n, n+1]$, $0 \le n$, and their union T, called a "telescope," see Figure A-1. We can describe a point of T by (x, t) where $x \in K^{(n)}$ if $n \le t < n+1$. There is the map $\theta\colon T \to K$ taking (x, t) to x. We claim that this is a homotopy equivalence. Choose a base point $x_0 \in K^{(0)}$. For any $0 \le t_0 < \infty$ we will identify $\pi_*(T, \{x_0, t_0\})$ with $\pi_*(T, \{x_0, 0\})$ via the path $\{x_0\} \times [0, t_0]$. This being said, we can disregard base points.

For any cellular map $f\colon \mathbf{S}^n \to K$ we know that its image is in $K^{(n)}$ and thus $g(x) = (f(x), n) \in K^{(n)} \times \{n\} \subset T$ is defined. Clearly $[f] = \theta_{\#}[g]$, so that $\theta_{\#}$ is onto. Any cellular map $g\colon \mathbf{S}^n \to T$ has image in

$$K^{(0)} \times [0,1] \cup K^{(1)} \times [1,2] \cup \cdots \cup K^{(n)} \times [n, n+1] \cup K^{(n+1)} \times [n+1, \infty)$$

and thus is homotopic, along $\{x_0\} \times [0, \infty)$, to a map into $K^{(n+1)} \times \{n+1\}$. If $\theta \circ g$ is homotopically trivial in K then it is so in $K^{(n+1)}$ (by a cellular homotopy). But that homotopy can be pulled back to $K^{(n+1)} \times \{n+1\}$, showing that g is homotopically trivial in $K^{(n+1)} \times \{n+1\} \subset T$. These two facts imply that $\theta_{\#}: \pi_n(T) \to \pi_n(K)$ is an isomorphism for all n. By Corollary 11.14 of Chapter VII, $\theta: T \to K$ is a homotopy equivalence.

Also note that θ restricts to

$$\theta_n: T_n = K^{(0)} \times [0,1] \cup \cdots \cup K^{(n-1)} \times [n-1, n] \cup K^{(n)} \times \{n\} \to K^{(n)}$$

and this restriction is clearly a homotopy equivalence because of the telescoping of T_n onto $K^{(n)} \times \{n\}$. We have the commutative diagram

$$\begin{array}{ccc} H_p(T_n) & \xrightarrow[\approx]{\theta_{n*}} & H_p(K^{(n)}) \\ \big\downarrow & & \big\downarrow \\ H_p(T) & \xrightarrow[\approx]{\theta_*} & H_p(K) \end{array}$$

where the verticals are induced by inclusion. This induces

$$\begin{array}{ccc} \varinjlim_n H_p(T_n) & \xrightarrow{\approx} & \varinjlim_n H_p(K^{(n)}) \\ \big\downarrow & & \big\downarrow \\ H_p(T) & \xrightarrow{\approx} & H_p(K). \end{array}$$

Our desired result is that the vertical map on the right is an isomorphism. Thus it suffices to prove that the vertical map on the left is an isomorphism. We will prove a slightly more general result.

Consider a sequence of spaces $X_0, X_1, \ldots$ and maps $f_n: X_n \to X_{n+1}$ for all $n \geq 0$. Let Y_n be the mapping cylinder of f_n but with parameter values in $[n, n+1]$. That is, Y_n is the quotient space of $X_n \times [n, n+1] \cup X_{n+1}$ by the relation $(x, n+1) \sim f_n(x)$. We will consider X_n as embedded as the "top" of Y_n and X_{n+1} as the "bottom." Then we can form the telescopic union $T = Y_0 \cup Y_1 \cup \cdots$; see Figure A-1. Again, we can describe points in T by pairs

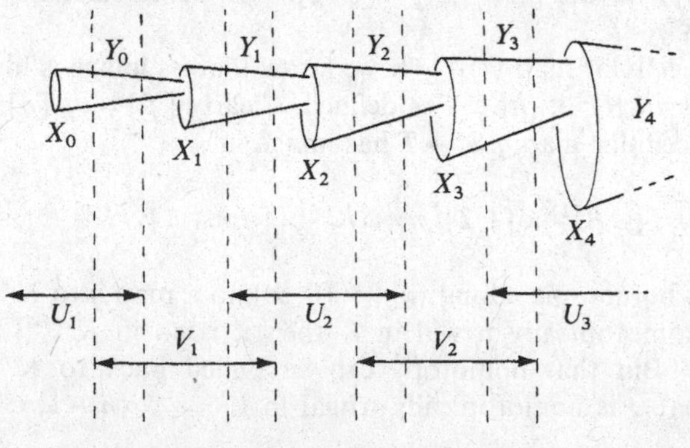

Figure A-1. A telescope.

(y, t) where $y \in X_n$ when $n \leq t < n+1$. Put

$$U_n = \{(y,t) \in T \mid 2(n-1) - \tfrac{2}{3} < t < 2(n-1) + \tfrac{2}{3}\},$$
$$V_n = \{(y,t) \in T \mid 2n - \tfrac{5}{3}t < t < 2n - \tfrac{1}{3}\},$$

for $n \geq 1$. Then the U_n are disjoint open sets in T homotopy equivalent to $X_{2(n-1)}$. The V_n are disjoint open sets in T homotopy equivalent to X_{2n-1}. Let $U = \bigcup U_n$ and $V = \bigcup V_n$. Then $U \cap V$ is the disjoint union $U_1 \cap V_1 + V_1 \cap U_2 + U_2 \cap V_2 + \cdots$ which is homotopically equivalent to the disjoint union $X_0 + X_1 + \cdots$. By the Additivity Axiom, $H_p(U \cap V)$ can be described as

$$H_p(U \cap V) = \{(a_0, a_1, \ldots) \mid a_i \in H_p(X_i),\ a_i = 0 \text{ for large } i\}.$$

Similarly,

$$H_p(U) = \{(a_0, 0, a_2, 0, \ldots) \mid a_{2i} \in H_p(X_{2i}),\ a_{2i} = 0 \text{ for large } i\},$$
$$H_p(V) = \{(0, a_1, 0, a_3, \ldots) \mid a_{2i+1} \in H_p(X_{2i+1}),\ a_{2i+1} = 0 \text{ for large } i\}.$$

Also, it is clear that the homomorphism $H_p(U \cap V) \to H_p(U)$ takes

$$(a_0, a_1, \ldots) \mapsto (a_0, 0, a_2 + f_{1*}(a_1), 0, a_4 + f_{3*}(a_3), 0, \ldots)$$

and $H_p(U \cap V) \to H_p(V)$ takes

$$(a_0, a_1, \ldots) \mapsto (0, a_1 + f_{0*}(a_0), 0, a_3 + f_{2*}(a_2), 0, \ldots).$$

Taking the direct sum of the first of these maps with the negative of the second gives the map

$$\phi\colon H_p(U \cap V) \to H_p(U) \oplus H_p(V)$$

described by

$$\phi(a_0, a_1, \ldots) = (a_0, -a_1 - f_{0*}(a_0), a_2 + f_{1*}(a_1), -a_3 - f_{2*}(a_2), \ldots).$$

Put $\phi(a_0, a_1, \ldots) = (b_0, b_1, \ldots)$. If the b_i are all 0 then so are the a_i as is seen by an induction. Thus ϕ is a monomorphism. The map ϕ is one of the maps in the Mayer–Vietoris sequence for (U, V). (See Problem 4 of Section 18 of Chapter IV.) It follows that $H_*(T) = H_*(U \cup V)$ is the cokernel of ϕ.

The image of ϕ is generated by the

$$\phi(0, 0, \ldots, a_i, 0, \ldots) = (0, \ldots, \pm a_i, \mp f_{i*}(a_i), 0, \ldots).$$

These are exactly the relations defining the direct limit of the factors $H_p(X_n)$. Consequently, we have proved that the maps $H_p(X_n) \to H_p(T)$ induce an isomorphism $\varinjlim_n H_p(X_n) \approx H_p(T)$, which is our desired result.

The same proof works for pairs of CW-complexes, but the result for such pairs also follows immediately from the 5-lemma.

For more on this topic see Milnor [2].

B. Background in Set Theory

Intuitively, a "set" is a collection of objects called "members" of the set. In mathematics the notion of a "set" is taken as undefined, as is the relation of membership, and axioms are put down for these to follow. We will not do so in this "naive" treatment. Of course, it is well known that an undisciplined approach easily leads to logical difficulties such as the "set of all sets that do not contain themselves." These problems are handled in set theory by careful treatment of the axioms. But that is not the purpose of this appendix. We merely intend to set down terminology and notation that the reader must already have a feeling for, or he would not be studying this book. We will briefly discuss some "obvious" concepts and results, and will then prove some things that are not so obvious.

The terms "collection" or "family" are synonyms of "set," although the term "family" is usually used only for somewhat complicated sets such as a family of subsets of a set or a family of functions. The term "class" is often used as a synonym for "set," but in axiomatic set theory, it is used for a more encompassing concept: a "set" is a class that is a member of another class. A "proper class" is a class that is not a "set." The phrase "the class of all sets that do not contain themselves" is meaningful, but "the class of all classes that do not contain themselves" is not. We will not worry about such things, but we will avoid the use of the term "class" when we mean a "set." (An exception to this is the use of "class" in the phrase "equivalence class" which is traditional.)

We shall use the logical symbols $\exists$ to mean "there exists", $\exists 1$ to mean "there exists a unique," $\forall$ to mean "for all," $\ni$ to mean "such that," $\Rightarrow$ to mean "implies," $\Leftarrow$ to mean "is implied by," and $\Leftrightarrow$ to mean "if and only if."

If an object x is a member of a set S then we write $x \in S$. If not then we write $x \notin S$. If $P(x)$ is a statement about objects x which can be true or false for a given object x, then $\{x \mid P(x)\}$ stands for the set of all objects for which $P(x)$ is true, provided this does in fact define a set. If S is a set then $\{x \in S \mid P(x)\}$ is the same as $\{x \mid x \in S \text{ and } P(x)\}$.

If S and T are sets then we say S is contained in T, or S is a "subset" of T if $x \in S \Rightarrow x \in T$. This is denoted by $S \subset T$ or $T \supset S$. The statement $S \subset S$ is true for all sets S. If $S \subset T$ is false then we write $S \not\subset T$.

The "empty set" $\emptyset$ is the unique set with no objects, i.e., $x \in \emptyset$ is false for all objects x. The statement $\emptyset \subset S$ is true for all sets S.

The "union" of two sets S and T is $S \cup T = \{x \mid x \in S \text{ or } x \in T\}$. The "or" here is always inclusive, i.e., in the previous sentence it means $x \in S$ or $x \in T$ or both $x \in S$ and $x \in T$. The "intersection" of two sets S and T is $S \cap T = \{x \mid x \in S \text{ and } x \in T\}$. The "difference" of two sets is $S - T = \{s \in S \mid s \notin T\}$.

If A is a collection of sets then $\bigcup \{S \mid S \in A\} = \{x \mid \exists S \in A \ni x \in S\}$ and $\bigcap \{S \mid S \in A\} = \{x \mid \forall S \in A, x \in S\}$. If $\{S_\alpha \mid \alpha \in A\}$ is an "indexed" family of sets, we also use the notation $\bigcup S_\alpha = \{x \mid \exists \alpha \in A \ni x \in S_\alpha\}$ and $\bigcap S_\alpha = \{x \mid \forall \alpha \in A, x \in S_\alpha\}$.

Unions, intersections and differences follow these laws:

$$S \cup T = T \cup S, \qquad S \cap T = T \cap S,$$
$$R \cup (S \cup T) = (R \cup S) \cup T, \qquad R \cap (S \cap T) = (R \cap S) \cap T,$$
$$R \cup (S \cap T) = (R \cup S) \cap (R \cup T), \qquad R \cap (S \cup T) = (R \cap S) \cup (R \cap T),$$
$$R \cup \bigcap S_\alpha = \bigcap (R \cup S_\alpha), \qquad R \cap \bigcup S_\alpha = \bigcup (R \cap S_\alpha),$$
$$(X - S) \cup (X - T) = X - (S \cap T), \qquad (X - S) \cap (X - T) = X - (S \cup T),$$
$$\bigcup (X - S_\alpha) = X - \bigcap S_\alpha, \qquad \bigcap (X - S_\alpha) = X - \bigcup S_\alpha,$$
$$\left(\bigcup S_\alpha\right) \cap \left(\bigcup T_\beta\right) = \bigcup (S_\alpha \cap T_\beta), \qquad \left(\bigcap S_\alpha\right) \cup \left(\bigcap T_\beta\right) = \bigcap (S_\alpha \cup T_\beta).$$

The "cartesian product," or simply the "product" of two sets S and T is the set of ordered pairs $S \times T = \{\langle s, t \rangle | s \in S, t \in T\}$. We sometimes use (s, t) instead of $\langle s, t \rangle$ to denote an ordered pair.

A "relation" R between two sets S and T is a set of ordered pairs $R \subset S \times T$. We usually write $s\,R\,t$ to mean $\langle s, t \rangle \in R$. For example, $\in$ is a relation between a set of objects and a collection of sets. Another example is the relation $x \leq y$ between the set $\mathbf{R}$ of real numbers and itself.

The "domain" of a relation $R \subset S \times T$ is $\{t | \exists s \in S \ni s\,R\,t\}$ and the "range" of R is $\{s | \exists t \in T \ni s\,R\,t\}$.

A "function" f from the set X to the set Y is a relation $f \subset Y \times X$ with domain X such that $(x \in X, y \in Y, y' \in Y, yfx$ and $y'fx) \Rightarrow y = y'$. One writes $y = f(x)$ to mean yfx. We also use $f: X \to Y$, and variants of this to mean that f is a function from X to Y. The notation $x \mapsto y$ is also used for $y = f(x)$.

A function $f: X \to Y$ is said to be "injective" or "one–one into" if $f(a) = f(b) \Rightarrow a = b$. It is said to be "surjective" or "onto" if $y \in Y \Rightarrow \exists x \in X \ni y = f(x)$. It is said to be "bijective" or a "one–one correspondence" if it is both injective and surjective.

The identity function on X taking every member of X to itself is denoted by 1_X, or simply by 1 when that is not ambiguous.

If R and S are relations (in particular, if they are functions) then we define the "composition" of R and S to be

$$R \circ S = \{\langle a, c \rangle | \exists b \ni a\,R\,b \text{ and } b\,S\,c\},$$

and the "inverse" of R to be

$$R^{-1} = \{\langle a, b \rangle | b\,R\,a\}.$$

It is easy to see that $(R \circ S)^{-1} = S^{-1} \circ R^{-1}$. It is also elementary that $g \circ f$ is a function when f and g are both functions.

If $R \subset Y \times X$ is a relation and $A \subset X$ then we put $R(A) = \{y \in Y | \exists a \in A \ni y\,R\,a\}$. Note that, for a function $f: X \to Y, f(A) \subset Y$ is defined for $A \subset X$ and $f^{-1}(B) \subset X$ is defined for $B \subset Y$.

If $f: X \to Y$ and $A \subset X$ then let $f|_A = f \cap (Y \times A)$, the "restriction" of f to A.

B.1. Definition. A relation $R \subset X \times X$ is an *equivalence relation* on X if:

(1) (reflexive) $\quad x\,R\,x \quad$ for all $x \in X$,

(2) (symmetric) $x\,R\,y \;\Rightarrow\; y\,R\,x$,
(3) (transitive) $x\,R\,y$ and $y\,R\,z \;\Rightarrow\; x\,R\,z$.

B.2. Definition. If R is an equivalence relation on X then we put

$$[x]=\{y\in X \mid x\,R\,y\}.$$

This is called the *equivalence class* of x.

B.3. Proposition. *If R is an equivalence relation, then $[x]=[y]\Leftrightarrow x\,R\,y$. Also $[x]\cap[y]\neq\varnothing\Leftrightarrow[x]=[y]$.* □

In other words the equivalence classes $[x]$ partition X into disjoint subsets whose union is X.

B.4. Definition. If R is an equivalence relation on X then the set of equivalence classes $\{[x]\mid x\in X\}$ is denoted by X/R. There is the canonical surjection $\phi: X\to X/R$ given by $\phi(x)=[x]$.

B.5. Definition. If X is a set then its *power set* is $\mathscr{P}(X)=\{A\mid A\subset X\}$. Also let $\mathscr{P}_0(X)=\mathscr{P}(X)-\{\varnothing\}$.

B.6. Definition. If X and Y are sets, put $Y^X=\{f\mid f: X\to Y\}$.

B.7. Proposition. *If 2 denotes the set $\{0,1\}$ of two elements then the correspondence $A\leftrightarrow\chi_A$ between $\mathscr{P}(X)$ and 2^X given by*

$$\chi_A(x)=\begin{cases}0 & \text{if } x\notin A,\\ 1 & \text{if } x\in A,\end{cases}$$

is a bijection. □

B.8. Definition. A *partial ordering* on a set X is a relation $\leq$ on X such that:

(1) (reflexive) $a\leq a$ for all $a\in X$,
(2) (antisymmetric) $a\leq b$ and $b\leq a \;\Rightarrow\; a=b$,
(3) (transitive) $a\leq b$ and $b\leq c \;\Rightarrow\; a\leq c$.

A set together with a partial ordering is called a *partially ordered set* or a *poset*.

B.9. Definition. A poset X is said to be *totally ordered* (or *simply ordered* or *linearly ordered* or a *chain*) if $a, b\in X\Rightarrow$ either $a\leq b$ or $b\leq a$.

B.10. Definition. A function $f: X\to X$ on a poset is called *isotone* if $x\leq y\Rightarrow f(x)\leq f(y)$.

B.11. Definition. If $(X,\leq)$ is a poset and $A\subset X$ then $x\in X$ is an *upper bound* for A if $a\in A\Rightarrow a\leq x$. The element x is a *least upper bound* or *lub* for A if it is an upper bound and x' an upper bound for $A\Rightarrow x\leq x'$. Similarly for *lower* *bound* and *greatest lower bound* or *glb*. Also, *supremum* = *least upper bound* and *infimum* = *greatest lower bound*, and *sup* and *inf* are abbreviations of these.

B.12. Definition. A *lattice* is a poset such that every two element subset has an lub and a glb. It is a *complete lattice* if every subset has an lub and a glb.

B.13. Proposition. *If S is a set then $\mathcal{P}(S)$ is partially ordered by inclusion (i.e., by $\subset$) and is a complete lattice.* □

B.14. Proposition. *If X is a complete lattice and $f: X \to X$ is isotone, then f has a fixed point, i.e., $\exists x \in X \ni f(x) = x$.*

Proof. Let $Y = \{x \in X \mid f(x) \geq x\}$ and put $y_0 = \sup(Y)$. Note that $y \in Y \Rightarrow f(y) \geq y \Rightarrow f(f(y)) \geq f(y) \Rightarrow f(y) \in Y$. Also $y \in Y \Rightarrow y \leq y_0 \Rightarrow y \leq f(y) \leq f(y_0) \Rightarrow f(y_0)$ is an upper bound for $Y \Rightarrow f(y_0) \geq y_0 \Rightarrow y_0 \in Y \Rightarrow f(y_0) \in Y \Rightarrow f(y_0) \leq y_0$. Since we had the opposite inequality, we conclude that $f(y_0) = y_0$. □

B.15. Proposition. *Let $f: X \to Y$ and $g: Y \to X$ be functions. Then there are sets $A \subset X$ and $B \subset Y$ such that $f(A) = B$ and $g(Y - B) = X - A$.*

Proof. Consider the power set $\mathcal{P}(X)$ ordered by inclusion. It is a complete lattice by Proposition B.13. If $S \subset X$ then let $h(S) \in \mathcal{P}(X)$ be $h(S) = X - g(Y - f(S))$.

If $S \subset T$ then it is easy to see that $h(S) \subset h(T)$, so that h is isotone. By Proposition B.14 there is a subset $A \subset X$ such that $h(A) = A$. Let $B = f(A)$. Then $g(Y - B) = g(Y - f(A)) = X - h(A) = X - A$. □

B.16. Definition. A totally ordered set X is said to be *well ordered* if every nonempty subset has a least element. That is, $\emptyset \neq A \subset X \Rightarrow \exists a \in A \ni (b \in A \Rightarrow a \leq b)$. (Of course, the least element of A is $\text{glb}(A)$.) If $x \in X$ then its *initial segment* is

$$\text{IS}(x) = \{y \in X \mid y < x\},$$

and its *weak initial segment* is

$$\text{WIS}(x) = \{y \in X \mid y \leq x\}.$$

Also, if $x \in X$ and is not the least upper bound of X (which may not exist) then we put

$$\text{succ}(x) = \text{glb}\{y \in X \mid y > x\},$$

the *successor* of x.

Note that every subset of a well ordered set is well ordered.

B.17. Lemma. *Let X be a poset such that every well ordered subset has an lub in X. If $f: X \to X$ is such that $f(x) \geq x$ for all $x \in X$, then f has a fixed point.*

PROOF. Pick an element $x_0 \in X$. Let **S** be the collection of subsets $Y \subset X$ such that:

(1) Y is well ordered with least element x_0 and successor function $f|_{Y-\{\text{lub}\,Y\}}$.
(2) $x_0 \neq y \in Y \Rightarrow \text{lub}_X(\text{IS}_Y(y)) \in Y$.

For example, $\{x_0\} \in \mathbf{S}$, $\{x_0, f(x_0)\} \in \mathbf{S}$, etc. We need the following sublemmas (A) and (B):

(A) If $Y \in \mathbf{S}$ and $Y' \in \mathbf{S}$, then Y is an initial segment of Y' or vice versa.

To prove (A) let $V = \{x \in Y \cap Y' \mid \text{WIS}_Y(x) = \text{WIS}_{Y'}(x)\}$. Suppose first that V has a last element v. If v is not the last element of Y then $\text{succ}_Y(v) = f(v)$. If v is not the last element of Y' then $\text{succ}_{Y'}(v) = f(v)$. Hence if neither of Y, Y' is an initial segment of the other then $f(v) \in V$, whence $f(v) = v$ and we are done.

If, on the contrary, V has no last element, let $z = \text{lub}_X(V)$. If $Y \neq V \neq Y'$ then it follows from (2) that $z \in Y \cap Y'$ (because if $y = \inf(Y - V)$ then $V = \text{IS}_Y(y)$ and therefore $z = \text{lub}_X(\text{IS}_Y(y)) \in Y$ by (2)). Therefore, $z \in V$, a contradiction, proving (A).

(B) The set $Y_0 = \bigcup\{Y \mid Y \in \mathbf{S}\}$ is in **S**.

To prove (B) note that if $y_0 \in Y \in \mathbf{S}$ then it follows from (A) that $\{y \in Y_0 \mid y < y_0\} = \text{IS}_Y(y_0)$ and so this subset is well ordered with successor function f. This implies immediately that Y_0 is well ordered and satisfies (1). Also $\text{lub}_X(\text{IS}(y_0)) \in Y \subset Y_0$ which gives condition (2) for Y_0. Thus (B) is proved.

Now we complete the proof of Lemma B.17. Let $y_0 = \text{lub}_X(Y_0)$. If $y_0 \notin Y_0$ then $Y_0 \cup \{y_0\} \in \mathbf{S}$ and so $y_0 \in Y_0$ after all. If $f(y_0) > y_0$ then $Y_0 \cup \{f(y_0)\} \in \mathbf{S}$ contrary to the definition of Y_0. Thus $f(y_0) = y_0$ as desired. □

B.18. Theorem. *The following statements are equivalent:*

(A) *For each set X, there is a function $f: \mathcal{P}_0(X) \to X$ such that $f(S) \in S$ for all $\varnothing \neq S \subset X$.*
(B) *If X is a poset such that every well ordered subset has an* lub *in X then X contains a maximal element, i.e., an element $a \in X \ni a' \geq a \Rightarrow a' = a$.*
(C) *(Maximal Chain Theorem.) If X is a poset then X contains a maximal chain, i.e., a chain not properly contained in any other chain in X.*
(D) *(Maximality Principle.) If X is a poset such that every chain in X has an upper bound, then X has a maximal element.*
(E) *(Zermelo, Well-Ordering Theorem.) Every set can be well ordered.*
(F) *If $f: X \to Y$ is surjective then there is a section $g: Y \to X$ of f, i.e., an injection $g: Y \to X$ such that $f \circ g = 1_Y$.*
(G) *(Axiom of Choice.) If $\{S_\alpha \mid \alpha \in A\}$ is an indexed family of nonempty sets S_α then there exists a function $f: A \to \bigcup S_\alpha$ such that $f(\alpha) \in S_\alpha$ for all $\alpha \in A$.*

PROOF. (A)⇒(B): Assume (B) is false. Then let $X_a = \{x \in X \mid x > a\}$. By assumption $X_a \neq \varnothing$ for all $a \in X$. Let $g: \mathcal{P}_0(X) \to X$ be a choice function. Define

$f: X \to X$ by $f(a) = g(X_a) > a$. Then $f(x) > x$ for all $x \in X$ contrary to Lemma B.17.

(B)$\Rightarrow$(C): Let $\mathbf{S}$ be the collection of all chains in X ordered by inclusion. If $\mathbf{C} \subset \mathbf{S}$ is a chain of chains (i.e., $Y_1, Y_2 \in \mathbf{C} \Rightarrow Y_1 \subset Y_2$ or $Y_2 \subset Y_1$) then $\bigcup\{Y \mid Y \in \mathbf{C}\}$ is a chain. Therefore *every* chain in $\mathbf{S}$ has a lub. By (B) there is a maximal element of $\mathbf{S}$, i.e., a maximal chain.

(C)$\Rightarrow$(D): Pick a maximal chain C and note that if x is an upper bound of C, then x is maximal.

(D)$\Rightarrow$(E): Consider the collection W of elements of the form $(U, \ll)$ where $U \subset X$ and $\ll$ is a well ordering on U. Order these by $(U, \ll) \leq (V, \ll') \Leftrightarrow$ they are equal or $(U, \ll)$ is an initial segment of $(V, \ll')$ and $\ll$ is the restriction of $\ll'$ to $U \times U$.

As in the proof of Lemma B.17 we see that every chain in W has a (least) upper bound, namely, the union of its elements. Thus (D) implies that there exists a maximal (with respect to $\leq$) well ordering, say $(U, \ll)$.

We claim that $U = X$. If not, let $x \in X - U$ and define $(U \cup \{x\}, \ll')$ where $\ll' = \ll \cup (U \times \{x\})$, i.e., make x larger than anything in U. This contradicts maximality of $(U, \ll)$.

(E)$\Rightarrow$(F): Well order X and let $g(y)$ be the first element of $f^{-1}(y)$. Then $f \circ g(y) = y$.

(F)$\Rightarrow$(G): Let $S = \bigcup S_\alpha$ and $X = \{\langle s, \alpha\rangle \in S \times A \mid x \in S_\alpha\}$. Let $p_s: X \to S$ and $p_A: X \to A$ be the projections $p_S\langle s, \alpha\rangle = s$ and $p_A\langle s, \alpha\rangle = \alpha$. Then p_A is onto since each $S_\alpha \neq \emptyset$. Thus there is a section $g: A \to X$ for p_A; i.e., $g(\alpha) = \langle s, \alpha\rangle$ for some $s \in S_\alpha$. Let $f = p_S \circ g: A \to S$. Then f is a choice function since $f(\alpha) = p_S g(\alpha) = p_S\langle s, \alpha\rangle = s$ for some $s \in S_\alpha$, all $\alpha \in A$.

(G)$\Rightarrow$(A): For $T \in \mathscr{P}_0(X)$ define $S_T = T$. Then $\mathscr{P}_0(X) = \{S_T \mid T \in \mathscr{P}_0(X)\}$ is an indexed collection of nonempty sets. Note that $\bigcup S_T = X$ since, for any $x \in X, x \in \{x\} = S_{\{x\}}$. By (G) there is a function $f: \mathscr{P}_0(X) \to \bigcup S_T = X$ such that $f(T) \in S_T = T$ for any $\emptyset \neq T \subset X$. □

The Maximality Principle (D) is often inappropriately referred to as "Zorn's Lemma." It is actually due, independently, to R.L. Moore and Kuratowski, a dozen years before Zorn.

The only numbered results in this appendix that depend on the Axiom of Choice are Theorems B.28 and B.26(d). (The latter requires only a countable number of arbitrary choices, and so is relatively innocuous.)

B.19. Definition. Two sets X and Y are said to have the *same cardinal number* if there exists a one–one correspondence between them.

Given a set $\mathbf{S}$ of sets, this relation is an equivalence relation on $\mathbf{S}$. If $X \in \mathbf{S}$ we denote the equivalence class of X by $\operatorname{card}(X)$. We also write $\operatorname{card}(X) \leq \operatorname{card}(Y)$ if there exists an injection $f: X \to Y$.

B.20. Theorem (Schroeder–Bernstein). *If* $\operatorname{card}(X) \leq \operatorname{card}(Y)$ *and* $\operatorname{card}(Y) \leq \operatorname{card}(X)$ *then* $\operatorname{card}(X) = \operatorname{card}(Y)$.

Proof. (Note that this proof does not use the Axiom of Choice.) By hypothesis there exist injections $f: X \to Y$ and $g: Y \to X$. By Proposition B.15 there exist subsets $A \subset X$ and $B \subset Y$ such that $f(A) = B$ and $g(Y - B) = X - A$. Therefore $f|_A: A \leftrightarrow B$ and $g|_{Y-B}: Y - B \leftrightarrow X - A$ are one–one correspondences. Put them together. □

B.21. Corollary. *The ordering* $\leq$ *on the cardinals is a partial ordering.* □

It is not hard to see that, assuming the Axiom of Choice in the guise of the Well-Ordering Theorem, the cardinals are well ordered by $\leq$. This is, in fact, equivalent to the Axiom of Choice.

B.22. Theorem. *For any* X, $\mathrm{card}(X) < \mathrm{card}(\mathscr{P}(X))$.

Proof. The relation $\mathrm{card}(X) \leq \mathrm{card}(\mathscr{P}(X))$ holds because of the injection $x \mapsto \{x\}$. Let $f: X \to \mathscr{P}(X)$ be any function. Put $A = \{x \in X \mid x \notin f(x)\}$. We claim that there can be no $y \in X$ with $A = f(y)$. If there is such a y then

$$y \in A \quad \Rightarrow \quad y \notin f(y) = A$$

and

$$y \notin A \quad \Rightarrow \quad y \in f(y) = A,$$

so neither possibility is tenable. Thus there never exists a surjection $f: X \to \mathscr{P}(X)$. □

The symbol ω is used to denote the set of nonnegative integers with the usual ordering. Let $\omega' = \omega \cup \{\omega\}$, tacking on a last element. Note that $\mathrm{card}(\omega)$ is the least infinite cardinal.

B.23. Definition. *A set* X *is said to be* countable *if there exists an injection* $f: X \to \omega$.

B.24. Lemma. *The product* $\omega \times \omega$ *is countable.*

Proof. The function $f: \omega \times \omega \to \omega$ given by $f(n, k) = (2n + 1)2^k - 1$ is a bijection. □

B.25. Lemma. *If* $f: X \to Y$ *is an injection with* $X \neq \varnothing$ *then there exists a surjection* $g: Y \to X$ *such that* $g \circ f = 1_X$.

Proof. For some $x_0 \in X$ let $g(y)$ be x_0 for $y \notin f(X)$ and $g(y) = f^{-1}(y)$ for $y \in f(X)$. □

B.26. Theorem.

(a) *If* X *is countable and* $f: X \to Y$ *is onto then* Y *is countable.*
(b) *A subset of a countable set is countable.*

(c) *X, Y countable $\Rightarrow X \times Y$ countable.*
(d) *A countable union of countable sets is countable.*

PROOF. For (a) let $g: X \to \omega$ be injective and define $h(y) = \inf g(f^{-1}(y))$. Then $h: Y \to \omega$ is an injection.

Part (b) is trivial.

For (c), if $f: X \to \omega$ and $g: Y \to \omega$ are injections then the composition of $f \times g: X \times Y \to \omega \times \omega$ with the injection $\omega \times \omega \to \omega$, given by Lemma B.24, gives an injection $X \times Y \to \omega$.

For (d), suppose that X_α is a countable set defined for $\alpha \in A \neq \varnothing$ which is countable. Then let $f_\alpha: \omega \to X_\alpha$ be a surjection and $g: \omega \to A$ a surjection. Let $h: \omega \times \omega \to \bigcup \{X_\alpha | \alpha \in A\}$ be given by $h(n, k) = f_{g(n)}(k)$. Then h is surjective and so $\bigcup X_\alpha$ is countable by (a). □

In general, it can be shown that if X, Y are nonempty, and not both finite, then $\text{card}(X \cup Y) = \max(\text{card}(X), \text{card}(Y)) = \text{card}(X \times Y)$. The consequence that $\text{card}(X \times X) = \text{card}(X)$, whenever X is infinite, is equivalent to the Axiom of Choice.

B.27. Theorem. *If $\mathbf{R}$ is the set of reals then* $\text{card}(\mathbf{R} - \{0\}) = \text{card}(\mathscr{P}(\omega)) > \text{card}(\omega)$.

PROOF. Let $\mathbf{R}_+$ denote the nonnegative reals. The injection $\mathbf{R}_+ \hookrightarrow \mathbf{R}$ and the injection $\mathbf{R} \to \mathbf{R}_+$ given by $x \mapsto e^x$ show that $\text{card}(\mathbf{R}) = \text{card}(\mathbf{R}_+)$. Similarly, $\text{card}(\mathbf{R}) = \text{card}(\mathbf{R} - \{0\})$. We shall exhibit a bijection $\mathbf{R}_+ \leftrightarrow \mathscr{P}(\omega)$.

First write each positive real number r in its continued fraction expansion

$$r = a_0 + \cfrac{1}{a_1 + \cfrac{1}{a_2 + \cfrac{1}{a_3 + \cdots}}}$$

where the a_i are integers with $a_0 \geq 0$ and $a_i > 0$ for $i > 0$. We shall denote this expansion by $r = [a_0, a_1, a_2, \ldots]$. A terminating (rational) continued fraction will be written in the form

$$r = a_0 + \cfrac{1}{a_1 + \cfrac{1}{\ddots \; a_{s-1} + \cfrac{1}{a_s + 1}}}$$

and condensed to $[a_0, a_1, \ldots, a_s]$. In particular, an integer $n > 0$ is written as $n = (n-1) + 1 = [n-1]$. With this understanding, a continued fraction representing r is uniquely determined by r. Thus this determines a one–one correspondence of the positive reals r with the sequences, infinite or finite, $[a_0, a_1, \ldots]$. Also let the real 0 correspond to the empty sequence. Finally, let a sequence $[a_0, a_1, \ldots]$ correspond to the subset $\{a_0, a_0 + a_1, a_0 + a_1 + a_2, \ldots\}$ of ω. This is easily seen to be a one–one correspondence between the non-negative reals and subsets of ω, as claimed. The real number 0 corresponds to the empty subset of ω. □

It should be noted that, in the above correspondence, the rationals correspond to the finite subsets of ω. Thus the set of all finite subsets of ω is countable. (I believe the foregoing proof is due to A. Gleason.)

B.28. Theorem. *There exists an uncountable well-ordered set Ω' with last element Ω such that $x < \Omega \Rightarrow \mathrm{IS}(x)$ is countable.*

Proof. Well order the reals and put on an extra element x_0 at the end. Then let Ω be the least element in the ordering such that $\mathrm{IS}(\Omega)$ is uncountable. This exists since $\mathrm{IS}(x_0)$ has cardinality that of $\mathbf{R}$ which is greater than that of ω. Then $\Omega' = \mathrm{WIS}(\Omega)$ is the desired set. Note that by an equivalence, one can regard Ω' as $\Omega \cup \{\Omega\}$. □

We shall refer to Ω, as in Theorem B.28, as "the least uncountable ordinal" and to other elements of Ω' as "countable ordinal numbers."

B.29. Theorem. *If* $\mathrm{card}(X) = \mathrm{card}(X \times Y)$ *then* $\mathrm{card}(\mathscr{P}_0(X)) = \mathrm{card}(\mathscr{P}_0(X) \times \mathscr{P}_0(Y))$.

Proof. By assumption there is a one–one correspondence $f: X \times Y \to X$. This induces a one–one correspondence $F: \mathscr{P}(X \times Y) \to \mathscr{P}(X)$ by $F(S) = \{f(x, y) | \langle x, y \rangle \in S\}$; i.e., $F(S) = f(S)$. But $g: \mathscr{P}_0(X) \times \mathscr{P}_0(Y) \to \mathscr{P}_0(X \times Y)$ given by $g(\langle S, T \rangle) = S \times T$ is an injection, and so $F \circ g: \mathscr{P}_0(X) \times \mathscr{P}_0(Y) \to \mathscr{P}_0(X)$ is also an injection. There is also an injection $\mathscr{P}_0(X) \to \mathscr{P}_0(X) \times \mathscr{P}_0(Y)$ (unless $Y = \varnothing$, in which case the result is trivial) and so the contention follows from the Schroeder–Bernstein Theorem (Theorem B.20). □

B.30. Corollary. *For any positive integer n we have* $\mathrm{card}(\mathbf{R}^n) = \mathrm{card}(\mathbf{R})$.

Proof. By Lemma B.24, $\mathrm{card}(\omega \times \omega) = \mathrm{card}(\omega)$. By Theorem B.27, $\mathrm{card}(\mathbf{R}) = \mathrm{card}(\mathbf{R} - \{0\}) = \mathrm{card}(\mathscr{P}_0(\omega))$, so Theorem B.29 implies that $\mathrm{card}(\mathbf{R}^2) = \mathrm{card}(\mathbf{R} \times \mathbf{R}) = \mathrm{card}(\mathbf{R})$. If we know that $\mathrm{card}(\mathbf{R}^n) = \mathrm{card}(\mathbf{R})$ then $\mathrm{card}(\mathbf{R}^{n+1}) = \mathrm{card}(\mathbf{R} \times \mathbf{R}^n) = \mathrm{card}(\mathbf{R} \times \mathbf{R}) = \mathrm{card}(\mathbf{R})$ and so an induction finishes the proof. □

As mentioned before, the Axiom of Choice implies similar facts for arbitrary infinite cardinals, but Theorem B.29 and Corollary B.30 do not depend on the Axiom of Choice.

In this book, we shall often make use of the Axiom of Choice without explicit mention. In cases where use of the axiom is known to be crucial, we do mention that. There are probably many other places where the axiom is crucial but where that fact is not definitely known to the author.

C. Critical Values

The purpose of this appendix is to give a proof of Sard's Theorem:

C.1. Theorem. *If M^n is a smooth manifold and $f: M^n \to \mathbf{R}^k$ is smooth and if $C = \operatorname{crit}(f) \subset M^n$ is the critical set of f then the set $f(C)$ of critical values has measure zero in $\mathbf{R}^k$.*

The proof does not require knowledge of measure theory. (The definition of "measure zero" does not presume a definition of "measure," and it is an invariant of smooth manifolds, and so is not really a measure-theoretic concept.) Our proof of Theorem C.1 is partly based on that of Holm [1].

C.2. Definition. A set $K \subset \mathbf{R}^n$ is said to have *measure zero* if, for any $\epsilon > 0$, there exists a sequence of open cubes Q_i with $K \subset \bigcup Q_i$ and $\sum_i \operatorname{vol}(Q_i) < \epsilon$.

This is equivalent to saying that K can be covered by countably many balls of arbitrarily small total volume, since the ratio of the volumes of a cube and the circumscribed ball depends only on n, and vice versa. Similarly, one could use rectangles, etc.

For an open set $U \subset \mathbf{R}^n$ let $\operatorname{vol}(U) = \inf\{\sum_i \operatorname{vol}(Q_i)\}$ where $\{Q_i\}$ is a sequence of cubes covering U. This may be infinite. The definition of "measure zero" can then be rephrased as the existence of open sets of arbitrarily small volume and containing the given set.

C.3. Proposition.

(a) *A countable union of sets of measure zero has measure zero.*

(b) *If every point $x \in K \subset \mathbf{R}^n$ has a neighborhood N with $K \cap N$ of measure zero then K has measure zero.*

Proof. If $K = K_1 \cup K_2 \cup \ldots$, and if $K_i \subset U_i$, an open set of volume $< \epsilon/2^i$ then $U = \bigcup U_i \supset K$ and $\operatorname{vol} U \le \sum_i \operatorname{vol} U_i < \sum_i \epsilon/2^i = \epsilon$, proving (a). For (b), let $\{W_i\}$ be a countable basis for the topology of $\mathbf{R}^n$. If $x \in K$, let N_x be a neighborhood of x with $K \cap N_x$ of measure zero. Let $W_{i(x)} \subset N_x$ with $x \in W_{i(x)}$. Then $K \cap W_{i(x)}$ has measure zero and $K = \bigcup K \cap W_{i(x)}$ is (really) a countable union of sets of measure zero. □

The following lemma shows that "measure zero" is a "smooth invariant":

C.4. Lemma. *If K has measure zero, where $K \subset U$ and $U \subset \mathbf{R}^n$ is open, and if $g: U \to \mathbf{R}^n$ is differentiable, then $g(K)$ has measure zero.*

Proof. Let $Q \subset U$ be a cubical neighborhood of $x \in K$. Let B be a bound for all $|\partial g_i/\partial x_j|$ on Q. If N is an open ball of radius r about a point of $K \cap Q$ then the Mean Value Theorem (Theorem 1.1 of Chapter II) implies that $g(Q \cap N)$ is contained in a ball of radius Br. Thus, if $Q \cap K$ is covered by balls of total volume $< \epsilon$, then $g(Q \cap K)$ is covered by balls of total volume $< B^n \epsilon$. □

Now we proceed with the proof of Sard's Theorem. Let $f: M^n \to \mathbf{R}^k$ be smooth and put $f = (f_1, \ldots, f_k)$. The proof will be by induction on n, and the case $n = 0$ is true. Let C be the critical set of f, which is closed in M^n. Let $D \subset C$ be the set of points where the differential of f vanishes. We shall divide the proof into the cases of showing that $f(D)$ has measure zero, and $f(C - D)$ has measure zero. Note that D is closed in C and hence in M.

C.5. Lemma. *The set $f(D)$ has measure zero in $\mathbf{R}^k$.*

Proof. If the differential f_* of f is zero at x then so is the differential of f_1. Thus, if E is the critical set of f_1 (which equals the set where the differential of f_1 vanishes) then $f(D) \subset f_1(E) \times \mathbf{R}^{k-1}$. Since $\mathbf{R}^{k-1}$ can be covered by a countable set of cubes, each of volume < 1, $f_1(E) \times \mathbf{R}^{k-1}$ has measure zero in $\mathbf{R}^k$ if $f_1(E)$ has measure zero in $\mathbf{R}$. Hence, it suffices to prove Lemma C.5 for $k = 1$, i.e., when $f: U \to \mathbf{R}$ is a real valued function, U open in $\mathbf{R}^n$.

Let $D_i = \{x \in U \mid \text{all partial derivatives of } f \text{ of order } \leq i \text{ vanish}\}$. Then we have that $D = D_1 \supset D_2 \supset \cdots \supset D_n$, and all are closed.

Case n. Proof that $f(D_n)$ has measure zero:

It suffices to show that $f(D_n \cap Q)$ has measure zero for any closed cube $Q \subset U$. Let s be the length of the sides of Q. Let m be an integer and partition Q into m^n cubes of side s/m, hence of diameter $s\sqrt{n}/m$. Let $\bar{x} \in Q \cap D_n$ and let Q' be one of the small cubes containing the point $\bar{x}$. Since the partial derivatives of f through order $n+1$ are bounded on Q there is a constant B, independent of m, such that

$$x \in Q' \quad \Rightarrow \quad |f(x) - f(\bar{x})| \leq B \| x - \bar{x} \|^{n+1} \leq B \cdot (s\sqrt{n}/m)^{n+1}$$

by Taylor's Theorem. Thus $f(Q')$ is contained in an interval of length A/m^{n+1}, where A is a constant independent of m. Hence $f(Q \cap D_n)$ is contained in a union of intervals of total length $\leq Am^n/m^{n+1} = A/m$. Since $A/m \to 0$ as $m \to \infty$, $f(Q \cap D_n)$ has measure zero, finishing Case n.

Case $i < n$. Proof that $f(D_i - D_{i+1})$ has measure zero:

Since D_{i+1} is closed in U, we may as well throw it out of U and so assume that $D_{i+1} = \emptyset$. At a point $\bar{x} \in D_i$, then, all partial derivatives of order $\leq i$ vanish and there is a partial derivative of order $i+1$ which does not vanish. Let g be such an ith order derivative of f, whose differential is nonzero at $\bar{x}$. We may pass to a smaller open neighborhood of $\bar{x}$ on which the differential of g is nonzero. Then $g=0$ on D_i and 0 is a regular value for g. Let $V^{n-1} = g^{-1}(0)$. Then $D_i \subset \text{crit}(f|_V)$ since the differential of $f|_V$ vanishes on D_i. By the inductive assumption, $f(D_i) = f|_V(D_i)$ has measure zero, as claimed.

Putting Case n together with Cases $1 \leq i < n$ proves the lemma. □

C.6. Lemma. *The set $f(C-D)$ has measure zero.*

Proof. Since D is closed we can remove it from M^n as far as the proof goes, and so we can assume that $D = \emptyset$. Since the differential of f does not vanish at $\bar{x}$, there is a coordinate projection, say the last coordinate, $g: \mathbf{R}^k \to \mathbf{R}$ such that the differential of $g \circ f$ does not vanish at $\bar{x}$. We can restrict attention to a neighborhood U of $\bar{x}$ where $g \circ f$ has nonzero differential; i.e., where all values are regular for $g \circ f: U \to \mathbf{R}$. For $t \in \mathbf{R}$, $V_t^{n-1} = (g \circ f)^{-1}(t)$ is then a smooth $(n-1)$-manifold. Put $f_t = f|_V: V_t^{n-1} \to g^{-1}(t) = \mathbf{R}^{k-1} \times \{t\}$. Since the differential of f at any $x \in V_t$ maps some vector to a vector not in $\mathbf{R}^{k-1} \times \{t\}$ (i.e., going nontrivially to $\mathbf{R}$ under g), it follows that x is critical for f if and only if it is critical for f_t. By the inductive hypothesis, the set of critical values of f_t is of measure zero in $\mathbf{R}^{k-1} \times \{t\}$ for each t. Since it suffices to show that the image of $C \cap Q$ has measure zero for each cube $Q \subset U$, it suffices to show that a compact set in $\mathbf{R}^k$ has measure zero if its intersection with each hyperplane $\mathbf{R}^{k-1} \times \{t\}$ has measure zero. This follows from the Fubini Theorem in measure theory. An elementary proof of it is given in the next lemma. □

This completes the proof of Sard's Theorem (Theorem C.1). □

As promised, we now give an elementary proof of the consequence of Fubini's Theorem used in the proof of Lemma C.6. Although Fubini's Theorem is part of every mathematician's education, a large number of students will not have seen it prior to the time of studying this book. In any case, the proof of the following lemma is very easy.

C.7. Lemma. *If $K \subset \mathbf{R}^n$ is a compact set whose intersection with each hyperplane $\mathbf{R}^{n-1} \times \{t\}$ has measure zero in the hyperplane, then K has measure zero in $\mathbf{R}^n$.*

Proof. We may as well assume that $K \subset \mathbf{I}^n$. For any closed set $S \subset \mathbf{I}^n$ let

$$\mu(S) = \inf\{\text{vol } U \mid U \supset S \text{ open}\}.$$

This clearly has the properties

$$\mu(S) \leq 1,$$
$$S \subset T \Rightarrow \mu(S) \leq \mu(T),$$
$$\mu(S \cup T) \leq \mu(S) + \mu(T).$$

Define a function $f: \mathbf{I} \to \mathbf{R}$ by

$$f(x) = \mu(K \cap \mathbf{I}^{n-1} \times [0, x]).$$

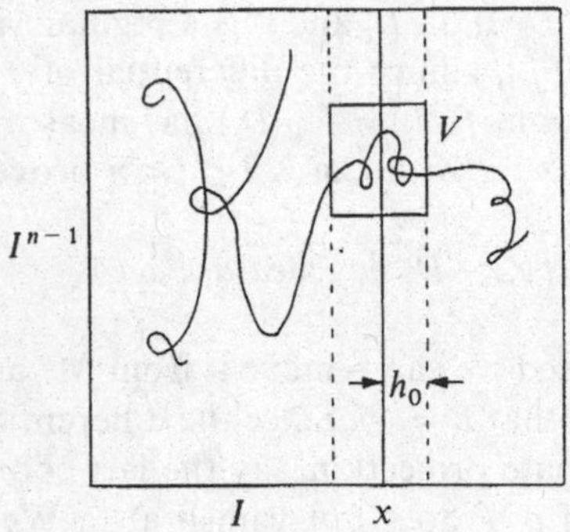

Figure C-1. Piccolo Fubini.

For any $x \in \mathbf{I}$, and given any $\epsilon > 0$, there is an open set V in $\mathbf{I}^{n-1}$ such that $V \times \{x\} \supset K \cap (\mathbf{I}^{n-1} \times \{x\})$ and with vol $V < \epsilon$. By compactness of K, there is a number $h_0 > 0$ such that $V \times [x - h_0, x + h_0] \supset K \cap (\mathbf{I}^{n-1} \times [x - h_0, x + h_0])$. (See Figure C-1.) Then, for any number h with $0 < h < h_0$, $(K \cap \mathbf{I}^{n-1} \times [0, x + h]) \subset (K \cap \mathbf{I}^{n-1} \times [0, x]) \cup (V \times [x, x + h])$ can be covered by an open set of volume less than $f(x) + \epsilon h$. That is,

$$f(x+h) \leq f(x) + \epsilon h \quad \text{for} \quad 0 \leq h < h_0.$$

Similarly $(K \cap \mathbf{I}^{n-1} \times [0, x]) \subset (K \cap \mathbf{I}^{n-1} \times [0, x - h]) \cup (V \times [x - h, x])$, so that

$$f(x) \leq f(x-h) + \epsilon h \quad \text{for} \quad 0 \leq h < h_0.$$

Therefore

$$0 \leq \frac{f(x+h) - f(x)}{h} \leq \epsilon$$

for all $|h| < h_0$. It follows that f is differentiable at x with derivative zero. Also $f(0) = 0$. Hence $f \equiv 0$ by elementary calculus. Consequently, $0 = f(1) = \inf\{\text{vol}\, U \mid K \subset U \text{ open}\}$, which is the desired conclusion. □

D. Direct Limits

In this appendix we discuss the algebraic notion of direct limits. This is used in the discussion of duality in Chapter VI, and to a lesser nonessential degree in some other parts of the book. It is also used in Appendices A and E.

D.1. Definition. Let D be a directed set and let G_α be an abelian group defined for each $\alpha \in D$. Suppose we are given homomorphisms $f_{\beta,\alpha}: G_\alpha \to G_\beta$ for each $\beta > \alpha$ in D. Assume that for all $\gamma > \beta > \alpha$ in D, we have $f_{\gamma,\beta} f_{\beta,\alpha} = f_{\gamma,\alpha}$. (Such a system is called a *direct system* of abelian groups.) Then $G = \varinjlim G_\alpha$ is defined to be the quotient group of the direct sum $G = \bigoplus G_\alpha$ modulo the relations $f_{\beta,\alpha}(g) \sim g$ for all $g \in G_\alpha$ and all $\beta > \alpha$.

Since any element of the direct sum involves only a finite number of G_α, it is equivalent to an element in a single G_α. Thus the direct sum could be replaced by the disjoint union, but the present definition makes the group structure transparent.

The inclusions $G_\alpha \hookrightarrow \bigoplus G_\alpha$ induce homomorphisms $i_\alpha: G_\alpha \to \varinjlim G_\alpha$ and it is clear that $i_\beta \circ f_{\beta,\alpha} = i_\alpha$. Moreover, for any $g \in G$ there is a $g_\alpha \in G_\alpha$, for some α, such that $g = i_\alpha(g_\alpha)$. Also, for any index $\alpha \in D$, and element $g_\alpha \in G_\alpha, i_\alpha(g_\alpha) = 0 \Leftrightarrow \exists \beta \geq \alpha \ni f_{\beta,\alpha}(g_\alpha) = 0$. In fact, these two properties characterize the direct limit:

D.2. Proposition. *Suppose given a direct system $\{G_\alpha, f_{\beta,\alpha}\}$ of abelian groups. Let A be an abelian group and $h_\alpha: G_\alpha \to A$ homomorphisms such that $\beta > \alpha \Rightarrow h_\beta \circ f_{\beta,\alpha} = h_\alpha$. Then there is a unique homomorphism $h: \varinjlim G_\alpha \to A$ such that $h \circ i_\alpha = h_\alpha$ for all α. Moreover:*

(1) $\operatorname{im}(h) = \{a \in A \mid a = h_\alpha(g)$ *for some* g *and* $\alpha\} = \bigcup \operatorname{im}(h_\alpha)$; *and*
(2) $\ker(h) = \{g \in \varinjlim G_\alpha \mid \exists \alpha$ *and* $g_\alpha \in G_\alpha \ni g = i_\alpha(g_\alpha)$ *and* $h_\alpha(g_\alpha) = 0\} = \bigcup i_\alpha(\ker h_\alpha)$.

Proof. Define $h(g_\alpha) = h_\alpha(g_\alpha)$. This defines a homomorphism on $\bigoplus G_\alpha$ and it respects the equivalence relation defining the direct limit, so it is well defined. Uniqueness is obvious as is property (1). The equivalence class g of $g_\alpha \in G_\alpha$ is taken to 0 by $h \Leftrightarrow h_\alpha(g_\alpha) = 0$ in A which is just another way of writing property (2). □

D.3. Corollary. *In the situation of Proposition* D.2, *$h: \varinjlim G_\alpha \to A$ is an isomorphism if and only if the following two statements hold true:*

(i) $\forall a \in A, \exists \alpha \in D$ *and* $g_\alpha \in G_\alpha \ni h_\alpha(g_\alpha) = a$; *and*
(ii) *if* $h_\alpha(g_\alpha) = 0$ *then* $\exists \beta > \alpha \ni f_{\beta,\alpha}(g_\alpha) = 0$. □

D.4. Theorem. *The direct limit is an exact functor. That is, if we have direct systems $\{A'_\alpha\}, \{A_\alpha\}$, and $\{A''_\alpha\}$ based on the same directed set, and if we have an exact sequence $A'_\alpha \to A_\alpha \to A''_\alpha$ for each α, where the maps commute with those defining the direct systems, then the induced sequence*

$$\varinjlim A'_\alpha \to \varinjlim A_\alpha \to \varinjlim A''_\alpha$$

is exact.

Proof. This is a very easy diagram chase in the diagram made up of all the original exact sequences and the limit sequence, using Corollary D.3. □

D.5. Theorem. *Suppose given two directed sets D and E. Define an order on $D \times E$ by $(\alpha, \beta) \geq (\alpha', \beta') \Leftrightarrow \alpha \geq \alpha'$ and $\beta \geq \beta'$. Suppose $G_{\alpha,\beta}$ is a direct system* *based on* $D \times E$. *Then the maps* $G_{\alpha,\beta} \to \varinjlim_\beta G_{\alpha,\beta} \to \varinjlim_\alpha (\varinjlim_\beta G_{\alpha,\beta})$ *induce an isomorphism*

$$\varinjlim_{\alpha,\beta} G_{\alpha,\beta} \xrightarrow{\approx} \varinjlim_\alpha (\varinjlim_\beta G_{\alpha,\beta}).$$

PROOF. This is just a matter of (easy) verification of (i) and (ii) of Corollary D.3. For (i), note that any element of the iterated limit comes from an element of some $\varinjlim_\beta G_{\alpha,\beta}$, but that, in turn, comes from some $G_{\alpha,\beta}$ implying it comes from the double limit. For (ii), if some element of $G_{\alpha,\beta}$ maps to zero in the iterated limit then it must already map to zero in some $\varinjlim_\beta G_{\alpha',\beta}$. But, similarly, that implies it must already map to zero in some $G_{\alpha',\beta'}$. □

E. Euclidean Neighborhood Retracts

In this appendix we answer the question of which subsets of euclidean space are retracts of some neighborhood there. Such a set is called an ENR (euclidean neighborhood retract). We shall see momentarily that the existence of such a retraction does not depend on the embedding. Thus this is an intrinsic property of the embeddable space and should be describable in terms of that space alone. Since a retract X of an open set $U \subset \mathbf{R}^n$ is closed in U, it is locally compact. Moreover, since $\mathbf{R}^n$ is locally contractible, it is easy to see that X is also locally contractible, meaning that any neighborhood U of a point $x \in X$ contains a smaller neighborhood V of x such that there is a homotopy $F: V \times I \to U$ starting at the inclusion and ending at a constant map. These, together with embeddability, are precisely the conditions needed.

E.1. Lemma. *Suppose* $X \subset \mathbf{R}^n$ *is a retract of an open neighborhood* U *of* X *in* $\mathbf{R}^n$. *Let* K *be a metrizable space and let* $Y \subset K$ *be homeomorphic to* X. *Then* Y *is the retract of some neighborhood* V *of* Y *in* K.

PROOF. Let $f: X \to Y$ be a homeomorphism and $r: U \to X$ the given retraction. Since Y is locally compact, it is the intersection of an open neighborhood W of Y in K with $\bar{Y}$, by Proposition 11.7 of Chapter I. Thus Y is closed in W. The composition of f^{-1} with any coordinate projection $X \subset \mathbf{R}^n \to \mathbf{R}$ can be extended to W by Tietze's Theorem (Theorem 10.4 of Chapter I). These combine to give a map $h: W \to \mathbf{R}^n$ extending f^{-1}. Let $V = h^{-1}(U)$. Then $f \circ r \circ h: V \to Y$ is the required retraction. □

E.2. Lemma. *Let* $X \subset \mathbf{R}^n$ *be locally compact. Then there is an embedding of* X *in* $\mathbf{R}^{n+1}$ *as a closed subset.*

PROOF. Since X is locally compact and hence locally closed, $X = U \cap \bar{X}$ for some open $U \subset \mathbf{R}^n$. Let $C = \bar{X} - X = \bar{X} - U$, which is closed. Let $f: \mathbf{R}^n \to \mathbf{R}$ be $f(x) = \operatorname{dist}(x, C)$ which is easily seen to be continuous. Then we claim that

$\phi: X \to \mathbf{R}^n \times \mathbf{R} = \mathbf{R}^{n+1}$, given by $\phi(x) = (x, 1/f(x))$, is an embedding of X onto a closed set. To see this, let $\{x_i\}$ be a sequence in X such that $\lim \phi(x_i) = (y_1, y_2)$ exists in $\mathbf{R}^n \times \mathbf{R}$. Then $\lim(x_i) = y_1$ and $\lim(1/f(x_i)) = y_2$, so that $\operatorname{dist}(x_i, C) = f(x_i) \not\to 0$. This implies that $y_1 = \lim(x_i) \in \bar{X} - C = X$ and so $\lim \phi(x_i) = \phi(y_1) \in \phi(X)$, as claimed. □

Note that the "$n+1$" in Lemma E.2 cannot be improved to "n" as is shown by the example of the open Möbius band $M \subset \mathbf{R}^3$. If M could be embedded as a closed subset in $\mathbf{R}^3$ then adding the point at infinity would give an embedding of $\mathbf{P}^2 = M^+$ in $\mathbf{S}^3$ contrary to Corollary 8.9 of Chapter VI.

E.3. Theorem (Borsuk). *If X is locally compact and locally contractible then any embedding of X in any $\mathbf{R}^n$ is a retract of some neighborhood there.*

Proof. By the lemmas we can assume that X is closed in $\mathbf{R}^n$. First, we will briefly outline the idea of the proof which is very simple. We divide $\mathbf{R}^n - X$ into cells which get small as they approach X. This is easiest done with cubes as cells. We then attempt to build a map to X on each cell by induction on the dimension of the cell. If a map is given on the boundary of a cell whose image is in a set which contracts in X then the map extends to the cell by putting the contraction along radii from the center of the cell. In order for this to produce a neighborhood retraction, we need two things:

(1) it must be continuous with the identity on X as cells approach X; and
(2) it must be defined on enough cells to provide a neighborhood of X.

Both these things are guaranteed by arranging that small cells don't map too far away.

Now the details. Divide $\mathbf{R}^n$ into cubes by hyperplanes parallel to the coordinate planes and of integer distance from the origin. Define a set C_1 as the union of those closed cubes which do not touch X. Note that any point a with $\operatorname{dist}(a, X) > \sqrt{n}$ is contained in a cube in C_1 since $\sqrt{n}$ is the diameter of an n-cube of side 1. Now divide the complement of $\operatorname{int}(C_1)$ into cubes of side $\frac{1}{2}$ whose sides are integer or half integer distance from the coordinate planes and let C_2 be C_1 together with all the smaller cubes not touching X. Continue this with cubes of side $\frac{1}{4}$, etc., giving sets $C_1, C_2, C_3, \ldots$. By the remark that C_1 contains all points a with $\operatorname{dist}(a, X) > \sqrt{n}$, and its analogues for C_i, any point $a \notin X$ is in only a finite number of such cubes. Therefore $C = \bigcup C_i$ is a locally finite CW-complex structure on $\mathbf{R}^n - X$.

Now we attempt the construction of the retraction r. For any 0-cell a of C let $r_0(a)$ be some point of X such that

$$\operatorname{dist}(a, r_0(a)) < 2 \cdot \inf\{\operatorname{dist}(a, x) \mid x \in X\}. \tag{1}$$

For any cell σ of C and map $f: \sigma \to X$ define

$$\rho(f) = \max\{\operatorname{dist}(x, f(x)) \mid x \in \sigma\}.$$

Suppose, by induction, that we have defined a map r_i on the union A_i of *some* of the i-cells of C to X. Then we define r_{i+1} as follows. Let σ be an $(i+1)$-cell of C for which r_i is defined on $\partial\sigma$. If there *exists* an extension of $r_i|_{\partial\sigma}$ to a map $\sigma \to X$ then let f be such an extension so that

$$\rho(f) < 2 \cdot \inf\{\rho(g) \mid g: \sigma \to X, g = r_i \text{ on } \partial\sigma\}. \tag{2}$$

These maps f fit together to give the desired map r_{i+1} on a union A_{i+1} of some of the $(i+1)$-cells of C, to X.

Let $A = \bigcup A_i \cup X$ and $r: A \to X$ the function which is r_i on A_i and the identity on X. We claim that r is the desired retraction. We must show that r is continuous, and that A is a neighborhood of X. Both of these will be proved by the same reasoning. Consider any point $p \in X$ and number $\epsilon_0 > 0$. Then we can find numbers $\epsilon_1, \epsilon_2, \ldots, \epsilon_{2n}$ such that

$$3\epsilon_{i+1} < \epsilon_i,$$

and

$$B_{i+1} \text{ is deformable to a point inside } B_i,$$

where $B_i = X \cap B_{\epsilon_i}(p)$. Let $U = B_{\epsilon_{2n}/4}(p)$. Then $r(U \cap A_0) \subset B_{2n}$ by (1). Assume inductively that r is defined on each i-cell σ of C such that $\sigma \subset U$ and that $r(\sigma) \subset B_{2n-2i}$. Let σ be an $(i+1)$-cell of C inside U. Since $r(\partial\sigma) \subset B_{2n-2i}$ and B_{2n-2i} contracts inside $B_{2n-2i-1}$, there exists an extension of $r|_{\partial\sigma}$ to $f: \sigma \to B_{2n-2i-1}$. Consequently, r is defined on σ. If $a \in \sigma$ then $\text{dist}(a, f(a)) \leq \text{dist}(a, p) + \text{dist}(p, f(a)) < (\frac{1}{4})\epsilon_{2n} + \epsilon_{2n-2i-1}$. By (2), $\text{dist}(a, r(a)) < (\frac{1}{2})\epsilon_{2n} + 2\epsilon_{2n-2i-1}$, for any $a \in \sigma$, and so

$$\begin{aligned}
\text{dist}(p, r(a)) &\leq \text{dist}(p, a) + \text{dist}(a, r(a)) \\
&< (\tfrac{1}{4})\epsilon_{2n} + (\tfrac{1}{2})\epsilon_{2n} + 2\epsilon_{2n-2i-1} \\
&< \epsilon_{2n} + 2\epsilon_{2n-2i-1} \\
&< 3\epsilon_{2n-2i-1} \\
&< \epsilon_{2n-2i-2},
\end{aligned}$$

whence $r(\sigma) \subset B_{2n-2i-2}$. This completes the induction and shows that r is defined on each cell in U with values in $B_{\epsilon_0}(p)$. This implies both that r is defined in a neighborhood of p and that it is continuous at p. □

E.4. Corollary. *If a topological manifold M^m is embedded in $\mathbf{R}^n$ then it is the retract of some neighborhood there.* □

E.5. Corollary. *If M is a compact topological manifold then $H_*(M)$ and $H^*(M)$ are finitely generated.*

Proof. The first part of the proof of Theorem 10.7 of Chapter II applies to topological manifolds and shows that M can be embedded in some $\mathbf{R}^n$. It is then a retract of some neighborhood there. There is a smaller neighborhood of M that is a finite CW-complex (e.g., a union of cubes). Thus there exists a finite CW-complex K, an inclusion $i: X \hookrightarrow K$ and a retraction $r: K \to X$. Since the identity map $1: X \to X$ factors as $1 = r \circ i$ we have that

$$1_* = r_* \circ i_*: H_*(X) \to H_*(K) \to H_*(X).$$

Since $H_*(K)$ is finitely generated, so is $H_*(X)$. A similar argument works for cohomology, or one can use the Universal Coefficient Theorem for that. □

E.6. Corollary. *If the manifold* M^m *(or any* ENR*) is embedded in* $\mathbf{R}^n$ *then*

$$\check{H}^*(M) = \varinjlim \{H^*(K)\} \to H^*(M)$$

is an isomorphism, where K *ranges over the neighborhoods of* M *in* $\mathbf{R}^n$.

PROOF. The proof has two disparate parts. First, we will show that any neighborhood $U \subset M$ contains a smaller neighborhood V which strongly deforms to M through U. Then we will show that this implies the direct limit statement.

For the deformation statement, it suffices to produce, for some neighborhood W of M, a deformation $F: W \times I \to \mathbf{R}^n$, with $F(m, t) = m$ for all $m \in M$ and $t \in I$, $f(w, 0) = w$, and $f(w, 1) \in M$, since, for any $U, F^{-1}(U)$ is an open set containing $M \times I$, and hence containing an open set of the form $V \times I$ by the compactness of I. But if $r: W \to M$ is a retraction of a neighborhood of M in $\mathbf{R}^n$, then $F: W \times I \to \mathbf{R}^n$ defined by $F(w, t) = tw + (1 - t)r(w)$ is such a deformation.

For the second part, let U be a neighborhood of M in $\mathbf{R}^n$ so small that there is a retraction of U to M. Then the restriction map (induced by inclusion) $H^*(U) \to H^*(M)$ is onto, and is split by the map induced by the retraction. Suppose that $M \subset V \subset U$ with $F: V \times I \to U$ a deformation as above. Let $r = F(\cdot, 1): V \times \{1\} \to M$, a retraction. Let $i: V \times \{0\} \to U$ be the inclusion $(v, 0) \mapsto v \in V \subset U$, and $j: M \hookrightarrow U$, the inclusion. Consider the commutative diagram

$$\begin{array}{ccc}
H^*(M) & \xrightarrow{r^*} & H^*(V \times \{1\}) \\
{\scriptstyle j^*}\uparrow & & \uparrow{\scriptstyle \approx} \\
H^*(U) & \xrightarrow{F^*} & H^*(V \times I) \\
& {\scriptstyle i^*}\searrow & \downarrow \\
& & H^*(V \times \{0\}).
\end{array}$$

We see that $j^*(\alpha) = 0 \Rightarrow F^*(\alpha) = 0 \Rightarrow i^*(\alpha) = 0$. Thus any element $\alpha \in H^*(U)$ going to 0 in $H^*(M)$ already goes to 0 in $H^*(V)$. This, together with the surjectivity pointed out before, proves the result, by Corollary D.3. □

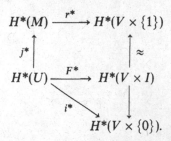

E.7. Corollary. *Let $f: X \to Y$ where X and Y are* ENRs *and X is compact. Then the mapping cylinder M_f and the mapping cone C_f (unreduced) are* ENRs.

PROOF. Local contractibility is clear, and so it suffices to produce embeddings in euclidean space. Assume that $X \subset \mathbf{R}^n$ and $Y \subset \mathbf{R}^m$. Then the map $(x, t) \mapsto (tx, (1-t)f(x), t) \in \mathbf{R}^n \times \mathbf{R}^m \times \mathbf{R}$ and $y \mapsto (0, y, 0)$ induce an embedding of M_f. Changing the tx to $t(1-t)x$ gives an embedding of C_f. □

E.8. Corollary. *A finite CW-complex is an* ENR. □

Bibliography

Artin, E. and Braun, H.
[1] *Introduction to Algebraic Topology*. Merrill, 1969.

Atiyah, M.
[1] *K-Theory*. Benjamin, 1967.

Bredon, G.
[1] On homogeneous cohomology spheres. *Ann. of Math.* **73** (1961), 556–565.
[2] The Künneth formula and functorial dependence in algebraic topology. *Amer. J. Math.* **40** (1968), 522–527.
[3] Some examples for the fixed point property. *Pacific J. Math.* **38** (1971), 571–575.
[4] *Introduction to Compact Transformation Groups*. Academic Press, 1972.

Bredon, G. and Wood, J.W.
[1] Non-orientable surfaces in orientable 3-manifolds. *Invent. Math.* **7** (1969), 83–110.

Brody, E.J.
[1] The topological classification of lens spaces. *Ann. of Math.* **71** (1960), 163–184.

Brown, M.
[1] A proof of the generalized Schoenflies theorem. *Bull. Amer. Math. Soc.* **66** (1960), 74–76.
[2] Locally flat imbeddings of topological manifolds. *Ann. of Math.* **75** (1962), 331–341.

Brown, R.F.
[1] *The Lefschetz Fixed Point Theorem*. Scott Foresman, 1971.

Bullett, S.R. and MacDonald, I.G.
[1] On the Adem relations. *Topology* **21** (1982), 329–332.

Dieudonné, J.
[1] *A History of Algebraic and Differential Topology*, 1900–1960. Birkhäuser. 1989.

Dold, A.
[1] *Lectures on Algebraic Topology*. Springer-Verlag, 1972.

Dugundji, J.
[1] *Topology*. Allyn and Bacon, 1966.

Eilenberg, S. and Steenrod, N.
[1] *Foundations of Algebraic Topology*. Princeton University Press, 1952.

Francis, G.K.
[1] *A Topological Picturebook*. Springer-Verlag, 1987.

Greenberg, M. and Harper, J.
[1] *Algebraic Topology: A First Course*. Benjamin/Cummings, 1981.

Hilton, P.J. and Wylie, S.
[1] *Homology Theory*. Cambridge University Press, 1960.

Hirzebruch, H.F., and Mayer, K.H.
[1] **O**(*n*)-*Mannigfaltigkeiten, Exotische Sphären und Singularitäten*. Springer-Verlag, 1968.

Hochschild, G.
[1] *The Structure of Lie Groups*. Holden-Day, 1965.

Hocking, J.G. and Young, G.S.
[1] *Topology*. Addison-Wesley, 1961.

Holm, P.
[1] The theorem of Brown and Sard. *Enseign. Math.* **33** (1987), 199–202.

Hu, S.-T.
[1] *Homotopy Theory*. Academic Press, 1959.

Kelley, J.L.
[1] *General Topology*. Van Nostrand, 1955.

Kervaire, M. and Milnor, J.
[1] Groups of homotopy spheres, I. *Ann. of Math.* **77** (1963), 503–537.

Kosinski, A.
[1] *Differential Manifolds*. Academic Press, 1992.

Lundell, A. and Weingram, S.
[1] *The Topology of CW Complexes*, Van Nostrand Reinhold, 1969.

Mazur, B.
[1] On embeddings of spheres. *Bull. Amer. Math. Soc.* **65** (1959), 59–65.

Mac Lane, S.
[1] *Homology*. Academic Press/Springer-Verlag, 1963.

Massey, W.S.
[1] *Algebraic Topology: An Introduction*. Springer-Verlag, 1967.

Milnor, J.
[1] On spaces having the homotopy type of a CW-complex. *Trans. Amer. Math. Soc.* **90** (1959), 272–280.
[2] On axiomatic homology theory. *Pacific J. Math.* **12** (1962), 337–341.
[3] *Topology from the Differentiable Viewpoint*. University Press of Virginia, 1965.
[4] Whitehead torsion. *Bull. Amer. Math. Soc.* **72** (1966), 358–426.

Milnor, J. and Stasheff, J.
[1] *Characteristic Classes*. Princeton University Press, 1974.

Montgomery, D. and Zippin, L.
[1] *Topological Transformation Groups*. Wiley, 1955; Krieger, 1974.

Morse, M.
[1] A reduction of the Schoenflies extension problem. *Bull. Amer. Math. Soc.* **66** (1960), 113–115.

Mosher, R. and Tangora, M.
[1] *Cohomology Operations and Applications in Homotopy Theory*. Harper & Row, 1968.

Nehari, Z.
[1] *Conformal Mapping*. McGraw-Hill, 1952.

Novikov, P.S.
[1] On the algorithmic unsolvability of the word problem in group theory. *Amer. Math. Soc. Trans.* (2) **9** (1958), 1–122.

Rushing, T.B.
[1] *Topological Embeddings.* Academic Press, 1973.

Seifert, H. and Threlfall, W.
[1] *A Textbook of Topology* (translation). Academic Press, 1980.

Singer, I.M. and Thorpe, J.A.
[1] *Lecture Notes on Elementary Topology and Geometry.* Springer-Verlag, 1967.

Smale, S.
[1] Generalized Poincaré conjecture in dimensions greater than four. *Ann. of Math.* **74** (1961), 391–406.

Spanier, E.H.
[1] *Algebraic Topology.* McGraw-Hill, 1966.

Steenrod, N.E., and Epstein, D.B.A.
[1] *Cohomology Operations.* Princeton University Press, 1962.

Stone, A.H. and Tukey, J.W.
[1] Generalized "sandwich" theorems. *Duke Math. J.* **9** (1942), 356–359.

Thom, R.
[1] Quelques propriétés globales des variétés différentiables, *Comment. Math. Helv.*, **28** (1954), 17–86.

Vick, J.W.
[1] *Homology Theory.* Academic Press, 1973.

Wall, C.T.C.
[1] On the exactness of interlocking sequences. *Enseign. Math.* **12** (1966), 95–100.

Weyl, H.
[1] *The Classical Groups*, 2nd ed. Princeton University Press, 1946.

Whitehead, G.
[1] *Elements of Homotopy Theory.* Springer-Verlag, 1978.

Whitehead, J.H.C.
[1] On incidence matrices, nuclei and homotopy types. *Ann. of Math.* **42** (1941), 1197–1239.
[2] Combinatorial homotopy I. *Bull. Amer. Math. Soc.* **55** (1949), 213–245.

Whitney, H.
[1] *Geometric Integration Theory.* Princeton University Press, 1957.

Index of Symbols

$\mathbf{Z}$	integers, 43
$\mathbf{Z}_p$	$\mathbf{Z}/p\mathbf{Z}$, 131
$\mathbf{Q}$	rational numbers, 215
$\mathbf{R}$	real numbers, 1
$\mathbf{C}$	complex numbers, 53
$\mathbf{H}$	quaternions, 53
$\mathbf{R}^n$	real euclidean n-space, 1
$\mathbf{S}^n$	unit sphere in $\mathbf{R}^{n+1}$, 40
$\mathbf{D}^n$	unit disk in $\mathbf{R}^n$, 40
$\mathbf{P}^n$ or $\mathbf{RP}^n$	real projective n-space, 43, 55
$\mathbf{CP}^n$	complex projective n-space, 55, 197
$\mathbf{QP}^n$	quaternionic projective n-space, 206
$\mathbf{I}$ or I	unit interval $[0, 1]$, 6
$\mathbf{I}^n$	n-cube, 198
$\mathbf{T}^n$	n-torus, 43
$\mathbf{K}^2$	Klein bottle, 43
$\mathbf{Gl}(n, \mathbf{F})$	general linear group, 53
$\mathbf{Sl}(n, \mathbf{F})$	special linear group, 53
$\mathbf{O}(n)$	orthogonal group, 53
$\mathbf{U}(n)$	unitary group, 53
$\mathbf{Sp}(n)$	symplectic group, 53
A^t	transpose of A, 53
$V_{n,k}$	Stiefel manifold, 54
$G_{n,k}$	Grassmann manifold, 463
$L(p, q)$	classical lens space, 86
$L(p; q_1, \ldots, q_n)$	generalized lens space, 151
Y^X	space of maps $X \to Y$, 23, 437
$T_p(M)$	tangent space at p of M, 76

$E(\xi)$	total space of vector bundle ξ, 108
$B(\xi)$	base space of vector bundle ξ, 108
$N(H)$	normalizer of H, 148
$\pi_n(X)$	nth homotopy group of X, 119, 130
χ	Euler class, 378
$\chi(X)$	Euler characteristic of X, 153, 215
$C_p(X)$	cellular/simplicial chain group of X, 203, 247
Δ_n	standard n-simplex, 169
$\Delta_p(X)$	singular chain group of X, 170
$\Delta^p(X)$	singular cochain group of X, 273
ϵ or ϵ_*	augmentation, 172
Z_p	group of p-cycles, 171
B_p	group of p-boundaries, 171
$H_p(X)$	pth homology group of X, 171
$H^p(X)$	pth cohomology group of X, 273
$\check{H}^p(X)$	pth Čech cohomology group of X, 348
$\tilde{H}_p(X)$	pth reduced homology group of X, 181, 184
$H^p_\Omega(M)$	pth de Rham cohomology group of M, 263
$K^{(n)}$	n-skeleton of K, 195
$K^{[n]}$	nth barycentric subdivision of K, 251
$\mathrm{mesh}(K)$	mesh of K, 251
$\mathrm{carr}(x)$	carrier of x, 251
$\mathrm{St}_K(v)$	open star in K of v, 252
$A^p(V)$	alternating p-forms on V, 260
$\omega \wedge \eta$	exterior product of ω and η, 260
$\Omega^p(M)$	differential p-forms on M, 262
$d\omega$	exterior derivative of ω, 262
β or β_0	Bockstein homomorphism, 181
$A \otimes B$	tensor product of A and B, 271
$\mathrm{Hom}(A, B)$	group of homomorphisms $A \to B$, 272
$\mathrm{Ext}(A, B)$	(extensions), derived functor of Hom, 275
$A * B$ or $\mathrm{Tor}(A, B)$	torsion product of A and B, 278
$\phi_{\sigma,\tau}: \mathbf{S}^n \to \mathbf{S}^n$	induced map from cellular map, 209
$X \times Y$	cartesian product of X and Y, 22
$\times X_\alpha$	cartesian product of the X_α, 22
$X + Y$	topological sum of X and Y, 24
$+ X_\alpha$	topological sum of the X_α, 24
$u \times v$	cross product of u and v, 220, 322
$u \cup v$	cup product of u and v, 326
$\alpha \cap a$	cap product of α and a, 334
$\langle \alpha, a \rangle$	Kronecker product of α and a, 322
θ	Eilenberg–Zilber map, 316
ϑ	orientation class, 348
$[M]$	fundamental class of manifold M, 335
$a \bullet b$	intersection product of a and b, 367
$[K] \cdot [N]$	intersection number of K and N, 380

τ	Thom class, 368
$f^!$ and $f_!$	transfer map, 368
D_M	(inverse of) Poincaré duality map, 367
Sq^i	Steenrod square, 404
$\mathscr{P}^i$	Steenrod cyclic reduced power, 410
w_i	ith Stiefel–Whitney class, 421
$K(\pi, n)$	Eilenberg–Mac Lane space, 484
c_f^{n+1}	obstruction class, 499
$d^n(f, g)$	difference obstruction class, 500
$\mathrm{dist}(x, y)$	distance from x to y, 1
$B_\epsilon(x)$	ϵ-ball about x, 2
$\mathrm{diam}(A)$	diameter of A, 28
M_f	mapping cylinder of f, 42
C_f	mapping cone of f, 42
∂c	boundary of c, 9, 170
δf	coboundary of f, 271, 321
$X \vee Y$	one-point union of X and Y, 44
$\bigvee X_\alpha$	one-point union of the X_α, 200
$X \wedge Y$	smash product of X and Y ($= X \times Y/X \vee Y$), 199
∇f	gradient of f, 80
Δ	deck transformation group, 147
Δ	diagonal approximation, 327
$[X, Y]$	Lie bracket of vector fields X and Y, 88
$[X; Y]$	homotopy classes of maps $X \to Y$, 48
$[\sigma{:}\, \tau]$	incidence number between cells σ and τ, 204
$[\pi, \pi]$	commutator subgroup of π, 173
$[\alpha, \beta]$	Whitehead product of α and β, 461
$G * H$	free product of groups G and H, 158
$\bigstar G_\alpha$	free product of groups G_α, 158
$f * g$	concatenation of paths, loops, or homotopies, 46
ΣX	unreduced suspension of X, 190
SX	reduced suspension of X, 128
CX	cone on X, 430
f_σ	characteristic map for σ, 195
$f_{\partial\sigma}$	attaching map for σ, 194
$p_\sigma{:}\, K^{(n)} \to \mathbf{S}^n$	collapsing projection, 200
$\gamma_n{:}\, \mathbf{I}^n \to \mathbf{S}^n$	canonical collapsing map, 199
$f_\#$	induced map on homotopy groups, 131
f_Δ	induced chain map, 175
f^Δ	induced cochain map, 286
f_*	induced map on homology, 176
f^*	induced map on cohomology, 273
Υ	subdivision operator, 223
$L(f)$	Lefschetz fixed point number, 254
$L(f, g)$	Lefschetz coincidence number, 394
tr	trace, 253

$f \pitchfork g$	f is transverse to g, 84, 114
TA or $T(A)$	torsion subgroup of A, 255, 283
PX	path space of X, 456
ΩX	loop space of X, 441
T_u	induced equivalence by characteristic element u, 491
$\approx$	isomorphism, homeomorphism, or diffeomorphism, 5
$\simeq$	homotopy or homotopy equivalence, 45, 219
$\sim$	homologous, 171
$[f]$	homotopy class of f, 124, 129
$[\![f]\!]$	homology class of f, 171
$a \mapsto b$	element a goes to element b, 523
$A \hookrightarrow X$	inclusion map of A in X, 46
$\bar{A}$	closure of A, 8
$\text{int}(A)$	interior of A, 8
G_x	isotropy subgroup of G at x, 54
$[v_0, \ldots, v_n]$	affine simplex, 169
$\langle v_0, \ldots, v_n \rangle$	simplicial simplex as a chain, 246
$(v_0, \ldots, v_n)$	simplicial simplex as a space, 246
$f \,\underline{\times}\, g$	the composition $(f \times g) \circ d: N \to N \times N \to K \times M$, 380
∇	codiagonal map $X \vee X \to X$, 441
$f \,\underline{\vee}\, g$	the composition $\nabla \circ (f \vee g): X \vee Y \to Z \vee Z \to Z$, 441
$\varinjlim$	direct limit, 535
$\varprojlim$	inverse limit, 505

Index

Action
effective 54
free 154
group 54, 86, 150
monodromy 146, 150
properly discontinuous 150–154, 219
smooth 306
Acyclic models (*see* Models, acyclic)
Adem relations 405, 408, 411, 412, 466
Adjoint 495
Affine independence 245
Affine simplex (*see* Simplex, affine)
Algebra
cohomology 263, 329
Lie 305
Alexander duality (*see* Duality, Alexander)
Alexander horned disk 231–232
Amalgamation 159
Amoeba 472
Anchoring 470–471
Antipodal map (*see* Map, antipodal)
Approximation
CW 488
diagonal 327–328, 334, 335, 413
simplicial 251–253
smooth 96–97
Atlas 69, 76, 347–348
Attaching 41, 482–484
Augmentation 172, 181, 220, 224, 335, 356
Axiom of Choice 18, 526–531
Barycenter 224, 251
Barycentric coordinates 169
Base point 118, 128
nondegenerate 435
Base ring (*see* Ring, base)
Basis
neighborhood 4, 6, 8, 13, 24, 32, 437
of a topology 5–6, 9, 437
Betti number (*see* Number, Betti)
Bicollaring 235
Bidegree 409, 412
Bijection 523
Bockstein homomorphism 181, 334, 338, 363, 404, 410, 417–418, 492–493
Bott periodicity 467, 469
Boundary 9, 71, 170–171, 177, 204, 213, 221, 247, 315, 321, 355*ff*, 360–361, 366
Bouquet (*also see* Union, one-point), 489
Braid diagram (*see* Diagram, braid)
Bundle 106–114, 455
disk 109, 367*ff*, 426–429
euclidean 109
induced 111–114
line 294, 389, 424, 426, 515, 517
normal 93, 99, 110, 113, 378*ff*, 515
orientation 340–348
principal 111
sphere 109, 368, 390–393, 408, 412
tangent 88–89, 109, 113, 378*ff*, 391, 420
vector 108, 378*ff*, 420–426, 511–517

Card (*see* Number, cardinal)
Carrier 251
Cartan formula 404, 410, 419–420
Category (first and second) 57, 81
Cell 194*ff*
Cell attachment 482–484
Cellular approximation (*see* Theorem, cellular approximation)
Chain 168*ff*, 524
maximal 526
Chain contraction 316
Characteristic, Euler 153, 215–217, 256, 285, 321, 355, 360, 366, 378, 380, 383–389, 391, 398, 515
Chart 68–75, 82–84, 347–348
flat 83
Class
characteristic (*also see* Class, Stiefel–Whitney *and* Class, Chern), 491*ff*
Chern 294
difference 500–501, 507, 509–511
equivalence 524
Euler 379*ff*, 421, 514–515
fundamental 199, 339, 355, 377
obstruction 500, 507, 509, 512–515
orientation (*also see* Class, fundamental) 301
self-intersection 382
Stiefel–Whitney 420–426
Thom 367–372, 377, 379–384, 514–515
Wu 423
Classification problem 487
Closure 8
Cobordism 120–126
Cochain 271*ff*
Coefficients 180, 184
Coexact 445–448
Cofibration 430–436
Cogroup 441–442
Cohomology
Čech 348*ff*
de Rham 263–271, 286–297, 304–314
sheaf 373
Cohomology operation (*see* Operation, cohomology)
Cohomology ring (*see* Algebra, cohomology)
Coincidence 393–404
Collapsing 40, 433–436, 445–447
Collar 267, 355
Compactification
one-point 32–33
Stone–Čech 34–35
Compactness (*see* Space, compact)
Completion 29
Complex
chain 177
CW 194–218, 246
relative CW 483
simplicial 245–254
Component 10–12, 31, 234–239, 353, 355
Component, arc 12
Concatenation 46, 127, 441
Cone 219, 223, 287, 430
mapping 42–43, 48–51, 433, 445–448
Connectivity (*see* Space, connected)
Continuity 1–7, 15, 437–439
Convergence, uniform 6, 440
Convex 56
Coproduct 442
Corner 383, 427
Countable
first 6
second 6
Covering 138–158, 198, 216, 341–348
orientable 348
regular 149–151
universal 145, 155
Crosscap 163
Curl 269, 291
CW-complex (*see* Complex, CW-)
CW pair 450
Cycle 168*ff*
Cylinder, mapping 42–43, 46–50, 432–436, 457, 481, 485

Decision procedure 487
Deck transformation (*see* Transformation, deck)
Deformation retract (*see* Retract, deformation)
Degree 124, 142–143, 186–187, 190–194, 244, 256, 300–304, 315, 333, 359, 403, 409, 449, 470
Dense 9
Derivation 77
Derivative, exterior 262
Diagonal 24
Diagonal approximation (*see* Approximation, diagonal)
Diagram, braid 188–189, 230
Diameter 28
Dictionary order 6
Diffeomorphism 63, 70
Difference, primary 508–509
Differential 78
Direct system 535

Directional derivative 76
Distance 1
Divergence (div) 269, 291
Dodecahedral space (*see* Space, dodecahedral)
Doubling 59, 370, 371, 402
Dual cell structure 339
Duality
  Alexander 351–353
  cup product 357–358
  Lefschetz 351–353
  Poincaré 338–340, 348*ff*, 367
  Whitney 422
Duality pairing 358
Dunce cap 50, 206, 257, 479

Embedding 27, 353, 362
  smooth 79, 89–92
ENR 255–256, 536–540
Equivalence
  bundle 108
  chain 219
  homotopy (*see* Homotopy equivalence)
  n- 485
  weak homotopy 485
Euler characteristic (*see* Characteristic, Euler)
Euler number (*see* Number, Euler)
Evaluation (*also see* Map, evaluation), 322
Exactness (*see* Sequence, exact)
Excision 183, 223–228, 285, 482
Exponential law (*see* Law, exponential)
Ext 274*ff*
Extension problem 430, 498*ff*

Face 169, 328
Fattened manifold (*see* Manifold, fattened)
Fiber 107, 146, 450
  homotopy 505
Fiber homotopic (*see* Homotopy, fiber)
Fibration 450–457
  Hopf (*see* Map, Hopf)
  Hurewicz 450
  induced 456
  path-loop 456–457, 484
  principal 498
  Serre 450
Final 16
Finite intersection property 19
Fixed point property 257–259, 294, 412
Flow 86–88, 386
Form
  alternating 260
  closed 263, 291
  differential 261*ff*, 304–314
  exact 263, 291
  induced 263
  integration of 265*ff*
  invariant 306, 308
  left-invariant 305
  symmetric 310
Fox–Artin wild arc 231
Fratricide 487
Function
  closed 7
  continuous 4
  isotone 524
  nowhere differentiable 60–62
  open 7
  periodic 291
  semicontinuous 62
  transition 70
Functional structure 69
  induced 72–74
Functor 131, 176
  exact 535
  left exact 272
  right exact 272
Fundamental group (*also see* Group, fundamental)
  of arbitrary complex 487
  of circle 142, 149
  of CW-complex 211
  of 4-manifold 164, 487
  of graph 153, 488
  of Klein bottle 161
  of knot complement 164
  of product space 137
  of projective space 143, 147, 149
  of surface 162
  of topological group 138
  of torus 143, 149
  of union 159–164

Gauss map (*see* Map, Gauss)
General position (*see* Transversality)
Genus 162, 516
Germ 76
Glb 525
Gradient (grad) 80, 93, 269, 291
Graph 152–153, 239, 240, 380*ff*, 427, 488
Grassmann manifold (*see* Manifold, Grassmann)

Group
classical 53–55, 101–106, 463–469
covering 158, 311
divisible 274
free 152–155, 158–161, 488
fundamental 132*ff*, 172–175, 211
graded 177
H- 441
homotopy 127–132, 393, 443*ff*
icosahedral 353–354
injective 274
isotropy 54, 146
Lie 53–55, 101–106, 110–111, 291, 304–314, 388–389, 463–467
one-parameter 102, 106, 312–314
orthogonal 53–55, 103–104, 164–167, 191, 354, 464–467
presentation of 164
projective 274
semisimple 311
simple 311
structure 107
symplectic 53–55, 103–104, 464–467
topological 51–56, 138, 158, 412
torsion free 279
unitary 53–55, 103–104, 393, 412, 463–469
Weyl 389

Haar integral 306
Heine–Borel property 19
Hom 270–281
Homeomorphism 5
Homology
axioms for 183
Borel–Moore 373
Čech 169, 184
cellular 200*ff*
computation of 204–206, 258
reduced 181, 184
relative 180
singular 168–182, 219*ff*
smooth singular 269–271, 286–291
uniqueness of 210
with coefficients 180
Homology groups (*and*/*or* Cohomology groups)
of CW-complex 200*ff*
of dunce cap 206
of figure eight 175
of Klein bottle 175, 205, 250, 284
of lens spaces 205–206, 363, 366
of Lie groups 304–314
of manifolds 338*ff*
of one-point union 190–191
of plumbing 427–429
of products 211–214, 320, 325, 374
of projective spaces 175, 193, 204, 206, 217–218, 240–241, 249, 283, 292–296, 329–330, 338, 359, 375–377, 411–412
of simplicial complex 247
of sphere 175, 185
of sphere bundle 390–393
of spherical complement 233
of Stiefel manifolds 392
of suspensions 190
of torus 175, 205–206, 250, 331
of union 228–231
of unitary group 393
Homomorphism, connecting 178, 183
Homotopy 44–51, 115*ff*, 127*ff*, 430*ff*
chain 219, 275, 317–318
constant 46
fiber 457
free 136, 457–461
inverse of 47
Homotopy equivalence 45, 486, 488
Homotopy extension property 430–431
Homotopy fiber (*see* Fiber, homotopy)
Homotopy groups
of attachment 482–484
of classical groups 463–469
of H-spaces 443
of one-point unions 457, 482, 488
of products 454
of projective spaces 457, 464, 480
of spheres 121–126, 131, 295, 453–454, 464–466
Homotopy inverse 45
Homotopy lifting property 450*ff*
Homotopy rel 46
Homotopy type 45
Hopf invariant 409, 412
Hopf map (*see* Map, Hopf)
H-space (*see* Space, H-)

Imbedding (*see* Embedding)
Immersion 79, 424–425
Index (*also see* Signature) 385–388
Indivisible 378
Initial segment 525
Inequalities, Morse 217
Infimum (inf) 525
Initial segment 525
Injection 523
Interior 8
Intersection matrix 428–429

Intersection product (*see* Product, intersection)
Isometry 29
Isomorphism, Thom 368–370
Isotopy 116, 120, 122
Isotropy group (*see* Group, isotropy)

Jello 165
Join 409, 468, 480

Killing homotopy groups 418, 483–484
Klein bottle 43, 88, 89, 111, 138, 140, 142, 151, 158, 161, 163, 175, 197, 205, 214, 250, 284, 361, 426, 450, 462
Knot 163–164, 516
K(π, n) (*see* Space, Eilenberg–Mac Lane)

Ladder 177, 227, 288, 480–481
Lattice 525
Law, exponential 438, 444
Lemma
  bootstrap 345, 351, 369, 370
  five 181–182
  Lebesgue 28
  Poincaré 288
  reparametrization 46
  Sperner's 253
  Urysohn's 29
  Zorn's 527
Lens space (*see* Space, lens)
Lie algebra (*see* Algebra, Lie)
Lie bracket 88, 263, 305
Lifting problem 143, 498 *ff*
Limit 14
  direct 231, 534–536
  inverse 505
  of net 14–18
Line integral 134
Linking number (*see* Number, linking)
Locally compact (*see* Space, locally compact)
Locally finite collection 35
Loop 132 *ff*
Loop space (*see* Space, loop)
Lub 524

Manifold
  closed 359
  differentiable 68 *ff*
  fattened 119
  framed 121
  Grassmann 463
  orientable 71
  oriented 70
  parallelizable 89, 389, 410, 469, 517
  product 75
  smooth 68 *ff*
  Stiefel 54, 391–393, 464
  strange 59–60
  topological 59, 68, 341 *ff*, 429
  unsmoothable 429
  with boundary 71
Map 4
  antipodal 40, 43, 163, 187, 217–218, 240–245, 253, 256, 362
  attaching 194
  bundle 108
  cellular 207
  chain 176, 177
  characteristic 195
  differentiable 70
  discrete valued 10
  equivalence to a fibration 457
  equivariant 240–245, 364, 487
  evaluation 437
  exponential 101–106, 314
  Gauss 384, 385, 389
  Hopf 131, 295, 406–409, 453, 464–469
  identification 39, 43
  isometric 29
  pointed 128, 434 *ff*
  proper 20, 22, 33
  restriction 455
  simple 501, 507
  simplicial 250
  smooth 70, 71
  splitting 179–180
Mapping cone (*see* Cone, mapping)
Mapping cylinder (*see* Cylinder, mapping)
Mayer–Vietoris (*see* Sequence, Mayer–Vietoris)
Maximality principle (*see* Principle, maximality)
Mesh 251
Metric, Riemannian 109
Metric space
  complete 25
  separable 10
  totally bounded 25
Metrizable 26–31, 36, 38, 69
Möbius band 537
Models, acyclic 221, 287, 317, 323, 327, 337, 413
Monodromy action (*see* Action, monodromy)
Moore–Postnikov decomposition 501

Neighborhood 4*ff*
cubic 267
symmetric 51
tubular 92–101
Net 14–18, 21, 23, 24, 440
universal 17, 21
Nowhere dense 9
Number
Betti 172, 258–259, 398, 401
cardinal 527–530
Cayley 469
coincidence 393*ff*
Euler 382–383
incidence 204, 246, 297
intersection 340, 380–382, 428
Lefschetz fixed point 254, 294, 381, 384, 389, 393, 394, 398
linking 117–118, 366
ordinal 530
self-intersection 382

Obstruction 297–298, 379, 497–517
difference 500–501, 507–509
primary 507–509
One-parameter subgroup (*see* Group, one-parameter)
One-sided 376
Operation, cohomology 328, 363, 393, 404–420, 492–497
decomposable 407
Orbit 54
Order
linear 524
partial 524
simple 524
total 524
well 525
Ordinal 6, 530
Orientation 71, 199, 205, 267, 301, 338–348, 355, 370, 471

Partition of unity 35–37, 89
Plane, projective (*also see* Space, projective) 40, 50, 74, 139, 142, 143, 147, 158, 162, 163, 193, 204, 214, 230, 248–250, 257, 283, 284, 300, 304, 320, 321, 329–332, 338, 361, 365, 375, 376, 401, 403, 425, 426, 436, 480, 493, 517, 537
Plumbing 426–429
Poincaré conjecture 354, 429
Poincaré duality (*see* Duality, Poincaré)
Pointed (*see* Map, pointed)
Polyhedron 246
Poset (*see* Order, partial)
Postnikov decomposition 501
Principle, maximality 17, 18, 37, 38, 275, 526–527
Product
cap 323, 334–338, 349, 358
cartesian 523
cross 220–223, 321–338
cup 326–337, 413–414
exterior 260
free 158–159
intersection 366–378, 395
Kronecker 322, 338
smash 435
tensor 180, 271–281, 315
torsion 278
wedge 260, 435
Whitehead 461–463
Pullback 111, 455, 484, 498

Quadratic residue 365
Quantum 167
Quasi-component 11, 12, 31

Refinement 35
Relation, equivalence 523
Representation, adjoint 308–314
Residual set (*see* Set, residual)
Resolution
injective 275
projective 277
Restriction (*see* Map, restriction)
Retract 42, 51, 186, 245, 257, 412, 431, 454
deformation 45, 432, 451
neighborhood 536–540
of disk 98, 186
Ring
base 280
cohomology (*see* Algebra, cohomology)
truncated polynomial 330

Saturation 41
Section 342, 368, 378–379, 382–383, 391, 514–515, 526
Semicontinuity 62
Separated sets 223, 345
Sequence
Barratt–Puppe 447
Cauchy 25
exact 178
fibration 453

Gysin 390–393
homology 180*ff*
homotopy 445–450, 463
Mayer–Vietoris 228–230, 285–287
Smith 240, 387
split 179–180
Set
cellular 237–239
clopen 10
closed 3
countable 528
dense 9, 18
directed 14
elementary 139
evenly covered 139
fixed point 253*ff*, 307, 378*ff*
open 3
power 524
residual 57–62, 81
symmetric 51
uncountable 530
Sheet 139, 147, 149, 216
Shriek (*see* Transfer)
Shrinking 37, 90
Signature 361, 517
Simple (*see* Group, simple *or* Map, simple *or* Space, simple)
Simplex 169, 170
affine 169, 223–226, 246
smooth 269, 286–291
Skeleton 195
Space
acyclic 182, 332, 334
arcwise connected 12
comb 134–135
compact 18–21
complete metric 25
completely regular 26, 30, 34–35
connected 10–12, 25
contractible 45, 51, 99, 185, 211, 219, 436, 479, 480
covering 138–158, 198, 216, 457
dodecahedral 353–354, 429, 480
Eilenberg–Mac Lane 484, 488–493
fiber 450–457
H- 333, 412, 441–443, 463
Hausdorff 13, 15, 24
irreducible 10
lens 86, 151, 206, 219, 363–366, 392, 486–487, 492
locally compact 31–34, 57, 437–441
locally connected 12
locally relatively simply connected 155
loop 441–445, 456–457
metric 1–3, 6, 14, 25–29, 57–62
metrizable 26, 28, 38–39, 69
n-connected 207, 449
n-simple 459
normal 13, 20, 30, 36, 40
orbit 150, 219
paracompact 35–39
path 456–457
pointed 44, 128, 199, 435
projective (*also see* Plane, projective) 55, 114, 147, 149, 175, 190, 197, 198, 206, 216, 217–218, 241, 245, 257, 292–297, 304, 359–360, 362, 377, 403, 411–412, 424–426, 457, 464, 481, 492
quotient 39–44
regular 13
semilocally 1-connected 155
separable metric 9
σ-compact 37–39
simple 459, 501
simply connected 132
topological 3*ff*
totally bounded metric 25
well-pointed 212, 435–436, 457, 493, 497
Zariski 10, 14
Sphere
exotic 429
homology 354, 428–429, 480, 488
Kervaire 429
Milnor 429
noncontractibility of 99
Split (*see* Sequence, split)
Stability 131
Star 252, 339
Steenrod squares (*see* Operation, cohomology)
Stereographic projection 73
Stiefel manifold (*see* Manifold, Stiefel)
Subdivision 223–228, 339
Submanifold 79
Submersion 79
Subnet 16
Subspace 8
locally closed 33
of compact space 34–35
of completely regular space 28
of Hausdorff space 14
of normal space 60
of regular space 13
Successor 525
Sum
connected 158, 358, 361, 366
topological 24, 183, 286
Whitney (*see* Whitney sum)

Support 36, 90, 265, 342, 355
Supremum (sup) 525
Surjection 523
Suspension 124–126, 128–129, 190, 212, 332, 406, 417, 435, 442, 448, 464, 486, 493–494

Tangent bundle (*see* Bundle, tangent)
Tangent field (*also see* Vector field) 87–88, 188, 256, 383–388, 515
Tangent space 76
Tangent vector (*see* Vector, tangent)
Telescope 519–521
Theorem
- Banach 64
- Borsuk 537
- Borsuk–Ulam 240–245, 362
- Brouwer fixed point 98, 186, 402
- cellular approximation 208
- classification of covering spaces 154
- classification of deck transformations 149
- covering homotopy 140
- de Rham 271, 286–291
- divergence 269
- Eilenberg–Zilber 318
- Euler–Poincaré 215
- exponential (*see* Law, exponential)
- Freudenthal suspension 126
- Fubini 533
- fundamental, of abelian groups 215
- fundamental, of algebra 81–82, 142, 194, 388
- Gauss' 269
- Green's 269
- Gysin's 390
- ham sandwich 242–243
- homotopy addition 469–475
- Hopf classification 124, 297–304, 497, 509
- Hurewicz 174, 475–480, 488
- implicit function 65–66
- invariance of dimension 235
- invariance of domain 235
- inverse function 67–68, 82–86
- Jordan–Brouwer 234
- Jordan curve 230–240, 353
- Künneth 318–320, 325
- Lefschetz–Hopf fixed point 253–259, 381, 384, 389
- lifting 143
- Lusternik–Schnirelmann 243
- Mayer–Vietoris (*see* Sequence, Mayer–Vietoris)
- maximal chain 526
- mean value 63
- metrization 28
- path lifting 140
- Poincaré duality (*see* Duality, Poincaré)
- Poincaré–Hopf 387
- Sard's 80–82, 531–534
- Schoenflies 235–239
- Schroeder–Bernstein 527
- Seifert–Van Kampen 158–164
- Serre 492
- simplicial approximation 252
- smooth approximation 96, 97
- Stokes' 267–269
- Stone–Tukey 243
- Thom isomorphism 368, 370
- Thom–Pontryagin 122
- Thom realizability 377
- Tietze extension 30
- tubular neighborhood 92–100
- Tychonoff 23
- universal coefficient 281–283
- Urysohn 28
- well-ordering 526
- Whitehead 481
- Whitney duality 422
- Whitney embedding 91, 92
- Wu 423

Theory, homology 183
Topological group (*see* Group, topological)
Topology
- coarsest 7
- compact-open 437–441
- discrete 5
- finest 7
- generated 5
- half open interval 6
- identification 39
- induced 39
- largest 7
- order 6, 7, 14, 18
- product (Tychonoff) 22, 214, 441
- quotient 39
- relative 8
- smallest 7
- strongest 7
- subspace 8
- trivial 5
- weak (CW) 194–195, 214
- weakest 7

Tor 274–280
Torsion 279
Torsion subgroup 255, 283, 346, 357, 366

Torus 43, 50, 72, 75, 143, 158, 163, 175, 197, 205, 206, 210, 216, 250, 304, 331, 334, 373, 378, 388–389, 400–401, 436
  maximal 388
Trace formula 254
Transfer 240, 368, 394–395
Transformation, deck 147
Transgression 297
Translation (left and right) 51, 468
Transversality 84–86, 91, 114–118, 371, 381
Tree 152–153, 427, 488
Triangulation 169, 246
Trivialization 107, 426
Tubular neighborhood (*see* Neighborhood, tubular)
Type, finite 215
Type (π, n) (*see* Space, Eilenberg–Mac Lane)

Union
  disjoint 24
  one-point 44, 50, 153, 199, 206, 257, 488
  separated 345
Unity, partition of (*see* Partition of Unity)
Universal net (*see* Net, universal)

Value
  critical 80, 531–534
  regular 80, 531–534
Vector
  probability 100
  tangent 76
Vector bundle (*see* Bundle, vector)
Vector field (*also see* Tangent field) 86–88, 268–269, 291, 305
  incompressible 291
  index of 385–388
  left-invariant 305*ff*

Well-pointed (*see* Space, well-pointed)
Whisker 436
Whitney sum 113, 422
Word 158–162, 487

Zariski space (*see* Space, Zariski)

Graduate Texts in Mathematics

(continued from page ii)

64 EDWARDS. Fourier Series. Vol. I. 2nd ed.
65 WELLS. Differential Analysis on Complex Manifolds. 2nd ed.
66 WATERHOUSE. Introduction to Affine Group Schemes.
67 SERRE. Local Fields.
68 WEIDMANN. Linear Operators in Hilbert Spaces.
69 LANG. Cyclotomic Fields II.
70 MASSEY. Singular Homology Theory.
71 FARKAS/KRA. Riemann Surfaces. 2nd ed.
72 STILLWELL. Classical Topology and Combinatorial Group Theory. 2nd ed.
73 HUNGERFORD. Algebra.
74 DAVENPORT. Multiplicative Number Theory. 3rd ed.
75 HOCHSCHILD. Basic Theory of Algebraic Groups and Lie Algebras.
76 IITAKA. Algebraic Geometry.
77 HECKE. Lectures on the Theory of Algebraic Numbers.
78 BURRIS/SANKAPPANAVAR. A Course in Universal Algebra.
79 WALTERS. An Introduction to Ergodic Theory.
80 ROBINSON. A Course in the Theory of Groups. 2nd ed.
81 FORSTER. Lectures on Riemann Surfaces.
82 BOTT/TU. Differential Forms in Algebraic Topology.
83 WASHINGTON. Introduction to Cyclotomic Fields. 2nd ed.
84 IRELAND/ROSEN. A Classical Introduction to Modern Number Theory. 2nd ed.
85 EDWARDS. Fourier Series. Vol. II. 2nd ed.
86 VAN LINT. Introduction to Coding Theory. 2nd ed.
87 BROWN. Cohomology of Groups.
88 PIERCE. Associative Algebras.
89 LANG. Introduction to Algebraic and Abelian Functions. 2nd ed.
90 BRØNDSTED. An Introduction to Convex Polytopes.
91 BEARDON. On the Geometry of Discrete Groups.
92 DIESTEL. Sequences and Series in Banach Spaces.
93 DUBROVIN/FOMENKO/NOVIKOV. Modern Geometry—Methods and Applications. Part I. 2nd ed.
94 WARNER. Foundations of Differentiable Manifolds and Lie Groups.
95 SHIRYAEV. Probability. 2nd ed.
96 CONWAY. A Course in Functional Analysis. 2nd ed.
97 KOBLITZ. Introduction to Elliptic Curves and Modular Forms. 2nd ed.
98 BRÖCKER/TOM DIECK. Representations of Compact Lie Groups.
99 GROVE/BENSON. Finite Reflection Groups. 2nd ed.
100 BERG/CHRISTENSEN/RESSEL. Harmonic Analysis on Semigroups: Theory of Positive Definite and Related Functions.
101 EDWARDS. Galois Theory.
102 VARADARAJAN. Lie Groups, Lie Algebras and Their Representations.
103 LANG. Complex Analysis. 3rd ed.
104 DUBROVIN/FOMENKO/NOVIKOV. Modern Geometry—Methods and Applications. Part II.
105 LANG. $SL_2(\mathbf{R})$.
106 SILVERMAN. The Arithmetic of Elliptic Curves.
107 OLVER. Applications of Lie Groups to Differential Equations. 2nd ed.
108 RANGE. Holomorphic Functions and Integral Representations in Several Complex Variables.
109 LEHTO. Univalent Functions and Teichmüller Spaces.
110 LANG. Algebraic Number Theory.
111 HUSEMÖLLER. Elliptic Curves.
112 LANG. Elliptic Functions.
113 KARATZAS/SHREVE. Brownian Motion and Stochastic Calculus. 2nd ed.
114 KOBLITZ. A Course in Number Theory and Cryptography. 2nd ed.
115 BERGER/GOSTIAUX. Differential Geometry: Manifolds, Curves, and Surfaces.
116 KELLEY/SRINIVASAN. Measure and Integral. Vol. I.
117 SERRE. Algebraic Groups and Class Fields.
118 PEDERSEN. Analysis Now.
119 ROTMAN. An Introduction to Algebraic Topology.
120 ZIEMER. Weakly Differentiable Functions: Sobolev Spaces and Functions of Bounded Variation.
121 LANG. Cyclotomic Fields I and II. Combined 2nd ed.
122 REMMERT. Theory of Complex Functions. *Readings in Mathematics*
123 EBBINGHAUS/HERMES et al. Numbers. *Readings in Mathematics*

124 DUBROVIN/FOMENKO/NOVIKOV. Modern Geometry—Methods and Applications. Part III
125 BERENSTEIN/GAY. Complex Variables: An Introduction.
126 BOREL. Linear Algebraic Groups. 2nd ed.
127 MASSEY. A Basic Course in Algebraic Topology.
128 RAUCH. Partial Differential Equations.
129 FULTON/HARRIS. Representation Theory: A First Course. *Readings in Mathematics*
130 DODSON/POSTON. Tensor Geometry.
131 LAM. A First Course in Noncommutative Rings. 2nd ed.
132 BEARDON. Iteration of Rational Functions.
133 HARRIS. Algebraic Geometry: A First Course.
134 ROMAN. Coding and Information Theory.
135 ROMAN. Advanced Linear Algebra.
136 ADKINS/WEINTRAUB. Algebra: An Approach via Module Theory.
137 AXLER/BOURDON/RAMEY. Harmonic Function Theory. 2nd ed.
138 COHEN. A Course in Computational Algebraic Number Theory.
139 BREDON. Topology and Geometry.
140 AUBIN. Optima and Equilibria. An Introduction to Nonlinear Analysis.
141 BECKER/WEISPFENNING/KREDEL. Gröbner Bases. A Computational Approach to Commutative Algebra.
142 LANG. Real and Functional Analysis. 3rd ed.
143 DOOB. Measure Theory.
144 DENNIS/FARB. Noncommutative Algebra.
145 VICK. Homology Theory. An Introduction to Algebraic Topology. 2nd ed.
146 BRIDGES. Computability: A Mathematical Sketchbook.
147 ROSENBERG. Algebraic *K*-Theory and Its Applications.
148 ROTMAN. An Introduction to the Theory of Groups. 4th ed.
149 RATCLIFFE. Foundations of Hyperbolic Manifolds.
150 EISENBUD. Commutative Algebra with a View Toward Algebraic Geometry.
SILVERMAN. Advanced Topics in Arithmetic of Elliptic Curves.
GLER. Lectures on Polytopes.
LTON. Algebraic Topology: A rst Course.
154 BROWN/PEARCY. An Introduction to Analysis.
155 KASSEL. Quantum Groups.
156 KECHRIS. Classical Descriptive Set Theory.
157 MALLIAVIN. Integration and Probability.
158 ROMAN. Field Theory.
159 CONWAY. Functions of One Complex Variable II.
160 LANG. Differential and Riemannian Manifolds.
161 BORWEIN/ERDÉLYI. Polynomials and Polynomial Inequalities.
162 ALPERIN/BELL. Groups and Representations.
163 DIXON/MORTIMER. Permutation Groups.
164 NATHANSON. Additive Number Theory: The Classical Bases.
165 NATHANSON. Additive Number Theory: Inverse Problems and the Geometry of Sumsets.
166 SHARPE. Differential Geometry: Cartan's Generalization of Klein's Erlangen Program.
167 MORANDI. Field and Galois Theory.
168 EWALD. Combinatorial Convexity and Algebraic Geometry.
169 BHATIA. Matrix Analysis.
170 BREDON. Sheaf Theory. 2nd ed.
171 PETERSEN. Riemannian Geometry.
172 REMMERT. Classical Topics in Complex Function Theory.
173 DIESTEL. Graph Theory. 2nd ed.
174 BRIDGES. Foundations of Real and Abstract Analysis.
175 LICKORISH. An Introduction to Knot Theory.
176 LEE. Riemannian Manifolds.
177 NEWMAN. Analytic Number Theory.
178 CLARKE/LEDYAEV/STERN/WOLENSKI. Nonsmooth Analysis and Control Theory.
179 DOUGLAS. Banach Algebra Techniques in Operator Theory. 2nd ed.
180 SRIVASTAVA. A Course on Borel Sets.
181 KRESS. Numerical Analysis.
182 WALTER. Ordinary Differential Equations.
183 MEGGINSON. An Introduction to Banach Space Theory.
184 BOLLOBAS. Modern Graph Theory.
185 COX/LITTLE/O'SHEA. Using Algebraic Geometry.
186 RAMAKRISHNAN/VALENZA. Fourier Analysis on Number Fields.

187 HARRIS/MORRISON. Moduli of Curves.
188 GOLDBLATT. Lectures on the Hyperreals: An Introduction to Nonstandard Analysis.
189 LAM. Lectures on Modules and Rings.
190 ESMONDE/MURTY. Problems in Algebraic Number Theory.
191 LANG. Fundamentals of Differential Geometry.
192 HIRSCH/LACOMBE. Elements of Functional Analysis.
193 COHEN. Advanced Topics in Computational Number Theory.
194 ENGEL/NAGEL. One-Parameter Semigroups for Linear Evolution Equations.
195 NATHANSON. Elementary Methods in Number Theory.
196 OSBORNE. Basic Homological Algebra.
197 EISENBUD/HARRIS. The Geometry of Schemes.
198 ROBERT. A Course in p-adic Analysis.
199 HEDENMALM/KORENBLUM/ZHU. Theory of Bergman Spaces.
200 BAO/CHERN/SHEN. An Introduction to Riemann–Finsler Geometry.
201 HINDRY/SILVERMAN. Diophantine Geometry: An Introduction.
202 LEE. Introduction to Topological Manifolds.
203 SAGAN. The Symmetric Group: Representations, Combinatorial Algorithms, and Symmetric Functions.
204 ESCOFIER. Galois Theory.
205 FÉLIX/HALPERIN/THOMAS. Rational Homotopy Theory. 2nd ed.
206 MURTY. Problems in Analytic Number Theory.
Readings in Mathematics
207 GODSIL/ROYLE. Algebraic Graph Theory.
208 CHENEY. Analysis for Applied Mathematics.
209 ARVESON. A Short Course on Spectral Theory.
210 ROSEN. Number Theory in Function Fields.